INTERMEDIATE
HEAT
TRANSFER

INTERMEDIATE HEAT TRANSFER

AHMAD FAKHERI

CRC Press
Taylor & Francis Group
Boca Raton London New York

CRC Press is an imprint of the
Taylor & Francis Group, an **informa** business

CRC Press
Taylor & Francis Group
6000 Broken Sound Parkway NW, Suite 300
Boca Raton, FL 33487-2742

First issued in paperback 2019

© 2014 by Taylor & Francis Group, LLC
CRC Press is an imprint of Taylor & Francis Group, an Informa business

No claim to original U.S. Government works

ISBN-13: 978-1-4398-1936-4 (hbk)
ISBN-13: 978-0-367-37981-0 (pbk)

Library of Congress Cataloging-in-Publication Data

Fakheri, Ahmad.
  Intermediate heat transfer / author, Ahmad Fakheri.
      pages cm
  Includes bibliographical references and index.
  ISBN 978-1-4398-1936-4 (hardback)
  1. Heat--Transmission. I. Title.

QC320.F356 2013
536'.2--dc23
2013010833

Visit the Taylor & Francis Web site at
http://www.taylorandfrancis.com

and the CRC Press Web site at
http://www.crcpress.com

To my alma mater, the University of Illinois

Contents

Preface

With any new book, there are the questions of why there is need for another and what its additional contributions are. Much has changed in the field of fluid mechanics and heat transfer and how and what is taught. Advanced thermal science courses tended also to be advanced math courses, where a great deal of time had to be spent teaching the requisite advanced analytical techniques. An important change has been a shift away from advanced analytical techniques to more of a reliance on numerical solutions, which has also broadened the topics that are covered in these courses. This book is a response to these changes and is intended for an advanced undergraduate or first-year graduate heat transfer course. It also fills the gap between the undergraduate heat transfer course and specialized advanced courses such as conduction, convection, radiation, and mass transfer.

The new tools of choice for solving heat transfer problems are numerical methods through computational fluid dynamics (CFD) packages. This, however, has not diminished the importance of analytical solutions, which are needed not only to understand the fundamental concepts but also to verify numerical solutions. Where appropriate, this book presents detailed step-by-step analytical solutions, including many classical problems, while many other problems are solved numerically. This approach significantly broadens the scope and depth of the topics covered and allows for the presentation of a much wider array of solutions to realistic problems.

Obtaining an accurate solution to realistic transport problems requires a clear understanding of the fundamental topics, the ability to think critically and creatively in order to include what is important and ignore what can be neglected, the mastery of the available tools for arriving at a solution, and the ability to critically examine the results. This book emphasizes formulating problems, starting from fundamental basics, and obtaining and analyzing their solutions. Nondimensionalization is greatly emphasized throughout the text as a tool for simplifying the governing equations, developing additional insights into the physics of the problems, identifying the relevant parameters, and arriving at general solutions.

Obtaining correct numerical solutions requires an additional skill set, independent of the specific tool used. The tool used for obtaining numerical solutions of the resulting differential and partial differential equations is spreadsheets. Spreadsheets are powerful, easily available, have built-in graphics capabilities, and, most importantly, students already have a good understanding of their use, which saves valuable class time. They are generally easy to follow, eliminating the need for debugging complicated codes. Students not only solve different problems and display the results, but they can also perform parametric studies and see the impact on the results immediately and develop better insights into the physics of the problems. Students actually program a wide variety of problems and prove the validity of many typically invoked, simplifying assumptions. Numerical solution using spreadsheets helps students develop a clearer perspective on how the problems are formulated, simplified, and solved numerically, providing an effective way of developing the skills needed in using commercial CFD software. Students also take this skill with them, as they have access to spreadsheets even after they complete their studies.

The numerical solutions presented in this book are sufficiently detailed and can also be easily obtained using other tools, including equation solvers and computing environments or direct programming using languages such as Fortran or C. These specialized softwares suffer from the same shortcoming as that of advanced mathematical techniques: one has to devote valuable class time teaching the particular tool since none has yet been accepted as a standard tool by the engineering community, and instructors cannot assume that all students are sufficiently familiar with their preferred software.

The chapter on CFD provides the necessary background and skills for obtaining numerical solutions. It includes a number of step-by step tutorials for solving more complicated problems, both to show how CFD codes are used and as a further check of some of the more commonly used assumptions, such as the boundary layer approximation and negligibility of axial diffusion in the entrance of pipes, among others.

The companion website www.ihtbook.com contains additional information, including the source code for many cases presented in the book, supplementary examples, additional end-of-chapter problems, solutions to the problems, and lecture PowerPoint presentations of the selected chapters. This book has been under development for over 20 years, having started as notes for a course with the same title. The field of heat transfer is a relatively mature field, thanks to the contributions of the thousands of researchers who have driven advances in this field and from whose work this book derives totally. I have only cited a small fraction of them in the pages of this book, but I want to acknowledge them all here.

I further want to thank a few specific people for their contributions to this book as a whole: my students for their feedback and corrections over the years; Wikipedia and its broad community of authors for the huge body of knowledge they have made freely available to the world; Professor Majid Molki, who coauthored the chapter on computational fluid dynamics, and Matt West, who wrote the second tutorial in that chapter; my dear colleague and friend, the late Professor Yahya Safdari, for his suggestions and encouragement; and, finally, my wife, Farzaneh, and my son, Rustin, for their love and support, without which this work would not have been possible.

Author

Dr. Ahmad Fakheri is a professor of mechanical engineering at Bradley University, Peoria, Illinois. He received his undergraduate and graduate degrees, all in mechanical engineering, from the University of Illinois in Urbana-Champaign. His teaching and research interests are in the area of thermal sciences. Since 1985, he has taught many different undergraduate and graduate level courses; developed four new ones, including an intermediate heat transfer course; and published over 50 refereed papers. Dr. Fakheri has served in a number of leadership positions at Bradley University and in the American Society of Mechanical Engineers (ASME) and is a fellow of the ASME.

1

Basic Concepts

1.1 Introduction to Heat Transfer

The human race's improvements in standard of living and the incredible expansion of available goods and services are directly related to our ability to harness energy. Early humans' primary energy need was thermal to keep them warm, and perhaps to cook. Progress increased the need for mechanical energy. "All ancient civilizations, no matter how enlightened or creative, rested on slavery and on grinding human labor, because human and animal muscle power were the principal forms of energy available for mechanical work. The discovery of ways to use less expensive sources of energy than human muscles made it possible for men to be free" [1]. Harnessing mechanical energy from thermal energy, made possible by James Watt's invention of the steam engine in 1765, started the Industrial Revolution. It is estimated that two thirds of mechanical energy used in the United States in the 1850s was provided by horses [2].

Energy is an abstract concept, difficult to define, yet we all have a good feel for what it is or at least what it can do. As stated by the Nobel Laureate late professor Richard Feynman [3], the law of conservation of energy "states that there is a certain quantity, which we call energy, that does not change in manifold changes which nature undergoes. That is a most abstract idea, because it is a mathematical principle; it says that there is a numerical quantity, which does not change when something happens. It is not a description of a mechanism, or anything concrete; it is just a strange fact that we can calculate some number, and when we finish watching nature go through her tricks and calculate the number again, it is the same."

The importance of this number is that humans' ability to harness what it represents has brought about the tremendous improvements in the standard of living, particularly since the industrial revolution.

Figure 1.1 is a plot of annual energy consumption globally by type and their projections until the year 2035 [4]. In 2011, of the total amount of energy used, 5.6E + 20 J, 33% came from liquid fuels, 28% from coal, 22% from natural gas, 11% from renewable sources of the energy, and 5% from nuclear. This level of energy consumption averages to about 219 MJ used per day by each one of the world's 7 billion inhabitants, with a large variation geographically—1000 MJ/day in the United States, 50 in Africa. To survive, we need 12.54 MJ (3000 kcal/day) from food.

Proper utilization of all these different forms of energy requires an understanding of energy, work, and heat transfer and the laws that govern their behavior.

Heat transfer is a mechanism by which energy can be transferred across the boundaries of a system and thus its study is intimately related to the general topic of energy. Energy

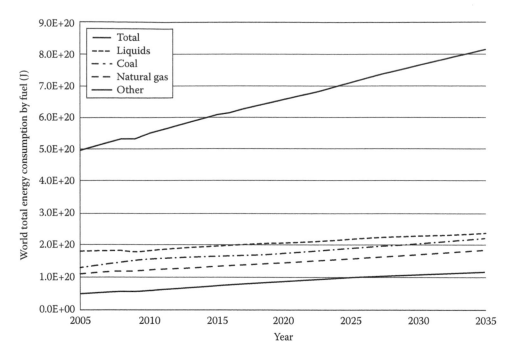

FIGURE 1.1
World energy use by fuel type.

can also cross the boundaries of a system as a result of work and mass transfer. Heat transfer is thus one of the three ways by which energy can be added or removed from a system. It is the transfer of energy due to a temperature difference.

The reason that temperature difference results in heat transfer is explained by the second law of thermodynamics. In its most basic version, the second law states that if a system is not in equilibrium with its surroundings, nature requires it to try to reach equilibrium. If the system and its surroundings are not at the same temperature, that is, they are not in thermal equilibrium, and are free to interact, then energy will be transferred in the form of heat until they reach thermal equilibrium, that is, reach the same temperature.

Another consequence of the second law of thermodynamics is the concept of entropy. Drawing a parallel with Professor Feyman's definition of energy, there is a quantity that we call entropy. It is not a description of a mechanism, or anything concrete; it is just a strange fact that we can calculate some number, and when we finish watching nature go through her tricks and calculate the number again, it has increased. The physical relevance of this number is that it is a measure of the randomness, or the universe's order continuously decreases.

Heat and work transfer are both mechanisms for transfer of energy and are therefore related through the law of conservation of energy. From a second law perspective, the difference between heat transfer and work transfer is that the former increases entropy, making it an irreversible process, while work transfer does not, or is reversible. Heat transfer is a disorderly mechanism for transfer of energy, whereas work is an orderly manner by which energy can be transferred across the boundaries of the system.

1.2 Conservation of Energy

In the absence of nuclear reactions, energy is conserved. Therefore, energy transferred to a system will increase its energy by the exact amount of energy transferred. The laws of nature take different mathematical forms depending on whether the system is a control mass or a control volume.

1.2.1 Conservation of Energy for a Control Mass

A control mass is a system where no mass crosses its boundaries, thus its mass remains fixed. There are two ways to transfer energy to a control mass; therefore, mathematically, the law of conservation of energy for a control mass undergoing a process from any initial state i to any final state f can be stated as

$$_iQ_f + _iW_f + Q_g = E_f - E_i \tag{1.1}$$

The first term on the left-hand-side (LHS) of Equation 1.1, $_iQ_f$, is the net rate of heat transfer to the system from the surroundings, $_iW_f$ is the net rate of work done on (transferred to) the system by the surroundings, and Q_g is the total amount of energy released in the system (generated) during the process. This energy could come as a result of chemical reactions, which would release some of the internal energy in the form of thermal energy, nuclear reactions, the absorption of electromagnetic (microwave) radiation, or internal resistance to the flow of electricity. The energy generation term is sometimes included in the work term and sometimes as a separate term as in Equation 1.1.

The term on the right-hand-side (RHS), E, is the total energy of the system. A system can have three forms of energy: kinetic $\left(\dfrac{mV^2}{2g_c}\right)$, potential $\left(\dfrac{mgz}{g_c}\right)$, and internal ($U = mu$), where u is internal energy per unit mass, specific internal energy. Note that g_c is a constant that depends on the system of units used and is equal to 1 in SI and 32.2 in the conventional system.

In many applications, the changes in the kinetic and potential energies are negligible, reducing the law of conservation of energy to

$$_iQ_f + Q_g + _iW_f = U_f - U_i = m(u_f - u_i) \tag{1.2}$$

It takes a finite amount of time for heat and work to be transferred to the system across its boundaries. Theoretically, it takes infinitely long for a process to reach the final equilibrium state, although depending on the process the time for it to reach very close to the final equilibrium is finite, and in many practical situations, very short.

If both sides of the aforementioned equation are divided by Δt and the limit is taken as Δt goes to zero, then we end up with the first law of thermodynamics on a rate basis;

$$q + q_g + \dot{W} = m\frac{du}{dt} \tag{1.3}$$

Instead of using $\dot{Q}$ to signify the net rate of heat transfer, the more accepted terminology is q, which is heat transfer per unit time and is expressed in Watts or Btu/h. The rate of energy generation is designated by q_g. There is a fundamental difference between Equations 1.2 and 1.3. Equation 1.2 is an equality between the total amounts during the process, from the initial state to the final state, regardless of how long it takes for the process to go to completion, whereas Equation 1.3 is an equality between the rates.

For incompressible substances (solids and liquids) as well as for ideal gases, the internal energy is only a function of temperature or

$$du = d(cT) \tag{1.4}$$

where c is the specific heat. Assuming constant specific heat, Equation 1.3 can be written as

$$q + q_g + \dot{W} = mc\frac{dT}{dt} \tag{1.5}$$

which states that the net rate at which energy is transferred as heat into the system, plus the rate at which energy is generated within the system, plus the net rate that energy is transferred as work to the system is equal to the rate of increase of the internal energy of the system.

Note that q represents the net rate of heat transfer to the control mass, which is the difference between the rate of heat transfer into the system from a source at a higher temperature and the rate of heat transfer out of the system into its surroundings, which are at a temperature lower than the system temperature.

$$q = q_{in} - q_{out} \tag{1.6}$$

Substituting Equation 1.6 into Equation 1.5 results in

$$q_{in} - q_{out} + q_g + \dot{W} = mc\frac{dT}{dt} \tag{1.7}$$

A steady process is one during which the state of the system, as defined by its properties, does not change with time, or all system properties, such as temperature and pressure, remain constant with respect to time, thus $\frac{dT}{dt} = 0$, simplifying Equation 1.7 to

$$q + q_g + \dot{W} = 0 \tag{1.8}$$

which is the first law of thermodynamics for a control mass undergoing a steady process.

1.2.2 Conservation of Energy for a Control Volume

A control volume is a system where mass does cross its boundaries. The first law of thermodynamics takes a different form for a control volume since we also have to account for the transfer of energy as a result of mass entering and leaving the system. The control volume formulation of the first law is

$$q + q_g + \dot{W} + \sum_{in} \dot{m}\left(h + \frac{V^2}{2g_c} + \frac{gz}{g_c}\right) = \frac{dE_{c.v}}{dt} + \sum_{out} \dot{m}\left(h + \frac{V^2}{2g_c} + \frac{gz}{g_c}\right) \tag{1.9}$$

The summation terms account for the rate of energy transfer into and out of the control volume as a result of the flow of mass in and out of the system. The first term on the RHS accounts for the rate of energy storage in the system, and h is enthalpy per unit mass, specific enthalpy, given by $h = u + Pv$. Typically, the kinetic and potential energy changes are negligible simplifying Equation 1.9 to

$$q + q_g + \dot{W} + \sum_{in} \dot{m}h = \frac{dE_{c.v}}{dt} + \sum_{out} \dot{m}h \qquad (1.10)$$

Example 1.1

A car engine has an efficiency of 33%. The fuel has a heating value of 42,000 kJ/kg and the engine rejects 30% of the energy generated as a result of combustion through the radiator and the rest is rejected through the exhaust. The car generates 75 hp, and the air fuel ratio is 17. Find the fuel consumption rate and the exhaust temperature.

The efficiency is defined as energy sought, the work output of the engine, divided by the energy that costs, which is the fuel energy:

$$\eta = \frac{\dot{W}}{q_g} = \frac{\dot{W}}{\dot{m}_f HV}$$

$$\dot{m}_f = \frac{\dot{W}}{\eta HV} = \frac{75 \times 735.6}{0.33 \times 42,000 \times 1,000} = 0.00398 \, kg/s = 3.98 \, g/s$$

$$\dot{m}_a = 17 \times 0.00398 = 0.0677 \, kg/s = 67.7 \, g/s$$

$$q_g = \dot{m}_f HV = 1.67 \times 10^5 \, W$$

Thirty percent of the energy generated by the combustion process is rejected in the radiator; therefore, the first law for the engine becomes

$$q + q_g + \dot{W} + \sum_{in} \dot{m}h = \sum_{out} \dot{m}h$$

$$q_{rad} + q_g + \dot{W} + (\dot{m}_a h_a + \dot{m}_f h_f)_i = (\dot{m}_a + \dot{m}_f)h_e$$

Assuming the fuel and air to have the same specific heat and enter the engine at the same temperature, then

$$T_e = T_i + \frac{q_{rad} + q_g + \dot{W}}{(\dot{m}_a + \dot{m}_f)c_p} = T_i + \frac{-0.3q_g + q_g - \dot{W}}{c_p(\dot{m}_a + \dot{m}_f)} = 300 + \frac{0.7 \times 1.67 \times 10^5 - 75 \times 735.6}{1005(0.00398 + 0.0677)} = 1159 \, K$$

Equations 1.7 and 1.10 are used to analyze the average behavior of the control masses and control volumes. When using these equations, we neglect the variation of properties inside the system and only deal with average behavior. Often, the average behavior is insufficient for the proper design and analysis of the systems, and the local values must be determined. This requires the knowledge of the energy transfer mechanisms not only at the boundaries but also at every point in the system. Studying heat transfer allows the determination of the local state of a system or provides the local values of system properties.

There are three main mechanisms by which heat can be transferred in a system: conduction, convection, and radiation. Most real problems involve a combination of all three modes of heat transfer.

1.3 Conduction

Conduction is energy transfer from a high-temperature region to a low-temperature region within a solid or fluid that is at rest. Temperature is a measure of the internal energy, which is a measure of the translational, vibrational, rotational, and other forms of energy the system has at the microscopic level. Conduction of heat is transfer of these forms of energy at the microscopic level. In liquids and gases, conduction is caused by the motion and collision of atoms and molecules.

Consider a cube 1 m on each side containing an ideal gas. The temperature of the gas is related to the average velocity of its particles through

$$T = \frac{1}{3k}mV^2 \tag{1.11}$$

where
    m is the mass of the particles
    V is the average velocity
    $k = 1.3805 \times 10^{-23}$ J/K is the Boltzmann constant

Solving for the average velocity of air molecules, from (1.11), and noting that $m = M/L$, where M is the molar mass (kg/kmol) and L is the Avogadro's number, which is $L = 6.02214 \times 10^{26}$ particles/kmol, then for air at room temperature,

$$V = \sqrt{\frac{3kT}{m}} = \sqrt{\frac{3kT}{M/L}} = \sqrt{\frac{3 \times 1.3805 \times 10^{-23} \times 300}{\dfrac{29}{6.02214 \times 10^{26}}}} = 508\,\text{m/s}$$

or the oxygen and nitrogen molecules travel with a velocity of around 500 m/s or bounce off the walls around 500 times every second. The mean free path is defined as the average distance that an atom or a molecule travels between successive collisions and is given by

$$\lambda = \frac{kT}{\sqrt{2}\pi d^2 P} \tag{1.12}$$

where
    d is the particle diameter
    P is the pressure

Although the diameters of atoms are difficult to define, the diameters of nitrogen and oxygen molecules can be approximated to be about 3.1 Å (3.1×10^{-10} m). Therefore, at room temperature and pressure,

$$\lambda = \frac{kT}{\sqrt{2}\pi d^2 P} = \frac{1.3805 \times 10^{-23} \times 300}{\sqrt{2}\pi (3.1 \times 10^{-10})^2 101,325} = 9.57 \times 10^{-8}\ m$$

or one molecule will collide with other molecules around 10 million times as it moves from one side of the container to the other side, and the molecules travel this distance on average about 500 times every second or molecules on the average experience 5 billion collisions per second. If one of the container walls is at a higher temperature, the molecules colliding with that wall absorb additional energy and leave the surface with higher kinetic energy. Through their numerous collisions with other molecules, the more energetic molecules will transfer their energy and effectively transfer energy from one of the container walls to the other walls. Therefore, in fluids that are at rest, conduction occurs a result of the random motion of the molecules.

In materials that conduct electricity, like metals, conduction is primarily caused by the motion of free electrons that carry kinetic energy from the regions of higher kinetic energy to the lower one, in a sense similar to what happens in liquids and gases. In addition, in solids, the atoms are arranged in the form of a crystal lattice. When the molecules in the lattice vibrate, they generate waves that propagate through the lattice. In classical mechanics, the vibrations are wavelike, and the frequency at which the lattice oscillates uniformly is known as the normal mode. Heat conduction in crystalline structures can be explained by phonons [5] that are wave-like phenomena, which sometimes are easier to study, by assuming that they have particle-like properties. Wave-particle duality is a concept similar to electromagnetic waves and photons. Unlike photons that are massless, phonons have mass. Therefore, the mechanism for conduction of heat in solids is the motion of phonons, and the thermal conductivity is primarily dependent on the speed by which phonons travel in nonconducting solids.

1.3.1 One-Dimensional Steady Conduction

A problem is classified as one dimensional when temperature changes in only one spatial direction and therefore heat is transferred in that same direction. Figure 1.2 shows a wall with thickness L, whose left surface is maintained at a constant temperature T_1, while the right surface is T_2, with $T_1 > T_2$. We want to determine the rate at which heat is transferred at any point in the wall. Since heat transfer occurs as a result of temperature difference, we must first determine the temperature distribution within the wall. In this case, the temperature will continuously drop from T_1 on the left surface to T_2 on the right surface, and our first objective is to determine the function that describes this change. To determine the function, we need to come up with the statement of the first law that is valid at a point, keeping in mind that Equation 1.7 is valid for a control mass that has a finite size.

Consider a control mass (CM) inside the wall with a small finite thickness Δx. We start with the first law of thermodynamics for a control mass, Equation 1.7,

$$q_{in} - q_{out} + q_g + \dot{W} = mc\frac{dT}{dt} \tag{1.13}$$

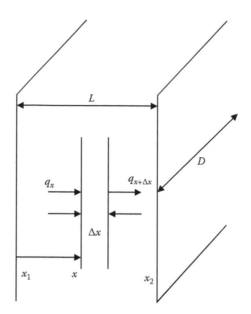

FIGURE 1.2
Energy balance for steady 1-D conduction.

Since the differential control mass is inside a solid, the only mechanism for heat transfer across the boundaries of the system is conduction. Assuming heat to be transferred from left to right, then q_x is the rate that heat is transferred into the control mass from the surroundings through the left boundary of the system and $q_{x+\Delta x}$ is the rate that heat is transferred from the control mass through the right boundary to the surroundings, or the net rate of heat transfer to the CM is

$$q_{in} - q_{out} = q_x - q_{x+\Delta x} \tag{1.14}$$

In Equation 1.13, $\dot{W}$ is the net rate that work is done on the differential control mass by all mechanisms. In this case, there is no change in volume (PV work), no other work is done on the CM, or the rate of work done is zero. Assume $\dot{q}$ is the rate by which energy is generated per unit volume, then q_g, which is the rate at which energy is generated in the entire control mass, becomes

$$q_g = \dot{q}\Delta x A \tag{1.15}$$

where A is the cross-sectional area of the control mass, which is the area perpendicular to the x-axis. For this case, we assume there is no energy generation ($\dot{q} = 0$). The term on the RHS of Equation 1.13 is the rate of energy storage in the control mass, which for a steady state process is zero, and therefore Equation 1.13 becomes

$$q_x = q_{x+\Delta x} \tag{1.16}$$

Here, we have one equation and two unknowns—the values of function q at two locations x and $x + \Delta x$.

As you may recall from calculus, if the values of a function and all of its derivatives are known at a point x, then the value of the function at another point located $x + \Delta x$ can be calculated from the Taylor series;

$$f(x+\Delta x) = \sum_{n=0}^{\infty} \frac{\Delta x^n}{n!} f^n(x) = f(x) + \frac{\Delta x}{1!}\frac{df}{dx}\bigg|_x + \frac{\Delta x^2}{2!}\frac{d^2 f}{dx^2}\bigg|_x + \frac{\Delta x^3}{3!}\frac{d^3 f}{dx^3}\bigg|_x + \cdots \tag{1.17}$$

where $f^n(x)$ is the nth derivative of f with respect to x, evaluated at x

$$f^n(x) = \frac{d^n f}{dx^n}\bigg|_x \tag{1.18}$$

Note that all the terms on the RHS are evaluated at x.

Example 1.2

Consider the following are known about a function and all its derivatives at $x = 0$:

$$f(0) = 0, \quad \frac{df}{dx}\bigg|_0 = 0, \quad \frac{d^2 f}{dx^2}\bigg|_0 = 2, \quad \frac{d^3 f}{dx^3}\bigg|_0 = \frac{d^4 f}{dx^4}\bigg|_0 = \cdots = 0 \tag{1.19}$$

Then from the Taylor series, the value of the function at a point that is Δx away from this point (zero) is given by

$$f(0 + \Delta x) = 0 + 0\frac{\Delta x}{1!} + 2\frac{(\Delta x)^2}{2!} + 0\frac{(\Delta x)^3}{3!} + 0 = (\Delta x)^2 \tag{1.20}$$

Therefore, the value of the unknown function at any location is equal to the square of its distance from zero (origin) or

$$f(x) = x^2 \tag{1.21}$$

since x is the distance from the origin. It is easy to verify that function (1.21) satisfies all the relations given by Equations 1.19 at zero or is the unknown function.

Returning to our analysis of the one-dimensional, steady conduction in the wall, since q is a function of x, then using the Taylor series, $q_{x+\Delta x}$ can be expressed in terms of q_x and its derivatives at location x

$$q_{x+\Delta x} = q_x + \frac{dq_x}{dx}\Delta x + \frac{d^2 q_x}{dx^2}\frac{\Delta x^2}{2!} + \text{H.O.T} \tag{1.22}$$

where H.O.T stands for higher order terms. From the law of conservation of energy, Equation 1.16, Equation 1.22 simplifies to

$$0 = \frac{dq_x}{dx}\Delta x + \frac{d^2 q_x}{dx^2}\frac{\Delta x^2}{2!} + \text{H.O.T} \tag{1.23}$$

This is a differential equation for one unknown, but unfortunately has infinitely many terms and is still not solvable. This equation is applicable to a control mass located at location x and extending to the point $x + \Delta x$. Dividing both sides by Δx

$$\frac{dq_x}{dx} = -\left[\frac{d^2q_x}{dx^2}\frac{\Delta x}{2!} + \text{H.O.T}\right] \tag{1.24}$$

Since Δx can take any small value, the aforementioned equation is also valid in the limit as Δx approaches zero or as the size of the control mass approaches zero and becomes a point. In this limit, all the terms on the RHS of Equation 1.24 approach zero since they all are multiplied by at least Δx to the first power, which results in

$$\frac{dq_x}{dx} = 0 \tag{1.25}$$

This equation is the mathematical representation of the law of conservation of energy, valid for all points of a solid or a fluid at rest, if heat is transferred in one direction. For one-dimensional conduction, the first law necessitates that the change in the rate of heat transfer with respect to the direction of its transfer (x) be zero or the rate of heat transfer in the x direction is constant

$$q_x = \text{constant} \tag{1.26}$$

For one-dimensional conduction, the rate of heat transfer is the same at any given point inside the wall, which is not surprising, since heat transfer is the only mechanism for transfer of energy and there are no generation or storage of energy. The first law shows that the rate of heat transfer is constant, but it does not show its value. To answer that question, we must express q_x in terms of measurable quantities, since in general heat transfer is difficult to measure directly.

Since heat is transferred across the temperature gradient, then x is also the direction over which the temperature changes. Experiments have shown that the rate of heat transfer by conduction is proportional to the temperature difference between the two points, is inversely proportional to the *distance* that the heat has to travel, and is also proportional to the total area available for the transfer of heat. Mathematically, these relationships become

$$q_x \propto A_x \frac{dT}{dx} \tag{1.27}$$

The proportionality constant is the thermal conductivity k; therefore,

$$q_x = -kA_x \frac{dT}{dx} \tag{1.28}$$

This is known as the Fourier's law of heat conduction, named after the French mathematician Joseph Fourier (1768–1830). In this equation, A_x is the area perpendicular to the direction of heat flow or is the area across which heat flows.

The thermal conductivity is a material property and represents the rate of heat transfer per unit length per unit temperature difference between two points and has units of W/(m K) in SI. Note that the K is for one Kelvin difference, which is the same as 1°C difference, or an equivalent unit is W/(m °C) in SI. The higher the thermal conductivity,

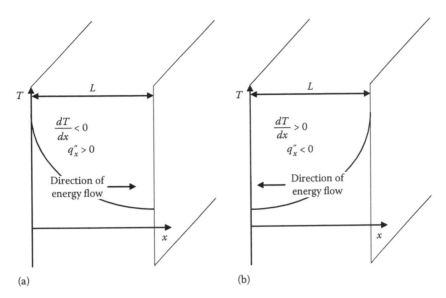

FIGURE 1.3
Sign convention for heat transfer.

the better conductor the material is. The thermal conductivity is 400 for copper, 1.4 for glass, 0.61 for water, and 0.026 W/m K for air at room temperature.

The negative sign in the Fourier's law of heat conduction (Equation 1.28) is to conform with the convention that positive q indicates that heat is transferred in the positive x-direction. To demonstrate this, consider Figure 1.3a, where temperature decreases in the x-direction and therefore heat is being transferred in the x-direction. In this case, $\dfrac{dT}{dx}$ is negative and since A_x and k are positive, the rate of heat transfer, q, becomes positive. Conversely, in Figure 1.3b, temperature increases in the x direction and $\dfrac{dT}{dx}$ is positive and the multiplication with a negative sign results in a negative value for q, which means that heat will flow in the negative x-direction, which is what happens in reality. This convention is often used along the concept of energy in and out of the system, and remaining consistent will be very important, as will soon become more clear.

Substituting the Fourier law of heat conduction (1.28) into the first law of thermodynamics (1.25) gives us

$$\frac{d}{dx}\left(-kA_x\frac{dT}{dx}\right) = 0 \tag{1.29}$$

and by multiplying both sides by −1, we get

$$\frac{d}{dx}\left(kA_x\frac{dT}{dx}\right) = 0 \tag{1.30}$$

In Cartesian coordinates, the area in the x-direction is constant and does not depend on x. If we assume the thermal conductivity to also be constant, that is, independent of temperature and therefore x, then k and A_x can be taken out of the derivative and both sides can be divided by them:

$$\frac{d^2T}{dx^2} = 0 \tag{1.31}$$

This equation resulted from the application of the law of conservation of energy, when heat is transferred in one direction, and the area through which heat is transferred does not change in that direction. The first law of thermodynamics requires the temperature in the medium to satisfy Equation 1.31, which is a second-order differential equation. To solve this equation, two boundary conditions are needed.

If the wall thickness is L, then the boundaries are located at 0 and L and the boundary conditions are

$$x = 0 \quad T = T_1$$
$$x = L \quad T = T_2 \tag{1.32}$$

Integrating the energy equation twice leads to

$$\frac{dT}{dx} = C_1 \tag{1.33}$$

$$T = C_1 x + C_2 \tag{1.34}$$

Applying the boundary conditions to determine the constants results in

$$T_1 = C_1 0 + C_2 \quad C_2 = T_1 \tag{1.35}$$

$$T_2 = C_1 L + T_1 \quad C_1 = \frac{T_2 - T_1}{L} \tag{1.36}$$

which when substituted in Equation 1.34 results in

$$T = \frac{T_2 - T_1}{L} x + T_1 \tag{1.37}$$

Equation 1.37 shows that temperature changes linearly from T_1 on the left face to T_2 on the right face. We now have the temperature variation in the wall that allows us to calculate the rate of heat transfer, which is what we are generally seeking. The two are related through the Fourier law of heat conduction

$$q_x = -kA_x \frac{dT}{dx} \tag{1.38}$$

From Equations 1.33 and 1.36 or by differentiating Equation 1.37

$$\frac{dT}{dx} = \frac{T_2 - T_1}{L} \tag{1.39}$$

and substituting into Equation 1.38 yields

$$q_x = -kA_x \frac{T_2 - T_1}{L} \tag{1.40}$$

which can be rewritten in the following form:

$$q_x = \frac{T_1 - T_2}{\dfrac{L}{kA_x}} \tag{1.41}$$

For one-dimensional conduction, it helps to draw an analogy between the flow of heat and the flow of electric current. Electric current flows to counteract a difference in electric potential; similarly heat flows to counteract a temperature difference. Both these effects are required from the second law of thermodynamics that states that if a system is not in equilibrium, it seeks it. Flow of current is to equalize the potentials, the flow of heat is to equalize the temperatures. What opposes the flow of electricity is the material's electrical resistance, and therefore we can define conduction resistance as

$$R_{cond} = \frac{L}{kA_x} \tag{1.42}$$

so

$$q_x = \frac{T_1 - T_2}{R_{cond}} \tag{1.43}$$

This looks like the equation for an electric circuit $I = \dfrac{V}{R}$ and is solved the same way, that is, equivalent resistances can be calculated when they are in series and parallel, as explained later in this chapter.

1.4 Convection

Convection heat transfer is a result of both molecular motion and material motion or it is conduction that is enhanced by the motion of the medium. When a fluid moves over the surface of a hot body, fluid particles at the interface stick momentarily to the surface and therefore their velocity becomes zero. At this interface, the energy is transferred by conduction into the fluid, and the energy transferred to the fluid is carried to other points by both random molecular motion and bulk fluid motion. The heat transfer by convection can be very complex.

There are two types of convection, differentiated by the cause of the fluid's bulk motion. Forced convection describes a situation in which the fluid motion is caused by an external agent, such as by a pump or a fan, or when the object moves in the fluid, as stirring coffee with a spoon. Free convection describes a situation in which the fluid motion is brought about by some naturally occurring force, which in most practical applications is the force of buoyancy.

1.4.1 Heat Transfer Coefficient

The concept of the heat transfer coefficient, h, which is also known as the film conductance coefficient, is used to facilitate calculating heat transfer by convection. If q'' is the heat flux (the rate of heat transfer per unit area) transferred from a surface at T_w to a fluid whose temperature far from the surface is T_∞, then the local value of the heat transfer coefficient is a number that when multiplied by the temperature difference provides the local heat flux or

$$q'' = h(T_w - T_\infty) \tag{1.44}$$

The heat transfer coefficient, h, incorporates fluid flow, fluid properties, and geometry and in general is not uniform over a surface. The total heat transfer from the surface is determined by using an average heat transfer coefficient

$$\bar{h} = \frac{1}{A} \int_A h \, dA \tag{1.45}$$

The total heat transfer from an isothermal surface is

$$q = \int_A q'' dA = \int_A h(T_w - T_\infty) dA = A(T_w - T_\infty) \frac{1}{A} \int_A h \, dA \tag{1.46}$$

$$q = \bar{h} A (T_w - T_\infty) \tag{1.47}$$

We can also define the thermal resistance for convection by rearranging Equation 1.47

$$q = \frac{(T_w - T_\infty)}{\dfrac{1}{\bar{h}A}} = \frac{(T_w - T_\infty)}{R_{conv}} \tag{1.48}$$

where

$$R_{conv} = \frac{1}{\bar{h}A} \tag{1.49}$$

Consider the wall of Figure 1.4, exposed to a convective environment. To determine the temperature distribution, it is easier to use the electrical analogy. There are two resistances, one conduction and one convection. The amount of heat transfer can be calculated from any of the three equations given next:

$$q = \frac{(T_1 - T_2)}{\dfrac{L}{kA}} = \frac{(T_2 - T_\infty)}{\dfrac{1}{hA}} = \frac{(T_1 - T_\infty)}{\dfrac{L}{kA} + \dfrac{1}{hA}} \tag{1.50}$$

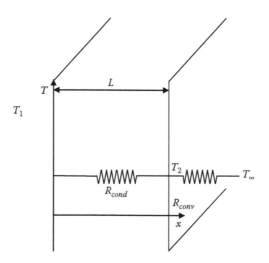

FIGURE 1.4
Conduction and convection resistances in series.

However, in this case, T_2 is unknown and therefore the last equation is to be used:

$$q = \frac{(T_1 - T_\infty)}{\dfrac{L}{kA} + \dfrac{1}{hA}} \tag{1.51}$$

Once heat transfer is calculated, T_2 can be calculated by using either of the two remaining equations.

1.4.2 Biot Number

As will be seen throughout this book, the behavior of fluid mechanics and heat transfer problems is largely determined not by any individual parameter or variable, rather by non-dimensional groupings of variables and parameters. We now introduce one such grouping by rearranging the middle equation in Equation 1.50:

$$\frac{T_1 - T_2}{T_2 - T_\infty} = \frac{\dfrac{L}{kA}}{\dfrac{1}{hA}} \tag{1.52}$$

This is a dimensionless equation. The LHS is the ratio of the temperature differences in the wall and in the fluid, and the RHS is the ratio of conduction resistance to convection resistance. The dimensionless group on the RHS is called the Biot number:

$$Bi = \frac{R_{cond}}{R_{conv}} = \frac{hL}{k} \tag{1.53}$$

It is important to note that the thermal conductivity appearing in the definition of Biot number is that of the solid and L characterizes the distance that heat travels by conduction. Therefore, Biot number is the ratio of conduction resistance to convection resistance and is also a measure of temperature variation in the solid to temperature variation in the fluid:

$$\frac{\Delta T_{solid}}{\Delta T_{fluid}} = Bi \tag{1.54}$$

The variation of the temperature in the wall and in the fluid is shown in Figure 1.5 for different Biot numbers. If the Biot number is small, the conduction resistance is small requiring a small temperature gradient in the solid for the heat to be transferred. The barrier to the transfer of heat or the bottle neck will be the resistance in the fluid. Once the energy reaches the solid boundary, it encounters a large convection resistance in the fluid requiring a large temperature difference between the surface of the solid and fluid. Therefore, the temperature variation in the solid is small compared to the temperature variation in the fluid, and as an approximation, the solid can be assumed to be isothermal. For $Bi < 0.1$, the temperature variation in the wall is less than 10%.

When the Biot number is very large (much greater than one), the resistance to the flow of heat by conduction, R_{cond}, is large. Therefore, in order to transfer a given amount of heat, a large ΔT will develop in the body. Also, the temperature variation in the fluid is much smaller than that in the solid. This means that the surface temperature of the solid exposed to the fluid will be approximately equal to the fluid temperature far from the solid. In the limit as the Biot number approaches infinity, the wall temperature will be at a constant value equal to the free stream temperature of the fluid.

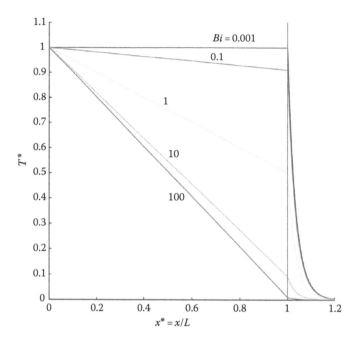

FIGURE 1.5
Variation of the temperature in the wall and in the fluid for different Biot numbers.

1.5 Radiation

Radiation is the energy emitted by a substance due to its absolute temperature. All substances continuously emit electromagnetic waves (photons) if their temperature is above absolute zero. Unlike conduction and convection, which require a medium (solid and/or fluid), radiation is most efficient in a vacuum. Radiation is the mechanism by which the sun's energy reaches us and is directly or indirectly the source of all renewable energy on Earth.

The sun is 1.5×10^{11} m from the earth and emits 3.85×10^{26} W of thermal energy, that is, every second it emits 700,000 more energy than the 2011 annual global energy consumption. By the time this energy reaches earth, it is distributed over the surface of a sphere having a radius equal to the distance from the sun to earth, providing a solar flux of 1369 W/m² known as the solar constant. The amount of energy intercepted by the earth is 1.76×10^{17} W, which is the solar constant times the area of a disk having a diameter equal to that of earth (1.28×10^{7} m). Less than an hour of sunshine on earth provides sufficient energy to meet the 2011 annual global energy consumption. This energy is then distributed over the surface of the earth, which is spherical; therefore, the average solar radiation on the surface of the earth is

$$\text{Solar flux} = \frac{1.76 \times 10^{17}}{\pi (1.28 \times 10^{7})^2} = 342 \text{ W/m}^2 \tag{1.55}$$

For earth not to warm up, the same amount of energy, 341.3 W/m² on the average, must be radiated back to the outer space. What happens to the solar energy reaching the earth is shown in Figure 1.6, which is a 2009 update of the earth's global annual mean energy budget based on new observations and analyses, that have also found the surface flux to be 341.3 W/m² [6].

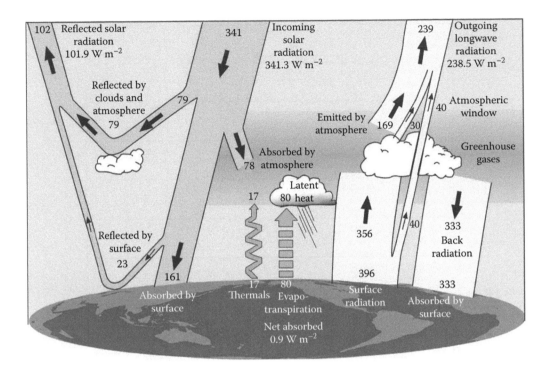

FIGURE 1.6
Earth's energy balance. (From Trenberth, K.E., Fasullo, J.T., and Kiehl, J., *Bulletin of the American Meteorological Society*, 90, 311, 2009.)

Of this amount, 79 W is reflected from the clouds, and another 78 W is absorbed by the atmosphere; therefore, about 184 W/m² reaches the ground. Of this amount, 23 is reflected. Multiplying this by the earth's area ($D = 1.28 \times 10^7$ m) shows that the earth receives 2.99×10^{24} J of energy annually. As indicated earlier, the total global energy consumption in 2009 was 5.17×10^{20} J, or 0.017% of the total solar energy received on the surface of the earth.

There is also around 333 W of back radiation at longer wavelengths from the atmosphere that can also be utilized as an energy source, for a total of 517 W/m² of direct and indirect thermal energy.

The solar energy, or in general radiation, travels in the form of electromagnetic waves (EM) over all wavelengths. Depending on the wavelength, the electromagnetic radiation has different names, as shown in Figure 1.7. For example, light is the EM radiation that

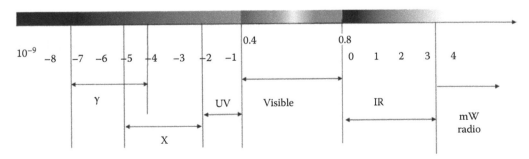

FIGURE 1.7
The electromagnetic spectrum.

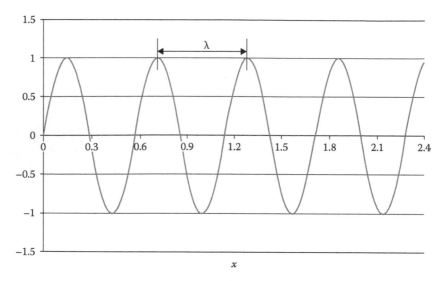

FIGURE 1.8
A continuous wave traveling in the x direction with the speed of light.

has wavelengths between 0.4 and about 0.8 μm, which are the EM waves that our eyes can detect, and send to the brain to process the visual stimuli, creating our perceptual experience of the world around us.

Electromagnetic waves travel with the speed of light, which is approximately $C = 3 \times 10^8$ m/s. Figure 1.8 is the plot of a wave having a particular wavelength traveling at the speed of light, which means that every point on the wave is traveling at the speed of light. The wavelength is defined as the distance between two corresponding points on two wave fronts. For example, if the plot corresponds to that for yellow light having a wavelength of approximately 0.570 μm, then the period (τ) which is defined as the time that it takes for one wave to travel the distance of one wavelength is

$$\tau = \frac{\lambda}{C} = \frac{0.57 \times 10^{-6}}{3 \times 10^8} = 1.9 \times 10^{-15} \text{ s} \tag{1.56}$$

and the frequency is

$$v = \frac{1}{\tau} = 5.26 \times 10^{14} \text{ s}^{-1} \text{ or Hz} \tag{1.57}$$

1.5.1 Blackbody

A convenient concept in radiation heat transfer is that of a blackbody. A blackbody emits radiation over all wavelengths in all directions. At a given temperature and wavelength, the radiation flux from a blackbody is given by

$$e_{\lambda,b}(\lambda) = \frac{3.742 \times 10^8}{\lambda^5 \left[e^{\frac{1.439 \times 10^4}{\lambda T}} - 1 \right]} \text{ W/m}^2 \tag{1.58}$$

This is also the maximum amount of radiation that can be emitted from any surface at a given T and λ.

The second characteristic of a blackbody is that it absorbs all radiant energy incident upon it. It represents an ideal case, not achieved physically. Although no surface behaves like a true blackbody, the concept provides a convenient tool for assessing the properties of real surfaces. Since by definition a blackbody absorbs all radiation incident upon it, it will reflect no radiation. Note that the objects that appear black to the eye absorb all the wavelengths in the visible range of 0.4–0.8 μm. There is much more to the spectrum than "meets the eyes" (the visible wavelengths). Snow, for example, is "black" at longer wavelengths of the infrared. Also, a blackbody that has a high temperature will emit enough in the visible range to appear as white.

The total energy flux emitted by a blackbody is obtained by integrating Equation 1.58 over all wavelengths, resulting in

$$\left(\frac{q}{A}\right)_b = e_b = \sigma T^4 \tag{1.59}$$

where $\sigma = 5.67 \times 10^{-8}$ W/m² K⁴ is the Stefan–Boltzmann constant. Equation 1.59 is the maximum amount of energy per unit area that can be emitted from a surface that is at temperature T and can only be emitted by a black surface. This energy is uniformly distributed over all directions. All real surfaces emit an amount of energy less than e_b. As mentioned earlier, sun emits 3.85×10^{26} W of energy. The sun's diameter is 1.39×10^9 m. Therefore, assuming sun to be a blackbody, its effective surface temperature can be calculated from

$$\frac{3.86 \times 10^{26}}{\pi (1.39 \times 10^9)^2} = 5.67 \times 10^{-8} T^4 \tag{1.60}$$

to be 5779 K.

1.5.2 Radiation Exchange between Gray Surfaces

A large class of surfaces can be classified as gray surfaces. A gray surface emits a constant fraction of blackbody radiation at each wavelength. Implicit in this definition is that properties of gray surfaces are independent of wavelength. The heat flux emitted by a gray surface can be expressed as

$$e = \varepsilon e_b = \varepsilon \sigma T^4 \tag{1.61}$$

Calculating radiation exchange between two surfaces is somewhat more complicated. Radiation depends not only on the surface properties but also on the relative orientation of the surfaces. Consider the relatively simple case of radiation exchange between two surfaces shown in Figure 1.9. The problem can be more easily solved by resorting to electrical

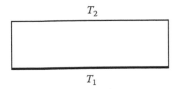

FIGURE 1.9
An enclosure made from two gray surfaces.

network analogy, although the electrical resistance network for this case is more complicated than those for conduction or convection. For radiation exchange between gray surfaces, we need to define two types of resistances to come up with the network. Associated with each surface, there is a surface resistance, which is

$$R_{su} = \frac{1-\varepsilon}{A\varepsilon} \tag{1.62}$$

Therefore, there will be two surface resistances since we have two surfaces. For each pair of surfaces that can exchange energy radiatively with one another, there is also a space resistance

$$R_{sp} = \frac{1}{A_1 F_{12}} \tag{1.63}$$

where F is the shape factor or "view factor." F_{12} is defined as the fraction of radiative energy that leaves surface 1 and reaches surface 2. The shape factor is a function of the geometry or the relative positions. It can be shown that

$$A_i F_{ij} = A_j F_{ji} \tag{1.64}$$

which is known as reciprocity relation. The equivalent electrical network is shown in Figure 1.10 and from the electrical network for this case

$$q_1 = -q_2 = \frac{\sigma\left(T_1^4 - T_2^4\right)}{\frac{1-\varepsilon_1}{A_1\varepsilon_1} + \frac{1}{A_1 F_{12}} + \frac{1-\varepsilon_2}{A_2\varepsilon_2}} \tag{1.65}$$

Following the same approach as that for conduction and convection,

$$q = \frac{T_1 - T_2}{R_{rad}} \tag{1.66}$$

$$R_{rad} = \frac{\dfrac{1-\varepsilon_1}{A_1\varepsilon_1} + \dfrac{1}{A_1 F_{12}} + \dfrac{1-\varepsilon_2}{A_2\varepsilon_2}}{\sigma(T_1 + T_2)\left(T_1^2 + T_2^2\right)} \tag{1.67}$$

If the surfaces are black, then

$$R_{rad} = \frac{1}{\sigma A_1 F_{12}(T_1 + T_2)\left(T_1^2 + T_2^2\right)} \tag{1.68}$$

R_{rad} has limited use since it is highly temperature dependent.

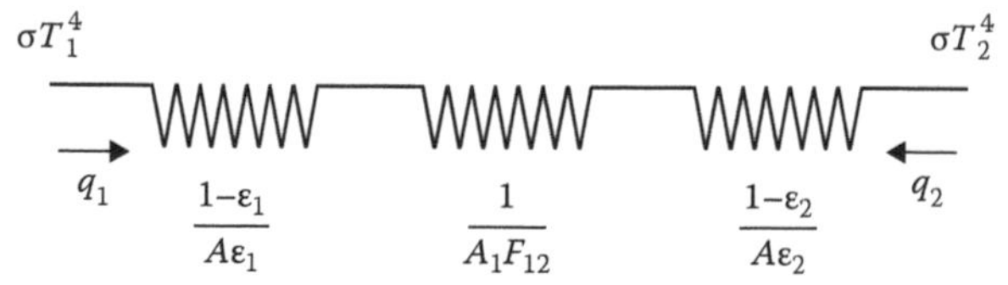

FIGURE 1.10
Electrical network for a two-surface enclosure.

For the case shown in Figure 1.9, surface 1 does not see itself, meaning that none of the energy leaving surface 1 is incident on surface 1; therefore, $F_{12} = 1$:

$$q_1 = \frac{\sigma\left(T_1^4 - T_2^4\right)}{\dfrac{1}{A_1\varepsilon_1} + \dfrac{1}{A_2\varepsilon_2} - \dfrac{1}{A_2}} \tag{1.69}$$

In addition, if surface 2 is either infinitely large or is black, then

$$q_1 = A_1\varepsilon_1\sigma\left(T_1^4 - T_2^4\right) \tag{1.70}$$

Example 1.3

Determine the highest heat transfer coefficient at which dew will form on the surface of a grass blade. Assume the emissivity of grass to be 0.6, the ambient temperature to be 20°C, the relative humidity to be 50%, and the sky temperature to be absolute zero.

Neglecting conduction from the ground and assuming the grass to be at a uniform temperature, then the energy loss to the night sky by radiation must be equal to the energy gain from the ambient air (Figure 1.11):

$$q_{rad} = q_{con}$$

Assuming both sides of grass exchange radiation with sky, and sky having a much larger area than the grass, then the heat transfer by radiation is given by Equation 1.70:

$$A\varepsilon_g\sigma\left(T_g^4 - T_s^4\right) = Ah(T_\infty - T_g)$$

which can be solved for the heat transfer coefficient

$$h = \varepsilon_g\sigma\frac{T_g^4 - T_s^4}{T_\infty - T_g}$$

Note that the calmer the air (the lower heat transfer coefficient), the less heat will be transferred from the ambient to the grass, and therefore the higher the ambient temperature at which dew will form. Also, higher relative humidity has the same effect on the dew formation.

FIGURE 1.11
Energy balance for grass blade.

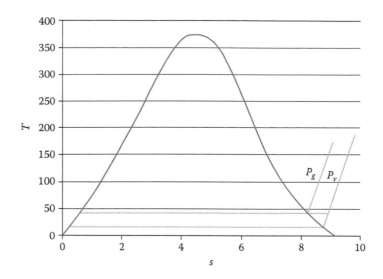

FIGURE 1.12
Vapor dome.

For the dew to form, the grass temperature, T_g, must be equal to the saturation temperature of water. The relative humidity is defined as

$$\phi = \frac{P_v}{P_g}$$

where
P_v is the partial pressure of water
P_g is the saturation pressure at ambient temperature, as shown in Figure 1.12

From steam tables, at 20°C, $P_g = 2.339$ kPa, therefore,

$$P_v = 0.5 \times 2.339 = 1.1695$$

At this pressure, the saturation temperature is 9.23°C, or the grass temperature must drop to this value. Therefore, the heat transfer coefficient must be

$$h = 0.6 \times 5.67 \times 10^{-8} \frac{(273.15 + 9.23)^4 - 0^4}{20 - 9.23} = 20.1 \text{ W/m}^2\text{C}$$

1.6 General Heat Diffusion Equation

Conduction is the mechanism by which heat is transferred within and among solids and fluids that are at rest. In Section 1.3.1, we used the first law of thermodynamics and Taylor series to derive an equation that described temperature distribution when heat is conducted in one direction and the properties are independent of time. We are now going to consider the general case in which heat is transferred in three dimensions, properties are a

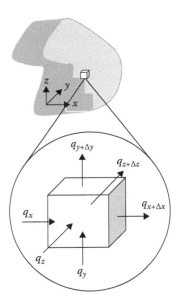

FIGURE 1.13
Energy balance on a differential control mass in Cartesian coordinates.

function of time, and energy can be generated within the medium. This is the most general form of the first law of thermodynamics for energy transfer by conduction; however, the process used is the same as that for the case of 1D conduction.

Consider an object of arbitrary shape as shown in Figure 1.13. Inside the object we consider an elemental control mass with dimensions Δx, Δy, and Δz. The control mass is selected by starting in a given position (x, y, z) in the medium and incrementing in the three directions. The control mass is magnified to show the details of the derivation.

After the system is selected, we apply the relevant physical law, which in this case is the first law of thermodynamics to the elemental system. Note that since the control mass is a solid, the only mechanism by which heat can be transferred to and from the system is conduction. Heat is conducted in and out of the control mass in x, y, and z directions, and there is no work, reducing Equation 1.13 to

$$q_x + q_y + q_z + \dot{q}(\Delta x\,\Delta y\,\Delta z) = \frac{\partial E}{\partial t} + q_{x+\Delta x} + q_{y+\Delta y} + q_{z+\Delta z} \tag{1.71}$$

This is one equation with seven unknowns. We can eliminate three unknowns by expanding the q_{x+dx}, q_{y+dy}, and q_{z+dz} terms using the Taylor series:

$$q_{x+\Delta x} = q_x + \frac{\partial q_x}{\partial x}\Delta x + \frac{\partial^2 q_x}{\partial x^2}\frac{\Delta x^2}{2!} + \text{H.O.T.} \tag{1.72}$$

$$q_{y+\Delta y} = q_y + \frac{\partial q_y}{\partial y}\Delta y + \frac{\partial^2 q_y}{\partial y^2}\frac{\Delta y^2}{2!} + \text{H.O.T.} \tag{1.73}$$

$$q_{z+\Delta z} = q_z + \frac{\partial q_z}{\partial z}\Delta z + \frac{\partial^2 q_z}{\partial z^2}\frac{\Delta z^2}{2!} + \text{H.O.T.s} \tag{1.74}$$

The change in internal energy is related to temperature and is given by

$$\frac{\partial E}{\partial t} = \frac{\partial (me)}{\partial t} = \frac{\partial \left[\rho(\Delta x \, \Delta y \, \Delta z)cT \right]}{\partial t} \tag{1.75}$$

where
    e is the internal energy per unit mass
    ρ is the density
    c is the specific heat
    T is the temperature

Substituting Equations 1.72 through 1.75 into Equation 1.71 results in

$$\dot{q}(\Delta x \, \Delta y \, \Delta z) = (\Delta x \, \Delta y \, \Delta z)\frac{\partial [\rho cT]}{\partial t} + \frac{\partial q_x}{\partial x}\Delta x + \frac{\partial q_y}{\partial y}\Delta y + \frac{\partial q_z}{\partial z}\Delta z + \text{H.O.T.} \tag{1.76}$$

This is now one equation with four unknowns. The next step is to relate the rates of heat transfer to temperature by Fourier's law of heat conduction:

$$q_x = -kA_x\frac{\partial T}{\partial x} = -k\Delta y \, \Delta z \frac{\partial T}{\partial x} \tag{1.77}$$

$$q_y = -kA_y\frac{\partial T}{\partial y} = -k\Delta x \, \Delta z \frac{\partial T}{\partial y} \tag{1.78}$$

$$q_z = -kA_z\frac{\partial T}{\partial z} = -k\Delta x \, \Delta y \frac{\partial T}{\partial z} \tag{1.79}$$

Then,

$$\dot{q}(\Delta x \, \Delta y \, \Delta z) = (\Delta x \, \Delta y \, \Delta z)\frac{\partial [\rho cT]}{\partial t} + \frac{\partial}{\partial x}\left[-k\Delta y \, \Delta z \frac{\partial T}{\partial x} \right]\Delta x + \frac{\partial}{\partial y}\left[-k\Delta x \, \Delta z \frac{\partial T}{\partial y} \right]\Delta y$$
$$+ \frac{\partial}{\partial z}\left[-k\Delta x \, \Delta y \frac{\partial T}{\partial z} \right]\Delta z + \text{H.O.T.} \tag{1.80}$$

Equation 1.80 has only one unknown, the temperature. This equation still is not in its final form since it has infinitely many terms and is valid over a region having dimensions Δx, Δy, and Δz. Since the size of the control volume is arbitrary, we can look at the limit of the size going to zero. This allows the elimination of all the higher-order terms and leads to an equation that is valid at every point inside the medium.

Dividing both sides of Equation 1.80 by $\Delta V = \Delta x \, \Delta y \, \Delta z$ results in

$$\dot{q} = \rho c \frac{\partial T}{\partial t} + \frac{\partial}{\partial x}\left[-k\frac{\partial T}{\partial x}\right] + \frac{\partial}{\partial y}\left[-k\frac{\partial T}{\partial y}\right] + \frac{\partial}{\partial z}\left[-k\frac{\partial T}{\partial z}\right] + \text{H.O.T.} \tag{1.81}$$

If Δx, Δy, and Δz (or ΔV) approach zero, then all higher order terms will go to zero, since they all have at least one of the increments, yielding

$$\frac{\partial}{\partial x}\left(k\frac{\partial T}{\partial x}\right) + \frac{\partial}{\partial y}\left(k\frac{\partial T}{\partial y}\right) + \frac{\partial}{\partial z}\left(k\frac{\partial T}{\partial z}\right) + \dot{q} = \frac{\partial(\rho c T)}{\partial t} \tag{1.82}$$

This is the general heat diffusion equation in Cartesian coordinates. If properties are constant, then

$$\frac{\partial^2 T}{\partial x^2} + \frac{\partial^2 T}{\partial y^2} + \frac{\partial^2 T}{\partial z^2} + \frac{\dot{q}}{k} = \frac{\rho c}{k}\frac{\partial T}{\partial t} \tag{1.83}$$

The fraction $\dfrac{\rho c}{k}$ on the RHS of Equation 1.83 is a combination of material properties and, therefore, is a property itself. Inverse of this ratio is called thermal diffusivity:

$$\alpha = \frac{k}{\rho c} \tag{1.84}$$

Thermal diffusivity, α, has the dimensions of m^2/s and is a measure of how quickly heat flows or diffuses through the material. Since heat transfer is proportional to temperature difference, α is also a measure of how quickly a change in temperature at a point in the material propagates throughout the material. The higher the thermal diffusivity of a substance, the faster a temperature change propagates. The heat diffusion equation in Cartesian coordinates with constant k is

$$\frac{\partial^2 T}{\partial x^2} + \frac{\partial^2 T}{\partial y^2} + \frac{\partial^2 T}{\partial z^2} + \frac{\dot{q}}{k} = \frac{1}{\alpha}\frac{\partial T}{\partial t} \tag{1.85}$$

In vector notation,

$$\nabla^2 T + \frac{\dot{q}}{k} = \frac{1}{\alpha}\frac{\partial T}{\partial t} \tag{1.86}$$

where

$$\nabla^2 = \frac{\partial^2}{\partial x^2} + \frac{\partial^2}{\partial y^2} + \frac{\partial^2}{\partial z^2} \tag{1.87}$$

is the Laplacian operator and is a short-hand notation for the sum of three second-order derivatives.

1.7 Coordinate Systems

In general, as shown in Figure 1.14, a point P in space is specified by the vector $\overrightarrow{OP}$ that extends from the origin to the point. In Cartesian coordinate system, the position of the point is specified by the x, y, and z components of vector OP. The vectors $\hat{x}$, $\hat{y}$, $\hat{z}$ are the unit vectors in x, y, and z directions. In the cylindrical coordinate system, the location of a point in space is defined by the point's distance (r) from a reference axis, (z), the point's distance (z) from a plane perpendicular to the reference axis (x–y plane), and the azimuth angle ϕ, which is the rotation around the z-axis, or is the angle between the projection of the vector OP on the reference plane and the x axis. Sometimes the symbol θ is used instead of ϕ for the azimuth angle. The differential control mass is shown in Figure 1.15.

The volume of the differential element is the product of the length of its three sides:

$$dV = rd\phi drdz \tag{1.88}$$

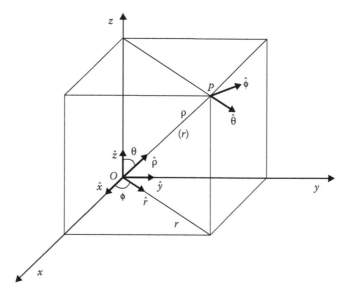

FIGURE 1.14
Specification of a point in different coordinate systems.

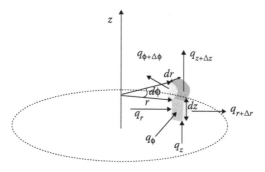

FIGURE 1.15
Energy balance on a differential control mass in Cylindrical coordinates.

Using a similar approach, the energy equation in cylindrical and spherical coordinate systems can be obtained. By doing an energy balance on the differential control mass shown in Figure 1.15, the heat conduction equation in cylindrical coordinates becomes

$$\frac{1}{r}\frac{\partial}{\partial r}\left(kr\frac{\partial T}{\partial r}\right)+\frac{1}{r^2}\frac{\partial}{\partial \phi}\left(k\frac{\partial T}{\partial \phi}\right)+\frac{\partial}{\partial z}\left(k\frac{\partial T}{\partial z}\right)+\dot{q}=\rho c\frac{\partial T}{\partial t} \tag{1.89}$$

In the spherical coordinate system shown in Figure 1.16, the position of a point is specified by the radial distance of that point from the origin (r), its inclination or zenith angle (θ), which is the angle measured from the z-axis, and the azimuth angle (ϕ) which is the rotation around the z-axis. The differential control mass is shown in Figure 1.16, and its volume is the product of the length of its three sides:

$$dV = r\sin\theta\,d\phi\,rd\theta\,dr = r^2\sin\theta\,d\theta\,d\phi\,dr \tag{1.90}$$

By doing an energy balance on the differential control mass shown in Figure 1.16, the heat conduction equation in spherical coordinates becomes

$$\frac{1}{r^2}\frac{\partial}{\partial r}\left(kr^2\frac{\partial T}{\partial r}\right)+\frac{1}{r^2\sin^2\theta}\frac{\partial}{\partial \phi}\left(k\frac{\partial T}{\partial \phi}\right)+\frac{1}{r^2\sin\theta}\frac{\partial}{\partial \theta}\left(k\sin\theta\frac{\partial T}{\partial \theta}\right)+\dot{q}=\rho c\frac{\partial T}{\partial t} \tag{1.91}$$

Again, the notations used for θ and ϕ are sometimes interchanged.

Although the different forms of the energy equation are partial differential equations (PDEs) and may appear intimidating, they just represent the first law of thermodynamics and are differential forms of Equation 1.7. For example, in Equation 1.83, the first three terms represent the net rate of heat transfer by conduction in the x, y, and z directions per unit volume. The fourth term is the rate at which energy is generated per unit volume, and the term on the RHS is the rate at which energy is stored per unit volume.

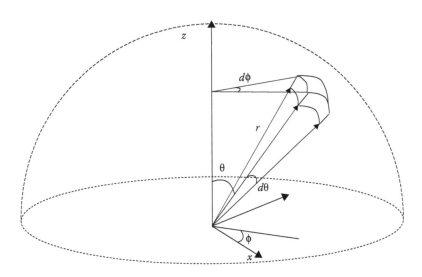

FIGURE 1.16
Energy balance on a differential control mass in Spherical coordinates.

Considering the heat diffusion equation in the three coordinate systems, we can see that for one-dimensional conduction the energy equation can be written in the general form

$$\frac{1}{r^n}\frac{\partial}{\partial r}\left(r^n\frac{\partial T}{\partial r}\right)+\frac{\dot{q}}{k}=\frac{\rho c}{k}\frac{\partial T}{\partial t} \tag{1.92}$$

where $n = 0, 1, 2$ for Cartesian, cylindrical, and spherical coordinates, respectively.

1.8 Initial and Boundary Conditions

A differential equation is valid over certain values of each of its independent variables; these values comprise the domain, and the limits of the domain are called boundaries. As such, the solution of a differential equation is obtained over the domain defined by the boundaries. The boundary conditions are the limitations on the dependent variable or its derivatives on the boundaries of the domain. The number of boundary conditions needed on each independent variable is equal to the order of the highest derivative in the equation with respect to that variable. For example, the temperature distribution for unsteady one-dimensional conduction in a wall of thickness $2L$ is given by the partial differential equation

$$\frac{\partial T}{\partial t}=\alpha\frac{\partial^2 T}{\partial x^2} \tag{1.93}$$

The independent variables are x and t. The equation is valid from $x = -L$ to $x = L$ and for $t > 0$. To solve this PDE, two boundary conditions are needed in the x-direction and one boundary condition (initial condition) in time. Therefore, we need one condition for T or one of its derivatives at $x = -L$, one at $x = L$, and one for $t = 0$. The boundary conditions for the energy equation are essentially determined through one of two methods:

1. Specified surface temperature: The simplest boundary condition to use is the constant surface temperature, or isothermal surface, where the temperature at the boundary is maintained at a known value. This requires transferring energy into or out of the surface, from or to an external source, in order to maintain the temperature at the specified value. If steam were condensing on the outer surface of a thin pipe, the temperature of the pipe would be constant and equal to the saturation temperature of the steam.

2. Surface energy balance: If the surface temperature is not explicitly specified, then the boundary condition is obtained by applying the law of conservation of energy to the surface defining the boundary. As indicated earlier, the first law of thermodynamics for a control mass is expressed as

$$q_{in}-q_{out}+q_g+\dot{W}=mc\frac{dT}{dt} \tag{1.94}$$

A surface does not have any volume (mass); therefore, it cannot store any energy and the storage term or the RHS of the equation goes to zero. Therefore,

$$q_{in}-q_{out}+q_g+\dot{W}=0 \tag{1.95}$$

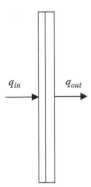

FIGURE 1.17
Surface energy balance.

In most cases, no energy is generated on the surface, exceptions include solid fuel combustion and catalytic surface reactions. The work done on a surface is also generally zero, with the work performed by friction being one exception. For most cases, therefore, the mathematical representation of the first law at a surface reduces to

$$q_{in} = q_{out} \tag{1.96}$$

To find the boundary condition, we must find all the ways by which heat is transferred to and from the surface and set the total heat transferred to the surface equal to the total heat transferred from the surface, as shown in Figure 1.17.

It is important to note that Equation 1.96 establishes a direction for the flow of heat. Energy that flows toward the surface is energy into the surface, and the energy that flows away from the surface is energy leaving the system. This is in addition to the conventions that we have established for the transfer of heat by conduction and convection. The Fourier law of heat conduction

$$q_x = -kA_x \frac{\partial T}{\partial x} \tag{1.97}$$

represents the heat that is conducted in the positive x-direction. Depending on the orientation of the surface, the heat conducted could be q_{in} or q_{out}. Similarly, the heat transfer by convective is given by

$$q = hA(T - T_\infty) \tag{1.98}$$

and describes the heat that is convected (transferred) from the surface to the fluid; in other words, Equation 1.98 is the energy leaving the surface. On the other hand, if the convective heat transfer is written as

$$q = hA(T_\infty - T) \tag{1.99}$$

it would describe the heat that is transferred from the fluid to the surface, representing the energy entering the surface. The interpretation of energy into and out of the system is independent of whether the surface temperature is more or less than the ambient temperature.

Although the law of conservation of energy requires the total energy in to be equal to the total energy out, since both sides of a surface have the same area, we can divide both sides of the surface energy balance equation by the area and do a surface flux balance, that is, the total energy flux in equals total energy flux out. Some of the more commonly used boundary conditions are described in the following sections.

1.8.1 Surface Heat Flux

This boundary condition is used when the surface receives heat from another source and all of the heat is absorbed by the surface. Consider the wall shown in Figure 1.18. Using the coordinate system shown, we perform an energy balance on surface 1. q'' is the energy that arrives at surface 1 (q_{in}). Excluding convection and radiation at the left side of surface 1, the only other way by which energy can be transferred to surface 1 is by conduction from its RHS. The amount of energy conducted is given by $q = -kA_x\left(\dfrac{dT}{dx}\right)$, and because of the convention we adopted, it is energy that is moving in the positive x-direction, representing the energy that is moving out of surface 1 (q_{out}). The surface energy balance requires that the total energy in equal to the total energy out (net energy transfer of zero)

$$q''A_x = -kA_x\frac{dT}{dx} \tag{1.100}$$

$$q'' = -k\frac{dT}{dx} \tag{1.101}$$

Since $q'' > 0$, then $\dfrac{dT}{dx}$ must be negative at $x = 0$; or the temperature is high at the left wall and decreases with increasing x near the left wall.

For surface 2, the heat flux again is energy that is coming in. The heat transfer by conduction is also pointing in the positive x direction and therefore they both represent energy input to surface 2:

$$-k\frac{dT}{dx} + q'' = 0. \tag{1.102}$$

This may appear unsettling, in that energy cannot just go to a surface and not leave, since surfaces cannot store energy, or if energy enters a surface, it must also leave the surface. A closer examination of Equation 1.102 shows that since q'' is positive, $\dfrac{dT}{dx}$ must also be positive at $x = L$ to ensure the first term in Equation 1.102 is negative, requiring the temperature

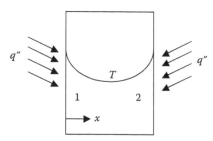

FIGURE 1.18
Surface energy balance for uniform heat flux.

to increase in the positive x-direction close to the right wall. From the aforementioned discussion, as one moves toward the center from both surfaces, the temperature decreases, or one expects the temperature to be minimum at the center of the wall.

1.8.2 Insulated or Adiabatic Boundary Condition

A special case is an adiabatic (insulated) wall, in which the heat transfer out of the surface is zero. Physically this boundary condition approximates the cases where insulation is placed on the surface or at sharp corners where there is no area for heat transfer. The mathematical representation of this boundary condition is

$$-kA_x \frac{dT}{dx} = 0 \quad \text{or} \quad \frac{dT}{dx} = 0 \tag{1.103}$$

which is obtained by setting $q'' = 0$ in Equation 1.102.

1.8.3 Convection Surface Condition

This boundary condition is encountered when the surface exchanges energy with a fluid by convection. By performing an energy balance on the surface at $y = 0$ in Figure 1.19, the amount of heat transfer from the surface by convection is

$$q'' = h(T - T_\infty) \tag{1.104}$$

This is the heat that is transferred from the surface, which is at T, to the environment, which is at T_∞. The conduction heat flux is

$$q'' = -k \frac{dT}{dy} \tag{1.105}$$

which, because of the adopted sign, is the energy that is transferred in the y-direction, that is, energy transferred away from the surface located at $y = 0$, making it heat transfer out. Therefore,

$$0 = h[T(x,0) - T_\infty] - k \frac{dT}{dy}\bigg|_{y=0} \tag{1.106}$$

It should be clear that T and dT/dy are evaluated at $y = 0$.

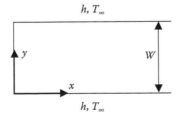

FIGURE 1.19
Surface energy balance for convection boundary condition.

The boundary condition at $y = W$ is obtained by doing an energy balance on that surface. At the top surface, conduction heat transfer is an energy "in" and convection is an energy "out." The energy balance can therefore become

$$-k\frac{dT}{dy}\bigg|_{y=L} = h[T(x,W) - T_\infty] \tag{1.107}$$

Alternatively, at the upper surface, the convective heat flux from the fluid to the wall is

$$q''_{conv} = h(T_\infty - T) \tag{1.108}$$

making the convection heat transfer also an energy input, and the surface energy balance becomes

$$q''_{in} = h(T_\infty - T) - k\frac{dT}{dy} = q''_{out} = 0 \tag{1.109}$$

which is the same as Equation 1.107. This shows that the mathematical form of the boundary conditions will be the same regardless of the relative magnitudes of the wall and ambient temperatures as long as a convention for energy in and out is chosen and consistently applied to the first law.

Equation 1.107 describes the balance between the energy conducted into the surface and the energy convected out of the surface. The amount of heat conducted into the surface is finite, so as h gets larger, the difference between the surface and ambient temperature decreases. With h approaching infinity, the surface temperature approaches the ambient temperature. This shows that the constant temperature boundary condition is a special case of the convective boundary, when the heat transfer coefficient approaches infinity. In practice, one way of achieving an isothermal condition is to have the solid in contact with a fluid undergoing a phase change (condensation or evaporation).

1.8.4 Radiative Boundary Condition

We considered the case where the boundary exchanges radiation only with one other surface, so that the amount of heat transfer is given by Equation 1.65. The situation where radiation is with more than one surface is discussed in Chapter 12.

Example 1.4

An electric strip heater, shown in Figure 1.20, loses heat by convection and radiation from its outer surfaces. Write the boundary conditions.

Assuming that the radiation exchange is with the surrounding surfaces that are also at T_∞, and are also much larger than the boundary, then the energy balance on surface 1 becomes

$$-k\frac{dT}{dx}\bigg|_1 + \varepsilon\sigma\left(T_1^4 - T_\infty^4\right) + h(T_1 - T_\infty) = 0$$

with all three forms being heat transfer from the wall. The energy balance on surface 2 is

$$-k\frac{dT}{dx}\bigg|_2 = \varepsilon_2\sigma\left(T_2^4 - T_\infty^4\right) + h(T_2 - T_\infty)$$

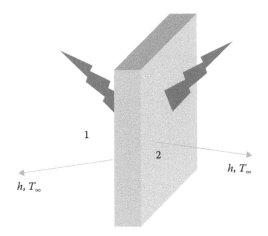

FIGURE 1.20
Radiation and convection boundary conditions.

1.8.5 Surface Energy Generation

As mentioned earlier, in most applications, the energy generation on the surface is zero. Some of the exceptions include surface chemical reactions (solid fuel combustion, heterogeneous catalytic combustion), evaporation and condensation, and frictional heating. In case of chemical reactions, in addition to the energy equation, one must also solve equations for the conservation of the different species (molecules) that make up the mixture, which are beyond the present discussion and will be dealt with in more detail in Chapter 15.

Example 1.5

Consider two disks that are in contact, with the top one spinning at a constant angular velocity as shown in Figure 1.21. Write the boundary conditions at the interface.

At the interface, the two solids must be at the same temperature; therefore,

$$T_1 = T_2$$

also the surface energy balance yields

$$q_{in} - q_{out} + \dot{W} = 0.$$

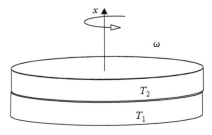

FIGURE 1.21
Boundary condition for surface energy generation.

applying this equation to a differential area (a disk of thickness dr)

$$-k_1 \frac{dT_1}{dx}(2\pi r dr) - \left(-k_2 \frac{dT_2}{dx}\right)(2\pi r dr) + F.V = 0.$$

$$V = r\omega$$

$$-k_1 \frac{dT_1}{dx}(2\pi r dr) - \left(-k_2 \frac{dT_2}{dx}\right)(2\pi r dr) + Fr\omega = 0.$$

The tangential component of force is equal to the normal force applied multiplied by the friction coefficient, so

$$-k_1 \frac{dT_1}{dx}(2\pi r dr) - \left(-k_2 \frac{dT_2}{dx}\right)(2\pi r dr) + \mu F_N r\omega = 0.$$

$$-k_1 \frac{dT_1}{dx} + k_2 \frac{dT_2}{dx} + \mu \frac{F_N}{2\pi r dr} r\omega = 0.$$

$$-k_1 \frac{dT_1}{dx} + k_2 \frac{dT_2}{dx} + \mu P(r) r\omega = 0.$$

To use these equations, one must know the distribution of the normal pressure between the two disks with respect to r. Generally, pressure is assumed to be constant and equal to the applied force divided by the area of the disk.

1.9 One-Dimensional Conduction without Energy Generation

Heat transfer by conduction between two points is proportional to the temperature difference and inversely proportional to the distance between the points. Therefore, if one dimension of the object is much larger than its other dimensions, the rate of heat transfer in that direction is generally much lower than that in the other directions and can be neglected. In one-dimensional conduction, one of the object's dimensions is typically much smaller than the other two, and temperature variation and therefore heat transfer are important only in that one direction.

1.9.1 Cartesian Coordinates

This case was considered in Section 1.3.1. Here, we start with the general heat diffusion equation (Equation 1.85), which for steady one-dimensional conduction without energy generation in Cartesian coordinates reduces to

$$\frac{d^2 T}{dx^2} = 0 \qquad\qquad (1.110)$$

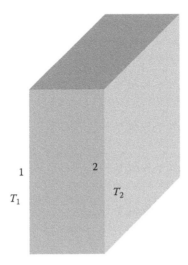

FIGURE 1.22
Steady 1D conduction in a plane wall.

which is the same as Equation 1.31 obtained by doing an energy balance on an element of volume having a thickness of Δx. The solution needs two boundary conditions. As mentioned earlier, boundary conditions are conditions that are imposed on the dependent variable (T) at the boundaries (the specified values of the independent variable, x). For a wall with length L, subject to isothermal conditions, Figure 1.22, the solution is given by Equation 1.37

$$T = \frac{T_2 - T_1}{L}x + T_1 \tag{1.111}$$

And the heat transfer rate is given by Equation 1.41

$$q_x = \frac{T_1 - T_2}{\dfrac{L}{kA_x}} \tag{1.112}$$

Equation 1.111 shows that temperature changes linearly from T_1 on the left face to T_2 on the right face. Rearranging the equation yields

$$\frac{T - T_1}{T_2 - T_1} = \frac{x}{L} \tag{1.113}$$

This equation is dimensionless. The LHS is a nondimensional variable that is the ratio of temperatures and the RHS is a dimensionless variable that is the ratio of lengths. Working with nondimensional variables and equations is an important topic in heat transfer that will be discussed later and utilized throughout this book.

Example 1.6

A kiln is a chamber, oven, used to dry materials like lumber, or harden clay or ceramics. A kiln used for the firing of pottery operates at a temperature of 950 K, and is made of fire brick 10 cm thickness, having thermal conductivity of $k = 1.2$ W/m K, the inside heat transfer coefficient being $h_i = 25$ W/m² K. The outer surface has an emissivity of $\varepsilon = 0.7$, and exposed to air at 30°C and $h_2 = 5$ W/m² K. Determine the inside and outside surface temperatures of the oven.

Solution

The kiln and the electrical circuit for the problem are shown in Figure 1.23. Heat is convected from the hot air in the oven to the inner surface of the oven. It is then conducted through the wall. At the outer surface, a portion of the heat is convected to the outside air and the rest is radiated to the surroundings, which are assumed to be at T_3. The radiation exchange is between two surfaces, one of which, the outer surface of the oven, does not see itself. Summing the currents at surface 2,

$$\frac{T_i - T_2}{\dfrac{1}{h_i A_i} + \dfrac{L}{k A_i}} = \frac{T_2 - T_\infty}{\dfrac{1}{h_2 A_2}} + \frac{\sigma T_2^4 - \sigma T_3^4}{\dfrac{1-\varepsilon_2}{\varepsilon_2 A_2} + \dfrac{1}{A_2 F_{2-3}} + \dfrac{1-\varepsilon_3}{\varepsilon_3 A_3}}$$

The ambient is taken to be a black surface ($\varepsilon_3 = 1$) and $F_{2-3} = 1$. Assuming the inside and outside areas of the kiln to be the same, then

$$\frac{T_i - T_2}{\dfrac{1}{h_i} + \dfrac{L}{k}} = \frac{T_2 - T_\infty}{\dfrac{1}{h_2}} + \frac{\sigma T_2^4 - \sigma T_3^4}{\dfrac{1}{\varepsilon_2}}$$

This is one equation for one unknown T_2, which can be solved iteratively:

$$T_2 = \frac{\dfrac{T_1}{\dfrac{1}{h_i} + \dfrac{L}{k}} + \dfrac{T_\infty}{\dfrac{1}{h_2}} + \dfrac{\sigma T_3^4}{\dfrac{1}{\varepsilon_2}}}{\dfrac{1}{\dfrac{1}{h_i} + \dfrac{L}{k}} + \dfrac{1}{\dfrac{1}{h_2}} + \dfrac{\sigma T_2^3}{\dfrac{1}{\varepsilon_2}}} = 515.3 \text{ K}$$

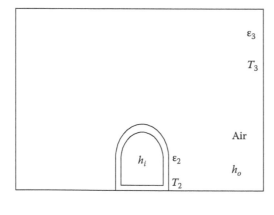
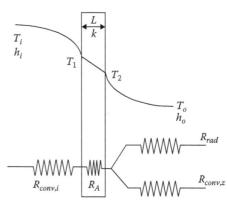

FIGURE 1.23
Kiln and the equivalent electrical circuit.

Knowing T_2, heat transfer can be calculated from

$$q'' = \frac{T_i - T_2}{\dfrac{1}{h_i} + \dfrac{L}{k}} = 3524.5 \text{ W/m}^2$$

which can then be used to calculate the inner surface temperature of the kiln (Figure 1.23):

$$T_1 = T_i - \frac{q''}{h_i} = 809.0 \text{ K}$$

1.9.2 Steady One-Dimensional Cylindrical Conduction

Consider the flow of a hot fluid in a pipe whose outside surface is exposed to the ambient. The heat transfer through the pipe wall can be approximated by conduction in a hollow cylinder whose inner and outer walls are exposed to different thermal conditions. In general, temperature will be a function of r, θ, and z. If the conditions are uniform in the θ and z directions, then the heat transfer and therefore the temperature variation in those two directions are zero and conduction is only in the radial direction (Figure 1.24).

The general heat diffusion equation in cylindrical coordinates, or Equation 1.89, for steady one-dimensional case with no energy generation simplifies to

$$\frac{\partial}{\partial r}\left(r \frac{\partial T}{\partial r} \right) = 0 \tag{1.114}$$

Integrating Equation 1.114 results in

$$r \frac{dT}{dr} = C_1 \tag{1.115}$$

Dividing by r to separate the variables and integrating again yields

$$T = C_1 \ln r + C_2 \tag{1.116}$$

where C_1 and C_2 are to be determined from the boundary conditions. For isothermal boundary conditions, where inside and outside temperatures are maintained at a constant value,

$$r = r_i \quad T = T_i \quad T_i = C_1 \ln(r_i) + C_2 \tag{1.117}$$

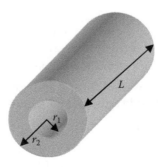

FIGURE 1.24
Steady one-dimensional conduction in a cylinder.

$$r = r_o \quad T = T_o \quad T_o = C_1 \ln(r_o) + C_2 \tag{1.118}$$

Solving for C_1 and C_2 and substituting in Equation 1.116 results in

$$T = \frac{T_i - T_o}{\ln \frac{r_i}{r_o}} \ln r + T_i - \frac{T_i - T_o}{\ln \frac{r_i}{r_o}} \ln r_i \tag{1.119}$$

which provides the temperature distribution in the pipe wall. This equation can be rearranged into

$$\frac{T - T_o}{T_i - T_o} = \frac{\ln \frac{r}{r_o}}{\ln \frac{r_i}{r_o}} \tag{1.120}$$

The LHS is a nondimensional temperature and the RHS is also nondimensional. If we define the nondimensional temperature and nondimensional radial distance as

$$T^* = \frac{T - T_o}{T_i - T_o} \tag{1.121}$$

and

$$r^* = \frac{r}{r_o} \tag{1.122}$$

Then, Equation 1.120 becomes

$$T^* = \frac{\ln r^*}{\ln r_i^*} \tag{1.123}$$

where

$$r_i^* = \frac{r_i}{r_o} \tag{1.124}$$

The temperature, T, varies between T_i and T_o, and r changes between r_i and r_o. However, working with the nondimensional form of the variables, the nondimensional temperature varies between zero and one and the nondimensional distance varies between $r_i^* = \frac{r_i}{r_o}$ and one. The solution in the nondimensional form is independent of the actual values of the parameters such as inside and outside radii and temperatures. It only depends on the ratio of inside and outside radii $\left(r_i^* = \frac{r_i}{r_o} \right)$.

Figure 1.25 is a plot of nondimensional temperature variation as a function of the nondimensional radial distance, for different values of the ratio of the inside to outside radii (r_i^*). Figure 1.25 shows the nondimensional solution to all the steady one-dimensional

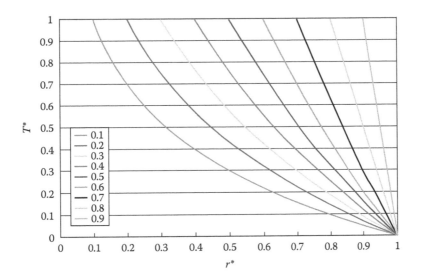

FIGURE 1.25
The radial variation of nondimensional temperature for different values of r_i^*.

conduction problems in cylindrical coordinates and is independent of the imposed temperatures or the material thermal conductivity. It also shows the power of nondimensionalization in allowing the determination of the general solutions. It can also be seen that as the r_i^* increases, which means that the cylinder's thickness decreases, the temperature distribution becomes very close to being linear or the solution approaches that of a plane wall.

Returning to the temperature distribution, our goal is to find the heat transfer. From Fourier's law,

$$q_n = -kA_n \frac{dT}{dn} \tag{1.125}$$

$$A_r = 2\pi r L \tag{1.126}$$

$$q_r = -kA_r \frac{dT}{dr} \tag{1.127}$$

$$\frac{dT}{dr} = \frac{T_i - T_o}{\ln \frac{r_i}{r_o}} \frac{1}{r} \tag{1.128}$$

$$q_r = -k2\pi r L \frac{T_i - T_o}{\ln \frac{r_i}{r_o}} \frac{1}{r} \tag{1.129}$$

$$q_r = \frac{T_i - T_o}{\dfrac{\ln \dfrac{r_o}{r_i}}{2\pi k L}} \tag{1.130}$$

Therefore, we can define the thermal resistance for a radial system as

$$R_{cond,radial} = \frac{\ln \frac{r_o}{r_i}}{2\pi k L}$$

(1.131)

And therefore the rate of heat transfer becomes

$$q_r = \frac{T_i - T_o}{R_{cond,radial}}$$

(1.132)

Note that the rate of heat transfer is constant and independent of the radial position.

1.10 Composite Systems

So far we have considered one-dimensional conduction in media where the thermal conductivity was uniform and constant. There are practical cases where heat is transferred from one material to another. If the heat transfer can be considered one dimensional, then the system can be analyzed more conveniently using the equivalent electrical system.

1.10.1 Composite Systems in Planar Coordinates

Consider the composite wall shown in Figure 1.26 separating a fluid at temperature T_i, having a heat transfer coefficient h_i, from another fluid at temperature T_o, having a heat transfer coefficient h_o.

The wall is composed of three layers of different materials. For a single layer, we showed that the heat transfer by conduction is

$$q = \frac{\Delta T}{\frac{L}{kA}}$$

(1.133)

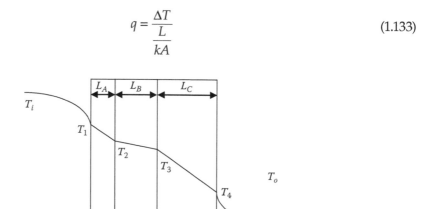

FIGURE 1.26
Composite wall.

The heat transfer q is the same in each layer, that is, whatever heat is traveling through layer A will be transferred by conduction at the interface between A and B to layer B, etc. Also, at the right-hand face of layer C, the same amount of heat will be convected to the outside fluid. The amount of heat transfer by convection is

$$q = \frac{\Delta T}{\dfrac{1}{hA}} \tag{1.134}$$

In terms of an equivalent network,

$$q = \frac{\Delta T}{R_{Total}} = \frac{T_i - T_o}{R_{Total}} \tag{1.135}$$

where

$$R_{Total} = \frac{1}{h_i A} + \frac{L_A}{k_A A} + \frac{L_B}{k_B A} + \frac{L_C}{k_C A} + \frac{1}{h_o A} \tag{1.136}$$

Since A is a constant for a planar medium,

$$q'' = \frac{T_i - T_o}{\dfrac{1}{h_i} + \dfrac{L_A}{k_A} + \dfrac{L_B}{k_B} + \dfrac{L_C}{k_C} + \dfrac{1}{h_o}} \tag{1.137}$$

Note that if another temperature, besides T_o, is given (e.g., T_4), then R_{total} will be the total resistance between the two given Ts, not the overall total resistance between T_i and T_o. Therefore,

$$q = \frac{T_\alpha - T_\beta}{R_{\alpha-\beta}} \tag{1.138}$$

where T_α and T_β are the two known temperatures and $R_{\alpha-\beta}$ is the resistance between the locations of T_α and T_β.

Once q is determined from Equation 1.138, the temperature at any other point can be found by noting that q is the same in all layers and known. Equation 1.138 can be applied again, this time between the location whose temperature is to be found (T_γ) and another location with known temperature T_α, using $R_{\gamma-\alpha}$ the resistance between the two locations.

1.10.2 Overall Heat Transfer Coefficient and *R* Value

Often the rate of heat transfer, q, is expressed in terms of an overall heat transfer coefficient, that is,

$$q = UA\Delta T \tag{1.139}$$

where U is the overall heat transfer coefficient. Note that

$$UA = \frac{1}{R} \tag{1.140}$$

or

$$U = \frac{1}{\dfrac{1}{h_i} + \dfrac{L_A}{k_A} + \dfrac{L_B}{k_B} + \dfrac{L_C}{k_C} + \dfrac{1}{h_o}} \tag{1.141}$$

In heat exchanger industry, U is used as a measure of the heat exchanger's ability to transfer heat.

In construction industry, the R-value is used to characterize the insulating value of insulating materials and is a measure of a material's capacity to impede heat transfer. The R-value is defined as conduction resistance times the area:

$$R_{value} = R_{cond} A \tag{1.142}$$

or conduction resistance can be considered as the R-value per unit area. Note that

$$q'' = \frac{\Delta T}{R_{Value}} \tag{1.143}$$

In Cartesian coordinates, the conduction resistance is given by Equation 1.42; therefore, the R-value for a single layer is

$$R_{value} = \frac{L}{k} \tag{1.144}$$

The higher the R-value, the better insulator the material is. The units of R-value are provided in ft^2 h°F/BTU (or m^2-K/W), and the commercial values quoted are in the conventional system (ft^2 h°F/BTU). In planar coordinates, like conduction resistances, the R-values of insulators in series can be added to come up with the R-value of the series.

Example 1.7

Determine the heat losses per unit area for windows and walls of a building and consider the improvement that can be made by using double pane window. The wall, shown in Figure 1.27, is made of brick, which is 3.5 in. wide, followed by a 0.5 in. air gap, 0.5 in. plywood, 2 × 4 studs (1.5 × 3.5 actual) that are 16 in. on center and are filled with fiber glass insulation with an R-value of 13. The window glass is 1/4 in. thick, the gap between the two glass panes is 1/2 in., and the inside and outside temperatures are 76°F and 20°F respectively.

The electrical network is also shown in Figure 1.27. Neglecting the heat transfer through the studs, from Equation 1.135

$$q'' = \frac{T_i - T_o}{\dfrac{1}{h_i} + \dfrac{L_A}{k_A} + \dfrac{L_B}{k_B} + \dfrac{L_C}{k_C} + \dfrac{L_D}{k_D} + \dfrac{L_E}{k_E} + \dfrac{1}{h_o}}$$

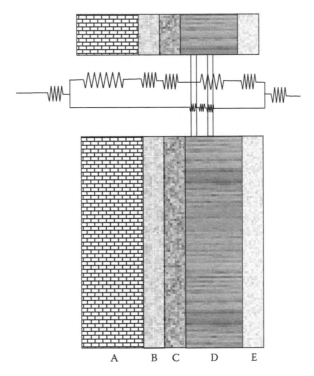

FIGURE 1.27
Details of the building wall.

Note that except the first and the last terms, the other terms are the *R*-value of the material.

Material	*k* (Btu/h-ft²-°F)
Brick (face)	9.3
Air	0.01457
Plywood	0.8
Gypsum (dry wall)	5.6
Glass	0.4

$$q'' = \frac{76 - 20}{\dfrac{1}{10} + \dfrac{3.5/12}{9.3} + \dfrac{0.5/12}{0.01457} + \dfrac{0.5/12}{0.8} + 13 + \dfrac{0.5/12}{5.6} + \dfrac{1}{20}}$$

$$q'' = \frac{76 - 20}{0.1 + 0.031 + 2.86 + 0.052 + 13 + 0.007 + 0.02} = \frac{56}{16.101}$$

$$q'' = 3.48 \text{ Btu/h-ft}^2 = 10.96 \text{ W/m}^2$$

For a single pane glass

$$q'' = \frac{T_i - T_o}{\dfrac{1}{h_i} + \dfrac{L_g}{k_g} + \dfrac{1}{h_o}} = \frac{76 - 20}{\dfrac{1}{10} + \dfrac{0.25/12}{0.4} + \dfrac{1}{20}} = \frac{56}{0.1 + 0.0037 + 0.05}$$

$$q'' = 364.3 \text{ Btu/h-ft}^2 = 1149 \text{ W/m}^2$$

For a double pane window

$$q'' = \frac{T_i - T_o}{\dfrac{1}{h_i} + \dfrac{L_g}{k_g} + \dfrac{L_a}{k_a} + \dfrac{L_g}{k_g} + \dfrac{1}{h_o}} = 18.56 \text{ Btu/h-ft}^2 = 58.5 \text{ W/m}^2$$

In the aforementioned example, if we consider a portion of the wall where there is a window, then the heat will be transferred from the inside to the outside both through the wall and the window. The portion of the heat transferred through the wall is first convected to the inner surface, passes through five resistances, and then is transferred by convection to the outside. The rest of the heat is convected to the first glass pane, then conducted through three resistance, and is convected to the outside.

Figure 1.26 also shows the equivalent electrical resistance network. The five resistances in the top can be combined into one as can the three on the bottom, forming two parallel resistances. The equivalent resistance can be added to the two convective resistances to find the total resistance. This is a quasi-one-dimensional problem.

This analysis is based on heat being transferred only in one direction and therefore requires that the temperature variation in the *y*-direction to be small. This would be the case only if the resistances in two branches are approximately equal. If the resistances are not too far apart, the temperature variation across the wall and across the window will be approximately the same and therefore the temperature variation and thus the heat transfer in the *y*-direction will be small, making the heat flow approximately one dimensional.

1.10.3 Composite Radial Systems

Consider the pipe shown in Figure 1.28, which is insulated on its outer wall and loses heat by convection to the environment at T_∞ and receive heat by convection at its inner surface. Recall that for a single tube

$$q_r = \frac{T_i - T_o}{\dfrac{\ln \dfrac{r_o}{r_i}}{2\pi k L}} \tag{1.145}$$

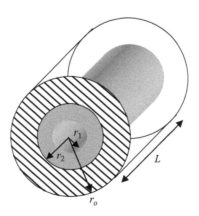

FIGURE 1.28
Composite radial system.

and we define the thermal resistance for a radial system as

$$R_{cond,radial} = \frac{\ln\frac{r_o}{r_i}}{2\pi k L} \tag{1.146}$$

then

$$q_r = \frac{T_i - T_o}{R_{tot}} \tag{1.147}$$

where

$$R_{Total} = R_{conv\,i} + R_{cond\,tube} + R_{cond\,insu} + R_{conv\,o} \tag{1.148}$$

$$R_{cond,radial} = \frac{1}{h_i 2\pi r_1 L} + \frac{\ln\frac{r_2}{r_1}}{2\pi k_p L} + \frac{\ln\frac{r_o}{r_2}}{2\pi k_l L} + \frac{1}{h_o 2\pi r_o L} \tag{1.149}$$

$$q_r = \frac{T_i - T_o}{\dfrac{1}{h_i 2\pi r_1 L} + \dfrac{\ln\frac{r_2}{r_1}}{2\pi k_p L} + \dfrac{\ln\frac{r_o}{r_2}}{\pi k_l L} + \dfrac{1}{h_o 2\pi r_o L}} \tag{1.150}$$

1.10.4 Critical Radius of Insulation

Consider an insulated pipe. The rate of heat transfer from the pipe is

$$q_r = \frac{T_i - T_o}{\dfrac{1}{h_i 2\pi r_i L} + \dfrac{\ln\frac{r_2}{r_i}}{2\pi k_p L} + \dfrac{\ln\frac{r_o}{r_2}}{2\pi k_l L} + \dfrac{1}{h_o 2\pi r_o L}} \tag{1.151}$$

What happens as we add more and more insulation? As r_o increases, the third term in the denominator, which is the resistance of the insulating material, increases. This results in a reduction in the rate of heat transfer. However, as r_o increases, more and more area will be available for transfer of heat by convection from the outer surface of the insulation and thus the last term in the denominator, which is the convection resistance, decreases. It is possible for the heat transfer to go through a maximum. This maximum amount of heat transfer happens at an outer r value called the critical radius of insulation (Figures 1.28 and 1.29). To find r_{crit}, Equation 1.151 must be differentiated with respect to r_o and set equal to zero

$$\frac{dq_r}{dr_o} = \frac{-(T_i - T_o)\left[\dfrac{1}{r_o}\dfrac{1}{2\pi k_l L} - \dfrac{1}{h_o 2\pi r_o^2 L}\right]}{\left[\dfrac{1}{h_i 2\pi r_i L} + \dfrac{\ln\frac{r_2}{r_i}}{2\pi k_p L} + \dfrac{\ln\frac{r_o}{r_2}}{2\pi k_l L} + \dfrac{1}{h_o 2\pi r_o L}\right]^2} = 0 \tag{1.152}$$

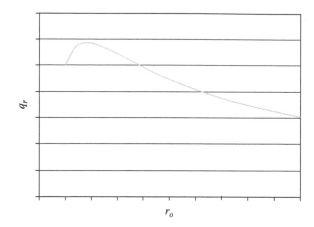

FIGURE 1.29
Heat transfer variation with increased insulation thickness.

resulting in

$$r_{crit} = \frac{k_I}{h_o} \tag{1.153}$$

If an insulating material is being considered and the heat transfer coefficient is known, then the critical radius of insulation can be determined from Equation 1.153. If the outer radius of the pipe is less than the r_{crit}, then adding the insulation will actually increase the heat transfer, and thus a different material with a smaller thermal conductivity needs to be considered.

1.11 Contact Resistance

The analysis of composite systems presented earlier was based on the assumption of the continuity of temperature across the interface between the materials, which strictly holds, if there is perfect contact between the two surfaces, without any gaps. This in general is not the case, due to the surface roughness resulting in less than ideal contact, leaving gaps that are either filled by air or bonding agent, if used to attach the surfaces. The temperature of the two materials at the interface can no longer be assumed to be continuous, and there is a temperature change across the gap. The continuity of the heat flux still applies, as the gap will not store much energy, and thus at the interface

$$-k_L \frac{dT_L}{dr} = -k_r \frac{dT_r}{dr} \tag{1.154}$$

However, the continuity of the temperature does not hold and there is a temperature drop across the interface that needs to be accounted for. The actual heat transfer process at the interface will be three dimensional because of the irregularity of the geometry and point

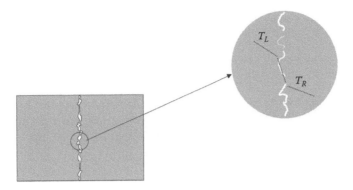

FIGURE 1.30
Temperature discontinuity due to imperfect contact.

contacts (Figure 1.30). The analysis can be simplified by assuming one-dimensional heat transfer and using the concept of resistance

$$q'' = \frac{T_L - T_R}{R_c} = \frac{(T_L - T_R)}{\dfrac{1}{h_c}} \tag{1.155}$$

where
  R_c is the thermal contact resistance, which is resistance per unit area
  h_c is the thermal contact conductance and is the value typically reported in the literature ranging between 2×10^3 and 2×10^5 W/m^2 K

1.12 Steady One-Dimensional Conduction with Heat Generation

There are many instances where energy is generated inside a solid and transferred as a result of conduction. Examples include a conductor dissipating heat due to its resistance to the flow of an electric current through it, chemical reactions within a material (e.g., setting of concrete), nuclear reaction within a material (reactor fuel rods), or molecular resistance to changing electric field (microwave oven). In these situations, the material is heated from within and the energy released is then conducted through and transferred out across the boundaries.

1.12.1 Planar Systems

For steady one-dimensional conduction with energy generation, assuming constant properties, the general heat diffusion equation

$$\frac{\partial^2 T}{\partial x^2} + \frac{\partial^2 T}{\partial y^2} + \frac{\partial^2 T}{\partial z^2} + \frac{\dot{q}}{k} = \frac{\rho c}{k} \frac{\partial T}{\partial t} \tag{1.156}$$

simplifies to

$$\frac{d^2T}{dx^2} + \frac{\dot{q}}{k} = 0 \qquad (1.157)$$

This is the general equation for determining temperature distribution. To find $T(x)$, we need to know how $\dot{q}$ depends on x to solve the differential equation. We must also have boundary conditions to solve for the constants of integration. Integrating Equation 1.157 twice

$$T = -\frac{\dot{q}}{k}\frac{x^2}{2} + C_1 x + C_2 \qquad (1.158)$$

Assuming isothermal wall boundary conditions

$$\begin{aligned} x = 0 \quad T = T_1 \\ x = L \quad T = T_2 \end{aligned} \qquad (1.159)$$

the constants of integration become

$$C_2 = T_1 \qquad (1.160)$$

$$C_1 = \frac{T_2 - T_1 + \dfrac{\dot{q}}{k}\dfrac{L^2}{2}}{L} \qquad (1.161)$$

Substituting and simplifying results in

$$T - T_1 = \frac{\dot{q}L^2}{2k}\left[\frac{x}{L} - \frac{x^2}{L^2}\right] + (T_2 - T_1)\frac{x}{L} \qquad (1.162)$$

which can be rearranged as

$$\frac{T - T_1}{(T_2 - T_1)} = \frac{\dot{q}L^2}{2k(T_2 - T_1)}\left[\frac{x}{L} - \frac{x^2}{L^2}\right] + \frac{x}{L} \qquad (1.163)$$

The heat flux is

$$q'' = -k\frac{dT}{dx} = -\frac{\dot{q}L}{2}\left[1 - \frac{2x}{L}\right] - (T_2 - T_1)\frac{k}{L} \qquad (1.164)$$

The heat flux at the left surface is

$$q''\Big|_{x=0} = -k\frac{dT}{dx}\Big|_{x=0} = (T_1 - T_2)\frac{k}{L}\left[1 - \frac{\dot{q}L^2}{2k(T_1 - T_2)}\right] \qquad (1.165)$$

and at the right surface

$$q''|_{x=L} = -k\frac{dT}{dx}\bigg|_{x=L} = (T_1 - T_2)\frac{k}{L}\left[1 + \frac{\dot{q}L^2}{2k(T_1 - T_2)}\right] \tag{1.166}$$

Thus if

$$\frac{\dot{q}L^2}{2k(T_1 - T_2)} > 1 \tag{1.167}$$

which will happen for large enough value of the heat generation, the heat flow is negative at $x = 0$ or heat will be transferred to the surroundings, both from the left and right surfaces of the slab. The term $q''|_{x=0} = -k\dfrac{dT}{dx}\bigg|_{x=0}$ is the heat that is transferred in the x-direction in the wall at $x = 0$. This is the heat that is transferred to the wall. The heat transferred from the wall is therefore $-q''|_{x=0} = k\dfrac{dT}{dx}\bigg|_{x=0}$. Similarly, the term $q''|_{x=L} = -k\dfrac{dT}{dx}\bigg|_{x=L}$ is the heat that is transferred in the x-direction in the wall at $x = L$. This is the heat that is transferred from the wall. Therefore, the total heat transfer *from* the wall will be

$$q''|_{tot} = q''|_{x=L} - q''|_{x=0} = (T_1 - T_2)\frac{k}{L}\left[1 + \frac{\dot{q}L^2}{2k(T_1 - T_2)}\right] - (T_1 - T_2)\frac{k}{L}\left[1 - \frac{\dot{q}L^2}{2k(T_1 - T_2)}\right] = \dot{q}L \tag{1.168}$$

The total heat transfer from the wall is the product of heat flux and area; therefore,

$$q = Aq''|_{tot} = A\dot{q}L = \dot{q}V \tag{1.169}$$

or the total heat transfer from the wall is equal to the total energy generated in the wall, as required by the first law.

1.12.2 Radial Systems: Solid Rod with Generation

Consider a solid cylinder in which energy is generated, subject to a convective boundary. This could represent a nuclear reactor fuel rod. Starting with Equation 1.89 or 1.92 and assuming steady, constant thermal conductivity, and uniform energy generation, the energy equation simplifies to

$$\frac{1}{r}\frac{\partial}{\partial r}\left(r\frac{\partial T}{\partial r}\right) + \frac{\dot{q}}{k} = 0 \tag{1.170}$$

The temperature distribution is expected to be symmetric with respect to r; also, one expects to have finite temperature at the center of the rod. At the outer surface, the rod is exposed to a convective boundary. Therefore, the boundary conditions are

$$r = 0. \quad \frac{dT}{dr} = 0. \tag{1.171}$$

$$r = \frac{D}{2} \quad -k\frac{dT}{dr} = h(T - T_\infty) \tag{1.172}$$

Integrating Equation 1.170 twice

$$T = -\frac{\dot{q}}{k}\frac{r^2}{4} + C_1 \ln r + C_2 \qquad (1.173)$$

Using the boundary conditions

$$C_1 = 0 \qquad (1.174)$$

$$C_2 = T_\infty + \frac{\dot{q}D}{4h} + \frac{\dot{q}}{k}\frac{D^2}{16} \qquad (1.175)$$

and the temperature distribution becomes

$$T = -\frac{\dot{q}}{k}\frac{r^2}{4} + T_\infty + \frac{\dot{q}D}{4h} + \frac{\dot{q}D^2}{16k} \qquad (1.176)$$

which can be rearranged into a dimensionless form

$$\frac{T - T_\infty}{\dot{q}\frac{D^2}{4k}} = \left(\frac{1}{4} - \frac{r^2}{D^2} + \frac{k}{hD} \right) \qquad (1.177)$$

or

$$T^* = \left(\frac{1}{4} - r^{*2} + \frac{1}{Bi} \right) \qquad (1.178)$$

The maximum value of temperature occurs at the center and is equal to

$$T_o^* = \frac{1}{4} + \frac{1}{Bi} \qquad (1.179)$$

and the rod surface temperature is

$$T_w^* = \frac{1}{Bi} \qquad (1.180)$$

Therefore, the temperature variation across the rod is 1/4. Figure 1.31 is a plot of nondimensional temperature distribution as a function of the nondimensional radial distance for different values of the Biot number. For small values of Biot number, $1/Bi \gg 1/4$ and thus the temperature of the rod is approximately uniform equal to $1/Bi$. Also for very large values of Bi, the surface temperature of the rod approaches that of the fluid-free stream temperature. Therefore, the limit of a very large Biot number corresponds to that of a cylinder whose surface temperature is maintained at a constant value.

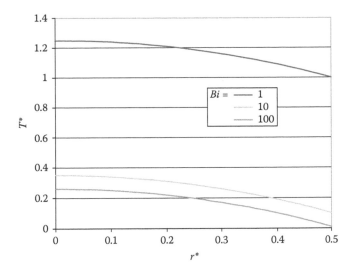

FIGURE 1.31
Radial variation of nondimensional temperature for different Biot numbers.

The heat transfer at any radial location in the rod is

$$q = -kA_r \frac{dT}{dr} = -k(2\pi r L)\left(-\frac{2r}{D^2}\right)\frac{\dot{q}D^2}{4k} = \pi r^2 L\dot{q} \tag{1.181}$$

The rate of heat transfer from the rod is obtained by evaluating Equation 1.181 at $r = D/2$

$$q = \frac{\pi}{4}D^2 L\dot{q} = V\dot{q} \tag{1.182}$$

which as expected is equal to the total amount of energy generated inside the rod.

Example 1.8

Calculate the maximum current that a copper wire with a 2 mm diameter can carry if the maximum temperature is not to exceed 100°C. Take resistance to be 3×10^{-3} Ω/m, $h_\infty = 5$ W/m^2 K, $T_\infty = 25$°C, and $k = 400$ W/m K.
   From Equation 1.178,

$$T^* = \left(\frac{1}{4} - r^{*2} + \frac{1}{Bi}\right)$$

The maximum temperature will happen at the center:

$$T^* = \frac{1}{4} + \frac{1}{Bi}$$

$$Bi = \frac{hD}{k} = \frac{5 \times 0.002}{400} = 2.5E-3$$

$$T^* = 40{,}000 = \frac{T_w - T_\infty}{\dfrac{\dot{q}D^2}{4k}}$$

$$40{,}000 = \frac{100 - 25}{\dfrac{\dot{q} \times 0.002^2}{4 \times 400}}$$

$$\dot{q} = 7.5 \times 10^5 \, \text{W/m}^3$$

$$\text{Power} = I^2 R = \dot{q}V = \dot{q}\frac{\pi}{4}R^2 L$$

$$I^2 = \frac{\dot{q}\dfrac{\pi}{4}D^2 L}{R} = \dot{q}\frac{\pi}{4\Omega}D^2$$

where $\Omega = R/L$ is the wire electrical resistance per unit length.

$$I = D\sqrt{\dot{q}\frac{\pi}{4\Omega}} = 2 \times 10^{-3}\sqrt{7.5 \times 10^5 \frac{\pi}{4 \times 3 \times 10^{-3}}} = 28\,\text{A}$$

1.13 Concluding Remarks

The objective of this chapter was to introduce some basic heat transfer concepts and lay the foundations for the topics that will be covered in the rest of this book. This chapter introduced the different heat transfer mechanisms, covered the approach by which the basic laws of nature are used to arrive at mathematical formulations of heat transfer problems, and presented nondimensionalization as an analytical tool.

Problems

1.1 By doing an energy balance on a differential control mass, derive the energy equation for steady one-dimensional conduction in cylindrical coordinates.

1.2 By doing an energy balance on a differential control mass, derive the energy equation for steady one-dimensional conduction in spherical coordinates with energy generation.

1.3 Determine the temperature distribution in the geometry shown next.

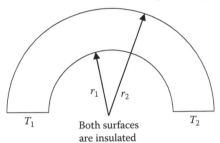

1.4 Derive an expression for the ambient temperature at which frost will form on the top surface of a car. Assume the ambient temperature to be T_∞, the effective sky temperature to be T_0, and the outside and inside heat transfer coefficients to be h_o and h_i, respectively.

1.5 Assuming steady state, one-dimensional conduction in the truncated cone shown next, determine the heat transfer.

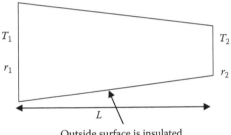

1.6 Simplify the heat diffusion equation to come up with the governing equation and also provide the needed boundary conditions for heat transfer in a large plane wall of thickness L, where the two sides are maintained at constant temperatures T_1 and T_2.

1.7 Simplify the heat diffusion equation to come up with the governing equation and also provide the needed boundary conditions for heat transfer in a large plane wall of thickness L, where the two sides are exposed to a convective medium at different temperatures.

1.8 Simplify the heat diffusion equation to come up with the governing equation and also provide the needed boundary conditions for heat transfer in a short cylinder of radius R and length L exposed to a convective medium.

1.9 Simplify the heat diffusion equation to come up with the governing equation and also provide the needed boundary conditions for heat transfer in a rectangular fin, losing heat by convection. Assume 2-D heat transfer and temperature distribution.

1.10 Write the energy equation and the boundary conditions for heat transfer in two large plane walls of thickness L_1, L_2, and thermal conductivities k_1, k_2, attached together over one of their faces, with the other two faces in contact with convective media at T_∞ and h.

1.11 A fuel plate is made from 3.5 mm 1.5% enriched uranium for which the rate of energy generation is 2.3×10^9 W/m³. The cladding is 0.25 mm 304 stainless steel and its outside surface temperature is 300°C. Determine the temperature at the center of the plate and at the interface of the plate and the cladding.

1.12 A uranium oxide fuel rod has a diameter of D and the clad thickness is t, and energy is generated at the rate of $\dot{q}$ W/m³. The outside heat transfer coefficient is h W/m²-K and the water temperature surrounding the fuel rod is at T_∞. Determine the temperature at the center of the rod and the heat flux at the fuel surface.

1.13 A uranium oxide fuel rod has a diameter of 1 cm and is clad with 0.25 mm 304 stainless steel. The rod is to develop 16.4 kW/m. The outside heat transfer coefficient is 7400 W/m²-K, and the water temperature surrounding the fuel rod is at 288°C. Determine the temperature at the center of the rod and the heat flux at the fuel surface.

1.14 Derive the governing and boundary conditions for heat transfer through a glass window of thickness L that receives heat flux at the rate of q'' on one side and loses heat by convection on both sides. As radiation passes through the glass, a portion gets absorbed at a rate of

$$q = q'' \alpha e^{-\alpha x}$$

where α is the absorption coefficient.

1.15 A thin rectangular plate receives uniform heat flux q'' on one side from a high-temperature source and loses energy by convection on both sides. The two ends of the plate are maintained at temperature T_0. Derive the governing equation and boundary conditions for the temperature distribution. Begin from the basic elemental control volume analysis for the temperature distribution.

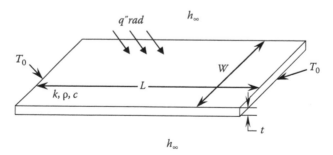

1.16 Model a stake placed on a grill.

1.17 Model an egg being boiled in a pan.

1.18 A gas pool heater is used to heat a 20,000 gal pool. The heater is rated at 120,000 BTU/h and has an efficiency of 80%. The water is being pumped through the heater at a rate of 50 gal/min. How long will it take to heat the pool from 75°F to 87°F. Assume the pool area to be 500 ft² and heat transfer coefficient between the pool surface and the ambient is 20 W/m² K.

1.19 A stainless steel pan is used to boil water on a gas stove. The pan is 3 mm thick, has a base diameter of 25 cm, a height of 10 cm, and receives 4 kW of thermal energy over a 2 cm ring of outer diameter 18 cm. The inside water temperature is 100°C and the heat transfer coefficient is 500 W/m² K. The ambient temperature is 25°C and the heat transfer coefficient is 10 W/m² K. Assume that the pan can be approximated as a flat disk. Write the governing equation and the boundary conditions for the heat transfer in the pan.

1.20 Consider a hollow cylinder with energy generation whose inside and outside surfaces are maintained at constant temperature. What are the governing equation and the boundary conditions?

1.21 Consider the maximum current that an insulated copper wire with a 2 mm diameter can carry if the maximum temperature is not to exceed 100°C. Take resistance to be 3×10^{-3} Ω/m, $h_\infty = 5$ W/m² K, $T_\infty = 25$°C, and $k = 400$ W/m K. The insulation on the wire 0.5 mm is Teflon.

1.22 A circular electric heater ($k = 40$ W/m K) has a diameter of 25 cm and a surface emissivity of 0.9. The surface of the heater is at 250°C, the heat transfer coefficient is 7 W/m² K, and the ambient is at 25°C. Determine the maximum temperature.

1.23 Determine the temperature distribution for one-dimensional conduction in a plane wall if the thermal conductivity of the wall is a linear function of temperature.

$$k = a_0 + a_1 T$$

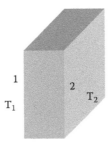

1.24 A man can be modeled as a vertical cylinder at an average temperature of 34°C. For a convection heat transfer coefficient of 20 W/m²°C, determine the rate of heat loss from this man by convection in an environment at 18°C.

1.25 An insulated travel coffee mug shown in the figure has an inner diameter D and length L, is made of a stainless steel inner cup of thickness t_s, and air gap of thickness t_a, and a plastic outer shell of thermal conductivity k_i and thickness t_i. The coffee loses heat by convection to the ambient, that is at T_∞, and h. Arrive at an expression for steady state temperature distribution, assuming 1-D heat transfer and the temperature of coffee is constant at T_c.

1.26 Repeat Problem 1.25, assuming the gap between the inner and outer shells is evacuated.

1.27 A thermocouple is used to measure the temperature of a gas flowing in a duct. The duct wall is at a temperature lower than the gas temperature. Determine the difference between the actual temperature of the gas (T_g) and the temperature that the thermocouple reads (T_c).

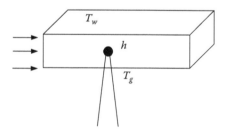

1.28 Liquid nitrogen at 100 K is stored in a stainless steel spherical vessel having a diameter of 5 m and a thickness of 2 cm. The vessel is insulated by a material having a thermal conductivity of 0.01 W/m K. The outside temperature is 20°C and the heat transfer coefficient is 30 W/m² K. A safety valve maintains the pressure constant by letting nitrogen vapor to escape. Determine the insulation thickness needed to keep the rate of nitrogen evaporation below 30 kg/h. The nitrogen enthalpy of evaporation at 100 K is 160.68 kJ/kg.

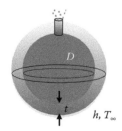

1.29 Consider a long circular cylinder with energy generation given by $\dot{q} = a + b\dfrac{r}{R}$ where a and b are two known constants, r is measured from the center of the cylinder, and R is the radius. Write the governing equations and the boundary conditions for the transient temperature distribution.

1.30 In metal casting, cores are forms, usually made of sand, that are placed into a mold cavity to form the interior surfaces of the castings. Thus, the void space between the core and mold–cavity surface is what eventually becomes the casting. In a particular casting process, cores are heated in a microwave oven for drying and curing. Heating in a microwave oven is equivalent to an internal energy generation $\dot{q}$ that is a function of location in the core. Assuming a transient one-dimensional model.

   a. Modeling the core as an infinitely long wall with energy generation inside and heat loss by convection to the surroundings, write the governing equation and boundary conditions, if $\dot{q} = ae^{bx}$ where a and b are two known constants and x is measured from the center of the wall.

   b. Obtain the *steady state* solution analytically.

1.31 Consider a long circular cylinder with energy generation given by $\dot{q} = a + b\dfrac{r}{R}$, where a and b are two known constants, r is measured from the center of the cylinder, and R is the radius. Obtain the *steady state* solution analytically.

References

1. Revelle, R. (1976) Energy use in rural India, *Science* 192, 969.
2. Beretta, G. P. (2007) World energy consumption and resources: An outlook for the rest of the century, *International Journal of Environmental Technology and Management* 7(1–2), 99–112. SCOPUS. Web. 28 May 2012.
3. Feynman, R. (1964) *The Feynman Lectures on Physics*, Vol. 1. Reading, MA: Addison Wesley.
4. U.S. Energy Information Administration. (2011) International energy outlook 2011, Report Number: DOE/EIA-0484(2011). http://www.eia.gov/forecasts/ieo/index.cfm
5. Sobhan, C. B. and Peterson, G. P. (2008) *Microscale and Nanoscale Heat Transfer: Fundamentals and Engineering Applications*, Boca Raton: CRC Press.
6. Trenberth, K. E., Fasullo, J. T., and Kiehl, J. (2009) Earth's global energy budget, *Bulletin of the American Meteorological Society* 90(3), 311–323. http://www.cgd.ucar.edu/cas/Trenberth/trenberth.papers/BAMSmarTrenberth.pdf

2

Basic Means

Mathematical representation of the physical laws often leads to ordinary differential equations (ODE) or partial differential equations (PDE), for example, conservation of energy derived in Chapter 1. Many of these equations cannot be solved analytically and one has to resort to approximate or numerical solutions. Often we are interested in gaining as much information as possible about the nature of the problem, before actually solving the equations. The studies of fluid mechanics and heat transfer are greatly facilitated by a number of tools. This chapter covers three of the more important ones, the nondimensionalization, analytical methods, and numerical methods, which will be used throughout this book for the analysis and solution of a broad range of problems.

2.1 Nondimensionalization

Since obtaining solutions to the governing equations is typically not a trivial matter, it is desirable to arrive at solutions that are as generally applicable as possible. Nondimensionalization of governing equations and boundary conditions is a tool that greatly helps these goals. There are a number of texts that provide excellent coverage of the topic [1,2].

Nondimensionalization is scaling the dependent and independent variables, the equations, the boundary conditions, and the parameters. This is done by selecting a new reference and a new scale for each variable and expressing the dimensional variables, equations, and parameters in terms of the nondimensional ones. The nondimensional variables are constructed from the scales and references intrinsic to the problem under consideration. The nondimensionalization results in the elimination of units from an equation, by replacing the variables in the equation by the equivalent nondimensional ones.

Consider the steady 1D conduction in a wall subject to isothermal boundaries. The temperature distribution in the wall is given by

$$T = T_1 + (T_2 - T_1)\frac{x}{L} \tag{2.1}$$

Without loss of generality, assume $T_2 > T_1$. In this case, the temperature in the wall depends on three parameters (T_1, T_2, L) and the independent variable, x. To determine temperature at a particular location, all of these must be known.

The parameters and the variables are expressed in terms of a consistent set of units, for example, SI or conventional system. For example, $L = 10$ in. means that the wall thickness is measured relative to a reference of zero using an arbitrary scale called 1 in. (and we all know how big 1 in. is), and the wall is 10 times thicker than this arbitrary scale. If the right wall temperature is at 75°F, it means that the temperature is 43 units (degree Fahrenheit) more than the freezing temperature of water that has been assigned the arbitrary value of 32. Now consider rearranging Equation 2.1:

$$\frac{T - T_1}{T_2 - T_1} = \frac{x}{L} \tag{2.2}$$

If we define two new variables

$$x^* = \frac{x}{L} \quad \text{and} \quad T^* = \frac{T - T_1}{T_2 - T_1} \tag{2.3}$$

then the temperature distribution is given as

$$T^* = x^* \tag{2.4}$$

which says that temperature varies between 0 and 1 over the width of the wall that also varies between 0 and 1. Here the nondimensional temperature only depends on the nondimensional-independent variable and all the parameters have been eliminated from the solution, having been incorporated in the nondimensional variables. In effect what we have done is rather than using a standard system of units, we have set up a new measuring system that uses the references and scales that are derived from the problem at hand and are more relevant to it.

Instead of measuring the width of the wall using arbitrary reference of zero and scale of 1 in., we are using a reference of zero and a scale equal to the difference between the maximum and minimum wall coordinates, or we are using the width of the wall as the scale for this problem. Similarly for temperature, instead of using the freezing point of the water we are using the minimum wall temperature as the reference and instead of degree Fahrenheit we are using the difference between maximum and minimum temperatures for the temperature scale. Using these scales and references, both the nondimensional temperature and the wall thickness vary between zero and one. From the nondimensional solution, for example, we see that in the middle of the wall ($x^* = 0.5$), the temperature is 0.5, which means at any chosen system on units, the center of the plate is at a temperature equal to the minimum wall temperature plus half the difference between the maximum and minimum temperatures in the walls, in the chosen system. Therefore, once the nondimensional temperature is known, it can be converted to any system of units desired, using the definitions used for the nondimensional variables. For example, if the wall thickness is 12 in., then the middle of the wall is at $x = 6$ in. at which $T^* = 0.5$.

Using a standard system of units has the advantage that the references and scales used are universally known and need no definition. The disadvantage is that the standard units may not be the most appropriate or general approach. The nondimensional form of the equations and boundary conditions reveal the relevant nondimensional parameters and also lend themselves to parametric investigations, resulting in more general solutions.

2.1.1 Nondimensionalization Process

The nondimensionalization of a differential or PDE and its associated boundary conditions involves the following steps:

i. Define new nondimensional-dependent and -independent variables using the general form:

$$z^* = \frac{z - z_r}{z_s} \tag{2.5}$$

The superscript * is used to denote a dimensionless variable; therefore, if z is a dimensional variable (dependent or independent), z^* is the nondimensional form of z, defined by Equation 2.5 where z_r is the new reference and z_s is the new scale for variable z.

ii. Once the nondimensional variables have been defined, then the original equation and boundary conditions need to be expressed in terms of the new nondimensional variables. The following simple formula, which results from the application of the variable change and chain rule, can be very helpful in relating the nth derivative of the dependent variable z with respect to the independent variable x to the nondimensional form of the two variables defined by Equation 2.5:

$$\frac{d^n z}{dx^n} = \frac{z_s}{(x_s)^n} \frac{d^n z^*}{dx^{*n}} \tag{2.6}$$

iii. Divide both sides of the equation and the boundary conditions by the dimensional coefficients of one of the terms, typically, but not always, the highest order derivative term, to rewrite them in terms of the new dimensionless variables, and dimensionless parameters.

In step i, the choice of z_r and z_s are quite arbitrary; the only limitations on them being that they should be constant and have the same dimension as z. The nondimensionalization process becomes more effective, if a few rules are followed in selecting z_r and z_s:

1. If possible, choose z_r and z_s so that z^* would vary between 0 and 1. This means that value of z_r must be the minimum (or maximum) expected value of z and the value of z_s must be the difference between the maximum and minimum values of z. For example if the minimum value of r is 0 and the maximum value is R, then

$$r^* = \frac{r-0}{R-0} = \frac{r}{R} \tag{2.7}$$

If a variable (dependent or independent) appears directly in the governing equation or boundary conditions, then the reference should be whatever is subtracted from it, if any. For example for convective boundary,

$$-k\frac{\partial T}{\partial x} = h(T - T_\infty) \tag{2.8}$$

the appropriate reference for temperature is T_∞.

2. When maximum or minimum values of z are not known or do not exist (for example, time), it is not possible to force z^* to vary between 0 and 1. In such cases, scales and references are generally selected from the values that directly appear in the governing equations or boundary conditions.

3. If the variable is raised to a power other than 1, then the appropriate reference is generally zero. For example in a convective radiative boundary $-k\left(\dfrac{dT}{dx}\right)=h(T-T_\infty)+\varepsilon\sigma(T^4-T_\infty^4)$, the appropriate reference for temperature is zero, since T is raised to the fourth power. If T_∞, is selected, the nondimensional form of the nonlinear term would become complicated and awkward.

4. When maximum or minimum values of z are not known or do not exist, or initially it is not clear what to use, leave the scales and references as unknowns to be determined later, and proceed with the nondimensionalization, and select them at the end. As a general rule, z_r and z_s must be selected so that the resulting nondimensional equations and the boundary conditions would have the minimum number of dimensionless parameters.

5. Generally, once a scale is chosen for a variable, the same scale is used for similar variables. This is analogous to measuring the size of a room using the same unit for length, width, and height. We typically do not use meter for sides, and inches for height. There are rare cases where you may want to use different scales for similarly dimensioned variables in order to simplify the resulting nondimensional equation.

Example 2.1

Consider steady 1D conduction in cylindrical coordinates with energy generation subject to convective boundary

$$\frac{1}{r}\frac{\partial}{\partial r}\left(r\frac{\partial T}{\partial r}\right)+\frac{\dot{q}}{k}=0 \tag{2.9}$$

$$r=0,\quad \frac{dT}{dr}=0 \tag{2.10}$$

$$r=R,\quad -k\frac{dT}{dr}=h(T-T_\infty) \tag{2.11}$$

We start by defining the nondimensional variables. In this case, since r directly appears in the governing equation, the reference used must be what is subtracted from it in the governing equation, which is zero. Therefore, the reference must be zero and the scale the maximum value of r, to ensure dimensionless r is less than 1, or

$$r^*=\frac{r}{R} \tag{2.12}$$

Temperature also appears directly in the second boundary condition and T_∞ is subtracted from it. Therefore, the temperature reference must be T_∞. Since we do not know the maximum value of the temperature, the nondimensional temperature is defined as

$$T^*=\frac{T-T_\infty}{T_s} \tag{2.13}$$

where T_s is yet to be determined. In terms of the nondimensional variables, with the help of Equation 2.6, Equation 2.9 can be written as

$$\frac{1}{Rr^*}\frac{1}{R}\frac{\partial}{\partial r^*}\left(Rr^*\frac{T_s}{R}\frac{\partial T^*}{\partial r^*}\right)+\frac{\dot{q}}{k}=0 \tag{2.14}$$

Dividing both sides by T_s/R^2 results in

$$\frac{1}{r^*}\frac{\partial}{\partial r^*}\left(r^*\frac{\partial T^*}{\partial r^*}\right)+\frac{\dot{q}R^2}{kT_s}=0 \tag{2.15}$$

The first term is nondimensional; therefore, the second term must also be nondimensional. We still have to decide what T_s should be. Since we do not know T_{max}, then we follow rule 4, and choose T_s so that the governing equation would have the minimum number of parameters. There are two limitations on T_s: it must have dimensions of temperature and it must be constant. Other than those, it can be anything. For example we can take it to be 100°C, but that does not have much relevance. Since the second term is nondimensional, then the quantity

$$\frac{\dot{q}R^2}{k} \tag{2.16}$$

must have dimension of temperature. It is also a constant, i.e., it is independent of r. Therefore by taking $T_s=\frac{\dot{q}R^2}{k}$, then

$$T^*=\frac{T-T_\infty}{\dfrac{\dot{q}R^2}{k}} \tag{2.17}$$

The second term on the LHS becomes equal to 1

$$\frac{\dot{q}R^2}{kT_s}=1 \tag{2.18}$$

and we manage to scale all these parameters out of the governing equation.
    The nondimensionalize boundary conditions become

$$r^*=0,\quad \frac{T_s}{R}\frac{dT^*}{dr^*}=0 \quad \text{or} \quad \frac{dT^*}{dr^*}=0 \tag{2.19}$$

$$r^*=1,\quad -k\frac{T_s}{R}\frac{dT^*}{dr^*}=hT_sT^* \quad \text{or} \quad -\frac{dT^*}{dr^*}=\frac{hR}{k}T^*=BiT^* \tag{2.20}$$

Therefore, the nondimensional form of the governing equation and the boundary conditions become

$$\frac{1}{r^*}\frac{\partial}{\partial r^*}\left(r^*\frac{\partial T^*}{\partial r^*}\right)+1=0 \tag{2.21}$$

$$r^*=0,\quad \frac{dT^*}{dr^*}=0 \tag{2.22}$$

$$r^*=1,\quad \frac{dT^*}{dr^*}=-BiT^* \tag{2.23}$$

Comparing Equations 2.21 through 2.23 to Equations 2.9 through 2.11, it is clear that the nondimensional form of the equations is far more general, in that the nondimensional temperature is a function of the nondimensional variable, r^*, and a nondimensional group, Biot number, or $T^* = T^*(r^*, Bi)$, whereas in the dimensional form temperature depends on one variable and five parameters $T = T(r, \dot{q}, k, R, h, T_\infty)$.

The solution is obtained by integrating the equation twice and evaluating the constants using the boundary conditions:

$$T^* = \frac{1}{4} - \frac{r^{*2}}{4} + \frac{1}{2Bi} \tag{2.24}$$

This equation provides the solution to all 1D conduction problems with energy generation in cylindrical coordinates. Through nondimensionalization, we ended up with the result that the nondimensional temperature is a function of the nondimensional distance and a parameter called Biot number. All the other physical parameters are incorporated in the nondimensional variables and the parameter. Note that this equation is slightly different than Equation 1.178 in Chapter 1, because r^* in that equation is defined as r/D.

Substituting for T^* and r^* from Equations 2.12 and 2.17, the dimensional form of Equation 2.24 reduces to Equation 1.176:

$$T = T_\infty + \frac{\dot{q}D^2}{16k} - \frac{\dot{q}r^2}{4k} + \frac{\dot{q}D}{4h} \tag{2.25}$$

2.2 Analytical Solutions

As seen in Chapter 1, second-order differential equations are often encountered in the formulation of physical phenomenon. In this section, we review the solution of some basic differential equations.

A differential equation is an equation that contains the derivatives or differentials. If the equation contains only one independent variable, it is an ODE and if it has two or more independent variables, it is a PDE. There are numerous texts devoted to the subject [3–5] and a brief overview of the solution to some commonly encountered differential equations are given as follows.

2.2.1 First-Order Ordinary Differential Equations

The general form of a first-order ODE is

$$\frac{dy}{dx} = f(x, y) \tag{2.26}$$

or

$$A(x, y)dx + B(x, y)dy = 0 \tag{2.27}$$

The general solution to this equation is given by the implicit function

$$F(x, y) = C \tag{2.28}$$

where

$$\frac{\partial F}{\partial x} = \mu A \tag{2.29}$$

$$\frac{\partial F}{\partial y} = \mu B \tag{2.30}$$

and function $\mu(x, y)$, if it exists, is the integration factor that satisfies

$$\frac{\partial(\mu A)}{\partial y} = \frac{\partial(\mu B)}{\partial x} \tag{2.31}$$

It may not always be possible to obtain a closed-form solution to a first-order differential equation. There are a number of cases where the solution procedure simplifies greatly.

2.2.1.1 Linear Equation

If the equation is linear, i.e., it can be expressed in the general form

$$\frac{dy}{dx} + A(x)y = B(x) \tag{2.32}$$

then the solution is given by

$$y = \frac{1}{\mu(x)} \left[\int \mu(x)B(x)dx + C \right] \tag{2.33}$$

where the integration factor is obtained from

$$\mu(x) = e^{\int A(x)dx} \tag{2.34}$$

2.2.1.2 Separable Equations

If the equation is such that all the terms containing the dependent variable y can be moved to one side of the equation and all the terms containing the independent variable x to the other side, then the equation is said to be separable. The solution to a separable equation is obtained by integrating each side with respect to the variable that it depends on. A separable equation has the general form

$$\frac{dy}{dx} = A(x)B(y) \tag{2.35}$$

and the solution is given by

$$\int \frac{dy}{B(y)} = \int A(x)dx + C \tag{2.36}$$

Example 2.2

The energy equation for unsteady temperature change in an object with low Biot number is given by

$$\frac{dT}{dt} + m(T - T_\infty) = 0$$

$$t = 0, \quad T = T_0$$

Obtain the solution.

This is a separable first-order differential equation and thus the solution can be obtained by separation of variables:

$$\frac{dT}{(T - T_\infty)} = -mdt$$

$$\ln(T - T_\infty) = -mt + C$$

Using the initial condition, the constant of integration can be determined as

$$\ln(T_0 - T_\infty) = C$$

and the solution is

$$\frac{T - T_\infty}{T_0 - T_\infty} = e^{-mt}$$

The same problem can be solved using integration factor. The differential equation can be written as

$$\frac{dT}{dt} + mT = mT_\infty$$

The integration factor is

$$\mu = e^{\int mdt} = e^{mt}$$

and the solution becomes

$$T = \frac{1}{e^{mt}} \left[\int e^{mt} mT_\infty dt + C \right]$$

$$T = \frac{1}{e^{mt}} [T_\infty e^{mt} + C]$$

Using the initial condition results in $C = T_0 - T_\infty$, and the solution becomes

$$T = T_\infty + (T_0 - T_\infty)e^{-mt}$$

which is the same as that obtained by separation of variables.

2.2.2 Second-Order Constant Coefficient Ordinary Differential Equations

The equation

$$A\frac{d^2y}{dx^2} + B\frac{dy}{dx} + Cy = D \tag{2.37}$$

is the general form of a second-order nonhomogeneous constant coefficient (A, B, C, and D are constants) differential equation. The first step in the solution of this equation is to turn it into a homogeneous differential equation by defining a new variable

$$z = y - \frac{D}{C} \tag{2.38}$$

Note that

$$\frac{dz}{dx} = \frac{dy}{dx} \tag{2.39}$$

Then in terms of the new variable, the original equation turns into the following homogeneous equation:

$$A\frac{d^2z}{dx^2} + B\frac{dz}{dx} + Cz = 0 \tag{2.40}$$

The characteristic equation for this differential equation is

$$A\lambda^2 + B\lambda + C = 0 \tag{2.41}$$

which has two roots

$$\lambda_1 = \frac{-B + \sqrt{B^2 - 4AC}}{2A} \tag{2.42}$$

$$\lambda_2 = \frac{-B - \sqrt{B^2 - 4AC}}{2A} \tag{2.43}$$

If

$$B^2 - 4AC > 0 \tag{2.44}$$

then the solution to the differential equation is

$$z = c_1 e^{\lambda_1 x} + c_2 e^{\lambda_2 x} \tag{2.45}$$

The solution can also be expressed in terms of hyperbolic functions, and therefore

$$z = c_1[\sinh(\lambda_1 x) + \cosh(\lambda_1 x)] + c_2[\sinh(\lambda_2 x) + \cosh(\lambda_2 x)] \tag{2.46}$$

where

$$\sinh(x) = \frac{e^x - e^{-x}}{2}$$

$$\cosh(x) = \frac{e^x + e^{-x}}{2} \tag{2.47}$$

Equations 2.45 and 2.46 are equivalent. If the domain of interest is finite, the form given by Equation 2.46 is generally easier to work with, whereas the exponential solution is preferable if the domain extends to infinity. When dealing with hyperbolic functions, the following identities may prove useful:

$$\cosh(x \pm y) = \cosh(x)\cosh(y) \pm \sinh(x)\sinh(y) \tag{2.48}$$

$$\sinh(x \pm y) = \sinh(x)\cosh(y) \pm \cosh(x)\sinh(y) \tag{2.49}$$

Returning back to the solution, substituting for y, the result is

$$y = c_1 e^{\lambda_1 x} + c_2 e^{\lambda_2 x} + \frac{D}{C} = c_1[\sinh(\lambda_1 x) + \cosh(\lambda_1 x)] + c_2[\sinh(\lambda_2 x) + \cosh(\lambda_2 x)] + \frac{D}{C} \tag{2.50}$$

If

$$B^2 - 4AC = 0 \tag{2.51}$$

then the solution is

$$z = c_1 e^{\lambda_1 x} + c_2 x e^{\lambda_1 x} \tag{2.52}$$

Finally for

$$B^2 - 4AC < 0 \tag{2.53}$$

the solution is

$$z = c_1 e^{\alpha x} \cos(\beta x) + c_2 e^{\alpha x} \sin(\beta x) \tag{2.54}$$

where

$$\alpha = \frac{-B}{2A}$$

$$\beta = \frac{\sqrt{4AC - B^2}}{2A} \tag{2.55}$$

Example 2.3

As will be shown later, the nondimensional energy equation for a constant area fin is

$$\frac{d^2T^*}{dx^{*2}} - m^{*2}T^* = 0$$

Obtain the solution.

$$B^2 - 4AC = 0^2 + 4m^{*2} > 0$$

$$\lambda_1 = \frac{-0 + 2m^*}{2} = +m^*$$

$$\lambda_2 = \frac{-0 - 2m^*}{2} = -m^*$$

$$T^* = c_1 e^{m^*x} + c_2 e^{-m^*x}$$

2.2.3 Second-Order Variable Coefficient Ordinary Differential Equations

The second-order variable coefficient differential equation

$$x^2 \frac{d^2y}{dx^2} + (2k+1)x\frac{dy}{dx} + (\alpha^2 x^{2r} + \beta^2)y = 0 \tag{2.56}$$

also appears frequently, particularly when dealing with cylindrical coordinates. The general solution to Equation 2.56 is

$$y = x^{-\kappa}\left[C_1 J_{\kappa/r}\left(\frac{\alpha x^r}{r}\right) + C_2 Y_{\kappa/r}\left(\frac{\alpha x^r}{r}\right)\right] \quad \text{where } \kappa = \sqrt{k^2 - \beta^2} \tag{2.57}$$

In Equation 2.57, J_n is the Bessel function of the first kind of order n, which is defined as

$$J_n(x) = \sum_{m=0}^{\infty} \frac{(-1)^m}{m!\,\Gamma(m+n+1)}\left(\frac{x}{2}\right)^{2m+n} \tag{2.58}$$

and Y_n is Bessel function of the second kind of order n defined as

$$Y_n(x) = \frac{J_n(x)\cos(n\pi) - J_{-n}(x)}{\sin(n\pi)} \tag{2.59}$$

The Bessel functions of the first and second kind are plotted in Figures 2.1 and 2.2, and as can be seen they are periodic functions, similar to sine and cosine functions. The Bessel

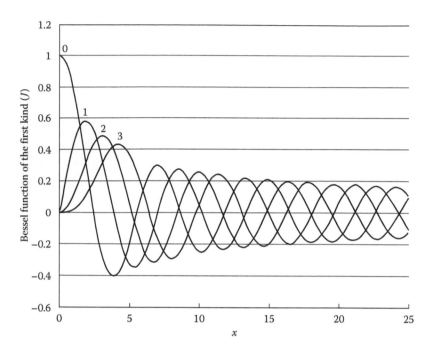

FIGURE 2.1
Bessel functions of the first kind of different orders.

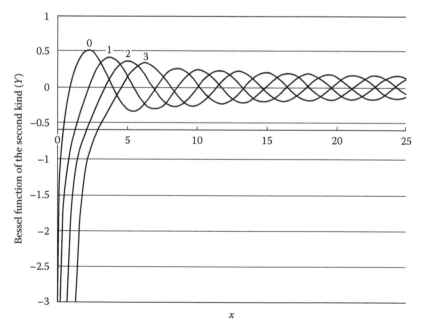

FIGURE 2.2
Bessel functions of the second kind of different orders.

functions are also valid when the variable is a complex number. In these cases, the modified Bessel functions are more useful. The modified Bessel functions of the first and second kind are defined as

$$I_n(x) = \sum_{m=0}^{\infty} \frac{1}{m!\,\Gamma(m+n+1)}\left(\frac{x}{2}\right)^{2m+n} \tag{2.60}$$

$$K_n(x) = \frac{\pi}{2}\frac{I_{-n}(x) - I_n(x)}{\sin(n\pi)} \tag{2.61}$$

and are plotted in Figures 2.3 and 2.4.

In the earlier equations, $\Gamma(z)$ is the gamma function, which is a generalization of the factorial function, defined for nonnegative integers, to real and complex numbers. If z is an integer, then

$$\Gamma(z) = (z-1)! \tag{2.62}$$

and for real and complex numbers

$$\Gamma(z+1) = z\Gamma(z) \tag{2.63}$$

and can be calculated from the definite integral

$$\Gamma(z) = \int_0^{\infty} t^{z-1}e^{-t}dt \quad \text{for } z \geq 0 \tag{2.64}$$

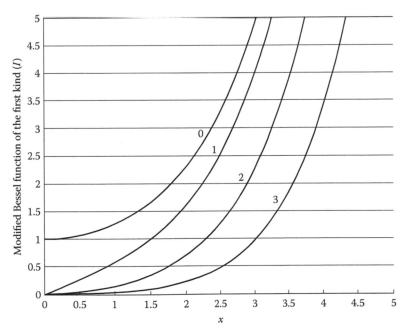

FIGURE 2.3
Modified Bessel functions of the first kind of different orders.

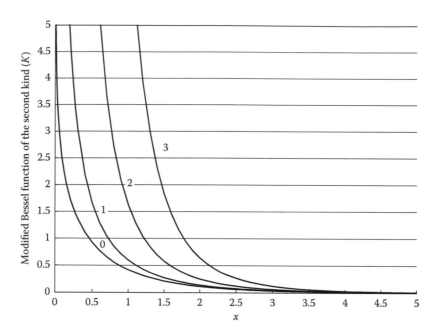

FIGURE 2.4
Modified Bessel functions of the second kind of different orders.

The gamma function can also be calculated from the infinite series

$$\Gamma(z) = \lim_{n \to \infty} \frac{n^x n!}{x(x+1)(x+2)\cdots(x+n)} \quad (2.65)$$

and is plotted in Figure 2.5.

The following equations are useful when dealing with Bessel functions:

$$J_n(iu) = i^n I_n(u) \quad (2.66)$$

$$Y_n(iu) = i^n K_n(u) \quad (2.67)$$

$$\frac{d}{dx}[x^{\pm n} J_n(x)] = \pm x^n J_{n\mp 1}(x) \quad (2.68)$$

$$\frac{d}{dx}[x^{\pm n} Y_n(x)] = \pm x^n Y_{n\mp 1}(x) \quad (2.69)$$

$$\frac{dI_n}{dx} = I_{n-1} - \frac{n}{x} I_n = I_{n+1} + \frac{n}{x} I_n \quad (2.70)$$

$$\frac{dK_n}{dx} = -K_{n-1} - \frac{n}{x} K_n = -K_{n+1} + \frac{n}{x} K_n \quad (2.71)$$

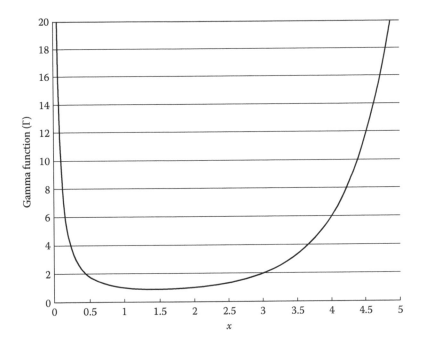

FIGURE 2.5
Gamma function.

2.3 Numerical Methods

The analytical solutions presented in the previous section are the exception, in that the ODEs and PDEs that describe the behavior of physical processes typically do not have analytical solution. Until recently, many such phenomena were poorly understood due to lack of solution techniques. ODEs and PDEs can now be solved readily due to the availability of computers and the use of numerical methods. Numerical methods are a collection of very powerful technique for solution and analysis of many problems. Numerical solutions [6,7] are an alternative to analytical solutions and often the only option for obtaining solutions.

There are fundamental differences between analytical and numerical solutions. The first difference is that numerical solutions are an approximation and always different than the exact solution. Furthermore, numerical solutions are only obtained at discrete points, whereas an analytical solution allows the determination of the dependent variable at any point or the solution is continuously defined in the domain of interest. Consider the first-order ODE

$$\frac{dy}{dx} - 2x = 0 \qquad (2.72)$$

$$y(0) = 0 \qquad (2.73)$$

The analytical solution of this ODE is $y = x^2$. This means that at any point (x), one can determine y from the solution. Numerical solutions provide solution only at discret preselected points in the domain of interest. The solution at other points is obtained by interpolation.

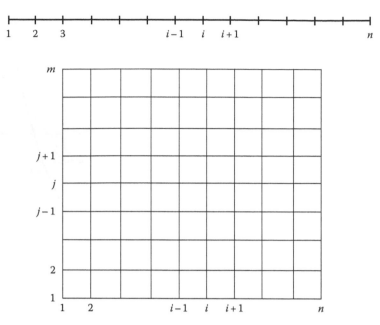

FIGURE 2.6
Discretization of 1D and 2D domain.

Numerical methods is an expanding field and covers many subjects including finite element, finite difference, and finite voume techniques. Here, we consider finite difference technique for the solution of ODEs and PDEs. In Chapter 15, the finite volume method is covered in great detail.

In finite difference solution, the region of interest is divided into discrete points and we try to find values that approximately satisfy the differential equation and the boundary conditions at these points. Therefore, the first step in obtaining the finite difference solution is to divide the region of interest into a series of smaller regions. This is known as the discretization process, and Figure 2.6 shows examples of discretization of 1D and 2D domains.

The second step is to replace the derivatives that appear in the ODE or the PDE with algebraic equations approximating the derivatives. Usually, the approximations to the derivatives are arrived at by using Taylor series.

The final step is to solve the resulting algebraic equations for the values of the dependent variable at the discrete values of the independent variable.

2.3.1 Finite Difference Approximation of Derivatives

As discussed in Chapter 1, Taylor series relates the value of a function at a point to the value of the function at a different point, or

$$f(x+\Delta x)=f(x)+\frac{\Delta x}{1!}\frac{df}{dx}\bigg|_x+\frac{\Delta x^2}{2!}\frac{d^2f}{dx^2}\bigg|_x+\frac{\Delta x^3}{3!}\frac{d^3f}{dx^3}\bigg|_x+\cdots \tag{2.74}$$

Different Taylor series expansions of a function can be used to arrive at different approximations to the derivatives of different orders. Some of the more commonly used ones are derived as follows.

2.3.1.1 First-Order Derivatives

If f_i denotes the value of the function f at a point x_i, then we use f_{i+1} to denote the value of the function f at point $x_{i+1} = x_i + \Delta x$ and we use f_{i-1} to denote the value of the function at point $x_{i-1} = x_i - \Delta x$, etc., then

$$f_{i+1} = f_i + \frac{\Delta x}{1!} \left.\frac{df}{dx}\right|_i + \frac{\Delta x^2}{2!} \left.\frac{d^2 f}{dx^2}\right|_i + \frac{\Delta x^3}{3!} \left.\frac{d^3 f}{dx^3}\right|_i + \cdots \tag{2.75}$$

$$f_{i-1} = f_i - \frac{\Delta x}{1!} \left.\frac{df}{dx}\right|_i + \frac{\Delta x^2}{2!} \left.\frac{d^2 f}{dx^2}\right|_i - \frac{\Delta x^3}{3!} \left.\frac{d^3 f}{dx^3}\right|_i + \cdots \tag{2.76}$$

$$f_{i+2} = f_i + \frac{2\Delta x}{1!} \left.\frac{df}{dx}\right|_i + \frac{4\Delta x^2}{2!} \left.\frac{d^2 f}{dx^2}\right|_i + \frac{8\Delta x^3}{3!} \left.\frac{d^3 f}{dx^3}\right|_i + \cdots \tag{2.77}$$

$$f_{i-2} = f_i - \frac{2\Delta x}{1!} \left.\frac{df}{dx}\right|_i + \frac{4\Delta x^2}{2!} \left.\frac{d^2 f}{dx^2}\right|_i - \frac{8\Delta x^3}{3!} \left.\frac{d^3 f}{dx^3}\right|_i + \cdots \tag{2.78}$$

Solving from Equation 2.75 for the first derivative of f

$$\left.\frac{df}{dx}\right|_i = \frac{f_{i+1} - f_i}{\Delta x} - \frac{\Delta x}{2!} \left.\frac{d^2 f}{dx^2}\right|_i - \frac{\Delta x^2}{3!} \left.\frac{d^3 f}{dx^3}\right|_i + \cdots \tag{2.79}$$

then an approximate expression for the first derivative of f is given by

$$\left.\frac{df}{dx}\right|_i \approx \frac{f_{i+1} - f_i}{\Delta x} \tag{2.80}$$

The truncation error which is the error introduced by neglecting of the higher-order terms of the infinite series is of the order of $O(\Delta x)$. This representation of the first derivative is called the forward differencing in which the value of the derivative of function the f at point i is related to the value of the function f at point i and its value at point $(i + 1)$.

Solving from Equation 2.76 for the first derivative of function f

$$\left.\frac{df}{dx}\right|_i = \frac{f_i - f_{i-1}}{\Delta x} + \frac{\Delta x}{2!} \left.\frac{d^2 f}{dx^2}\right|_i - \frac{\Delta x^2}{3!} \left.\frac{d^3 f}{dx^3}\right|_i + \cdots \tag{2.81}$$

another approximate expression for the first derivative can be found as

$$\left.\frac{df}{dx}\right|_i \approx \frac{f_i - f_{i-1}}{\Delta x} \tag{2.82}$$

Again, the truncation error is of the order of $O(\Delta x)$. This is called the backward difference representation of the first derivative in which the value of the derivative of function f at point i is related to the value of the function f at point i and its value at point $(i - 1)$.

Subtracting Equation 2.76 from Equation 2.75 results in

$$f_{i+1} - f_{i-1} = \frac{2\Delta x}{1!} \frac{df}{dx}\bigg|_i + \frac{2\Delta x^3}{3!} \frac{d^3 f}{dx^3}\bigg|_i + \cdots \tag{2.83}$$

from which

$$\frac{df}{dx}\bigg|_i = \frac{f_{i+1} - f_{i-1}}{2\Delta x} - \frac{\Delta x^2}{3!} \frac{d^3 f}{dx^3}\bigg|_i + \cdots \tag{2.84}$$

or

$$\frac{df}{dx}\bigg|_i \approx \frac{f_{i+1} - f_{i-1}}{2\Delta x} \tag{2.85}$$

This is called the central difference representation of the first derivative in which the value of the derivative of function f at point i is related to the value of the function f at point $(i + 1)$ and its value at point $(i - 1)$. This method is more accurate since its truncation error is of the order of $O(\Delta x)^2$.

We can also come up with higher-order approximation to the first derivative by taking additional terms, for example,

$$\frac{df}{dx} = \frac{3f_i - 4f_{i-1} + f_{i-2}}{2\Delta x} \tag{2.86}$$

is the second-order backward approximation to the first derivative, obtained by using Equations 2.76 and 2.78.

2.3.1.2 Second-Order Derivatives

In a similar fashion, one can arrive at different representations of the second derivatives. Multiplying Equation 2.75 by 2 and subtracting Equation 2.77 from it results in

$$2f_{i+1} - f_{i+2} = f_i - \frac{2\Delta x^2}{2!} \frac{d^2 f}{dx^2}\bigg|_i - \frac{6\Delta x^3}{3!} \frac{d^3 f}{dx^3}\bigg|_i + \cdots \tag{2.87}$$

Solving for the second derivative from Equation 2.87 and neglecting the higher-order terms, we get

$$\frac{d^2 f}{dx^2}\bigg|_i \approx \frac{f_{i+2} + f_i - 2f_{i+1}}{\Delta x^2} \tag{2.88}$$

This is the forward difference representation of the second derivative and the truncation error is of the order of $O(\Delta x)$. Similarly, multiplying Equation 2.76 by 2 and subtracting Equation 2.78 from it results in

$$\frac{d^2 f}{dx^2}\bigg|_i \approx \frac{f_{i-2} - 2f_{i-1} + f_i}{\Delta x^2} \tag{2.89}$$

This is the backward difference representation of the second derivative with a truncation error of the order of $O(\Delta x)$.

Adding Equations 2.75 and 2.76 results in the central differencing formulation of the second derivative given by

$$\frac{d^2 f}{dx^2}\bigg|_i \approx \frac{f_{i+1} - 2f_i + f_{i-1}}{\Delta x^2} \qquad (2.90)$$

with a truncation error of the order of $O(\Delta x)^2$. This is the form most frequently used for the second derivative approximation.

The earlier formulations are based on assuming uniform spacing between the nodes. There are situations where it is advantageous to use variable spacing. Using the same process, we can come up with the finite difference formulation when the spacing is not uniform. Rewriting Equations 2.75 and 2.76

$$f_{i+1} = f_i + \frac{\Delta x^+}{1!} \frac{df}{dx}\bigg|_i + \frac{\Delta x^{+2}}{2!} \frac{d^2 f}{dx^2}\bigg|_i + \frac{\Delta x^{+3}}{3!} \frac{d^3 f}{dx^3}\bigg|_i + \cdots \qquad (2.91)$$

$$f_{i-1} = f_i - \frac{\Delta x^-}{1!} \frac{df}{dx}\bigg|_i + \frac{\Delta x^{-2}}{2!} \frac{d^2 f}{dx^2}\bigg|_i - \frac{\Delta x^{-3}}{3!} \frac{d^3 f}{dx^3}\bigg|_i + \cdots \qquad (2.92)$$

where $\Delta x^+ = x_{i+1} - x_i$ and $\Delta x^- = x_i - x_{i-1}$. As earlier, rearranging (2.91) and (2.92) results in the following for the central difference approximations of the first and second derivatives

$$\frac{df}{dx} = \frac{\dfrac{f_{i+1}}{\Delta x^{+2}} - \left(\dfrac{1}{\Delta x^{+2}} - \dfrac{1}{\Delta x^{-2}}\right) f_i - \dfrac{f_{i-1}}{\Delta x^{-2}}}{\dfrac{1}{\Delta x^+} + \dfrac{1}{\Delta x^-}} \qquad (2.93)$$

$$\frac{d^2 f}{dx^2} = \frac{\dfrac{f_{i+1}}{\Delta x^+} - \left(\dfrac{1}{\Delta x^+} + \dfrac{1}{\Delta x^-}\right) f_i + \dfrac{f_{i-1}}{\Delta x^-}}{\dfrac{1}{2}(\Delta x^+ + \Delta x^-)} \qquad (2.94)$$

2.3.2 Spreadsheet Solution

A drawback of numerical solutions is the need for either the knowledge of a programming language or using specialized software. The techniques and solution methodologies covered throughout this book can easily be adopted to any programming language, including equation solvers, and there is no requirement to use spreadsheets. Spreadsheets are increasingly used as an effective tool for the solution of many engineering problems. They are on almost all personal computers, and most people have a basic knowledge of how to use them. They also have powerful calculation and graphic capabilities. Spreadsheets provide an option that at least eliminates the need for learning a programming language. They provide a natural environment for numerical solutions, as each cell becomes a node. The built-in plotting capabilities are also great help in presenting and understanding the results.

Example 2.4

Obtain the solution to the following ODE using finite difference and compare the results to the exact solution

$$\frac{d^2y}{dx^2} + y = 0 \tag{2.95}$$

subject to

$$y(0) = 0$$
$$\left.\frac{dy}{dx}\right|_0 = 1 \tag{2.96}$$

The exact solution of the earlier equation is $y = \sin x$.

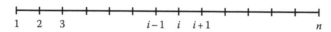

The finite difference approximation is

$$\frac{y_{i+1} - 2y_i + y_{i-1}}{\Delta x^2} + y_i = 0 \tag{2.97}$$

which is then rearranged into

$$y_{i+1} + y_i[\Delta x^2 - 2] + y_{i-1} = 0 \tag{2.98}$$

From the boundary conditions,

$$y_1 = 0 \tag{2.99}$$

and the finite difference approximation to the second boundary condition results in

$$\frac{y_2 - y_1}{\Delta x} = 1 \tag{2.100}$$

or

$$y_2 - y_1 = \Delta x \tag{2.101}$$

Assuming $\Delta x = 0.1$, the first four y value calculations are shown in Table 2.1.

TABLE 2.1

The Finite Difference and Exact Calculations for the First Four y Values

x	Finite Difference	Exact Analytical
0.0	$y_1 = 0.0$	$y_1 = 0.0$
0.1	$y_2 = y_1 + 0.1 = 0.1$	$y_2 = \sin(0.1) = 0.0998$
0.2	$y_3 = -y_2[0.1^2 - 2] - y_1 = 0.1990$	$y_3 = \sin(0.2) = 0.1987$
0.3	$y_4 = -y_3[0.1^2 - 2] - y_2 = 0.2960$	$y_3 = \sin(0.3) = 0.2955$

	A	B	C	D	E	F	G
1	dx	=0.1	i	x	y Finite Difference	y Exact Analytical	Error
2			1	=0	0	=SIN(D2)	
3			2	=D2+dx	=E2+dx	=SIN(D3)	=ABS(E3-F3)/F3
4			3	=D3+dx	=-E3*(dx^2-2)-E2	=SIN(D4)	=ABS(E4-F4)/F4
5			4	=D4+dx	=-E4*(dx^2-2)-E3	=SIN(D5)	=ABS(E5-F5)/F5
6			5	=D5+dx	=-E5*(dx^2-2)-E4	=SIN(D6)	=ABS(E6-F6)/F6
7			6	=D6+dx	=-E6*(dx^2-2)-E5	=SIN(D7)	=ABS(E7-F7)/F7
8			7	=D7+dx	=-E7*(dx^2-2)-E6	=SIN(D8)	=ABS(E8-F8)/F8
9			8	=D8+dx	=-E8*(dx^2-2)-E7	=SIN(D9)	=ABS(E9-F9)/F9
10			9	=D9+dx	=-E9*(dx^2-2)-E8	=SIN(D10)	=ABS(E10-F10)/F10
11			10	=D10+dx	=-E10*(dx^2-2)-E9	=SIN(D11)	=ABS(E11-F11)/F11
12			11	=D11+dx	=-E11*(dx^2-2)-E10	=SIN(D12)	=ABS(E12-F12)/F12
13			12	=D12+dx	=-E12*(dx^2-2)-E11	=SIN(D13)	=ABS(E13-F13)/F13
14			13	=D13+dx	=-E13*(dx^2-2)-E12	=SIN(D14)	=ABS(E14-F14)/F14
15			14	=D14+dx	=-E14*(dx^2-2)-E13	=SIN(D15)	=ABS(E15-F15)/F15
16			15	=D15+dx	=-E15*(dx^2-2)-E14	=SIN(D16)	=ABS(E16-F16)/F16
17			16	=D16+dx	=-E16*(dx^2-2)-E15	=SIN(D17)	=ABS(E17-F17)/F17
18			17	=D17+dx	=-E17*(dx^2-2)-E16	=SIN(D18)	=ABS(E18-F18)/F18
19			18	=D18+dx	=-E18*(dx^2-2)-E17	=SIN(D19)	=ABS(E19-F19)/F19
20			19	=D19+dx	=-E19*(dx^2-2)-E18	=SIN(D20)	=ABS(E20-F20)/F20
21			20	=D20+dx	=-E20*(dx^2-2)-E19	=SIN(D21)	=ABS(E21-F21)/F21
22			21	=D21+dx	=-E21*(dx^2-2)-E20	=SIN(D22)	=ABS(E22-F22)/F22
23			22	=D22+dx	=-E22*(dx^2-2)-E21	=SIN(D23)	=ABS(E23-F23)/F23
24			23	=D23+dx	=-E23*(dx^2-2)-E22	=SIN(D24)	=ABS(E24-F24)/F24
25			24	=D24+dx	=-E24*(dx^2-2)-E23	=SIN(D25)	=ABS(E25-F25)/F25
26			25	=D25+dx	=-E25*(dx^2-2)-E24	=SIN(D26)	=ABS(E26-F26)/F26
27			26	=D26+dx	=-E26*(dx^2-2)-E25	=SIN(D27)	=ABS(E27-F27)/F27
28			27	=D27+dx	=-E27*(dx^2-2)-E26	=SIN(D28)	=ABS(E28-F28)/F28
29			28	=D28+dx	=-E28*(dx^2-2)-E27	=SIN(D29)	=ABS(E29-F29)/F29
30			29	=D29+dx	=-E29*(dx^2-2)-E28	=SIN(D30)	=ABS(E30-F30)/F30
31			30	=D30+dx	=-E30*(dx^2-2)-E29	=SIN(D31)	=ABS(E31-F31)/F31
32							

FIGURE 2.7
Spreadsheet for solving Example 2.4.

Figure 2.7 shows the formulas that are entered in the spreadsheet to solve the earlier differential equation using finite difference and compares the numerical results with the exact results. Figure 2.8 shows the results of the spreadsheet's calculations. Note that in the program $\Delta x = 0.1$. As can be seen, the finite difference results and the exact analytical solutions are not identical; however, the two are close and the error can be reduced by even taking smaller Δx.

We have also defined a variable called dx in cell B1, which is equivalent to naming cell B1 as dx. This is done by inserting the cursor in cell B1 and then Insert → Name → Define then enter dx in the dialog box of the Define Name window and OK, as shown in Figure 2.9.

Alternatively select cell B1 and then click inside the "Name" box that is normally on the formula bar on your toolbar at the top of your screen. When you click inside this box, you see the cell name (B1) is highlighted and moves to the left of the box, as shown in Figure 2.10.

While B1 highlighted, type dx and hit Enter (Figure 2.11).

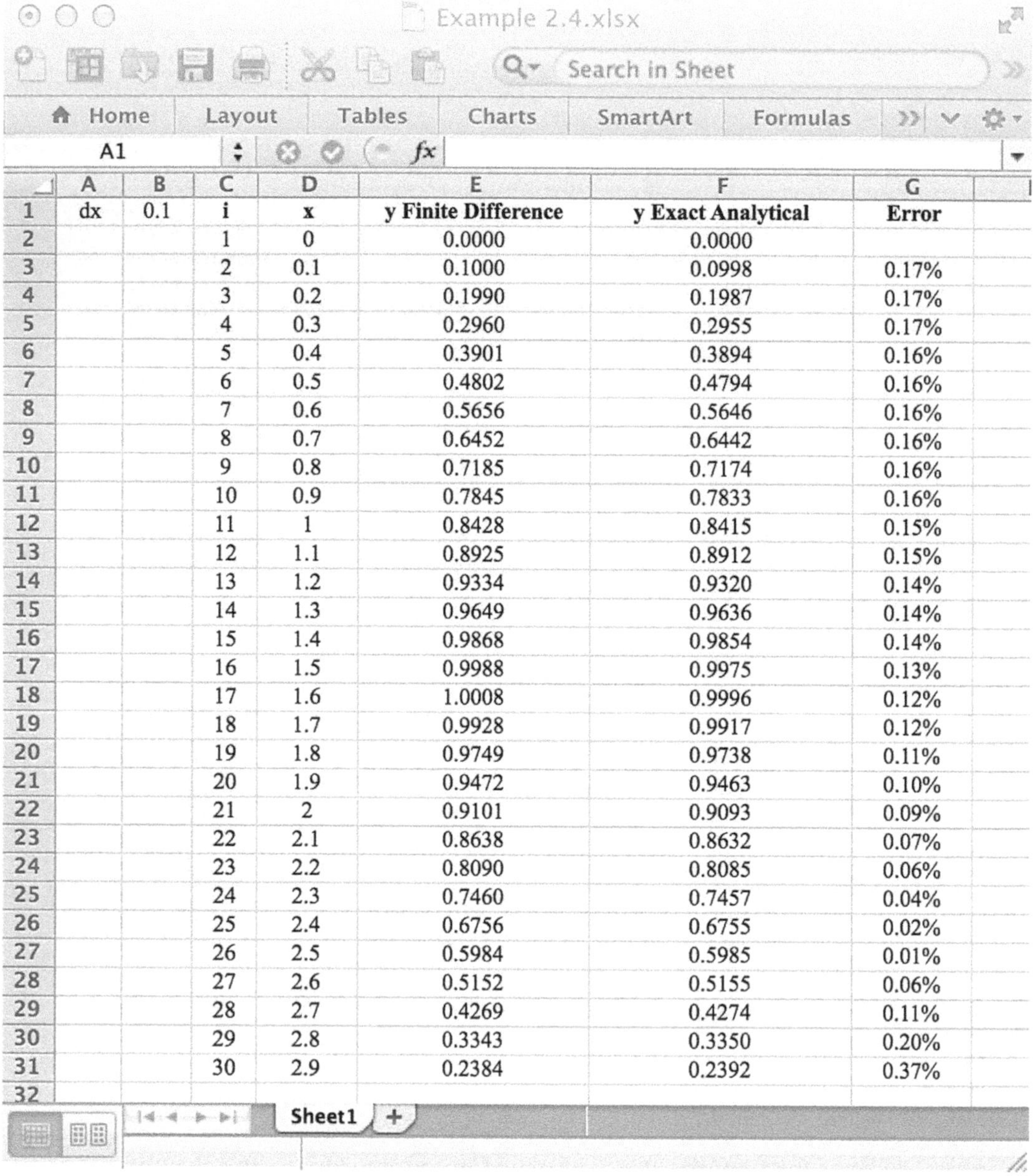

	A	B	C	D	E	F	G
1	dx	0.1	i	x	y Finite Difference	y Exact Analytical	Error
2			1	0	0.0000	0.0000	
3			2	0.1	0.1000	0.0998	0.17%
4			3	0.2	0.1990	0.1987	0.17%
5			4	0.3	0.2960	0.2955	0.17%
6			5	0.4	0.3901	0.3894	0.16%
7			6	0.5	0.4802	0.4794	0.16%
8			7	0.6	0.5656	0.5646	0.16%
9			8	0.7	0.6452	0.6442	0.16%
10			9	0.8	0.7185	0.7174	0.16%
11			10	0.9	0.7845	0.7833	0.16%
12			11	1	0.8428	0.8415	0.15%
13			12	1.1	0.8925	0.8912	0.15%
14			13	1.2	0.9334	0.9320	0.14%
15			14	1.3	0.9649	0.9636	0.14%
16			15	1.4	0.9868	0.9854	0.14%
17			16	1.5	0.9988	0.9975	0.13%
18			17	1.6	1.0008	0.9996	0.12%
19			18	1.7	0.9928	0.9917	0.12%
20			19	1.8	0.9749	0.9738	0.11%
21			20	1.9	0.9472	0.9463	0.10%
22			21	2	0.9101	0.9093	0.09%
23			22	2.1	0.8638	0.8632	0.07%
24			23	2.2	0.8090	0.8085	0.06%
25			24	2.3	0.7460	0.7457	0.04%
26			25	2.4	0.6756	0.6755	0.02%
27			26	2.5	0.5984	0.5985	0.01%
28			27	2.6	0.5152	0.5155	0.06%
29			28	2.7	0.4269	0.4274	0.11%
30			29	2.8	0.3343	0.3350	0.20%
31			30	2.9	0.2384	0.2392	0.37%
32							

FIGURE 2.8
Spreadsheet results.

2.3.2.1 Boundary Value Problems

The boundary value problems are differential equations whose boundary conditions are specified at more than one point over the domain. Most ODEs and PDEs encountered in mechanical engineering are boundary value problems. The numerical solution of these equations results in a set of coupled algebraic equations. There are two methods for solving boundary value problems, iteratively or directly through matrix inversion.

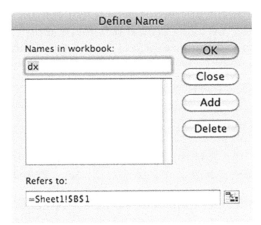

FIGURE 2.9
Defining variables.

FIGURE 2.10
Original cell name, B1.

FIGURE 2.11
New cell name *dx* (cell B1 can now also be referred to as *dx*).

In the iterative approach, an initial guess is made to start the iteration process and the solution is improved on in successive iterations until it converges, i.e., the difference between successive iterations at all points becomes less than a specified value. In the direct approach, the solution to the set of algebraic equations is obtained by inversion of the coefficient matrix. Both approaches are shown through an example.

Consider the nondimensional energy equation and the boundary conditions for conduction in a cylinder with energy generation, subject to a convective boundary

$$\frac{1}{r^*}\frac{d}{dr^*}\left(r^*\frac{dT^*}{dr^*}\right)+1=\frac{d^2T^*}{dr^{*2}}+\frac{1}{r^*}\frac{dT^*}{dr^*}+1=0 \tag{2.102}$$

$$r^*=0.,\quad \frac{dT^*}{dr^*}=0$$

$$r^*=1.0,\quad \frac{dT^*}{dr^*}=-BiT^* \tag{2.103}$$

The exact solution is given by

$$T^*=\frac{1}{4}-\frac{r^{*2}}{4}+\frac{1}{2Bi} \tag{2.104}$$

To numerically solve Equation 2.102, the first step is to divide the domain into $(N-1)$ divisions using N nodes. Then, by using central difference approximation for the first and second derivatives, the finite difference form of the energy equation becomes

$$\frac{T_{i-1}-2T_i+T_{i+1}}{\Delta r^2}+\frac{1}{r_i}\frac{T_{i+1}-T_{i-1}}{2\Delta r}+1=0 \tag{2.105}$$

which can be rearranged into

$$T_{i-1}\left[\frac{1}{\Delta r^2}-\frac{1}{2r_i\Delta r}\right]+T_i\left[\frac{-2}{\Delta r^2}\right]+T_{i+1}\left[\frac{1}{\Delta r^2}+\frac{1}{2r_i\Delta r}\right]=-1 \tag{2.106}$$

for $1 < i < N$. The finite difference form of the boundary conditions results in

$$T_1 = T_2 \tag{2.107}$$

$$-T_{N-1}+(1+Bi\Delta r)T_N = 0 \tag{2.108}$$

2.3.2.1.1 Iterative Solution of Boundary Value Problems

The iterative solution can be easily obtained using the built-in iterative capabilities of Excel®. To use Excel to solve the earlier equations iteratively, we solve for T_i from the finite difference equation (2.106) to get

$$T_i = \frac{T_{i+1}\left[\dfrac{1}{\Delta r^2}+\dfrac{1}{2\Delta rr_i}\right]+T_{i-1}\left[\dfrac{1}{\Delta r^2}-\dfrac{1}{2\Delta rr_i}\right]+1}{\dfrac{2}{\Delta r^2}} \tag{2.109}$$

If we consider each cell in a spreadsheet column as one of the finite difference nodes, then in the cell corresponding to node i, we should enter the RHS of Equation 2.109, that is, the content of cell i is equal to the content of the cell immediately below it $(i + 1)$ times the quantity in the first square bracket plus the content of the cell immediately above cell i, cell $(i - 1)$, times the quantity in the second square bracket, plus one, divided by the term in the denominator.

The formulas needed for the spreadsheet implementation of the earlier procedure are shown in Figure 2.12. Note that the formulas in the columns D and E are repeating except for the first and last rows. Therefore, the equations in each column have to be written only once in one of the rows, for example E3 ($i = 2$), and then copied in the rest of the cells in the column by selecting the cell, then placing the cursor on the lower right-hand corner of the cell, making sure that the cursor has changed to a plus sign and dragging to the desired cell, in this case E51. The first and last row values are specified by the boundary conditions. The formula for node 1 comes from Equation 2.107 which requires the contents of cell E2 (which is temperature at node 1) to be equal to cell E3 (which is temperature at node 2). For node N, the formula to be used comes from Equation 2.108 which can be solved for T_N:

$$T_N = \frac{T_{N-1}}{(1+Bi\Delta r)} \tag{2.110}$$

and entered in cell E52.

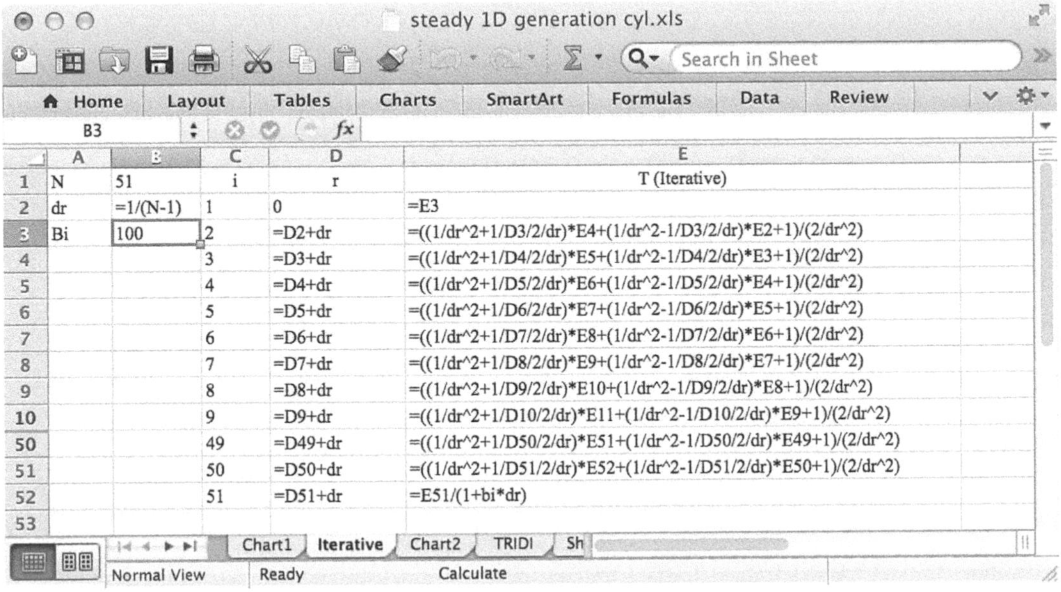

	A	B	C	D	E
1	N	51	i	r	T (Iterative)
2	dr	=1/(N-1)	1	0	=E3
3	Bi	100	2	=D2+dr	=((1/dr^2+1/D3/2/dr)*E4+(1/dr^2-1/D3/2/dr)*E2+1)/(2/dr^2)
4			3	=D3+dr	=((1/dr^2+1/D4/2/dr)*E5+(1/dr^2-1/D4/2/dr)*E3+1)/(2/dr^2)
5			4	=D4+dr	=((1/dr^2+1/D5/2/dr)*E6+(1/dr^2-1/D5/2/dr)*E4+1)/(2/dr^2)
6			5	=D5+dr	=((1/dr^2+1/D6/2/dr)*E7+(1/dr^2-1/D6/2/dr)*E5+1)/(2/dr^2)
7			6	=D6+dr	=((1/dr^2+1/D7/2/dr)*E8+(1/dr^2-1/D7/2/dr)*E6+1)/(2/dr^2)
8			7	=D7+dr	=((1/dr^2+1/D8/2/dr)*E9+(1/dr^2-1/D8/2/dr)*E7+1)/(2/dr^2)
9			8	=D8+dr	=((1/dr^2+1/D9/2/dr)*E10+(1/dr^2-1/D9/2/dr)*E8+1)/(2/dr^2)
10			9	=D9+dr	=((1/dr^2+1/D10/2/dr)*E11+(1/dr^2-1/D10/2/dr)*E9+1)/(2/dr^2)
50			49	=D49+dr	=((1/dr^2+1/D50/2/dr)*E51+(1/dr^2-1/D50/2/dr)*E49+1)/(2/dr^2)
51			50	=D50+dr	=((1/dr^2+1/D51/2/dr)*E52+(1/dr^2-1/D51/2/dr)*E50+1)/(2/dr^2)
52			51	=D51+dr	=E51/(1+bi*dr)
53					

FIGURE 2.12
Spreadsheet formulas for iterative solution (note rows 11–49 are hidden).

Three parameters have been defined in the spreadsheet, and used in the formulas. They are *N*, the number of nodes, *dr*, which is Δr, and the other is *Bi*, which is the Biot number, which is set to 100 in this example. Solution for different Biot numbers can be obtained by changing the value in cell B4.

Since the solution in cell E4 depends on the value of the cell E3 and vice versa, the spreadsheet cannot be calculated and an error will be issued indicating circular reference, which is a hint that the solution has to be obtained iteratively. To enable the iterative capabilities of Excel, go to **Preferences → Calculations** and check **Iterations** and set the **Maximum iterations** to 10,000 and **Maximum Change** to 1e–9. After several thousand iterations, the solution will converge to the values shown in Figure 2.13.

2.3.2.1.2 *Direct Solution of Boundary Value Problems Thomas Algorithm*

One drawback of using spreadsheets for numerical solution is the user's lack of much control over spreadsheets' built-in iterative schemes. This results in the slow convergence rates and limits their use to solving only simple problems. The finite difference formulation of the governing equation and the boundary conditions can be written as

$$T_1 - T_2 = 0 \tag{2.111}$$

$$T_{i-1}\left[\frac{1}{\Delta r^2} - \frac{1}{2r_i\Delta r}\right] + T_i\left[\frac{-2}{\Delta r^2}\right] + T_{i+1}\left[\frac{1}{\Delta r^2} + \frac{1}{2r_i\Delta r}\right] = -1 \tag{2.112}$$

$$-T_{N-1} + (1 + Bi\Delta r)T_N = 0 \tag{2.113}$$

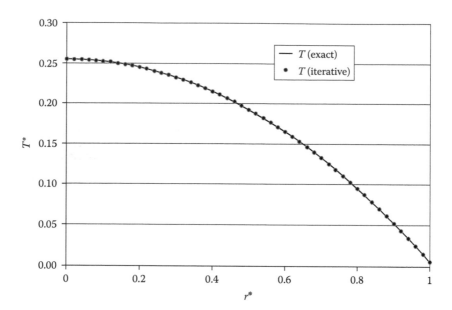

FIGURE 2.13
Comparison of exact and iterative solutions.

Equations 2.111 through 2.113 can be written in matrix form as

$$
\begin{bmatrix}
1 & -1 \\
\left[\dfrac{1}{\Delta r^2} - \dfrac{1}{2r_2\Delta r}\right] & \left[\dfrac{-2}{\Delta r^2}\right] & \left[\dfrac{1}{\Delta r^2} + \dfrac{1}{2r_2\Delta r}\right] \\
& & & & \ddots \\
& & & \left[\dfrac{1}{\Delta r^2} - \dfrac{1}{2r_{n-1}\Delta r}\right] & \left[\dfrac{-2}{\Delta r^2}\right] & \left[\dfrac{1}{\Delta r^2} + \dfrac{1}{2r_{n-1}\Delta r}\right] \\
& & & & -1 & 1 + Bi\Delta r
\end{bmatrix}
\begin{bmatrix}
T_1 \\ T_2 \\ T_3 \\ \vdots \\ T_{n-1} \\ X_n
\end{bmatrix}
=
\begin{bmatrix}
0 \\ -1 \\ \\ \\ -1 \\ 0
\end{bmatrix}
$$

$$(2.114)$$

or more generally as

$$
\begin{bmatrix}
B_1 & C_1 \\
A_2 & B_2 & C_2 \\
& A_3 & B_3 & C_3 \\
& & & \ddots \\
& & & A_{n-1} & B_{n-1} & C_{n-1} \\
& & & & A_n & B_n
\end{bmatrix}
\begin{bmatrix}
X_1 \\ X_2 \\ X_3 \\ \vdots \\ X_{n-1} \\ X_n
\end{bmatrix}
=
\begin{bmatrix}
R_1 \\ R_2 \\ R_3 \\ \vdots \\ R_{n-1} \\ R_n
\end{bmatrix}
$$

$$(2.115)$$

The algebraic Equation 2.115 is a set of N equations for the N unknowns. The coefficient matrix is a tri-diagonal matrix. A tri-diagonal matrix is one with nonzero elements only on the main diagonal and on the diagonals immediately above and immediately below the main one. Linear algebraic equations with tri-diagonal coefficient matrix frequently arise in the finite difference solutions. The inversion of the coefficient matrix for this case is relatively easy. The coefficient matrix can be transformed into a lower triangular matrix, where all the elements on the main diagonal are equal to 1, and the lower diagonal is non-zero. This can be done by dividing the last equation by B_n and the $(n-1)$ equation by C_{n-1}, subtracting the last equation from the $(n-1)$ equation, and dividing the resulting equation by the coefficient of X_{n-1} term resulting in

$$
\begin{bmatrix}
B_1 & C_1 & & & & \\
A_2 & B_2 & C_2 & & & \\
& A_3 & B_3 & C_3 & & \\
& & & & & \\
& & & & A'_{n-1} & 1 \\
& & & & & A'_n & 1
\end{bmatrix}
\begin{bmatrix}
X_1 \\
X_2 \\
X_3 \\
\\
X_{n-1} \\
X_n
\end{bmatrix}
=
\begin{bmatrix}
R_1 \\
R_2 \\
R_3 \\
\\
R'_{n-1} \\
R'_n
\end{bmatrix}
\tag{2.116}
$$

This process can now be repeated until the set of algebraic equations becomes a lower triangular matrix:

$$
\begin{bmatrix}
1 & & & & & \\
A'_1 & 1 & & & & \\
& A'_2 & 1 & & & \\
& & & & & \\
& & & & A'_{n-1} & 1 \\
& & & & & A'_n & 1
\end{bmatrix}
\begin{bmatrix}
X_1 \\
X_2 \\
X_3 \\
\\
X_{n-1} \\
X_n
\end{bmatrix}
=
\begin{bmatrix}
R'_1 \\
R'_2 \\
R'_3 \\
\\
R'_{n-1} \\
R'_n
\end{bmatrix}
\tag{2.117}
$$

From the first equation, X_1 is determined which can then be substituted in the next equation to get X_2, etc.

This is known as the Thomas algorithm, and its implementation in FORTRAN is shown in Figure 2.14, providing an efficient method for solving set of algebraic equations whose coefficient matrix is tri-diagonal.

This code is extensively used in CFD, as the encountered ODEs or PDEs are predominantly second order and their numerical approximation results in sets of algebraic equations whose coefficient matrices are or can be made to be tri-diagonal. Note that the values of $A(1)$ and $C(N)$ are never used and can be set to any value.

This function is not one of the built-in functions of Excel. However, Excel has the capability for the users to define their own functions. Visual Basic Applications (VBA) has been integrated in Excel, except 2008 version for Mac. Visual Basic is a high-level programming language that allows development of among other things User-Defined Functions in Excel. As described by Fakheri and Naraghi [8], Figure 2.15 is the VBA implementation of Thomas algorithm.

```
C PROGRAM TRIDI SOLVES A SET OF EQUATIONS
C WITH TRIDIAGONAL COEFFICIENT MATRIX
      SUBROUTINE TRIDI(A,B,C,X,R,N)
      REAL A(100),B(100),C(100),R(100),X(100)
      A(N)=A(N)/B(N)
      R(N)=R(N)/B(N)
      DO 1 I=2,N
      II=-I+N+2
      BN=1/(B(II-1)-A(II)*C(II-1))
      A(II-1)=A(II-1)*BN
   1  R(II-1)=(R(II-1)-C(II-1)*R(II))*BN
      X(1)=R(1)
      DO 2 I=2,N
   2  X(I)=R(I)-A(I)*X(I-1)
      RETURN
      END
```

FIGURE 2.14
FORTRAN code for solving a tri-diagonal matrix.

```
Option Base 1
Function TRIDI(ByVal Ac As Range, ByVal Bc As Range, ByVal Cc As Range, _ ByVal Rc As Range) As Variant
   Dim BN As Single
   Dim i As Integer
   Dim II As Integer
   Dim A() As Single, B() As Single, C() As Single, R() As Single, X() As Single

   N = Ac.Rows.Count
   ReDim A(N), B(N), C(N), R(N), X(N)
   For i = 1 To N
      A(i) = Ac.Parent.Cells(Ac.Row + i - 1, Ac.Column)
      B(i) = Bc.Parent.Cells(Bc.Row + i - 1, Bc.Column)
      C(i) = Cc.Parent.Cells(Cc.Row + i - 1, Cc.Column)
      R(i) = Rc.Parent.Cells(Rc.Row + i - 1, Rc.Column)
   Next i
      A(N) = A(N)/B(N)
      R(N) = R(N)/B(N)
   For i = 2 To N
      II = -i + N + 2
      BN = 1/(B(II - 1) - A(II)*C(II - 1))
      A(II - 1) = A(II - 1)*BN
      R(II - 1) = (R(II - 1) - C(II - 1)*R(II))*BN
   Next i
      X(1) = R(1)
   For i = 2 To N
      X(i) = R(i) - A(i) * X(i - 1)
   Next i
      TRIDI = Application. WorksheetFunction.Transpose(X)
End Function
```

FIGURE 2.15
VBA code for solving a tri-diagonal matrix using Excel®.

As described in Ref. [9] to turn this algorithm into a user-defined Excel function:

1. Start a new workbook
2. Enter the VBA environment (Press Alt + F11) or Tools → Macro → Visual Basic Editor
3. Insert a new module (Insert → Module)
4. Type all the text in Figure 2.15
5. Exit VBA (Press Alt + Q)
6. Save the spreadsheet and give a name like MYTRIDI
7. The function can be used Insert → Function → User Defined → TRIDI
8. To use TRIDI in more than the workbook just created, you need to save your function as a custom add-in. File → Save As → Excel Add-In file (.xlam)
9. Load the Add-In (Tools → Add-Ins → Select → xlam file you just saved)

Steps 1–6 create a spreadsheet that contains the function TRIDI. Steps 8 and 9 will make TRIDI a user-defined function that is available to all new sheets started on the user's computer. However, if you open a spreadsheet that uses TRIDI on another computer that has not had TRIDI installed on it, the function will not be recognized. If you think that you may want to share your calculations, always start with a sheet that already has TRIDI, like the one saved in step 6. Alternatively, you can download a copy from the Book's companion website www.ihtbook.com.

The function TRIDI is an array function that returns N values for X_1 through X_N. Because TRIDI is an array function, you need to select N cells in a column then enter the function in the first cell, and then press Control–Shift–Enter (Command–Enter on Mac).

The syntax of TRIDI is

$$= \text{TRIDI}(\text{a1:an}, \text{b1:bn}, \text{c1:cn}, \text{r1:rn})$$

where a1, b1, c1, and r1 are addresses of the first cells in the columns containing arrays A, B, C, and R, respectively, and an, bn, cn, and rn are addresses of the last cells in the columns containing arrays A, B, C, and R, respectively.

Example 2.5

Solve the problem of heat conduction in a cylindrical bar with energy generation, using subroutine TRIDI.

From Equations 2.111 through 2.113

$$B_1 = 1, \quad C_1 = -1, \quad R_1 = 0$$

$$A_i = \frac{1}{\Delta r^2} - \frac{1}{2r_i \Delta r}, \quad B_i = -\frac{2}{\Delta r^2}, \quad C_i = \frac{1}{\Delta r^2} + \frac{1}{2r_i \Delta r}, \quad R_i = -1$$

$$A_N = -1, \quad B_N = 1 + Bi\Delta r, \quad R_N = 0$$

As mentioned earlier, A_1 and C_N are not used and are assigned a value of 1. Figure 2.16 shows the spreadsheet use of TRIDI for the solution of Example 2.5.

There are 51 nodes, Δr is 0.01, and Biot number is 100. Column C contains the node numbering, column D calculates the values of r, and the columns E, F, G, and H contain steady 1D generation cyl.xls

100% ▾   Q▾ Search in Sheet

🏠 Home   Layout   Tables   Charts   SmartArt   Formulas   Data   Review

A3 | fx

	A	B	C	D	E	F	G	H	I
1	N	=51	i	r	A	B	C	R	T (TRIDI)
2	dr	=1/(N-1)	1	=0	1	1	-1	0	=TRIDI(E2:E52,F2:F52,G2:G52,H2:H52)
3	Bi	100	2	=D2+dr	=1/dr^2-1/D3/2/dr	=-2/dr^2	=1/dr^2+1/D3/2/dr	=-1	=TRIDI(E2:E52,F2:F52,G2:G52,H2:H52)
4			3	=D3+dr	=1/dr^2-1/D4/2/dr	=-2/dr^2	=1/dr^2+1/D4/2/dr	=-1	=TRIDI(E2:E52,F2:F52,G2:G52,H2:H52)
5			4	=D4+dr	=1/dr^2-1/D5/2/dr	=-2/dr^2	=1/dr^2+1/D5/2/dr	=-1	=TRIDI(E2:E52,F2:F52,G2:G52,H2:H52)
6			5	=D5+dr	=1/dr^2-1/D6/2/dr	=-2/dr^2	=1/dr^2+1/D6/2/dr	=-1	=TRIDI(E2:E52,F2:F52,G2:G52,H2:H52)
7			6	=D6+dr	=1/dr^2-1/D7/2/dr	=-2/dr^2	=1/dr^2+1/D7/2/dr	=-1	=TRIDI(E2:E52,F2:F52,G2:G52,H2:H52)
8			7	=D7+dr	=1/dr^2-1/D8/2/dr	=-2/dr^2	=1/dr^2+1/D8/2/dr	=-1	=TRIDI(E2:E52,F2:F52,G2:G52,H2:H52)
9			8	=D8+dr	=1/dr^2-1/D9/2/dr	=-2/dr^2	=1/dr^2+1/D9/2/dr	=-1	=TRIDI(E2:E52,F2:F52,G2:G52,H2:H52)
10			9	=D9+dr	=1/dr^2-1/D10/2/dr	=-2/dr^2	=1/dr^2+1/D10/2/dr	=-1	=TRIDI(E2:E52,F2:F52,G2:G52,H2:H52)
50			49	=D49+dr	=1/dr^2-1/D50/2/dr	=-2/dr^2	=1/dr^2+1/D50/2/dr	=-1	=TRIDI(E2:E52,F2:F52,G2:G52,H2:H52)
51			50	=D50+dr	=1/dr^2-1/D51/2/dr	=-2/dr^2	=1/dr^2+1/D51/2/dr	=-1	=TRIDI(E2:E52,F2:F52,G2:G52,H2:H52)
52			51	=D51+dr	=-1	=(1+bi*dr)	1	=0	=TRIDI(E2:E52,F2:F52,G2:G52,H2:H52)

◄ ◄ ► ►| Chart1   Iterative   Chart2   **TRIDI**   Sheet3   +

Normal View   Ready   Calculate   Sum=0   ▾

FIGURE 2.16
Using function TRIDI in spreadsheet.

values of *A*, *B*, *C*, and *R*. To use the TRIDI, select the cells that will contain the values of the unknown *T*, in this case I2 to I52 and type

$$= TRIDI(E2{:}E52, F2{:}F52, G2{:}G52, H2{:}H52)$$

in cell I2, then press Control–Shift–Enter (Command–Enter on Mac) to complete the array function. In the formula bar, Excel places { } characters around the function to indicate that it is an array function and computes the values of *T* and puts them in cells I2:I52.

Figure 2.17 is a comparison between the numerical solution using TRIDI function and the exact solution. As can be seen the solution is very close to the exact solution, as is the solution obtained iteratively. The advantage of TRIDI is that the solution is obtained directly without any need for iteration.

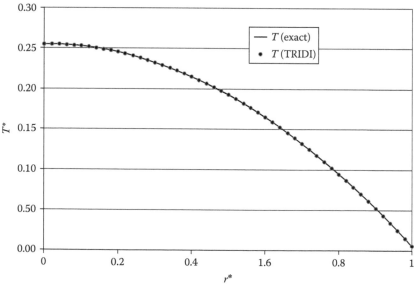

FIGURE 2.17
Comparison of exact and direct (TRIDI) solutions.

Problems

2.1 Consider steady 1D conduction in a plane wall with constant thermal conductivity. If both sides are exposed to a convective environment having heat transfer coefficient h, with the medium on the LHS at T_L and the medium on the RHS at T_R:
   a. Nondimensionalize the governing equation and boundary conditions
   b. Numerically obtain the solution
   c. Compare the numerical results with the exact solution

2.2 Consider steady 1D conduction in a plane wall with variable thermal conductivity $k = aT + b$. If both sides are exposed to an environment at T_∞ and h:
   a. Nondimensionalize the governing equation and boundary conditions
   b. Obtain the solution numerically

2.3 Simplify the heat diffusion equation to come up with the governing equation and also provide the needed boundary conditions for steady, 2D heat transfer in a rectangular plate, where all the sides are maintained at constant temperature T_1, except the bottom surface which is at T_2. Nondimensionalize the governing equation and the boundary conditions.

2.4 Simplify the heat diffusion equation to come up with the governing equation and also provide the needed boundary conditions for heat transfer in a rectangular plate, losing heat by convection from top, bottom, and the right face, while the left surface is maintained at temperature T_b. Nondimensionalize the governing equation and the boundary conditions.

2.5 The energy equation for a constant area fin becomes

$$\frac{d^2T}{dx^2} - \frac{hp}{kA}(T - T_\infty) = 0$$

$$x = 0, \quad T = T_b$$

$$x = L, \quad -k\frac{dT}{dx} = h(T - T_\infty)$$

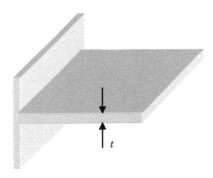

   a. Nondimensionalize the governing equation and the boundary conditions.
   b. Obtain the analytical solution.

2.6 The energy equation for transient conduction for an object having volume V, surface area A, whose temperature (T) does not change with location and loses heat by convection (h, T_∞) is

$$\rho c V \frac{\partial T}{\partial t} = -hA(T - T_\infty)$$

$$t = 0, \quad T = T_i$$

where
  ρ is the density
  c is the specific heat of the object

  a. Nondimensionalize the governing equation and the boundary conditions.
  b. Determine the solution.

2.7 The energy equation for 1D transient conduction in a wall is

$$\frac{\partial^2 T}{\partial x^2} = \frac{1}{\alpha} \frac{\partial T}{\partial t}$$

$$t = 0, \quad T = T_i$$

$$x = 0, \quad \frac{dT}{dx} = 0$$

$$x = L, \quad -k\frac{dT}{dx} = h(T - T_\infty)$$

Nondimensionalize the governing equation and the boundary conditions.

2.8 Derive the governing equation and the boundary conditions for steady 1D conduction in a semitransparent medium, if the rate of energy absorption is given as

$$\gamma = \mu I_{in} e^{-\mu x}$$

where
  μ is the absorption coefficient
  I_{in} is the intensity of the incoming radiation

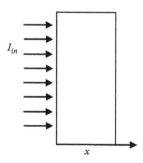

The left surface is exposed to environment at T_∞ and h and the right surface is maintained at T_s. Nondimensionalize the governing equation and the boundary conditions and obtain the solution numerically.

2.9 Consider steady 1D conduction with energy generation in a long cylinder exposed to an environment at T_∞ and h. The thermal conductivity and energy generation are not constant and are given by

$$k = aT + b$$

$$\dot{q} = cT + d$$

   a. Nondimensionalize the governing equation and boundary conditions.
   b. Obtain the solution numerically for a reasonable set of parameters.

2.10 Numerically solve for the temperature distribution and heat transfer in the truncated cone shown whose sides are insulated. Assume that $D = D_1 e^x$, the left end diameter is D_1 and the two ends are at T_1 and T_2, and the length is L. Compare the results with the exact solution.

2.11 Consider a long circular cylinder with energy generation given by $\dot{q} = a + b\dfrac{r}{R}$, where a and b are two known constants, r is measured from the center of the cylinder, and R is the radius:

   a. Nondimensionalize the governing equation and boundary conditions.
   b. Obtain the solution analytically.

2.12 For Problem 2.11, obtain the solution numerically and compare with the analytical solution, for $b/a = -1$, and $Bi = 100$.

2.13 A spherical nuclear fuel element generates energy at the constant rate of $\dot{q}$ W/m³. The heat transfer coefficient depends on the surface temperature and is given by

$$h = CT_s^m$$

where C and m are constants

a. Derive the governing equation for temperature distribution in the fuel, assuming *h* varies according to the expression given earlier.
b. Nondimensionalize the governing equation and the boundary conditions.
c. Obtain the solution numerically.

2.14 A hot wire anemometer is a device used for measuring velocity of a fluid. It is made up of a thin wire carrying a current. As the velocity of the fluid flowing over the wire changes, the temperature of the wire and therefore the resistance of the wire change. If the wire conductivity and resistance are a linear function of temperature, $k = k_0(1 + a_1 T)$, $R = R_0(1 + b_1 T)$:
a. Derive the governing equation.
b. Nondimensionalize the governing equation and boundary conditions.
c. Solve the problem numerically, for an arbitrary set of parameters.

2.15 Derive the transient formulation of Problem 2.14, and nondimensionalize the governing equation and boundary conditions.

2.16 A nuclear fuel rod generates energy at the constant rate of $\dot{q}$ W/m^3 and is submerged in fluid at temperature T_∞ and the heat transfer coefficient is *h*. There is a sudden loss of coolant, dropping the heat transfer coefficient to 1/10 of the original value:

a. Derive the governing equations and boundary conditions for temperature distribution in the fuel.
b. Nondimensionalize the governing equation and the boundary conditions.

2.17 A spherical nuclear fuel element generates energy at the constant rate of $\dot{q}$ W/m^3. It has a cladding of thickness *t* and is suddenly submerged in fluid at temperature T_∞ and the heat transfer coefficient is *h*:

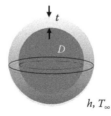

a. Derive the governing equations and boundary conditions for temperature distribution in the fuel, and cladding.
b. Nondimensionalize the governing equation and the boundary conditions.

2.18 A cylindrical rod receives a nonuniform heat flux over the top half of its circumference and loses heat by convection over its entire circumference

$$q''(\theta) = \begin{array}{ll} q_0'' \sin(\theta) & 0 \le \theta \le \pi \\ 0 & \pi \le \theta \le 2\pi \end{array}$$

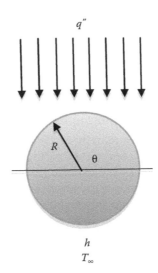

a. Write the governing equation and boundary conditions (20 pts).
b. Nondimensionalize the governing equation and boundary conditions (20 pts).

2.19 The aluminum frying pan has a thickness of 0.75 cm, an outer diameter of 40 cm, and is used to boil water on a gas stove. The pan receives 4 kW of thermal energy from the bottom over an area with a diameter of 30 cm. The inside water temperature is 100°C and the heat transfer coefficient is 500 W/m² K. The ambient temperature is 25°C and the heat transfer coefficient is 10 W/m² K. The pan analysis can be greatly simplified by approximating the pan as a round thin disk having a diameter of 66 cm, losing heat from the top to water and receiving heat over an area with a diameter of 30 cm on the bottom surface and losing heat to the ambient over the rest:

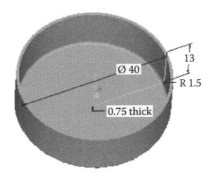

a. Derive the governing equation and the boundary conditions for the temperature distribution in the pan.
b. Nondimensionalize the governing equation and the boundary conditions.

2.20 The energy equation for fully developed flow between two parallel plates that are $2L$ apart is given by distribution in

$$1.5U_{in}\left[1-\left(\frac{y}{L}\right)^2\right]\frac{\partial T}{\partial x} = \alpha\frac{\partial^2 T}{\partial y^2}$$

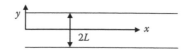

a. What are the boundary conditions if the plates are subjected to a uniform heat flux? Assume at $x = 0$, $u = U_{in}$, $T = T_{in}$.

b. Nondimensionalize the governing equation and the boundary conditions.

2.21 In metal casting, cores are forms, usually made of sand, which are placed into a mold cavity to form the interior surfaces of the castings. Thus the void space between the core and mold-cavity surface is what eventually becomes the casting. In a particular casting process, cores are heated in a microwave oven for drying and curing. Heating in a microwave is equivalent to an internal energy generation $\dot{q}$ that is a function of location in the core. Assuming a transient 1D model

a. Modeling the core as an infinitely long wall with energy generation inside and heat loss by convection to the surroundings, write the governing equation and boundary conditions, if $\dot{q} = ae^{bx}$, where a and b are two known constants and x is measured from the center of the wall.

b. Nondimensionalize the governing equation and the boundary condition.

c. Obtain the steady-state solution analytically.

d. Obtain the steady-state solution numerically for a reasonable set of parameters.

e. Discuss the results.

References

1. Lin, C. C. and Segel, L. A. (1974) *Mathematics Applied to Deterministic Problems in the Natural Sciences*. New York: Macmillan.
2. Fowler, A. C. (1997) *Mathematical Models in the Applied Sciences*. Cambridge: Cambridge University Press.
3. Arfken, G. B. (1985) *Mathematical Methods for Physicists*. New York: Academic Press.
4. Boyce, W. E. and DiPrima, R. C. (2009) *Elementary Differential Equations*. Hoboken, NJ: John Wiley & Sons.
5. Spiegel, M. R. (1971) *Schaum's Outline of Theory and Problems of Advanced Mathematics for Engineers and Scientists*. New York: McGraw-Hill.
6. Ferziger, J. H. and Perić, M. (1996) *Computational Methods for Fluid Dynamics*. Berlin, Germany: Springer.
7. Jaluria, Y. and Torrance, K. E. (2003) *Computational Heat Transfer*. New York: Taylor & Francis.
8. Fakheri, A. and Naraghi, M. H. (2009) Non-iterative solution of ordinary and partial differential equations using spreadsheets. *Computers in Education Journal* 19(3), 28–37.
9. Wittwer, J. (2004) How to create custom Excel functions. User-defined function (UDF) examples for Excel. Retrieved September 4, 2011, from http://www.vertex42.com/ExcelArticles/user-defined-functions.html

3

Fins

3.1 Introduction

Fins or extended surfaces are used to increase the effective surface area available for the transfer of heat and therefore increase the rate of heat transfer from a surface by convection or radiation. The convection heat transfer from a surface is calculated from

$$q = hA(T_s - T_\infty) \tag{3.1}$$

There are three options for increasing the rate of heat transfer.

1. The first is to increase $(T_s - T_\infty)$. Surface temperature, T_s, is often set by other design considerations, and T_∞ is the ambient temperature and mostly beyond our control. There are cases that this option is used, for example, cooling the CPU of a computer with liquid nitrogen to improve performance.
2. The second option is to increase the heat transfer coefficient h. This also requires major design modifications, either to use a fan to move the fluid faster (needs power), to use a different fluid, or to use a fluid that goes through phase change.
3. The third alternative is to increase the area available for the transfer of heat. The projected area is usually fixed, but the surface area available for heat transfer can be increased, by adding fins.

Similar arguments hold for heat transfer by radiation or combined convection and radiation. Fins come in many configurations some of which are shown in Figures 3.1 and 3.2. They can be of the same material as the surface (base) they are attached to or they may be different. Usually, economics and manufacturability determine fin material and construction. Fins are particularly useful in enhancing heat transfer to gases that typically have small heat transfer coefficient, particularly in cases involving free convection. Optimizing fin shape for a given application is an important consideration in fin design.

The fin face attached to the surface is called the fin base, and heat is conducted from the surface that it is attached to through the fin base. In the fin, heat is conducted perpendicular to the base, as well as in the transverse direction (e.g., x and y) and eventually is convected and/or radiated to the ambient from the surfaces of the fin. Conduction in the axial direction (x) is the mechanism responsible for the transfer of heat away from the base and

(a)   (b)

(c)   (d)

FIGURE 3.1
(a) Internally and externally finned tubes, (b) pin fin, (c) miscellaneous fins, and (d) CPU heat sink.

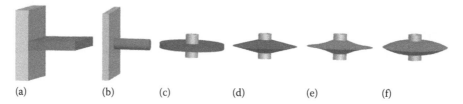

(a)   (b)   (c)   (d)   (e)   (f)

FIGURE 3.2
Different profile fins (a) straight, (b) pin, (c) annular, (d) annular triangular, and (e) and (f) annular parabolic.

along the fin, and conduction in the y direction is the mechanism that primarily carries the heat to the fin surfaces, allowing it to be transferred to the environment. Even though heat transfer in both directions must be considered, it may be possible to neglect temperature variation across the fin compared to that along the fin, making fins a practical example of what are known as quasi one-dimensional conduction problems.

3.2 Quasi One-Dimensional Heat Transfer

The quasi one-dimensional heat transfer problems are those where heat is transferred in two directions, but temperature changes are important only in one. We next examine when temperature variation across the fin can be neglected. Consider the heat transfer in the y direction for the fin shown in Figure 3.3. As discussed in Section 1.4.2, Biot number signifies the ratio of conduction to convection resistances. It is also the ratio of temperature change in the transverse direction in the solid to the temperature difference in the fluid, and for the fin shown in Figure 3.3 is given by

$$\frac{T_c - T_s}{T_s - T_\infty} \approx \frac{ht/2}{k} = Bi \tag{3.2}$$

where
T_c is the center temperature
T_s is the surface temperature
$t/2$ is the characteristic distance in the y direction, which is the distance that heat has to travel

In general, fins are thin and have high thermal conductivity, and as long as h is not very large, the Biot number would be very small. This makes the difference between the center and surface temperatures much smaller than the difference between surface and ambient temperatures, or the temperature variation across the fin will be very small compared to the temperature difference between the fin and the ambient.

We can also perform an order of magnitude analysis, which will be more fully covered in Chapter 6, to estimate the magnitude of the different terms, without actually solving the equations. The rate of heat transfer in the x and y directions by conduction are

$$q_x = -kA_x \frac{\partial T}{\partial x} \tag{3.3}$$

$$q_y = -kA_y \frac{\partial T}{\partial y} \tag{3.4}$$

Assume that the rate of heat transfer in both directions is of the same order of magnitude, i.e., one rate is not much different than the other one, since as argued earlier, both are needed for the fin to function; therefore,

$$-kA_x \frac{\partial T}{\partial x} \approx -kA_y \frac{\partial T}{\partial y} \tag{3.5}$$

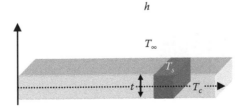

FIGURE 3.3
Axial and lateral temperature change in constant area fin.

Using a first-order linear approximation for the derivatives,

$$-ktW\frac{(\Delta T)_x}{L} \approx -kLW\frac{(\Delta T)_y}{\dfrac{t}{2}} \tag{3.6}$$

$$\frac{(\Delta T)_y}{(\Delta T)_x} \approx \left(\frac{t}{L}\right)^2 \tag{3.7}$$

For fins, t is usually much smaller than L; therefore, for the heat transfers to be of the same order of magnitude, the temperature variation in the y direction must be much smaller than the temperature variation in the x direction, which is the case for $Bi \ll 1$, as shown earlier.

3.3 Energy Equation

To arrive at the governing equation for the general case of a variable cross-sectional area fin, shown in Figure 3.4, we assume the fin thickness to be much smaller than the fin length, making each cross section approximately isothermal. An energy balance on the differential control mass in the fin shown in Figure 3.4 results in

$$q_{cond,x} = q_{cond,x+\Delta x} + dq_{conv} \tag{3.8}$$

Expanding the first term on the right-hand side using Taylor series

$$q_{cond,x} = q_{cond,x} + \frac{dq_{cond,x}}{dx}\Delta x + \frac{dq_{cond,x}}{dx}\frac{\Delta x^2}{2!} + \cdots + dq_{conv} \tag{3.9}$$

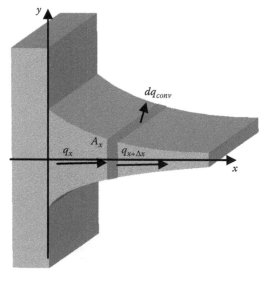

FIGURE 3.4
Energy balance for a differential control mass in a variable area fin.

which simplifies to

$$\frac{dq_{cond,x}}{dx}\Delta x + \frac{dq_{cond,x}}{dx}\frac{\Delta x^2}{2!} + \cdots + dq_{conv} = 0 \qquad (3.10)$$

The heat transfer by convection is

$$dq_{conv} = hdA_s(T - T_\infty) \qquad (3.11)$$

where A_s is the total fin surface area (from fin base to position x) through which heat is transferred by convection. Substituting from (3.11) into (3.10) and dividing both sides of Equation 3.10 by Δx, and letting $\Delta x \to 0$, all higher-order terms will approach zero resulting in

$$\frac{dq_{cond,x}}{dx} + h\frac{dA_s}{dx}(T - T_\infty) = 0 \qquad (3.12)$$

This equation is a statement of the first law and basically says that the net energy out of the system by conduction plus the net heat transfer out by convection must be zero, or energy in by conduction must be equal to energy out by convection, since the problem is steady. From Fourier law of heat conduction,

$$q_x = -kA_x\frac{\partial T}{\partial x} \qquad (3.13)$$

where A_x cross-sectional area at position x. Substituting Equation 3.13 into Equation 3.12 results in the general form of the energy equation for any fin:

$$\frac{d}{dx}\left(kA_x\frac{\partial T}{\partial x}\right) - h\frac{dA_s}{dx}(T - T_\infty) = 0 \qquad (3.14)$$

Assuming constant thermal conductivity,

$$\frac{d}{dx}\left(A_x\frac{\partial T}{\partial x}\right) - \frac{h}{k}\frac{dA_s}{dx}(T - T_\infty) = 0 \qquad (3.15)$$

The first term in this equation is the net heat transfer by conduction in the x direction and the second term is the heat loss by convection. As shown in Figure 3.5a, for a straight fin,

$$\Delta A_s = p\Delta s \qquad (3.16)$$

where p is the circumference of the fin at location x and Δs is the differential length along the surface of the fin. For an annular fin (Figure 3.5b), accounting for the top and bottom areas,

$$\Delta A_s = 2p\Delta s = 4\pi r\Delta s \qquad (3.17)$$

Also

$$\Delta s = \sqrt{(\Delta x)^2 + (\Delta y)^2} \qquad (3.18)$$

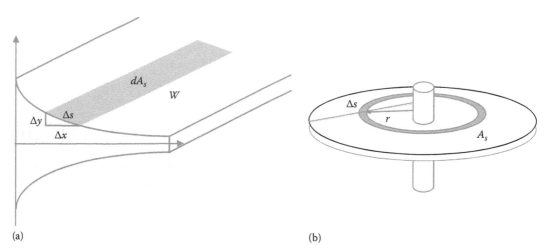

FIGURE 3.5
Variable area (a) straight and (b) annular fins.

Therefore,

$$\frac{dA_s}{dx} = np\sqrt{1 + \left(\frac{dy}{dx}\right)^2} \tag{3.19}$$

where $n = 1$ for straight, and $n = 2$ for annular fins. Integrating Equation 3.19 results in the fin surface area from the fin base to any location x,

$$A_s = \int_0^x np\sqrt{1 + \left(\frac{dy}{dx}\right)^2}\, dx \tag{3.20}$$

Expanding the first term in Equation 3.15 and dividing both sides by A_x, we get an alternative form

$$\frac{d^2T}{dx^2} + \frac{1}{A_x}\frac{dA_x}{dx}\frac{dT}{dx} - \frac{h}{k}\frac{1}{A_x}\frac{dA_s}{dx}(T - T_\infty) = 0 \tag{3.21}$$

This is a second differential equation, requiring two boundary conditions. The boundaries are located at $x = 0$ and $x = L$. At the base of the fin,

$$x = 0, \quad T = T_b \tag{3.22}$$

For the boundary condition at the tip, we consider four different cases. If the fin is long, the fin tip temperature approaches the ambient temperature, or it can be considered to be an infinitely long fin:

$$x \to \infty \quad \text{and} \quad T \to T_\infty \tag{3.23}$$

If the fin loses heat by convection from the tip, then by doing an energy balance at the fin tip,

$$x = L, \quad -k\frac{dT}{dx} = h_L(T - T_L) \tag{3.24}$$

Note that h_L and T_L are the heat transfer coefficient and the ambient temperature at the fin tip respectively, which may be different than the similar quantities at the sides. A special case of convective boundary at the tip is an insulated-tip fin

$$x = L, \quad \frac{dT}{dx} = 0 \tag{3.25}$$

where the heat loss from the fin tip is zero, or the fin tip is insulated. Practical applications of this boundary condition include sharp tip fins or fins that experience a local minimum temperature along their length. Also to simplify the analytical solutions, actual fins are often replaced by an equivalent fin whose tip is insulated. The equivalent fin is slightly longer than the actual one to allow the tip heat loss to be accounted for by the losses from the added fin area to the sides.

The final boundary condition considered is that of an isothermal tip fin, where the fin tip is maintained at a constant known temperature

$$x = L, \quad T = T_L \tag{3.26}$$

3.4 Nondimensionalization

Before solving it, we first nondimensionalize the energy equation and the boundary conditions by defining nondimensional variables

$$T^* = \frac{T - T_\infty}{T_b - T_\infty}, \quad x^* = \frac{x}{L}, \quad y^* = \frac{y}{L}, \quad A_x^* = \frac{A_x^*}{A_{scale}}, \quad A_s^* = \frac{A_s^*}{A_{scale}} \tag{3.27}$$

where A_{scale} is an appropriately defined area (e.g., the fin cross-sectional area at the base) whose definition will not change the form of the equation. In terms of the nondimensional variables, Equation 3.21 becomes

$$\frac{d^2T^*}{dx^{*2}} + \frac{1}{A_x^*}\frac{dA_x^*}{dx^*}\frac{dT^*}{dx^*} - \frac{hL}{k}\frac{1}{A_x^*}\frac{dA_s^*}{dx^*}T^* = 0 \tag{3.28}$$

The nondimensional boundary condition at the base is

$$x^* = 0, \quad T^* = 1 \tag{3.29}$$

and the nondimensional tip boundary condition for the four cases of infinitely long fin, fin with convective, insulated, and isothermal tip, discussed earlier, are

$$x^* \to \infty \quad T^* \to 0 \tag{3.30}$$

$$x^* = 1$$

$$\frac{dT^*}{dx^*} = -Bi\left[T^* - T_L^*\right] \tag{3.31}$$

$$\frac{dT^*}{dx^*} = 0 \tag{3.32}$$

$$T^* = T_L^* \tag{3.33}$$

respectively, where

$$Bi = \frac{h_L L}{k} \tag{3.34}$$

and

$$T_L^* = \frac{T_L - T_\infty}{T_b - T_\infty} \tag{3.35}$$

Figures shown in Table 3.1 are the three commonly encountered variable-area fins. The different terms in Equation 3.21 are also derived and shown in Table 3.1 for the three cases. For straight and pin fins, x is the distance from the base, and for annular fins, x must be replaced with r. For the three cases, Equation 3.21 can be written in the following general form:

$$\frac{d^2T}{dx^2} + \left(\frac{a}{x} + \frac{b}{y}\frac{dy}{dx}\right)\frac{dT}{dx} - \frac{h}{k}\frac{c}{y}\sqrt{1 + \left(\frac{dy}{dx}\right)^2}(T - T_\infty) = 0 \tag{3.36}$$

and in dimensionless form, it becomes

$$\frac{d^2T^*}{dx^{*2}} + \left(\frac{a}{x^*} + \frac{b}{y^*}\frac{dy^*}{dx^*}\right)\frac{dT^*}{dx^*} - \frac{hL}{k}\frac{c}{y^*}\sqrt{1 + \left(\frac{dy^*}{dx^*}\right)^2}T^* = 0 \tag{3.37}$$

where the values of the constants are also listed in Table 3.1.

TABLE 3.1

Parameters in the Energy Equation for Variable Area Fins

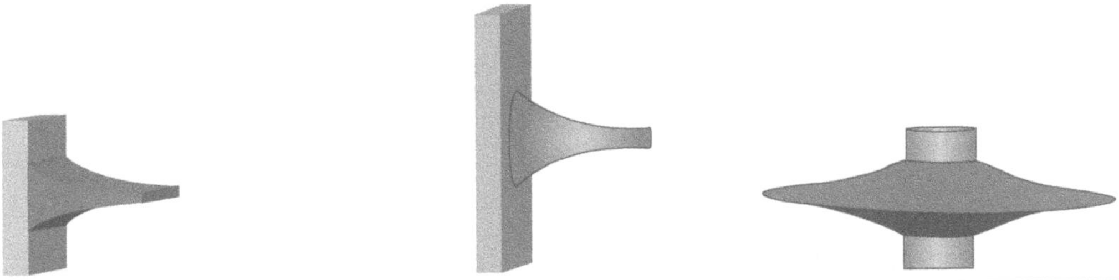

Variable-Thickness Straight Fins	Variable-Thickness Pin Fins	Variable-Thickness Annular Fins

$A_x = 2yW$

$\dfrac{1}{A_x}\dfrac{dA_x}{dx} = \dfrac{1}{y}\dfrac{dy}{dx}$

$A_s = \displaystyle\int_0^x p\sqrt{1+\left(\dfrac{dy}{dx}\right)^2}\,dx$

$\dfrac{dA_s}{dx} = p\sqrt{1+\left(\dfrac{dy}{dx}\right)^2}$

$p = 2(2y + W)$

a	b	c
0	1	1

$A_x = \pi r^2$

$\dfrac{1}{A_x}\dfrac{dA_x}{dx} = \dfrac{2}{r}\dfrac{dr}{dx}$

$A_s = \displaystyle\int_0^x p\sqrt{1+\left(\dfrac{dy}{dx}\right)^2}\,dx$

$\dfrac{dA_s}{dx} = p\sqrt{1+\left(\dfrac{dy}{dx}\right)^2}$

$p = 2\pi r$

a	b	c
0	2	2

$A_r = 4\pi yr$

$\dfrac{1}{A_r}\dfrac{dA_r}{dr} = \dfrac{1}{r}+\dfrac{1}{y}\dfrac{dy}{dr}$

$A_s = 2\displaystyle\int_0^r p\sqrt{1+\left(\dfrac{dy}{dr}\right)^2}\,dr$

$\dfrac{dA_s}{dr} = 2p\sqrt{1+\left(\dfrac{dy}{dr}\right)^2}$

$p = 2\pi r$

a	b	c
1	1	1

$$\frac{d^2T}{dx^2}+\left(\frac{a}{x}+\frac{b}{y}\frac{dy}{dx}\right)\frac{dT}{dx}-\frac{h}{k}\frac{c}{y}\sqrt{1+\left(\frac{dy}{dx}\right)^2}\,(T-T_\infty)=0$$

$$\frac{d^2T^*}{dx^{*2}}+\left(\frac{a}{x^*}+\frac{b}{y^*}\frac{dy^*}{dx^*}\right)\frac{dT^*}{dx^*}-\frac{hL}{k}\frac{c}{y^*}\sqrt{1+\left(\frac{dy^*}{dx^*}\right)^2}\,T^*=0$$

3.5 Constant-Area Fins

Constant-area fins refer to the extended surfaces whose cross section is constant along the fin. Figure 3.6 shows two examples of straight and pin fins. For constant-area fins, y is a constant; therefore, $\dfrac{dy}{dx}=0$, and from Table 3.1, for both cases of straight and pin fins, $a = 0$; therefore, Equation 3.37 simplifies to

$$\frac{d^2T^*}{dx^{*2}}-\frac{hL}{k}\frac{c}{y^*}T^*=0 \tag{3.38}$$

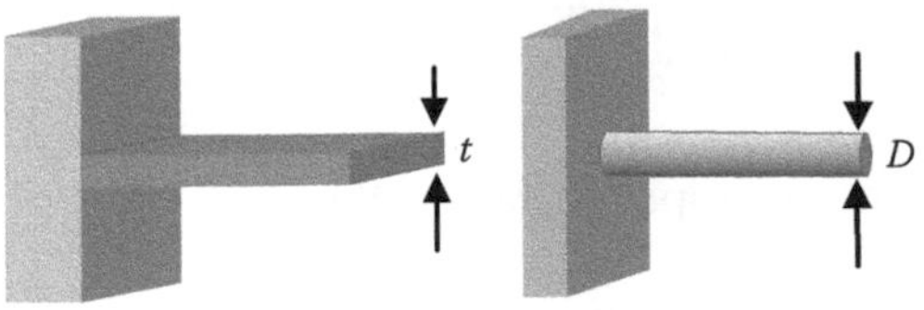

FIGURE 3.6
Straight and pin fins.

TABLE 3.2

Parameters in the Energy Equation for Constant
Area Fins

Constant-Area Straight Fin	Constant-Area Pin Fin
$y = \dfrac{t}{2} \rightarrow y^* = \dfrac{t}{2L}$	$y = \dfrac{D}{2} \rightarrow y^* = \dfrac{D}{2L}$
$c = 1$	$c = 2$
$m^{*2} = \dfrac{2hL^2}{kt}$	$m^{*2} = \dfrac{4hL^2}{kD}$

Further simplification of Equation 3.38 is shown in Table 3.2, and for both cases, the energy
equation simplifies to

$$\frac{d^2 T^*}{dx^{*2}} - m^{*2} T^* = 0 \tag{3.39}$$

3.5.1 Alternative Derivation of the Fin Equation

We can also derive Equation 3.39 for a straight fin by starting with the general heat diffusion equation

$$\frac{\partial^2 T}{\partial x^2} + \frac{\partial^2 T}{\partial y^2} + \frac{\partial^2 T}{\partial z^2} + \frac{\dot{q}}{k} = \frac{1}{\alpha} \frac{\partial T}{\partial t} \tag{3.40}$$

which for steady two-dimensional conduction without energy generation reduces to

$$\frac{\partial^2 T}{\partial x^2} + \frac{\partial^2 T}{\partial y^2} = 0 \tag{3.41}$$

Integrating this equation from $y = 0$ to $y = t/2$

$$\int_0^{t/2} \frac{\partial^2 T}{\partial x^2} dy + \int_0^{t/2} \frac{\partial^2 T}{\partial y^2} dy = 0 \tag{3.42}$$

Since the order of integration and differentiation can be interchanged,

$$\frac{\partial^2}{\partial x^2} \int_0^{t/2} T dy + \frac{\partial T}{\partial y}\bigg|_{t/2} - \frac{\partial T}{\partial y}\bigg|_0 = 0 \tag{3.43}$$

Doing an energy balance at the upper surface of the fin,

$$-k \frac{\partial T}{\partial y}\bigg|_{t/2} = h(T_s - T_\infty) \tag{3.44}$$

and also due to symmetry,

$$\frac{\partial T}{\partial y}\bigg|_0 = 0 \tag{3.45}$$

If we also multiply and divide the first term by $t/2$, then Equation 3.43 becomes

$$\frac{\partial^2}{\partial x^2}\left\{(t/2)\left[\frac{1}{t/2}\int_0^{t/2} Tdy\right]\right\} - \frac{h}{k}(T_s - T_\infty) = 0 \tag{3.46}$$

By definition, the term inside the square bracket is the average temperature across the fin

$$\overline{T} = \frac{1}{t/2}\int_0^{t/2} Tdy \tag{3.47}$$

Therefore, Equation 3.46 becomes

$$\frac{d^2\overline{T}}{dx^2} - \frac{h}{k(t/2)}(T_s - T_\infty) = 0 \tag{3.48}$$

So far, we have not made any approximations, and this equation is as exact as Equation 3.41. However, we have one equation for two unknowns $\overline{T}$ and T_s. As shown earlier, for fins, the temperature variation in the y direction is negligible; therefore, the average temperature is very close to the surface temperature and thus

$$\overline{T} = T_s = T \tag{3.49}$$

and Equation 3.48 becomes

$$\frac{d^2T}{dx^2} - \frac{2h}{kt}(T - T_\infty) = 0 \tag{3.50}$$

which, if nondimensionalized, reduces to Equation 3.39.

3.5.2 Fin Efficiency

The performance of fins is usually expressed in terms of fin efficiency. Fin efficiency is defined as the ratio of the actual heat transfer from the fin and the maximum heat that can ideally be transferred from the fin. Since temperature decreases along the fin, so does the rate of heat transfer. The maximum rate of heat transfer occurs, if the entire fin were at a constant temperature equal to its base temperature. Therefore, the fin efficiency is

$$\eta = \frac{\text{Actual heat transfer from the fin}}{\text{Heat transfer if the fin is at the base temperature}} = \frac{q}{hA_s(T_b - T_\infty)} \tag{3.51}$$

The fin efficiency for a constant-area fin becomes

$$\eta = \frac{q}{(hpL + h_L A_L)(T_b - T_\infty)} \tag{3.52}$$

To calculate the heat transfer from the fin, we note that all the heat that is transferred from the fin surface by convection must have reached the fin through its base by conduction. Therefore, the heat transfer from the fin is equal to the heat conducted into its base and is given as

$$q = -kA_x \left.\frac{dT}{dx}\right|_{x=0} = -kA_x \frac{(T_b - T_\infty)}{L} \left.\frac{dT^*}{dx^*}\right|_{x=0} \tag{3.53}$$

3.5.3 Analytical Solution

For a constant-area fin, the energy equation is

$$\frac{d^2 T^*}{dx^{*2}} - m^{*2} T^* = 0 \tag{3.54}$$

which is a second-order constant coefficient differential equation, whose solution was given in Chapter 2. The general solution is of the form

$$T^* = C_1 e^{-m^* x^*} + C_2 e^{m^* x^*} = D_1 \sinh(m^* x^*) + D_2 \cosh(m^* x^*) \tag{3.55}$$

The two solution forms in Equation 3.55 are equivalent. Generally, if the domain is infinitely long, the first form is preferable, and if the domain is finite, the second form is easier to work with. The constants are determined from the boundary conditions, and therefore we need two boundary conditions.

3.5.3.1 Very Long Fins

The first limiting case considered is that of an infinitely long fin for which the fin temperature at the tip will approach the ambient temperature. Therefore, the boundary conditions are

$$
\begin{aligned}
x^* = 0 \quad & T^* = 1 \\
x^* \to \infty \quad & T^* \to 0
\end{aligned} \tag{3.56}
$$

Note that in this case, the scale used to nondimensionalize length cannot be the fin length, since it approaches infinity, and some other length must be used whose choice is immaterial. Starting with the exponential form of Equation 3.55, and applying the boundary conditions

$$T^* = C_1 e^{-m^* x^*} + C_2 e^{m^* x^*} \tag{3.57}$$

$$
\begin{aligned}
C_1 &= 1 \\
C_2 &= 0
\end{aligned} \tag{3.58}
$$

The temperature distribution in the fin is then given by

$$T^* = e^{-m^* x^*}$$

(3.59)

In terms of dimensioned variables,

$$\frac{T - T_\infty}{T_b - T_\infty} = e^{-mx} = e^{-\sqrt{\frac{hp}{kA_c}}\,x}$$

(3.60)

where $A_x = A_c$ is the fin cross section. This function is plotted in Figure 3.7, and as can be seen for $mx > 5$, the tip temperature becomes very close to the ambient temperature, thus a constant cross-sectional area fins can be assumed to behave like infinitely long fins, if $mL > 5$.

Therefore, from Equation 3.53, the heat transfer from the fin becomes

$$q = -kA_x \frac{dT}{dx}\bigg|_{x=0} = -kA_c(T_b - T_\infty)\left(-\sqrt{\frac{hp}{kA_c}}\,e^{-\sqrt{\frac{hp}{kA}}\,x}\right)\bigg|_{x=0} = kA_c(T_b - T_\infty)\sqrt{\frac{hp}{kA_c}}$$

(3.61)

$$q = (T_b - T_\infty)\sqrt{hpkA_c}$$

(3.62)

Since there is no heat loss from the fin tip, this must also be equal to the amount of heat that is lost from the fin surface to the environment that can be obtained by integrating the convective heat loss over the fin surface

$$q = \int_0^\infty h(T - T_\infty)dA_s = \int_0^\infty hp(T - T_\infty)dx = \int_0^\infty hp(T_b - T_\infty)e^{-\sqrt{\frac{hp}{kA_c}}\,x}dx = hp(T_b - T_\infty)\int_0^\infty e^{-\sqrt{\frac{hp}{kA_c}}\,x}dx$$

(3.63)

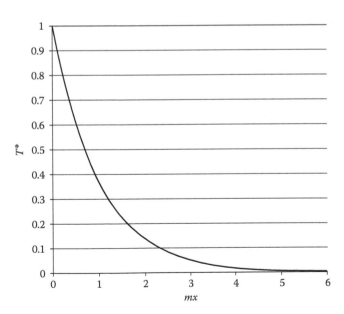

FIGURE 3.7
Temperature variation in an infinitely long fin.

thus

$$q = hp(T_b - T_\infty)\frac{1}{-\sqrt{\dfrac{hp}{kA_c}}}e^{-\sqrt{\frac{hp}{kA_c}}x}\Bigg|_0^\infty = hp(T_b - T_\infty)\frac{1}{\sqrt{\dfrac{hp}{kA_c}}} = (T_b - T_\infty)\sqrt{hpkA_c} \qquad (3.64)$$

which, as expected, is the same as Equation 3.62.

For an infinitely long fin, the efficiency is

$$\eta = \frac{q}{hA_s(T_b - T_\infty)} = \underset{L\to\infty}{Lim}\frac{(T_b - T_\infty)\sqrt{hpkA_c}}{hpL(T_b - T_\infty)} = 0 \qquad (3.65)$$

Note that even though the fin efficiency is zero, the fin still transfers a finite amount of heat, since its surface approaches infinity.

3.5.3.2 Finite-Length Fin with Tip Heat Loss

The most general solution is given for the convective boundary condition at the tip, and as will be shown next, the other three conditions are special cases of this general solution and can be derived from it. The nondimensional form of the boundary conditions for a convective boundary is

$$x^* = 0, \quad T^* = 1 \qquad (3.66)$$

$$x^* = 1, \quad \frac{dT^*}{dx^*} = -Bi[T^* - T_L^*] \qquad (3.67)$$

where

$$Bi = \frac{h_L L}{k} \qquad (3.68)$$

$$T_L^* = \frac{T_L - T_\infty}{T_b - T_\infty} \qquad (3.69)$$

Here we have assumed that the tip heat transfer coefficient (h_L) and the fluid ambient temperature at the tip (T_L) are not necessarily the same as the values on the top and bottom of the fin (h and T_∞).

The general form of the solution from Equation 3.55 is

$$T^* = D_1 \sinh(m^* x^*) + D_2 \cosh(m^* x^*) \qquad (3.70)$$

and the constants are found from the boundary conditions

$$1 = D_1 \sinh(m^* 0) + D_2 \cosh(m^* 0) = D_2 \qquad (3.71)$$

$$D_1 m^* \cosh(m^*) + m^* \sinh(m^*) = -Bi[D_1 \sinh(m^*) + \cosh(m^*) - T_L^*] \qquad (3.72)$$

Solving for D_1,

$$D_1 = -\frac{m^* \sinh(m^*) + Bi[\cosh(m^*) - T_L^*]}{m^* \cosh(m^*) + Bi \sinh(m^*)} \tag{3.73}$$

Substituting D_1 and D_2 in the general solution,

$$T^* = -\frac{m^* \sinh(m^*) + Bi[\cosh(m^*) - T_L^*]}{m^* \cosh(m^*) + Bi \sinh(m^*)} \sinh(m^* x^*) + \cosh(m^* x^*) \tag{3.74}$$

which can be written as

$$T^* = \frac{\cosh(m^*)\cosh(m^* x^*) - \sinh(m^*)\sinh(m^* x^*)}{\cosh(m^*) + \dfrac{Bi}{m^*}\sinh(m^*)}$$

$$+ \frac{\dfrac{Bi}{m^*}[\sinh(m^*)\cosh(m^* x^*) - \cosh(m^*)\sinh(m^* x^*)]}{\cosh(m^*) + \dfrac{Bi}{m^*}\sinh(m^*)} + \frac{T_L^* \dfrac{Bi}{m^*}\sinh(m^* x^*)}{\cosh(m^*) + \dfrac{Bi}{m^*}\sinh(m^*)} \tag{3.75}$$

Using hyperbolic function identities given by Equations 2.48 and 2.49 and using some algebra, Equation 3.75 can be rewritten as

$$T^* = \frac{\cosh[m^*(1 - x^*)] + \dfrac{Bi}{m^*}\{\sinh[m^*(1 - x^*)] + T_L^* \sinh(m^* x^*)\}}{\cosh(m^*) + \dfrac{Bi}{m^*}\sinh(m^*)} \tag{3.76}$$

This is the general solution for temperature distribution in a constant-area fin. The heat transfer from the fin can be determined from Equation 3.53, which simplifies to

$$q = \sqrt{hpkA_c}\,(T_b - T_\infty)\frac{\sinh(m^*) + \dfrac{Bi}{m^*}[\cosh(m^*) - T_L^*]}{\cosh(m^*) + \dfrac{Bi}{m^*}\sinh(m^*)} \tag{3.77}$$

This is the total amount of heat transfer from a fin. This must also be equal to the amount of heat that is lost from the fin surface to the environment by convection, which can be determined by integrating the convective heat loss over the fin surface

$$q = \int_{A_s} h(T - T_\infty)dA_s \tag{3.78}$$

Note that A_s includes the tip surface. Generally, calculating heat transfer is much easier using Equation 3.53.

For the case when the fluid at the tip is also at T_∞, then $T_L^* = 0$, and the solution simplifies to

$$T^* = \frac{\cosh[m^*(1-x^*)] + \dfrac{Bi}{m^*}\sinh[m^*(1-x^*)]}{\cosh(m^*) + \dfrac{Bi}{m^*}\sinh(m^*)} \qquad (3.79)$$

and the heat transfer from the fin is

$$q = \sqrt{hpkA_c}\,(T_b - T_\infty)\,\frac{\sinh(m^*) + \dfrac{Bi}{m^*}\cosh(m^*)}{\cosh(m^*) + \dfrac{Bi}{m^*}\sinh(m^*)} \qquad (3.80)$$

The solution for the infinitely long fin arrived at in Section 3.5.3.1 can also be obtained from this general solution. In this case, the fin temperature at the tip will approach the ambient temperature. Therefore, we need to examine Equation 3.76 for the limiting case when

$$m^* \to \infty, \qquad (3.81)$$
$$T_L^* \to 0$$

which simplifies to

$$T^* = \lim_{\substack{m^* \to \infty \\ T_L^* \to 0}} \frac{\cosh[m^*(1-x^*)] + \dfrac{Bi}{m^*}\{\sinh[m^*(1-x^*)] + T_L^*\sinh(m^*x^*)\}}{\cosh(m^*) + \dfrac{Bi}{m^*}\sinh(m^*)} = \lim_{m^* \to \infty} \frac{\cosh[m^*(1-x^*)]}{\cosh(m^*)} \qquad (3.82)$$

$$T^* = \lim_{m^* \to \infty} \frac{e^{m^*(1-x^*)} + e^{-m^*(1-x^*)}}{e^{m^*} + e^{-m^*}} = \lim_{m^* \to \infty} \frac{e^{(m^*)}e^{(-m^*x^*)}}{e^{m^*}} = \lim_{L \to \infty} e^{(-m^*x^*)} = e^{(-mx)} \qquad (3.83)$$

Therefore, the temperature distribution for an infinitely long fin becomes

$$T^* = e^{-mx} \qquad (3.84)$$

which is the same as the results obtained before.

3.5.3.3 Finite-Length Fin with Prescribed Tip Temperature

This corresponds to the case where Biot number approaches infinity in Equation 3.76, and the tip temperature assumes a constant value equal to that of the ambient to which the tip is exposed to. The solution is obtained by finding the limit of Equation 3.76 as $Bi \to \infty$, which becomes

$$T^* = \frac{T_L^*\sinh(m^*\,x^*) + \sinh[m^*(1-x^*)]}{\sinh(m^*)} \qquad (3.85)$$

For this case, it is possible for the fin temperature to reach a minimum along the fin, or the fin can receive heat not only from the base, but also from the tip, and therefore the total heat loss from the fin is equal to the heat transferred to the fin from the base and from the tip. The heat transfer from the fin is then equal to the heat transferred by conduction at the base minus the heat transferred by conduction at the tip.

$$q = q(0) - q(L) \tag{3.86}$$

$$q = -kA \frac{T_b - T_\infty}{L} \frac{dT^*}{dx^*} = -kA_c m^* \frac{T_b - T_\infty}{L} \frac{T_L^* \cosh(m^* x^*) - \cosh(m^* (1 - x^*))}{\sinh m^*} \tag{3.87}$$

$$q(0) = -kA_c m^* \frac{T_b - T_\infty}{L} \frac{T_L^* - \cosh m^*}{\sinh m^*} \tag{3.88}$$

$$q(L) = -kA_c m^* \frac{T_b - T_\infty}{L} \frac{T_L^* \cosh m^* - 1}{\sinh m^*} \tag{3.89}$$

$$q = \sqrt{hPkA_c} \frac{T_b - T_\infty}{\sinh m^*} (\cosh m^* - 1)\left(1 + T_L^*\right) \tag{3.90}$$

Alternatively, the heat loss from the fin can be calculated by calculating convection heat transfer from the surface of the fin:

$$q = \int h(T - T_\infty) dA_s \tag{3.91}$$

Substituting from Equation 3.85 for temperature and performing the integration result in an expression identical to Equation 3.90. The fin efficiency is then given by

$$\eta = \frac{q}{q_{max}} = \frac{q}{hpL(T_b - T_\infty) + h_L A_c (T_b - T_L)} \tag{3.92}$$

3.5.3.4 Insulated Tip Fin

The insulated-tip boundary condition means that there is no heat loss from the tip of the fin. This can happen if the tip is actually insulated (no reason to do so), or if the tip area is zero (tapered tip) or if the temperature in the fin reaches a minimum along the fin (attaching the two ends of the fin to two isothermal surfaces). For this case, the tip boundary condition is

$$x^* = 1, \quad \frac{dT^*}{dx^*} = 0 \tag{3.93}$$

which corresponds to the case of $h_L = 0$ ($Bi = 0$) in Equation 3.76, which reduces to

$$T^* = \frac{\cosh[m^*(1-x^*)]}{\cosh(m^*)} \tag{3.94}$$

and the heat transfer rate from Equation 3.77 becomes

$$q = \sqrt{hpkA_c}\,(T_b - T_\infty)\tanh(m^*) \tag{3.95}$$

From Equation 3.94, the temperature at the tip of the fin is

$$T^* = \frac{1}{\cosh(m^*)} \tag{3.96}$$

The fin efficiency is

$$\eta = \frac{q}{q_{max}} = \frac{\sqrt{hpkA_c}\,(T_b - T_\infty)\tanh(m^*)}{hpL(T_b - T_\infty)} = \frac{\tanh(m^*)}{\sqrt{\dfrac{hp}{kA_c}}\,L} \tag{3.97}$$

which simplifies to

$$\eta = \frac{\tanh(m^*)}{m^*} \tag{3.98}$$

The efficiency of insulated-tip fin is only a function of one nondimensional variable

$$m^* = \sqrt{\frac{hpL^2}{kA_c}} \tag{3.99}$$

Figure 3.8 is a plot of fin efficiency as a function of m^*, which shows that the fin efficiency decreases with increasing m^*. Therefore, the short fins with low heat transfer coefficient have higher efficiency, although they are not a particularly good choice.

Figure 3.9 is a comparison of the fin axial temperature variation for the four boundary conditions for a given value of $m^* = 2$. As can be seen, in all cases, temperature drops almost exponentially with the infinitely long approximation representing the worst case. The prescribed tip temperature can produce interesting result in that if the tip temperature is high enough, the fin can reach a minimum temperature at an axial location along the fin.

Example 3.1

A thin constant-area cylindrical bar is attached to two surfaces maintained at temperatures T_1 and T_2. Determine the location of minimum temperature.

We can assume the thin bar to behave like two insulated-tip fins, one whose base is at T_1 having a length z, and the second having a length $(L - z)$ whose base is maintained at T_2, as shown in Figure 3.10.

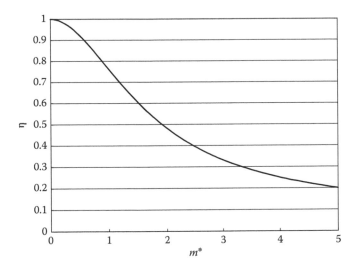

FIGURE 3.8
Efficiency of constant area insulated tip fin.

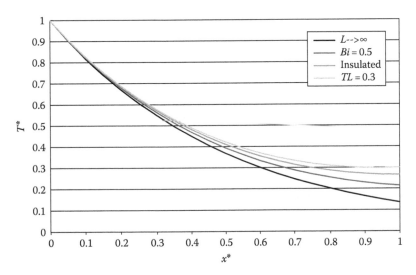

FIGURE 3.9
Fin axial temperature variation for the four boundary conditions ($m^* = 2$).

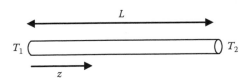

FIGURE 3.10
Fin having minimum temperature along its length.

For an insulated-tip fin, the tip temperature is given by

$$T^* = \frac{1}{\cosh(m^*)}$$

which can be solved for T

$$T = T_\infty + \frac{T_b - T_\infty}{\cosh(m^*)}$$

The tip temperature for the two fins must be equal or

$$T_\infty + \frac{T_1 - T_\infty}{\cosh(m_1^*)} = T_\infty + \frac{T_2 - T_\infty}{\cosh(m_2^*)}$$

$$\frac{T_1 - T_\infty}{T_2 - T_\infty} = \frac{\cosh(m_1^*)}{\cosh(m_2^*)} = \frac{\cosh\left(\sqrt{\dfrac{hp}{kA}}\,z\right)}{\cosh\left(\sqrt{\dfrac{hp}{kA}}(L-z)\right)}$$

$$\frac{T_1 - T_\infty}{T_2 - T_\infty} = \frac{\cosh\left(\sqrt{\dfrac{hpL^2}{kA}}\,\dfrac{z}{L}\right)}{\cosh\left(\sqrt{\dfrac{hpL^2}{kA}}\left[1 - \dfrac{z}{L}\right]\right)} = \frac{\cosh(m^* z^*)}{\cosh(m^*(1 - z^*))}$$

The unknown in this equation is z^*, which is the location where the temperatures are equal. Expanding the term in the denominator using cosh identities

$$\frac{T_1 - T_\infty}{T_2 - T_\infty} = \frac{\cosh(m^* z^*)}{\cosh(m^*)\cosh(m^* z^*) - \sinh(m^*)\sinh(m^* z^*)} = \frac{1}{\cosh(m^*) - \sinh(m^*)\tanh(m^* z^*)}$$

which can be solved for z^* to yield

$$z^* = \frac{1}{m^*}\tanh^{-1}\left[\frac{\cosh(m^*) - \dfrac{T_2 - T_\infty}{T_1 - T_\infty}}{\sinh(m^*)}\right]$$

Since the argument of inverse hyperbolic function must be less than one, then a minimum will exist as long as $m^* > \ln\left[\dfrac{T_1 - T_\infty}{T_2 - T_\infty}\right]$. In order for the minimum to be in the fin, $z^* < 1$.

For example, if $T_1 = T_2$, then

$$z^* = \frac{1}{m^*}\tanh^{-1}\left[\frac{\cosh(m^*) - 1}{\sinh(m^*)}\right]$$

using the identity

$$\tanh x = \frac{\cosh(2x) - 1}{\sinh(2x)}$$

then

$$z^* = \frac{1}{2}$$

which as expected means that the minimum is in the middle of the fin.

Alternatively, each fin can be considered as one with a fixed tip temperature; however, in this case, in addition to being constant, the tip is also the location of the minimum temperature. Therefore, differentiating Equation 3.85

$$\frac{dT^*}{dz^*} = \frac{T_L^* m^* \cosh(m^* z^*) - m^* \cosh[m^*(1 - z^*)]}{\sinh(m^*)} = 0$$

$$T_L^* = \frac{\cosh[m^*(1 - z^*)]}{\cosh(m^* z^*)}$$

For both fins, the tip temperature occurs at $z^* = 1$; therefore,

$$T_L^* = \frac{1}{\cosh(m^*)}$$

which is the same as the result obtained earlier.

Example 3.2

The solar collector plate shown in Figure 3.11 receives a uniform radiation heat flux on the top, while the bottom surface is well insulated. Tubes attached to the collector are $2L$ apart and carry a working fluid at temperature T_b. The top surface is exposed to a fluid at h_∞, T_∞.

1. Derive the governing equation and boundary conditions.
2. Nondimensionalize the equation and boundary conditions.
3. Obtain the temperature distribution and heat transfer to the working fluid.

For the collector, starting with Equation 3.41,

$$\frac{\partial^2 T}{\partial x^2} + \frac{\partial^2 T}{\partial y^2} = 0$$

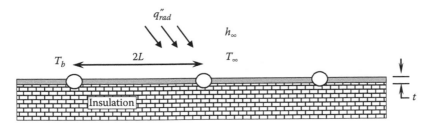

FIGURE 3.11
Solar collector.

Integrating this equation from $y = 0$ to $y = t$

$$\int_0^t \frac{\partial^2 T}{\partial x^2} dy + \int_0^t \frac{\partial^2 T}{\partial y^2} dy = 0$$

$$\frac{\partial^2}{\partial x^2} \int_0^t T dy + \frac{\partial T}{\partial y}\bigg|_t - \frac{\partial T}{\partial y}\bigg|_0 = 0$$

$$\frac{\partial^2}{\partial x^2} \left\{ (t) \left[\frac{1}{t} \int_0^t T dy \right] \right\} + \frac{\partial T}{\partial y}\bigg|_t - \frac{\partial T}{\partial y}\bigg|_0 = 0$$

By definition, the term inside the square bracket is the average temperature of the fin across the fin

$$\bar{T} = \frac{1}{t} \int_0^t T dy$$

$$\frac{\partial^2 \bar{T}}{\partial x^2} + \frac{1}{t} \frac{\partial \bar{T}}{\partial y}\bigg|_t = 0$$

Doing an energy balance on the top surface of collector, assuming it to absorb all the incident radiation,

$$-k \frac{\partial T}{\partial y} + q'' = h(T - T_\infty)$$

Substituting for the temperature gradient,

$$\frac{\partial^2 \bar{T}}{\partial x^2} + \frac{1}{tk} \left[q'' - h(T - T_\infty) \right] = 0$$

Assuming the surface and mean temperatures to be the same,

$$\frac{d^2 T}{dx^2} - \frac{h}{tk} \left[(T - T_\infty) - \frac{q''}{h} \right] = 0$$

Defining the nondimensional temperature and distance along the collector as

$$T^* = \frac{T - T_\infty - \dfrac{q''}{h}}{T_b - T_\infty - \dfrac{q''}{h}}$$

$$x^* = \frac{x}{L}$$

then the energy equation reduces to

$$\frac{d^2T^*}{dx^{*2}} - m^{*2}T^* = 0$$

where

$$m^{*2} = \frac{hL^2}{kt}$$

and the boundary conditions are

$$x^* = 0 \quad T^* = 1$$

$$x^* = 1 \quad \frac{dT^*}{dx^*} = 0$$

The solution to this equation is given by Equation 3.94

$$T^* = \frac{\cosh[m^*(1-x^*)]}{\cosh(m^*)}$$

or

$$T = T_\infty + \frac{q''}{h} + \left[T_b - T_\infty - \frac{q''}{h}\right] \frac{\cosh\left[\sqrt{\frac{hL^2}{kt}}\left(1-\left(\frac{x}{L}\right)\right)\right]}{\cosh\left(\sqrt{\frac{hL^2}{kt}}\right)}$$

The amount of heat transferred from the solar collector to the fluid flowing in the pipe from collector areas on both sides of the pipe is

$$q = 2ktW\frac{dT}{dx}\bigg|_{x=0} = 2ktW\left[\frac{q''}{h} - T_b + T_\infty\right] \frac{-\sqrt{\frac{h}{kt}}\sinh\left[\sqrt{\frac{hL^2}{kt}}\left(1-\frac{x}{L}\right)\right]}{\cosh\left(\sqrt{\frac{hL^2}{kt}}\right)}$$

The heat transfer to the fluid per unit depth of the collector is therefore

$$\frac{q}{W} = 2\left[\frac{q''}{h} + T_\infty - T_b\right]\sqrt{hkt}\,\tanh\left[\sqrt{\frac{hL^2}{kt}}\right]$$

As can be seen, the reduction of heat loss by convection (h) is very important for efficient functioning of the collector, and that is the reason for using glass covers on solar collectors to allow the radiation in, but increase the conduction resistance due to glass and the air gap between the collector and the glass.

3.5.4 Practical Considerations

The fin efficiency expression obtained analytically for constant-area insulated-tip case has a simple form compared with the more realistic case of a fin with heat loss at the tip. Generally, $h_L \approx h$; therefore, we can account for the heat loss at the tip of the fin by replacing the actual fin with an equivalent fin whose total area is equal to that of the actual fin. This equivalent fin can then be analyzed using the efficiency results for insulated-tip fin. Figure 3.12 shows an actual fin and its equivalent insulated-tip fin, whose length is slightly more. For a straight fin, the corrected length is

$$L_c = L + \frac{t}{2} \tag{3.100}$$

and for a pin fin, the corrected length is

$$\pi dL + \frac{\pi}{4}d^2 = \pi dL_c \tag{3.101}$$

$$L_c = L + \frac{\pi}{4}d \tag{3.102}$$

One should not confuse high efficiency with high heat transfer rate. Even though increasing the fin length decreases the efficiency, it increases the area and thus the rate of heat transfer. This can be readily shown for an insulated-tip pin fin, for which the rate of heat transfer from Equation 3.95 becomes

$$q = h\pi dL(T_b - T_\infty)\frac{\tanh\left(\sqrt{\frac{4h}{kd}}L\right)}{\sqrt{\frac{4h}{kd}}L} = \frac{\pi}{2}(T_b - T_\infty)\sqrt{hkd^3}\tanh\left(\sqrt{\frac{4h}{kd}}L\right) \tag{3.103}$$

From this equation, the rate of heat transfer is proportional to the fin length. The heat transfer from the fin is also proportional to heat transfer coefficient and temperature difference.

Figure 3.13 shows the impact of fin diameter, fin length, heat transfer coefficient, and thermal conductivity on the heat transfer from the fin (Equation 3.103). As seen from the

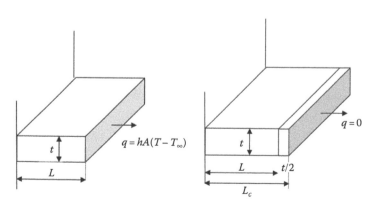

FIGURE 3.12
Equivalent insulated tip fin.

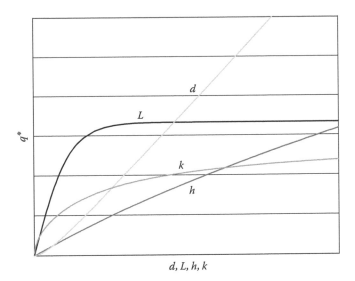

FIGURE 3.13
Impact of different parameters on the heat transfer from the fin.

chart, increasing all of these parameters increases the heat transfer rate from the fin. The heat transfer from the fin is an almost linear function of the diameter and heat transfer coefficient. Increasing the fin length initially increases the heat transfer rate, but then additional increase in length has no appreciable effect on the heat transfer rate, since past some length, the fin temperature becomes close to the ambient temperature, and further increase in length will not help heat transfer. The fin thermal conductivity also shows a similar behavior in that past some value, additional increase in thermal conductivity does not increase the heat transfer rate appreciably; hence, depending on the application, a lower-thermal conductivity material may be almost as effective as a material with substantially higher thermal conductivity and cost.

Another practical concern is maximizing the heat transfer for a given material cost (weights). For example, the mass of a pin fin is

$$m = \rho \frac{\pi}{4} d^2 L \tag{3.104}$$

Eliminating L from Equation 3.103 with the help of (3.104) and noting that $V = m/\rho$,

$$q = \frac{\pi}{2}(T_b - T_\infty)\sqrt{hkd^3} \tanh\left(\sqrt{\frac{h}{k}} \frac{8V}{\pi d^{2.5}}\right) \tag{3.105}$$

As can be seen, Equation 3.105 is the product of two terms, one that increases with d and the other that decreases with d, and thus there is a possibility of an optimum solution.

Figure 3.14 is a plot of this function's dependence on d, for a particular set of parameters, and shows that there is an optimum solution. The optimum fin diameter can be obtained by differentiating Equation 3.105 with respect to diameter, and setting it equal to zero results in

$$\sinh\left(2\sqrt{\frac{h}{k}} \frac{8V}{\pi} d^{-5/2}\right) = \frac{10}{3}\sqrt{\frac{h}{k}} \frac{8V}{\pi} d^{-5/2} \tag{3.106}$$

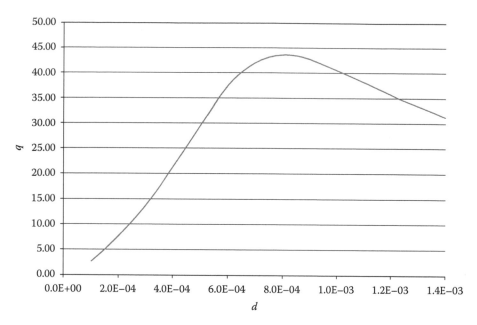

FIGURE 3.14
Existence of optimum diameter for maximum heat transfer for a given fin material cost (weights).

which can be solved iteratively, and the root is

$$\sqrt{\frac{h}{k}}\frac{8V}{\pi}d^{-5/2} = 0.9193 \tag{3.107}$$

from which optimum diameter can be determined to be

$$d = \left(7.6727\frac{h}{k}V^2\right)^{1/5} \tag{3.108}$$

Substituting for volume in terms of diameter and length of the fin, Equation 3.108 simplifies to

$$d = 4.733\frac{h}{k}L^2 \tag{3.109}$$

which shows the relationship between the fin diameter and fin length at optimum point, noting that d and L must also satisfy Equation 3.104 for the given mass.

To determine the efficiency of the fin at the optimum point, m^* needs to be calculated. Rearranging Equation 3.109 results in

$$\sqrt{\frac{4h}{kd}L^2} = m^* = 0.9193 \tag{3.110}$$

The efficiency for the maximum amount of heat transfer is a constant value at

$$\eta_m = \frac{\tanh(m^*)}{m^*} = 78.9\% \tag{3.111}$$

These are very interesting results in that regardless of the weight of the fin, the optimum design diameter and length are only a function of the volume of the fin, the heat transfer coefficient, and thermal conductivity, and regardless of the size, the fin has the same efficiency.

Example 3.3

Aluminum pin fins, with the total weight not exceeding 40 g, are to be attached to a 10 × 10 cm surface at 100°C and the ambient is at 25°C. Design the fins, i.e., determine the diameter, length, and the number of the fins, for the maximum amount of heat transfer, if $h = 10$ W/m^2 K, as long as there is a minimum 2 mm gap between the fins.

The aluminum thermal conductivity is 200 W/m K, and its density is 2700 kg/m^3.

If M is the number of fins per row, then there are $M - 1$ gaps and therefore

$$Md + 0.002(M-1) = 0.1 \tag{a}$$

and the total number of fins will be $N = M^2$. The total amount of heat transfer can be obtained by multiplying heat transfer from each fin by the number of fins. The heat transfer for each fin is given by Equation 3.105

$$q = N\frac{\pi}{2}(T_b - T_\infty)\sqrt{hkd^3} \tanh\left(\sqrt{\frac{h}{k}\frac{8m}{N\rho\pi d^{2.5}}}\right) \tag{b}$$

Figure 3.15 is a plot of heat transfer as a function of diameter and the number of fins for this particular case, and as can be seen, there is an optimum for a fin diameter of around 0.5 mm. To find the values at optimum point, from Equation 3.109,

$$d = 4.733\frac{10}{200}L^2 = 0.2366L^2 \tag{c}$$

$$m = N\rho\frac{\pi}{4}d^2L$$

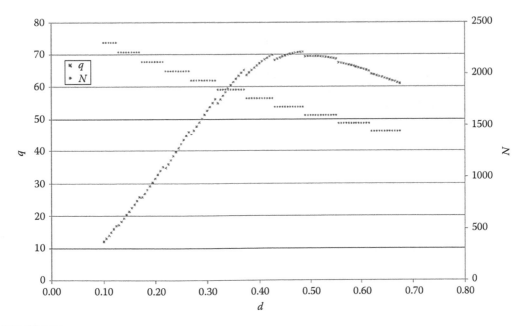

FIGURE 3.15
Optimum fin array.

Substituting the values

$$0.04 = N2700\frac{\pi}{4}d^2L \tag{d}$$

d, L, and N can be obtained from Equations a, c, and d, and the results are

$$d = 0.485 \text{ mm}$$

$$L = 45.3 \text{ mm}$$

$$N = 1681$$

Substituting these in Equation b, the total amount of heat transfer from the fins is

$$q = 70.83 \text{ W}$$

3.6 Variable-Area Fins: Analytical Solutions

The earlier discussion was limited to the fins whose cross-sectional area is constant in the direction where heat is transferred. Many fins used in actual applications are those whose cross-sectional area changes in the direction of heat transfer. The variable-area fins also come in many different configurations, and we consider two more commonly encountered ones in the following text. The analysis for all variable-area fins is very similar, and not that different from that of constant-area fins, except that the math is a little different since we deal with variable-coefficient second-order differential equations, whose solutions involve Bessel functions (instead of the exponential) that were reviewed in Chapter 2.

3.6.1 Triangular Fins

A triangular profile fin is shown in Figure 3.16. This is a variable-area straight fin, and from Table 3.1, for this case, $a = 0$, $b = 1$, and $c = 1$, and the energy equation becomes

$$\frac{d^2T^*}{dx^{*2}} + \frac{1}{y^*}\frac{dy^*}{dx^*}\frac{dT^*}{dx^*} - \frac{hL}{k}\frac{1}{y^*}\sqrt{1+\left(\frac{dy^*}{dx^*}\right)^2}\,T^* = 0 \tag{3.112}$$

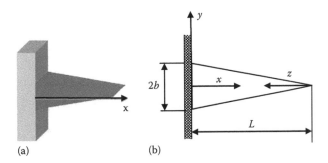

FIGURE 3.16
Triangular fin.

From Figure 3.16b, it can be seen that

$$y = b\left[1 - \frac{x}{L}\right]$$ (3.113)

which can be nondimensionalized into

$$y^* = b^*[1 - x^*]$$ (3.114)

where

$$y^* = \frac{y}{L} \quad \text{and} \quad b^* = \frac{b}{L}$$ (3.115)

Therefore, Equation 3.112 becomes

$$(1 - x^*)\frac{d^2T^*}{dx^{*2}} - \frac{dT^*}{dx^*} - \frac{hL}{k}\sqrt{1 + \frac{1}{b^{*2}}}T^* = 0$$ (3.116)

since $b^* \ll 1$ then $1 + \left(\dfrac{1}{b^{*2}}\right) \approx \dfrac{1}{b^{*2}}$, and Equation 3.116 becomes

$$(1 - x^*)\frac{d^2T^*}{dx^{*2}} - \frac{dT^*}{dx^*} - \frac{hL^2}{kb}T^* = 0$$ (3.117)

Defining a new independent variable z,

$$z = L - x$$ (3.118)

Then

$$z^*\frac{d^2T^*}{dz^{*2}} + \frac{dT^*}{dz^*} - m^{*2}T^* = 0$$ (3.119)

where

$$m^{*2} = \frac{hL^2}{kb} \tag{3.120}$$

Multiplying Equation 3.119 with z^*, and comparing it with Equation 2.56,

$$k = 0, \quad \beta = 0, \quad r = \frac{1}{2}, \quad \alpha = im^*, \quad \kappa = 0 \tag{3.121}$$

and therefore the solution is

$$T^* = \left[C_1 J_0(2i\,m^* z^{*1/2}) + C_2 Y_0(2i\,m^* z^{*1/2}) \right] \tag{3.122}$$

which can be expressed in terms of modified Bessel functions, as

$$T^* = C_1 I_0(2m^* z^{*1/2}) + C_2 K_0(2m^* z^{*1/2}) \tag{3.123}$$

The integration constants are determined from the boundary conditions

$$z^* = 0 \quad \frac{dT^*}{dz^*} = 0$$

$$\tag{3.124}$$

$$z^* = 1 \quad T^* = 1$$

$$\frac{dT^*}{dz^*} = \left[C_1 I_1(2m^* z^{*1/2}) + C_2 K_1(2m^* z^{*1/2}) \right] m^* z^{*-1/2} \tag{3.125}$$

$$0 = \frac{[C_1 I_1(0) + C_2 K_1(0)]}{0^{*1/2}} m^* \tag{3.126}$$

since $K_1(0) \to \infty$, then $C_2 = 0$ \hfill (3.127)

$$1 = C_1 I_0(2m^*) + C_2 K_0(2m^*) \tag{3.128}$$

$$C_1 = \frac{1}{I_0(2m^*)} \tag{3.129}$$

Therefore, the temperature distribution in a triangular fin is given by

$$T^* = \frac{I_0(2m^* z^{*1/2})}{I_0(2m^*)} = \frac{I_0(2m^* \sqrt{1 - x^*})}{I_0(2m^*)} \tag{3.130}$$

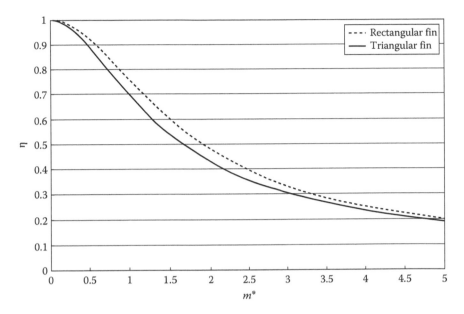

FIGURE 3.17
Efficiencies of triangular and rectangular fins.

and the heat transfer by

$$q = kA \frac{dT}{dz}\bigg|_{z=L} = kA \frac{(T_b - T_\infty)}{L} \frac{dT^*}{dz^*}\bigg|_{z^*=1} = kA \frac{(T_b - T_\infty)}{L} \frac{I_1(2m^*)}{I_0(2m^*)} \tag{3.131}$$

and the fin efficiency can be calculated from

$$\eta = \frac{q}{q_{max}} = \frac{k2bW \dfrac{(T_b - T_\infty)}{L} m^* \dfrac{I_1(2m^*)}{I_0(2m^*)}}{2WLh(T_b - T_\infty)} = \frac{kb}{L^2 h} m^* \frac{I_1(2m^*)}{I_0(2m^*)} = \frac{1}{m^*} \frac{I_1(2m^*)}{I_0(2m^*)} \tag{3.132}$$

which is plotted in Figure 3.17. On the same figure, there is also a plot of the efficiency of a constant-area rectangular fin (Equation 3.98), and as can be seen for the same base thickness ($t = 2b$), same fin length, same heat transfer coefficient, and same material, a rectangular fin has a higher efficiency compared to a triangular fin; at the same time, the triangular fin has a slightly higher area, which increases the heat transfer rate somewhat. However the more important consideration may be that for the same base thickness and length, the triangular fin weighs half that of a rectangular fin, and therefore, when material cost or weight is an issue, triangular fins provide a viable alternative.

3.6.2 Constant-Thickness Annular Fin

Annular fins are typically placed on tubes and provide for an effective and inexpensive method of enhancing heat transfer. For a constant-thickness annular fin, shown in Figure 3.18, $r^* = \dfrac{r}{r^2}$, and $y^* = \dfrac{t/2}{r^2}$ and

$$\frac{dy^*}{dx^*} = \frac{dy^*}{dr^*} = 0 \tag{3.133}$$

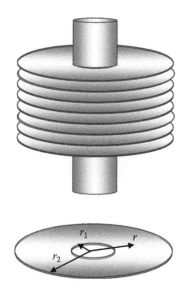

FIGURE 3.18
Constant area annular fin.

Then from Table 3.1, the energy equation becomes

$$\frac{d^2T^*}{dr^{*2}} + \frac{1}{r^*}\frac{dT^*}{dr^*} - \frac{2hr_2^2}{kt}T^* = 0 \tag{3.134}$$

The boundary conditions are

$$r^* = \frac{r_1}{r_2} = r_1^*, \quad T^* = 1 \tag{3.135}$$

$$r^* = 1 \quad \frac{dT^*}{dr^*} = 0 \tag{3.136}$$

Multiplying Equation 3.134 by r^{*2}

$$r^{*2}\frac{d^2T^*}{dr^{*2}} + r^*\frac{dT^*}{dr^*} - m^{*2}r^{*2}T^* = 0 \tag{3.137}$$

and comparing it with Equation 2.56, $k = 0$, $\beta = 0$, $r = 1$, $\alpha = im^*$, $\kappa = 0$ and therefore

$$T^* = C_1J_0(im^*r^*) + C_2Y_0(im^*r^*) = C_1I_0(m^*r^*) + C_2K_0(m^*r^*)$$

Applying the boundary conditions

$$1 = C_1I_0(m^*r_i^*) + C_2K_0(m^*r_i^*) \tag{3.138}$$

$$\frac{dT^*}{dr^*} = C_1 m^* I_1(m^*) - C_2 m^* K_1(m^*) = 0 \tag{3.139}$$

results in

$$C_1 = -\frac{K_1(m^*)}{[K_0(m^* r_1^*)I_1(m^*) + K_1(m^*)I_0(m^* r_1^*)]} \tag{3.140}$$

$$C_2 = \frac{I_1(m^*)}{K_0(m^* r_1^*)I_1(m^*) + K_1(m^*)I_0(m^* r_1^*)} \tag{3.141}$$

Therefore, the temperature distribution in the fin is given by

$$T^* = \frac{K_0(m^* r^*)I_1(m^*) + K_1(m^*)I_0(m^* r^*)}{K_0(m^* r_1^*)I_1(m^*) + K_1(m^*)I_0(m^* r_1^*)} \tag{3.142}$$

Differentiating Equation 3.142 with respect to r^* and evaluating at r_1^*, the efficiency of an annular fin becomes

$$\eta = \frac{2}{m^*\left(\dfrac{1}{r_1^*} - r_1^*\right)} \frac{K_1(m^* r_1^*)I_1(m^*) - I_1(m^* r_1^*)K_1(m^*)}{I_0(m^* r_1^*)K_1(m^*) + K_0(m^* r_1^*)I_1(m^*)} \tag{3.143}$$

3.7 Numerical Solution

We now consider the numerical solution of the fin equation, using the general nondimensional form of the energy equation, given by Equation 3.37. The first step is to discretize the domain into $N - 1$ intervals by choosing N nodes, as shown in Figure 3.19. In this case, the domain extends from zero to one, and therefore, the length of each interval is

$$\Delta x^* = \frac{1}{N-1} \tag{3.144}$$

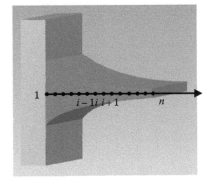

FIGURE 3.19
Discretization of fin.

Replacing the derivatives with their central finite-difference approximation

$$\frac{T_{i+1}^* - 2T_i^* + T_{i-1}^*}{\Delta x^{*2}} + \left(\frac{a}{x_i^*} + \frac{b}{y_i^*}\frac{dy^*}{dx^*}\Big|_i\right)\frac{T_{i+1}^* - T_{i-1}^*}{2\Delta x^*} - \frac{hL}{k}\frac{c}{y_i^*}\sqrt{1 + \left(\frac{dy^*}{dx^*}\Big|_i\right)^2} T_i^* = 0 \qquad (3.145)$$

and collecting the terms, Equation 3.145 can be expressed as a set of algebraic equations of the form

$$A_i T_{i-1}^* + B_i T_i^* + C_i T_{i+1}^* = R_i \qquad (3.146)$$

where

$$A_i = \frac{1}{\Delta x^{*2}} - \left(\frac{a}{x_i^*} + \frac{b}{y_i^*}\frac{dy^*}{dx^*}\Big|_i\right)\frac{1}{2\Delta x^*} \qquad (3.147)$$

$$B_i = -\frac{2}{\Delta x^{*2}} - \frac{hL}{k}\frac{c}{y_i^*}\sqrt{1 + \left(\frac{dy^*}{dx^*}\Big|_i\right)^2} \qquad (3.148)$$

$$C_i = \frac{1}{\Delta x^{*2}} + \left(\frac{a}{x_i^*} + \frac{b}{y_i^*}\frac{dy^*}{dx^*}\Big|_i\right)\frac{1}{2\Delta x^*} \qquad (3.149)$$

$$R_i = 0 \qquad (3.150)$$

The solution can be obtained iteratively, by solving for T_i from Equation 3.146

$$T_i^* = \frac{R_i - \left(A_i T_{i-1}^* + C_i T_{i+1}^*\right)}{B_i} \qquad (3.151)$$

or directly using the subroutine TRIDI. Additional details can be found in Ref. [1].

3.7.1 Constant-Area Fin

For constant-area fin from Table 3.1, $a = 0$, $b = 1$, $c = 1$, $\dfrac{dy^*}{dx^*} = 0$, and $y_i^* = \dfrac{t/2}{L}$, therefore

$$A_i = \frac{1}{\Delta x^{*2}} \qquad (3.152)$$

$$B_i = -\frac{2}{\Delta x^{*2}} - \frac{2hL^2}{kt} = -\frac{2}{\Delta x^{*2}} - m^{*2} \qquad (3.153)$$

$$C_i = \frac{1}{\Delta x^{*2}} \qquad (3.154)$$

The tip of the fin is assumed to be insulated. The details of how to set up the spreadsheet and the iterative solution are shown in Figure 3.20. To facilitate parametric studies, we have

◇	A	B	C	D	E	
1	k	30	i	x*	A	
2	h	60	1	0	1	1
3	L	0.02	2	=D2+Dx	=(1/Dx)^2	=
4	w	0.04	3	=D3+Dx	=(1/Dx)^2	=
5	t	=0.001	4	=D4+Dx	=(1/Dx)^2	=
6	N	101	5	=D5+Dx	=(1/Dx)^2	=
7	Dx	0.01	6	=D6+Dx	=(1/Dx)^2	=
8	ms	=SQRT(2*h*L^2/(k*t))	7	=D7+Dx	=(1/Dx)^2	=
9			8	=D8+Dx	=(1/Dx)^2	=

C	D	E	F	G	H	I	J
i	x*	A	B	C	R	T	Exact
1	0	1	1	0	1	1	=COSH(ms*(1-D2))/COSH(ms)
2	=D2+Dx	=(1/Dx)^2	=-(2*(1/Dx)^2+ms^2)	=1/(Dx)^2	0	=-(E3*I2+I4*G3)/F3	=COSH(ms*(1-D3))/COSH(ms)
3	=D3+Dx	=(1/Dx)^2	=-(2*(1/Dx)^2+ms^2)	=1/(Dx)^2	0	=-(E4*I3+I5*G4)/F4	=COSH(ms*(1-D4))/COSH(ms)
4	=D4+Dx	=(1/Dx)^2	=-(2*(1/Dx)^2+ms^2)	=1/(Dx)^2	0	=-(E5*I4+I6*G5)/F5	=COSH(ms*(1-D5))/COSH(ms)
5	=D5+Dx	=(1/Dx)^2	=-(2*(1/Dx)^2+ms^2)	=1/(Dx)^2	0	=-(E6*I5+I7*G6)/F6	=COSH(ms*(1-D6))/COSH(ms)
6	=D6+Dx	=(1/Dx)^2	=-(2*(1/Dx)^2+ms^2)	=1/(Dx)^2	0	=-(E7*I6+I8*G7)/F7	=COSH(ms*(1-D7))/COSH(ms)
7	=D7+Dx	=(1/Dx)^2	=-(2*(1/Dx)^2+ms^2)	=1/(Dx)^2	0	=-(E8*I7+I9*G8)/F8	=COSH(ms*(1-D8))/COSH(ms)
8	=D8+Dx	=(1/Dx)^2	=-(2*(1/Dx)^2+ms^2)	=1/(Dx)^2	0	=-(E9*I8+I10*G9)/F9	=COSH(ms*(1-D9))/COSH(ms)

FIGURE 3.20
Spreadsheet implementation of the numerical solution.

defined the variable named k for thermal conductivity, h for heat transfer coefficient, L for fin length, W for fin width, and t for fin thickness, and ms for m^*, which for the parameters chosen is 1.26 ($m^* = 1.26$), and dx for the interval length. There are 101 nodes along the fin ($\Delta x = 0.01$). As mentioned in Chapter 2, to define a parameter, select the cell that contains the value of the parameter, for example, B8 and enter the variable name (ms) in the box that normally contains the location of the cell (the box to the left of the formula bar) and make sure to hit return.

Figure 3.21 is a comparison of the temperature distribution obtained numerically using Excel®'s iterative capabilities and the exact analytical solution given by Equation 3.94. As can be seen, there is an excellent agreement between the two solutions.

3.7.2 Triangular Fin

For a straight triangular profile fin, $a = 0$, $b = 1$, $c = 1$, $y^* = b^*[1 - x^*]$, and $\dfrac{dy^*}{dx^*} = -b^* = -\dfrac{b}{L}$; therefore,

$$A_i = \frac{1}{\Delta x^{*2}} + \frac{1}{2\Delta x^*[1 - x_i^*]} \tag{3.155}$$

$$B_i = -\frac{2}{\Delta x^{*2}} - \frac{hL^2}{kb}\frac{1}{[1 - x_i^*]} \tag{3.156}$$

$$C_i = \frac{1}{\Delta x^{*2}} - \frac{1}{2\Delta x^*[1 - x_i^*]} \tag{3.157}$$

$$R_i = 0 \tag{3.158}$$

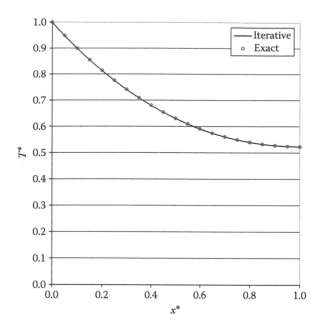

FIGURE 3.21
Comparison of the numerical and exact temperature distributions in a constant area fin.

The solution can be obtained using the previous spreadsheet, by changing the expressions for A, B, and C from the earlier equations and solving the equations iteratively. The equations can also be solved directly using the function TRIDI.

Figure 3.22 is again a comparison of the analytical solution given by Equation 3.130 and the numerical results.

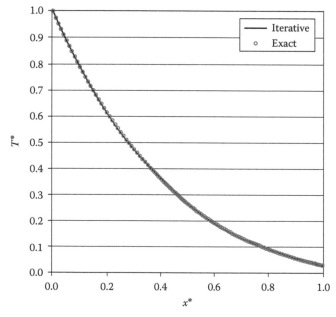

FIGURE 3.22
Comparison of the numerical and exact temperature distributions in a triangular fin.

3.7.3 Annular Fin

As was shown in Section 3.6.2, for a constant-thickness annular fin, $a = 1$, $b = 1$, $c = 2$, and $L = r_2$, and

$$y^* = \frac{t/2}{r_2} \tag{3.159}$$

$$\frac{dy^*}{dx^*} = 0 \tag{3.160}$$

$$A_i = \frac{1}{\Delta r^{*2}} - \frac{1}{r_i^*}\frac{1}{2\Delta r^*} \tag{3.161}$$

$$B_i = -\frac{2}{\Delta r^{*2}} - \frac{2hr_2^2}{kt} \tag{3.162}$$

$$C_i = \frac{1}{\Delta r^{*2}} + \frac{1}{r_i^*}\frac{1}{2\Delta r^*} \tag{3.163}$$

$$R_i = 0 \tag{3.164}$$

In this case, the solution is obtained using the function TRIDI, and the details are shown in Figure 3.23.

◇	A	B	C	D	E	F
1	t	0.001		I	r*	A
2	k	50		1	0.2	1
3	rl	0.02		2	0.22	3761.6
4	N	51		3	0.23	3771.6
5	h	10		4	0.25	3780.2
6	ro	0.1		5	0.26	3787.9
7	rls	0.2		6	0.28	3794.6
8	dr	0.016		7	0.3	3800.7
9	ms	2.00		8	0.31	3806.1
10				9	0.33	3811.0

◇	D	E	F	G	H	I	J
1	I	r*	A	B	C	R	T direct
2	1	=rls	1	1	0	1	=TRIDI(F2:F52,G2:G52,H2:H52,I2:I52)
3	2	=E2+dr	=1/dr^2-1/(2*dr*E3)	=-(2/dr^2+M)	=1/dr^2+1/(2*dr*E3	=0	=TRIDI(F2:F52,G2:G52,H2:H52,I2:I52)
4	3	=E3+dr	=1/dr^2-1/(2*dr*E4)	=-(2/dr^2+M)	=1/dr^2+1/(2*dr*E4	=0	=TRIDI(F2:F52,G2:G52,H2:H52,I2:I52)
50	49	=E49+dr	=1/dr^2-1/(2*dr*E50)	=-(2/dr^2+M)	=1/dr^2+1/(2*dr*E5	=0	=TRIDI(F2:F52,G2:G52,H2:H52,I2:I52)
51	50	=E50+dr	=1/dr^2-1/(2*dr*E51)	=-(2/dr^2+M)	=1/dr^2+1/(2*dr*E5	=0	=TRIDI(F2:F52,G2:G52,H2:H52,I2:I52)
52	51	=E51+dr	1	-1	0	=0	=TRIDI(F2:F52,G2:G52,H2:H52,I2:I52)

FIGURE 3.23
Spreadsheet for direct numerical solution of an annular fin.

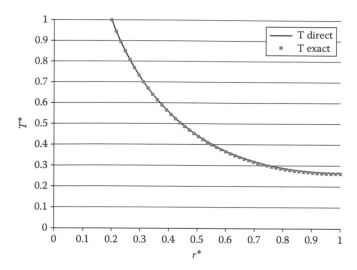

FIGURE 3.24
Comparison of the direct (TRIDI) and exact temperature distributions in an annular fin.

Figure 3.24 is a plot of radial temperature distribution and its comparison with the exact solution is given by Equation 3.142. As can be seen, again there is an excellent agreement between the two, although a slight difference is seen near the tip. The reason for the difference is explained as follows.

3.7.4 Numerical Calculation of Heat Transfer and Fin Efficiency

Once the temperature distribution is obtained, the heat transfer can be found by calculating the heat transfer by conduction at the base. For example, for the annular fin,

$$q = -k2\pi r_i t \frac{(T_b - T_\infty)}{r_o} \left. \frac{dT^*}{dr^*} \right|_{r_i^*} \tag{3.165}$$

Alternatively, heat transfer from the fin can be determined by calculating the convection heat transfer from each element and adding them all up:

$$q = \sum_{j=2}^{N} h2\pi \left(r_j^2 - r_{j-1}^2 \right) \left(\frac{T_j + T_{j-1}}{2} - T_\infty \right) \tag{3.166}$$

This method is more accurate compared to Equation 3.165, since it uses all the calculated temperatures and not just the first two.

To determine the fin efficiency, the maximum amount of heat that can be transferred from the annular fin must be calculated. For the annular fin, it is given by

$$q_{max} = h2\pi \left(r_o^2 - r_i^2 \right) (T_b - T_\infty) \tag{3.167}$$

TABLE 3.3

Efficiency of Annular Fin

Method	Efficiency (%)
Equation 3.168	$\eta = 35.6$
Equation 3.143	$\eta = 37.8$
Equation 3.169	$\eta = 37.4$

then

$$\eta = \frac{-k2\pi r_i t \dfrac{(T_b - T_\infty)}{r_o}}{h2\pi(r_o^2 - r_i^2)(T_b - T_\infty)} \frac{dT^*}{dr^*} = -\frac{\left(\dfrac{t}{r_o}\right)^2}{\dfrac{ht}{k}\left(\dfrac{1}{r_i^*} - r_i^*\right)} \frac{dT^*}{dr^*} \approx -\frac{\left(\dfrac{t}{r_o}\right)^2}{\dfrac{ht}{k}\left(\dfrac{1}{r_i^*} - r_i^*\right)} \frac{T_2^* - T_1^*}{\Delta r^*} \tag{3.168}$$

Alternatively, the efficiency can be determined by calculating the convection heat transfer from each element and adding them all up and dividing by the maximum heat transfer:

$$\eta = \frac{\displaystyle\sum_{j=2}^{N} h2\pi\left(r_j^2 - r_{j-1}^2\right)\left(\dfrac{T_j + T_{j-1}}{2} - T_\infty\right)}{h2\pi\left(r_o^2 - r_i^2\right)(T_b - T_\infty)} = \frac{\displaystyle\sum_{j=2}^{N} \left(r_j^{*2} - r_{j-1}^{*2}\right)\left(\dfrac{T_j^* + T_{j-1}^*}{2}\right)}{\left(1 - r_i^{*2}\right)} \tag{3.169}$$

The exact solution for efficiency of an annular fin is given by Equation 3.143. For the case in Figure 3.24, the efficiency values calculated from the three approaches are given in Table 3.3. As expected, Equation 3.169 is much closer to the exact solution of Equation 3.143 compared with Equation 3.168

The accuracy can be improved by better approximation of the tip boundary condition. The finite-difference approximation of the insulated boundary condition at the tip results in

$$T_N^* = T_{N-1}^* \tag{3.170}$$

Of course this is an approximation, and one does not expect temperature to be constant over a distance Δx close to the tip. A better approximation of the boundary condition can be obtained by assuming that there is a fictitious node $N + 1$, making node N an interior one, which means that Equation 3.146 is also applicable to node N, or

$$A_N T_{N-1}^* + B_N T_N^* + C_N T_{N+1}^* = R_N \tag{3.171}$$

For

$$\frac{dT^*}{dx^*} = 0 \tag{3.172}$$

node N must be the location of a local minimum, and therefore, it is reasonable to assume

$$T_{N-1}^* = T_{N+1}^* \tag{3.173}$$

simplifying Equation 3.171 to

$$(A_N + C_N)T^*_{N-1} + B_N T^*_N = R_N \tag{3.174}$$

which can be used for the boundary condition at N. Using this boundary condition, the agreement with exact solution becomes even better, and the efficiency calculated from Equation 3.169 is within 0.01% of the exact solution.

3.8 Fin Arrays

Fins are often used in an array as shown in Figure 3.25. For fin arrays, the heat transfer per unit area of the finned surface is the quantity of interest, which is the sum of heat transfer from the finned and unfinned areas. Consider a more general case of heat transfer across a wall whose outside surface is finned and assume a one-dimensional analysis. Heat is transferred by convection from the inside gas, which is at T_1, to the inner wall, which is at T_2, conducted through the wall and then convected from the outer surface as well as the fin surfaces to the ambient:

$$q = q_{finned} + q_{unfinned} \tag{3.175}$$

$$q = \eta A_f h(T_b - T_\infty) + A_u h(T_b - T_\infty) \tag{3.176}$$

$$q = [\eta A_f + A_u] h(T_b - T_\infty) \tag{3.177}$$

$$q = \frac{T_b - T_\infty}{\dfrac{1}{h(\eta A_f + A_u)}} \tag{3.178}$$

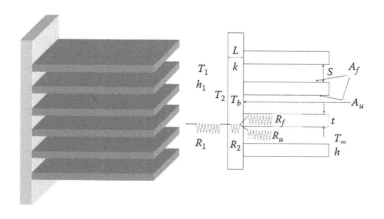

FIGURE 3.25
Fin array.

If A is the total area available for the transfer heat, A_f the total fin area, and A_u the total area of the unfinned portion, then

$$A = A_f + A_u \tag{3.179}$$

$$\eta A_f + A_u = \eta A_f + A - A_f = A\left(1 - \frac{A_f}{A}(1-\eta)\right) \tag{3.180}$$

Surface efficiency is defined as

$$\eta_o = 1 - \frac{A_f}{A}(1-\eta) \tag{3.181}$$

Therefore, the total heat transfer from a finned surface is

$$q = hA\eta_o(T_b - T_\infty) \tag{3.182}$$

and for the case shown in Figure 3.25, the total heat transfer, including the other two resistances, is

$$q = \frac{T_1 - T_\infty}{\dfrac{1}{h_1 A_1} + \dfrac{L}{kA_1} + \dfrac{1}{hA\eta_0}} \tag{3.183}$$

where the last term in the denominator is the resistance of the finned surface. The heat transfer per unit area of the wall is

$$q'' = \frac{T_1 - T_\infty}{\dfrac{1}{h_1} + \dfrac{L}{k} + \dfrac{A_1}{hA\eta_0}} \tag{3.184}$$

In fin arrays, an often used quantity is fin density, which represents the number of fins per unit length. Fin density is therefore as follows:

$$\text{Fin density} = \frac{N}{L} = \frac{1}{S+t} \tag{3.185}$$

Example 3.4

Find the heat transfer per unit area for stainless steel triangular and rectangular fins that have the same base area and volume (i.e., same material cost), $k_{st} = 15.1$ W/m K for the conditions shown in the figure.

The solution details are shown in the spreadsheet. Even though the triangular fin has a lower efficiency, it transfers more heat per unit length because it has more area, for the same weight.

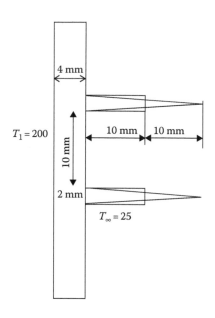

		Rectangular		Triangular
L		0.01		0.02
T_1		200		200
h_1		70		70
L_w		0.004		0.004
k		15.1		15.1
T_∞		25		25
W	Assume fin depth of 1 m	1		1
Fin spacing		0.01		0.01
h		10		10
t		0.002		
b				0.001
L_c	$L_c = L + \dfrac{t}{2}$	0.011	$L_c = L$	0.02
m^*	$m^* = \sqrt{\dfrac{2h}{kt}} L_c$	0.283	$m^* = \sqrt{\dfrac{h}{kb}} L_c$	0.515
η	$\eta = \dfrac{\tan h(m^*)}{m^*}$	0.974	$\eta = \dfrac{1}{m^*} \dfrac{I_1(2m^*)}{I_0(2m^*)}$	0.887
A_f	$A = 2WL_c$	0.022	$A = 2W\sqrt{L^2 + b^2}$	0.040
Au		0.010		0.010
A		0.032		0.050
A_1		0.012		0.012
η_0	$\eta_0 = 1 - \dfrac{A_f}{A}(1 - \eta)$	0.982	$\eta_0 = 1 - \dfrac{A_f}{A}(1 - \eta)$	0.910
q	$q'' = \dfrac{T_1 - T_\infty}{\dfrac{1}{h_1} + \dfrac{L}{k} + \dfrac{A_1}{hA\eta_0}}$	**3319**		**4278**

3.8.1 Finned Tube Arrays

Consider the heat exchanger shown in Figure 3.26, which is made up a fin array and also an array of tubes. The overall heat transfer from a single tube is given by

$$q = \frac{T_i - T_o}{\dfrac{1}{h_i A_i} + \left(\dfrac{\ln \dfrac{r_o}{r_i}}{2\pi k L}\right) + \dfrac{1}{h_o A \left(1 - \dfrac{A_f}{A}(1 - \eta)\right)}} \tag{3.186}$$

Other resistances, such as fouling or contact resistances may also have to be added to the terms in the denominator. To calculate the heat transfer, the values of the inside and outside heat transfer coefficients must be known, and their calculations are shown in the subsequent chapters.

For situations like that shown in Figure 3.26, an approximate solution for fin efficiency can be obtained by replacing the actual fin attached to each tube with an annular fin whose area is equal to the area of the fin strip divided by the number of tubes passing through it. The efficiency of this fin is then used in Equation 3.143.

A more accurate approach is to divide the fin into hexagonal-shaped fins as shown in Figure 3.27. Stewart [2] has shown that the efficiency of the hexagonal fins:

$$\eta = \frac{\tanh(m^*)}{m^*} \cos(0.1 m^*) \tag{3.187}$$

where

$$m^* = \sqrt{\frac{2h}{kt}} R_e \phi \tag{3.188}$$

$$R_e = 1.27 B \sqrt{\frac{H}{B} - 0.3} \tag{3.189}$$

$$\phi = \left(\frac{R_e}{r_t} - 1\right)\left(1 + 0.35 \ln\left(\frac{R_e}{r_t}\right)\right) \tag{3.190}$$

FIGURE 3.26
Fin and tube compact heat exchanger.

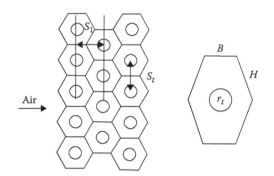

FIGURE 3.27
Dividing the fin strip into hexagons.

$$B = S_l \qquad \text{if } S_l < S_t/2$$
$$B = S_t/2 \quad \text{if } S_l \geq S_t/2 \tag{3.191}$$

$$H = \frac{1}{2}\sqrt{\left(\frac{S_t}{2}\right)^2 + S_l^2} \tag{3.192}$$

and r_t is the outer radius of the pipe.

3.9 Radiation Fins

A more challenging problem is radiation fins, whose formulation results in nonlinear differential equations. Examples of such fins are given in Refs. [3,4]. Consider a constant-area fin that receives energy by radiation and also loses energy by radiation. If G is the energy incident on the surface per unit area, then assuming that the radiation heat loss from the fin surface is to ambient at 0 K, then an energy balance on a differential control volume results in

$$q_{cond,x} + dA_s G\alpha = q_{cond,x+\Delta x} + \varepsilon\sigma dA_s T^4 \tag{3.193}$$

$$\frac{dq_{cond,x}}{dx} + \frac{dA_s}{dx}\varepsilon\sigma\left(T^4 - \frac{G\alpha}{\varepsilon\sigma}\right) = 0 \tag{3.194}$$

$$\frac{d}{dx}\left(kA_x\frac{\partial T}{\partial x}\right) - \varepsilon\sigma\frac{dA_s}{dx}\left(T^4 - \frac{\alpha G}{\varepsilon\sigma}\right) = 0 \tag{3.195}$$

which for a constant-area fin becomes

$$\frac{d^2T}{dx^2} - \frac{\varepsilon\sigma}{kt}\left(T^4 - \frac{\alpha G}{\varepsilon\sigma}\right) = 0 \tag{3.196}$$

Equation 3.196 is the same as that given by Chapman [3], where, ε and α are total surface emissivity and absorptivity, G is the external irradiation (e.g., solar irradiation), k is the fin conductivity, and t is the thickness of the fin. The equilibrium temperature that a surface would achieve if it is insulated and is subject to irradiation G is given by

$$T_s^4 = \frac{\alpha G}{\varepsilon \sigma} \tag{3.197}$$

Using the equilibrium temperature and dimensionless variables

$$T^* = \frac{T}{T_b}, \quad T_s^* = \frac{T_s}{T_b}, \quad x^* = \frac{x}{L}, \quad \text{and} \quad \lambda = \frac{\varepsilon \sigma T_b^4 L^2}{kt} \tag{3.198}$$

leads to the following differential equation:

$$\frac{d^2 T^*}{dx^{*2}} - \lambda \left(T^{*4} - T_s^{*4} \right) = 0 \tag{3.199}$$

The boundary conditions to be satisfied are

$$T^* = 1 \quad \text{at } x^* = 0 \tag{3.200}$$

$$\frac{dT^*}{dx^*} = 0 \quad \text{at } x^* = 1 \tag{3.201}$$

The solution to Equation 3.199 needs to be obtained numerically. Using central difference approximation for the second derivative

$$\frac{T_{i+1} - 2T_i + T_{i-1}}{\Delta x^2} - \lambda \left(T_i^4 - T_s^{*4} \right) = 0 \tag{3.202}$$

and simplifying result in

$$T_{i+1} - \left(\lambda \Delta x^2 T_i^3 + 2 \right) T_i + T_{i-1} = -\lambda \Delta x^2 T_s^{*4} \tag{3.203}$$

which again can be written in the form of a set of algebraic equations having a tridiagonal coefficient matrix

$$A_i T_{i-1}^* + B_i T_i^* + C_i T_{i+1}^* = R_i \tag{3.204}$$

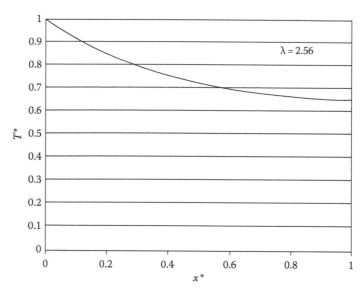

FIGURE 3.28
Temperature distribution in a radiative fin.

where

$$A_i = 1 \tag{3.205}$$

$$B_i = -\left(\lambda \Delta x^2 T_i^3 + 2\right) \tag{3.206}$$

$$C_i = 1 \tag{3.207}$$

$$R_i = -\lambda \Delta x^2 T_s^{*4} \tag{3.208}$$

These equations are nonlinear, since B_i depends on T_i, the temperature that is the unknown, and need to be solved iteratively. The solution is obtained using the function TRIDI in Excel and Excel's iterative capabilities. Taking $\Delta x^* = 0.01$ for $\lambda = 2.56$, the solution converges quickly after around 30 iterations, and the temperature distribution is shown in Figure 3.28.

The efficiency of radiative fins is expressed by Chapman [3]:

$$\eta = \frac{q}{\left(\varepsilon \sigma T_0^4 - \alpha G\right)L} = \frac{q}{\varepsilon \sigma L\left(T_0^4 - T_s^4\right)} \tag{3.209}$$

Substituting for q and making it dimensionless, the earlier equation becomes

$$\eta = -\frac{1}{\lambda\left(1 - T_s^4\right)}\left(\frac{dT}{dx}\right)_{\xi=0} \tag{3.210}$$

Figure 3.29 shows a graph of radiative fin efficiencies versus $\sqrt{\lambda}$ with T_s as the parameter. This graph is a replica of what is given in heat transfer textbook by Chapman [3], which is based on a relatively complex numerical approach given in a NASA report [4].

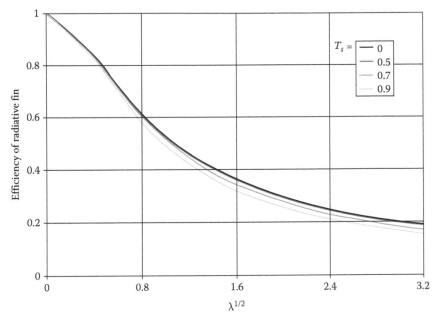

FIGURE 3.29
Efficiency of radiative fin for different values of λ.

3.10 Foam Fins

Metal foams or metal sponges, or porous metals shown in Figure 3.30, are interconnected metal cells, roughly homogeneous, randomly oriented, and often shaped like geodesic domes or buckyballs. They provide more surface area per volume and low weight compared to conventional fins and are being used in applications where high heat transfer is required. The foams are usually manufactured from high conducting solid metals like aluminum, and copper or their alloys.

Foam fins have been the subject of many investigations [5–10] focused on different aspects of the problem, like determining the effective thermal conductivity, friction factor, permeability, etc. One approach in modeling metal foams is to treat them as porous media. Another approach is to treat the foam as connected metal struts, which is presented as follows.

Macroscopically, foams are defined by pore density (PPI) and porosity (ε). Pore density is defined as the number of pores per linear inch (PPI). Pore density is measured microscopically over different lengths of the foam and averaged out. Porosity is defined as the ratio of volume of void spacing to the total bulk volume:

$$\varepsilon = \frac{V_T - V_s}{V_T} \tag{3.211}$$

where V_s is the total volume of the solid and V_t is the volume of the cubical lattice.

Microscopically, two parameters are used to characterize the foam. The strut diameter (d_f) is the average diameter of the foam strands, and strut length (d_p) is the average spacing between the struts, as shown in Figure 3.31.

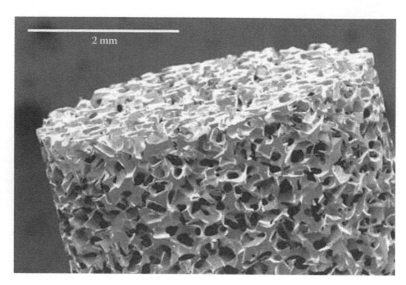

FIGURE 3.30
SEM image of a metal foam. (From Brothers, A.H. and Dunand, D.C., *Advanced Materials*, 17, 484, 2005.)

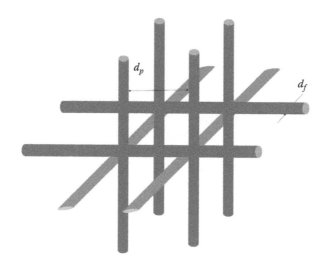

FIGURE 3.31
Continuous cubic structure of metallic foam.

3.10.1 Analytical Model

The basic approach used in modeling the metal foam can be better understood by considering Figure 3.31. The metal foam is considered to have been made up of a simple continuous cubical lattice with slender tubes of diameter (d_f) and length (d_p) [5,11,12]. The struts are modeled as pin fins with heat transferred in the struts by conduction and convected away from them to the fluid that is flowing over the foam. The temperature of the fluid increases as it passes over the foam as it gains the energy lost by the foam.

The first step is to relate the strut diameter (d_f) and pore size (d_p) to porosity (ε) and pore density (PPI). Consider the cubical lattice structure in Figure 3.31 with side equal to dp and diameter d_f. From the definition of porosity [12],

$$V_s = 12\frac{\pi}{4}d_f^2 d_p + (16 - 8\sqrt{2})\frac{d_f^3}{8} \tag{3.212}$$

and

$$V_T = (d_p + d_f)^3 \tag{3.213}$$

Substituting Equations 3.212 and 3.213 in Equation 3.211 results in Ref. [11]

$$\varepsilon = 1 - \frac{9.42r^2 + 0.586r^3}{(1+r)^3} \tag{3.214}$$

where

$$r = \frac{d_f}{d_p} \tag{3.215}$$

As stated earlier, pore density (PPI) is the number of pores per linear inch. If the dimensions are expressed in millimeters, then

$$PPI = \frac{25.4}{d_f + d_p} \tag{3.216}$$

Rearranging Equation 3.216 results in

$$d_p = \frac{25.4}{PPI(1+r)} \tag{3.217}$$

For a known porosity (ε), and PPI, Equations 3.214 and 3.220 provide two equations for calculating strut diameter (d_f) to strut length (d_p). Mahajan and Calmidi [5] provided the following relationship between d_f, d_p, and ε:

$$\frac{d_f}{d_p} = \frac{1.18\sqrt{\left(\dfrac{1-\varepsilon}{3\pi}\right)}}{1 - e^{\frac{-(1-\varepsilon)}{0.04}}} \tag{3.218}$$

Figure 3.32 is a comparison of the manufacturer's data, the model presented by Mahajan and Calmidi [5], and the model presented here for foams with different PPI. In all cases, the foam porosity was 0.90. As can be seen, the two models closely match with the manufacturer's data [12].

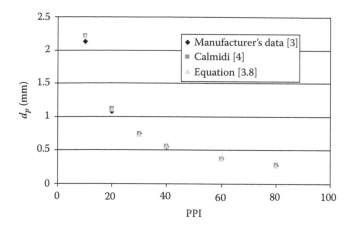

FIGURE 3.32
Comparison of pore size (d_p) models with the manufacturer's data.

3.10.2 Temperature Distribution

The heat transfer model is based on an analogy to a three-dimensional finite-difference mesh [12]. At the junction where struts intersect (nodes), heat is conducted from the base to the struts that are attached to the base and then conducted in the struts and convected to the fluid. The flow of fluid is assumed to be in the y-direction through the foam along its width (W) and with a constant velocity. The base of the fin is maintained at a constant base temperature (T_b).

Figure 3.33 shows the strut intersection at nodes i, j, and k, where six pin fins (struts) intersect. Energy balance is done at this particular node in order to arrive at the governing equation for the foam and fluid. The foam temperature at nodes i, j, and k is represented by $T_{i,j,k}$ and fluid temperature is assumed to be at T_∞. Each strut is assumed to be a pin fin with the base at temperature $T_{i,j,k}$, and the tip at a specified temperature at $i + 1$, or $j + 1$, or $k + 1$. From the first law, sum of all the energy leaving nodes i, j, and k must be zero. Substituting

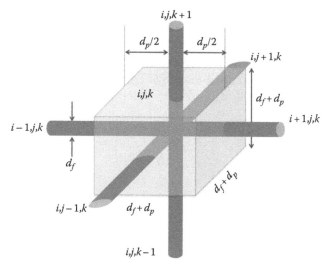

FIGURE 3.33
Model used for the foam fin.

for the different rates of heat transfer and after much algebra and simplifications, the governing equation for the temperature distribution in foam becomes [12]

$$
T^*_{i+1,j,k} + T^*_{i-1,j,k} + T^*_{i,j+1,k} + T^*_{i,j-1,k} - T^*_{i,j,k}\left[4\coth(m^*) + 2\tanh\left(\frac{m^*}{2}\right)\right]\sinh(m^*)
$$

$$
= -6T^*_{f,i,j,k} \cdot \tanh\left(\frac{m^*}{2}\right)\sinh m^* \tag{3.219}
$$

where

$$
T^* = \frac{T - T_\infty}{T_b - T_\infty} \tag{3.220}
$$

$$
m^* = \sqrt{\frac{hP}{k_f A_c}}d_p \tag{3.221}
$$

and heat transfer coefficient h is calculated from that of flow over a horizontal cylinder, for example,

$$
Nu = 0.8Re_{d_f}^{0.43} \cdot Pr^{0.33} \tag{3.222}
$$

This equation is coupled to the fluid temperature, so we need to model the fluid as well. The details of the analysis can be found in Ref. [12]. For a constant fluid temperature, and a known m^*, the set of algebraic equations given by Equations 3.219 can be solved numerically. The details of the analysis and comparison with experimental results can be found in the work of Mudunuri et al. [12].

Problems

3.1 A straight aluminum fin ($k = 200$) has a base thickness of 2 mm and is 30 mm long. The heat transfer coefficient is 30 W/m² K and the ambient is at 300°C. How many fins are needed to remove 1 kW from the environment?

3.2 Starting with the general heat diffusion equation in cylindrical coordinates, derive the energy equation for a straight pin fin. Nondimensionalize the equation and show that it reduces to Equation 3.39.

3.3 Determine the fin efficiency of a constant-area straight fin, having different heat transfer coefficients on the top and the bottom.

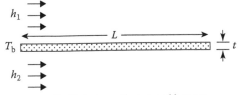

Coefficients on the top and bottom

3.4 Determine the efficiency of a hollow pin fin, having a thickness t, length L, and different inside and outside heat transfer coefficients.

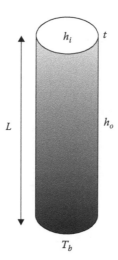

3.5 The solar collector plate shown receives a uniform radiation heat flux on the top, while the bottom surface is well insulated. The top surface is also exposed to a fluid at h_∞ T_∞. The fluid flowing in the pipe is at temperature T_0, and the tube side resistances can be neglected. Determine the heat transfer to the fluid.

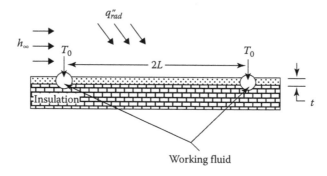

Working fluid

3.6 A thin rectangular solar collector receives uniform heat flux q'' on the top and loses energy by convection also from the top. The two ends of the plate are maintained at temperature T_0, and the other two ends and the bottom are insulated.

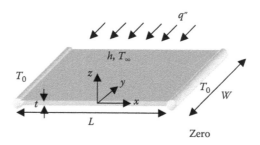

a. Provide a justification as to why temperature variation in the y direction is zero.
b. Starting from the basic elemental control volume analysis, derive the governing equation and the boundary conditions for the steady-state temperature distribution in the collector.

c. Nondimensionalize the governing equation and the boundary conditions.

d. Write the finite-difference form of the governing equation and the boundary conditions and discuss how they would be solved.

3.7 A two-stroke 4 HP lawn mower engine has a cast aluminum alloy 195 cylinder 5 cm high and 5 cm outside diameter. The engine has a thermal efficiency of 40% and is to be cooled by air. Rectangular or hyperbolic fins, not exceeding 3 cm in length, are to be used to cool the cylinder whose outside surface temperature is 400°C. The ambient temperature is 25°C and the heat transfer coefficient is 50 W/m² K. Design the fins.

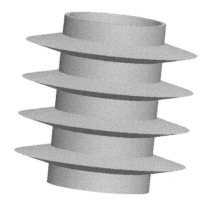

3.8 A two-stroke 4 HP lawn mower engine has a cast aluminum alloy 195 cylinder 5 cm high and 5 cm outside diameter. The engine has a thermal efficiency of 40% and is to be cooled by air. Rectangular longitudinal fins, not exceeding 3 cm in length, are to be used to cool the cylinder whose outside surface temperature is 400°C. The ambient temperature is 25°C and the heat transfer coefficient is 50 W/m² K. Compare the heat loss from each fin, with and without radiation heat loss and comment on the impact of radiation on your design.

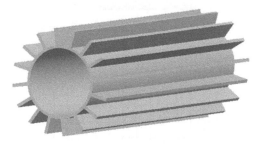

3.9 Numerically, obtain the solution for an annular triangular profile fin.

3.10 Determine the temperature distribution for a pin fin of triangular profile and show that the fin efficiency is

$$\eta = \frac{2}{m^*} \frac{I_2(2m^*)}{I_1(2m^*)}$$

3.11 Solve Problem 3.7 numerically and compare the temperature distribution to the analytical results.

3.12 Consider a truncated triangular fin of length L whose tip thickness is one-third of its base thickness. Compare the heat transfer when the smaller or larger ends are considered the base.

3.13 Obtain the solution to the earlier problem numerically.

3.14 Calculate the heat transfer per unit mass for the two fins in Problems 3.7 and 3.8 and compare those to heat transfer per unit mass of a constant-area rectangular fin.

3.15 The local heat transfer coefficient for natural convection from an isothermal vertical plate is given by

$$h = Cx^{-\left(\frac{1}{4}\right)}$$

where C is a constant.

Derive the governing equation for temperature distribution in the fin, assuming that h varies according to the earlier expression given.

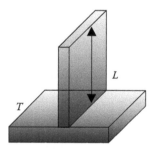

3.16 Consider the wire mesh shown in the following diagram, for which the grid dimensions are $a \times b$. Assuming that each section can be treated as a pin fin of diameter d, determine the temperature distribution in the mesh.

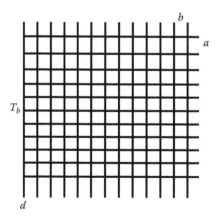

3.17 Determine the temperature distribution for a pin fin losing heat by convection and radiation.

3.18 The aluminum frying pan shown in the following has a thickness of 0.75 cm, an outer diameter of 40 cm, and is used to boil water on a gas stove. The pan receives 4 kW of thermal energy from the bottom over an area with a diameter of 30 cm. The inside water temperature is 100°C and the heat transfer coefficient is 100 W/m² K. The ambient temperature is 25°C, and the outside heat transfer coefficient is 10 W/m² K. The pan analysis can be greatly simplified by approximating the pan as a round thin disk having the same weight as that of the pan and 10% thicker, losing heat from the top to water and receiving heat over an area with a diameter of 30 cm on the bottom surface and losing heat to the ambient over the rest.

What is the diameter of the equivalent disk?

Can the disk be treated as a fin?

Assuming it can, derive the governing equation and the boundary conditions for the temperature distribution in the pan. Nondimensionalize the governing equations and the boundary conditions. Determine the temperature distribution in the pan.

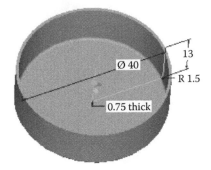

3.19 Derive the governing equation for the temperature distribution in the pan cover's handle. If the handle is made of aluminum and is 11.5 cm wide and 3 mm thick, and the spacing between the two screws is 7 cm and the height is 4 cm, for water boiling in the pan, determine the minimum temperature in the handle, if the heat transfer coefficient is 10 W/m²K.

3.20 An arc welding electrode has a diameter d, carries a current I, and has a coating with an outer diameter D. Derive the governing equation and the boundary conditions and solve for temperature distribution, neglecting the length change.

3.21 Determine the heat transfer from a circular copper ring having a circular cross section, soldered to a surface at temperature T.

3.22 Water at a rate of 0.1 kg/s is to be cooled from 80°C to 40°C in a finned copper pipe having an outside diameter of 10 mm, by transferring heat to the environment at 25°C by natural convection. The fins are constant-thickness annular ones with an outer radius R. Assume the heat transfer coefficient is given by the following correlation:

$$h = \left[0.01 \frac{R^2}{S^6} + .003R^{0.5} \right]^{-0.5}$$

where S is the fin spacing, R is the fin radius in mm, and h is W/m² K. Varying the fin radius (<30 mm), fin thickness (>0.1 mm), and fin spacing, design the lightest system.

3.23 A thin rectangular plate receives uniform heat flux q'' on one side from a high-temperature source and loses energy by convection on both sides. The two ends of the plate are maintained at a temperature T_0. Derive the governing equation and boundary conditions for the temperature distribution. Begin from the basic elemental control volume analysis for the temperature distribution.

3.24 Coolant enters the radiator at 90°C and a velocity of 0.1 m/s and is to be cooled to 30°C. The radiator tube is flat, 1.5 cm wide, 1 mm thick, and the inside height is 2 mm. It is made of aluminum, and the tubes are spaced 2 cm apart. Find the number of fins and their thickness. Assume $h = 150$ W/m² K.

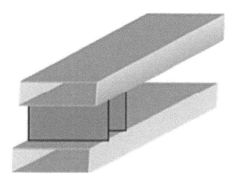

3.25 The solar collector plate shown has a glass cover and receives a uniform radiation heat flux on the top, while the bottom surface is well insulated. The top surface of glass is exposed to a fluid at $h_\infty T_\infty$. The base temperature is T_b.

a. Derive the governing equation and boundary conditions for plate and glass.

b. Nondimensionalize the equations and boundary conditions.

c. Numerically obtain the temperature distribution.

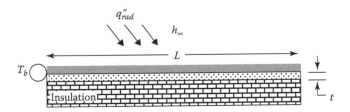

3.26 A constant-area pin fin is initially in thermal equilibrium with the ambient at T_∞. The base temperature is suddenly raised to a temperature change T_b.

a. Derive the governing equation and boundary conditions for the UNSTEADY temperature distribution. Begin from the basic elemental control volume analysis for the transient temperature distribution. Take the fin properties, dimensions (area A, length L, perimeter P), temperatures T_b, T_∞, and the heat transfer coefficient h as all being constants. The length is finite and the tip is not insulated.

b. Nondimensionalize the governing equation and the boundary conditions.

References

1. Fakheri, A. and Naraghi, M. H. (2009) Non-iterative solution of ordinary and partial differential equations using spreadsheets, *Computers in Education Journal* 19(3), 28–37.
2. Stewart, S. W. (2003) Enhanced finned-tube condense design and optimization, PhD dissertation, Georgia Institute of Technology, Atlanta, GA.
3. Chapman, A. J. (1987) *Fundamentals of Heat Transfer*. New York: Macmillan Publishing Company.
4. Lieblein, S. (1959) Analysis of temperature distribution and radiant heat transfer along rectangular fin of uniform thickness, NASA Technical Note D-196, Washington, DC.
5. Mahajan, R. L. and Calmidi, V. V. (2000) Forced convection in high porosity metallic foams, *Journal of Heat Transfer* 122, 557–565.
6. Boomsma, K., Poulikakos, D., and Zwick, F. (2003) Metal foams as compact high performance heat exchangers, *Mechanics of Materials* 35, 1161–1176.
7. Brothers, A. H. and Dunand, D. C. (2005) Ductile bulk metallic glass foams. *Advanced Materials*, 17(4), 484–486.
8. Bhattacharya, A., Mahajan, R. L., and Calmidi, V. V. (2002) Thermo physical properties of high porosity metallic foams, *International Journal of Heat and Mass Transfer* 45, 1017–1031.
9. Bhattacharya, A. and Mahajan, R. L. (2002) Finned metal foam heat sinks for electronics cooling in forced convection, *Journal of Electronic Packing* 124, 155–163.

10. Salas, K. I. and Waas, A. M. (2007) Convective heat transfer in open cell metal foams, *Journal of Heat Transfer* 129, 1217–1229.
11. Ghosh, I. (2008) Heat transfer analysis of high porosity open cell metal foams, *Journal of Heat Transfer* 130(3).
12. Mudunuri, V. R., Sudhini, G., and Fakheri, A. (2010) Analytical and experimental investigations of high porosity metallic foams, *Proceedings of the 2010 ASME International Mechanical Engineering Conference and Exposition*, Vancouver, British Columbia, Canada. IMECE2010-39749.

4

Multidimensional Conduction

Most practical conduction problems involve temperature variation in more than one spatial direction. For example, for a fin, if Biot number, based on the half thickness, is not small, then we need to include the temperature variation in both directions. This makes the governing equation, a partial differential one. There are not many exact solutions available for multidimensional conduction problems, and the few available ones are for simple geometries and even they require relatively advanced mathematical techniques. Yet, the analytical solutions are important in providing insights into the behavior of problems. In this chapter, we will first present a number of different analytical solutions to multidimensional conduction problems, and then examine the numerical solution of more complex problems.

4.1 Steady 2D Conduction

We first examine the case where temperature varies in two directions. Starting with the heat diffusion equation

$$\frac{\partial^2 T}{\partial x^2} + \frac{\partial^2 T}{\partial y^2} + \frac{\partial^2 T}{\partial z^2} + \frac{\dot{q}}{k} = \frac{\rho c}{k}\frac{\partial T}{\partial t} \tag{4.1}$$

for steady 2D conduction without energy generation, it simplifies to

$$\frac{\partial^2 T}{\partial x^2} + \frac{\partial^2 T}{\partial y^2} = 0 \tag{4.2}$$

which is a second-order PDE, with two independent variables. For the rectangular bar shown in the Figure 4.1, the boundary conditions are

$$\begin{aligned}
x &= 0, & T &= T_1 \\
y &= 0, & T &= T_1 \\
x &= L, & T &= T_1 \\
y &= W, & T &= T_1 + T_m f(x)
\end{aligned} \tag{4.3}$$

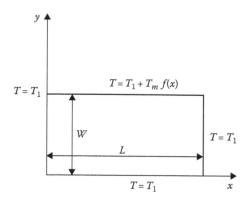

FIGURE 4.1
Steady 2D conduction in a rectangular bar.

4.1.1 Separation of Variables

For this case, the governing equation is homogeneous but the boundary conditions are nonhomogeneous. A differential, or partial differential, or boundary condition is said to be homogeneous provided that if it is satisfied by a function y, then it is also satisfied by Cy where C is a constant. This requires that every term in the differential/partial differential equation or boundary conditions to contain the dependent variable or one of its derivatives. In Equation 4.3, the RHS of all boundary conditions does not involve only T or its derivatives therefore they are nonhomogeneous. To reduce the number of nonhomogeneous terms, we define nondimensional variables like

$$T^* = \frac{T - T_1}{T_m} \tag{4.4}$$

which reduces the governing equations and the boundary conditions

$$\frac{\partial^2 T^*}{\partial x^{*2}} + \frac{\partial^2 T^*}{\partial y^{*2}} = 0 \tag{4.5}$$

$$
\begin{aligned}
&x^* = 0, \quad T^* = 0 \\
&y^* = 0, \quad T^* = 0 \\
&x^* = 1, \quad T^* = 0 \\
&y^* = \frac{W}{L} = W^*, \quad T^* = f(x^*)
\end{aligned}
\tag{4.6}
$$

Figure 4.2 shows the nondimensional problem. This equation is linear and homogeneous (RHS is zero). Three of the boundary conditions are homogeneous with one nonhomogeneous boundary condition in the y direction at $y^* = W^*$.

With only one nonhomogeneity, this equation can be solved by the method of separation of variables, i.e., the solution can be written as the product of two functions, one only a function of x^* and one only a function of y^*

$$T^*(x^*, y^*) = X(x^*)Y(y^*) \tag{4.7}$$

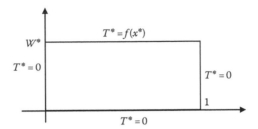

FIGURE 4.2
The nondimensionalized steady 2D conduction in a rectangular bar.

Substituting in Equation 4.5, and noting that X only depends on x^* and Y on y^*, then

$$Y\frac{\partial^2 X}{\partial x^{*2}} + X\frac{\partial^2 Y}{\partial y^{*2}} = 0 \tag{4.8}$$

Dividing both sides of the equation by XY,

$$\frac{1}{X}\frac{\partial^2 X}{\partial x^{*2}} = -\frac{1}{Y}\frac{\partial^2 Y}{\partial y^{*2}} \tag{4.9}$$

In this equation, the LHS is only a function of x^* and the RHS is only a function of y^*. Since x^* and y^* are independent variables, they can be changed independent of each other, e.g., change x^* while keeping y^* constant, and Equation 4.9 must still hold. This requires that X and $\dfrac{\partial^2 X}{\partial x^{*2}}$ change with x^* in a manner that will keep $\dfrac{1}{X}\dfrac{\partial^2 X}{\partial x^{*2}}$ constant, since it must continue to remain equal to the RHS of Equation 4.9 which will remain invariant since y^*, which is independent of x^*, is kept constant.

Therefore the only way that Equation 4.9 can hold, for all values of x^* and y^*, is for both sides of the equation to be equal to a constant. The constant can be positive, negative, or zero, therefore

$$\frac{1}{X}\frac{\partial^2 X}{\partial x^{*2}} = -\frac{1}{Y}\frac{\partial^2 Y}{\partial y^{*2}} = \begin{cases} \lambda^2 \\ 0 \\ -\lambda^2 \end{cases} \tag{4.10}$$

and in what follows, we consider all three possibilities, although as shown in Section 4.2, the correct sign can be immediately determined without the need for examining all three possibilities.

4.1.1.1 Constant Is Zero

Assuming that both sides are equal to zero,

$$\frac{1}{X}\frac{\partial^2 X}{\partial x^{*2}} = -\frac{1}{Y}\frac{\partial^2 Y}{\partial y^{*2}} = 0 \tag{4.11}$$

Then

$$\frac{\partial^2 X}{\partial x^{*2}} = 0 \tag{4.12}$$

and the solution is

$$X = C_1 x^* + C_2 \tag{4.13}$$

and

$$\frac{\partial^2 Y}{\partial y^{*2}} = 0 \tag{4.14}$$

the solution becomes

$$Y = C_3 y^* + C_4 \tag{4.15}$$

The temperature distribution becomes

$$T^* = (C_1 x^* + C_2)(C_3 y^* + C_4) \tag{4.16}$$

The constants must be determined from the boundary conditions:

$$x^* = 0, \quad T^* = 0 = (C_1 0 + C_2)(C_3 y^* + C_4) \rightarrow C_2 = 0 \tag{4.17}$$

$$x^* = 1, \quad T^* = 0 = (C_1)(C_3 y^* + C_4) \rightarrow C_1 = 0 \tag{4.18}$$

Therefore, we get the trivial solution or $T^* = 0$, which will not satisfy the last boundary condition. Therefore, the constant cannot be zero.

4.1.1.2 Constant Is Positive

If the constant is assumed to be positive

$$\frac{1}{X}\frac{\partial^2 X}{\partial x^{*2}} = -\frac{1}{Y}\frac{\partial^2 Y}{\partial y^{*2}} = \lambda^2 \tag{4.19}$$

The step-by-step solution is given in Table 4.1.

TABLE 4.1

Separation of Variable

$\dfrac{d^2 X}{dx^{*2}} - X\lambda^2 = 0$ (4.20a)		$\dfrac{d^2 Y}{dy^{*2}} + Y\lambda^2 = 0$ (4.20b)	
$X = C_1 e^{\lambda x^*} + C_2 e^{-\lambda x^*}$ (4.21a)		$Y = C_3 \cos(\lambda y^*) + C_4 \sin(\lambda y^*)$ (4.21b)	

$$T^* = (C_1 e^{\lambda x^*} + C_2 e^{-\lambda x^*})[C_3 \cos(\lambda y^*) + C_4 \sin(\lambda y^*)] \tag{4.22}$$

$$y^* = 0, \quad T^* = 0 = T^* = (C_1 e^{\lambda x^*} + C_2 e^{-\lambda x^*})(C_3 + C_4 0) \rightarrow C_3 = 0 \tag{4.23}$$

$$x^* = 0, \quad T^* = 0 = T^* = (C_1 + C_2)[C_4 \sin(\lambda y^*)] \rightarrow C_1 = -C_2 \tag{4.24}$$

$$T^* = C_1(e^{\lambda x^*} - e^{-\lambda x^*})(C_4 \sin(\lambda y^*)) = C_5(e^{\lambda x^*} - e^{-\lambda x^*})\sin(\lambda y^*) \tag{4.25}$$

where $C_5 = C_1 C_4$

Equation 4.25 still has two unknowns that need to be determined by the remaining two boundary conditions. Applying the third boundary condition

$$x = 1, \quad T^* = 0 = C_5(e^\lambda - e^{-\lambda})\sin(\lambda y^*) \tag{4.26}$$

which is only equal to zero, if C_5 is equal to zero, in which case we again end up with the trivial solution. Therefore, if the separation of variables is to lead to a solution, the constant must be negative.

4.1.1.3 Constant Is Negative

$$\frac{1}{X}\frac{\partial^2 X}{\partial x^{*2}} = -\frac{1}{Y}\frac{\partial^2 Y}{\partial y^{*2}} = -\lambda^2 \tag{4.27}$$

The step-by-step solution is given in the below table.

$\frac{d^2 X}{dx^{*2}} + X\lambda^2 = 0$	(4.28a)	$\frac{d^2 Y}{dy^{*2}} - Y\lambda^2 = 0$	(4.28b)
$X = C_1\cos(\lambda x^*) + C_2\sin(\lambda x^*)$	(4.29a)	$Y = C_3 e^{\lambda y^*} + C_4 e^{-\lambda y^*}$	(4.29b)
$x^* = 0, \quad X = 0 = C_1 + C_2 0 \rightarrow C_1 = 0$	(4.30a)	$y^* = 0, \quad Y = 0 = (C_3 + C_4) \rightarrow C_3 = -C_4$	(4.30b)
$X = C_2 \sin(\lambda)$	(4.31a)	$Y = C_3(e^{\lambda y^*} - e^{-\lambda y^*})$	(4.31b)
$x^* = 1, \quad X = 0 = C_2 \sin(\lambda)$	(4.32a)	$Y = 2C_3 \dfrac{e^{\lambda y^*} - e^{-\lambda y^*}}{2}$	(4.32b)
$\sin(\lambda) = 0 \rightarrow \lambda = n\pi$	(4.33)		
$X = C_n\sin(\lambda_n x^*)$	(4.34a)	$Y = 2C_3\sinh(\lambda_n y^*)$	(4.34b)
$T^* = c_n\sin(\lambda_n x^*)\sinh(\lambda_n y^*)$ where $c_n = 2C_3 C_n$ (4.35)			

The general solution is given by Equation 4.35 and represents infinitely many solutions all of which satisfy the governing partial differential equation (Equation 4.5) and the first three boundary conditions. These solutions require λ to have the specific values defined by Equation 4.33. The values of λ that satisfy this equation are called the characteristic values or eigenvalues.

The individual solutions given by Equation 4.35 do not satisfy the fourth boundary condition. Since the governing equation, Equation 4.5, is linear and the boundary conditions are also linear, then any linear combination of the solutions will also satisfy the governing equation and three of the four boundary conditions or the general solution can be written as

$$T^* = \sum_{n=1}^{\infty} c_n\sin(n\pi x^*)\sinh(n\pi y^*) = \sum_{n=1}^{\infty} c_n\sin\left(\frac{n\pi x}{L}\right)\sinh\left(\frac{n\pi y}{L}\right) \tag{4.36}$$

To find the solution to the problem at hand, we must find the form that satisfies the fourth boundary condition

$$y = W, \quad T^* = f(x) = T^* = \sum_{n=1}^{\infty} c_n\sin\left(\frac{n\pi x}{L}\right)\sinh\left(\frac{n\pi W}{L}\right) \tag{4.37}$$

We need to find the constants c_n that would satisfy Equation 4.37. This requires that we learn about orthogonal functions.

4.1.2 Orthogonal Functions

Two functions $f(x)$ and $g(x)$ are said to be orthogonal in an interval L with respect to a weighting function $w(x)$, if

$$\int_0^L w(x)f(x)g(x)dx = 0 \tag{4.38}$$

as long as f and g are not identical functions. Consider the general second-order differential equation

$$\frac{d^2y}{dx^2} + f_1(x)\frac{dy}{dx} + [f_2(x) + \lambda^2 f_3(x)]y = 0 \tag{4.39}$$

Subject to the following boundary conditions at the two boundaries a and b

$$x = a, \quad a_1\frac{dy}{dx} + a_2y = 0$$

$$x = b, \quad b_1\frac{dy}{dx} + b_2y = 0 \tag{4.40}$$

The differential equation given by Equation 4.39 is a linear differential equation. For a linear differential equation, if Y_1 and Y_2 are two solutions of the equation, then any linear combination of the two solutions, Y_3

$$Y_3 = c_1Y_1 + c_2Y_2 \tag{4.41}$$

is also a solution, satisfying the differential equation and the boundary conditions. This is shown as follows:

$$\frac{d^2(c_1Y_1 + c_2Y_2)}{dx^2} + f_1(x)\frac{d(c_1Y_1 + c_2Y_2)}{dx} + [f_2(x) + \lambda^2 f_3(x)](c_1Y_1 + c_2Y_2) = 0 \tag{4.42}$$

Expanding and collecting the terms

$$c_1\left\{\frac{d^2Y_1}{dx^2} + f_1(x)\frac{dY_1}{dx} + [f_2(x) + \lambda^2 f_3(x)]Y_1\right\} + c_2\left\{\frac{d^2Y_2}{dx^2} + f_1(x)\frac{dY_2}{dx} + [f_2(x) + \lambda^2 f_3(x)]Y_2\right\} = 0 \tag{4.43}$$

Since Y_1 and Y_2 are solutions, then the terms inside the curly brackets are zero, or Y_3 satisfies the solution. It is easy to show that Y_3 also satisfies the boundary conditions.

The boundary value problem, given by Equations 4.39 and 4.40 only has nontrivial solutions for specific values of λ. These values of λ are called the eigenvalues or characteristic values and the corresponding solutions are called the eigenfunctions or characteristic functions. As an example, consider Equation 4.28a

$$\frac{d^2X}{dx^2} + X\lambda^2 = 0 \tag{4.44}$$

subject to

$$x = 0, \quad X = 0$$

$$x = 1, \quad X = 0$$

We showed that functions

$$\phi = \sin(\lambda x) \tag{4.45}$$

satisfy this equation as long as $\lambda = n\pi$ for $n = 1, 2, 3, \dots$

Therefore, $\pi, 2\pi, 3\pi, \dots$ are the eigenvalues and $\sin(\pi x)$, $\sin(2\pi x)$, $\sin(3\pi x), \dots$ are eigenfunctions.

Let ϕ_m and ϕ_n ($m \neq n$) be two different characteristic functions (eigenfunctions) of Equation 4.39 and λ_m and λ_n be the two corresponding characteristic values (eigenvalues). It can be proven [1] that the characteristic functions are orthogonal with respect to a weighting function $w(x)$ over the interval from a to b, i.e.,

$$\int_a^b w\phi_m\phi_n \, dx = 0 \quad \text{for } m \neq n \tag{4.46}$$

where

$$w = f_3 e^{\int f_1 dx} \tag{4.47}$$

As an example comparing Equation 4.44 with 4.39

$$f_1(x) = 0$$

$$f_2(x) = 0 \tag{4.48}$$

$$f_3(x) = 1$$

and therefore $w = 1$. We now actually prove that for this particular case, the eigenfunctions $\sin\left(\dfrac{m\pi x}{L}\right)$ and $\sin\left(\dfrac{n\pi x}{L}\right)$ are orthogonal with respect to the function $w = 1$ over the interval 0 to L, or we have to show

$$\int_0^L \sin\left(\frac{m\pi x}{L}\right)\sin\left(\frac{n\pi x}{L}\right)dx = 0 \quad m \neq n \tag{4.49}$$

Using trigonometric identities, Equation 4.49 can be written as

$$\int_0^L \frac{1}{2}\left[\cos\frac{\pi x}{L}(m-n) - \cos\frac{\pi x}{L}(m+n)\right]dx = \frac{1}{2}\left[\frac{L}{\pi(m-n)}\sin\pi(m-n) + \frac{L}{\pi(m+n)}\sin\pi(m+n)\right]$$

$$\tag{4.50}$$

For all values of m and n, the second term is always zero. The first term is also zero, unless $n = m$. Assuming $s = \pi(m - n)$ and using L'Hopital's rule to find the limit

$$= \underset{m \to n}{\text{Lim}} \frac{1}{2} \left[\frac{L}{\pi(m - n)} \sin\pi(m - n) \right] = \underset{s \to 0}{\text{Lim}} \frac{L}{2} \frac{\sin(s)}{s} = \frac{L}{2} \tag{4.51}$$

Therefore

$$\int_0^L \sin\left(\frac{m\pi x}{L}\right) \sin\left(\frac{n\pi x}{L}\right) dx = \begin{cases} 0 & m \neq n \\ \dfrac{L}{2} & m = n \end{cases} \tag{4.52}$$

4.1.3 Final Solution

Going back to Equation 4.37, to determine the constants we take advantage of the orthogonality of the sine and cosine functions. Multiplying both side of Equation 4.37 by $\sin\left(\dfrac{m\pi x}{L}\right)$ and integrating from 0 to L

$$\int_0^L f(x)\sin\left(\frac{m\pi x}{L}\right) dx = \sum_{n=1}^\infty c_n \sinh\left(\frac{n\pi W}{L}\right) \int_0^L \sin\left(\frac{n\pi x}{L}\right) \sin\left(\frac{m\pi x}{L}\right) dx \tag{4.53}$$

The RHS has infinitely many terms all of which are zero except when $m = n$, for which

$$\int_0^L \sin\left(\frac{n\pi x}{L}\right) \sin\left(\frac{m\pi x}{L}\right) dx = \frac{L}{2} \tag{4.54}$$

Therefore, Equation 4.53 becomes

$$\int_0^L f(x)\sin\left(\frac{n\pi x}{L}\right) dx = c_n \frac{L}{2} \sinh\left(\frac{n\pi W}{L}\right) \tag{4.55}$$

which can be solved for the constants c to yield

$$c_n = \frac{2}{L} \frac{\displaystyle\int_0^L f(x)\sin\left(\frac{n\pi x}{L}\right) dx}{\sinh\left(\dfrac{n\pi W}{L}\right)} \tag{4.56}$$

Therefore, the solution is given by

$$T^* = \sum_{n=1}^\infty c_n \sin\left(\frac{n\pi x}{L}\right) \sinh\left(\frac{n\pi y}{L}\right) \tag{4.57}$$

where c_n is given by Equation 4.56 which must be evaluated for each value of n. As an example, consider a rectangular bar shown in Figure 4.3, whose three surfaces are at temperature T_1 and the top surface is at a different constant value T_0, or

$$y = W, \quad T = T_0 = T_1 + T_0 - T_1 \tag{4.58}$$

Therefore,

$$T_m = T_0 - T_1 \quad \text{and} \quad f(x) = 1 \tag{4.59}$$

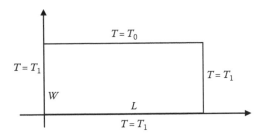

FIGURE 4.3
Steady 2D conduction in a rectangular bar with three sides at one temperature and the top surface at a different temperature.

and

$$c_n = \frac{2}{L} \frac{1}{\sinh\left(\frac{n\pi W}{L}\right)} \int_0^L \sin\left(\frac{n\pi x}{L}\right) dx = \frac{2}{L} \frac{1}{\sinh\left(\frac{n\pi W}{L}\right)} \frac{1}{n\pi} \left[-\cos\left(\frac{n\pi x}{L}\right)\right]_0^L \tag{4.60}$$

$$c_n = \frac{2}{n\pi} \frac{1}{\sinh\left(\frac{n\pi W}{L}\right)} [1 - \cos(n\pi)] = \frac{2}{n\pi} \frac{[1 - (-1)^n]}{\sinh\left(\frac{n\pi W}{L}\right)} \tag{4.61}$$

Therefore, the solution for temperature distribution in a 2D rectangular bar becomes

$$T^* = \frac{2}{\pi} \sum_{n=1}^{\infty} \frac{1 + (-1)^{n+1}}{n} \sin(n\pi x^*) \frac{\sinh(n\pi y^*)}{\sinh(n\pi W^*)} \tag{4.62}$$

and in terms of the dimensional variables

$$T = T_1 + (T_0 - T_1)\frac{2}{\pi} \sum_{n=1}^{\infty} \frac{1 + (-1)^{n+1}}{n} \sin\left(\frac{n\pi x}{L}\right) \frac{\sinh\left(\frac{n\pi y}{L}\right)}{\sinh\left(\frac{n\pi W}{L}\right)} \tag{4.63}$$

Figure 4.4 is a plot of temperature distribution in the x direction at different values of y^* by taking the first 50 nonzero terms of the infinite series. Note that the even terms in Equation 4.63 are all zero.

The heat flux in each direction is

$$q_x = -k\frac{\partial T}{\partial x} = (T_0 - T_1)\frac{2}{\pi} \sum_{n=1}^{\infty} \frac{1 + (-1)^{n+1}}{n} \frac{n\pi}{L} \cos\left(\frac{n\pi x}{L}\right) \frac{\sinh\left(\frac{n\pi y}{L}\right)}{\sinh\left(\frac{n\pi W}{L}\right)} \tag{4.64}$$

$$q_y = -k\frac{\partial T}{\partial y} = (T_0 - T_1)\frac{2}{\pi} \sum_{n=1}^{\infty} \frac{1 + (-1)^{n+1}}{n} \frac{n\pi}{L} \sin\left(\frac{n\pi x}{L}\right) \frac{\cosh\left(\frac{n\pi y}{L}\right)}{\sinh\left(\frac{n\pi W}{L}\right)} \tag{4.65}$$

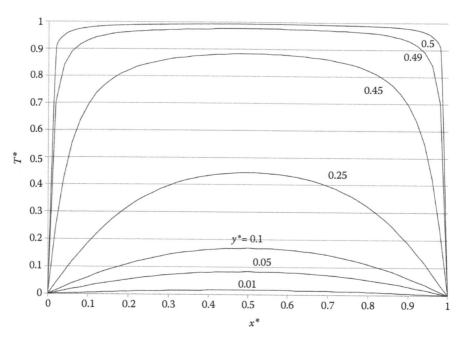

FIGURE 4.4
Temperature distribution in the rectangular bar.

and the heat transfer from each face is given by integrating the heat flux along that face. For example, the total amount of heat transfer in the x direction is

$$q_x = \int_0^W -k\frac{\partial T}{\partial x}dy = \int_0^W -k\left[(T_0-T_1)\frac{2}{\pi}\sum_{n=1}^{\infty}\frac{1+(-1)^{n+1}}{n}\frac{n\pi}{L}\cos\left(\frac{n\pi x}{L}\right)\frac{\sinh\left(\dfrac{n\pi y}{L}\right)}{\sinh\left(\dfrac{n\pi W}{L}\right)}\right]dy \qquad (4.66)$$

which simplifies to

$$\frac{q_x}{k(T_0-T_1)} = -2\left[\sum_{n=1}^{\infty}\frac{1+(-1)^{n+1}}{n\pi}\frac{\cos\left(\dfrac{n\pi x}{L}\right)}{\sinh\left(\dfrac{n\pi W}{L}\right)}\left(\cosh\left(\frac{n\pi W}{L}\right)-1\right)\right] \qquad (4.67)$$

And similarly the heat transfer in the y direction

$$\frac{q_y}{k(T_0-T_1)} = -2\left[\sum_{n=1}^{\infty}\frac{1+(-1)^{n+1}}{n\pi}\frac{\cosh\left(\dfrac{n\pi y}{L}\right)}{\sinh\left(\dfrac{n\pi W}{L}\right)}(1-\cos(n\pi))\right] \qquad (4.68)$$

From an overall energy perspective, the net heat transfer in must be equal to heat transfer out:

$$(q_{x=0}+q_{y=0})-(q_{x=L}+q_{y=W})=0 \qquad (4.69)$$

Evaluating the heat transfers at each boundary results in

$$-2\left[\sum_{n=1}^{\infty}\frac{1+(-1)^{n+1}}{n\pi}\frac{1}{\sinh\left(\dfrac{n\pi W}{L}\right)}\left(\cosh\left(\dfrac{n\pi W}{L}\right)-1\right)\right]$$

$$-2\left[\sum_{n=1}^{\infty}\frac{1+(-1)^{n+1}}{n\pi}\frac{1}{\sinh\left(\dfrac{n\pi W}{L}\right)}(1-\cos(n\pi))\right]$$

$$+2\left[\sum_{n=1}^{\infty}\frac{1+(-1)^{n+1}}{n\pi}\frac{\cos(n\pi)}{\sinh\left(\dfrac{n\pi W}{L}\right)}\left(\cosh\left(\dfrac{n\pi W}{L}\right)-1\right)\right]$$

$$+2\left[\sum_{n=1}^{\infty}\frac{1+(-1)^{n+1}}{n\pi}\frac{1}{\tanh\left(\dfrac{n\pi W}{L}\right)}(1-\cos(n\pi))\right]=0 \qquad (4.70)$$

By combining the sums, it is easy to show that the LHS of the equation adds up to zero.

To demonstrate the limits of infinite solutions, Figure 4.5 is the heat transfer at the four surfaces, each term of Equation 4.70, as a function of the number of terms of the infinite series evaluated. As can be seen, even after taking over 1100 terms the three of the faces ($x = 0$, $x = L$, and $Y = W$) have not yet converged even though the net heat transfer is zero, regardless of the number of terms chosen.

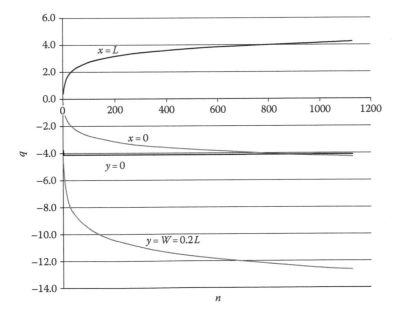

FIGURE 4.5
Heat transfer from the surfaces of the rectangular bar.

4.2 Superposition

To obtain the earlier solution, we tried all the three choices of the signs for the constant, and it turned out that the solution can only exist if the constant is negative. There is no need to check all three possibilities. The sign of the constant should be chosen such that the solution in the direction where both boundary conditions are homogeneous be in the form of periodic functions like sine and cosine or modified Bessel functions, or alternatively the solution in the direction with the nonhomogeneous boundary condition to be in the form of exponential functions.

The method of separation variables works when the problem has one nonhomogeneous term. Consider the problem shown in Figure 4.6, where there are two nonhomogeneous boundary conditions:

$$\frac{\partial^2 T^*}{\partial x^{*2}} + \frac{\partial^2 T^*}{\partial y^{*2}} = 0 \tag{4.71}$$

We will use the fact that the governing equation is linear and the boundary conditions are also linear and show that the solution can be represented as sum of two simpler solutions, each having one nonhomogeneous boundary condition. Each of these problems can be solved by the method of separation of variables and the two solutions added together provide the solution to the original problem.

Assume the solution can be expressed in terms of sum of two solutions:

$$T^* = T_1^* + T_2^* \tag{4.72}$$

Substituting for T^* in the governing equation and expanding results in

$$\frac{\partial^2 T_1^*}{\partial x^{*2}} + \frac{\partial^2 T_1^*}{\partial y^{*2}} + \frac{\partial^2 T_2^*}{\partial x^{*2}} + \frac{\partial^2 T_2^*}{\partial y^{*2}} = 0 \tag{4.73}$$

with the boundary conditions

$$\begin{aligned}
x^* = 0, \quad & T^* = 0 = T_1^* + T_2^* \\
y^* = 0, \quad & T^* = 0 = T_1^* + T_2^* \\
x^* = 1, \quad & T^* = g(y) = T_1^* + T_2^* \\
y^* = W^*, \quad & T^* = f(x) = T_1^* + T_2^*
\end{aligned} \tag{4.74}$$

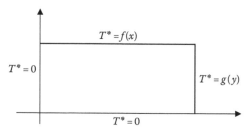

FIGURE 4.6
Superposition solution for rectangular bar.

TABLE 4.2

Simplification of a Problem with Two Nonhomogeneous Boundary Conditions
into Two Problems Each Having One

First Problem	Second Problem

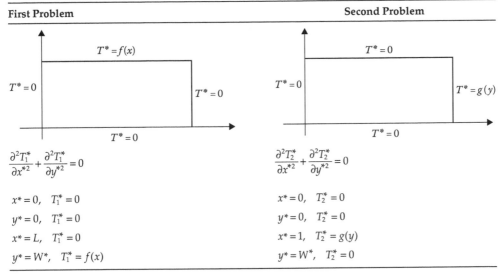

$$\frac{\partial^2 T_1^*}{\partial x^{*2}} + \frac{\partial^2 T_1^*}{\partial y^{*2}} = 0 \qquad\qquad \frac{\partial^2 T_2^*}{\partial x^{*2}} + \frac{\partial^2 T_2^*}{\partial y^{*2}} = 0$$

$$x^* = 0, \quad T_1^* = 0 \qquad\qquad\qquad x^* = 0, \quad T_2^* = 0$$

$$y^* = 0, \quad T_1^* = 0 \qquad\qquad\qquad y^* = 0, \quad T_2^* = 0$$

$$x^* = L, \quad T_1^* = 0 \qquad\qquad\qquad x^* = 1, \quad T_2^* = g(y)$$

$$y^* = W^*, \quad T_1^* = f(x) \qquad\qquad y^* = W^*, \quad T_2^* = 0$$

Now we can construct two problems, which when added together, become the original problem as shown in Table 4.2.

Both of these problems are the same type as the one considered earlier.

Example 4.1

If $f(x) = g(x) = 1$, what is the temperature distribution in the bar.
The solution is given in the table.

$$T_1^* = \frac{2}{\pi} \sum_{n=1}^{\infty} \frac{1+(-1)^{n+1}}{n} \sin(n\pi x^*) \frac{\sinh(n\pi y^*)}{\sinh(n\pi W^*)} \qquad T_2^* = \frac{2}{\pi} \sum_{n=1}^{\infty} \frac{1+(-1)^{n+1}}{n} \sin\left(\frac{n\pi}{W^*} y^*\right) \frac{\sinh\left(\frac{n\pi}{W^*} x^*\right)}{\sinh\left(\frac{n\pi}{W^*}\right)}$$

$$T^* = \frac{2}{\pi} \sum_{n=1}^{\infty} \frac{1+(-1)^{n+1}}{n} \left[\sin(n\pi x^*) \frac{\sinh(n\pi y^*)}{\sinh(n\pi W^*)} + \sin\left(\frac{n\pi}{W^*} y^*\right) \frac{\sinh\left(\frac{n\pi}{W^*} x^*\right)}{\sinh\left(\frac{n\pi}{W^*}\right)} \right]$$

4.2.1 Cylindrical Coordinates

Consider the heat transfer in the cylinder shown in Figure 4.7. The nondimensional form of the governing equation and the boundary conditions are

$$\frac{1}{r^*} \frac{\partial}{\partial r^*}\left(r^* \frac{\partial T^*}{\partial r^*}\right) + \frac{\partial^2 T^*}{\partial z^{*2}} = 0 \qquad\qquad (4.75)$$

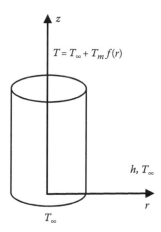

FIGURE 4.7
2D conduction in a cylinder.

$$r* = 0, \quad \frac{\partial T^*}{\partial r^*} = 0$$

$$r* = 1, \quad \frac{\partial T^*}{\partial r^*} = -BiT^* \qquad (4.76)$$

$$z* = 0, \quad T^* = 0$$

$$z* = H^* = \frac{H}{R}, \quad T^* = f(r^*)$$

where

$$r* = \frac{r}{R}, \quad z* = \frac{z}{R}, \quad T^* = \frac{T - T_\infty}{T_m}, \quad Bi = \frac{hR}{k} \qquad (4.77)$$

The boundary condition at $z = 0$ corresponds to having a very large heat transfer coefficient at the bottom surface. Since we have one nonhomogeneous boundary condition, the solution can be obtained by separation of variables. Assume

$$T^*(r^*, z^*) = [\Gamma^*(r^*)][Z^*(z^*)] \qquad (4.78)$$

Substituting in Equation 4.75, separating the terms, and dropping the asterisks

$$\frac{1}{\Gamma} \frac{1}{r} \frac{\partial}{\partial r}\left(r \frac{\partial \Gamma}{\partial r}\right) = -\frac{1}{Z} \frac{\partial^2 Z}{\partial z^2} = -\lambda^2 \qquad (4.79)$$

Since the nonhomogeneity is in the z direction, we want to have the exponential solution in that direction; therefore, the constant must be negative, hence

$$r^2 \frac{\partial^2 \Gamma}{\partial r^2} + r \frac{\partial \Gamma}{\partial r} + \lambda^2 r^2 \Gamma = 0 \qquad (4.80)$$

$$\frac{\partial^2 Z}{\partial z^2} - \lambda^2 Z = 0 \qquad (4.81)$$

From Equation 2.56,

$$\Gamma = C_1 J_0(\lambda r) + C_2 Y_0(\lambda r) \tag{4.82}$$

$$Z = C_3 e^{\lambda z} + C_4 e^{-\lambda z} \tag{4.83}$$

$$T = [C_1 J_0(\lambda r) + C_2 Y_0(\lambda r)][C_3 e^{\lambda z} + C_4 e^{-\lambda z}] \tag{4.84}$$

Applying the boundary condition at $z = 0$ results in

$$T = [b_1 J_0(\lambda r) + b_2 Y_0(\lambda r)] \sin h(\lambda z) \tag{4.85}$$

where b_1 and b_2 are the new constants. Applying the boundary condition at $r = 0$

$$\frac{dT}{dr} = [-b_1 \lambda J_1(\lambda r) - b_2 Y_0(\lambda r)] \sin h(\lambda z) \tag{4.86}$$

$$0 = [-b_1 \lambda J_1(\lambda 0) - b_2 Y_0(\lambda 0)] \sin h(\lambda z) \tag{4.87}$$

$$b_2 = 0 \tag{4.88}$$

Therefore

$$T = b_1 J_0(\lambda r) \sin h(\lambda z) \tag{4.89}$$

Using the third homogeneous boundary condition

$$\frac{dT}{dr} = -BiT \tag{4.90}$$

results in the eigenfunctions

$$\lambda_n J_1(\lambda_n) + Bi J_0(\lambda_n) = 0 \tag{4.91}$$

and therefore the general solution becomes

$$T = \sum_{n=0}^{\infty} b_n J_0(\lambda_n r) \sin h(\lambda_n z) \tag{4.92}$$

Consider Equation 4.80

$$\frac{\partial^2 \Gamma}{\partial r^2} + r^{-1} \frac{\partial \Gamma}{\partial r} + \lambda^2 \Gamma = 0 \tag{4.93}$$

Comparing with Equation 4.39

$$f_1(x) = r^{-1}$$
$$f_2(x) = 0 \tag{4.94}$$
$$f_3(x) = 1$$

Therefore, the weighting function becomes

$$w = f_3 e^{\int f_1 dx} = e^{\int \frac{1}{2} dr} = e^{\ln r} = r \tag{4.95}$$

Since Equation 4.93 is an eigenvalue problem, then $J_0(\lambda_n r)$ are orthogonal functions with respect to weighting function r and therefore

$$\int_0^1 r J_0(\lambda_n r) J_0(\lambda_m r) dr = 0 \tag{4.96}$$

as long as m and n are different. The final boundary condition is

$$f(r) = \sum_{n=0}^{\infty} b_n J_0(\lambda_n r) \sin h(\lambda_n H) \tag{4.97}$$

multiplying both sides of the equation by $r J_0(\lambda_m r)$

$$r f(r) J_0(\lambda_m r) = \sum_{n=0}^{\infty} b_n r J_0(\lambda_m r) J_0(\lambda_n r) \sin h(\lambda_n H) \tag{4.98}$$

and integrating from 0 to 1, all the terms on the RHS are zero, unless $m = n$, in which case the integral is 1, therefore

$$b_n = \frac{\int_0^1 r f(r) J_0(\lambda_n r) dr}{\sin h(\lambda_n H)} \tag{4.99}$$

Therefore, the solution is given by

$$T = \sum_{n=0}^{\infty} b_n J_0(\lambda_n r) \sin h(\lambda_n z) \tag{4.100}$$

with coefficients calculated from Equation 4.99.

As an example, let us determine the temperature distribution if the top surface is maintained at a constant temperature T_1. In this case, $f(r) = 1$ and assuming $s = \lambda_n r$

$$b_n = \frac{\int_0^1 r J_0(\lambda_n r) dr}{\sin h(\lambda_n H)} = \frac{1}{\lambda_n \lambda_n} \frac{\int_0^1 r \lambda_n J_0(\lambda_n r) d(\lambda_n r)}{\sin h(\lambda_n H)} = \frac{1}{\lambda_n \lambda_n} \frac{\int_0^{\lambda_n} s J_0(s) ds}{\sin h(\lambda_n H)} \tag{4.101}$$

Using the Bessel function identity

$$\frac{d}{dx}[x^{\pm n} J_n(x)] = \pm x^n J_{n \mp 1}(x) \tag{4.102}$$

introduced in Chapter 2, then

$$\frac{d}{dx}[xJ_1(x)] = xJ_0(x) \tag{4.103}$$

therefore

$$\int_0^{\lambda_n} sJ_0(s)ds = \lambda_n J_1(\lambda_n) \tag{4.104}$$

which simplifies Equation 4.101 to

$$b_n = \frac{J_1(\lambda_n)}{\lambda_n \sin h(\lambda_n H)} \tag{4.105}$$

or the series solution becomes

$$T^* = \sum_{n=0}^{\infty} \frac{J_1(\lambda_n)J_0(\lambda_n r^*)}{\lambda_n} \frac{\sin h(\lambda_n z^*)}{\sin h(\lambda_n H^*)} \tag{4.106}$$

where

$$\lambda_n J_1(\lambda_n) + Bi J_0(\lambda_n) = 0 \tag{4.107}$$

4.3 Conduction Shape Factors

The concept conduction shape factor can be used to calculate the rate of heat transfer in certain 2D or 3D conduction problems, when heat is transferred between only two surfaces that are at different temperatures. The amount of heat transfer between the two surfaces is assumed to be given by

$$q = kS\Delta T \tag{4.108}$$

where S is the conduction shape factor. For 1D heat transfer, the conduction resistance was defined as

$$q = \frac{DT}{R_{cond}} \tag{4.109}$$

Thus

$$S = \frac{1}{k R_{cond}} \tag{4.110}$$

For example, for a plane wall

$$S = \frac{1}{kR_{cond}} = \frac{1}{k\frac{L}{kA}} = \frac{A}{L} \tag{4.111}$$

The conduction shape factor with several other geometries is given in Table 4.3 [2,3].

TABLE 4.3

Conduction Shape Factors

An isothermal cylinder of length L buried in a semi-infinite medium $L \gg D$, $z > 1.5D$	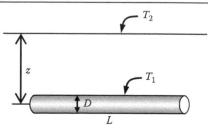	$S = \dfrac{2\pi L}{\ln \dfrac{4z}{D}}$ (4.112)
Two parallel isothermal cylinders of length L buried in an infinitely large medium $L \gg D_1$, D_2, W	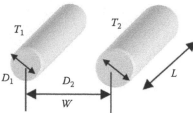	$S = \dfrac{2\pi L}{\cosh^{-1}\left(\dfrac{4W^2 - D_1^2 - D_2^2}{D_1 D_2}\right)}$ (4.113)
Edge of two adjoining walls of equal thickness	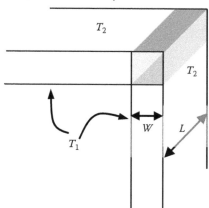	$S = 0.54L$ (4.114)
A square hollow passage of length L	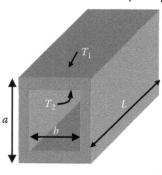	$S = \dfrac{2\pi L}{0.93\ln\left(0.948\dfrac{a}{b}\right)}$ (4.115)
An isothermal cylinder of length L placed in the midplane of an infinite wall, $z > 0.5D$	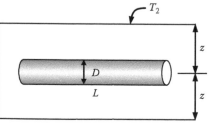	$S = \dfrac{2\pi L}{\ln\left(\dfrac{8z}{\pi D}\right)}$ (4.116)

TABLE 4.3 (continued)

Conduction Shape Factors

A vertical isothermal cylinders of length L buried in a semi-infinite medium $L \gg D$	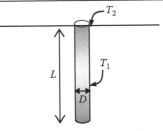	$S = \dfrac{2\pi L}{\ln\left(\dfrac{4L}{D}\right)}$ (4.117)
A row of equally spaced parallel isothermal cylinders of length L diameter D, buried in a semi-infinite medium $L \gg D$, z, and $w > 1.5D$	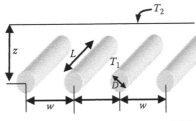	$S = \dfrac{2\pi L}{\cosh^{-1}\left(\dfrac{4W^2 - D_1^2 - D_2^2}{D_1 D_2}\right)}$ (4.118)
Corner at the intersection of three adjoining walls of equal thickness	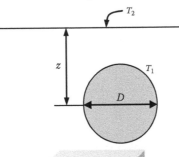	$S = 0.15L$ (4.119)
An isothermal sphere buried in a semi-infinite medium	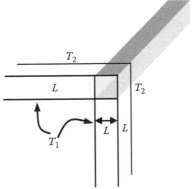	$S = \dfrac{2\pi D}{1 - \dfrac{D}{4z}}$ (4.120)
A circular isothermal cylinder of length L placed at the center of a square bar of length L	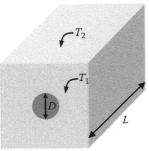	$S = \dfrac{2\pi L}{\ln\left(1.08\,\dfrac{w}{D}\right)}$ (4.121)

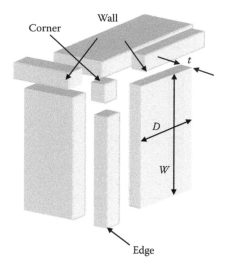

FIGURE 4.8
Heat transfer analysis from the walls of a room.

For example, for heat transfer from a room to outside, shown in Figure 4.8, the wall can be decomposed into its basic elements:

$$S_{wall} = \frac{A}{L} = \frac{DW}{t}$$

$$S_{edge} = 0.54W$$

$$S_{corner} = 0.15t$$

Example 4.2

A spherical shell is used to store radioactive material buried in the ground. The surface temperature of the shell is 100°C, and the ground surface temperature is 20°C. Find the rate of heat generation:

$$q = kS\Delta T = k\frac{2\pi D}{1 - \dfrac{D}{4z}}\Delta T$$

$$q = 0.8\frac{2\pi 0.3}{1 - \dfrac{0.3}{4 \times 2}}80 = 125.34 \text{ W}$$

which is also equal to the total heat generated in the sphere

$$q = \dot{q}\frac{\pi}{6}D^3 = 125.34 \text{ W}$$

$$\dot{q} = \frac{125.34}{\dfrac{\pi}{6}0.3^3} = 8866 \text{ W/m}^3$$

Once this is known, from 1D conduction results, the maximum temperature in the sphere can be calculated.

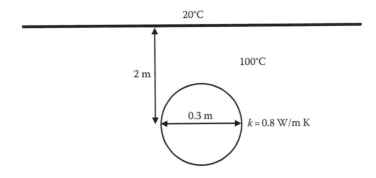

4.4 Numerical Solution

In general, multidimensional conduction problems do not have analytical solutions and the solution to realistic problems need to be found numerically. The available analytical solutions are still invaluable for providing insight into the physics of more realistic problems as well as for checking accuracy of numerical solutions.

We now examine the numerical solution to the problem considered in Section 4.1.3

$$\frac{\partial^2 T^*}{\partial x^{*2}} + \frac{\partial^2 T^*}{\partial y^{*2}} = 0. \tag{4.122}$$

$$x^* = 0, \quad T^* = 0$$

$$y^* = 0, \quad T^* = 0$$

$$x^* = 1, \quad T^* = 0 \tag{4.123}$$

$$y^* = W^* = \frac{W}{L}, \quad T^* = 1$$

Dropping the superscript *, the finite difference form of Equation 4.122 becomes

$$\frac{T_{i+1,j} - 2T_{i,j} + T_{i+1,j}}{\Delta x^2} + \frac{T_{i,j+1} - 2T_{i,j} + T_{i,j+1}}{\Delta y^2} = 0 \tag{4.124}$$

which simplifies to

$$T_{i,j-1} + T_{i,j}\left(-2 - 2\frac{\Delta y^2}{\Delta x^2}\right) + T_{i,j+1} = -(T_{i-1,j} + T_{i+1,j})\frac{\Delta y^2}{\Delta x^2} \tag{4.125}$$

The solution can again be obtained iteratively, by solving for $T_{i,j}$ from Equation 4.125 or more directly using subroutine TRIDI. For using TRIDI, the coefficients are

$$A_j = 1, \quad B_j = -2 - 2\frac{\Delta y^2}{\Delta x^2}, \quad C_j = 1, \quad R_j = -(T_{i-1,j} + T_{i+1,j})\frac{\Delta y^2}{\Delta x^2} \tag{4.126}$$

The solution at nodes $i = 1$ are known from the $x = 0$ boundary condition, and there-fore, the solution starts at nodes with $i = 2$. At this i location, all temperatures are obtained for nodes $j = 1, M$ using subroutine TRIDI. However since the R values need temperature at the downstream nodes, the values of T at $(i + 1)$ are guessed and thus the solution is obtained iteratively, until the difference between the consecutive val-ues of T at all locations is less than a predetermined value, specified by the iterative capabilities of Excel®. This is an example of using a combination of direct and iterative solutions, where the solution in the y direction is found directly and the solution in the x direction iteratively.

Figure 4.9 shows the comparison between the analytical and numerical solutions along the centerline. As can be seen, there is close agreement between the two.

We next consider the numerical solution to heat transfer from the square furnace shown in Figure 4.10 and compare the results with those obtained by using conduction shape factor

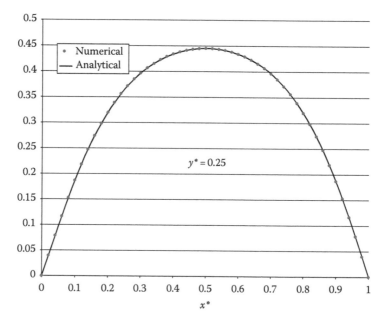

FIGURE 4.9
The comparison between the analytical and numerical solutions along the centerline of a rectangular bar.

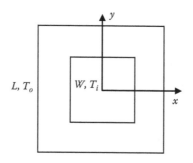

FIGURE 4.10
Square furnace.

correlations. Using L as the length scale, the governing equation is given by Equation 4.122 and the boundary conditions are

$$
\begin{aligned}
x^* &= \pm\frac{W^*}{2} & -\frac{W^*}{2} \le y^* \le \frac{W^*}{2}, & \quad T^* = 1 \\
y^* &= \pm\frac{W^*}{2} & -\frac{W^*}{2} \le x^* \le \frac{W^*}{2}, & \quad T^* = 1 \\
x^* &= \pm\frac{1}{2} & -\frac{1}{2} \le y^* \le \frac{1}{2}, & \quad T^* = 0 \\
y^* &= \pm\frac{1}{2} & -\frac{1}{2} \le x^* \le \frac{1}{2}, & \quad T^* = 0
\end{aligned}
\tag{4.127}
$$

The finite difference of the energy equation is given by Equation 4.126. Due to the irregular geometry, the number of equations in the y direction is not constant, although TRIDI can still be used, the iterative solution is perhaps more convenient.

Solving for temperature

$$
T_{i,j} = \frac{T_{i,j-1} + T_{i,j+1} + (T_{i-1,j} + T_{i+1,j})\dfrac{\Delta y^2}{\Delta x^2}}{2 + 2\dfrac{\Delta y^2}{\Delta x^2}}
\tag{4.128}
$$

This can be solved easily and we can also take advantage of the symmetry and solve only for 1/4 of the solution domain. Figure 4.11 shows the temperature distribution for the case $W^* = 1/2$ using 101 nodes along each exterior side. Once the temperature

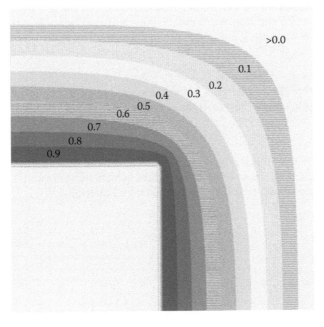

FIGURE 4.11
Temperature distribution in a square furnace for $W^* = 1/2$.

distribution is obtained, then heat transfer along each side, for example horizontal sides, can be calculated from

$$q = \int_0^L -kD\frac{\partial T}{\partial y} dx \tag{4.129}$$

where D is the depth of the furnace. Nondimensionalizing Equation 4.129 and evaluating the integral numerically using the trapezoidal rule result in

$$\frac{q}{kD(T_1 - T_2)} = \int_0^1 -\frac{\partial T^*}{\partial y^*} dx^* = -\left[\sum_{i=1}^n \left. \frac{\partial T^*}{\partial y^*} \right|_i - \frac{1}{2} \left(\left. \frac{\partial T^*}{\partial y^*} \right|_1 + \left. \frac{\partial T^*}{\partial y^*} \right|_n \right) \right] \Delta x^* \tag{4.130}$$

A similar expression can be obtained for heat transfer along the vertical sides. The numerical solution for nondimensional heat transfer results in a value of 10.25 along the outside surfaces and 9.84 along the inside surface for a difference of about 4%. Using conduction shape factor for this geometry

$$q = \frac{2\pi D}{0.93\ln(0.948L/W)} k(T_1 - T_2) \tag{4.131}$$

$$\frac{q}{k(T_1 - T_2)D} = \frac{2\pi}{0.93\ln\left(\dfrac{0.948}{W^*}\right)} = 10.56 \tag{4.132}$$

which is close to the numerical solution.

Example 4.3

The spreadsheet used for the calculating heat transfer from the rectangular furnace can be easily modified for the case of a circular isothermal cylinder of length L placed at the center of a square bar of length L, as shown in Table 4.3. The inner boundary should be replaced by a circle which is defined by

$$x^{*2} + y^{*2} = D^{*2}$$

For the case where the diameter is ½ the sides, the temperatures at the nodes whose coordinates closest satisfy the equation

$$x^{*2} + y^{*2} = \frac{1}{4}$$

should be set equal to 1. From the numerical solution, the nondimensional heat transfer from the outside surfaces of the cube is 8.14, which compares well with the value obtained by using conduction shape factor given by Equation 4.121

$$S = \frac{2\pi L}{\ln\left(1.08\dfrac{w}{D}\right)}$$

which becomes

$$\frac{q}{k(T_1 - T_2)D} = \frac{2\pi}{\ln\left(\dfrac{1.08}{D^*}\right)} = \frac{2\pi}{\ln\left(\dfrac{1.08}{0.5}\right)} = 8.16$$

The temperature distribution is shown in Figure 4.12.

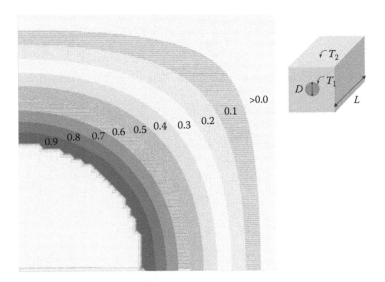

FIGURE 4.12
Temperature distribution in a square bar of length L, with an isothermal circular hole (isothermal cylinder) at its center.

4.4.1 2D Fin

Consider the fin shown in Figure 4.13, which has a finite depth W. We are interested in determining if temperature variation is important in the y direction. Starting with 3D heat diffusion equation

$$\frac{\partial^2 T}{\partial x^2} + \frac{\partial^2 T}{\partial y^2} + \frac{\partial^2 T}{\partial z^2} = 0 \tag{4.133}$$

Following the same procedure as in Chapter 3, when we derived the fin equation from the energy equation, we integrate this equation from $z = 0$ to $z = t/2$

$$\int_0^{t/2} \frac{\partial^2 T}{\partial x^2}\,dz + \int_0^{t/2} \frac{\partial^2 T}{\partial y^2}\,dz + \frac{\partial T}{\partial z}\bigg|_{t/2} - \frac{\partial T}{\partial z}\bigg|_0 = 0 \tag{4.134}$$

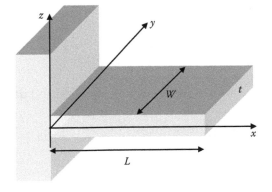

FIGURE 4.13
2D fin.

$$\frac{2}{t}\frac{\partial^2 T}{\partial x^2} + \frac{2}{t}\frac{\partial^2 T}{\partial y^2} - \frac{h}{k}(T - T_\infty) = 0 \tag{4.135}$$

Equation 4.135 is valid for constant or variable h. The boundary conditions are

$$x = 0, \quad T = T_b$$

$$x = L, \quad -k\frac{\partial T}{\partial x} = h_s(T - T_\infty)$$

$$y = 0, \quad k\frac{\partial T}{\partial y} = h_s(T - T_\infty) \tag{4.136}$$

$$y = W, \quad -k\frac{\partial T}{\partial y} = h_s(T - T_\infty)$$

The heat transfer coefficient on the sides and tip, h_s, does not necessarily have to be assumed to be the same as the one on the top and bottom. Assuming constant h and nondimensionalizing Equations 4.135 and 4.136

$$\frac{\partial^2 T^*}{\partial x^{*2}} + \frac{\partial^2 T^*}{\partial y^{*2}} - m^{*2}\, T^* = 0 \tag{4.137}$$

where

$$m^{*2} = \frac{2hL^2}{kt} \tag{4.138}$$

$$x^* = 0, \quad T^* = 1$$

$$x^* = 1, \quad \frac{\partial T^*}{\partial x^*} = -BiT^*$$

$$y^* = 0, \quad \frac{\partial T^*}{\partial y^*} = BiT^* \tag{4.139}$$

$$y^* = W^*, \quad \frac{\partial T^*}{\partial y^*} = -BiT^*$$

$$\text{where} \quad Bi = \frac{h_s L}{k}$$

writing the finite difference form of the governing equation and dropping the superscript $*$

$$\frac{T_{i+1,j} - 2T_{i,j} + T_{i-1,j}}{\Delta x^2} + \frac{T_{i,j+1} - 2T_{i,j} + T_{i,j-1}}{\Delta y^2} - m^2 T_{i,j} = 0 \tag{4.140}$$

$$T_{i-1,j} + \left(-2 - 2\frac{\Delta x^2}{\Delta y^2} - m\Delta x^2\right)T_{i,j} + T_{i+1,j} = -(T_{i,j-1} + T_{i,j+1})\frac{\Delta x^2}{\Delta y^2} \tag{4.141}$$

$$x^* = 0, \quad T_{1,j} = 1 \tag{4.142}$$

$$x^* = 1, \quad T_{N,j} = \frac{1}{1 + Bi\Delta x} T_{N-1,j} \tag{4.143}$$

$$y^* = 0, \quad T_{i,1} = \frac{1}{1 + Bi\Delta y} T_{i,2} \tag{4.144}$$

$$y^* = W^*, \quad T_{i,M} = \frac{1}{1 + Bi\Delta y} T_{i,M-1} \tag{4.145}$$

The solution can be obtained by using function TRIDI,

$$A_1 = 1, \quad B_1 = 1, \quad C_1 = 0, \quad R_1 = 1$$

$$A_i = 1, \quad B_i = -2 - 2\frac{\Delta x^2}{\Delta y^2} - m^2\Delta x^2, \quad C_i = 1, \quad R_i = -(T_{i,j+1} + T_{i,j+1})\frac{\Delta x^2}{\Delta y^2} \quad 1 < i < N \tag{4.146}$$

$$A_N = -\frac{1}{1 + Bi\Delta x}, \quad B_N = 1, \quad C_N = 1, \quad R_N = 0$$

Equation 4.141 is valid for rows $j = 2$ through $j = M - 1$. The temperatures at rows $j = 1$ and $j = M$ are given by Equations 4.144 and 4.145.

The solution can also be obtained through direct iteration by solving for $T_{i,j}$ from Equation 4.141:

$$T_{i,j} = \frac{T_{i-1,j} + T_{i+1,j} + (T_{i,j-1} + T_{i,j+1})\frac{\Delta x^2}{\Delta y^2}}{2 + 2\frac{\Delta x^2}{\Delta y^2} + m^2\Delta x^2} \tag{4.147}$$

Figure 4.14 provides the temperature distribution at three different y locations for the parameters given in the table. The results presented are obtained both iteratively and by using TRIDI for the parameters given in the table.

h	20.0	W/m^2 K
h_s	10.00	W/m^2 K
t	0.0010	m
k	50.0	W/m K
L	0.02	m
W	0.01	m
m^{*2}	0.32	
Bi	0.004	

The solutions are also compared to the exact results for 1D temperature distribution in fins given by

$$T^* = \frac{\cos h[m^*(1 - x^*)] + \dfrac{Bi}{m^*}\sin h[m^*(1 - x^*)]}{\cos h(m^*) + \dfrac{Bi}{m^*}\sin h(m^*)} \tag{4.148}$$

The TRIDI and iterative results are obtained after over 3000 iterations.

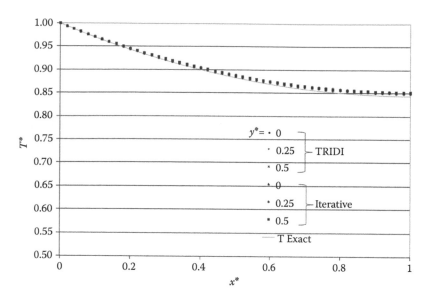

FIGURE 4.14
The temperature distribution at three different y locations in a 2D fin and comparison with 1D solution.

Since Biot number is small, the amount of heat transfer from the edges is negligible and the heat transfer and therefore temperature variation in the y direction are small, and the results should be close to the exact solution for 1D fin. As can be seen, both TRIDI and iterative methods provide the same results and that the temperature variation in the y direction is in fact negligible, and both are very close to the quasi 1D solution.

4.4.2 Variable Heat Transfer Coefficient

We next consider the case where h is not constant over the surface of the fin. Assume that the air is flowing over the surface of the fin in the y direction leading to a variable h in that direction. We assume h to be constant in the x direction

$$\frac{\partial^2 T^*}{\partial x^{*2}} + \frac{\partial^2 T^*}{\partial y^{*2}} - \frac{hy}{k_f} \frac{2L^2 k_f}{ykt} T^* = 0 \tag{4.149}$$

In general,

$$Nu = CRe^n Pr^m = C\left(\frac{Uy}{\nu}\right)^n Pr^m \tag{4.150}$$

For example for laminar flow over a flat plate

$$Nu = \frac{hy}{k_f} = 0.662 Re^{\frac{1}{2}} Pr^{\frac{1}{3}} \tag{4.151}$$

$$\frac{\partial^2 T^*}{\partial x^{*2}} + \frac{\partial^2 T^*}{\partial y^{*2}} - C\left(\frac{Uy}{\nu L}\right)^n Pr^m \frac{2L^{2+n} k_f}{ykt} T^* = 0 \tag{4.152}$$

which simplifies to

$$\frac{\partial^2 T^*}{\partial x^{*2}} + \frac{\partial^2 T^*}{\partial y^{*2}} - \beta y^{*n-1} T^* = 0 \tag{4.153}$$

where

$$\beta = C(Re_L)^n Pr^m \frac{2k_f}{kt^*} \tag{4.154}$$

Assuming that the heat transfer from the edges to be negligible, the insulated boundary condition on all edges except the base can be used to obtain the temperature distribution.

4.4.3 Louvered Fins

Since the heat transfer coefficient decreases in the flow direction, one method of enhancing heat transfer from fins is to cut slits into them, as shown in Figure 4.15. These are known as louvered fins and are used extensively in automotive and air-conditioning applications. Assuming fin to be thin so that temperature variation in the z direction can be neglected, it was shown that the governing equation for the temperature distribution is

$$\frac{\partial^2 T}{\partial x^2} + \frac{\partial^2 T}{\partial y^2} - \frac{2h}{kt}(T - T_\infty) = 0 \tag{4.155}$$

Unlike the 3D fin case, the temperature variation in the y direction may no longer be negligible. However, as seen in the previous example, in general, we can neglect the heat transfer from all edges that have thickness t, also taking advantage of the symmetry, except at the base, the temperature gradient around all external edges is zero.

To simplify the analysis and just to focus on the impact of the slit on temperature distribution, rather than taking variable h, we assume constant h, taking into consideration that the impact of the slits is to increase the effective value of the average heat transfer coefficient over the fin. The nondimensional form of the energy equation for the fin is

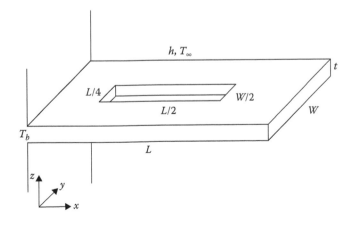

FIGURE 4.15
Louvered fin.

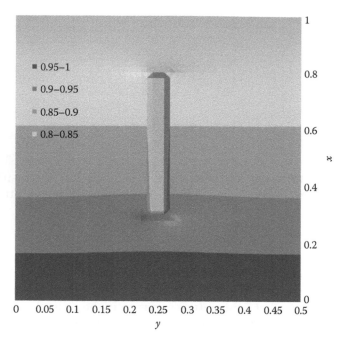

FIGURE 4.16
Temperature distribution in a louvered fin.

given Equation 4.147 and the boundary conditions are Equations 4.142 through 4.145 for the insulated boundary ($Bi = 0$).

To account for the impact of the slit, we set the temperature along the slit's edge equal to the adjacent cell in the fin, to satisfy the condition that there is negligible heat transfer from the edge of the fin. The result of the numerical simulation is shown in Figure 4.16, for constant heat transfer coefficient, to show the ease by which such solutions can be obtained. As can be seen, in this case there is some variation in temperature in the y direction. However, to better assess the impact of the cuts, variable heat transfer coefficient needs to be used, as was done in the previous section.

4.4.4 2D Conduction in Cylindrical Coordinates

Consider a cylinder of radius R and length L placed in an ambient which is at T_∞ and h where the base ($x = 0$) temperature is T_b. The dimensionless energy equation is

$$\frac{1}{r^*}\frac{\partial}{\partial r^*}\left(r^*\frac{\partial T^*}{\partial r^*}\right) + \frac{\partial^2 T^*}{\partial z^{*2}} = 0 \tag{4.156}$$

which can be written as

$$\frac{\partial^2 T^*}{\partial z^{*2}} + \frac{\partial^2 T^*}{\partial r^{*2}} + \frac{1}{r^*}\frac{\partial T^*}{\partial r^*} = 0 \tag{4.157}$$

The finite difference form of Equation 4.157 becomes

$$\frac{T_{i+1,j}-2T_{i,j}+T_{i-1,j}}{\Delta z^2}+\frac{T_{i,j+1}-2T_{i,j}+T_{i,j-1}}{\Delta r^2}+\frac{1}{r_j}\frac{T_{i,j+1}-T_{i,j-1}}{2\Delta r}=0 \tag{4.158}$$

which simplifies to

$$T_{i-1,j}+\left(-2-2\frac{\Delta z^2}{\Delta r^2}\right)T_{i,j}+T_{i+1,j}=-\frac{\Delta z^2}{\Delta r^2}\left[\left(1+\frac{\Delta r}{2r_j}\right)T_{i,j+1}+\left(1-\frac{\Delta r}{2r_j}\right)T_{i,j-1}\right] \tag{4.159}$$

The boundary condition at $r^* = R^* = R/L$, and $z^* = 1$ become

$$T_{i,M} = \frac{T_{i,M-1}}{1+\Delta rBi} \tag{4.160}$$

$$T_{N,j} = \frac{T_{N-1,j}}{1+\Delta zBi} \tag{4.161}$$

The solution is obtained using 51 nodes in r and z directions, with $D = L = 0.1$ m, $k = 40$ W/m K, and $h = 100$ W/m^2 K and $Bi = hL/k = 0.25$. The results of temperature change at the center, surface, and halfway radially are shown in Figure 4.17, and as can be seen, there is little temperature variation in the radial direction. This should not be surprising since Biot number based on the radial distance $Bi = \dfrac{hR}{k} = 0.125$ and therefore temperature distribution should be close to that of a pin fin, which is also plotted in Figure 4.17.

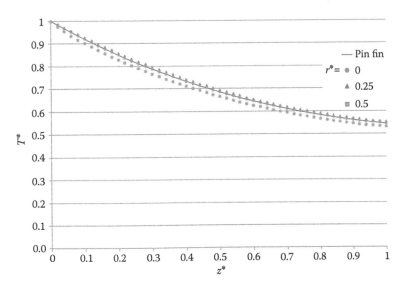

FIGURE 4.17
The temperature distribution at three different r locations in a circular bar and comparison with fin solution.

4.4.5 Heat Exchanger Fins

As was discussed in Chapter 3, compact heat exchangers have a large area-to-volume ratio. Often times, this is achieved by addition of fins. The method for calculating the efficiency of finned tube heat exchangers was discussed there as well, by dividing the fin into hexagonal-shaped segments. We can now directly solve for the temperature distribution and fin efficiency. The energy equation for 2D fin was derived earlier as

$$\frac{\partial^2 T^*}{\partial x^{*2}} + \frac{\partial^2 T^*}{\partial y^{*2}} - m^{*2} T^* = 0 \tag{4.162}$$

where

$$m^{*2} = \frac{2hS_l^2}{kt} \tag{4.163}$$

$$r_t^* = \frac{r_t}{S_l}$$

$$S_t^* = \frac{S_t}{S_l} \tag{4.164}$$

Taking advantage of the symmetry, the computational domain is highlighted in Figure 4.18. The temperature distribution everywhere is given by Equation 4.162 and the gradient along all the boundaries are equal to zero, except at the tubes

$$x^{*2} + y^{*2} - R^{*2} < 0 \quad \text{or} \quad (x^* - 1)^2 + (y^* - S_t^*)^2 - R^{*2} < 0 \tag{4.165}$$

where temperature is 1.

For $d = 0.01$ m, $S_l = 0.05$ m, $S_t = 0.04$ m, $h = 20$ W/m² K, $t = 0.001$ m, $k = 50$ W/m K, using 51 nodes in each direction, the temperature variation with x for several values of y are shown

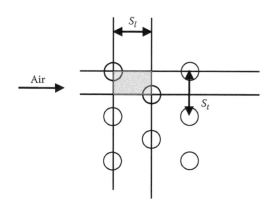

FIGURE 4.18
Heat exchanger fin.

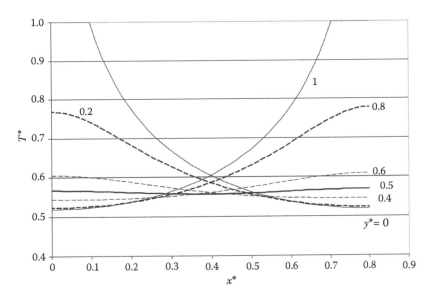

FIGURE 4.19
The temperature variation with x at different y locations.

in Figure 4.19. The temperature is showing the correct behavior. Once the temperature distribution is found, the heat transfer can be evaluated numerically from

$$q = \frac{1}{4}hLW(T_b - T_\infty)\left[\sum_{i=1}^{N-1}\sum_{j=1}^{M-1}\frac{1}{N-1}\frac{1}{M-1}(T_{i,j}^* + T_{i+1,j}^* + T_{i,j+1}^* + T_{i+1,j+1}^*) - 2\pi\frac{r^{*2}}{W^*}\right] \quad (4.166)$$

which is then used to determine fin efficiency from

$$\eta = \frac{\dfrac{4\sum_{i=1}^{N}T_{i,j}^* - \left(\sum_{j=1}^{M}T_{1,j}^* + \sum_{j=1}^{M}T_{N,j}^* + \sum_{i=1}^{N}T_{i,1}^* + \sum_{i=1}^{N}T_{i,M}^*\right)}{4(N-1)(M-1)} - \dfrac{\pi}{2}\dfrac{r^{*2}}{W^*}}{1 - \dfrac{\pi}{2}\dfrac{r^{*2}}{W^*}} \quad (4.167)$$

For this case, efficiency is 0.6 from the numerical solution. A crude approximate solution can be obtained by considering a straight insulated tip fin having a length $S_l/2 = 0.025$ m. For this fin, the efficiency is

$$\eta = \frac{\tan h\left(\sqrt{\dfrac{2\times20}{50\times0.001}}\times0.025\right)}{\left(\sqrt{\dfrac{2\times20}{50\times0.001}}\times0.025\right)} = 0.86 \quad (4.168)$$

This efficiency overestimates the actual efficiency, since for this case the base which is $S_t/2 = 0.02$ m is maintained at the base temperature, whereas for the actual fin, only a length equal to $\pi D/4 = 0.0078$ or only 39% of the base can be approximately considered to be at base temperature. Nevertheless, the 1D results can be used to guide the modification of the parameters in order to improve the overall heat transfer rate from the finned tubes.

Problems

4.1 Analytically solve for temperature distribution in a long rectangular bar $L \times W$, where the left and right surfaces are maintained at constant temperature T_1, the bottom surface is insulated, and the tight side temperature is given by

$$T = T_1 \left[1 + \sin\left(\pi \frac{x}{L} \right) \right]$$

Verify that the net heat transfer to the bar is zero.

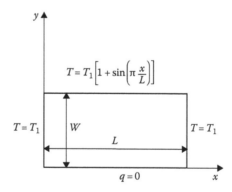

4.2 Analytically solve for temperature distribution in a long rectangular bar $L \times W$, where the left and right surfaces are maintained at a constant temperature and the top and bottom are losing heat by convection.

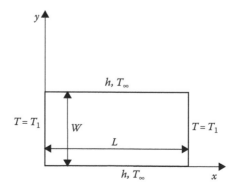

4.3 Mathematically prove that the temperature distribution in a cylinder of $D \times H$, where the top and bottom are maintained at constant temperature and the sides that are losing heat by convection are given by the product of an infinitely long cylinder and a plane wall.

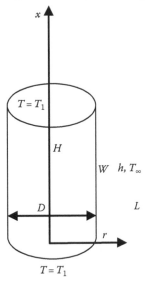

4.4 Analytically solve for temperature distribution in a cylinder of $D \times H$, where the top and bottom are maintained at constant temperature and the sides are losing heat by convection.

4.5 Analytically solve for temperature distribution in a cylinder of $D \times H$, where the top and bottom are loosing heat by convection and the sides are maintained at a constant temperature.

4.6 Solve Problem 4.1 numerically.

4.7 Solve Problem 4.2 numerically.

4.8 Solve Problem 4.3 numerically.

4.9 Solve Problem 4.4 numerically.

4.10 For fins, it is assumed that the base can be considered to be at a constant temperature. Verify the accuracy of this assumption by numerically solving for the temperature distribution for the case shown later if $t/L = 0.02$, for $0.01 < ht/k < 1$. Take advantage of the symmetry.

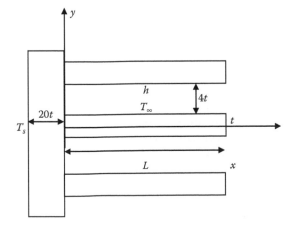

4.11 Numerically verify Equation 4.120.

4.12 Numerically verify Equation 4.121.

4.13 To reduce the heat loss from a thin copper solar collector, it is covered with a sheet of glass. Assuming 2D temperature distribution:
   a. Determine the governing equation and boundary conditions.
   b. Determine the temperature and heat loss from the collector numerically.

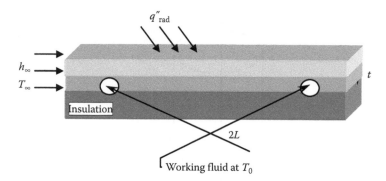

4.14 A 2 cm long microchannel heat exchanger having a circular cross section of 0.5 mm wide is used to condense Refrigerant-134a at 70°C. The channel wall thickness is 1 mm. Determine the temperature distribution in the channel, if the outside is assumed to be at a constant temperature of 25°C.

4.15 Numerically solve for temperature distribution in a rectangular heating element loosing heat to a medium by convection. The exact solution for the large heat transfer coefficient is given by Arpaci as

$$\frac{T - T_\infty}{\dfrac{\dot{q}L^2}{k}} = \frac{1}{2}\left[1 - \left(\frac{x}{L}\right)^2\right] - 2\sum_{n=0}^{\infty}\frac{(-1)^n}{\left(\dfrac{(2n+1)\pi}{2}\right)^3}\left[\cos\left(\frac{(2n+1)\pi}{2}\frac{x}{L}\right)\frac{\cos h\left(\dfrac{(2n+1)\pi}{2}\dfrac{y}{L}\right)}{\cos h\left(\dfrac{(2n+1)\pi}{2}\dfrac{W}{L}\right)}\right]$$

Use this to check the accuracy of your numerical solution.

4.16 A cylindrical rod receives a uniform heat flux over half of its circumference and looses heat by convection over from its outer surface. Numerically solve for temperature distribution in the rod for

$$q''(\theta) = \begin{cases} q_0'' \sin(\theta) & 0 \leq \theta \leq \pi \\ 0 & \pi \leq \theta \leq 2\pi \end{cases}$$

and compare with the exact solution

$$\frac{T-T_\infty}{\frac{q_0''}{h}} = \frac{1}{\pi} + \frac{\frac{r}{R}}{2\left(1+\frac{k}{hR}\right)}\sin(\theta) - \frac{2}{\pi}\sum_{n=1}^{\infty}\frac{\left(\frac{r}{R}\right)^{2n}}{(4n^2-1)\left(1+\frac{2nk}{hR}\right)}\cos(2n\theta)$$

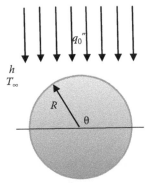

4.17 Numerically solve for temperature distribution in a semi-infinite cylindrical bar whose base is at a uniform temperature and looses heat to the environment by convection. Compare your solution to the exact solution for high heat transfer coefficient given by Arpaci

$$\frac{T-T_\infty}{T_b-T_\infty} = 2\sum_{n=1}^{\infty}\frac{e^{-\lambda_n x}J_0(\lambda_n r)}{(\lambda_n R)J_1(\lambda_n R)}$$

where $J_0(\lambda_n R) = 0$.

4.18 Numerically solve for temperature distribution in a triangle whose three sides are maintained at a constant temperature.

4.19 A cylindrical pipe of diameter d and thickness t receives a uniform heat flux over half of its circumference from a hemispherical solar reflector of diameter D. The pipe looses heat by convection over its outer surface and the inner surface is at a constant temperature T_s. Numerically solve for the temperature distribution in the pipe and determine the amount of heat transferred to the fluid inside the pipe, assuming a reasonable set of parameters.

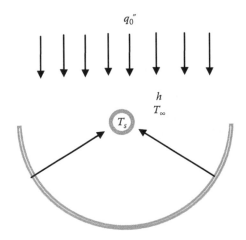

4.20 Stream is flowing in an underground pipe. What is the daily heat loss?

4.21 A computer chip in the shape of a disk (D = 10 mm) is flush mounted in a substrate as shown having a $2D$ diameter. The rate of energy generation in the chip is 10^8 W/m^3 and its thermal conductivity is 30 W/m K. The ambient is at 25°C and the heat transfer coefficient is 100 W/m^2 K and the outer edge and the bottom of substrate are insulated. Numerically determine the maximum temperature in the chip. What substrate (k = 80 W/m K) diameter is needed to ensure that the chip maximum temperature will be below 80°C.

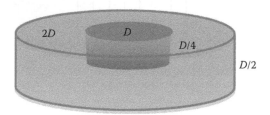

4.22 In metal casting, cores are forms, usually made of sand, which are placed into a mold cavity to form the interior surfaces of the castings. Thus the void space between the core and mold-cavity surface is what eventually becomes the casting. In a particular casting process, cores are dipped in a refractory material and then heated in a microwave oven for drying and curing. Heating in a microwave is equivalent to an internal energy generation $\dot{q}(x, y)$ that is a function of location in the core:

     a. Assuming a 2D geometry (L by H), model the process for steady-state case.

     b. Nondimensionalize the governing equation and the boundary conditions.

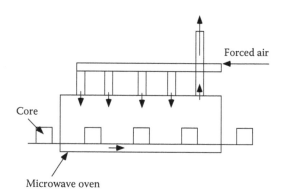

4.23 One method of enhancing heat transfer from fins is to cut slits into them. These are known as louvered fins and are used extensively in automotive applications. Assuming fin to be thin so that temperature variation in the z direction can be neglected, starting with the heat diffusion equation or basic elemental control volume analysis, show that the governing equation for the temperature distribution is

$$\frac{\partial^2 T}{\partial x^2} + \frac{\partial^2 T}{\partial y^2} - \frac{2h}{kt}(T - T_\infty) = 0$$

a. What are the boundary conditions?
b. Nondimensionalize the governing equation and the boundary conditions.
c. Solve for the temperature distribution using finite difference method, for the parameters in the table.

k	200	t	0.005
h	100	L	0.05
T_∞	25	W	0.03
T_b	100		

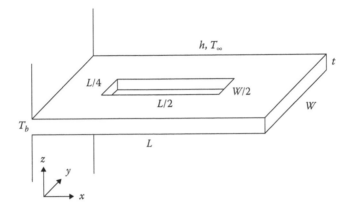

4.24 A cylindrical rod receives a nonuniform heat flux over the top half of its circumference and looses heat by convection over its entire circumference

$$q''(\theta) = \begin{cases} q_0'' \sin(\theta) & 0 \le \theta \le \pi \\ 0 & \pi \le \theta \le 2\pi \end{cases}$$

a. Write the governing equation and boundary conditions.
b. Nondimensionalize the governing equation and boundary conditions.
c. Obtain the solution numerically.

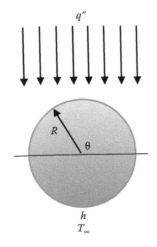

References

1. Arpaci, V. S. (1996) *Conduction Heat Transfer*. Reading, MA: Addison-Wesley.
2. Bergman, T. A., Lavine, S., Incropera, F., and Dewitt, D. P. (2011) *Fundamentals of Heat and Mass Transfer*, 7th edn. Hoboken, NJ: John Wiley.
3. Çengel, Y. A. (2007) *Heat and Mass Transfer: A Practical Approach*. Boston, MA: McGraw-Hill.

5

Transient Conduction

5.1 Introduction

In steady-state problems, the properties (temperature, density, etc.) do not change with respect to time, indicating no storage of energy within the system. We now consider the case where the net energy transfer to the control mass, heat and work transfer, is not zero, causing the internal energy to change with time which in turn causes the temperature to change with time, absent phase change. The temperature may also be a function of location. Biot number significantly influences the behavior of the transient problems. As shown in Chapter 1, Biot number is the ratio of resistance to heat transfer by conduction in the solid to the resistance to convection in the fluid surrounding the object. As was shown in Chapter 1, Biot number is also the ratio of the temperature change within the body (ΔT_s) to the temperature change in the fluid (ΔT_f), and is given by

$$\frac{\Delta T_s}{\Delta T_f} = Bi = \frac{hL_c}{k} \tag{5.1}$$

For small Biot number, the conduction resistance is small; thus a small temperature gradient in the solid is all that is needed to transfer the heat, or the solid will be approximately isothermal and its temperature can be assumed to be independent of location and only dependent on time. Spatial variation of temperature may also be neglected in a well-stirred fluid (Figure 5.1).

The length scale, L_c, is a dimension that characterizes the object and is generally taken as the ratio of the volume of the object (which is proportional to how much energy the object can store) over the surface area through which heat is transferred to the surroundings:

$$L_c = \frac{V}{A_s} \tag{5.2}$$

and is calculated as follows for several common geometries as shown in Figure 5.2:

$$L_{c,wall} = \frac{V}{A_s} = \frac{LA}{2A} = \frac{L}{2}$$

$$L_{c,long\,cylinder} = \frac{V}{A_s} = \frac{\frac{\pi}{4}D^2 h}{\pi D h} = \frac{D}{4} \tag{5.3}$$

$$L_{c,sphere} = \frac{V}{A_s} = \frac{\pi D^3/6}{\pi D^2} = \frac{D}{6}$$

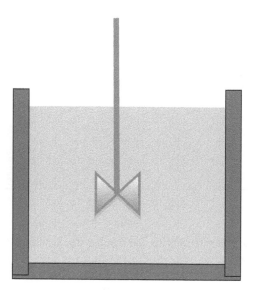

FIGURE 5.1
Well-stirred fluid.

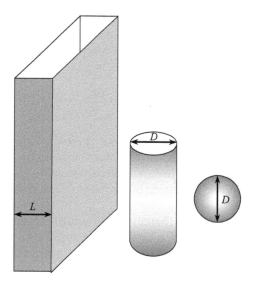

FIGURE 5.2
Basic geometries.

5.2 Lumped Heat Capacity System (Lumped Parameter Method)

Consider an arbitrary-shaped object having a surface area A_s and a volume V as shown in Figure 5.3. If the Bi number is small (<0.1), then we can assume that the entire body is at a uniform temperature at any given time. Starting with the first law of thermodynamics for the object

$$q + \dot{W} + \dot{q} = \frac{dE}{dt} \tag{5.4}$$

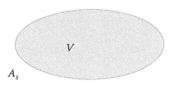

FIGURE 5.3
Cooling of an arbitrary shaped object.

Assuming that the rate of work done on the system and the rate of energy generation in the solid to be zero and that the heat transfer is by convection only, then Equation 5.4 becomes

$$\rho V c \frac{dT}{dt} = -hA_s(T - T_\infty) \tag{5.5}$$

subject to $t = 0$, $T = T_i$.

5.2.1 Nondimensionalization

Assume

$$T^* = \frac{T - T_\infty}{T_i - T_\infty} \quad \text{and} \quad t^* = \frac{t}{t_s}$$

Then the energy equation becomes

$$\frac{dT^*}{dt^*} = -\frac{t_s h A_s}{\rho V c} T^* \tag{5.6}$$

The number of parameters can be reduced significantly by assuming

$$\frac{t_s h A_s}{\rho V c} = 1 \rightarrow t_s = \frac{\rho V c}{h A_s} \tag{5.7}$$

and

$$t^* = \frac{h}{\rho c L_c} t \tag{5.8}$$

then Equation 5.6 and its initial condition reduce to

$$\frac{dT^*}{dt^*} = -T^* \tag{5.9}$$

$$t^* = 0, \quad T^* = 1 \tag{5.10}$$

whose solution is

$$T^* = e^{-t^*}$$ (5.11)

or the nondimensional temperature of the solid drops exponentially. In terms of the dimensional quantities

$$\frac{T - T_\infty}{T_i - T_\infty} = e^{-\frac{h}{\rho c L_c} t}$$ (5.12)

The nondimensional time

$$t^* = \frac{ht}{\rho L_c c} = \frac{h}{\rho c L_c} \frac{k}{k} \frac{L_c}{L_c} t = \frac{\alpha Bi}{L_c^2} t = BiFo$$ (5.13)

where

$$\alpha = \frac{k}{\rho c}$$ (5.14)

and *Fo* is the Fourier modulus which is the relevant nondimensional time in conduction defined as

$$Fo = \frac{\alpha t}{L_c^2}$$ (5.15)

Finally, the temperature variation can be written as

$$\frac{T - T_\infty}{T_i - T_\infty} = e^{-BiFo}$$ (5.16)

This result can be used to estimate how long it takes to heat up or cool down an object. For $t^* = 4.6$, the object's nondimensional temperature becomes 0.01 or if $Bi < 0.1$, then after

$$t \approx \frac{4.6}{Bi} \frac{L_c^2}{\alpha}$$ (5.17)

the object's temperature has reached essentially that of the ambient temperature.

Example 5.1

A can of soda at room temperature of 75°F is to be cooled to 37°F. How long does it take to cool it in the refrigerator which is at 36°F versus in the freezer which is at 0°F.

Neglecting the effect of the aluminum can and assuming properties of soda to be the same as water, then solution is obtained from Equation 5.16 and is shown in the table as follows:

		SI		USCS	
d	0.064	m	2.5	in.	
L	0.122	m	4.8	in.	
A	0.031	m²			
V	3.86E−04	m²			
L_c	0.013	m			
h	5.000	W/m² K			
k	0.580	W/m K			
ρ	1000.0	kg/m³			
c_p	4180.0	J/kg K			
α	1.39E−07	m²/s			
Bi	0.11				
T_i	23.9	°C	75	°F	
T	2.8	°C	37	°F	
T_∞	2.22	°C	36	°F	
t	10.7	h	Refrigerator		
T_∞	−17.8	°C	0	°F	
t	2.1	h	Freezer		

5.2.2 Convection and Radiation

If the initial solid temperature is high, and convection is with a gas, then heat transfer by radiation needs to be included and the energy equation, Equation 5.4 becomes,

$$\rho V c \frac{dT}{dt} = -hA_s(T - T_\infty) - \varepsilon\sigma A_s\left(T^4 - T_\infty^4\right) \tag{5.18}$$

When dealing with nonlinear equations, it is generally more convenient to use zero as the reference when defining nondimensional variables, therefore defining

$$T^* = \frac{T}{T_\infty} \tag{5.19}$$

and

$$t^* = \frac{t}{t_s} \tag{5.20}$$

Then the nondimensional form of Equation 5.17 becomes

$$\frac{dT^*}{dt^*} = -\frac{hA_s t_s}{\rho V c}(T^* - 1) - \frac{\varepsilon\sigma A_s T_\infty^3 t_s}{\rho V c}(T^{*4} - 1) \tag{5.21}$$

Taking

$$t_s = \frac{\rho V c}{h A_s} \tag{5.22}$$

$$N_r = \frac{\varepsilon \sigma T_\infty^3}{h} \tag{5.23}$$

then, Equation 5.21 simplifies to

$$\frac{dT^*}{dt^*} = -(T^* - 1) - N_r(T^{*4} - 1) \tag{5.24}$$

and the boundary conditions are

$$t^* = 0, \quad T^* = \frac{T_i}{T_\infty} = T_i^* \tag{5.25}$$

The solution to this equation is obtained numerically, using an implicit method in which the temperature on the RHS is evaluated at the present time level (T_k). The finite difference approximation is

$$\frac{T_k - T_{k-1}}{\Delta t} = -(T_k - 1) - N_r(T_k^4 - 1) \tag{5.26}$$

$$T_k = \frac{1 + \dfrac{T_{k-1}}{\Delta t} + N_r}{\dfrac{1}{\Delta t} + 1 + N_r T_k^3} \tag{5.27}$$

Since this is a nonlinear equation, the solution must be obtained iteratively. Figure 5.4 shows the temperature variation with time for different values of the radiation parameter.

The case $N_r = 0$ is for when the radiation is negligible and the heat transfer is by convection only. As expected, inclusion of radiation enhances the total rate of heat transfer, accelerating cooling.

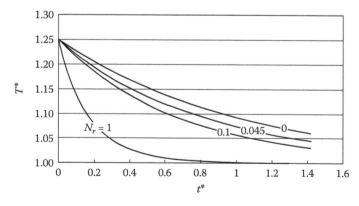

FIGURE 5.4
Transient temperature change for heat loss by convection and radiation.

5.3 General Unsteady Heat Conduction

When the Biot number, based on the characteristic length, is not small, the spatial variation of temperature must be included. The solution to most practical conduction problems
needs to be obtained numerically. The analytical solutions are only available for a limited
number of cases, some of which are presented as follows. Again, although these analytical
solutions have limited practical applications, they are invaluable in shedding light on the
basic behavior of transient systems.

As shown in Chapter 1, the energy equation for unsteady heat transfer in 1D can be
written as

$$\frac{1}{r^n}\frac{\partial}{\partial r}\left(r^n\frac{\partial T}{\partial r}\right)+\frac{\dot{q}}{k}=\frac{\rho c}{k}\frac{\partial T}{\partial t} \tag{5.28}$$

where $n = 0$, 1, or 2 corresponding to Cartesian, cylindrical, and spherical coordinates,
respectively.

5.3.1 Finite Solids (1D Unsteady)

Consider a wall of thickness $2L$, initially at temperature T_i, being exposed to an environment at $T = T_\infty$ with a heat transfer coefficient h as shown in Figure 5.5. The objective is to
determine how much heat is transferred from the solid to the fluid. Clearly, the rate of heat
transfer in this case is a function of time. Initially, the wall is at its highest temperature so
the transfer will be maximum. At the final state, the wall reaches the ambient temperature
and the rate of heat transfer will go to zero. To solve for heat transfer, we need to know the
temperature distribution. With no energy generation, the energy equation simplifies to

$$\frac{\partial^2 T}{\partial x^2}=\frac{1}{\alpha}\frac{\partial T}{\partial t} \tag{5.29}$$

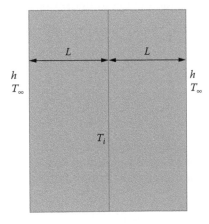

FIGURE 5.5
One dimensional transient conduction in plane wall.

The solution of Equation 5.29 requires two boundary conditions and one initial condition

$$x = -L, \quad 0 = -k\frac{\partial T}{\partial x} + h(T - T_\infty) \tag{5.30}$$

$$x = L, \quad -k\frac{\partial T}{\partial x} = h(T - T_\infty) \tag{5.31}$$

$$t = 0, \quad T = T_i \tag{5.32}$$

Recognizing that the temperature distribution is symmetric with respect to the y axis, instead of solving for the entire wall, we can get the temperature distribution between 0 and L, recognizing that the temperature distribution between $-L$ and 0 is the mirror image. In this case, our boundary condition would be between 0 and L and the boundary condition at 0 is

$$x = 0, \quad \frac{\partial T}{\partial x} = 0 \tag{5.33}$$

Because of the symmetry, the temperature will be maximum or minimum at the center. Defining

$$x^* = \frac{x}{L}, \quad T^* = \frac{T - T_\infty}{T_i - T_\infty}, \quad t^* = \frac{t}{t_s} \tag{5.34}$$

$$\frac{\partial^2 T^*}{\partial x^{*2}} = \frac{1}{\alpha}\frac{L^2}{t_s}\frac{\partial T^*}{\partial t^*} \tag{5.35}$$

The time scale is yet to be determined, keeping in mind that it can be any constant, having dimension of time. The term on the LHS is dimensionless and the last term on the RHS is also dimensionless; therefore, $\dfrac{L^2}{\alpha t_s}$ must also be nondimensional, which means that $\dfrac{L^2}{\alpha}$ must have dimension of time, and since it is also a constant, we can take

$$t_s = \frac{L^2}{\alpha} \tag{5.36}$$

which results in t^*, the nondimensional time, becoming the Fourier number, which as mentioned earlier is considered the relevant nondimensional time in conduction. This choice also eliminates all the parameters from the differential equation, simplifying to

$$\frac{\partial^2 T^*}{\partial x^{*2}} = \frac{\partial T^*}{\partial t^*} \tag{5.37}$$

The nondimensional form of the boundary conditions are

$$t^* = 0, \quad T^* = 1 \tag{5.38}$$

$$x^* = 0, \quad \frac{\partial T^*}{\partial x^*} = 0 \tag{5.39}$$

$$x^* = 1, \quad \frac{\partial T^*}{\partial x^*} = -\frac{hL}{k}T^* = -BiT^* \tag{5.40}$$

where Biot number is $Bi = \dfrac{hL}{k}$. For this problem, the characteristic length is half the wall thickness which also coincidentally is equal to the volume-to-area ratio.

The nondimensional temperature is a function of two nondimensional variables and one nondimensional parameter (Bi) or

$$T^* = T^*(x^*, t^*, Bi) \tag{5.41}$$

With one nonhomogeneity, the solution to Equation 5.37 can be obtained by the separation of variables, similar to the approach used for steady 2D conduction. Assume the solution is separable, i.e.,

$$T^*(x^*, t^*) = \chi(x^*)\tau(t^*) \tag{5.42}$$

then

$$\frac{1}{\chi}\frac{\partial^2 \chi}{\partial x^{*2}} = \frac{1}{\tau}\frac{\partial \tau}{\partial t^*} \tag{5.43}$$

The LHS is only a function of x^* and the RHS only a function t^* and for the equality to hold, both sides need to be equal to the same constant. Since the x direction is the direction with the homogeneous boundary condition, and time is the direction with the non-homogeneous boundary condition, the constant has to negative, in order to end up with the sine and cosine function in x (*orthogonal functions in the homogeneous direction*) and with exponential function in t (*nonhomogeneous direction*). Therefore,

$$\frac{1}{\chi}\frac{\partial^2 \chi}{\partial x^{*2}} = \frac{1}{\tau}\frac{\partial \tau}{\partial t^*} = -\lambda^2 \tag{5.44}$$

and the solution becomes

$$T^* = \left[c_1 \sin(\lambda x^*) + c_2 \cos(\lambda x^*)\right]e^{-\lambda^2 t^*} \tag{5.45}$$

From Equation 5.39 $c_1 = 0$, and from Equation 5.40

$$\lambda_n \tan(\lambda_n) = Bi \tag{5.46}$$

and the general solution becomes

$$T^* = \sum_{n=1}^{\infty} c_n \cos(\lambda_n x^*) e^{-\lambda_n^2 t^*} \tag{5.47}$$

To find the values of the constants c_n, we use the initial condition and take advantage of the orthogonality of the eigenfunctions

$$\int_0^1 \cos(\lambda_m x^*) dx^* = \sum_{n=1}^{\infty} c_n \int_0^1 \cos(\lambda_m x^*) \cos(\lambda_n x^*) dx^* \tag{5.48}$$

Since all the integrals on the RHS are equal to 0, unless $m = n$, then Equation 5.48 simplifies to

$$\int_0^1 \cos(\lambda_n x^*) dx^* = c_n \int_0^1 \cos^2(\lambda_n x^*) dx^* \tag{5.49}$$

Evaluating the integrals

$$\frac{1}{\lambda_n} \sin(\lambda_n x^*) \Big|_0^1 = c_n \int_0^1 \frac{1 + \cos(2\lambda_n x^*)}{2} dx^* = \frac{c_n}{2} \left[x^* + \frac{1}{2\lambda_n} \sin(2\lambda_n x^*) \right]_0^1 \tag{5.50}$$

solving for c_n

$$c_n = \frac{4 \sin \lambda_n}{2\lambda_n + \sin(2\lambda_n)} \tag{5.51}$$

Therefore, the temperature distribution is

$$T^* = \sum_{n=1}^{\infty} \frac{4 \sin \lambda_n}{2\lambda_n + \sin(2\lambda_n)} \cos(\lambda_n x^*) e^{-\lambda_n^2 t^*} \tag{5.52}$$

where λ_n, the eigenvalues, are the roots of Equation 5.46, which has infinitely many roots. Figure 5.6 shows the first five roots of Equation 5.46 for $Bi = 5$.

The same procedure can be followed for an infinitely long cylinder or a sphere, although the math is somewhat more involved. The analytical solutions for all three cases as well as the equations from which the eigenvalues can be calculated are presented in Table 5.1.

As can be seen from Figure 5.6, the values of λ_n are increasing function of n and every term in the infinite series is multiplied by exponential of $-\lambda_n^2 t^*$ which makes them decrease rapidly with increasing n and time. For $t^* > 0.2$, it turns out that the first term of the infinite series provides accurate results, and by taking the first term of the infinite series, the temperature distribution in the wall can be approximated by

$$T^* = A \cos(\lambda_1 x^*) e^{-\lambda_1^2 t^*} \tag{5.53}$$

where

$$A = \frac{4 \sin \lambda_1}{2\lambda_1 + \sin(2\lambda_1)} \tag{5.54}$$

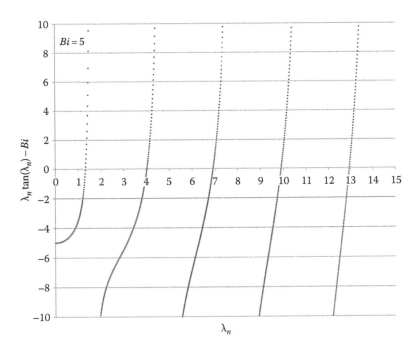

FIGURE 5.6
First five roots (eigenvalues) of the characteristic function for transient plane conduction.

is a constant that depends on Biot number. The centerline temperature is obtained by evaluating Equation 5.53 at $x^* = 0$:

$$T^*(0,t^*) = Ae^{-\lambda_1^2 t^*} \tag{5.55}$$

which is only a function of time and Bi. Therefore, the temperature at any point inside the wall at any time can be written as the product of the centerline temperature, which is only a function of time (and Bi), and another term that only depends on location (and Bi),

$$T^*(x^*,t^*) = T^*(0,t^*)\cos(\lambda_1 x^*) \tag{5.56}$$

Example 5.2

One side has a copper plate having a thickness of 2.5 cm is insulated. The plate is initially $T_i = 150°C$, and the ambient temperature $T_\infty = 40°C$, and $h = 570\dfrac{W}{m^2 K}$. Find the T at the insulation after 30 s.

Properties are evaluated at $\bar{T} = (150 + 40)/2 = 95°C$.
For copper

$$\rho = 8933\ \frac{kg}{m^3}$$

$$c = 385\ \frac{J}{kg\ K}$$

$$\alpha = 117 \times 10^{-6}\ \frac{m^2}{s}$$

$$k = 396\ \frac{W}{m\ K}$$

TABLE 5.1

Transient 1D Solutions

Exact	Geometry	Approximate
$T^* = \sum_{n=1}^{\infty} \dfrac{4\sin\lambda_n}{2\lambda_n + \sin(2\lambda_n)}\cos(\lambda_n x^*)e^{-\lambda_n^2 t^*}$ (5.57a) $\lambda_n \tan(\lambda_n) = Bi$	$x^* = \dfrac{x}{L}$ $Bi = \dfrac{hL}{k}$	$T^* = A\cos(\lambda_1 x^*)e^{-\lambda_1^2 t^*}$ (5.57b) $A = \dfrac{4\sin\lambda_1}{2\lambda_1 + \sin(2\lambda_1)}$
$T^* = \sum_{n=1}^{\infty} \dfrac{2J_1(\lambda_n)}{\lambda_n\left[J_0^2(\lambda_n) + J_1^2(\lambda_n)\right]}J_0(\lambda_n r^*)e^{-\lambda_n^2 t^*}$ (5.58a) $\lambda_n \dfrac{J_1(\lambda_n)}{J_0(\lambda_n)} = Bi$	$D = 2R$ $r^* = \dfrac{r}{R}$ $Bi = \dfrac{hR}{k}$	$T^* = AJ_0(\lambda_1 r^*)e^{-\lambda_1^2 t^*}$ (5.58b) $A = \dfrac{2J_1(\lambda_1)}{\lambda_1\left[J_0^2(\lambda_1) + J_1^2(\lambda_1)\right]}$
$T^* = \sum_{n=1}^{\infty} 2\dfrac{\sin\lambda_n - \lambda_n\cos\lambda_n}{\lambda_n - \sin\lambda_n\cos\lambda_n}\dfrac{\sin(\lambda_n r^*)}{\lambda_n r^*}e^{-\lambda_n^2 t^*}$ (5.59a) $\lambda_n \cot(\lambda_n) = 1 - Bi$	$D = 2R$ $r^* = \dfrac{r}{R}$ $Bi = \dfrac{hR}{k}$	$T^* = A\dfrac{\sin(\lambda_1 r^*)}{\lambda_1 r^*}e^{-\lambda_1^2 t^*}$ (5.59b) $A = 2\dfrac{\sin\lambda_1 - \lambda_1\cos\lambda_1}{\lambda_1 - \sin\lambda_1\cos\lambda_1}$

Lumped parameter solution:

$$L_c = \frac{V}{A_s} = \frac{0.025 A_s}{A_s} = 0.025$$

$$Bi = \frac{hL_c}{k_s} = \frac{570 \times 0.025}{396} = 0.036$$

$$Fo = \frac{\alpha t}{L_c^2} = \frac{117 \times 10^{-6} \times 30}{0.025^2} = 5.616$$

The lumped parameter method should be valid since $Bi < 0.1$

$$\frac{T - T_\infty}{T_i - T_\infty} = e^{-BiFo}$$

$$T = T_\infty + (T_i - T_\infty)e^{-BiFo} = 40 + (150 - 40)e^{-0.36 \times 5.616} = 129.9$$

Therefore, temperature is ~129.9°C throughout the wall.

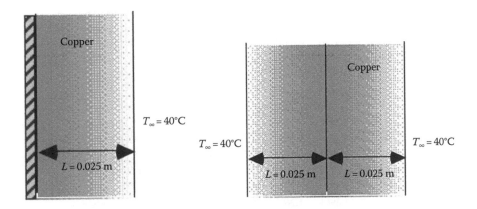

The solution to this problem is also the same as that for a wall with thickness 2L. For $Bi = 0.036$, the first eigenvalue is $\lambda_1 = 0.1886$

Therefore,

$$T^*(0, t^*) = \frac{4\sin(0.1886)}{(2 \times 0.1886) + \sin(2 \times 0.1886)} e^{-0.1886^2 \times 5.616} = 0.824$$

$$T = 40 + (150 - 40) \times 0.824 = 130.6$$

which is close to the results, using lumped capacitance approach.

The approximate solutions for the three cases are also given in Table 5.1 and the values of A and the first eigenvalues for different Biot numbers are given in Table 5.2.

TABLE 5.2

Coefficients for One Term Approximation
of 1D Transient Solutions

Bi	Wall λ_1	Wall A	Cylinder λ_1	Cylinder A	Sphere λ_1	Sphere A
0.01	0.10	1.0017	0.1412	1.00250	0.1730	1.0030
0.02	0.1410	1.0033	0.1995	1.00498	0.2445	1.0060
0.03	0.1723	1.0049	0.2440	1.00746	0.2991	1.0090
0.04	0.1987	1.0066	0.2814	1.00993	0.3450	1.0120
0.05	0.2218	1.0082	0.3143	1.01240	0.3854	1.0150
0.06	0.2425	1.0098	0.3439	1.01485	0.4239	1.0181
0.07	0.2615	1.0114	0.3709	1.01730	0.4603	1.0214
0.08	0.2791	1.0130	0.3960	1.01973	0.4877	1.0240
0.09	0.2956	1.0145	0.4192	1.02212	0.5189	1.0273
0.1	0.3111	1.0161	0.4417	1.02458	0.5423	1.0298
0.2	0.4328	1.0311	0.6172	1.04835	0.7576	1.0589
0.3	0.5218	1.0450	0.7465	1.07116	0.9209	1.0880
0.4	0.5932	1.0580	0.8515	1.09313	1.0527	1.1163
0.5	0.6533	1.0701	0.9406	1.11422	1.1656	1.1441
0.6	0.7051	1.0814	1.0184	1.13448	1.2644	1.1712
0.7	0.7506	1.0918	1.0872	1.15388	1.3525	1.1978
0.8	0.7910	1.1016	1.1490	1.17243	1.4320	1.2236
0.9	0.8273	1.1106	1.2048	1.19016	1.5044	1.2488
1	0.8603	1.1191	1.2558	1.20709	1.5708	1.2732
2	1.0769	1.1785	1.5995	1.33838	2.0288	1.4793
3	1.1924	1.2102	1.7887	1.41910	2.2889	1.6226
4	1.2645	1.2287	1.9081	1.46980	2.4556	1.7202
5	1.3137	1.2402	1.9897	1.50284	2.5703	1.7869
6	1.3496	1.2479	2.0490	1.52533	2.6537	1.8338
7	1.3766	1.2532	2.0938	1.54113	2.7165	1.8673
8	1.3978	1.2570	2.1287	1.55261	2.7654	1.8920
9	1.4149	1.2598	2.1567	1.56117	2.8044	1.9106
10	1.4289	1.2620	2.1795	1.56769	2.8363	1.9249
20	1.4961	1.2699	2.2881	1.59194	2.9857	1.9781
30	1.5202	1.2717	2.3261	1.59729	3.0372	1.9898
40	1.5325	1.2723	2.3455	1.59927	3.0632	1.9942
50	1.5400	1.2727	2.3572	1.60022	3.0788	1.9962
60	1.5451	1.2728	2.3651	1.60075	3.0893	1.9974
70	1.5487	1.2729	2.3707	1.60107	3.0967	1.9980
80	1.5514	1.2730	2.3750	1.60128	3.1023	1.9985
90	1.5535	1.2731	2.3783	1.60142	3.1067	1.9988
100	1.5552	1.2731	2.3809	1.60152	3.1102	1.9990
10,000	1.5706	1.2732	2.4046	1.60197	3.1413	2.0000

The time that it takes for any of the three objects earlier to reach a particular temperature T^* is

$$t = -\frac{1}{\lambda_1^2}\frac{L_c^2}{\alpha}\ln\left(\frac{T^*}{A}\right) \tag{5.60}$$

where
  $L_c = L$ for wall
  $L_c = R$ for cylinder and sphere

For example, the time that it takes for the nondimensional temperature at the center of the object to reach 1% of its final value ($T^* = 0.01$) is

$$t = \frac{L_c^2}{\alpha\lambda_1^2}\ln(100A) \tag{5.61}$$

From the values in Table 5.2, for $Bi > 0.5$, the term $\ln\dfrac{100A}{\lambda_1^2}$ varies between 0.54 and 6.4, or is of the order of 1; therefore, the time that it takes for an object to reach the ambient temperature is of the order of

$$t \approx \frac{L_c^2}{\alpha} \tag{5.62}$$

Or $Fo = 1$ provides the order of magnitude of the time that a change in temperature at the boundary has impacted a distance L_c into the object.

Example 5.3

An orange at room temperature of 24°C is to be cooled in the refrigerator which is at 2°C. How long does it take to cool it to 7°C, 5°C, and 3°C?
   Assuming properties of the orange to be the same as water, then the approximate solution is obtained from Equation 5.59b. The values used and the times for the three temperatures are given in the following table:

		SI
D	0.080	m
R	0.040	m
h	10.0	W/m² K
k	0.580	W/m K
ρ	1000.0	kg/m³
c_p	4180.0	J/kg K
α	1.39E-07	m²/s
Bi	0.690	
λ_1	1.343	
A	1.195	
T_1	24.0	°C
T_∞	2.0	°C
T	7.0	°C
t	2.95	h
T	5.0	°C
t	3.85	h
T	3.0	°C
t	5.80	h

5.3.2 Semi-Infinite Medium

Consider 1D unsteady conduction in a plane wall. From Equation 5.60, the centerline temperature will have been changed by <1% as long as t^* is less than

$$t^* < \frac{1}{\lambda_1^2} \ln\left(\frac{A}{0.99}\right)$$

Figure 5.7 is a plot of the variation of t^* for which the centerline temperature is within 99% of its original value as a function of Biot number from Equation 5.57b. For every Biot number, there is a finite time before the central line temperature begins to respond to the change that happened at the surface. For a broad range of Biot numbers, the nondimensional time that it takes for the change to penetrate is about 0.15 or the penetration depth is approximately

$$x \approx 2.5\sqrt{\alpha t} \tag{5.63}$$

or, for a given time, the depth that a change in conditions at the boundary has penetrated inside an object is of the order of $2.5\sqrt{\alpha t}$. For positions inside the object beyond this distance, the conditions remain unaffected.

A semi-infinite object, shown in Figure 5.8, is one where the object is large enough so that far from the boundary the object does not know that any change has occurred at the boundary. All real objects behave like a semi-infinite medium at short times, and thus the solution can be determined from

$$\frac{\partial^2 T}{\partial x^2} = \frac{1}{\alpha}\frac{\partial T}{\partial t} \tag{5.64}$$

subject to

$$t = 0, \quad T = T_i \tag{5.65}$$

$$x \rightarrow \infty, \quad T = T_i \tag{5.66}$$

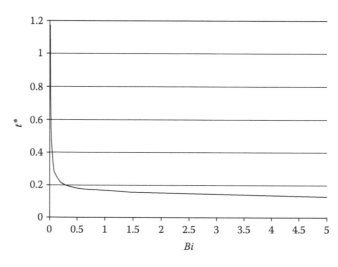

FIGURE 5.7
Dependence of the nondimensional time needed for the centerline temperature to change by 1% of its original value on *Bi*.

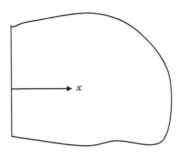

FIGURE 5.8
Schematic of a semi-infinite medium.

Three different boundary conditions are considered at $x = 0$,

$$T_w$$

$$x = 0, \quad -k\frac{\partial T}{\partial x} = h(T_\infty - T) \tag{5.67}$$

$$-k\frac{\partial T}{\partial x} = q'' $$

We first consider the first boundary condition in which the surface temperature is raised to a constant value and held at that value. Defining

$$x^* = \frac{x}{x_s}, \quad T^* = \frac{T - T_i}{T_w - T_i}, \quad t^* = \frac{t}{t_s} \tag{5.68}$$

$$\frac{\partial^2 T^*}{\partial x^2} = \frac{1}{\alpha}\frac{\partial T^*}{\partial t} \tag{5.69}$$

Solution to this problem can be obtained by the similarity method. If we define a new variable

$$\eta = \frac{x}{\sqrt{4\alpha t}} \tag{5.70}$$

$$\frac{\partial}{\partial x} = \frac{\partial}{\partial \eta}\frac{\partial \eta}{\partial x} = \frac{1}{\sqrt{4\alpha t}}\frac{\partial}{\partial \eta} \tag{5.71}$$

$$\frac{\partial^2}{\partial x^2} = \frac{\partial}{\partial x}\left[\frac{\partial}{\partial \eta}\frac{\partial \eta}{\partial x}\right] = \frac{\partial \eta}{\partial x}\frac{\partial}{\partial \eta}\left[\frac{\partial}{\partial \eta}\frac{\partial \eta}{\partial x}\right] = \left(\frac{\partial \eta}{\partial x}\right)^2\frac{\partial^2}{\partial \eta^2} = \frac{1}{4\alpha t}\frac{\partial^2}{\partial \eta^2} \tag{5.72}$$

$$\frac{\partial}{\partial t} = \frac{\partial}{\partial \eta}\frac{\partial \eta}{\partial t} = \frac{x}{\sqrt{4\alpha t}}\frac{-1}{2t}\frac{\partial}{\partial \eta} = -\frac{\eta}{2t}\frac{\partial}{\partial \eta} \tag{5.73}$$

$$\frac{1}{4\alpha t}\frac{\partial^2 T}{\partial \eta^2} = \frac{1}{\alpha}\frac{-\eta}{2t}\frac{\partial T}{\partial \eta} \tag{5.74}$$

$$\frac{\partial^2 T^*}{\partial \eta^2} = -2\eta\frac{\partial T^*}{\partial \eta} \tag{5.75}$$

Note that this equation only needs two boundary conditions, yet the original equation had two boundary conditions and an initial condition. Therefore, similarity solution exists only if the three original boundary conditions reduce to two in terms of the new similarity variable and are also only a function of the similarity variable. The boundary conditions are

$$t = 0, \quad \eta \to \infty, \quad T^* \to 0 \tag{5.76}$$

$$x \to \infty, \quad \eta \to \infty, \quad T^* \to 0 \tag{5.77}$$

$$x = 0, \quad \eta = 0, \quad T^* = 1 \tag{5.78}$$

As can be seen, the initial condition and the boundary condition at infinity are identical in terms of the similarity variable. Therefore, all three original conditions are satisfied by the two two in terms of the new similarity variable. To obtain the solution, assume

$$y = \frac{\partial T}{\partial \eta} \tag{5.79}$$

Then

$$\frac{dy}{d\eta} = -2\eta y \tag{5.80}$$

$$\ln y = -\eta^2 + c_1 \tag{5.81}$$

$$y = \frac{\partial T}{\partial \eta} = C_1 e^{-\eta^2} \tag{5.82}$$

$$T = C_1 \int_0^\eta e^{-\eta^2} d\eta + C_2 \tag{5.83}$$

From the boundary condition

$$1 = C_1 \int_0^0 e^{-\eta^2} d\eta + C_2 \Rightarrow C_2 = 1 \tag{5.84}$$

$$T^* = 1 + C_1 \int_0^\eta e^{-\eta^2} d\eta \tag{5.85}$$

The integral on the RHS shows up frequently in the solution of many problems, particularly in probability and statistics. A related function called error function is defined as

$$erf(\xi) = \frac{2}{\sqrt{\pi}} \int_0^\xi e^{-\xi^2} d\xi \tag{5.86}$$

Therefore,

$$T^* = 1 + C_1 \frac{\sqrt{\pi}}{2} erf(\eta) \qquad (5.87)$$

Using the boundary condition at infinity,

$$0 = 1 + C_1 \frac{\sqrt{\pi}}{2} 1 \qquad (5.88)$$

$$C_1 = -\frac{2}{\sqrt{\pi}} \qquad (5.89)$$

and the final solution becomes

$$T^* = 1 - erf(\eta) = erfc(\eta) \qquad (5.90)$$

where *erfc* is the complementary error function which is defined as $erfc(\eta) = 1 - erf(\eta)$. In dimensional form, Equation 5.90 becomes

$$\frac{T - T_i}{T_s - T_i} = erfc\left(\frac{x}{\sqrt{4\alpha t}}\right) \qquad (5.91)$$

and is plotted in Figure 5.9. As can be seen from Figure 5.9, for $\frac{x}{\sqrt{4\alpha t}} > 1.8$ or $x > 3.6\sqrt{\alpha t}$, the solid temperature is essentially at its initial temperature. This is close to the results given by Equation 5.63. Beyond a particular x location, as long as time is

$$t < \frac{x^2}{13\alpha} \qquad (5.92)$$

the temperature is within 1% of the initial temperature.

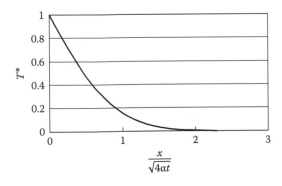

FIGURE 5.9
Transient response of a semi-infinite medium to temperature change.

The other two boundary conditions do not have similarity solution. Using Laplace transform, the solution for the convective boundary condition is

$$\frac{T-T_i}{T_\infty-T_i} = erfc\left(\frac{x}{\sqrt{4\alpha t}}\right) - \exp\left(2\frac{h\sqrt{\alpha t}}{k}\frac{x}{\sqrt{4\alpha t}} + \left(\frac{h\sqrt{\alpha t}}{k}\right)^2\right)\left[erfc\left(\frac{x}{\sqrt{4\alpha t}} + \frac{h\sqrt{\alpha t}}{k}\right)\right] \quad (5.93)$$

and the solution for the uniform heat flux condition is

$$\frac{T-T_i}{\frac{q''}{k}} = \sqrt{\frac{4\alpha t}{\pi}}\exp\left(-\frac{x^2}{4\alpha t}\right) - x\,erfc\left(\frac{x}{\sqrt{4\alpha t}}\right) \quad (5.94)$$

Analytical solution to a number of other transient conduction problems can be found in the classic text on conduction heat transfer by Arpaci [1].

5.3.3 Energy Loss during Cooling

In addition to temperature distribution, another quantity of interest is the amount of heat transfer, Q, between the initial time and any time t, which is the difference between the initial internal energy of the object and its internal energy at time t

$$Q = E(t=0) - E(t) \quad (5.95)$$

since the mass of the object is constant, then

$$Q = m^*\left[e(t=0) - e(t)\right] \quad (5.96)$$

$$Q = \int_V \rho C\left[T_i - T(x,t)\right]dV = \int_V \rho C\frac{T_i - T + T_\infty - T_\infty}{T_i - T_\infty}(T_i - T_\infty)dV \quad (5.97)$$

$$Q = \rho C(T_i - T_\infty)\int_V\left[1 - \frac{T-T_\infty}{T_i-T_\infty}\right]dV = \rho C(T_i-T_\infty)\frac{V}{V}\int_V[1-T^*]dV \quad (5.98)$$

The maximum amount of heat that can be transferred from the object is the difference between the initial internal energy and the final internal energy when the object's temperature reaches the ambient temperature:

$$Q_{max} = \rho VC(T_i - T_\infty) \quad (5.99)$$

Therefore,

$$\frac{Q}{Q_{max}} = \frac{1}{V}\int_V[1-T^*]dV \quad (5.100)$$

For a plane wall

$$V = Ax \tag{5.101}$$

$$\frac{Q}{Q_{max}} = \int_0^1 [1 - T^*]dx^* \tag{5.102}$$

The variation of temperature with space is known from the analytical solution. Assuming that the first term of the infinite series provides sufficient accuracy, then

$$\frac{Q}{Q_{max}} = \int_0^1 \left[1 - Ae^{-\lambda_1^2 t^*}\cos(\lambda_1 x^*)\right]dx^* \tag{5.103}$$

Evaluating the integral results in

$$\frac{Q}{Q_{max}} = 1 - Ae^{-\lambda_1^2 t^*}\frac{\sin(\lambda_1)}{\lambda_1} \tag{5.104}$$

which can be written as

$$\frac{Q}{Q_{max}} = 1 - T^*(0,\tau)\frac{\sin(\lambda_1)}{\lambda_1} \tag{5.105}$$

For all Biot numbers, $0.81 < A\dfrac{\sin(\lambda_1)}{\lambda_1} < 1$ and therefore,

$$\frac{Q}{Q_{max}} \approx 1 - e^{-\lambda_1^2 t^*} \tag{5.106}$$

5.4 Multidimensional System

Since the energy equation is linear, many transient multidimensional problems can be expressed in terms of the product of simpler 1D problems. The approach is demonstrated for transient conduction in a rectangular bar shown in Figure 5.10. The energy equation and the initial and boundary conditions are

$$\frac{\partial^2 T}{\partial x^2} + \frac{\partial^2 T}{\partial y^2} = \frac{1}{\alpha}\frac{\partial T}{\partial t} \tag{5.107}$$

$$t = 0, \quad T = T_i \tag{5.108}$$

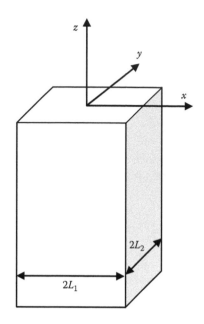

FIGURE 5.10
2D transient conduction in a rectangular bar.

Since because of the symmetry, the temperature will be maximum or minimum at the center, then

$$x = 0, \quad \frac{\partial T}{\partial x} = 0 \tag{5.109}$$

$$y = 0, \quad \frac{\partial T}{\partial y} = 0 \tag{5.110}$$

$$x = L_1, \quad -k\frac{\partial T}{\partial x} = h(T - T_\infty) \tag{5.111}$$

$$y = L_2, \quad -k\frac{\partial T}{\partial y} = h(T - T_\infty) \tag{5.112}$$

Defining

$$T^* = \frac{T - T_\infty}{T_i - T_\infty} \tag{5.113}$$

then

$$\frac{\partial^2 T^*}{\partial x^2} + \frac{\partial^2 T^*}{\partial y^2} = \frac{1}{\alpha}\frac{\partial T^*}{\partial t} \tag{5.114}$$

Assume the solution can be expressed as the product of two functions

$$T^*(x, y, t) = X(x, t)Y(y, t) \tag{5.115}$$

TABLE 5.3

Reduction of a Transient 2D to Two Transient 1D Problems

$$\frac{\partial^2 XY}{\partial x^2} + \frac{\partial^2 XY}{\partial y^2} = \frac{1}{\alpha}\frac{\partial XY}{\partial t} \tag{5.116}$$

$$Y\frac{\partial^2 X}{\partial x^2} + X\frac{\partial^2 Y}{\partial y^2} = \frac{1}{\alpha}Y\frac{\partial X}{\partial t} + X\frac{1}{\alpha}\frac{\partial Y}{\partial t} \tag{5.117}$$

$$\frac{1}{X}\frac{\partial^2 X}{\partial x^2} - \frac{1}{\alpha}\frac{1}{X}\frac{\partial X}{\partial t} = -\left[\frac{1}{Y}\frac{\partial^2 Y}{\partial y^2} - \frac{1}{Y}\frac{1}{\alpha}\frac{\partial Y}{\partial t}\right] = 0 \tag{5.118}$$

$$\frac{\partial^2 X}{\partial x^2} = \frac{1}{\alpha}\frac{\partial X}{\partial t} \qquad\qquad \frac{\partial^2 Y}{\partial y^2} = \frac{1}{\alpha}\frac{\partial Y}{\partial t}$$

$t = 0$	$X = 1$	$t = 0$	$Y = 1$
$x = 0$	$\dfrac{\partial X}{\partial x} = 0$	$y = 0$	$\dfrac{\partial Y}{\partial y} = 0$
$x = L_1$	$-k\dfrac{\partial X}{\partial x} = hX$	$y = L_2$	$-k\dfrac{\partial Y}{\partial y} = hY$

As shown in Table 5.3, the solution for the temperature distribution in the rectangular bar can be obtained as the product of the solution of two infinitely long walls. In a similar manner, solution to a short cylinder of diameter D and length $2L$ is the product of that of an infinitely long cylinder of diameter D and infinitely long wall of thickness $2L$.

5.5 Numerical Solution of Transient Conduction

As in other cases, for more complicated geometries and boundary conditions, the solution to transient problems needs to be obtained by numerical methods. The general form of transient 1D energy equation is

$$\frac{1}{r^{*n}}\frac{\partial}{\partial r^*}\left(r^{*n}\frac{\partial T^*}{\partial r^*}\right) = \frac{\partial T^*}{\partial t^*} \tag{5.119}$$

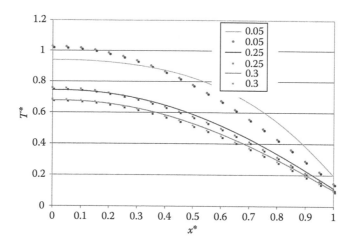

FIGURE 5.11
Comparison of numerical (lines) and analytical (points) solutions for transient conduction in a plane wall.

which can be expanded as

$$\frac{\partial^2 T^*}{\partial r^{*2}} + \frac{n}{r^*}\frac{\partial T^*}{\partial r^*} = \frac{\partial T^*}{\partial t^*} \tag{5.120}$$

$$\frac{T_{i+1}^k - 2T_i^k + T_{i-1}^k}{\Delta r^2} + \frac{n}{r_i}\frac{T_{i+1}^k - T_{i-1}^k}{2\Delta r} = \frac{T_i^k - T_i^{k-1}}{\Delta t} \tag{5.121}$$

$$T_{i+1}^k - 2T_i^k + T_{i-1}^k + \frac{n\Delta r}{2r_i}T_{i+1}^k - \frac{n\Delta r}{2r_i}T_{i-1}^k = \frac{\Delta r^2}{\Delta t}T_i^k - \frac{\Delta r^2}{\Delta t}T_i^{k-1} \tag{5.122}$$

$$\left(1 + \frac{n\Delta r}{2r_i}\right)T_{i+1}^k + \left(-2 - \frac{\Delta r^2}{\Delta t}\right)T_i^k + \left(1 - \frac{n\Delta r}{2r_i}\right)T_{i-1}^k = -\frac{\Delta r^2}{\Delta t}T_i^{k-1} \tag{5.123}$$

This equation can be solved iteratively or using TRIDI. Figure 5.11 is a comparison of the numerical and exact solutions using just the first term of the infinite series, Equation 5.53 for transient conduction in a plane wall ($n = 0$) for the case of $Bi = 10$, for the different values of Fourier number, t^*. Again, there is close agreement between the analytical and numerical solutions at large values of t^*. The agreement for $t^* = 0.05$ is not good, since as stated earlier the first term provides accurate results as long as $t^* > 0.2$.

5.6 Transient Fins

There are instances where a boundary may experience transient conditions. For example, the base temperature of the fins attached to a motorcycle engine experiences temperature oscillation. Arpaci [1] provides the analytical solution to a number of problems with periodic boundary conditions, including a semi-infinite medium and semi-infinite rod, whose

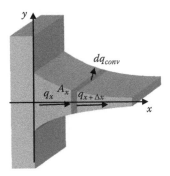

FIGURE 5.12
General variable area fin.

temperature at $x = 0$ oscillates as $T_0\cos(\omega t)$. Two distinct regions can be identified in the solution of such problems. During the initial period, the temperature changes periodically and also ramps up from the initial value, followed by a steady periodic solution, that although a function of time, follows a periodic pattern. The available analytical solutions are generally for the steady periodic part. The general solution to these problems can be obtained numerically.

The governing equation for transient fin can be determined by doing an energy balance on a differential control mass as shown in Figure 5.12

$$q_{cond,x} - q_{cond,x+\Delta x} - dq_{conv} = \frac{d(\rho \Delta V c T)}{\partial t} \tag{5.124}$$

which simplifies to

$$\frac{d}{dx}\left(kA_x \frac{\partial T}{\partial x}\right) - h\frac{dA_s}{dx}(T - T_\infty) = \rho c A_x \frac{dT}{\partial t} \tag{5.125}$$

This is the general transient fin equation. For a constant area fin, Equation 5.125 simplifies to

$$\frac{d^2 T}{dx^2} - \frac{hp}{kA}(T - T_\infty) = \frac{1}{\alpha}\frac{dT}{\partial t} \tag{5.126}$$

Consider a straight fin whose base temperature fluctuates sinusoidally between T_{max} and T_{min} with a specified period. For this particular case, the boundary conditions are

$$x = 0, \quad T = \frac{T_{max} + T_{min}}{2} + \frac{T_{max} - T_{min}}{2}\sin(\omega t) \tag{5.127}$$

$$x = L, \quad \frac{dT}{dx} = 0 \tag{5.128}$$

The governing equation and the boundary conditions are nondimensionalized next, by defining

$$T^* = \frac{T - T_\infty}{T_s}, \quad t^* = \frac{t}{t_s}, \quad x^* = \frac{x}{L} \tag{5.129}$$

$$\frac{d^2 T^*}{dx^{*2}} - \frac{hpL^2}{kA} T^* = \frac{1}{\alpha} \frac{L^2}{t_s} \frac{dT^*}{\partial t^*}$$ (5.130)

$$\frac{d^2 T^*}{dx^{*2}} - m^* T^* = \frac{\partial T^*}{\partial t^*}$$ (5.131)

where

$$t_s = \frac{L^2}{\alpha}$$ (5.132)

$$m^* = \frac{hpL^2}{kA}$$

assuming $T_s = \frac{(T_{max} - T_{min})}{2}$ then

The nondimensional forms of the boundary conditions are

$$x^* = 1, \quad \frac{dT^*}{dx^*} = 0$$ (5.133)

$$T^* = \bar{T}_b + \sin(\omega^* t^*)$$ (5.134)

where

$$\bar{T}_b = \frac{(T_{max} - T_\infty) + (T_{min} - T_\infty)}{(T_{max} - T_\infty) - (T_{min} - T_\infty)}$$ (5.135)

$$\omega^* = \omega \frac{L^2}{\alpha}$$ (5.136)

The solution is obtained numerically, and the finite difference forms of the equation and boundary conditions are

$$T_{i+1}^k + \left(-2 - m\Delta x^2 - \frac{\Delta x^2}{\Delta t}\right) T_i^k + T_{i-1}^k = -\frac{\Delta x^2}{\Delta t} T_i^{k-1}$$ (5.137)

$$T_1^k = \bar{T}_b + \sin(\omega^* t^*)$$ (5.138)

$$T_N = T_{N-1}$$ (5.139)

Figures 5.13 and 5.14 show the numerical solution for the variation of the fin temperature with location and time.

Figure 5.13 shows the temperature distribution along the fin at different times. Near the base, the fin temperature changes significantly with time, owing to the variation in the base temperature. However for $x^* > 0.2$, the temperature fluctuations with time are greatly dampened. This behavior is more clearly shown in Figure 5.14 that shows the variation of temperature with time at the base, middle, and tip of the fin. The temperature increases

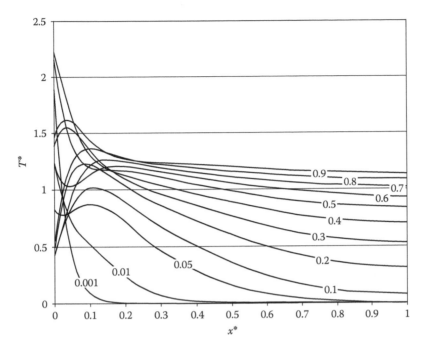

FIGURE 5.13
Temperature distribution along the fin at different times.

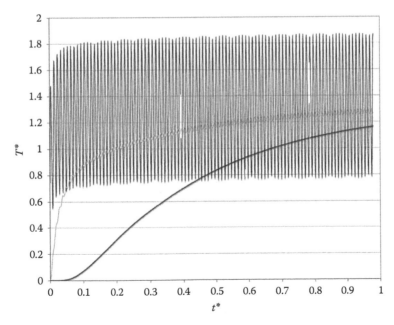

FIGURE 5.14
Variation of temperature with time at the base, middle, and tip of the fin.

with time and reaches a uniform value which is different than the steady case. Near the base, the temperature is fluctuating with a large amplitude; however, the fluctuations drop rapidly. In order to resolve the temperature field, the Δt was chosen to be 10^{-3}, and to reach steady state, the integration must be carried out to $t = 0.9$ and beyond or over 900 steps must be taken.

5.7 Moving Boundary Problem

In problems where melting or solidification takes place, the phase change boundary moves and therefore it poses a different constraint on the solution. The basic physics and the solution approach is demonstrated by an example. Consider a liquid which is initially at temperature T_s where T_s is the same as the solidification temperature (Figure 5.15). If the bottom surface's temperature is suddenly reduced to T_w which is less than T_s, then the liquid will freeze and the solidification front will move in the x direction.

Defining

$$T^* = \frac{T - T_w}{T_s - T_w} \tag{5.140}$$

$$x^* = \frac{x}{L}, \quad t^* = \frac{\alpha t}{L^2} \tag{5.141}$$

where L is a yet to be determined constant

$$\frac{\partial^2 T^*}{\partial x^{*2}} = \frac{\partial T^*}{\partial t^*} \tag{5.142}$$

$$t = 0, \quad T^* = 1 \tag{5.143}$$

$$x = s, \quad T^* = 1 \tag{5.144}$$

$$x = 0, \quad T^* = 0 \tag{5.145}$$

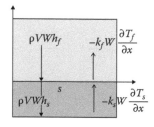

FIGURE 5.15
The solidification of a saturated liquid.

The unknown position of the freeze fronts is determined by doing an energy balance at the interface between solid and liquid. The front is moving up with a velocity V and therefore relative to the front, the liquid appears to move down, passes through the front, loses heat, and the solid exits from the back side. The heat that the liquid is losing as it solidifies is conducted in the solid in the negative x direction and eventually leaves the system as $x = 0$. Consider a section of interface having a width W, the fluid approaches with a velocity V at a high temperature relative, and then it solidifies. The amount of heat transferred from the liquid plus the energy released as the liquid solidifies must equal the amount of energy conducted away in the x direction in the solid. However, since the liquid is isothermal, the amount of heat transfer by conduction is zero, and therefore

$$\rho V W h_{fs} - kW \frac{\partial T}{\partial x} = 0 \tag{5.146}$$

since

$$V = \frac{dx}{dt} = \frac{ds}{dt} \tag{5.147}$$

then

$$\rho \frac{dx}{dt} h_{fs} - k \frac{\partial T}{\partial x} = 0 \tag{5.148}$$

In terms of the dimensionless variables

$$\frac{c(T_s - T_w)}{h_{fs}} \frac{\partial T^*}{\partial x^*} = \frac{dx^*}{dt^*} \tag{5.149}$$

The nondimensional group on the LHS is called Jacobs' number

$$Ja = \frac{c(T_s - T_w)}{h_{fs}} \tag{5.150}$$

$$Ja \frac{\partial T^*}{\partial x^*} = \frac{dx^*}{dt^*} \tag{5.151}$$

This problem eventually reaches a quasi steady solution, i.e., even though the temperature decreases with time, since the volume increasing the total system energy will remain constant. At this quasi steady condition, the temperature in the solid will vary linearly

$$Ja \frac{1}{s^*} = \frac{ds^*}{dt^*} \tag{5.152}$$

and the solution becomes

$$s^* = \sqrt{2Ja\, t^*} \tag{5.153}$$

which in terms of the dimensional variables is

$$s = \sqrt{2Ja\, \alpha t} \tag{5.154}$$

Also,

$$V = \frac{ds}{dt} = \sqrt{\frac{Ja\, \alpha}{2t}} \tag{5.155}$$

The results show that the phase change interface moves at a speed proportional to $\sqrt{\frac{\alpha}{t}}$. The finite difference form of the equation is

$$\frac{T_{i+1}^k + T_{i-1}^k + \dfrac{\Delta x^2}{\Delta t} T_i^{k-1}}{\left[2 + \dfrac{\Delta x^2}{\Delta t}\right]} = T_i^k \tag{5.156}$$

The numerical solution of this problem is complicated due to the fact that the boundary at which $T^* = 1$ is moving and its location is not known until the problem is completely solved. However if we use Equation 5.154 to determine the unknown location of the freeze front at each time level, then the temperature distribution can be found as shown in Figure 5.16. As can be seen, the solid temperature varies linearly, while the freeze front moves.

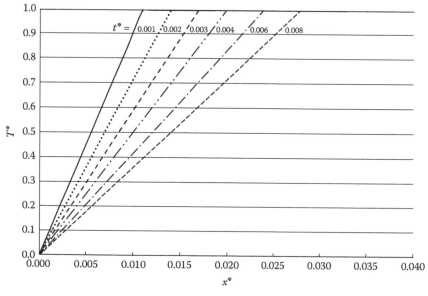

FIGURE 5.16
Transient temperature profile during solidification.

Problems

5.1 An egg may be modeled as a sphere having a diameter of 5 cm. The egg is initially at 4°C and is immersed in boiling water. Assuming egg's properties to be the same as water, determine the temperature at the center of the egg after
  a. 2 min
  b. 10 min

5.2 A hotdog can be treated as a cylinder 1.5 cm in diameter and 12 cm long, cooking in water at 100°C with a heat transfer coefficient of 1000 W/m² K. Analytically obtain the solution to how long it takes for the hotdog to cook, assuming it is cooked if its temperature everywhere is >95°C.

5.3 Show that the transient energy balance for an object with small internal resistance is

$$\rho c V \frac{dT}{dt} = -hA(T - T_\infty) - \varepsilon \sigma A\left(T^4 - T_\infty^4\right)$$

nondimensionalize the governing equation and solve for the temperature distribution numerically.

5.4 Show that the transient energy balance for an object with small internal resistance can be linearized into

$$\frac{dT^*}{dt^*} = -(T^*-1)[Bi + CNr]$$

where

$$C = \frac{\left(T_m^* + 1\right)^3}{2}$$

$$Bi = \frac{hL_c}{k_s}$$

$$Nr = \frac{\varepsilon \sigma L_c T_\infty^3}{k_s}$$

by noting that

$$2\frac{(T^{*2} + 1)}{(T^* + 1)^2} \approx 1$$

and

$$(T^* + 1)^3 \approx \left(\frac{T_i^* + 1}{2} + 1\right)^3 = (T_m^* + 1)^3$$

Compare the solution of the linearized equation with the numerical solution of the Problem 5.3.

5.5 Two large stainless steel plates, each 30 cm thick and insulated on one side, are initially at 25°C and 100°C. They are pressed together at their un-insulated surface. Numerically determine the temperature distribution.

5.6 Repeat problem 5.5 if the two sides are not insulated, rather exposed to environment at 20°C and heat transfer coefficient is 5 W/m² K.

5.7 The base temperature of a pin fin of triangular profile changes

$$T_b = T_\infty + 100 * \sin(t)$$

a. Derive the governing equation to the unsteady temperature distribution.
b. Obtain the solution numerically.

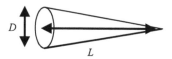

5.8 Copper tubing is joined to a solar collector plate of thickness t, and the working fluid enters the pipe at T_o. There is a net radiation heat flux to the top surface of the plate that varies with time according to

$$q'' = q_0'' \sin\left(2\pi \frac{t}{T} \right)$$

where
   q_0'' is the maximum flux
   T is the period

The bottom surface is well insulated, and the top surface is also exposed to a fluid at T_∞ that provides for a uniform convection coefficient. Determine the total heat transfer to the fluid.

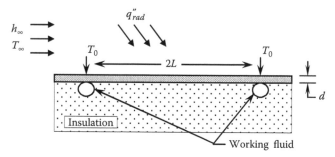

5.9 Consider the situation where the thermal conductivity is a function of temperature:

$$\frac{\partial}{\partial x}\left(k \frac{\partial T}{\partial x} \right) = \rho c \frac{\partial T}{\partial t}$$

$$k = a + bT$$

Solve the problem numerically and compare the results to the case of constant thermal conductivity.

5.10 A 3 cm thick brass plate is placed in an oven whose temperature changes from 500°C to 900°C in 10 min. Find the plate temperature as a function of time.

5.11 Estimate what laser power is needed to cut through a 0.5 cm steel sheet, if the beam focuses to a 2 mm disk.

5.12 Numerically obtain the solution for the time that it takes for the beam to start melting the steel, assuming the heat transfer coefficient is $h = 10 \times (T_s - T_\infty)^{1/2}$ W/m² K what laser power is needed to cut through a 0.5 cm steel sheet, if the beam focuses to a 2 mm disk.

5.13 A constant area annular fin is initially in thermal equilibrium with the ambient. The ambient temperature T_∞ and the heat transfer coefficient h are constant. The base temperature is suddenly raised to temperature T_b:

a. Starting with an elemental energy balance or heat diffusion equation, show that the energy equation for the UNSTEADY case is

$$\frac{\partial^2 T}{\partial r^2} + \frac{1}{r}\frac{\partial T}{\partial r} - \frac{2h}{kt}(T - T_\infty) = \frac{1}{\alpha}\frac{\partial T}{\partial t}$$

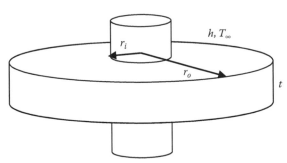

b. Nondimensionalize the governing equation and the boundary conditions.

c. For the values given in the following table, obtain the solution for the transient temperature distribution using finite difference.

h	200	W/m² K
k	80	W/m K
T_∞	25	°C
T_b	250	°C
t	0.001	m
r_i	0.01	m
r_o	0.03	m
α	1e–5	m²/s

5.14 A constant area fin is attached to a motorcycle engine initially in thermal equilibrium with the ambient at T_∞. The engine starts and the base temperature can be assumed to change $T = T_\infty + T_m^*(1 - e^{-\lambda t}) \sin(\omega t)$

a. Derive the governing equation and boundary conditions for the UNSTEADY temperature distribution. Begin from the basic elemental control volume analysis for the transient temperature distribution. Take the fin properties, dimensions (area A, length L, perimeter P), temperatures T_m, T_∞, λ, and the heat transfer coefficient, h, as all being constants. The length is finite and the tip is not insulated.

    b. Nondimensionalize the governing equation and the boundary conditions.

    c. Obtain the solution for the transient temperature distribution using finite difference.

5.15 A steel ball initially at 400°C is losing heat by convection and radiation as it cools down:

    a. Nondimensionalize the governing equation and the boundary conditions.

    b. For the conditions shown, plot the centerline temperature as a function of time.

h	200	W/m² K
k	40	W/m K
D	0.1	m
ε	0.5	
T_∞	25	°C
T_i	400	°C
α	1e–5	m²/s

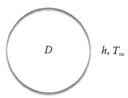

5.16 Solve Problem 5.2 numerically.

Reference

1. Arpaci, V. S. (1966) *Conduction Heat Transfer*. Reading, MA: Addison-Wesley Pub. Co.

6

Convection

6.1 Introduction

Convection is conduction enhanced by the additional energy transferred due to the flow of fluid. Conduction in the fluids is energy transfer due to microscopic motion of molecules (vibration, random translational motion). If the fluid moves, the bulk motion of the fluid further augments the transfer of thermal energy and is generally responsible for most of the heat transfer.

Most convection problems involve heat transfer between a solid and a fluid. As an example, consider the air flowing over a rectangular copper pipe carrying saturated stream. The inside surface of the pipe is at 100°C. The free electrons in the copper pipe carry the heat from inside surface of the pipe, where they have higher kinetic energy, toward the outer surface. Once the energy arrives at the outer surface, it must be transferred to the air. The first layer of the air molecules are stuck to the outer surface of the pipe and do not move. Therefore, the electrons' kinetic energy will be transferred to the air molecules on the surface by conduction only. As these molecules detach from the surface, they have higher kinetic energy compared to the other molecules they collide with, and transfer some of their energy to them. The transfer of heat as a result of molecular collision is conduction. Simultaneously, as a result of the fluid motion, all molecules are moving to other parts, carrying energy with them. Convection is the combination of the energy transfer due to molecular (microscopic) and bulk fluid (macroscopic) motions. The objective of convection heat transfer is the determination of the heat that is transferred to the fluid, in this case, from the outer surface of the pipe.

At the interface, between the solid and the fluid, the fluid molecules stick to the solid; therefore, the energy transfer between the solid and fluid at the interface is by conduction only. Doing an energy balance on the outer surface of the pipe results in

$$-k_s \left.\frac{\partial T_s}{\partial n}\right|_{n=0} = -k_f \left.\frac{\partial T_f}{\partial n}\right|_{n=0} \tag{6.1}$$

where
n is the coordinate normal to the surface, measured from the surface and stands for any coordinate (e.g., x, y, z in Cartesian or r, z, ϕ in cylindrical coordinates)
subscripts s and f refer to solid and fluid

Equation 6.1 also represents the heat transferred from the pipe to the air per unit area, or the heat flux:

$$-k_s \left.\frac{\partial T_s}{\partial n}\right|_{n=0} = -k_f \left.\frac{\partial T_f}{\partial n}\right|_{n=0} = \frac{Q}{A} = q'' \tag{6.2}$$

Equation 6.2 can be written in a more convenient-to-use form

$$-k_s \frac{\partial T_s}{\partial n}\bigg|_{n=0} = -k_f \frac{\partial T_f}{\partial n}\bigg|_{n=0} = q'' = h(T_w - T_\infty) \tag{6.3}$$

where h is called the local heat transfer coefficient. The heat transfer coefficient is a convenient concept defined to simplify the convection calculations. From Equation 6.3, h can be calculated as

$$h = \frac{-k_f \frac{\partial T_f}{\partial n}\bigg|_{n=0}}{T_w - T_\infty} \tag{6.4}$$

In fact, Equation 6.4 is the mathematical definition of heat transfer coefficient and how it is calculated. Evaluation of h requires the determination of the gradient of fluid temperature at the wall and therefore needs the knowledge of how temperature changes near the wall which generally requires the knowledge of the temperature in the entire flow field.

Therefore, to find h we need to know the temperature distribution in the fluid $T_f(x, y, z, t)$. Since the fluid is moving, one expects that the temperature of the fluid depends on how the fluid flows which also depends on the geometry, or $h = h(\text{fluid, flow})$. If h dependence with position on the surface is known, then the total heat transfer can be calculated from

$$Q = \int_A q'' dA = \int_A h(T_w - T_\infty) dA \tag{6.5}$$

and if the surface is isothermal

$$Q = (T_w - T_\infty) \int_A h \, dA \tag{6.6}$$

The average of any function, including h, over an area is mathematically defined as

$$\bar{h} = \frac{1}{A} \int_A h \, dA \tag{6.7}$$

thus the total heat transfer is given by

$$q = \bar{h} A (T_w - T_\infty) \tag{6.8}$$

The calculation of the heat transfer by convection from a surface boils down to knowing how h changes over the surface, which in turn requires the knowledge of how temperature of the fluid changes. Once h is known, it can be integrated over the area to find the average heat transfer coefficient and to determine q, using Equation 6.8.

If we define the nondimensional variables

$$T^* = \frac{T - T_\infty}{T_w - T_\infty}, \quad n^* = \frac{n}{L_c} \tag{6.9}$$

where L_c is a physical dimension that characterizes the object, then Equation 6.4 can be nondimensionalized as

$$h = \frac{-k_f}{L_c} \frac{\partial T^*}{\partial n^*}\bigg|_{n^*=0} \quad (6.10)$$

rearranging Equation 6.10

$$\frac{hL_c}{k_f} = -\frac{\partial T^*}{\partial n^*}\bigg|_{n^*=0} \quad (6.11)$$

The RHS of Equation 6.11 is dimensionless, therefore so should the LHS, and this nondimensional group is called Nusselt number, or

$$Nu = \frac{hL_c}{k_f} \quad (6.12)$$

Nusselt number is the nondimensional heat transfer coefficient. Therefore, to find Nusselt number we need to know how T^* changes in the flow field, i.e., we need to know $T^*(t^*, x^*, y^*, z^*)$ for the given fluid and the particular geometry. Once T^* is known, it can be differentiated in any direction and evaluated on any surface to determine Nu for that surface. A high value of h means a high temperature gradient near the surface, or a rapid change in the temperature as one moves away from the surface.

6.2 Conservation Equations

The temperature field in the fluid is to be determined from the first law of thermodynamics. The temperature distribution will be heavily impacted by how the fluid flows. Therefore, to find the temperature field, the fluid velocity must be known. The fluid velocity is determined by the laws of conservation of mass and conservation of momentum. We need to come up with equations that allow us to determine the three components of velocity as well as temperature at any point in the flow for all times.

All these equations are derived using the same procedure as we used for deriving the conduction equation, i.e., the conservation laws are applied to an elemental control volume containing fluid having dimensions Δx, Δy, and Δz containing fluid. Consider again the 3D flow of Figure 6.1. To derive the governing equations, the conservation laws are applied

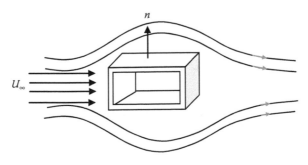

FIGURE 6.1
Flow over a rectangular duct.

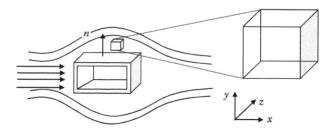

FIGURE 6.2
Elemental control volume.

to a control volume, which is a fluid cube having dimensions Δx, Δy, and Δz. The lower left-hand corner of the front face is located at point (x, y, z). This differential control volume has six surfaces. The three passing through the point (x, y, z) are taken to be inlets and the other three the outlets as shown in Figure 6.2.

6.2.1 Law of Conservation of Mass

The law of conservation of mass for a control volume is

$$\frac{dm}{dt} = \sum_{in} \dot{m} - \sum_{out} \dot{m} \tag{6.13}$$

where

$$\dot{m} = \rho A U \tag{6.14}$$

In Equation 6.14, U is the component of the average velocity normal to the area A. The mass flow rate at each inlet and exit is shown in Figure 6.2. The mass flow rate at x is the density times the area times the component of velocity perpendicular to the area, and therefore, the rate at which mass enters the control volume at the surface x is $(\rho \Delta y \Delta z u)_x$. The mass flow rate at other five faces is obtained in a similar way.

Substituting these in Equation 6.13

$$\frac{\partial(\rho \Delta x \Delta y \Delta z)}{\partial t}$$

$$= (\rho u \Delta y \Delta z)_x - (\rho u \Delta y \Delta z)_{x+\Delta x} + (\rho v \Delta x \Delta z)_y - (\rho v \Delta x \Delta z)_{y+\Delta y} + (\rho w \Delta x \Delta y)_z - (\rho w \Delta x \Delta y)_{z+\Delta z} \tag{6.15}$$

Using Taylor series to relate the terms on the RHS

$$(\rho u \Delta y \Delta z)_{x+\Delta x} = (\rho u \Delta y \Delta z)_x + \frac{\partial(\rho u \Delta y \Delta z)}{\partial x}\Delta x + \frac{\partial^2(\rho u \Delta y \Delta z)}{\partial x^2}\frac{\Delta x^2}{2!} + \cdots \tag{6.16}$$

Therefore, the rate of mass flow out minus mass flow in through the inlet and outlet surfaces that are perpendicular to the x-axis is

$$(\rho u \Delta y \Delta z)_{x+\Delta x} - (\rho u \Delta y \Delta z)_x = \frac{\partial(\rho u)}{\partial x}\Delta y \Delta z \Delta x + \frac{\partial^2(\rho u)}{\partial x^2}\Delta y \Delta z \frac{\Delta x^2}{2!} + \cdots \tag{6.17}$$

In the same manner, the rate of momentum increases as a result of flow in the z direction is

$$(\rho v \Delta x \Delta z)_{y+\Delta y} - (\rho v \Delta x \Delta z)_y = \frac{\partial(\rho v)}{\partial y} \Delta x \Delta z \Delta y + \frac{\partial^2(\rho v)}{\partial y^2} \Delta x \Delta z \frac{\Delta y^2}{2!} + \cdots \qquad (6.18)$$

$$(\rho w \Delta x \Delta y)_{z+\Delta z} - (\rho w \Delta x \Delta y)_z = \frac{\partial(\rho w)}{\partial z} \Delta x \Delta y \Delta z + \frac{\partial^2(\rho w)}{\partial z^2} \Delta x \Delta y \frac{\Delta z^2}{2!} + \cdots \qquad (6.19)$$

Substituting Equations 6.17 through 6.19 back into Equation 6.15 results in

$$\frac{\partial(\rho \Delta x \Delta y \Delta z)}{\partial t} = -\left[\frac{\partial(\rho u)}{\partial x} \Delta y \Delta z \Delta x + \frac{\partial^2(\rho u)}{\partial x^2} \Delta y \Delta z \frac{\Delta x^2}{2!} + \cdots + \frac{\partial(\rho v)}{\partial y} \Delta x \Delta z \Delta y + \frac{\partial^2(\rho v)}{\partial y^2} \Delta x \Delta z \frac{\Delta y^2}{2!} + \cdots \right.$$
$$\left. + \frac{\partial(\rho w)}{\partial z} \Delta x \Delta y \Delta z + \frac{\partial^2(\rho w)}{\partial z^2} \Delta x \Delta y \frac{\Delta z^2}{2!} + \cdots \right] \qquad (6.20)$$

Dividing both sides of the equation by the volume of the control volume, $\Delta x \Delta y \Delta z$

$$\frac{\partial(\rho)}{\partial t} = \frac{\partial(\rho u)}{\partial x} + \frac{\partial^2(\rho u)}{\partial x^2} \frac{\Delta x}{2!} + \frac{\partial(\rho v)}{\partial y} + \frac{\partial^2(\rho v)}{\partial y^2} \frac{\Delta y}{2!} + \frac{\partial(\rho w)}{\partial z} + \frac{\partial^2(\rho w)}{\partial z^2} \frac{\Delta z}{2!} + \cdots \qquad (6.21)$$

All the high-order terms on the RHS are multiplied by at least a Δx, Δy, or Δz and if we shrink the size of the control volume to zero, will all go to zero, and we end up with an equation (partial differential equation) which is valid at any point inside the fluid:

$$\frac{\partial(\rho)}{\partial t} + \frac{\partial(\rho u)}{\partial x} + \frac{\partial(\rho v)}{\partial y} + \frac{\partial(\rho w)}{\partial z} = 0 \qquad (6.22)$$

Equation 6.22 is the most general form of the law of conservation mass, valid for unsteady 3D compressible flow. For a steady flow, continuity equation becomes

$$\frac{\partial(\rho u)}{\partial x} + \frac{\partial(\rho v)}{\partial y} + \frac{\partial(\rho w)}{\partial z} = 0 \qquad (6.23)$$

and for incompressible flow it further simplifies to

$$\frac{\partial u}{\partial x} + \frac{\partial v}{\partial y} + \frac{\partial w}{\partial z} = 0 \qquad (6.24)$$

and finally for 2D incompressible flow, whether the flow is steady or unsteady, the conservation of mass becomes

$$\frac{\partial u}{\partial x} + \frac{\partial v}{\partial y} = 0 \qquad (6.25)$$

This equation states that if the fluid is accelerating in the x direction, its velocity in the y direction is decreasing (decelerating).

6.2.2 Vector Field Operators

The differential equations representing the velocity and temperature fields are often expressed in shorthand form using vector operators. For example, expanding the derivative terms in Equation 6.22

$$\frac{\partial(\rho)}{\partial t} + u\frac{\partial(\rho)}{\partial x} + v\frac{\partial(\rho)}{\partial y} + w\frac{\partial(\rho)}{\partial z} + \rho\frac{\partial(u)}{\partial x} + \rho\frac{\partial(v)}{\partial y} + \rho\frac{\partial(w)}{\partial z} = 0 \qquad (6.26)$$

The continuity equation can be expressed in a number of different ways. Equation 6.26 can be written in a more compact and general form by defining several operators. The first one is the substantial derivative operator which can be applied to a scalar or a vector and is defined as

$$\frac{D}{Dt} = \frac{\partial}{\partial t} + u\frac{\partial}{\partial x} + v\frac{\partial}{\partial y} + w\frac{\partial}{\partial z} \qquad (6.27)$$

The fluid velocity vector $\vec{V}$ is given by

$$\vec{V} = V_x\hat{i} + V_y\hat{j} + V_z\hat{k} = u\hat{i} + v\hat{j} + w\hat{k} \qquad (6.28)$$

where
$\hat{i}$ is the unit vector in x direction
$\hat{j}$ is the unit vector in y direction
$\hat{k}$ is the unit vector in z direction

The substantial time derivative of a property represents the rate of change of that property with time for a "particle" or "small packet" of fluid. For example, the substantial derivative of temperature represents the rate that the temperature of a small packet of fluid changes with time, if we followed that fluid packet as it moves with the rest of the flow.

If we insert a thermocouple at a point in a moving fluid and measure the temperature change with time, what we have measured is $\frac{\partial T}{\partial t}$ which is the temperature change at a particular location (x, y, z) with time. If we allow the thermocouple to move at the same velocity as that of the fluid, then measuring the temperature at two different times and dividing by Δt provide an approximation to the time derivative of the temperature. However this derivative is different, because it also takes into account the variation in position, since the probe is moving with the fluid. The additional terms, $u\frac{\partial T}{\partial x} + v\frac{\partial T}{\partial y} + w\frac{\partial T}{\partial z}$ account for the temperature change due to position change as a result of the flow of the fluid.

The gradient operator is applied to a scalar and results in a vector, and in Cartesian coordinates is defined as

$$\nabla = \frac{\partial}{\partial x}\hat{i} + \frac{\partial}{\partial y}\hat{j} + \frac{\partial}{\partial z}\hat{k} \qquad (6.29)$$

The physical significance of the gradient vector is that it points in the direction of greatest change of the scalar field. Using the earlier notations, the substantial derivative of a scalar S becomes

$$\frac{DS}{Dt} = \frac{\partial S}{\partial t} + \vec{V}\cdot\nabla S \qquad (6.30)$$

In Cartesian coordinates, the divergence of a vector $\vec{V}$

$$\vec{V} = V_x\hat{i} + V_y\hat{j} + V_z\hat{k} \tag{6.31}$$

is a scalar defined as

$$\nabla \cdot \vec{V} = \frac{\partial V_x}{\partial x} + \frac{\partial V_y}{\partial y} + \frac{\partial V_z}{\partial z} \tag{6.32}$$

Although $\vec{V}$ can be any vector, in the context of fluid mechanics and heat transfer, $\vec{V}$ is the fluid velocity vector.

The continuity equation can be written in a more compact form using these operators:

$$\frac{D\rho}{Dt} + \rho\nabla \cdot \vec{V} = 0 \tag{6.33}$$

For incompressible flow, density does not change with time or location and therefore $\dfrac{D\rho}{Dt} = 0$, simplifying Equation 6.33 to

$$\nabla \cdot \vec{V} = 0 \tag{6.34}$$

which for 2D flow results in Equation 6.25.

Using Equation 6.29, the continuity equation can also be written as

$$\frac{\partial\rho}{\partial t} + \vec{V} \cdot \nabla\rho + \rho\nabla \cdot \vec{V} = 0 \tag{6.35}$$

$$\frac{\partial\rho}{\partial t} + \nabla \cdot (\rho\vec{V}) = 0 \tag{6.36}$$

For steady compressible flow, the continuity equation simplifies to

$$\nabla \cdot (\rho\vec{V}) = 0 \tag{6.37}$$

Note that in general divergence of a vector is not equal to the dot product of the gradient operator and the vector, although it appears to work that way in Cartesian coordinates.

Another frequently used operator is the Laplacian operator which can be applied to a scalar as well as a vector and in Cartesian coordinates is defined as

$$\nabla^2 = \frac{\partial^2}{\partial x^2} + \frac{\partial^2}{\partial y^2} + \frac{\partial^2}{\partial z^2} \tag{6.38}$$

Laplacian is frequently encountered in the formulation of the physical laws.

These same operators can be defined in any curvilinear coordinate system, ξ, η, ζ. In this coordinate system, the gradient of a scalar S, the divergence of a vector V, and the Laplacian of a scalar S are given by

$$\nabla S = \frac{1}{h_\xi} \frac{\partial S}{\partial \xi} \hat{\xi} + \frac{1}{h_\eta} \frac{\partial S}{\partial \eta} \hat{\eta} + \frac{1}{h_\zeta} \frac{\partial S}{\partial \zeta} \hat{\zeta} \tag{6.39}$$

$$\nabla \cdot V = \frac{1}{h_\xi h_\eta h_\zeta} \left(\frac{\partial (h_\eta h_\zeta V_\xi)}{\partial \xi} + \frac{\partial (h_\xi h_\zeta V_\eta)}{\partial \eta} + \frac{\partial (h_\xi h_\eta V_\zeta)}{\partial \zeta} \right) \tag{6.40}$$

$$\nabla^2 S = \frac{1}{h_\xi h_\eta h_\zeta} \left[\frac{\partial}{\partial \xi} \left(\frac{h_\eta h_\zeta}{h_\xi} \frac{\partial S}{\partial \xi} \right) + \frac{\partial}{\partial \eta} \left(\frac{h_\xi h_\zeta}{h_\eta} \frac{\partial S}{\partial \eta} \right) + \frac{\partial}{\partial \zeta} \left(\frac{h_\xi h_\eta}{h_\zeta} \frac{\partial S}{\partial \zeta} \right) \right] \tag{6.41}$$

respectively, where

$$h_\xi = \left| \frac{\partial \vec{r}}{\partial \xi} \right| \tag{6.42}$$

$$h_\eta = \left| \frac{\partial \vec{r}}{\partial \eta} \right| \tag{6.43}$$

$$h_\zeta = \left| \frac{\partial \vec{r}}{\partial \zeta} \right| \tag{6.44}$$

and $\vec{r}$ is the position vector. In cylindrical coordinates, ξ, η, $\zeta = r$, ϕ, z, and the position vector is given by

$$\vec{r} = r \cos \phi \, \hat{i} + r \sin \phi \, \hat{j} + z \, \hat{k} \tag{6.45}$$

and for example

$$\frac{\partial \vec{r}}{\partial r} = \cos \phi \, \hat{i} + \sin \phi \, \hat{j} \tag{6.46}$$

and therefore

$$h_r = \left| \frac{\partial \vec{r}}{\partial r} \right| = \sqrt{\cos^2 \phi + \sin^2 \phi} = 1 \tag{6.47}$$

In spherical coordinates ξ, η, $\zeta = r$, ϕ, θ and the position vector is

$$\vec{r} = r \sin \theta \cos \phi \, \hat{i} + r \sin \theta \sin \phi \, \hat{j} + r \cos \theta \, \hat{k} \tag{6.48}$$

and for example

$$\frac{\partial \vec{r}}{\partial r} = \sin\theta\cos\phi\,\hat{i} + \sin\theta\sin\phi\,\hat{j} + \cos\theta\,\hat{k} \tag{6.49}$$

and

$$h_r = \left|\frac{\partial \vec{r}}{\partial r}\right| = \sqrt{(\sin\theta\cos\phi)^2 + (\sin\theta\sin\phi)^2 + \cos^2\theta} = 1 \tag{6.50}$$

Table 6.1 summarizes the different operators in the commonly used coordinate systems. For example, in cylindrical coordinates, the conservation of mass, or Equation 6.36 becomes

$$\frac{\partial\rho}{\partial t} + \frac{1}{r}\frac{\partial}{\partial r}(r\rho V_r) + \frac{1}{r}\frac{\partial(\rho V_\phi)}{\partial\phi} + \frac{\partial(\rho V_z)}{\partial z} = 0 \tag{6.51}$$

which for steady incompressible flow simplifies to

$$\frac{1}{r}\frac{\partial}{\partial r}(rV_r) + \frac{1}{r}\frac{\partial V_\phi}{\partial\phi} + \frac{\partial V_z}{\partial z} = 0 \tag{6.52}$$

In cylindrical coordinates, if the variation in ϕ is zero, then the problem is called an axisymmetric and if the variation with z is negligible, the problem is called polar.

TABLE 6.1

Vector Field Operators in Cartesian, Cylindrical, and Spherical Systems

Cartesian	Gradient of a scalar S	$\nabla S = \dfrac{\partial S}{\partial x}\hat{x} + \dfrac{\partial S}{\partial y}\hat{y} + \dfrac{\partial S}{\partial z}\hat{z}$
	Laplacian of a scalar S	$\nabla^2 S = \dfrac{\partial^2 S}{\partial x^2} + \dfrac{\partial^2 S}{\partial y^2} + \dfrac{\partial^2 S}{\partial z^2}$
	Divergence of a vector V	$\nabla\cdot\vec{V} = \dfrac{\partial V_x}{\partial x} + \dfrac{\partial V_y}{\partial y} + \dfrac{\partial V_z}{\partial z}$
Cylindrical	Gradient of a scalar S	$\nabla S = \dfrac{\partial S}{\partial r}\hat{r} + \dfrac{1}{r}\dfrac{\partial S}{\partial\phi}\hat{\phi} + \dfrac{\partial S}{\partial z}\hat{z}$
	Laplacian of a scalar S	$\nabla^2 S = \dfrac{1}{r}\dfrac{\partial}{\partial r}\left(r\dfrac{\partial S}{\partial r}\right) + \dfrac{1}{r^2}\dfrac{\partial^2 S}{\partial\phi^2} + \dfrac{\partial^2 S}{\partial z^2}$
	Divergence of a vector V	$\nabla\cdot\vec{V} = \dfrac{1}{r}\dfrac{\partial}{\partial r}(rV_r) + \dfrac{1}{r}\dfrac{\partial V_\phi}{\partial\phi} + \dfrac{\partial V_z}{\partial z}$
Spherical	Gradient of a scalar S	$\nabla S = \dfrac{\partial S}{\partial r}\hat{r} + \dfrac{1}{r}\dfrac{\partial S}{\partial\theta}\hat{\theta} + \dfrac{1}{r\sin\theta}\dfrac{\partial S}{\partial\phi}\hat{\phi}$
	Laplacian of a scalar S	$\nabla^2 S = \dfrac{1}{r^2}\dfrac{\partial}{\partial r}\left(r^2\dfrac{\partial S}{\partial r}\right) + \dfrac{1}{r^2\sin^2\theta}\dfrac{\partial^2 S}{\partial\phi^2} + \dfrac{1}{r^2\sin\theta}\dfrac{\partial}{\partial\theta}\left(\sin\theta\dfrac{\partial S}{\partial\theta}\right)$
	Divergence of a vector V	$\nabla\cdot\vec{V} = \dfrac{1}{r^2}\dfrac{\partial}{\partial r}(r^2 V_r) + \dfrac{1}{r\sin\theta}\dfrac{\partial}{\partial\theta}(\sin\theta V_\theta) + \dfrac{1}{r\sin\theta}\dfrac{\partial V_\phi}{\partial\phi}$

6.3 Conservation of Momentum

The conservation of linear momentum for an object (control mass) states that sum of all forces acting on the object must be equal to the rate of momentum increase or

$$\vec{F} = \frac{d(m\vec{V})}{dt} \tag{6.53}$$

This is a vector equation and has three components. If we apply this law to the control volume (CV) shown in Figure 6.3, the forces acting on the CV would increase the momentum inside the CV as well as momentum leaving the CV compared to the momentum entering it. The amount of mass in the control volume is $\rho\Delta x\Delta y\Delta z$ and this much mass has a velocity u in the x direction; therefore, the rate of increase in x momentum in the control volume with respect to time or the rate of momentum accumulation is

$$\frac{\partial(\rho\Delta x\Delta y\Delta zu)}{\partial t} \tag{6.54}$$

The mass flow rate at x is $(\rho\Delta y\Delta z\,u)_x$ and the fluid velocity is u in the x direction; therefore, the rate at which momentum enters the control volume at the surface x is $(\rho u\Delta y\Delta z\,u)_x$ which is obtained by multiplying the mass flow rate with the x component of velocity. Similarly, the rate of momentum leaving the control volume at $x + \Delta x$ is $(\rho u\Delta y\Delta z\,u)_{x+\Delta x'}$ which can be related to the momentum at x, using Taylor series:

$$(\rho u\Delta y\Delta z\,u)_{x+\Delta x} = (\rho u\Delta y\Delta z\,u)_x + \frac{\partial(\rho u\Delta y\Delta z\,u)}{\partial x}\Delta x + \frac{\partial^2(\rho u\Delta y\Delta z\,u)}{\partial x^2}\frac{\Delta x^2}{2!} + \cdots \tag{6.55}$$

Therefore, the rate of momentum out minus momentum in through the inlet and outlet surfaces that are perpendicular to the x axis is

$$(\rho u\Delta y\Delta z\,u)_{x+\Delta x} - (\rho u\Delta y\Delta z\,u)_x = \frac{\partial(\rho u\,u)}{\partial x}\Delta y\Delta z\Delta x + \frac{\partial^2(\rho u\,u)}{\partial x^2}\Delta y\Delta z\frac{\Delta x^2}{2!} + \cdots \tag{6.56}$$

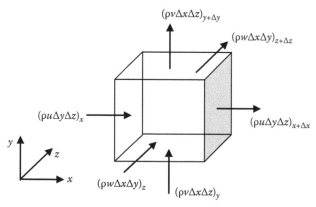

FIGURE 6.3
Mass balance for the elemental control volume.

The mass flow rate entering the control volume from the surface located at y is $(\rho\Delta x\Delta z\, v)_y$. This mass carries with it momentum into the control volume at the rate of $(\rho\Delta x\Delta z\, vV)_y$ in the direction of velocity vector V. The component in the x direction is therefore $(\rho\Delta x\Delta z\, vu)_y$. Similarly, the rate of momentum leaving the control volume at $y + \Delta y$ is $(\rho\Delta x\Delta z\, vu)_{y+\Delta y}$. This can be related to the momentum at y, again using Taylor series, and therefore the rate of momentum out minus momentum in through the inlet and outlet surfaces that are perpendicular to the y axis is

$$(\rho v\Delta x\Delta z\, u)_{y+\Delta y} - (\rho v\Delta x\Delta z\, u)_y = \frac{\partial(\rho v\, u)}{\partial y}\Delta x\Delta z\Delta y + \frac{\partial^2(\rho v\, u)}{\partial y^2}\Delta x\Delta z\frac{\Delta y^2}{2!} + \cdots \tag{6.57}$$

In the same manner, the rate of momentum increases as a result of flow in the z direction is

$$(\rho w\Delta x\Delta y\, u)_{z+\Delta z} - (\rho w\Delta x\Delta y\, u)_z = \frac{\partial(\rho w\, u)}{\partial z}\Delta x\Delta y\Delta z + \frac{\partial^2(\rho w\, u)}{\partial z^2}\Delta x\Delta y\frac{\Delta z^2}{2!} + \cdots \tag{6.58}$$

Substitute these in Equation 6.53, we get

$$F = \Delta x\Delta y\Delta z$$

$$\times\left[\frac{\partial(\rho u)}{\partial t} + \frac{\partial(\rho u\, u)}{\partial x} + \frac{\partial(\rho v\, u)}{\partial y} + \frac{\partial(\rho w\, u)}{\partial z} + \frac{\partial^2(\rho u\, u)}{\partial z^2}\frac{\Delta x}{2!} + \frac{\partial^2(\rho v\, u)}{\partial z^2}\frac{\Delta y}{2!} + \frac{\partial^2(\rho w\, u)}{\partial z^2}\frac{\Delta z}{2!} + \cdots\right] \tag{6.59}$$

We next need to determine all the forces acting on the control volume in the x direction. The forces acting on a fluid are shown in Figure 6.4. Depending on whether the forces act on the mass or surface of the control volume, they are divided into body forces and surface forces, respectively. The body forces are those forces that act on the mass of the fluid with the most commonly encountered body force being due to gravity:

$$\vec{F}_B = X\hat{i} + Y\hat{j} + Z\hat{k} \tag{6.60}$$

Rather than dealing with forces, it is more convenient to consider stresses which are forces per unit area. The surface forces act either normal to a surface or tangential to it. Normal stresses are broken down into two components: those that result from the fluid motion and viscosity and those that are present even in a stationary fluid, which is referred to as the thermodynamic pressure. Pressure is defined such that it acts against the surface, where the other normal stresses point away.

The tangential stresses are also a result of viscosity and fluid motion. The surface stresses are denoted by τ_{ij} where i designates the surface where the stress acts on and j designates

$$\text{Forces in the fluid} = \begin{cases} \text{Body} \\ \text{Surface} \begin{cases} \text{Normal} \begin{cases} \text{Pressure} \\ \text{Fluid motion} \end{cases} \\ \text{Tangential} - \text{Fluid motion} \end{cases} \end{cases}$$

FIGURE 6.4
Forces acting on a fluid.

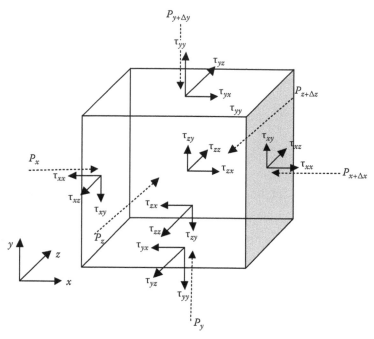

FIGURE 6.5
Normal and tangential stresses acting on the surface of the fluid element.

the direction. On a positive surface (a surface whose outward directed normal points in the positive coordinate direction) positive stress points in the positive coordinate direction and on a negative surface (a surface whose outward directed normal points in the negative coordinate direction) positive stress points in the negative coordinate direction, as shown in Figure 6.5. Sum of the forces acting on the fluid element in the x direction is

$$\sum F_x = \left[(\tau_{xx})_{x+\Delta x} - (\tau_{xx})_x + P_x - P_{x+\Delta x} \right] \Delta y \Delta z$$

$$+ \left[(\tau_{yx})_{y+\Delta y} - (\tau_{yx})_y \right] \Delta x \Delta z + \left[(\tau_{zx})_{z+\Delta z} - (\tau_{zx})_z \right] \Delta x \Delta y + X \Delta x \Delta y \Delta z \qquad (6.61)$$

Again using Taylor series

$$\sum F_x = -\left[\frac{\partial P}{\partial x} + \frac{\partial^2 P}{\partial x^2} \frac{\Delta x}{2} + \cdots \right] \Delta x \Delta y \Delta z + \left[\frac{\partial \tau_{xx}}{\partial x} + \frac{\partial^2 \tau_{xx}}{\partial x^2} \frac{\Delta x}{2} + \cdots \right] \Delta x \Delta y \Delta z$$

$$+ \left[\frac{\partial \tau_{yx}}{\partial y} + \frac{\partial^2 \tau_{yx}}{\partial y^2} \frac{\Delta y}{2} + \cdots \right] \Delta x \Delta y \Delta z + \left[\frac{\partial \tau_{zx}}{\partial z} + \frac{\partial^2 \tau_{zx}}{\partial z^2} \frac{\Delta z}{2} + \cdots \right] \Delta x \Delta y \Delta z + X \Delta x \Delta y \Delta z$$

$$\qquad (6.62)$$

Equating the terms on the RHS of Equations 6.59 and 6.62, dividing both sides by the volume of the control volume, $\Delta x \Delta y \Delta z$, and shrinking the size of the control volume to zero result in

$$\frac{\partial (\rho u)}{\partial t} + \frac{\partial (\rho u\, u)}{\partial x} + \frac{\partial (\rho v\, u)}{\partial y} + \frac{\partial (\rho w\, u)}{\partial z} = -\frac{\partial P}{\partial x} + \frac{\partial \tau_{xx}}{\partial x} + \frac{\partial \tau_{yx}}{\partial y} + \frac{\partial \tau_{zx}}{\partial z} + X \qquad (6.63)$$

For most commonly encountered fluids, it has been found experimentally that

$$\tau_{xx} = \mu\left(2\frac{\partial u}{\partial x} - \frac{2}{3}\nabla \cdot \vec{V}\right) \tag{6.64}$$

$$\tau_{yy} = \mu\left(2\frac{\partial v}{\partial y} - \frac{2}{3}\nabla \cdot \vec{V}\right) \tag{6.65}$$

$$\tau_{zz} = \mu\left(2\frac{\partial w}{\partial z} - \frac{2}{3}\nabla \cdot \vec{V}\right) \tag{6.66}$$

$$\tau_{yx} = \tau_{xy} = \mu\left(\frac{\partial u}{\partial y} + \frac{\partial v}{\partial x}\right) \tag{6.67}$$

$$\tau_{zx} = \tau_{xz} = \mu\left(\frac{\partial u}{\partial z} + \frac{\partial w}{\partial x}\right) \tag{6.68}$$

$$\tau_{zy} = \tau_{yz} = \mu\left(\frac{\partial v}{\partial z} + \frac{\partial w}{\partial y}\right) \tag{6.69}$$

note that summing the Equations 6.64 through 6.66 at a point results in

$$\tau_{xx} + \tau_{yy} + \tau_{zz} = 0 \tag{6.70}$$

which shows that thermodynamic pressure is the normal stresses acting at a point against the surface.

Substituting for stresses back in Equation 6.51

$$\frac{\partial(\rho u)}{\partial t} + \frac{\partial(\rho u\,u)}{\partial x} + \frac{\partial(\rho v\,u)}{\partial y} + \frac{\partial(\rho w\,u)}{\partial z}$$

$$= -\frac{\partial P}{\partial x} + \frac{\partial}{\partial x}\left[\mu\left(2\frac{\partial u}{\partial x} - \frac{2}{3}\nabla\cdot\vec{V}\right)\right] + \frac{\partial}{\partial y}\left[\mu\left(\frac{\partial u}{\partial y} + \frac{\partial v}{\partial x}\right)\right] + \frac{\partial}{\partial z}\left[\mu\left(\frac{\partial u}{\partial z} + \frac{\partial w}{\partial x}\right)\right] + X \tag{6.71}$$

Expanding the terms on the LHS

$$\rho\frac{\partial u}{\partial t} + u\frac{\partial \rho}{\partial t} + u\frac{\partial(\rho u)}{\partial x} + \rho u\frac{\partial(u)}{\partial x} + \rho v\frac{\partial(u)}{\partial y} + u\frac{\partial(\rho v)}{\partial y} + u\frac{\partial(\rho w)}{\partial z} + \rho w\frac{\partial(u)}{\partial z}$$

$$= \rho\frac{\partial u}{\partial t} + u\left[\frac{\partial \rho}{\partial t} + \frac{\partial(\rho u)}{\partial x} + \frac{\partial(\rho v)}{\partial y} + \frac{\partial(\rho w)}{\partial z}\right] + \rho u\frac{\partial(u)}{\partial x} + \rho v\frac{\partial(u)}{\partial y} + \rho w\frac{\partial(u)}{\partial z} \tag{6.72}$$

From continuity equation, the term inside the bracket is zero, simplifying Equation 6.72 to

$$\rho\left[\frac{\partial u}{\partial t}+u\frac{\partial u}{\partial x}+v\frac{\partial u}{\partial y}+w\frac{\partial u}{\partial z}\right]$$

$$=-\frac{\partial P}{\partial x}+\frac{\partial}{\partial x}\left[\mu\left(2\frac{\partial u}{\partial x}-\frac{2}{3}\nabla\cdot\vec{V}\right)\right]+\frac{\partial}{\partial y}\left[\mu\left(\frac{\partial u}{\partial y}+\frac{\partial v}{\partial x}\right)\right]+\frac{\partial}{\partial z}\left[\mu\left(\frac{\partial u}{\partial z}+\frac{\partial w}{\partial x}\right)\right]+X \qquad (6.73)$$

By applying the conservation of momentum in the y and z directions, similar equations for v and w components of velocity can be obtained.

For incompressible flow

$$\frac{D\rho}{Dt}=0 \qquad (6.74)$$

and therefore from continuity equation

$$\nabla\cdot\vec{V}=\frac{\partial u}{\partial x}+\frac{\partial v}{\partial y}+\frac{\partial w}{\partial z}=0 \qquad (6.75)$$

and the x momentum becomes

$$\rho\frac{Du}{Dt}=-\frac{\partial P}{\partial x}+\frac{\partial}{\partial x}\left[\mu\left(2\frac{\partial u}{\partial x}\right)\right]+\frac{\partial}{\partial y}\left[\mu\left(\frac{\partial u}{\partial y}+\frac{\partial v}{\partial x}\right)\right]+\frac{\partial}{\partial z}\left[\mu\left(\frac{\partial w}{\partial x}+\frac{\partial u}{\partial z}\right)\right]+X \qquad (6.76)$$

which for constant property fluids simplifies to

$$\rho\frac{Du}{Dt}=-\frac{\partial P}{\partial x}+\mu\left[\frac{\partial^2 u}{\partial x^2}+\frac{\partial^2 u}{\partial y^2}+\frac{\partial^2 u}{\partial x^2}\right]+\rho g_x \qquad (6.77)$$

6.4 Conservation of Energy Equation

The energy equation for a control volume, Equation 1.9 is

$$q+q_g+\dot{W}+\sum_{in}\dot{m}\left(h+\frac{V^2}{2}+gz\right)=\frac{dE_{c.v}}{dt}+\sum_{out}\dot{m}\left(h+\frac{V^2}{2}+gz\right) \qquad (6.78)$$

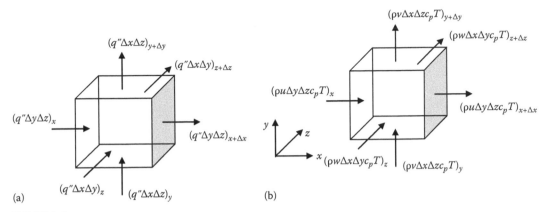

FIGURE 6.6
(a) Energy balance due to conduction and (b) mass flow on the differential control volume.

The first term on the LHS is the rate at which heat is transferred across the boundaries of the control volume. Applying this equation to the differential control volume, Figure 6.6a shows the heat transfer through the boundaries which is by conduction. Figure 6.6b shows the two summation terms in Equation 6.78. There are a number of ways by which work can be done on the control volume. The most significant one is the work due to the viscous forces, with ϕ representing this mode of work transfer per unit volume. There may also be energy generation that could be for example as a result of chemical reaction, with $\dot{q}$ representing the rate of energy generation per unit volume:

$$q''_x \Delta y \Delta z + q''_y \Delta x \Delta z + q''_z \Delta x \Delta y + \Phi \Delta x \Delta y \Delta z + \dot{q} \Delta x \Delta y \Delta z + (\rho u \Delta y \Delta z \, c_p T)_x + (\rho v \Delta x \Delta z \, c_p T)_y$$

$$+ (\rho w \Delta x \Delta y \, c_p T)_z = \frac{\partial (\rho \Delta x \Delta y \Delta z c_p T)}{\partial t} + q''_{x+\Delta x} \Delta y \Delta z + q''_{y+\Delta y} \Delta x \Delta z + q''_{z+\Delta z} \Delta x \Delta y + (\rho u \Delta y \Delta z \, c_p T)_{x+\Delta x}$$

$$+ (\rho v \Delta x \Delta z \, c_p T)_{y+\Delta y} + (\rho w \Delta x \Delta y \, c_p T)_{z+\Delta z} \tag{6.79}$$

Expanding the terms using Taylor series expansion, and dividing both sides by the volume of the CV, simplifies Equation 6.79 to

$$\left(-\frac{\partial q''_x}{\partial x} - \frac{\partial^2 q''_x}{\partial x^2}\frac{\Delta x}{2!} + \cdots \right) + \left(-\frac{\partial q''_y}{\partial y} - \frac{\partial^2 q''_y}{\partial y^2}\frac{\Delta y}{2!} + \cdots \right) + \left(-\frac{\partial q''_z}{\partial z} - \frac{\partial^2 q''_z}{\partial z^2}\frac{\Delta z}{2!} + \cdots \right) + \Phi + \dot{q}$$

$$= \frac{\partial (\rho c_p T)}{\partial t} + \frac{\partial (\rho u \, c_p T)}{\partial x} + \frac{\partial (\rho v \, c_p T)}{\partial y} + \frac{\partial (\rho w \, c_p T)}{\partial z} + \frac{\partial^2 (\rho u \, c_p T)}{\partial z^2}\frac{\Delta x}{2!} + \frac{\partial^2 (\rho v \, c_p T)}{\partial z^2}\frac{\Delta y}{2!}$$

$$+ \frac{\partial^2 (\rho w \, c_p T)}{\partial z^2}\frac{\Delta z}{2!} + \cdots \tag{6.80}$$

Taking the limit as the size of the CV is reduced to zero

$$\frac{\partial (\rho c_p T)}{\partial t} + \frac{\partial (\rho u \, c_p T)}{\partial x} + \frac{\partial (\rho v \, c_p T)}{\partial y} + \frac{\partial (\rho w \, c_p T)}{\partial z} = -\left[\frac{\partial q''_x}{\partial x} + \frac{\partial q''_y}{\partial y} + \frac{\partial q''_z}{\partial z} \right] + \Phi + \dot{q} \tag{6.81}$$

The heat flux by conduction is given by Fourier law, which says that the heat flux is proportional to the temperature difference and the inverse of the distance that heat has to travel. In Cartesian coordinate system,

$$q_x'' = -k\frac{\partial T}{\partial x} \tag{6.82}$$

$$q_y'' = -k\frac{\partial T}{\partial y} \tag{6.83}$$

$$q_z'' = -k\frac{\partial T}{\partial z} \tag{6.84}$$

$$\frac{\partial(\rho c_p T)}{\partial t} + \frac{\partial(\rho u\, c_p T)}{\partial x} + \frac{\partial(\rho v\, c_p T)}{\partial y} + \frac{\partial(\rho w\, c_p T)}{\partial z} = \frac{\partial}{\partial x}\left[k\frac{\partial T}{\partial x}\right] + \frac{\partial}{\partial y}\left[k\frac{\partial T}{\partial y}\right] + \frac{\partial}{\partial z}\left[k\frac{\partial T}{\partial z}\right] + \Phi + \dot{q} \tag{6.85}$$

Expanding the terms on the LHS

$$\rho\frac{\partial(c_p T)}{\partial t} + c_p T\left[\frac{\partial(\rho)}{\partial t} + \frac{\partial(\rho u)}{\partial x} + \frac{\partial(\rho v)}{\partial y} + \frac{\partial(\rho w)}{\partial z}\right] + \rho u\frac{\partial(c_p T)}{\partial x} + \rho v\frac{\partial(c_p T)}{\partial y} + \rho w\frac{\partial(c_p T)}{\partial z}$$

$$= \frac{\partial}{\partial x}\left[k\frac{\partial T}{\partial x}\right] + \frac{\partial}{\partial y}\left[k\frac{\partial T}{\partial y}\right] + \frac{\partial}{\partial z}\left[k\frac{\partial T}{\partial z}\right] + \Phi + \dot{q} \tag{6.86}$$

Then using continuity equation

$$\rho\frac{\partial(c_p T)}{\partial t} + \rho u\frac{\partial(c_p T)}{\partial x} + \rho v\frac{\partial(c_p T)}{\partial y} + \rho w\frac{\partial(c_p T)}{\partial z} = \frac{\partial}{\partial x}\left[k\frac{\partial T}{\partial x}\right] + \frac{\partial}{\partial y}\left[k\frac{\partial T}{\partial y}\right] + \frac{\partial}{\partial z}\left[k\frac{\partial T}{\partial z}\right] + \Phi + \dot{q} \tag{6.87}$$

Assuming constant properties, steady flow, without energy generation or viscous dissipation, the energy equation simplifies to

$$\rho c_p\left(\frac{\partial T}{\partial t} + u\frac{\partial T}{\partial x} + v\frac{\partial T}{\partial y} + w\frac{\partial T}{\partial z}\right) = k\left[\frac{\partial^2 T}{\partial x^2} + \frac{\partial^2 T}{\partial y^2} + \frac{\partial^2 T}{\partial z^2}\right] \tag{6.88}$$

The conservation equations and the constituent equations for the general 3D compressible flow of Newtonian fluids are listed in Tables 6.2 through 6.7 for Cartesian, cylindrical, and spherical coordinate systems [1]. For incompressible, constant property fluids, the Navier–Stokes (NS) and energy equations can be written in the general form

$$\nabla \cdot \vec{V} = 0 \tag{6.89}$$

TABLE 6.2

Conservation Equations for 3D Transient Compressible Flow in Cartesian Coordinates

$$\frac{\partial(\rho)}{\partial t} + \frac{\partial(\rho u)}{\partial x} + \frac{\partial(\rho v)}{\partial y} + \frac{\partial(\rho w)}{\partial z} = 0$$

$$\rho\left(\frac{\partial u}{\partial t} + u\frac{\partial u}{\partial x} + v\frac{\partial u}{\partial y} + w\frac{\partial u}{\partial z}\right) = -\frac{\partial P}{\partial x} + \frac{\partial \tau_{xx}}{\partial x} + \frac{\partial \tau_{yx}}{\partial y} + \frac{\partial \tau_{zx}}{\partial z} + \rho g_x$$

$$\rho\left(\frac{\partial v}{\partial t} + u\frac{\partial v}{\partial x} + v\frac{\partial v}{\partial y} + w\frac{\partial v}{\partial z}\right) = -\frac{\partial P}{\partial y} + \frac{\partial \tau_{xy}}{\partial x} + \frac{\partial \tau_{yy}}{\partial y} + \frac{\partial \tau_{yz}}{\partial z} + \rho g_y$$

$$\rho\left(\frac{\partial w}{\partial t} + u\frac{\partial w}{\partial x} + v\frac{\partial w}{\partial y} + w\frac{\partial w}{\partial z}\right) = -\frac{\partial P}{\partial z} + \frac{\partial \tau_{xz}}{\partial x} + \frac{\partial \tau_{yz}}{\partial y} + \frac{\partial \tau_{zz}}{\partial z} + \rho g_z$$

$$\rho c_v\left(\frac{\partial T}{\partial t} + u\frac{\partial T}{\partial r} + v\frac{\partial T}{\partial y} + w\frac{\partial T}{\partial z}\right) = -\left[\frac{\partial q_x''}{\partial x} + \frac{\partial q_y''}{\partial y} + \frac{\partial q_z''}{\partial z}\right] - T\left(\frac{\partial P}{\partial T}\right)_\rho\left(\frac{\partial u}{\partial x} + \frac{\partial v}{\partial y} + \frac{\partial w}{\partial z}\right) + \Phi + \dot q$$

TABLE 6.3

Stress and Strain Rate Relations and Viscous Dissipation for Newtonian Fluids in Cartesian Coordinates

$\tau_{xx} = \mu\left(2\dfrac{\partial u}{\partial x} - \dfrac{2}{3}\nabla\cdot\vec V\right)$	$q_x'' = -k\dfrac{\partial T}{\partial x}$
$\tau_{yy} = \mu\left(2\dfrac{\partial v}{\partial y} - \dfrac{2}{3}\nabla\cdot\vec V\right)$	$q_y'' = -k\dfrac{\partial T}{\partial y}$
$\tau_{zz} = \mu\left(2\dfrac{\partial w}{\partial z} - \dfrac{2}{3}\nabla\cdot\vec V\right)$	$q_z'' = -k\dfrac{\partial T}{\partial z}$
$\tau_{yx} = \tau_{xy} = \mu\left(\dfrac{\partial u}{\partial y} + \dfrac{\partial v}{\partial x}\right)$	
$\tau_{zx} = \tau_{xz} = \mu\left(\dfrac{\partial u}{\partial z} + \dfrac{\partial w}{\partial x}\right)$	
$\tau_{zy} = \tau_{yz} = \mu\left(\dfrac{\partial v}{\partial z} + \dfrac{\partial w}{\partial y}\right)$	

$$\Phi = \tau_{xx}\frac{\partial u}{\partial x} + \tau_{yy}\frac{\partial v}{\partial y} + \tau_{zz}\frac{\partial w}{\partial z} + \tau_{xy}\left(\frac{\partial u}{\partial y} + \frac{\partial v}{\partial x}\right) + \tau_{xz}\left(\frac{\partial u}{\partial z} + \frac{\partial w}{\partial x}\right) + \tau_{yz}\left(\frac{\partial v}{\partial z} + \frac{\partial w}{\partial y}\right)$$

$$\frac{\partial \vec V}{\partial t} + \vec V\cdot\nabla\vec V = -\nabla\cdot P + \nabla^2\vec V + \nabla\cdot\vec g \tag{6.90}$$

$$\frac{\partial T}{\partial t} + \vec V\cdot\nabla T = \nabla^2 T + \Phi \tag{6.91}$$

with the operators given in Table 6.1. The incompressible constant property form of the conservation equations in the three coordinate systems is given in Tables 6.8 through 6.10, and the 2D form of the equations in Cartesian and cylindrical coordinates is given in Table 6.11.

TABLE 6.4

Conservation Equations for 3D Transient Compressible Flow in Cylindrical Coordinates

$$\frac{\partial \rho}{\partial t} + \frac{1}{r}\frac{\partial}{\partial r}(r\rho u_r) + \frac{1}{r}\frac{\partial(\rho u_\phi)}{\partial \phi} + \frac{\partial(\rho u_z)}{\partial z} = 0$$

$$\rho\left(\frac{\partial u_r}{\partial t} + u_r\frac{\partial u_r}{\partial r} + \frac{u_\phi}{r}\frac{\partial u_r}{\partial \phi} - \frac{u_\phi^2}{r} + u_z\frac{\partial u_r}{\partial z}\right) = -\frac{\partial P}{\partial r} + \frac{1}{r}\frac{\partial(r\tau_{rr})}{\partial r} + \frac{1}{r}\frac{\partial(\tau_{r\phi})}{\partial \phi} - \frac{\tau_{\phi\phi}}{r} + \frac{\partial(\tau_{rz})}{\partial z} + \rho g_r$$

$$\rho\left(\frac{\partial u_\phi}{\partial t} + u_r\frac{\partial u_\phi}{\partial r} + \frac{u_\phi}{r}\frac{\partial u_\phi}{\partial \phi} + \frac{u_r u_\phi}{r} + u_z\frac{\partial u_\phi}{\partial z}\right) = -\frac{1}{r}\frac{\partial P}{\partial \phi} + \frac{1}{r^2}\frac{\partial\left(r^2\tau_{r\phi}\right)}{\partial r} + \frac{1}{r}\frac{\partial(\tau_{\phi\phi})}{\partial \phi} - \frac{\tau_{\phi\phi}}{r} + \frac{\partial(\tau_{\phi z})}{\partial z} + \rho g_\phi$$

$$\rho\left(\frac{\partial u_z}{\partial t} + u_r\frac{\partial u_z}{\partial r} + \frac{u_\phi}{r}\frac{\partial u_z}{\partial \phi} + u_z\frac{\partial u_z}{\partial z}\right) = -\frac{\partial P}{\partial z} + \frac{1}{r}\frac{\partial(r\tau_{rz})}{\partial r} + \frac{1}{r}\frac{\partial(\tau_{\phi z})}{\partial \phi} + \frac{\partial(\tau_{zz})}{\partial z} + \rho g_z$$

$$\rho c_v\left(\frac{\partial T}{\partial t} + u_r\frac{\partial T}{\partial r} + \frac{u_\phi}{r}\frac{\partial T}{\partial \phi} + u_z\frac{\partial T}{\partial z}\right) = -\left[\frac{1}{r}\frac{\partial(rq_r'')}{\partial r} + \frac{1}{r}\frac{\partial(q_\phi'')}{\partial \phi} + \frac{\partial(q_z'')}{\partial z}\right] - T\left(\frac{\partial P}{\partial T}\right)_\rho\left(\frac{1}{r}\frac{\partial}{\partial r}(ru_r) + \frac{1}{r}\frac{\partial u_\phi}{\partial \phi} + \frac{\partial u_z}{\partial z}\right) + \Phi + \dot{q}$$

TABLE 6.5

Stress and Strain Rate Relations and Viscous Dissipation for Newtonian Fluids in Cylindrical Coordinates

$$\tau_{rr} = \mu\left(2\frac{\partial u_r}{\partial r} - \frac{2}{3}\nabla\cdot\vec{V}\right) \qquad\qquad q_r'' = -k\frac{\partial T}{\partial r}$$

$$\tau_{\phi\phi} = \mu\left[2\left(\frac{1}{r}\frac{\partial u_\phi}{\partial \phi} + \frac{u_r}{r}\right) - \frac{2}{3}\nabla\cdot\vec{V}\right] \qquad\qquad q_\phi'' = -k\frac{1}{r}\frac{\partial T}{\partial \phi}$$

$$\tau_{zz} = \mu\left(2\frac{\partial u_z}{\partial z} - \frac{2}{3}\nabla\cdot\vec{V}\right) \qquad\qquad q_z'' = -k\frac{\partial T}{\partial z}$$

$$\tau_{r\phi} = \tau_{\phi r} = \mu\left[r\frac{\partial}{\partial r}\left(\frac{u_\phi}{r}\right) + \frac{1}{r}\frac{\partial u_r}{\partial \phi}\right]$$

$$\tau_{z\phi} = \tau_{\phi z} = \mu\left[\frac{\partial u_\phi}{\partial z} + \frac{1}{r}\frac{\partial u_z}{\partial \phi}\right]$$

$$\tau_{rz} = \tau_{zr} = \mu\left[\frac{\partial u_z}{\partial r} + \frac{\partial u_r}{\partial z}\right]$$

$$\Phi = \tau_{rr}\frac{\partial u_r}{\partial r} + \tau_{\phi\phi}\frac{1}{r}\left(\frac{\partial u_\phi}{\partial \phi} + u_r\right) + \tau_{zz}\frac{\partial u_z}{\partial z} + \tau_{r\phi}\left[r\frac{\partial}{\partial r}\left(\frac{u_\phi}{r}\right) + \frac{1}{r}\frac{\partial u_r}{\partial \phi}\right] + \tau_{rz}\left(\frac{\partial u_z}{\partial r} + \frac{\partial u_r}{\partial z}\right) + \tau_{\phi z}\left(\frac{1}{r}\frac{\partial u_z}{\partial \phi} + \frac{\partial u_\phi}{\partial z}\right)$$

6.5 Index Notation

Index notation is a shorthand method of writing long equations, like those encountered in fluid mechanics in a more compact form. In index form, the continuity can be written as

$$\frac{\partial}{\partial t}(\rho) + \frac{\partial}{\partial x_k}(\rho u_k) = 0 \tag{6.92}$$

TABLE 6.6

Conservation Equations for 3D Transient Compressible Flow in Spherical Coordinates

$$\frac{\partial \rho}{\partial t} + \frac{1}{r^2}\frac{\partial}{\partial r}\left(\rho r^2 u_r\right) + \frac{1}{r\sin\theta}\frac{\partial}{\partial\theta}(\rho u_\theta \sin\theta) + \frac{1}{r\sin\theta}\frac{\partial}{\partial\phi}(\rho u_\phi) = 0$$

$$\rho\left(\frac{\partial u_r}{\partial t} + u_r\frac{\partial u_r}{\partial r} + \frac{u_\theta}{r}\frac{\partial u_r}{\partial\theta} + \frac{u_\phi}{r\sin\theta}\frac{\partial u_r}{\partial\phi} - \frac{u_\theta^2 + u_\phi^2}{r}\right)$$

$$= -\frac{\partial P}{\partial r} + \frac{1}{r^2}\frac{\partial\left(r^2\tau_{rr}\right)}{\partial r} + \frac{1}{r\sin\theta}\frac{\partial(\tau_{r\theta}\sin\theta)}{\partial\theta} + \frac{1}{r\sin\theta}\frac{\partial(\tau_{r\phi})}{\partial\phi} - \frac{\tau_{\theta\theta} + \tau_{\phi\phi}}{r} + \rho g_r$$

$$\rho\left(\frac{\partial u_\theta}{\partial t} + u_r\frac{\partial u_\theta}{\partial r} + \frac{u_\theta}{r}\frac{\partial u_\theta}{\partial\theta} + \frac{u_\phi}{r\sin\theta}\frac{\partial u_\theta}{\partial\phi} + \frac{u_r u_\theta}{r} - \frac{u_\phi^2\cot\theta}{r}\right)$$

$$= -\frac{1}{r}\frac{\partial P}{\partial\theta} + \frac{1}{r^2}\frac{\partial\left(r^2\tau_{r\theta}\right)}{\partial r} + \frac{1}{r\sin\theta}\frac{\partial(\tau_{\theta\theta}\sin\theta)}{\partial\theta} + \frac{1}{r\sin\theta}\frac{\partial(\tau_{\theta\phi})}{\partial\phi} + \frac{\tau_{r\theta}}{r} - \frac{\tau_{\phi\phi}\cot\theta}{r} + \rho g_\theta$$

$$\rho\left(\frac{\partial u_\phi}{\partial t} + u_r\frac{\partial u_\phi}{\partial r} + \frac{u_\theta}{r}\frac{\partial u_\phi}{\partial\theta} + \frac{u_\phi}{r\sin\theta}\frac{\partial u_\phi}{\partial\phi} + \frac{u_r u_\phi}{r} + \frac{u_\theta u_\phi\cot\theta}{r}\right)$$

$$= -\frac{1}{r\sin\theta}\frac{\partial P}{\partial\phi} + \frac{1}{r^2}\frac{\partial\left(r^2\tau_{r\phi}\right)}{\partial r} + \frac{1}{r}\frac{\partial(\tau_{\theta\phi})}{\partial\theta} + \frac{1}{r\sin\theta}\frac{\partial(\tau_{\phi\phi})}{\partial\phi} + \frac{\tau_{r\phi}}{r} + \frac{2\tau_{\phi\theta}\cot\theta}{r} + \rho g_\phi$$

$$\rho c_v\left(\frac{\partial T}{\partial t} + u_r\frac{\partial T}{\partial r} + \frac{u_\theta}{r}\frac{\partial T}{\partial\theta} + \frac{u_\phi}{r\sin\theta}\frac{\partial T}{\partial\phi}\right) = -\left[\frac{1}{r^2}\frac{\partial\left(r^2 q_r''\right)}{\partial r} + \frac{1}{r\sin\theta}\frac{\partial(q_\theta''\sin\theta)}{\partial\theta} + \frac{1}{r\sin\theta}\frac{\partial q_\phi''}{\partial\phi}\right]$$

$$-T\left(\frac{\partial P}{\partial T}\right)_\rho\left(\frac{1}{r^2}\frac{\partial(r^2 u_r)}{\partial r} + \frac{1}{r\sin\theta}\frac{\partial(u_\theta\sin\theta)}{\partial\theta} + \frac{1}{r\sin\theta}\frac{\partial u_\phi}{\partial\phi}\right) + \Phi + \dot{q}$$

and the momentum equations become

$$\frac{\partial}{\partial t}(\rho u_i) + u_k\frac{\partial}{\partial x_k}(\rho u_i) = -\frac{\partial p}{\partial x_i} + \frac{\partial}{\partial x_k}\left(\mu\frac{\partial}{\partial x_k}(\rho u_i)\right) \tag{6.93}$$

The indices (i, j, k) appearing as subscripts assume the values of 1, 2, and 3, indicating the three coordinates. For example, $(x_1, x_2, x_3) = (x, y, z)$ and $u_1, u_2,$ and u_3 represent the velocity component in the $x, y,$ and z directions normally designated as $u, v,$ and w. If an index is repeated in a product, summation is implied over the repeated index. For example, $u_k\frac{\partial}{\partial x_k}(\rho u_i) = u\frac{\partial}{\partial x}(\rho u_i) + v\frac{\partial}{\partial y}(\rho u_i) + w\frac{\partial}{\partial z}(\rho u_i)$ where u_i stands for either of the three velocity components.

6.6 Streamlines and Stream Function

Fluid flow problems can be complex and visualizing the flow field is a useful tool for better understanding of the phenomenon. We are intuitively used to visualizing the fluid flow; leaves flowing over the surface of water, smoke coming out of a chimney, air, and vapor bubbles rising during boiling, sand, dust, or debris pickup by air or tornadoes are examples

TABLE 6.7

Stress and Strain Rate Relations and Viscous Dissipation for Newtonian Fluids in Spherical Coordinates

$$\tau_{rr} = \mu\left(2\frac{\partial u_r}{\partial r} - \frac{2}{3}\nabla\cdot\vec{V}\right)$$

$$q_r = -k\frac{\partial T}{\partial r}$$

$$\tau_{\theta\theta} = \mu\left[2\left(\frac{1}{r}\frac{\partial u_\theta}{\partial\theta} + \frac{u_r}{r}\right) - \frac{2}{3}\nabla\cdot\vec{V}\right]$$

$$q_\phi = -k\frac{1}{r}\frac{\partial T}{\partial\theta}$$

$$\tau_{\phi\phi} = \mu\left[2\left(\frac{1}{r\sin\theta}\frac{\partial u_\phi}{\partial\phi} + \frac{u_r}{r} + \frac{u_\theta\cot\theta}{r}\right) - \frac{2}{3}\nabla\cdot\vec{V}\right]$$

$$q_z = -k\frac{1}{r\sin\theta}\frac{\partial T}{\partial\phi}$$

$$\tau_{r\theta} = \tau_{\theta r} = \mu\left[r\frac{\partial}{\partial r}\left(\frac{u_\theta}{r}\right) + \frac{1}{r}\frac{\partial u_r}{\partial\theta}\right]$$

$$\tau_{\theta\phi} = \tau_{\phi\theta} = \mu\left[\frac{\sin\theta}{r}\frac{\partial}{\partial z}\left(\frac{u_\phi}{\sin\theta}\right) + \frac{1}{r\sin\theta}\frac{\partial u_\theta}{\partial\phi}\right]$$

$$\tau_{r\phi} = \tau_{\phi r} = \mu\left[\frac{1}{r\sin\theta}\frac{\partial u_r}{\partial\phi} + r\frac{\partial}{\partial r}\left(\frac{u_\phi}{r}\right)\right]$$

$$\Phi = \tau_{rr}\frac{\partial u_r}{\partial r} + \tau_{\phi\phi}\frac{1}{r}\left(\frac{\partial u_\phi}{\partial\phi} + \frac{u_r}{r}\right) + \tau_{\theta\theta}\left(\frac{1}{r}\frac{\partial u_\theta}{\partial\theta} + \frac{u_r}{r}\right) + \tau_{\phi\phi}\left(\frac{1}{r\sin\theta}\frac{\partial u_\phi}{\partial\phi} + \frac{u_r}{r} + \frac{u_\theta\cot\theta}{r}\right)$$

$$+ \tau_{r\theta}\left(\frac{\partial u_\theta}{\partial r} + \frac{1}{r}\frac{\partial u_r}{\partial\theta} - \frac{u_\theta}{r}\right) + \tau_{r\phi}\left[\frac{\partial u_\phi}{\partial r} + \frac{1}{r\sin\theta}\frac{\partial u_r}{\partial\phi} - \frac{u_\phi}{r}\right] + \tau_{\theta\phi}\left(\frac{1}{r}\frac{\partial u_\phi}{\partial\theta} + \frac{1}{r\sin\theta}\frac{\partial u_\theta}{\partial\phi} - \frac{u_\phi\cot\theta}{r}\right)$$

TABLE 6.8

Incompressible, Constant Property, Flow in Cartesian Coordinates

$$\frac{\partial u}{\partial x} + \frac{\partial v}{\partial y} + \frac{\partial w}{\partial z} = 0$$

$$\rho\left(\frac{\partial u}{\partial t} + u\frac{\partial u}{\partial x} + v\frac{\partial u}{\partial y} + w\frac{\partial u}{\partial z}\right) = -\frac{\partial P}{\partial x} + \mu\left[\frac{\partial^2 u}{\partial x^2} + \frac{\partial^2 u}{\partial y^2} + \frac{\partial^2 u}{\partial x^2}\right] + \rho g_x$$

$$\rho\left(\frac{\partial v}{\partial t} + u\frac{\partial v}{\partial x} + v\frac{\partial v}{\partial y} + w\frac{\partial v}{\partial z}\right) = -\frac{\partial P}{\partial y} + \mu\left[\frac{\partial^2 v}{\partial x^2} + \frac{\partial^2 v}{\partial y^2} + \frac{\partial^2 v}{\partial x^2}\right] + \rho g_y$$

$$\rho\left(\frac{\partial w}{\partial t} + u\frac{\partial w}{\partial x} + v\frac{\partial w}{\partial y} + w\frac{\partial w}{\partial z}\right) = -\frac{\partial P}{\partial z} + \mu\left[\frac{\partial^2 w}{\partial x^2} + \frac{\partial^2 w}{\partial y^2} + \frac{\partial^2 w}{\partial x^2}\right] + \rho g_z$$

$$\rho c_p\left(\frac{\partial T}{\partial t} + u\frac{\partial T}{\partial x} + v\frac{\partial T}{\partial y} + w\frac{\partial T}{\partial z}\right) = k\left[\frac{\partial^2 T}{\partial x^2} + \frac{\partial^2 T}{\partial y^2} + \frac{\partial^2 T}{\partial z^2}\right] + \Phi + \dot{q}$$

where $\Phi = \mu\left[2\left(\frac{\partial u}{\partial x}\right)^2 + 2\left(\frac{\partial v}{\partial y}\right)^2 + 2\left(\frac{\partial w}{\partial z}\right)^2 + \left(\frac{\partial u}{\partial y} + \frac{\partial v}{\partial x}\right)^2 + \left(\frac{\partial u}{\partial z} + \frac{\partial w}{\partial x}\right)^2 + \left(\frac{\partial v}{\partial z} + \frac{\partial w}{\partial y}\right)^2\right]$

TABLE 6.9

Incompressible, Constant Property, Flow in Cylindrical Coordinates

$$\frac{\partial u_z}{\partial z} + \frac{1}{r}\frac{\partial(ru_r)}{\partial r} + \frac{1}{r}\frac{\partial(u_\phi)}{\partial \phi} = 0$$

$$\rho\left(\frac{\partial u_z}{\partial t} + u_r\frac{\partial u_z}{\partial r} + \frac{u_\phi}{r}\frac{\partial u_z}{\partial \phi} + u_z\frac{\partial u_z}{\partial z}\right) = -\frac{\partial P}{\partial z} + \mu\left[\frac{\partial^2 u_z}{\partial z^2} + \frac{1}{r}\frac{\partial}{\partial r}\left(r\frac{\partial u_z}{\partial r}\right) + \frac{1}{r^2}\frac{\partial^2 u_z}{\partial \phi^2}\right] + \rho g_z$$

$$\rho\left(\frac{\partial u_r}{\partial t} + u_r\frac{\partial u_r}{\partial r} + \frac{u_\phi}{r}\frac{\partial u_r}{\partial \phi} + u_z\frac{\partial u_r}{\partial z} - \frac{u_\phi^2}{r}\right) = -\frac{\partial P}{\partial r} + \mu\left[\frac{\partial^2 u_r}{\partial z^2} + \frac{1}{r}\frac{\partial}{\partial r}\left(r\frac{\partial u_r}{\partial r}\right) + \frac{1}{r^2}\frac{\partial^2 u_r}{\partial \phi^2} - \frac{u_r}{r^2} - \frac{2}{r^2}\frac{\partial u_\phi}{\partial \phi}\right] + \rho g_r$$

$$\rho\left(\frac{\partial u_\phi}{\partial t} + u_r\frac{\partial u_\phi}{\partial r} + \frac{u_\phi}{r}\frac{\partial u_\phi}{\partial \phi} + u_z\frac{\partial u_\phi}{\partial z} + \frac{u_r u_\phi}{r}\right) = -\frac{1}{r}\frac{\partial P}{\partial \phi} + \mu\left[\frac{\partial^2 u_\phi}{\partial z^2} + \frac{1}{r}\frac{\partial}{\partial r}\left(r\frac{\partial u_\phi}{\partial r}\right) + \frac{1}{r^2}\frac{\partial^2 u_\phi}{\partial \phi^2} - \frac{u_\phi}{r^2} + \frac{2}{r^2}\frac{\partial u_r}{\partial \phi}\right] + \rho g_\phi$$

$$\rho c_p\left(\frac{\partial T}{\partial t} + u_r\frac{\partial T}{\partial r} + \frac{u_\phi}{r}\frac{\partial T}{\partial \phi} + u_z\frac{\partial T}{\partial z}\right) = k\left[\frac{1}{r}\frac{\partial}{\partial r}\left(r\frac{\partial T}{\partial r}\right) + \frac{1}{r^2}\frac{\partial^2 T}{\partial \phi^2} + \frac{\partial^2 T}{\partial z^2}\right] + \Phi + \dot{q}$$

where $\Phi = \mu\left[2\left(\frac{\partial u_r}{\partial r}\right)^2 + 2\frac{1}{r^2}\left(\frac{\partial u_\phi}{\partial \phi} + v_r\right)^2 + 2\left(\frac{\partial u_z}{\partial z}\right)^2 + \left(\frac{\partial u_\phi}{\partial z} + \frac{1}{r}\frac{\partial u_z}{\partial \phi}\right)^2 + \left(\frac{\partial u_z}{\partial r} + \frac{\partial u_r}{\partial z}\right)^2 + \left(\frac{1}{r}\frac{\partial u_r}{\partial \phi} + r\frac{\partial}{\partial r}\left(\frac{u_\phi}{r}\right)\right)^2\right]$

TABLE 6.10

Incompressible, Constant Property, Flow in Spherical Coordinates

$$\frac{1}{r^2}\frac{\partial}{\partial r}\left(r^2 u_r\right) + \frac{1}{r\sin\theta}\frac{\partial}{\partial \theta}(u_\theta\sin\theta) + \frac{1}{r\sin\theta}\frac{\partial}{\partial \phi}(u_\phi) = 0$$

$$\rho\left(\frac{\partial u_r}{\partial t} + u_r\frac{\partial u_r}{\partial r} + \frac{u_\theta}{r}\frac{\partial u_r}{\partial \theta} + \frac{u_\phi}{r\sin\theta}\frac{\partial u_r}{\partial \phi} - \frac{u_\theta^2 + u_\phi^2}{r}\right)$$

$$= -\frac{\partial P}{\partial r} + \frac{1}{r^2}\frac{\partial\left(r^2\tau_{rr}\right)}{\partial r} + \frac{1}{r\sin\theta}\frac{\partial(\tau_{r\theta}\sin\theta)}{\partial \theta} + \frac{1}{r\sin\theta}\frac{\partial(\tau_{r\phi})}{\partial \phi} - \frac{\tau_{\theta\theta} + \tau_{\phi\phi}}{r} + \rho g_r$$

$$\rho\left(\frac{\partial u_\theta}{\partial t} + u_r\frac{\partial u_\theta}{\partial r} + \frac{u_\theta}{r}\frac{\partial u_\theta}{\partial \theta} + \frac{u_\phi}{r\sin\theta}\frac{\partial u_\theta}{\partial \phi} + \frac{u_r u_\theta}{r} - \frac{u_\phi^2\cot\theta}{r}\right)$$

$$= -\frac{1}{r}\frac{\partial P}{\partial \theta} + \frac{1}{r^2}\frac{\partial\left(r^2\tau_{r\theta}\right)}{\partial r} + \frac{1}{r\sin\theta}\frac{\partial(\tau_{\theta\theta}\sin\theta)}{\partial \theta} + \frac{1}{r\sin\theta}\frac{\partial(\tau_{\theta\phi})}{\partial \phi} + \frac{\tau_{r\theta}}{r} - \frac{\tau_{\phi\phi}\cot\theta}{r} + \rho g_\theta$$

$$\rho\left(\frac{\partial u_\phi}{\partial t} + u_r\frac{\partial u_\phi}{\partial r} + \frac{u_\theta}{r}\frac{\partial u_\phi}{\partial \theta} + \frac{u_\phi}{r\sin\theta}\frac{\partial u_\phi}{\partial \phi} + \frac{u_r u_\phi}{r} + \frac{u_\theta u_\phi\cot\theta}{r}\right)$$

$$= -\frac{1}{r\sin\theta}\frac{\partial P}{\partial \phi} + \frac{1}{r^2}\frac{\partial\left(r^2\tau_{r\phi}\right)}{\partial r} + \frac{1}{r}\frac{\partial(\tau_{\theta\phi})}{\partial \theta} + \frac{1}{r\sin\theta}\frac{\partial(\tau_{\phi\phi})}{\partial \phi} + \frac{\tau_{r\phi}}{r} + \frac{2\tau_{\theta\phi}\cot\theta}{r} + \rho g_\phi$$

$$\rho c_p\left(\frac{\partial T}{\partial t} + u_r\frac{\partial T}{\partial r} + \frac{u_\theta}{r}\frac{\partial T}{\partial \theta} + \frac{u_\phi}{r\sin\theta}\frac{\partial T}{\partial \phi}\right) = k\left[\frac{1}{r^2}\frac{\partial}{\partial r}\left(r^2\frac{\partial T}{\partial r}\right) + \frac{1}{r^2\sin\theta}\frac{\partial}{\partial \theta}\left(\sin\theta\frac{\partial T}{\partial \theta}\right) + \frac{1}{r^2\sin^2\theta}\frac{\partial^2 T}{\partial \phi^2}\right] + \Phi + \dot{q}$$

where $\Phi = \mu\left[2\left(\frac{\partial u_r}{\partial r}\right)^2 + 2\left(\frac{1}{r}\frac{\partial u_\theta}{\partial \theta} + \frac{u_r}{r}\right)^2 + 2\left(\frac{1}{r\sin\theta}\frac{\partial u_\phi}{\partial \phi} + \frac{u_r}{r} + \frac{u_\theta\cot\theta}{r}\right)^2 + \left(r\frac{\partial}{\partial r}\left(\frac{u_\theta}{r}\right) + \frac{1}{r}\frac{\partial u_r}{\partial \theta}\right)^2 \\ + \left(\frac{1}{r\sin\theta}\frac{\partial u_r}{\partial \phi} + r\frac{\partial}{\partial r}\left(\frac{u_\phi}{r}\right)\right)^2 + \left(\frac{\sin\theta}{r}\frac{\partial}{\partial \theta}\left(\frac{u_\phi}{\sin\theta}\right) + \frac{1}{r\sin\theta}\frac{\partial u_\theta}{\partial \phi}\right)^2\right]$

TABLE 6.11

Governing Equations for Steady, 2D Incompressible, Constant Property Flow

Cartesian Coordinates	Cylindrical Coordinates
$\dfrac{\partial u}{\partial x}+\dfrac{\partial v}{\partial y}=0$	$\dfrac{\partial u_z}{\partial z}+\dfrac{1}{r}\dfrac{\partial(ru_r)}{\partial r}=0$
$\rho\left(u\dfrac{\partial u}{\partial x}+v\dfrac{\partial u}{\partial y}\right)=-\dfrac{\partial P}{\partial x}+\mu\left[\dfrac{\partial^2 u}{\partial x^2}+\dfrac{\partial^2 u}{\partial y^2}\right]+\rho g_x$	$\rho\left(u_z\dfrac{\partial u_z}{\partial z}+u_r\dfrac{\partial u_z}{\partial r}\right)=-\dfrac{\partial P}{\partial z}+\mu\left[\dfrac{\partial^2 u_z}{\partial z^2}+\dfrac{1}{r}\dfrac{\partial}{\partial r}\left(r\dfrac{\partial u_z}{\partial r}\right)\right]+\rho g_z$
$\rho\left(u\dfrac{\partial v}{\partial x}+v\dfrac{\partial v}{\partial y}\right)=-\dfrac{\partial P}{\partial y}+\mu\left[\dfrac{\partial^2 v}{\partial x^2}+\dfrac{\partial^2 v}{\partial y^2}\right]+\rho g_y$	$\rho\left(u_z\dfrac{\partial u_r}{\partial z}+u_r\dfrac{\partial u_r}{\partial r}\right)=-\dfrac{\partial P}{\partial r}+\mu\left[\dfrac{\partial^2 u_r}{\partial z^2}+\dfrac{1}{r}\dfrac{\partial}{\partial r}\left(r\dfrac{\partial u_r}{\partial r}\right)-\dfrac{u_r}{r^2}\right]+\rho g_r$
$\rho c_p\left(u\dfrac{\partial T}{\partial x}+v\dfrac{\partial T}{\partial y}\right)=k\left[\dfrac{\partial^2 T}{\partial x^2}+\dfrac{\partial^2 T}{\partial y^2}\right]$	$\rho c_p\left(u_z\dfrac{\partial T}{\partial z}+u_r\dfrac{\partial T}{\partial r}\right)=k\left[\dfrac{\partial^2 T}{\partial z^2}+\dfrac{1}{r}\dfrac{\partial}{\partial r}\left(r\dfrac{\partial T}{\partial r}\right)\right]$

of how flows are visualized. These are more qualitative measures, and to improve our understanding, we need to define more quantitative ones.

A very useful concept is that of streamlines for 2D flows or stream surfaces for 3D ones. The concepts can be applied to steady or unsteady, compressible or incompressible flows. Streamlines are curves that at a given point in time are tangent to the velocity vector. Streamlines provide a snapshot of the flow field, giving a clear picture of how each fluid particle is flowing at that instant. The fluid's velocity vector is tangent to the streamline, or the normal component of velocity on a streamline is zero, or no mass crosses a streamline; therefore, the amount of mass between two streamlines is constant. In the regions where the streamlines are close to each other, the velocity is high and in the areas where the streamlines diverge, the velocity slows down. Also, the normal (as well as tangential) components of velocity is zero on a solid surface, or no mass crosses a solid surface placed in a flow field, thus the surface of the object is a streamline.

A related concept to streamline is the stream function which for a 2D flow is defined by a function $\psi(x, y)$ such that its partial derivatives are the fluid's velocity components,

$$u = \frac{\partial \psi}{\partial y}, \quad v = -\frac{\partial \psi}{\partial x} \tag{6.94}$$

in Cartesian coordinates. The velocity field can be written as

$$\vec{V} = u\hat{i} + v\hat{j} = \frac{\partial \psi}{\partial y}\hat{i} - \frac{\partial \psi}{\partial x}\hat{j} \tag{6.95}$$

For 2D, incompressible flow, from the law of conservation of mass

$$\frac{\partial u}{\partial x} + \frac{\partial v}{\partial y} = 0 \tag{6.96}$$

and if we substitute for velocities in terms of the stream function

$$\frac{\partial}{\partial x}\left(\frac{\partial \psi}{\partial y}\right)+\frac{\partial}{\partial y}\left(-\frac{\partial \psi}{\partial x}\right)=0 \tag{6.97}$$

then the continuity equation is satisfied. Since stream function is a function of two variables, then the differential of ψ

$$d\psi = \frac{\partial \psi}{\partial x}dx+\frac{\partial \psi}{\partial y}dy \tag{6.98}$$

and from definition, the differential of the stream function becomes

$$d\psi = -vdx+udy \tag{6.99}$$

Along a line of constant stream function,

$$d\psi = -vdx+udy = 0 \tag{6.100}$$

or

$$\frac{dy}{dx}=\frac{v}{u} \tag{6.101}$$

The term $\dfrac{dy}{dx}$ is the slope of tangent along a constant ψ line. Next, consider a streamline shown in Figure 6.7. From the definition of streamline, the slope of velocity vector at any point along a streamline is also $\dfrac{v}{u}$. Therefore, the streamlines and lines of constant stream function have the same slope (Equation 6.101), and therefore the lines of constant stream function are streamlines, or streamlines are contours of stream function. The difference between values of two streamlines is proportional to the mass flow rate. The streamline passing through a solid surface is generally assigned the arbitrary value of zero.

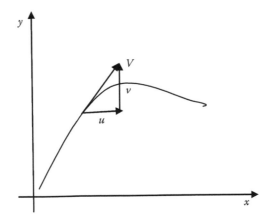

FIGURE 6.7
Relation between velocity, its components, and streamlines.

6.7 Nondimensionalization

Consider 2D transient, incompressible flow of a fluid at T_∞ and u_∞, over an object with characteristic length L, maintained at T_w. If we define the nondimensional variables

$$t^* = \frac{t}{t_s}, \quad x^* = \frac{x}{L}, \quad y^* = \frac{y}{L}, \quad u^* = \frac{u}{u_\infty}, \quad v^* = \frac{v}{u_\infty}, \quad P^* = \frac{P - P_{ref}}{P_{scale}}, \quad T^* = \frac{T - T_\infty}{T_w - T_\infty} \qquad (6.102)$$

Then the continuity equation in terms of nondimensional variables becomes

$$\frac{u_\infty}{L}\frac{\partial u^*}{\partial x^*} + \frac{u_\infty}{L}\frac{\partial v^*}{\partial y^*} = 0 \qquad (6.103)$$

which simplifies to

$$\frac{\partial u^*}{\partial x^*} + \frac{\partial v^*}{\partial y^*} = 0 \qquad (6.104)$$

The conservation of momentum equation in the x direction is

$$\rho\left(\frac{\partial u}{\partial t} + u\frac{\partial u}{\partial x} + v\frac{\partial u}{\partial y}\right) = -\frac{\partial P}{\partial x} + \mu\left[\frac{\partial^2 u}{\partial x^2} + \frac{\partial^2 u}{\partial y^2}\right] + \rho g_x \qquad (6.105)$$

which in terms of nondimensional variables is

$$\rho\left(\frac{u_\infty}{t_s}\frac{\partial u^*}{\partial t^*} + u_\infty u^*\frac{u_\infty}{L}\frac{\partial u^*}{\partial x^*} + u_\infty v^*\frac{u_\infty}{L}\frac{\partial u^*}{\partial y^*}\right) = -\frac{P_{scale}}{L}\frac{\partial P^*}{\partial x^*} + \mu\left(\frac{u_\infty}{L^2}\frac{\partial^2 u^*}{\partial x^{*2}} + \frac{u_\infty}{L^2}\frac{\partial^2 u^*}{\partial y^{*2}}\right) + \rho g_x$$

$$(6.106)$$

and simplifies to

$$\left(\frac{L}{u_\infty t_s}\frac{\partial u^*}{\partial t^*} + u^*\frac{\partial u^*}{\partial x^*} + v^*\frac{\partial u^*}{\partial y^*}\right) = -\frac{P_{scale}}{\rho u_\infty^2}\frac{\partial P^*}{\partial x^*} + \frac{\mu}{\rho u_\infty L}\left(\frac{\partial^2 u^*}{\partial x^{*2}} + \frac{\partial^2 u^*}{\partial y^{*2}}\right) + \frac{g_x}{u_\infty^2}L \qquad (6.107)$$

The scales are arbitrary, as long as they are constant and have dimensions the same as the variable that they are used to scale; therefore, the time and pressure scales can be set to

$$t_s = \frac{L}{u_\infty} \qquad (6.108)$$

and

$$P_{scale} = \rho u_\infty^2 \qquad (6.109)$$

which meet both requirements, and also simplify the equation greatly. The nondimensional group

$$Re = \frac{\rho u_\infty L}{\mu} \tag{6.110}$$

is the Reynolds number. Also the dimensionless group

$$Fr = \frac{u_\infty^2}{gL} \tag{6.111}$$

is the Froude number, which is a measure of inertial force to gravitational force. The nondimensional form of the momentum equation becomes

$$\frac{\partial u^*}{\partial t^*} + u^* \frac{\partial u^*}{\partial x^*} + v^* \frac{\partial u^*}{\partial y^*} = -\frac{\partial P^*}{\partial x^*} + \frac{1}{Re}\left(\frac{\partial^2 u^*}{\partial x^{*2}} + \frac{\partial^2 u^*}{\partial y^{*2}} \right) + \frac{1}{Fr} \tag{6.112}$$

The nondimensional y momentum equation will also have the same form. Consider laminar flow of water with a velocity of 0.1 m/s in a 1 cm circular pipe ($Re = 1800$). If the pipe is vertical, then the Froude number for this case

$$Fr = \frac{0.1^2}{9.81 \times 0.01} = 0.1$$

Or the gravitational effects may have to be considered in this case since most of the other terms are of the order of one. For $U = 5$ m/s ($Re = 91,400$). The flow becomes turbulent and $Fr = 255$, which means the gravitational effects will be negligible.

The energy equation is

$$\rho c_p \left(\frac{\partial T}{\partial t} + u \frac{\partial T}{\partial x} + v \frac{\partial T}{\partial y} \right) = k\left[\frac{\partial^2 T}{\partial x^2} + \frac{\partial^2 T}{\partial y^2} \right] + \mu\left[2\left(\frac{\partial u}{\partial x} \right)^2 + 2\left(\frac{\partial v}{\partial y} \right)^2 + \left(\frac{\partial u}{\partial y} + \frac{\partial v}{\partial x} \right)^2 \right] \tag{6.113}$$

which can be nondimensionalized

$$\rho c_p \left(\frac{\frac{T_w - T_\infty}{L}}{u_\infty} \frac{\partial T^*}{\partial t^*} + u_\infty u^* \frac{T_w - T_\infty}{L} \frac{\partial T^*}{\partial x^*} + u_\infty v^* \frac{T_\infty - T_w}{L} \frac{\partial T^*}{\partial y^*} \right)$$

$$= k\left(\frac{T_w - T_\infty}{L^2} \frac{\partial^2 T^*}{\partial x^{*2}} + \frac{T_w - T_\infty}{L^2} \frac{\partial^2 T^*}{\partial y^{*2}} \right) + \mu\left(\frac{u_\infty}{L} \right)^2 \left[2\left(\frac{\partial u^*}{\partial x^*} \right)^2 + 2\left(\frac{\partial v^*}{\partial y^*} \right)^2 + \left(\frac{\partial u^*}{\partial y^*} + \frac{\partial v^*}{\partial x^*} \right)^2 \right] \tag{6.114}$$

and simplifies to

$$\frac{\partial T^*}{\partial t^*} + u^* \frac{\partial T^*}{\partial x^*} + v^* \frac{\partial T^*}{\partial y^*}$$

$$= \frac{1}{RePr}\left\{\left(\frac{\partial^2 T^*}{\partial x^{*2}} + \frac{\partial^2 T^*}{\partial y^{*2}}\right) + \frac{\mu u_\infty^2}{k(T_w - T_\infty)}\left[2\left(\frac{\partial u^*}{\partial x^*}\right)^2 + 2\left(\frac{\partial v^*}{\partial y^*}\right)^2 + \left(\frac{\partial u^*}{\partial y^*} + \frac{\partial v^*}{\partial x^*}\right)^2\right]\right\} \quad (6.115)$$

The dimensionless group

$$Br = \frac{\mu u_\infty^2}{k(T_w - T_\infty)} \quad (6.116)$$

is the Brinkman number which is a measure of viscous dissipation to convection. For most fluids at moderate velocities, the viscous dissipation is generally not significant. For example for water flowing at 25°C with a velocity of 5 m/s over a surface maintained at 75°C, the Brinkman number becomes

$$Br = \frac{5.47 \times 10^{-4} \times 5^2}{0.644 \times 50} = 4.25 \times 10^{-4}$$

which is much smaller than one, which is of the same order of magnitude as the other terms in the energy equation, if important. Under the same conditions, Brinkman number becomes 0.5 for engine oil. Therefore, viscous heating may have to be considered for highly viscous fluids or high speed flows.

Problems

6.1 Starting with a differential control volume, derive the continuity equation for steady 2D flow in cylindrical coordinates.

6.2 Starting with a differential control volume, derive the continuity equation for steady 2D (no change in θ direction) flow in spherical coordinate.

6.3 Show that

$$\nabla \cdot (\rho \vec{V}) = \vec{V} \cdot \nabla\rho + \rho \nabla \cdot \vec{V}$$

6.4 Starting with a differential control volume, derive the continuity equation for steady 2D (no change in z direction) flow in cylindrical coordinates.

6.5 Starting with a differential control volume, derive the conservation of momentum equation in the r direction for steady 2D flow in cylindrical coordinates.

6.6 Simplify the continuity equation for flow between two infinitely long parallel plates, with the upper plate moving with a constant velocity V (Couette flow) and both plates are isothermal. Include the effect of viscous heating and come up with the boundary conditions.

6.7 Write the governing equations and the boundary conditions for cross flow over a long cylinder by simplifying the NS equations.

6.8 Simplify the NS and energy equation for flow between two infinitely long parallel plates, with the upper plate moving with a constant velocity V (Couette flow) and both plates are isothermal. Include the effect of viscous heating and come up with the boundary conditions.

6.9 Nondimensionalize the governing equations and boundary conditions for Problem 6.8.

6.10 Simplify the NS equation for flow between two infinitely long parallel plates, with the upper suddenly starts moving with a constant velocity V (unsteady Couette flow). List the boundary conditions.

6.11 Simplify the NS equation and provide the boundary conditions for temperature distribution in a fluid in contact with a long cylinder that is at a slightly higher temperature than the fluid, so that natural convection effects can be neglected.

6.12 Write the governing equations and the boundary conditions for cross flow over two long parallel cylinders by simplifying the NS equations.

6.13 Write the governing equations and the boundary conditions for a jet exiting a round nozzle into a quiescent fluid by simplifying the NS equations. Nondimensionalize the equations and boundary conditions.

6.14 Write the governing equations and the boundary conditions for a jet exiting a round nozzle into a quiescent fluid by simplifying the NS equations.

6.15 Write the governing equations and the boundary conditions for flow over a square cylinder.

6.16 Give the stream function for steady 2D incompressible axisymmetric flow.

6.17 Give the stream function for steady 2D compressible axisymmetric flow.

Reference

1. Bird, R. B., Stewart, W. E., and Lightfoot, E. N. (1960) *Transport Phenomena*. New York: Wiley.

7

External Flow

7.1 Boundary Layer Flows

External flows are those where fluid is not confined on all sides by a solid boundary. Examples include flow over an airplane or exiting its engines. In general, the external flow problems of practical interest are 3D and generally turbulent, requiring a combination of experimental and numerical analysis. Studying problems that focus on the basic characteristics of external flows provides great insights into their behavior and those of more realistic problems. In this chapter, a variety of basic external flow problems are considered, and analytical and numerical solutions are provided to gain an understanding of the basic features of external flows.

A large class of external flow problems falls under the category of boundary layer flows. Boundary layer flows are those where the fluid predominantly moves in one direction over a solid surface, as shown in Figure 7.1, and one of the velocity components is much larger than the others. For a boundary layer flow, the impact of the surface on the flow is confined to a narrow region near the surface, which becomes the focus of the study, instead of the entire flow field. When a fluid flows over a surface, the fluid that is in contact with the solid sticks to the surface, assuming its velocity. The fluid velocity changes with the distance from the surface, and the fluid layer across which the velocity increases from zero to its free stream value is called the boundary layer. The boundary layer thickness, δ, increases in the direction of flow. The velocity change results in the momentum change which requires a force. In boundary layer flows, viscous forces are responsible for the momentum change, and therefore, the boundary layer is also the region where the viscous forces are important. Outside of the boundary layer, the normal velocity gradient, and therefore the shear stress, is zero.

In boundary layer flows, the flow is divided into two regions: a narrow region near the solid where the viscous forces are important and the rest of the flow where the viscous forces are negligible. The velocity field inside the boundary layer is obtained by simplifying the Navier–Stokes (NS) equations for boundary layer flow, and the solution outside the boundary layer is obtained by neglecting the viscous terms in the NS equations, or they are obtained from the solution of inviscid flow problem or from the potential flow theory.

The heat transfer calculations are made in a similar manner. If the wall and the fluid are at different temperatures, the change from the surface to the free stream temperature occurs over a narrow fluid layer called the thermal boundary layer. Outside the boundary layer, the temperature is known, and therefore, solution only needs to be obtained over the narrow boundary layer regions where the change in the temperature occurs.

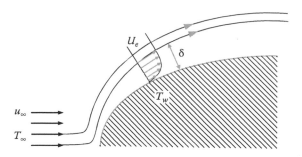

FIGURE 7.1
Boundary layer flow.

7.2 Boundary Layer Flow over a Flat Plate

Boundary layer flow over a flat plate is a classical problem used to demonstrate some of the fundamental concepts in fluid mechanics and heat transfer, and there are numerous books devoted to the subject, including the classic book by Schlichting [1]. Consider a fluid approaching a smooth flat plate with a uniform free stream velocity U_∞ and temperature T_∞ as shown in Figure 7.2. If the plate is at an angle with the free stream velocity, a portion of the flow will turn and predominantly move along the plate in the x direction. In this case, u, v, and T changes occur over a narrow region close to the plate, the boundary layer region. Outside of the boundary layer, the velocity and temperature fields are unaffected by the viscous force generated by the presence of the plate and are equal to their free stream values, obtained by neglecting viscous forces in the equations of motion. If the flow outside of the boundary layer is known, then what is left is obtaining the solution in the narrow boundary layer region. The continuity, momentum, and energy equations for steady 2D, incompressible laminar flow over an inclined flat plate reduce to

$$\frac{\partial u}{\partial x} + \frac{\partial v}{\partial y} = 0 \tag{7.1}$$

$$\rho\left(u\frac{\partial u}{\partial x} + v\frac{\partial u}{\partial y}\right) = -\frac{\partial P}{\partial x} + \mu\left(\frac{\partial^2 u}{\partial x^2} + \frac{\partial^2 u}{\partial y^2}\right) \tag{7.2}$$

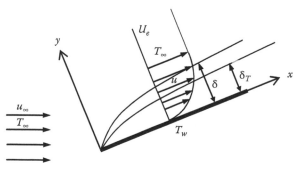

FIGURE 7.2
Boundary layer flow over an inclined flat plate.

$$\rho\left(u\frac{\partial v}{\partial x}+v\frac{\partial v}{\partial y}\right)=-\frac{\partial P}{\partial y}+\mu\left(\frac{\partial^2 v}{\partial x^2}+\frac{\partial^2 v}{\partial y^2}\right) \tag{7.3}$$

$$\rho c_p\left(u\frac{\partial T}{\partial x}+v\frac{\partial T}{\partial y}\right)=k\left(\frac{\partial^2 T}{\partial x^2}+\frac{\partial^2 T}{\partial y^2}\right) \tag{7.4}$$

$$
\begin{aligned}
&x=0, &&u=U_e, &&v=0, &&T=T_\infty \\
&y=0, &&u=0, &&v=0, &&T=T_w \\
&y\to\infty, &&u=U_e, &&v=0, &&T=T_\infty
\end{aligned} \tag{7.5}
$$

The terms on the left-hand side (LHS) of the momentum equations represent the rate of change of momentum per unit volume, and the terms on the right-hand side (RHS) are pressure and viscous forces per unit volume. Note that velocity outside of the boundary layer is $U_e(x)$ which in general can be a function of x, and u_∞ is a reference constant velocity that for the case shown in Figure 7.2 is the velocity of the fluid approaching the inclined plate.

7.2.1 Nondimensionalization

The equations are nondimensionalized first, by defining nondimensional-independent and -dependent variables as

$$x^*=\frac{x}{L},\quad y^*=\frac{y}{L},\quad u^*=\frac{u}{u_\infty},\quad v^*=\frac{v}{u_\infty},\quad P^*=\frac{P-P_{ref}}{P_{scale}},\quad T^*=\frac{T-T_w}{T_\infty-T_w} \tag{7.6}$$

Using Equation 2.6 and substituting the terms, the continuity equation can be expressed in terms of the dimensionless variables

$$\frac{u_\infty}{L}\frac{\partial u^*}{\partial x^*}+\frac{u_\infty}{L}\frac{\partial v^*}{\partial y^*}=0 \tag{7.7}$$

Dividing both sides by the constants that have dimensions simplifies the equation to

$$\frac{\partial u^*}{\partial x^*}+\frac{\partial v^*}{\partial y^*}=0 \tag{7.8}$$

Repeating the process for the x momentum equation

$$\rho\left(u_\infty u^*\frac{u_\infty}{L}\frac{\partial u^*}{\partial x^*}+u_\infty v^*\frac{u_\infty}{L}\frac{\partial u^*}{\partial y^*}\right)=-\frac{P_{scale}}{L}\frac{\partial P^*}{\partial x^*}+\mu\left(\frac{u_\infty}{L^2}\frac{\partial^2 u^*}{\partial x^{*2}}+\frac{u_\infty}{L^2}\frac{\partial^2 u^*}{\partial y^{*2}}\right) \tag{7.9}$$

and simplifying result in

$$\left(u^* \frac{\partial u^*}{\partial x^*} + v^* \frac{\partial u^*}{\partial y^*} \right) = -\frac{P_{scale}}{\rho u_\infty^2} \frac{\partial P^*}{\partial x^*} + \frac{\mu}{\rho u_\infty L} \left(\frac{\partial^2 u^*}{\partial x^{*2}} + \frac{\partial^2 u^*}{\partial y^{*2}} \right) \tag{7.10}$$

Pressure scale is arbitrary, as long as it is constant and has dimension of pressure. By setting

$$P_{scale} = \rho u_\infty^2 \tag{7.11}$$

which meets both requirements, we eliminate two parameters, making the dimensionless pressure as

$$P^* = \frac{P - P_{ref}}{\rho u_\infty^2} \tag{7.12}$$

and leaving only one dimensionless parameter, Reynolds number, defined as $Re = \frac{\rho u_\infty L}{\mu}$. The nondimensional form of the momentum equations becomes

$$u^* \frac{\partial u^*}{\partial x^*} + v^* \frac{\partial u^*}{\partial y^*} = -\frac{\partial P^*}{\partial x^*} + \frac{1}{Re} \left(\frac{\partial^2 u^*}{\partial x^{*2}} + \frac{\partial^2 u^*}{\partial y^{*2}} \right) \tag{7.13}$$

$$u^* \frac{\partial v^*}{\partial x^*} + v^* \frac{\partial v^*}{\partial y^*} = -\frac{\partial P^*}{\partial y^*} + \frac{1}{Re} \left(\frac{\partial^2 v^*}{\partial x^{*2}} + \frac{\partial^2 v^*}{\partial y^{*2}} \right) \tag{7.14}$$

Similarly, energy equation is nondimensionalized

$$\rho c_p \left(u_\infty u^* \frac{T_\infty - T_w}{L} \frac{\partial T^*}{\partial x^*} + u_\infty v^* \frac{T_\infty - T_w}{L} \frac{\partial T^*}{\partial y^*} \right) = k \left(\frac{T_\infty - T_w}{L^2} \frac{\partial^2 T^*}{\partial x^{*2}} + \frac{T_\infty - T_w}{L^2} \frac{\partial^2 T^*}{\partial y^{*2}} \right) \tag{7.15}$$

which simplifies to

$$\left(u^* \frac{\partial T^*}{\partial x^*} + v^* \frac{\partial T^*}{\partial y^*} \right) = \frac{1}{Re} \frac{1}{Pr} \left(\frac{\partial^2 T^*}{\partial x^{*2}} + \frac{\partial^2 T^*}{\partial y^{*2}} \right) \tag{7.16}$$

where Prandtl number $Pr = \frac{\nu}{\alpha} = \frac{\mu c_p}{k}$ is a property. Also, the product of Reynolds and Prandtl number is called Peclet number:

$$Pe = Re\, Pr = \frac{u_\infty L}{\nu} \frac{\nu}{\alpha} = \frac{u_\infty L}{\alpha} \tag{7.17}$$

The nondimensional equations for steady, incompressible 2D flow assuming constant properties in Cartesian coordinates become

$$\frac{\partial u^*}{\partial x^*} + \frac{\partial v^*}{\partial y^*} = 0 \tag{7.18}$$

$$u^* \frac{\partial u^*}{\partial x^*} + v^* \frac{\partial u^*}{\partial y^*} = -\frac{\partial P^*}{\partial x^*} + \frac{1}{Re}\left(\frac{\partial^2 u^*}{\partial x^{*2}} + \frac{\partial^2 u^*}{\partial y^{*2}} \right) \tag{7.19}$$

$$u^* \frac{\partial v^*}{\partial x^*} + v^* \frac{\partial v^*}{\partial y^*} = -\frac{\partial P^*}{\partial y^*} + \frac{1}{Re}\left(\frac{\partial^2 v^*}{\partial x^{*2}} + \frac{\partial^2 v^*}{\partial y^{*2}} \right) \tag{7.20}$$

$$u^* \frac{\partial T^*}{\partial x^*} + v^* \frac{\partial T^*}{\partial y^*} = \frac{1}{Re}\frac{1}{Pr}\left(\frac{\partial^2 T^*}{\partial x^{*2}} + \frac{\partial^2 T^*}{\partial y^{*2}} \right) \tag{7.21}$$

These equations are valid for any steady 2D incompressible laminar flow, implying that the dimensionless velocity components are a function of dimensionless coordinates, x^* and y^*, and the nondimensional parameter Reynolds number, and the nondimensional temperature distribution additionally depends on the Prandtl number. There is the possibility of introducing additional dimensionless parameters through the boundary conditions.

The next step is to determine if the above four partial differential equations can be simplified more for the case of boundary layer flow, i.e., any of the equations or terms can be neglected for the case under consideration. This is done by doing an order of magnitude analysis (OMA), which is the process by which one determines how large each term in an equation is expected to be and if some terms are much smaller than others and thus can be neglected.

7.2.2 Order of Magnitude Analysis

A basic issue in analyzing any problem is determining what is important and what can be neglected. The same applies when trying to solve a differential or partial differential equation. Are all of the terms in the governing equations important all of the times or are there conditions under which some of the terms can be neglected, thus simplifying the equations? In order to answer this question, we need to be able to estimate the magnitude of the terms. OMA estimates the magnitude of each term, within a factor of 10 or one order of magnitude. The approach is explained by considering the cubic function over the specified interval:

$$y = x^3, \quad 0 \le x \le 1 \tag{7.22}$$

The average value of the function can be evaluated as

$$\bar{y} = \frac{1}{L}\int_0^L y\,dx = \frac{1}{1}\int_0^1 x^3\,dx = \left.\frac{x^4}{4}\right|_0^1 = \frac{1^4 - 0^4}{4} = 0.25 \tag{7.23}$$

The cubic function changes from 0 to 1 and has an average value of 0.25 over the interval considered. The average value of the function can also be approximated by

$$\bar{y} \approx \frac{y(L) + y(0)}{2} = 0.5 \tag{7.24}$$

therefore, we can consider this average, which although is twice the actual average, to still be an approximation to the actual average since they are of the same order of magnitude. The average value of the function is also of the same order of magnitude (a factor of 10) as its maximum value, the value at L,

$$O(y) \approx O(\bar{y}) \approx y(L) \approx 1 \tag{7.25}$$

To estimate how large the derivatives are, consider the order of magnitude of the first derivative of a function, $O\left(\dfrac{dy}{dx}\right)$. The average value of the derivative of a function over an interval is

$$\overline{\frac{dy}{dx}} = \frac{1}{L}\int_0^L \frac{dy}{dx}\,dx = \frac{1}{L}\int_0^L dy = \frac{y(L)-y(0)}{L} \tag{7.26}$$

or

$$O\left(\frac{dy}{dx}\right) = \frac{\Delta y}{\Delta x} \tag{7.27}$$

where Δy is the change in y over the domain of interest (Δx). Similarly, the order of magnitude of the second derivative is

$$O\left(\frac{d^2 y}{dx^2}\right) = \frac{\Delta y}{(\Delta x)^2} \tag{7.28}$$

and in general for the nth derivative

$$O\left(\frac{d^n y}{dx^n}\right) = \frac{\Delta y}{(\Delta x)^n} \tag{7.29}$$

7.2.2.1 Order of Magnitude Analysis for Boundary Layer Flow

The boundary layer flow is based on the assumption that there is a thin region (δ) close to the solid where the velocity and temperature changes occur or where the shear forces and heat transfer are important. This region is small compared with the characteristic size of the object or the nondimensional boundary layer thickness $\delta^* = \dfrac{\delta}{L} \ll 1$ and $\delta_T^* = \dfrac{\delta_T}{L} \ll 1$. We first do an OMA on the continuity equation

$$\frac{\partial u^*}{\partial x^*} + \frac{\partial v^*}{\partial y^*} = 0 \tag{7.30}$$

$$\frac{\Delta u^*}{\Delta x^*} + \frac{\Delta v^*}{\Delta y^*} = 0$$

$$\frac{1-0}{1-0} \qquad \frac{v^*-0}{\delta^*-0} \tag{7.31}$$

The second term has to be important; otherwise, we will be dealing with uniform flow. In order for the second term to be important, it must also be of the same order of magnitude as the first term or of the order of 1, which requires the v^* to have the same order of magnitude as the nondimensional boundary layer thickness, or

$$v^* \approx \delta^* \tag{7.32}$$

Since the boundary layer approximation is valid when $\delta^* \ll 1$, then for boundary layer flows $v^* \ll 1$. Although small, v^* cannot be neglected. Next, we perform the OMA on the x momentum equation:

$$u^* \frac{\partial u^*}{\partial x^*} + v^* \frac{\partial u^*}{\partial y^*} = -\frac{\partial P^*}{\partial x^*} + \frac{1}{Re}\left(\frac{\partial^2 u^*}{\partial x^{*2}} + \frac{\partial^2 u^*}{\partial y^{*2}} \right)$$

$$1 \quad \frac{1}{1} \quad \delta^* \frac{1}{\delta^*} \quad = \quad -\frac{\partial P^*}{\partial x^*} + \frac{1}{Re}\left(\frac{1}{1^2} \quad \frac{1}{\delta^{*2}} \right) \tag{7.33}$$

The first term in the parenthesis on the RHS is of the order of 1. Since $\delta^* \ll 1$, then $\frac{1}{\delta^*} \gg 1$, making the second term of the order of a number which is much larger than one. Therefore we can neglect the first term compared to the second term, and the x momentum equation becomes

$$u^* \frac{\partial u^*}{\partial x^*} + v^* \frac{\partial u^*}{\partial y^*} = -\frac{\partial P^*}{\partial x^*} + \frac{1}{Re}\frac{\partial^2 u^*}{\partial y^{*2}} \tag{7.34}$$

Performing an OMA on the y momentum equation,

$$u^* \frac{\partial v^*}{\partial x^*} + v^* \frac{\partial v^*}{\partial y^*} = -\frac{\partial P^*}{\partial y^*} + \frac{1}{Re}\left(\frac{\partial^2 v^*}{\partial x^{*2}} + \frac{\partial^2 v^*}{\partial y^{*2}} \right)$$

$$1 \quad \frac{\delta^*}{1} \quad \delta^* \frac{\delta^*}{\delta^*} \quad = \quad -\frac{\partial P^*}{\partial y^*} + \frac{1}{Re}\left(\frac{\delta^*}{1^2} \quad \frac{\delta^*}{\delta^{*2}} \right) \tag{7.35}$$

Again, the first term in the parentheses is much smaller than the second term and therefore can be neglected. The forces and the momentum in the y direction are of the order of $\delta^* \ll 1$, whereas those in the x momentum equation are of the order of 1, or the forces and momentum in the y direction are much smaller than those in the x direction and can be neglected. This also implies that

$$\frac{\partial P^*}{\partial y^*} \approx \delta^* \tag{7.36}$$

or the pressure variation in the y direction (across the boundary layer) is small, or pressure does not change much in the y direction and can only be a function of x.

Performing an OMA on the energy equation, assuming that the thermal boundary layer is a small region near the wall where the temperature changes occur, then

$$u^* \frac{\partial T^*}{\partial x^*} + v^* \frac{\partial T^*}{\partial y^*} = \frac{1}{RePr} \left(\frac{\partial^2 T^*}{\partial x^{*2}} + \frac{\partial^2 T^*}{\partial y^{*2}} \right)$$

(7.37)

$$1 \quad \frac{1}{1} \qquad \delta^* \frac{1}{\delta_T^*} = \frac{1}{RePr} \left(\frac{1}{1^2} \qquad \frac{1}{\delta_T^{*2}} \right)$$

and simplifies to

$$u^* \frac{\partial T^*}{\partial x^*} + v^* \frac{\partial T^*}{\partial y^*} = \frac{1}{RePr} \frac{\partial^2 T^*}{\partial y^{*2}}$$

(7.38)

Equations 7.30, 7.34, and 7.38 are three equations for four unknowns (u, v, p, T). Clearly we need one more equation. The solution inside the boundary layer is coupled with the solution outside of the boundary layer, and the two solutions must match at the edge of the boundary layer, and this requirement provides the final equation needed. The x momentum equation is valid in the boundary layer, including at the edge of the boundary layer. At the edge of the boundary layer, velocity has reached its free stream and will not change in the y direction, beyond the boundary layer's edge

$$\frac{\partial u}{\partial y} = \frac{\partial^2 u}{\partial y^2} = 0$$

(7.39)

and the x momentum equation becomes

$$U_e^* \frac{dU_e^*}{dx^*} = -\frac{dP_e^*}{dx^*}$$

(7.40)

We showed that the pressure variation across the boundary layer is negligible at a given x location implying that pressure inside and outside the boundary layer are the same or

$$P^* = P_e^*$$

(7.41)

or

$$U_e^* \frac{dU_e^*}{dx^*} = -\frac{dP_e^*}{dx^*} = -\frac{dP^*}{dx^*}$$

(7.42)

We still have three equations for four unknowns (u, v, U_e, T) and thus need to find the velocity distribution outside the boundary layer. By definition, outside the boundary layer, the viscous forces are negligible, and thus the NS equations, without the viscous terms, can be solved to obtain the velocity distribution in that region, or the solution to velocity field outside of the boundary layer comes from the inviscid or potential flow solutions.

The above simplifications are valid as long as our original assumption of a thin boundary layer is satisfied. We now need to show the circumstance under which this assumption is valid. From the OMA, we concluded that all the terms on the LHS of the x momentum equation are of the order of 1, and therefore, if the viscous term is important, then

$$\frac{1}{Re} \frac{1}{\delta^{*2}} \approx 1 \tag{7.43}$$

or

$$\delta^* \approx \frac{1}{\sqrt{Re}} \ll 1 \tag{7.44}$$

Therefore, the boundary layer approximation will be valid as long as Re is large enough to satisfy 7.44 and ensuring the boundary layer is thin. This condition is clearly met for $Re > 1000$. It turns out that the approximation is valid at Reynolds numbers as small as 1. Expressing Equation 7.44 in terms of dimensional variable and replacing L, with x, distance from the leading edge

$$\delta \approx \sqrt{\frac{vx}{U_\infty}} \tag{7.45}$$

indicating that the thickness of the hydrodynamic boundary layer is proportional to the square root of the distance from the leading edge, and kinematic viscosity, and inversely proportional to the square root of the velocity. The actual solution shows that the boundary layer thickness for flow over a flat plate is different from Equation 7.45 only by a constant multiplier, or

$$\delta = \sqrt{\frac{Cvx}{U_\infty}} \tag{7.46}$$

where C is determined from the solution. This analysis shows the power of nondimensionalization, OMA, and insight that combined lead to reasonable estimate of the expected results, without solving any equations.

The NS equations for the boundary layer flow reduce to

$$\frac{\partial u^*}{\partial x^*} + \frac{\partial v^*}{\partial y^*} = 0 \tag{7.47}$$

$$u^* \frac{\partial u^*}{\partial x^*} + v^* \frac{\partial u^*}{\partial y^*} = U_e^* \frac{dU_e^*}{dx^*} + \frac{1}{Re} \frac{\partial^2 u^*}{\partial y^{*2}} \tag{7.48}$$

$$u^* \frac{\partial T^*}{\partial x^*} + v^* \frac{\partial T^*}{\partial y^*} = \frac{1}{RePr} \frac{\partial^2 T^*}{\partial y^{*2}} \tag{7.49}$$

This set of equations is significantly simpler than the original set, in that the use of boundary layer approximations

- Eliminates the y momentum equation.
- Means pressure variation across the boundary layer is negligible.
- Eliminates the second derivative in the axial direction, which turns the governing equations from elliptic to parabolic, requiring only one boundary condition in the axial direction.
- Results in parabolic equations (not having second-order derivative in one direction) that are simpler to solve, because their solution only depends on the solution upstream and not what happens downstream. This means that the numerical solution can proceed by marching in the x direction where the axial diffusion is negligible. The elliptic equations depend on the downstream boundary conditions, and as a result, the entire flow field must be solved simultaneously which means numerically they have to be solved iteratively. The solution only needs to be obtained over a very narrow region close to the object, the boundary layer, and not the entire flow field, as solution outside of this region is given by the potential flow solution which must be known a priori.

The boundary layer approximation can be used in cases where the flow predominantly moves in one direction, i.e., when one of the velocity components is much larger than the other ones. Obviously, anytime the boundary layer approximation can be used, it greatly simplifies arriving at a solution.

7.3 Boundary Layer Equations in Curvalinear Coordinates

For axisymmetric flows, rather than solving in the cylindrical coordinates, Boltze as indicated in Ref. [1] (p. 223) has shown that by using a coordinate system, shown in Figure 7.3, where x is measured along the curved surface and y is measured perpendicular to the surface at the current x location, the boundary layer equations reduce to

$$\frac{\partial u r^k}{\partial x} + \frac{\partial v r^k}{\partial y} = 0 \tag{7.50}$$

$$\rho\left(u\frac{\partial u}{\partial x} + v\frac{\partial u}{\partial y} \right) = -\frac{dP}{dx} + \frac{\partial \tau}{\partial y} \tag{7.51}$$

$$\tau = \mu\left(\frac{\partial u}{\partial y} \right)^n \tag{7.52}$$

$$\rho c_p\left(u\frac{\partial T}{\partial x} + v\frac{\partial T}{\partial y} \right) = k\frac{\partial^2 T}{\partial y^2} \tag{7.53}$$

where
  u and v are the velocity component parallel and perpendicular to the wall
  $k = 0$, 1, or 2 for Cartesian, cylindrical, and spherical coordinates, respectively
  $n = 1$ for a Newtonian fluid

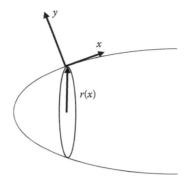

FIGURE 7.3
Axisymmetric boundary layer.

The contour of the axisymmetric object is given by $r(x)$ which is assumed to change smoothly, i.e., $\dfrac{d^2 r}{dx^2}$ is not very large. Also, for large values of r, the curvature effect can be neglected, and therefore r can be assumed not to be a strong function of x and y, and therefore both sides of the continuity equation can be divided by r, making the equation set the same as that for Cartesian coordinate system.

7.4 Boundary Layer over a Horizontal Flat Plate

For the flow of a Newtonian fluid over a horizontal flat plate, shown in Figure 7.4,

$$u_e^* = \frac{u_e}{u_\infty} = 1 \tag{7.54}$$

$$-\frac{dp^*}{dx^*} = u_e^* \frac{du_e^*}{dx^*} = 0 \tag{7.55}$$

Therefore, in terms of the nondimensional variables, the conservation equations and boundary conditions are

$$\frac{\partial u}{\partial x} + \frac{\partial v}{\partial y} = 0 \tag{7.56}$$

$$u\frac{\partial u}{\partial x} + v\frac{\partial u}{\partial y} = \nu \frac{\partial^2 u}{\partial y^2} \tag{7.57}$$

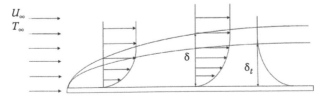

FIGURE 7.4
Boundary layer flow over a horizontal flat plate.

$$u\frac{\partial T}{\partial x} + v\frac{\partial T}{\partial y} = \alpha\frac{\partial^2 T}{\partial y^2} \tag{7.58}$$

$$x = 0, \quad u = U_\infty, \quad v = 0, \quad T = T_\infty$$

$$y = 0, \quad u = 0, \quad v = 0, \quad T = T_w \tag{7.59}$$

$$y \to \infty, \quad u = U_\infty, \quad v = 0, \quad T = T_\infty$$

The solution to these equations is next obtained using similarity method.

7.4.1 Similarity Solution

We used the idea of the similarity method to obtain the solution to transient conduction in an infinitely large medium. The velocity and temperature fields depend on two independent variables x and y. The similarity solution works since the velocity and temperature profiles appear to be geometrically similar. If we consider two different x locations along the plate, the nondimensional velocity increases from zero and reaches the value of 1 at the distance δ. The difference between the two locations is the thickness of the boundary layer. Therefore, it is plausible that if we define a new variable

$$\eta \propto \frac{y}{\delta} \tag{7.60}$$

then the velocity and temperature fields may only be a function of this single variable. This variable is a combination of our two original independent variables x and y, since δ is a function of x. From Equation 7.46, define

$$\eta = \frac{y}{\sqrt{Cvx/U_\infty}} = \sqrt{\frac{U_\infty}{Cv}}yx^{-1/2} = \sqrt{\frac{U_\infty}{Cvx}}y \tag{7.61}$$

If the similarity solution exists, then the velocity distribution must only be a function of the similarity variable. To find the similarity solution, we first eliminate the continuity equation by using the concept of the stream function.

Stream function is directly related to the concept of streamlines which are a family of curves that are instantaneously tangent to the velocity vector of the flow. Streamlines are a snapshot of the flow of the fluid that show the direction a fluid element will travel at any point in time. For 2D flows, streamlines can be defined by the contours of a two-variable function called stream function. The stream function is a function whose derivatives are related to the velocity components through

$$u = \frac{\partial \psi}{\partial y} \tag{7.62}$$

$$v = -\frac{\partial \psi}{\partial x} \tag{7.63}$$

Instead of finding velocity components, we can arrive at an equation for determining the stream function, which can then be used to determine the velocity components. Stream function automatically satisfies the continuity equation, eliminating the need for its solution. Substituting for the velocity components from the above two equations in the momentum and energy equations results in

$$\frac{\partial \psi}{\partial y}\frac{\partial^2 \psi}{\partial x \partial y} - \frac{\partial \psi}{\partial x}\frac{\partial^2 \psi}{\partial y^2} = \nu\frac{\partial^3 \psi}{\partial y^3} \tag{7.64}$$

$$\frac{\partial \psi}{\partial y}\frac{\partial T}{\partial x} - \frac{\partial \psi}{\partial x}\frac{\partial T}{\partial y} = \alpha\frac{\partial^2 T}{\partial y^2} \tag{7.65}$$

We now have two equations for two unknowns (ψ, T). The cost of eliminating the continuity equation is that the resulting equation for stream function is third order compared to the original second-order momentum equation. If the similarity solution exists, then u and T must be only a function of the similarity variable. Assume that there is a function $f(\eta)$, yet unknown, that is only a function of the single variable η, whose derivative with respect to η, $f'(\eta)$, is the nondimensional x component of velocity, i.e.,

$$u = U_\infty f'(\eta) = \frac{\partial \psi}{\partial y} \tag{7.66}$$

Integrating this equation

$$\psi = \int U_\infty f'(\eta)\partial y = \int U_\infty f'(\eta)\frac{\partial y}{\partial \eta}\partial\eta \tag{7.67}$$

$$\frac{\partial \eta}{\partial y} = \sqrt{\frac{U_\infty}{Cvx}} \tag{7.68}$$

$$\psi = \int U_\infty f'(\eta)\sqrt{\frac{Cvx}{U_\infty}}\partial\eta = \sqrt{U_\infty Cvx}\int f'(\eta)\partial\eta = \sqrt{U_\infty Cvx}\,f(\eta) \tag{7.69}$$

or

$$f(\eta) = \frac{\psi}{\sqrt{U_\infty Cvx}} \tag{7.70}$$

It can be seen that f is indeed the nondimensional stream function. Multiplying and dividing the term under the square root by the free stream velocity, from Equation 7.46, it can be seen that Equation 7.70 can be written as

$$f(\eta) = \frac{\psi}{U_\infty \delta} \tag{7.71}$$

or in general,

$$f(\eta) = \frac{\psi}{U_e \delta} \tag{7.72}$$

We now replace ψ with f and x and y with η in Equations 7.64 and 7.65. Using chain rule again,

$$\frac{\partial \eta}{\partial x} = -\frac{1}{2x}\sqrt{\frac{U_\infty}{Cvx}}y = -\frac{\eta}{2x} \tag{7.73}$$

$$\frac{\partial \psi}{\partial x} = \frac{\partial}{\partial x}\left[\sqrt{U_\infty Cvx}\,f(\eta)\right] = \frac{1}{2}\sqrt{\frac{U_\infty Cv}{x}}f(\eta) + \sqrt{U_\infty Cvx}\frac{\partial f(\eta)}{\partial x} = \frac{1}{2}\sqrt{\frac{U_\infty Cv}{x}}f(\eta) + \sqrt{U_\infty Cvx}\frac{\partial f(\eta)}{\partial \eta}\frac{\partial \eta}{\partial x} \tag{7.74}$$

$$u = \frac{\partial \psi}{\partial x} = \frac{1}{2}\sqrt{\frac{U_\infty Cv}{x}}\left[f - \eta f'\right] \tag{7.75}$$

$$v = -\frac{\partial \psi}{\partial x} = \frac{1}{2}\sqrt{\frac{U_\infty Cv}{x}}\left(\eta f'(\eta) - f(\eta)\right) \tag{7.76}$$

$$\frac{\partial^2 \psi}{\partial y^2} = \frac{\partial}{\partial y}\left[\frac{\partial \psi}{\partial y}\right] = \frac{\partial}{\partial y}\left[U_\infty f'(\eta)\right] = \frac{\partial \eta}{\partial y}\frac{\partial}{\partial \eta}\left[U_\infty f'(\eta)\right] = U_\infty f''(\eta)\sqrt{\frac{U_\infty}{Cvx}} \tag{7.77}$$

$$\frac{\partial^3 \psi}{\partial y^3} = \frac{\partial}{\partial y}\left[\frac{\partial^2 \psi}{\partial^2 y}\right] = \frac{\partial}{\partial y}\left[U_\infty f''(\eta)\sqrt{\frac{U_\infty}{Cvx}}\right] = \frac{\partial \eta}{\partial y}\frac{\partial}{\partial \eta}\left[U_\infty f''(\eta)\sqrt{\frac{U_\infty}{Cvx}}\right] = U_\infty f'''(\eta)\sqrt{\frac{U_\infty}{Cvx}}\sqrt{\frac{U_\infty}{Cvx}} \tag{7.78}$$

$$\frac{\partial^3 \psi}{\partial y^3} = U_\infty f'''(\eta)\frac{U_\infty}{Cvx} \tag{7.79}$$

$$\frac{\partial^2 \psi}{\partial x \partial y} = \frac{\partial}{\partial x}\left[\frac{\partial \psi}{\partial y}\right] = \frac{\partial}{\partial x}\left[U_\infty f'(\eta)\right] = \frac{\partial}{\partial \eta}\left[U_\infty f'(\eta)\right]\frac{\partial \eta}{\partial x} = -\frac{\eta}{2x}U_\infty f''(\eta) \tag{7.80}$$

Substituting all of these in Equation 7.64,

$$U_\infty f'(\eta)\left[-\frac{\eta}{2x}U_\infty f''(\eta)\right] - \left[\frac{1}{2}\sqrt{\frac{U_\infty Cv}{x}}f - \frac{\eta}{2x}\sqrt{U_\infty Cvx}f'\right]U_\infty f''\sqrt{\frac{U_\infty}{Cvx}} = vU_\infty f'''\frac{U_\infty}{Cvx} \tag{7.81}$$

which simplifies to

$$f''' + \frac{C}{2}ff'' = 0 \tag{7.82}$$

As can be seen, all the terms involving x and y drop out, and only terms that are function of the similarity variable remain, and therefore similarity solution exists. This is an ordinary third-order differential equation. This equation is known as the Blasius equation. In a similar manner, the energy equation reduces to

$$T^{*\prime\prime} + Pr\frac{C}{2}fT^{*\prime} = 0 \tag{7.83}$$

Note that if constant C is chosen so that Equation 7.60 is to hold exactly, then the boundary layer thickness would correspond to $\eta = 1$. Since this value of C is only known, after the solution is obtained, then the constant C can be assigned an arbitrary value, the choice of which would only result in the boundary layer thickness to correspond to a value of η different than 1. In the original derivation of Blasius, the constant C was assumed to be 1; however, to simplify the equation, we assume $C = 2$ making

$$\eta = \sqrt{\frac{U_\infty}{2vx}}y \tag{7.84}$$

and

$$f(\eta) = \frac{\psi}{\sqrt{2U_\infty vx}} \tag{7.85}$$

reducing the momentum and energy equations to

$$f''' + ff'' = 0 \tag{7.86}$$

$$T^{*\prime\prime} + Prf\,T^{*\prime} = 0 \tag{7.87}$$

To solve the momentum and energy equations, we need three and two boundary conditions, respectively. The solid walls are usually assumed to be streamlines having a value of zero ($\psi = 0$); therefore, the boundary conditions are

$$\eta = 0, \quad f = 0, \quad f' = 0, \quad T^* = 0$$
$$\eta \to \infty, \quad f' = 1, \quad T^* = 1 \tag{7.88}$$

The analytical solution to the Blasius equation can be obtained by series solution, asymptotic expansion, or numerically.

7.4.2 Numerical Solution of Blasius Equation

The Blasius equation (Equation 7.86) represents a third-order, nonlinear boundary value problem. To obtain the solution, we must first deal with the nonlinearity. Assume a new variable g where

$$g = f' \tag{7.89}$$

Then, Equation 7.86 can be written as

$$g'' + fg' = 0 \tag{7.90}$$

Using central differencing, the finite difference approximation to Equations 7.86 and 7.87 becomes

$$\left[\frac{1}{(\Delta\eta)^2} - f_i\frac{1}{2(\Delta\eta)}\right]g_{i-1} - \frac{2}{(\Delta\eta)^2}g_i + \left[\frac{1}{(\Delta\eta)^2} + f_i\frac{1}{2(\Delta\eta)}\right]g_{i+1} = 0 \tag{7.91}$$

$$\left[\frac{1}{(\Delta\eta)^2} - Prf_i\frac{1}{2(\Delta\eta)}\right]T_{i-1}^* - \frac{2}{(\Delta\eta)^2}T_i^* + \left[\frac{1}{(\Delta\eta)^2} + Prf_i\frac{1}{2(\Delta\eta)}\right]T_{i+1}^* = 0 \tag{7.92}$$

If f_is are known, then Equation 7.91 represents a set of N algebraic equations for the N unknown values of g_i. Similarly, Equation 7.92 represents a set of N algebraic equations for the N unknown values of T_i^*. These sets of algebraic equations can be easily solved using the built-in iterative capabilities of spreadsheets or TRIDI.

To be able to evaluate the values of g in Equation 7.91, values of f at each node are needed. Since g is the derivative of f, then f can be obtained from the Taylor series

$$f_i = f_{i-1} + f_{i-1}'(\Delta\eta) + f_{i-1}''\frac{(\Delta\eta)^2}{2} \tag{7.93}$$

replacing f' and f'' with g and its finite difference form

$$f_i = f_{i-1} + g_{i-1}(\Delta\eta) + \frac{g_i - g_{i-1}}{(\Delta\eta)}\frac{(\Delta\eta)^2}{2} \tag{7.94}$$

then we can arrive at an explicit equation relating the value of the stream function at a node to the velocity at the same node and one upstream

$$f_i = f_{i-1} + (g_{i-1} + g_i)\frac{(\Delta\eta)}{2} \tag{7.95}$$

To solve for f_i from Equation 7.95, the values of g_i must be known. However, the values of g_i from Equation 7.91 require the knowledge of the values of f_i or the two sets of equations 7.95 and 7.91 are coupled together and have to be solved iteratively. Regardless of whether Equation 7.91 is solved iteratively or using subroutine TRIDI, a second iteration to determine f from Equation 7.95 must be included. Once the values of f are obtained, Equation 7.92 can be solved iteratively to obtain the values of the nondimensional temperature.

The values of the nondimensional velocity, stream function, and temperature (f, f', and T^*) at different η values for $Pr = 0.1$, 1.0, and 10 are given in Table 7.1 and plotted in Figure 7.5 which also includes shear stress.

7.4.3 Boundary Layer Thickness

For external flows, one is interested in drag, lift, and heat transfer. For flow over a flat plate, pressure is constant and symmetric over top and bottom; therefore, there is not lift. The x component of velocity reaches 99% of its free stream value, $f' = \dfrac{u}{U_\infty} = 0.99$, at a specific

TABLE 7.1

Blasius Solution Results

η	f'	f	T Pr = 0.1	Pr = 1	Pr = 10	η	f'	f	T Pr = 0.1	Pr = 1	Pr = 10
0.00	0.00000	0.00000	0.00000	0.00000	0.00000	3.90	0.99651	2.63440	0.69308	0.99651	1.00000
0.05	0.02348	0.00059	0.00990	0.02348	0.05149	4.00	0.99741	2.73410	0.70642	0.99741	1.00000
0.10	0.04696	0.00235	0.01980	0.04696	0.10297	4.10	0.99810	2.83388	0.71940	0.99810	1.00000
0.15	0.07044	0.00528	0.02970	0.07044	0.15439	4.20	0.99862	2.93371	0.73202	0.99861	1.00000
0.20	0.09391	0.00939	0.03961	0.09391	0.20567	4.30	0.99900	3.03360	0.74427	0.99900	1.00000
0.25	0.11737	0.01467	0.04951	0.11737	0.25672	4.40	0.99929	3.13351	0.75616	0.99928	1.00000
0.30	0.14081	0.02113	0.05941	0.14081	0.30739	4.50	0.99950	3.23345	0.76768	0.99949	1.00000
0.35	0.16423	0.02875	0.06930	0.16423	0.35752	4.60	0.99965	3.33341	0.77883	0.99964	1.00000
0.40	0.18762	0.03755	0.07920	0.18761	0.40694	4.70	0.99975	3.43338	0.78962	0.99975	1.00000
0.45	0.21096	0.04751	0.08910	0.21096	0.45545	4.80	0.99983	3.53336	0.80004	0.99983	1.00000
0.50	0.23424	0.05864	0.09899	0.23424	0.50281	4.90	0.99988	3.63334	0.81010	0.99988	1.00000
0.60	0.28059	0.08439	0.11877	0.28059	0.59320	5.00	0.99992	3.73333	0.81981	0.99992	1.00000
0.70	0.32656	0.11475	0.13853	0.32655	0.67626	5.10	0.99995	3.83333	0.82916	0.99994	1.00000
0.80	0.37199	0.14968	0.15826	0.37199	0.75030	5.20	0.99996	3.93332	0.83815	0.99996	1.00000
0.90	0.41675	0.18912	0.17797	0.41675	0.81406	5.30	0.99998	4.03332	0.84680	0.99997	1.00000
1.00	0.46067	0.23300	0.19764	0.46067	0.86683	5.40	0.99998	4.13332	0.85511	0.99998	1.00000
1.10	0.50358	0.28122	0.21727	0.50358	0.90864	5.50	0.99999	4.23332	0.86308	0.99999	1.00000
1.20	0.54530	0.33367	0.23684	0.54529	0.94021	5.60	0.99999	4.33331	0.87072	0.99999	1.00000
1.30	0.58564	0.39023	0.25634	0.58564	0.96282	5.70	0.99999	4.43331	0.87803	0.99999	1.00000
1.40	0.62445	0.45074	0.27577	0.62444	0.97812	5.80	1.00000	4.53331	0.88503	0.99999	1.00000
1.50	0.66154	0.51505	0.29511	0.66154	0.98787	5.90	1.00000	4.63331	0.89172	0.99999	1.00000
1.60	0.67940	0.54858	0.30475	0.67939	0.99118	6.00	1.00000	4.73331	0.89811	1.00000	1.00000
1.70	0.71364	0.61824	0.32393	0.71364	0.99556	6.10	1.00000	4.83331	0.90420	1.00000	1.00000
1.80	0.74584	0.69123	0.34300	0.74583	0.99792	6.20	1.00000	4.93331	0.91000	1.00000	1.00000
1.90	0.77588	0.76733	0.36194	0.77588	0.99909	6.30	1.00000	5.03331	0.91553	1.00000	1.00000
2.00	0.80371	0.84632	0.38073	0.80370	0.99963	6.40	1.00000	5.13331	0.92078	1.00000	1.00000
2.10	0.82928	0.92798	0.39937	0.82927	0.99987	6.50	1.00000	5.23331	0.92577	1.00000	1.00000
2.20	0.85258	1.01209	0.41783	0.85258	0.99996	6.60	1.00000	5.33331	0.93051	1.00000	1.00000
2.30	0.87364	1.09842	0.43611	0.87364	0.99999	6.70	1.00000	5.43331	0.93500	1.00000	1.00000
2.40	0.89251	1.18674	0.45418	0.89251	1.00000	6.80	1.00000	5.53331	0.93925	1.00000	1.00000
2.50	0.90927	1.27684	0.47205	0.90926	1.00000	6.90	1.00000	5.63331	0.94327	1.00000	1.00000
2.60	0.92402	1.36852	0.48968	0.92401	1.00000	7.00	1.00000	5.73331	0.94708	1.00000	1.00000
2.70	0.93688	1.46157	0.50708	0.93688	1.00000	7.10	1.00000	5.83331	0.95067	1.00000	1.00000
2.80	0.94800	1.55583	0.52422	0.94799	1.00000	7.20	1.00000	5.93331	0.95405	1.00000	1.00000
2.90	0.95751	1.65111	0.54110	0.95750	1.00000	7.30	1.00000	6.03331	0.95725	1.00000	1.00000
3.00	0.96557	1.74727	0.55771	0.96557	1.00000	7.40	1.00000	6.13331	0.96025	1.00000	1.00000
3.10	0.97235	1.84418	0.57402	0.97234	1.00000	7.50	1.00000	6.23331	0.96308	1.00000	1.00000
3.20	0.97798	1.94170	0.59004	0.97797	1.00000	7.60	1.00000	6.33331	0.96573	1.00000	1.00000
3.30	0.98261	2.03974	0.60575	0.98261	1.00000	7.70	1.00000	6.43331	0.96823	1.00000	1.00000
3.40	0.98640	2.13819	0.62114	0.98639	1.00000	7.80	1.00000	6.53331	0.97057	1.00000	1.00000
3.50	0.98945	2.23699	0.63621	0.98945	1.00000	7.90	1.00000	6.63331	0.97276	1.00000	1.00000
3.60	0.99189	2.33606	0.65094	0.99189	1.00000	8.00	1.00000	6.73331	0.97480	1.00000	1.00000
3.70	0.99382	2.43535	0.66534	0.99382	1.00000	8.10	1.00000	6.83331	0.97672	1.00000	1.00000
3.80	0.99534	2.53481	0.67939	0.99533	1.00000	8.20	1.00000	6.93331	0.97851	1.00000	1.00000

(continued)

TABLE 7.1 (continued)

Blasius Solution Results

η	f'	f	Pr = 0.1	Pr = 1	Pr = 10	η	f'	f	Pr = 0.1	Pr = 1	Pr = 10
				T						T	
8.30	1.00000	7.03331	0.98018	1.00000	1.00000	9.70	1.00000	8.43331	0.99418	1.00000	1.00000
8.40	1.00000	7.13331	0.98173	1.00000	1.00000	9.80	1.00000	8.53331	0.99471	1.00000	1.00000
8.50	1.00000	7.23331	0.98318	1.00000	1.00000	9.90	1.00000	8.63331	0.99519	1.00000	1.00000
8.60	1.00000	7.33331	0.98453	1.00000	1.00000	10.00	1.00000	8.73331	0.99563	1.00000	1.00000
8.70	1.00000	7.43331	0.98578	1.00000	1.00000	10.10	1.00000	8.83331	0.99603	1.00000	1.00000
8.80	1.00000	7.53331	0.98695	1.00000	1.00000	10.20	1.00000	8.93331	0.99640	1.00000	1.00000
8.90	1.00000	7.63331	0.98802	1.00000	1.00000	10.30	1.00000	9.03331	0.99674	1.00000	1.00000
9.00	1.00000	7.73331	0.98902	1.00000	1.00000	10.40	1.00000	9.13331	0.99705	1.00000	1.00000
9.10	1.00000	7.83331	0.98995	1.00000	1.00000	10.50	1.00000	9.23331	0.99733	1.00000	1.00000
9.20	1.00000	7.93331	0.99080	1.00000	1.00000	10.60	1.00000	9.33331	0.99759	1.00000	1.00000
9.30	1.00000	8.03331	0.99159	1.00000	1.00000	10.70	1.00000	9.43331	0.99783	1.00000	1.00000
9.40	1.00000	8.13331	0.99232	1.00000	1.00000	10.80	1.00000	9.53331	0.99804	1.00000	1.00000
9.50	1.00000	8.23331	0.99300	1.00000	1.00000	10.90	1.00000	9.63331	0.99823	1.00000	1.00000
9.60	1.00000	8.33331	0.99361	1.00000	1.00000	11.00	1.00000	9.73331	0.99841	1.00000	1.00000

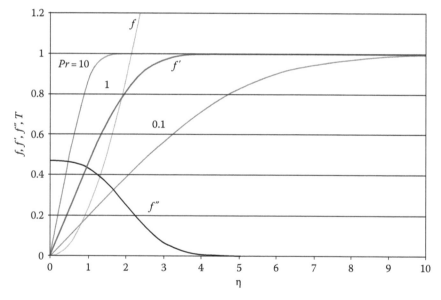

FIGURE 7.5
Blasius solution.

value of η designated by $\eta_{0.99}$. From Table 7.1, and by linear interpolation $\eta_{0.99} = 3.524$. The hydrodynamic boundary layer thickness is defined as the distance from the wall (y) where the velocity has reached 99% of its free stream value; therefore

$$\sqrt{\frac{U_\infty}{2vx}}\delta = \eta_{0.99} \tag{7.96}$$

$$\delta = \sqrt{2}\eta_{0.99}\sqrt{\frac{vx}{U_\infty}} \tag{7.97}$$

resulting in

$$\delta = 4.98\sqrt{\frac{\nu x}{U_\infty}} \qquad (7.98)$$

As expected, for $Pr = 1$, the velocity and temperature distributions are the same, as confirmed from the solution; therefore, the thermal boundary layer thickness is the same as hydrodynamic boundary layer thickness, or

$$\delta_T = 4.98\sqrt{\frac{\nu x}{U_\infty}} \qquad (7.99)$$

Similarly, from the results in Table 7.1, for $Pr = 0.1$ and 10, the thicknesses of the thermal boundary are

$$\delta_T = 13.02\sqrt{\frac{\nu x}{U_\infty}} \qquad (7.100)$$

$$\delta_T = 2.21\sqrt{\frac{\nu x}{U_\infty}} \qquad (7.101)$$

For $Pr > 1$, the thickness of thermal boundary layer is inversely proportional to $Pr^{1/3}$, or the thermal boundary layer thickness can be approximated by

$$\delta_T = \frac{4.98}{Pr^{1/3}Re^{1/2}} \qquad (7.102)$$

7.4.4 Reynolds Number

Reynolds number is generally considered to be the ratio of inertial to viscous forces. As mentioned earlier, the terms in the momentum equations are forces per unit volume. For boundary layer flow over a flat plate,

$$\frac{\text{Inertial force/volume}}{\text{Viscous force/volume}} = \frac{\rho u \dfrac{\partial u}{\partial x}}{\dfrac{\tau_w}{L}} \qquad (7.103)$$

The shear stress is $\tau_w = \mu\left(\dfrac{\partial u}{\partial y}\right)$ which is approximately $\tau_w \approx \mu\dfrac{u_\infty}{\delta}$, with $\delta \approx \sqrt{\dfrac{\nu L}{u_\infty}}$. Therefore

$$\frac{\text{Inertial force/volume}}{\text{Viscous force/volume}} \approx \frac{\rho u_\infty \dfrac{u_\infty}{L}}{\mu \dfrac{u_\infty}{\delta L}} \approx \frac{\rho}{\mu}u_\infty\sqrt{\frac{\nu L}{u_\infty}} = \sqrt{Re} \qquad (7.104)$$

Or for external flow, it is the square root of Reynolds number that is a measure of the inertial force to viscous force:

$$Re^{1/2} \approx \frac{\text{Inertial force/volume}}{\text{Viscous force/volume}} \qquad (7.105)$$

The inertial forces are those that sustain the external flow, and friction opposes or dampens it. The transitions to turbulence happen at Reynolds number of about 500,000, or when the inertial forces are about 700 times the friction forces. Bejan [2] argues that the square root of Reynolds number is important and is a geometric parameter indicating the slenderness ratio, which is the ratio of the distance along the wall to the boundary layer thickness.

7.4.5 Skin Friction Coefficient

The shear force exerted by the fluid on a solid is expressed in terms of the nondimensional quantity skin friction coefficient or friction coefficient. The local value of the friction coefficient is defined as

$$c_f = \frac{\tau_w}{\frac{1}{2}\rho U_\infty^2} = \frac{\mu \frac{\partial u}{\partial y}\Big|_{y=0}}{\frac{1}{2}\rho U_\infty^2} \tag{7.106}$$

In terms of the nondimensional velocity and the similarity variable,

$$c_f = \frac{\mu U_\infty \frac{\partial u^*}{\partial \eta}\Big|_0 \frac{\partial \eta}{\partial y}}{\frac{1}{2}\rho U_\infty^2} = \frac{\mu f''(0)\sqrt{\frac{U_\infty}{2\nu x}}}{\frac{1}{2}\rho U_\infty} = \sqrt{2}f''(0)Re_x^{-1/2} \tag{7.107}$$

The value of f'' at the wall can be calculated from Table 7.1,

$$f''(0) = \frac{0.02348 - 0}{0.05} = 0.4696 \tag{7.108}$$

$$c_{f,x} = 0.664 Re_x^{-1/2} \tag{7.109}$$

The average shear stress on the plate is given by

$$\overline{\tau}_w = \frac{1}{L}\int_0^L \tau_w dx = \frac{1}{L}\int_0^L 0.664 Re_x^{-1/2} \frac{1}{2}\rho U_\infty^2 dx = 0.664 \frac{\frac{1}{2}\rho U_\infty^2}{L}\left(\frac{U_\infty}{\nu}\right)^{-1/2}\int_0^L x^{-1/2}dx \tag{7.110}$$

and therefore, the average drag coefficient becomes

$$\overline{c}_f = \frac{\overline{\tau}_w}{\frac{1}{2}\rho U_\infty^2} = 1.328 Re_L^{1/2} \tag{7.111}$$

Note that the average drag coefficient from the plate edge to a location L is twice the local drag coefficient at L

$$\overline{c}_f = 2c_{f,L} \tag{7.112}$$

7.4.6 Nusselt Number

To find the Nusselt number,

$$Nu_x = \frac{hx}{k} = \frac{-k\frac{\partial T}{\partial y}x}{(T_w - T_\infty)k} = \frac{-(T_\infty - T_w)\frac{\partial T^*}{\partial y}x}{(T_w - T_\infty)} = \left.\frac{\partial T^*}{\partial \eta}\right|_0 \left.\frac{\partial \eta}{\partial y}\right.x = \left.\frac{\partial T^*}{\partial \eta}\right|_0 \sqrt{\frac{U_\infty}{2vx}}x = \frac{1}{\sqrt{2}}\left.\frac{\partial T^*}{\partial \eta}\right|_0 Re_x^{1/2}$$

(7.113)

Also, for this case,

$$\frac{\left.\dfrac{\partial T^*}{\partial \eta}\right|_{0,Pr=1}}{\left.\dfrac{\partial T^*}{\partial \eta}\right|_{0,Pr=10}} = \frac{\dfrac{0.0234-0}{0.05}}{\dfrac{0.0514-0}{0.05}} = 0.4553 \approx \frac{1}{\sqrt[3]{Pr}}$$

(7.114)

Therefore, the expression for determining the local value of the Nusselt number becomes

$$Nu_x = 0.332 Pr^{1/3} Re_x^{1/2}$$

(7.115)

The average heat transfer coefficient can be determined from

$$\bar{h} = \frac{1}{L}\int_0^L h\,dx = \frac{1}{L}\int_0^L 0.332 Pr^{1/3} Re_x^{1/2}\frac{k}{x}\,dx$$

(7.116)

resulting in

$$\overline{Nu} = \frac{\bar{h}L}{k} = 0.664 Pr^{1/3} Re_L^{1/2}$$

(7.117)

Also, similar to the drag coefficient, the average value of Nusselt number over a plate of length L is twice the local value of the Nusselt number at $x = L$:

$$\overline{Nu} = 2Nu_L$$

(7.118)

7.4.7 Reynolds–Coulborn Analogy and Stanton Number

Multiplying and dividing the RHS of Equation 7.115 by a factor of two, dividing both sides by Reynolds and Prandtl numbers and using Equation 7.104 yield

$$\frac{Nu}{PrRe_x} = \frac{1}{2}\frac{0.664}{Pr^{2/3}Re_x^{1/2}} = \frac{1}{2}\frac{c_f}{Pr^{2/3}}$$

(7.119)

The LHS of the equation is dimensionless and is called Stanton number. Expanding the terms,

$$St = \frac{Nu}{PrRe_x} = \frac{\dfrac{hx}{k}}{\dfrac{c_p\mu}{k}\dfrac{\rho U_\infty x}{\mu}} = \frac{h}{\rho c U_\infty}$$

(7.120)

Therefore, Equation 7.119 can be written as

$$StPr^{2/3} = \frac{C_f}{2} \tag{7.121}$$

This expression is exactly correct for laminar boundary layer flow over a flat plate. The LHS is the nondimensional heat transfer coefficient, and the RHS is the nondimensional shear stress. The RHS depends only on the velocity distribution and the LHS on temperature distribution. Equation 7.121 states that for this case, if friction factor is known, then the heat transfer coefficient can also be calculated, which is quite remarkable in that the need for the solution of energy equation is eliminated. This is known as the Reynolds–Colburn analogy between fluid flow and heat transfer. It turns out that the analogy is also approximately valid for many other flows, like flow over other geometries for both laminar and turbulent flow and developing flow in a pipe. One case where this equation is not valid is for laminar fully developed flow in pipes, although it is approximately valid for fully developed turbulent flow. It is also valid, not only for local values, but also for average values of Stanton and Nusselt numbers. In the absence of better correlations, if the drag coefficient is known, then the Reynolds–Colburn analogy can be used to arrive at a reasonable estimate of the heat transfer coefficient.

7.4.8 Small Prandtl Number Solution

As was seen earlier, the thermal boundary layer is proportional to $Pr^{-1/3}$, for $Pr > 1$. For liquid metals, the value of Prandtl number is much smaller than one. For this situation, an exact analytical solution can be arrived at. In the limit as Prandtl number approaches zero, the velocity boundary layer is much thinner than the thermal boundary layer, and therefore the velocity in the energy equation can be assumed to be uniform everywhere or $f' = 1$. It then follows that the nondimensional stream function will vary linearly

$$f = \eta \tag{7.122}$$

Substituting back in the energy equation,

$$\frac{\partial^2 T^*}{\partial \eta^2} + Pr\eta \frac{\partial T^*}{\partial \eta} = 0 \tag{7.123}$$

This equation can be integrated by separation of variables

$$\frac{\dfrac{\partial^2 T^*}{\partial \eta^2}}{\dfrac{\partial T^*}{\partial \eta}} = -Pr\eta \tag{7.124}$$

and integrating

$$\frac{\partial T^*}{\partial \eta} = Ce^{-(Pr/2)\eta^2} \tag{7.125}$$

taking

$$z = \sqrt{\frac{Pr}{2}}\eta \tag{7.126}$$

therefore

$$\frac{\partial T^*}{\partial z} = c e^{-z^2} \tag{7.127}$$

Integrating once more

$$T^* = c \int_0^z e^{-z^2} dz = K\, erf(z) \tag{7.128}$$

Since $T^*(\infty) = 1$, then

$$T^* = \frac{2}{\sqrt{\pi}} \int_0^z e^{-z^2} dz = erf\left(\sqrt{\frac{Pr}{2}}\,\eta \right) \tag{7.129}$$

The temperature gradient at the wall is

$$\frac{dT^*}{d\eta} = \sqrt{\frac{Pr}{2}}\,\frac{2}{\sqrt{\pi}}\,e^{-\frac{Pr}{2}\eta^2} = \sqrt{\frac{2Pr}{\pi}} \tag{7.130}$$

And Nusselt number becomes

$$Nu = \frac{hx}{k} = \frac{1}{\sqrt{2}}\,\frac{\partial T^*}{\partial \eta}\bigg|_0 Re_x^{1/2} = \frac{1}{\sqrt{2}}\sqrt{\frac{2}{\pi}}Pr^{1/2}Re_x^{1/2} \tag{7.131}$$

or

$$Nu = 0.564 Re_x^{1/2} Pr^{1/2} \tag{7.132}$$

Figure 7.6 is a comparison of $NuRe_x^{-1/2}$ from Equation 7.119 and the numerical results from Blasius solution, for a broad range of Prandtl numbers. The difference between the two

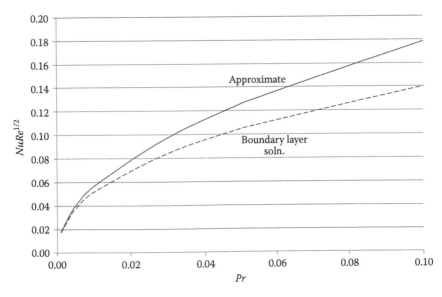

FIGURE 7.6
Low Pr number solutions for flow over a flat plate.

solutions diminishes at small values of Prandtl number. It is also important to note that for small values of Prandtl number, Nusselt number is proportional to Prandtl number to one-half power, whereas for $Pr > 1$, it is proportional to the one-third power.

7.4.9 Numerical Solution Using Primitive Variables

The solution to boundary layer equations can also be obtained using the original variables. As shown in the following, solution to the continuity, momentum, and energy equations in primitive variables is not much different from the similarity solution using spreadsheets. The solution to these equations provides the x and y components of the velocity and the temperature in the flow field above the plate. To obtain general solutions, the governing equations are nondimensionalized first by defining the nondimensional variables

$$x^* = \frac{x}{L}, \quad y^* = \frac{y}{L}\sqrt{Re}$$

$$u^* = \frac{u}{u_\infty}, \quad v^* = \frac{v}{u_\infty}\sqrt{Re}, \quad T^* = \frac{T - T_w}{T_\infty - T_w}$$

(7.133)

Note that different scales are used to nondimensionalize x and y and u and v. As shown in the following, this choice makes v^* and y^* to have the same order of magnitude as u^* and x^* (of the order 1), even though $v \ll u$ and $y \ll x$. The scaling employed also eliminates Reynolds number as a direct parameter in the solution of the problem, making one nondimensional solution valid for all values of Reynolds number in the laminar range. Also,

$$\eta = \frac{y^*}{\sqrt{2x^*}} = y\sqrt{\frac{u_\infty}{2vx}}$$

(7.134)

The nondimensional form of the boundary layer equations reduces to

$$\frac{\partial u^*}{\partial x^*} + \frac{\partial v^*}{\partial y^*} = 0$$

(7.135)

$$u^*\frac{\partial u^*}{\partial x^*} + v^*\frac{\partial u^*}{\partial y^*} = \frac{\partial^2 u^*}{\partial y^{*2}}$$

(7.136)

$$u^*\frac{\partial T^*}{\partial x^*} + v^*\frac{\partial T^*}{\partial y^*} = \frac{1}{Pr}\frac{\partial^2 T^*}{\partial y^{*2}}$$

(7.137)

$$x^* = 0, \quad u^* = 1, \quad v^* = 0, \quad T^* = 1$$

(7.138)

$$y^* = 0, \quad u^* = 0, \quad v^* = 0, \quad T^* = 0$$

(7.139)

$$y^* \to \infty, \quad u^* \to 1, \quad T^* \to 1$$

(7.140)

It is also interesting to note that the ratio of

$$\frac{y^*}{\sqrt{x^*}} = \frac{\frac{y}{L}\sqrt{Re}}{\sqrt{\frac{x}{L}}} = \frac{y}{\sqrt{x}}\frac{1}{\sqrt{L}}\sqrt{\frac{u_\infty L}{\nu}} = y\sqrt{\frac{u_\infty}{\nu x}} = \eta \tag{7.141}$$

is the similarity variable with $C = 1$ in Equation 7.61.

The finite difference approximation used is somewhat more involved than the central and backward differencing used for the similarity solution. Dropping the asterisks, and using the procedure outlined in Oosthuizen and Naylor [3], the finite difference form of the continuity Equation 7.134 becomes

$$\frac{1}{2}\left[\frac{u_{i,j} - u_{i-1,j}}{\Delta x} + \frac{u_{i,j-1} - u_{i-1,j-1}}{\Delta x}\right] + \frac{v_{i,j} - v_{i,j-1}}{\Delta y} = 0 \tag{7.142}$$

from which using the same procedure as earlier, we can solve for the y component of velocity at nodes i, j to get

$$v_{i,j} = v_{i,j-1} - \frac{\Delta y}{2\Delta x}[u_{i,j} - u_{i-1,j} + u_{i,j-1} - u_{i-1,j-1}] \tag{7.143}$$

The momentum equation can be discretized as

$$u_{i-1,j}\frac{u_{i,j} - u_{i-1,j}}{\Delta x} + v_{i-1,j}\frac{u_{i,j+1} - u_{i,j-1}}{2\Delta y} = \frac{u_{i,j+1} - 2u_{i,j} + u_{i,j-1}}{(\Delta y)^2} \tag{7.144}$$

and solved for $u_{i,j}$

$$u_{i,j} = \frac{1}{\frac{u_{i-1,j}}{\Delta x} + \frac{2}{(\Delta y)^2}}\left[\frac{u^2_{i-1,j}}{\Delta x} - u_{i,j-1}\left(\frac{-v_{i-1,j}}{2\Delta y} - \frac{1}{(\Delta y)^2}\right) - u_{i,j+1}\left(\frac{v_{i-1,j}}{2\Delta y} - \frac{1}{(\Delta y)^2}\right)\right] \tag{7.145}$$

The energy equation can be discretized as

$$u_{i-1,j}\frac{T_{i,j} - T_{i-1,j}}{\Delta x} + v_{i-1,j}\frac{T_{i,j+1} - T_{i,j-1}}{2\Delta y} = \frac{1}{Pr}\frac{T_{i,j+1} - 2T_{i,j} + T_{i,j-1}}{(\Delta y)^2} \tag{7.146}$$

and solved for $T_{i,j}$

$$T_{i,j} = \frac{1}{\left[\frac{u_{i-1,j}}{\Delta x} + \frac{2}{Pr(\Delta y)^2}\right]}\left[\frac{u_{i-1,j}T_{i-1,j}}{\Delta x} - T_{i,j-1}\left(\frac{-v_{i-1,j}}{2\Delta y} - \frac{1}{Pr(\Delta y)^2}\right) - T_{i,j+1}\left(\frac{+v_{i-1,j}}{2\Delta y} - \frac{1}{Pr(\Delta y)^2}\right)\right]$$

$$\tag{7.147}$$

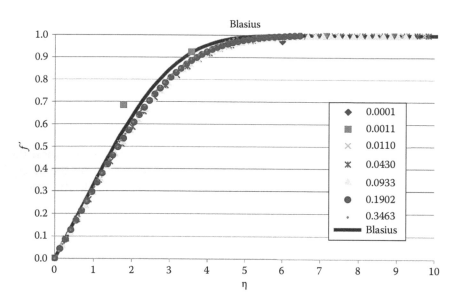

FIGURE 7.7
Comparison of similarity and primitive variables solutions.

Since each of the three variables u, v, and T are 2D, one worksheet for each, named for the unknowns (u, v, T), is needed to calculate and store their values at different x and y locations. In each worksheet, each cell is a node in the computational domain. Figure 7.7 is a plot of velocity distribution at several x locations and comparing them with the Blasius solution. The results at x and y locations correspond to a η value, and they should fall on the same curve, the Blasius solution.

7.5 Nonsimilar Boundary Layer Flow

The existence of similarity solution for boundary layer flow depends on the geometry and boundary conditions. For example, the similarity solution does not exist, if the wall instead of being isothermal is subjected to a uniform heat flux. Generalizing the boundary layer equations, for 2D ($k = 0$) and axisymmetric ($k = 1$) flows, the conservation equations are given by Equations 7.50 through 7.53

$$\frac{\partial u r^k}{\partial x} + \frac{\partial v r^k}{\partial y} = 0 \tag{7.148}$$

$$\rho\left(u\frac{\partial u}{\partial x} + v\frac{\partial u}{\partial y}\right) = -\frac{dP}{dx} + \frac{\partial}{\partial y}\left[\mu\left(\frac{\partial u}{\partial y}\right)^n\right] \tag{7.149}$$

$$\rho c_p\left(u\frac{\partial T}{\partial x} + v\frac{\partial T}{\partial y}\right) = k\frac{\partial^2 T}{\partial y^2} \tag{7.150}$$

where $n = 1$ for a Newtonian fluid and some other value for non-Newtonian ones. The boundary layer equations are singular at $x = y = 0$. Since the boundary layer thickness varies in the axial direction, we saw that the numerical solution using primitive variables is challenging, requiring either many nodes in the y direction or using variable spacing. The thickness of boundary layer does not change much if the governing equations are expressed using similarity variables, making numerical solutions easier to obtain for boundary layer flow problems. Using the similarity variable also removes the singularity. Defining a generalized stream function by [4]

$$u = \frac{1}{r^k} \frac{\partial r^k \psi}{\partial y} \tag{7.151}$$

$$v = -\frac{1}{r^k} \frac{\partial r^k \psi}{\partial x} = -\frac{\partial \psi}{\partial x} - \frac{\psi}{r^k} \frac{\partial r^k}{\partial x} \tag{7.152}$$

and the transformed x and y coordinates and the dimensionless stream function f are given by

$$\xi = \frac{1}{L} \int_0^x \left(\frac{r^k}{L^k}\right)^{n+1} \left(u_e^*\right)^{2n-1} dx \tag{7.153}$$

$$\eta = \left(\frac{r}{L}\right)^k \left(u_e^*\right) \left(\frac{Re}{2\xi}\right)^{\frac{1}{n+1}} \frac{y}{L} \tag{7.154}$$

$$\psi(x, y) = L U_\infty \left(\frac{2\xi}{Re}\right)^{\frac{1}{n+1}} f(\xi, \eta) \tag{7.155}$$

where

$$Re = \frac{\rho L^n U_\infty^{2-n}}{\mu} \tag{7.156}$$

and

$$u_e^* = \frac{U_e}{U_\infty} \tag{7.157}$$

In these earlier equations, U_∞ and L are appropriately defined velocity and length scales, and using these, the momentum equation becomes

$$\frac{\partial^3 f}{\partial \eta^3} + \frac{2}{n(n+1)} f \left(\frac{\partial^2 f}{\partial \eta^2}\right)^{2-n} + \frac{\beta}{n} \left(\frac{\partial^2 f}{\partial \eta^2}\right)^{1-n} \left(1 - \left(\frac{\partial f}{\partial \eta}\right)^2\right) = \frac{2\xi}{n} \left(\frac{\partial^2 f}{\partial \eta^2}\right)^{1-n} \left[\frac{\partial^2 f}{\partial \xi \partial \eta} \frac{\partial f}{\partial \eta} - \frac{\partial f}{\partial \xi} \frac{\partial^2 f}{\partial \eta^2}\right] \tag{7.158}$$

where

$$\beta = L \frac{2\xi \frac{dU_e^*}{dx}}{\left[\left(\frac{r}{L}\right)^k\right]^{n+1} \left(u_e^*\right)^{2n}} \tag{7.159}$$

The transformed energy equation becomes

$$\frac{\partial^2 T^*}{\partial \eta^2} + \frac{2}{n(n+1)} f \left(\frac{\partial T^*}{\partial \eta} \right)^{2-n} = \frac{2\xi}{n} \left(\frac{\partial^2 f}{\partial \eta^2} \right)^{1-n} \left[\frac{\partial f}{\partial \eta} \frac{\partial T^*}{\partial \xi} - \frac{\partial f}{\partial \xi} \frac{\partial T^*}{\partial \eta} \right] \qquad (7.160)$$

These equations are a generalization of the boundary layer equations, valid, regardless of whether the similarity solutions exist or not. The shear stress at the wall is given by

$$\tau_w = \frac{\mu U_\infty^n}{L^n} Re^{\frac{n}{n+1}} \left[\left(\frac{r}{L} \right)^k \right]^n \left(\frac{u_e}{u_\infty} \right)^{2n} (2\xi)^{\frac{-n}{n+1}} \left(\frac{\partial^2 f}{\partial \eta^2}(\xi,0) \right)^n \qquad (7.161)$$

For laminar Newtonian flow over a flat plate in Cartesian coordinates,

$$Re = \frac{\rho L u_\infty}{\mu} \qquad (7.162)$$

$$\xi = \frac{x}{L} \qquad (7.163)$$

$$\eta = \sqrt{\left(\frac{u_\infty}{2vx} \right)} y \qquad (7.164)$$

$$\beta = L \frac{2\xi \frac{du_e^*}{dx}}{\left[\left(\frac{r}{L} \right)^k \right]^{n+1} \left(u_e^* \right)^{2n}} = 0 \qquad (7.165)$$

and the momentum and energy equations become

$$\frac{\partial^3 f}{\partial \eta^3} + f \frac{\partial^2 f}{\partial \eta^2} = 2\xi \left[\frac{\partial^2 f}{\partial \xi \partial \eta} \frac{\partial f}{\partial \eta} - \frac{\partial f}{\partial \xi} \frac{\partial^2 f}{\partial \eta^2} \right] \qquad (7.166)$$

$$\frac{\partial^2 T^*}{\partial \eta^2} + f \frac{\partial T^*}{\partial \eta} = 2\xi \left[\frac{\partial f}{\partial \eta} \frac{\partial T^*}{\partial \xi} - \frac{\partial f}{\partial \xi} \frac{\partial T^*}{\partial \eta} \right] \qquad (7.167)$$

The shear stress at the wall is

$$\tau_w = \frac{\mu u_\infty}{L} Re^{1/2} \frac{1}{\sqrt{2\xi}} \frac{\partial^2 f}{\partial \eta^2}(\xi,0) \qquad (7.168)$$

which if rearranged results in

$$\tau_w = \frac{\mu u_\infty}{L} Re^{1/2} \frac{1}{\sqrt{2\xi}} \frac{\partial^2 f}{\partial \eta^2}(\xi,0) \qquad (7.169)$$

$$\frac{c_f}{\sqrt{Re_x}} = \sqrt{2} \frac{\partial^2 f(\xi,0)}{\partial \eta^2} \qquad (7.170)$$

7.5.1 Plate Subjected to a Uniform Heat Flux

We next consider the case where the plate is subjected to a uniform heat flux. The velocity profile is still similar and given by Blasius equation, and the nonsimilar form of the transformed energy equation becomes

$$\frac{1}{Pr}\frac{\partial^2 T^*}{\partial \eta^2} + f\frac{\partial T^*}{\partial \eta} = 2\xi\frac{\partial f}{\partial \eta}\frac{\partial T^*}{\partial \xi} \tag{7.171}$$

The wall boundary condition on the energy equation is

$$q_w = -k\frac{\partial T}{\partial y}\bigg|_0 = -kT_s\frac{\partial T^*}{\partial \eta}\bigg|_0\frac{\partial \eta}{\partial y} \tag{7.172}$$

$$\frac{\partial \eta}{\partial y} = \left(\frac{r}{L}\right)^k (u_e^*)\left(\frac{Re}{2\xi}\right)^{\frac{1}{n+1}}\frac{1}{L} = \sqrt{\frac{\rho u_\infty}{2\mu x}} \tag{7.173}$$

$$q_w = -k\frac{\partial T}{\partial y}\bigg|_0 = -kT_s\frac{\partial T^*}{\partial \eta}\bigg|_0\sqrt{\frac{\rho u_\infty}{2\mu x}} \tag{7.174}$$

since rearranging

$$\frac{\partial T^*}{\partial \eta}\bigg|_0 = -\frac{q_w}{kT_s}\sqrt{\frac{2\mu x}{\rho u_\infty}} = -\frac{q_w L}{kT_s\sqrt{Re}}\sqrt{2\xi} \tag{7.175}$$

Assuming

$$T_s = \frac{q_w L}{k\sqrt{Re}} \tag{7.176}$$

then

$$\frac{\partial T^*}{\partial \eta}\bigg|_0 = -\sqrt{2\xi} \tag{7.177}$$

$$Nu_x = \frac{hx}{k} = \frac{\dfrac{q_w}{T_w - T_\infty}x}{k} = \frac{-kT_s\dfrac{\partial T^*}{\partial \eta}\bigg|_0\sqrt{\dfrac{\rho u_\infty}{2\mu x}}x}{k\dfrac{T_w - T_\infty}{T_s}T_s} = \frac{-\dfrac{\partial T^*}{\partial \eta}\bigg|_0\sqrt{\dfrac{\rho u_\infty x}{2\mu}}}{T_w^*} = \frac{1}{\sqrt{2}}\frac{-\dfrac{\partial T^*}{\partial \eta}\bigg|_0}{T_w^*}\sqrt{\dfrac{\rho u_\infty x}{\mu}} \tag{7.178}$$

$$\frac{Nu_x}{\sqrt{Re_x}} = \frac{1}{\sqrt{2}}\frac{-\dfrac{\partial T^*}{\partial \eta}\bigg|_0}{T_w^*} = \frac{\sqrt{\xi}}{T_w^*} \tag{7.179}$$

Since the wall boundary condition on the energy equation is a function of ξ, for a uniform heat flux wall, the energy equation does not have similarity solution.

The analytical results are given by Kays and Crawford [5] (p. 179) who provide the following expression for the determination of the wall temperature for the general case of varying heat flux:

$$(T_w - T_\infty) = \frac{0.623}{k} \frac{1}{Re_x^{1/2} Pr^{1/3}} \int_0^x \left[1 - \left(\frac{\xi}{x} \right)^{3/4} \right]^{-2/3} q_w(\xi) d\xi \tag{7.180}$$

which for constant heat flux simplifies to

$$\frac{Nu_x}{Re_x^{1/2} Pr^{1/3}} = \frac{1}{0.623 \dfrac{1}{x} \displaystyle\int_0^x \left[1 - \dfrac{\xi}{(x)^{3/4}} \right]^{-2/3} d\xi} \tag{7.181}$$

To evaluate the integral on the RHS, define a new variable

$$z = 1 - \left(\frac{\xi}{x} \right)^{3/4} \tag{7.182}$$

$$dz = -\frac{3}{4} \left(\frac{1}{x} \right)^{3/4} (\xi)^{-1/4} d\xi \tag{7.183}$$

substituting and simplifying Equation 7.181 result in

$$\frac{Nu_x}{Re_x^{1/2} Pr^{1/3}} = \frac{1}{0.623 \dfrac{4}{3} \displaystyle\int_0^1 z^{-2/3} (1-z)^{1/3} dz} \tag{7.184}$$

To evaluate the integral, we take advantage of the identity

$$\int_0^1 z^{m-1} (1-z)^{n-1} dz = \beta(m,n) = \frac{\Gamma(m)\Gamma(n)}{\Gamma(m+n)} \tag{7.185}$$

where β and Γ are beta and gamma functions, respectively. Therefore,

$$\frac{Nu_x}{Re_x^{1/2} Pr^{1/3}} = \frac{1}{0.623 \dfrac{4}{3} \dfrac{\Gamma(1/3)\Gamma(4/3)}{\Gamma(5/3)}} = 0.453 \tag{7.186}$$

A comparison of the numerical and analytical results is shown in Figure 7.8, and as seen, the two match closely, slightly downstream of the leading edge.

7.5.2 Plate Subjected to a Varying Temperature

Consider flow over a flat plate whose temperature changes linearly and given by

$$T_w = T_\infty + a + bx \tag{7.187}$$

Assuming

$$T_w^* = \frac{T_w - T_\infty}{T_s} \tag{7.188}$$

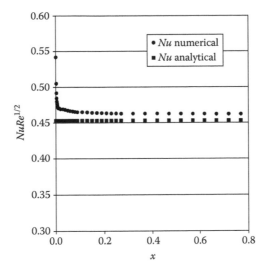

FIGURE 7.8
Comparison of the Nusselt number obtained analytically and numerically for flow over a flat plate subject to a uniform heat flux.

where

$$T_s = a, \quad \gamma = \frac{bL}{a} \tag{7.189}$$

then

$$T_w^* = 1 + \gamma\xi \tag{7.190}$$

and

$$Nu = \frac{hx}{k} = \frac{\dfrac{q_w}{T_w - T_\infty} x}{k} = \frac{q_w L \xi}{k \dfrac{T_w - T_\infty}{T_s} T_s} = -\frac{1}{T_w^*} \frac{\partial T^*}{\partial \eta}\bigg|_0 \sqrt{\frac{\rho u_\infty x}{2\mu}} \tag{7.191}$$

or

$$\frac{Nu_x}{\sqrt{Re_x}} = -\frac{\partial T^*}{\partial \eta_0} \frac{1}{\sqrt{2}} \frac{1}{T_w^*} \tag{7.192}$$

Kays and Crawford [5] give the following

$$\frac{Nu_x}{0.332 Re_x^{1/2} Pr^{1/3}} = \frac{a + 1.61bx}{a + bx} \tag{7.193}$$

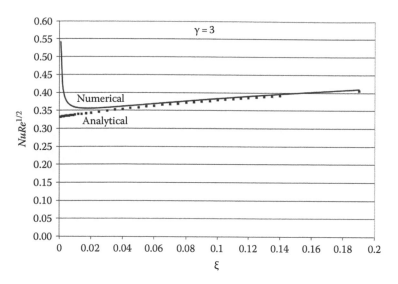

FIGURE 7.9
Comparison of the Nusselt number obtained analytically and numerically for flow over a flat plate subject to a linear surface temperature for $\gamma = 3$.

which can be written as

$$\frac{Nu_x}{Re_x^{1/2}Pr^{1/3}} = 0.332\left(1.61 - \frac{0.61}{1+\gamma\xi}\right) \tag{7.194}$$

Figure 7.8 is a comparison of the analytical and numerical solutions, and again, comparison improves away from the leading edge of the plate (Figure 7.9).

7.5.3 Flow over a Wedge

For flow of a Newtonian fluid ($n = 1$) over a wedge ($k = 0$), having a wedge angle of 2α radians, or for flow over a flat plate inclined at an angle α relative to the incoming fluid as shown in Figure 7.10, the free stream velocity is given by

$$U_e = Cx^m \tag{7.195}$$

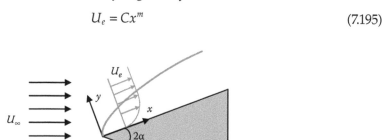

FIGURE 7.10
Falkner Skan flow.

where

$$m = \frac{\alpha}{\pi - \alpha} \tag{7.196}$$

and C is a constant. If negative, α is called the angle of attack.

For this case, the transformed variables become

$$\xi = \frac{1}{L} \int_0^x \frac{C}{U_\infty} x^m dx = \frac{1}{L} \frac{C}{U_\infty} \frac{1}{m+1} x^{m+1} = \frac{1}{m+1} \frac{x}{L} \frac{U_e}{U_\infty} \tag{7.197}$$

$$\eta = \sqrt{\frac{m+1}{2} \frac{U_e}{vx}} y \tag{7.198}$$

$$\psi(x,y) = \sqrt{\frac{2}{m+1} vxU_e} f(\xi, \eta) \tag{7.199}$$

$$\beta = \frac{L}{U_\infty} \frac{\frac{2}{m+1} \frac{x}{L} \frac{U_e}{U_\infty} mU_e}{\frac{U_e^2}{U_\infty} x} = \frac{2m}{m+1} = \frac{2\alpha}{\pi} \tag{7.200}$$

Therefore, the momentum equation becomes

$$\frac{\partial^3 f}{\partial \eta^3} + f \left(\frac{\partial^2 f}{\partial \eta^2} \right) + \frac{2\alpha}{\pi} \left(1 - \left(\frac{\partial f}{\partial \eta} \right)^2 \right) = 2\xi \left[\frac{\partial^2 f}{\partial \xi \partial \eta} \frac{\partial f}{\partial \eta} - \frac{\partial f}{\partial \xi} \frac{\partial^2 f}{\partial \eta^2} \right] \tag{7.201}$$

For flow over the wedge, the only ξ dependence is on the RHS for f. The boundary conditions are

$$\eta = 0, \quad f = f' = 0$$
$$\eta \to \infty \quad \text{or} \quad \xi = 0, \quad f' = 1 \tag{7.202}$$

The boundary conditions are independent of ξ, and furthermore two of the original boundary conditions (at $x = 0$, and $y \to \infty$) reduce to the same condition at $\eta \to \infty$.

These all point to the existence of similarity solution, i.e., the velocity distribution is only a function η (f is independent of ξ) or the RHS will be zero, simplifying the momentum equation to

$$\frac{\partial^3 f}{\partial \eta^3} + f \left(\frac{\partial^2 f}{\partial \eta^2} \right) + \beta \left(1 - \left(\frac{\partial f}{\partial \eta} \right)^2 \right) = 0 \tag{7.203}$$

where

$$\beta = \frac{2\alpha}{\pi} = \frac{2m}{m+1} \tag{7.204}$$

Note in some texts (Ref. [5], p. 97; Ref. [6], p. 187), the similarity variable and stream functions are defined as

$$\eta' = \sqrt{\frac{U_e}{\nu x}} y = \frac{\eta}{\sqrt{\frac{(m+1)}{2}}} \tag{7.205}$$

$$\psi(x,y) = \sqrt{\nu U_e x} F(\eta) \tag{7.206}$$

or

$$F(\eta) = \frac{f(\eta)}{\sqrt{\frac{(m+1)}{2}}} \tag{7.207}$$

resulting in momentum equation taking the following form

$$F''' + \frac{m+1}{2} FF'' + m(1 - F'^2) = 0 \tag{7.208}$$

The momentum equation (Equation 7.203) is solved numerically, by assuming

$$g = f' \tag{7.209}$$

$$g'' + fg' + \beta(1 - g^2) = 0 \tag{7.210}$$

$$\frac{g_{i+1} - 2g_i + g_{i-1}}{(\Delta\eta)^2} + f_i \frac{g_{i+1} - g_{i-1}}{2(\Delta\eta)} + \beta\left(1 - g_i^2\right) = 0 \tag{7.211}$$

$$A_i g_{i-1} + B_i g_i + C_i g_{i+1} = R_i \tag{7.212}$$

where

$$A_i = \frac{1}{(\Delta\eta)^2} - f_i \frac{1}{2(\Delta\eta)} \tag{7.213}$$

$$B_i = \frac{-2}{(\Delta\eta)^2} - \beta g_i \tag{7.214}$$

$$C_i = \frac{1}{(\Delta\eta)^2} + f_i \frac{1}{2(\Delta\eta)} \tag{7.215}$$

$$R_i = -\beta \tag{7.216}$$

From Equation 7.95,

$$f_i = f_{i-1} + (g_{i-1} + g_i)\frac{(\Delta\eta)}{2} \tag{7.217}$$

These equations can be solved numerically using spreadsheets [7]. Figure 7.11 shows the velocity (f'), stream function (f), and shear stress (f'') distribution for three different cases of flat plate parallel to the flow and at ±17.9°. In all three cases, the velocity increases to the free stream value. As the plate angle decreases from 17.9 to −17.9, the wall shear stress decreases and at −17.9 it reaches zero at the wall. For this angle, the maximum value of

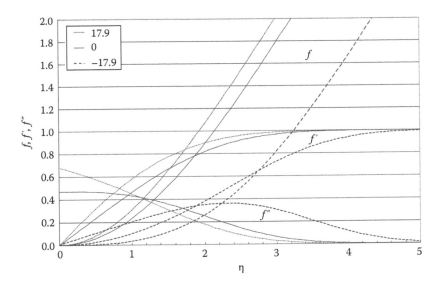

FIGURE 7.11
Impact of the plate angle.

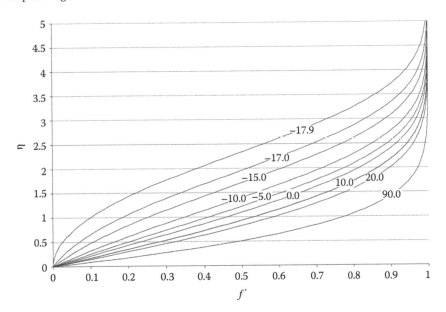

FIGURE 7.12
Velocity profile for flow over a wedge.

the shear stress occurs away from the wall in the fluid, before it drops to zero at the edge of the boundary layer. The velocity profiles for a broad range of plate angles, from 90, the stagnation point flow, to −17.9, which is the angle at which flow separates, are presented in Figure 7.12, and the values of shear stress at the wall are provided in Table 7.2.

As the plate angle decreases from 90 (stagnation point flow), the shear stress at the wall decreases, and the fluid eventually will not be able to follow the plate and will separate. The separation takes place at around −17.9° (107.9°). Past 107.9° the velocity gradient at the wall goes to zero, and the numerical solution diverges, and the flow is said to separate, i.e.,

TABLE 7.2

Shear Stress at the Wall for Different
Wedge Angles

$\alpha°$	m	β	$f'(0)$
90.00	1	1	1.2077
20.00	0.1250	0.2222	0.7016
10.00	0.0588	0.1111	0.5961
0.00	0.0000	0.0000	0.4696
−5.00	−0.0270	−0.0556	0.3934
−10.00	−0.0526	−0.1111	0.3017
−15.00	−0.0769	−0.1667	0.1764
−17.00	−0.0863	−0.1889	0.0962
−17.80	−0.0900	−0.1978	0.0335
−17.85	−0.0902	−0.1983	0.0247
−17.90	−0.0905	−0.1989	0.0054

in addition to the plate surface, there is a second point in the flow field, where the velocity is zero. This is further explained as follows, by considering the momentum equation at the edge of the boundary layer

$$\frac{dP}{dx} = -\rho U_e \frac{dU_e}{dx} = -\rho U_\infty^2 m x^{2m-1} \tag{7.218}$$

Positive values of m indicate that the plate slopes up relative to the incoming fluid and therefore, as indicated by the equation, the pressure gradient along the wedge is negative, or the pressure is decreasing in the direction of the fluid flow. This is known as the favorable pressure gradient, as the fluid prefers to flow in the direction of decreasing pressure.

For negatively sloped plate, m is negative and the fluid is therefore flowing in the direction of increasing pressure or the flow experiences adverse pressure gradient. The fluid's momentum carries it along the plate. This motion is always opposed by the drag and normally assisted by a favorable pressure gradient. For a sloping down plate, both drag and pressure oppose the flow of the fluid, and as the plate angle increases, eventually a point will be reached that the pressure and drag forces are more than the inertial forces and the fluid is no longer able to flow along the plate and it separates. This means that some of the fluid actually flows from the high pressure downstream region to the low pressure upstream region, along the plate. This fluid, as it moves upstream, encounters the main flow and therefore a point of zero velocity; actually a curve (streamline) is formed along which the velocity is zero. The $\psi = 0$ streamline, which normally is along the wall, extends into the fluid. Once the flow separates, the boundary layer approximation is no longer valid. In the separated region and near the wall, vortices are formed and the flow becomes turbulent. The separation of the flow would cause the incoming flow to perceive a different shaped object. This makes the potential flow solution obtained for the object not valid past the separation point, since it was based on a different geometry than the one perceived by the potential flow (Figure 7.13).

7.5.4 Flow over Curved Surfaces

Flow around curved objects is encountered commonly in practical applications like flow over a circular cylinder, a tube banks, a sphere, a car, wings of an airplane, turbine blades, and pump impeller vanes. Unless the body is streamlined, like an airfoil, and the angle of

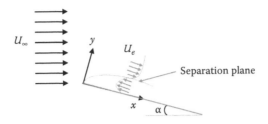

FIGURE 7.13
Flow separation for a negatively sloped flat plate.

attack is not too large, the flow will separate, similar to what we discussed for a sloping down flat plate. Unlike the sloping down plate, the flow is initially attached to the object, and the boundary layer assumption will provide a reasonable approximation to the flow field, up to the point where the flow separates. The flow around curved objects is also a strong function of Reynolds number: $Re = \dfrac{\rho U_\infty D}{\mu}$ which is based on the free stream velocity, upstream of the cylinder, and the cylinder diameter. The different flow regimes encountered for flow around a cylinder are shown in Table 7.3. For very small Reynolds numbers, the viscosity dominates the flow, and flow is very orderly and is slow enough to be able to follow the object, and no separation occurs. There are two stagnation points: one in front and one in the back. This is known as the creeping flow or Stokes flow. This flow is actually very similar to the ideal case

TABLE 7.3

Flow Patterns around a Cylinder

$0 < Re_D < 5$		No separation, creeping, or Stokes flow.
$5 < Re_D < 40$		A pair of symmetric vortices appears, counter-rotating and fixed in the wake, with their elongation growing with Re_D.
$40 < Re_D < 150$		A laminar boundary layer in front of the cylinder, with Von Kármán vortex street formed behind with vortices shed with a frequency f from the two sides. The flow behind cylinder is laminar and unsteady.
$150 < Re_D < 3 \times 10^5$		The boundary layer is laminar, past the separation point, the vortex street is turbulent.
$3 \times 10^5 < Re_D < 3.5 \times 10^6$		The boundary layer and the wake are turbulent. The separation point moves behind the cylinder, and the wake becomes narrower and disorganized.
$Re_D > 3.5 \times 10^6$		A turbulent vortex street is re-established, but it is narrower than that formed for $150 < Re_D < 3 \times 10^5$.

of inviscid flow. Once Reynolds number increases past 5, flow separates, and the separation occurs at around 80° from the front stagnation point. For $5 < Re < 40$, a pair of counter-rotating vortices will form behind the cylinder, as shown in Table 7.3, that is stable and symmetric with respect to the midplane. The size of the vortices grows with increasing Reynolds number, and past $Re = 40$, the vortices begin to separate from the cylinder, and the separation is alternating with one separating from the top, and at a later time another separating from the bottom (vortex shedding). The vortices move downstream and disappear due to friction. The flow becomes unsteady past separation, and asymmetrical, with the alternate shedding of vortices creating periodic lateral forces on the cylinder that although the average force perpendicular to the flow (lift) is zero, the shedding of vortices results in a reversing force perpendicular to the direction of the flow, which can result in flow-induced vibration and potentially failure.

The pattern of vortices formed is known as Von-Karman Vortex Street. The frequency by which the vortices are shed is expressed by a nondimensional group called Strouhal number, defined as

$$Sr = \frac{fD}{U_\infty} \tag{7.219}$$

where f is the frequency by which vortices are shed. Roshko [8] provided the following correlations between the Strouhal number and Reynolds number,

$$Sr = 0.212\frac{1-21.2}{Re} \quad 50 < Re < 150$$

$$\tag{7.220}$$

$$Sr = 0.212\frac{1-12.7}{Re} \quad 300 < Re < 2000$$

which is shown in Figure 7.14. Between $Re = 50$ and 150, Strouhal number increases almost linearly with Reynolds number from 0.12 to about 0.18 and then levels off. Past Reynolds

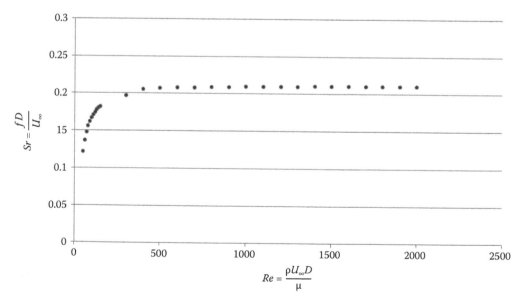

FIGURE 7.14
Strouhal number variation with the Reynolds number.

number of 300 and until transition to turbulent flow at *Re* of around 300,000, Strouhal number essentially remains constant at 0.2. In this range, the flow is laminar ahead and turbulent, past the separation point.

Past Reynolds number of about 300,000, the entire flow makes a transition to turbulent flow and the separation point moves further down along the cylinder to about an angle of about 130°, and the wake region narrows significantly, but becomes highly disorganized. The moving of the separation point means that the size of the low pressure region behind the cylinder decreases significantly, and even though the flow becomes fully turbulent, the total drag experienced by the cylinder actually decreases. For $Re_D > 3.5 \times 10^6$, the turbulent vortex street is reestablished, but it is narrower than that formed for $150 < Re_D < 3 \times 10^5$.

The *Re* ranges given are approximate and depend strongly on the surface roughness of the cylinder and the free stream turbulence.

7.5.5 Friction and Pressure Drag

The drag force is the net force acting on the object in the direction of the flow. The drag coefficient is defined as

$$C_D = \frac{\dfrac{F_D}{A}}{\dfrac{1}{2}\rho U_\infty^2} \tag{7.221}$$

where A is the projected area perpendicular to the flow direction. The total drag acting on an object can be the result of two forces: the shear or viscous force acting on the surface and the pressure force:

$$F_D = F_{friction} + F_{pressure} \tag{7.222}$$

Therefore, the drag coefficient can be written as

$$C_D = \frac{\dfrac{F_{friction}}{A}}{\dfrac{1}{2}\rho U_\infty^2} + \frac{\dfrac{F_{pressure}}{A}}{\dfrac{1}{2}\rho U_\infty^2} = C_{D,f} + C_{D,P} \tag{7.223}$$

where
$C_{D,f}$ is the drag coefficient due to friction
$C_{D,P}$ is the drag coefficient due to pressure, sometimes referred to as form drag, since it is caused by the object's shape

The drag coefficient due to friction, $C_{D,f}$, is based on the component of the shear force acting in the direction of the flow and is different from c_f that is based on the total drag experienced by the object. For flow over a horizontal flat plate, the two are equal. For separated flows, the pressure force is the dominant force contributing to the total drag.

At very low Reynolds numbers, in the absence of flow separation, Stokes showed that the total drag force acting on a sphere is given by

$$F_D = 3\pi\mu U D \tag{7.224}$$

or at low velocities, the drag is directly proportional to the velocity, the fluid viscosity, and the diameter of the sphere. Equation 7.224 can be rearranged to

$$C_D = \frac{24}{Re} \tag{7.225}$$

White [9] provides the following correlation for the drag coefficient on a sphere

$$C_{D,sphere} = \frac{24}{Re} + \frac{6}{1+\sqrt{Re}} + 0.4 \quad Re < 2\times 10^5 \tag{7.226}$$

that is accurate to within 10% until the critical Reynolds number, where the transition to turbulence happens. Based on the available experimental data, White [9] also provides

$$C_{D,cylinder} = 1 + 10Re^{-2/3}, \quad Re < 2\times 10^5 \tag{7.227}$$

for cylinders.

7.5.6 Ideal Flow over a Cylinder

From the potential flow theory, the stream function for inviscid flow around a cylinder is given by

$$\psi = U_\infty r \sin\theta \left(1 - \frac{R^2}{r^2}\right) \tag{7.228}$$

From the definition of stream function, then

$$u_\theta = -\frac{\partial \psi}{\partial r} = -U_\infty \sin\theta \left(1 + \frac{R^2}{r^2}\right) \tag{7.229}$$

$$u_r = \frac{1}{r}\frac{\partial \psi}{\partial \theta} = U_\infty \left(1 - \frac{R^2}{r^2}\right)\cos\theta \tag{7.230}$$

Close to the cylinder $u_r = 0$ and $u_\theta = -2U_\infty \sin\theta$, and the fluid velocity is

$$U = -2U_\infty \sin\theta \tag{7.231}$$

Once the velocity is known, the pressure for inviscid flow is obtained from the Bernoulli equation

$$\frac{P}{\rho} + \frac{U^2}{2} = \frac{P_\infty}{\rho} + \frac{U_\infty^2}{2} \tag{7.232}$$

The pressure coefficient is then

$$C_P = \frac{P - P_\infty}{\frac{1}{2}\rho U_\infty^2} = 1 - \frac{U^2}{U_\infty^2} \tag{7.233}$$

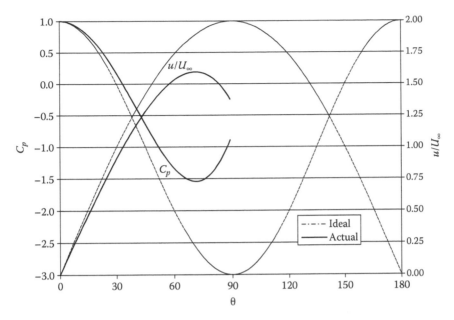

FIGURE 7.15
Pressure coefficient and velocity field for ideal and actual case.

and near the cylinder

$$C_P = 1 - 4\sin^2\theta \tag{7.234}$$

This expression shows that pressure is symmetric with respect to the x and y coordinates passing through the center of the cylinder, which means that according to the potential flow theory, there is no lift or drag.

The experimental measurements of flow velocity around a cylinder can be reasonably approximated by

$$\frac{U}{U_\infty} = 1.814\left(\frac{x}{R}\right) - 0.271\left(\frac{x}{R}\right)^3 - 0.0471\left(\frac{x}{R}\right)^5 \tag{7.235}$$

for $Re = 19,000$. The ideal and actual velocity distributions as well as the pressure coefficient based on them are plotted in Figure 7.15.

7.5.7 Numerical Solution of Boundary Layer Flow over a Cylinder

The boundary layer approximation should be valid up until around the separation point. The momentum equation for boundary layer flow over a circular cylinder becomes

$$\frac{\partial^3 f}{\partial\eta^3} + f\left(\frac{\partial^2 f}{\partial\eta^2}\right) + \beta\left(1 - \left(\frac{\partial f}{\partial\eta}\right)^2\right) = 2\xi\left[\frac{\partial^2 f}{\partial\xi\partial\eta}\frac{\partial f}{\partial\eta} - \frac{\partial f}{\partial\xi}\frac{\partial^2 f}{\partial\eta^2}\right] \tag{7.236}$$

where

$$\beta = R \frac{2\xi \dfrac{dU_e^*}{dx}}{\left(\dfrac{r}{R}\right)^2 U_e^{*2}} \tag{7.237}$$

For flow around a cylinder, the ideal velocity distribution is given by

$$U_e^* = \frac{U_e}{U_\infty} = 2\sin\theta = 2\sin\frac{x}{R} \tag{7.238}$$

This equation will remain valid, as long as the flow has not been separated. Therefore,

$$\frac{dU_e^*}{dx} = \frac{2}{R}\cos\frac{x}{R} \tag{7.239}$$

$$\frac{dP}{dx} = -\rho U_e \frac{dU_e}{\partial x} = -\rho \frac{2}{R} U_\infty^2 \sin 2\theta \tag{7.240}$$

$$\xi = \frac{1}{R}\int_0^x \left(\frac{r}{R}\right)^2 \left(U_e^*\right) dx = \frac{1}{R}\int_0^x U_e^* dx = \frac{1}{R}\int_0^x \left(2\sin\frac{x}{R}\right) dx = 2\left(1-\cos\frac{x}{R}\right) \tag{7.241}$$

In the boundary layer approximation, we have neglected variation of r with respect to x, since the thickness of boundary layer is small. Solving for x

$$\cos\frac{x}{R} = 1 - \frac{\xi}{2} \tag{7.242}$$

and

$$\beta = R \frac{2\xi\left(\dfrac{2}{R}\cos\dfrac{x}{R}\right)}{\left(\dfrac{r}{R}\right)^2 \left(2\sin\left(\dfrac{x}{R}\right)\right)^2} = \frac{(4-2\xi)}{(4-\xi)} \tag{7.243}$$

Before solving the momentum equation, let us explore its expected behavior. Since from Equation 7.243 β depends on ξ, the governing equations will not have similarity solution, and the terms on the RHS represent the deviation from similarity solution, which prior to separation are expected to be small. So if these terms are small, then momentum equation is the same as that for flow over a wedge, and therefore the separation is expected to happen, when

$$\beta = \frac{(4-2\xi)}{(4-\xi)} \approx -0.1989 \tag{7.244}$$

which corresponds to a location around the cylinder corresponding to

$$\xi = 2.18 \tag{7.245}$$

and

$$\theta_{sep} = \cos^{-1}\left(1 - \frac{2.18}{2}\right) = 95.2° \tag{7.246}$$

The experimental results show the flow separation to occur around 80°, and as is presented as follows, the numerical solution of the boundary layer equations shows that the flow separation happens at around 103.6°. To solve the problem numerically, as before defining $g = f'$, then

$$\frac{\partial^2 g}{\partial \eta^2} + f \frac{\partial g}{\partial \eta} = 2\xi \left[g \frac{\partial g}{\partial \xi} - \frac{\partial f}{\partial \xi} \frac{\partial g}{\partial \eta} \right] - \beta(1 - g^2) \tag{7.247}$$

Using central difference approximation for the derivatives in the η direction and backward difference for derivatives in the ξ direction, the finite difference form of the equation written as Equation 7.8

$$A_j g_{j-1} + B_j g_j + C_j g_{j+1} = R_j \tag{7.248}$$

where

$$A_j = \frac{1}{(\Delta \eta)^2} - \frac{f_{i,j}}{2\Delta \eta} \tag{7.249}$$

$$B_j = \frac{-2}{(\Delta \eta)^2} \tag{7.250}$$

$$C_j = \frac{1}{(\Delta \eta)^2} + \frac{f_{i,j}}{2\Delta \eta} \tag{7.251}$$

$$R_j = 2\xi \left[g_{i,j} \frac{g_{i,j} - g_{i-1,j}}{\Delta \xi} - \left(\frac{f_{i,j} - f_{i-1,j}}{\Delta \xi} \right) \frac{g_{i,j+1} - g_{i,j-1}}{2(\Delta \eta)} \right] - \beta \left(1 - g_{i,j}^2 \right) \tag{7.252}$$

In the form given, Equation 7.248 does not converge by using the traditional Gauss–Seidel iteration scheme. To obtain the solution, we use an iterative scheme based on Newton's method. The procedure is explained next, by defining N functions based on Equation 7.248 as

$$F_j(x_j) = A_j x_{j-1} + B_j x_j + C_j x_{j+1} - R_j(x_j) \tag{7.253}$$

Therefore, obtaining the solution to Equation 7.253 is equivalent to finding the values of x_j that make Equation 7.253 zero or finding the roots of Equation 7.253. The roots are obtained iteratively. If x_j^m is the value of x_j after m iterations, then using Taylor series,

$$F_i\left(x_j^{m+1}\right) = F_i\left(x_j^m\right) + \left(x_j^{m+1} - x_j^m\right) \frac{dF_j}{dx_j} \tag{7.254}$$

If x_j^{m+1} is the root of the equation, then substituting from Equation 7.254,

$$0 = A_j x_{j-1}^m + B_j x_j^m + C_j x_{j+1}^m - R_j(x_j) + \left(x_j^{m+1} - x_j^m\right) \left(B_j - \frac{dR_j}{dx_j}\right) \tag{7.255}$$

which simplifies to

$$A_j x_{j-1}^m + \left(B_j - \frac{dR_j}{dx_j}\right) x_j^{m+1} + C_j x_{j+1}^m = R_j(x_j) - \frac{dR_j}{dx_j} x_j^m \tag{7.256}$$

The difference between the values in two successive iterations is typically small; therefore Equation 7.256 can be written as

$$A_j x_{j-1}^{m+1} + \left(B_j - \frac{dR_j}{dx_j} \right) x_j^{m+1} + C_j x_{j+1}^{m+1} = R_j(x_j) - \frac{dR_j}{dx_j} x_j^m \tag{7.257}$$

Note that x_j in the above equations stands for $g_{i,j}$ and that after the solution converges, the terms involving the derivatives on both sides cancel out, and therefore the solution to Equation 7.257 is the same as 7.256. Applying this approach to Equation 7.252,

$$\frac{dR_j}{dg_{i,j}} = 2\xi \left[\frac{2g_{i,j} - g_{i-1,j}}{\Delta\xi} \right] + 2\beta g_{i,j} \quad \text{at } \xi = 0 \tag{7.258}$$

and at $\xi = 0$

$$R_j(g_{i,j}) = -\left(1 - g_{i,j}^2 \right) \tag{7.259}$$

$$\frac{dR_j}{dg_{i,j}} = (2g_{i,j}) \tag{7.260}$$

Again, these equations are solved using spreadsheets. Figure 7.16 is a plot of velocity profile at different angles along the cylinder, and as can be seen, the flow separation is predicted to occur at an angle of 103.6°. The series solution results presented in Schlichting [1] predict a separation angle of 108.8, and the numerical solutions of Schoenauer [1, p. 203] predict a value of 104.5°.

The experimental studies of flow over a circular cylinder, for example, those of Hiemenz as indicated in Ref. [1] (p. 162), have shown that the flow separation happens at around 81° for laminar flow. The earlier numerical results over predict separation angle by over 20°. A number of factors contribute to the discrepancy. For Reynolds number greater than 40, the flow actually becomes transient, i.e., vortices are formed and separate from the surface

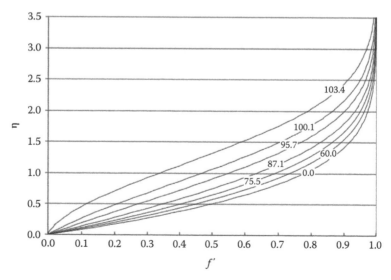

FIGURE 7.16
Velocity profiles using ideal flow outside the boundary layer.

of the cylinder. In the separated region, the flow becomes turbulent, and the axial diffusion becomes important, which means that the upstream conditions are now influenced by what happens downstream. The flow separation also changes the potential flow solution, as it effectively changes the shape of the cylinder seen by the inviscid flow. To more accurately predict flow separation, better estimate of velocity distribution outside of the boundary layer must be used.

The velocity distribution given by Equation 7.235 is a much more realistic approximation. Based on this velocity distribution,

$$\xi = \frac{1}{R}\int_0^x \left(\frac{U}{U_\infty}\right)dx = \frac{1.814}{2}\frac{x^2}{R^2} - \frac{0.271}{4}\frac{x^4}{R^4} - \frac{0.0471}{6}\frac{x^6}{R^6} \tag{7.261}$$

and

$$\frac{dU^*}{dx} = 1.814\left(\frac{1}{R}\right) - 3\times 0.271\frac{x^2}{R^3} - 5\times 0.0471\frac{x^4}{R^5} \tag{7.262}$$

note that $\theta = \dfrac{x}{R}$. These can then be substituted in Equation 7.237 to calculate β and improve the solution.

Figure 7.17 shows the velocity distribution obtained numerically using the more realistic free stream velocity distribution given by Equation 7.235. The solution converges until 78.8°, after which the solution oscillates and eventually diverges, indicating boundary layer separation taking place around 79°. This is close to the experimentally measured value of 81 or solution result of Smith and Clutter [10] of 80°.

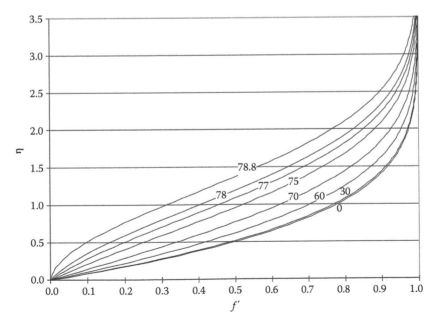

FIGURE 7.17
Velocity profiles using more realistic velocity field outside boundary layer.

7.6 Jet Flow

Jets are studied extensively [11–24] as they are used in numerous heat transfer applications like cooling of microelectronic equipments, annealing of metal and plastic sheets, and drying of textiles. They are also extensively encountered in combustion, both in diffusion flames where the fuel exits a nozzle and mixes with the surrounding oxidizer as in diesel engines, or in premixed flames where the fuel and oxidizer are mixed before they exit the nozzle as in gas welding. High-speed jets are also used as a cutting tool, including for cutting metals.

7.6.1 Planar Jet

Consider a narrow rectangular opening (Figure 7.18) of width $2b$ from which a fluid jet exits with a uniform velocity U into a similar fluid that is initially stationary. The jet applies a shear force to the surrounding fluid, dragging it along and therefore inducing a motion in the surrounding fluid, resulting in an expansion of the jet. As the jet expands and entrains more of the surrounding fluid, it slows down.

Since the jet is predominantly moving in the x direction, the boundary layer approximation can be invoked. Using this approximation, Schlichting [1] solved the momentum equation for 2D and axisymmetric jets using similarity solution. For boundary layer flows, the pressure variation across the boundary layer is negligible, and therefore the axial pressure variation inside and outside of the boundary layer is the same. For the case of jet flow, pressure outside of the boundary layer is constant, and therefore, pressure is constant everywhere, and therefore the governing equations are the same as those for flat plate:

$$\frac{\partial u}{\partial x} + \frac{\partial v}{\partial y} = 0 \tag{7.263}$$

$$\rho \left(u \frac{\partial u}{\partial x} + v \frac{\partial u}{\partial y} \right) = \mu \frac{\partial^2 u}{\partial y^2} \tag{7.264}$$

$$\rho c_p \left(u \frac{\partial T}{\partial x} + v \frac{\partial T}{\partial y} \right) = k \frac{\partial^2 T}{\partial y^2} \tag{7.265}$$

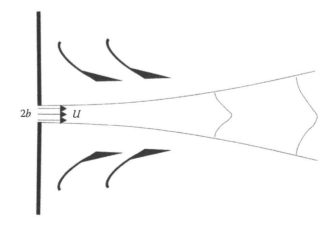

FIGURE 7.18
Planar jet flow.

The boundary conditions are

$$x = 0, \quad -b \le y \le b, \quad u = U$$

$$y = 0, \quad v = 0, \quad \frac{\partial u}{\partial y} = 0 \tag{7.266}$$

$$y \to \infty, \quad u \to 0$$

The solutions presented are for laminar flow. In practice, the flow becomes turbulent for Reynolds numbers as low as 30; however, as shown later, the turbulent flow solution is almost identical to the laminar flow one, and much insight can be gained by studying the laminar flow cases.

Writing the momentum equation in its conservative form and integrating it with respect to y from $-\infty$ to $+\infty$ give

$$\frac{\partial}{\partial x} \int_{-\infty}^{\infty} \rho u^2 dy + \rho v u \Big|_{\infty} - \rho v u \Big|_{-\infty} = \mu \left[\frac{\partial u}{\partial y} \Big|_{\infty} - \frac{\partial u}{\partial y} \Big|_{-\infty} \right] \tag{7.267}$$

Since

$$v \quad \text{and} \quad \frac{\partial u}{\partial y} \to 0 \quad \text{as} \quad y \to \pm\infty \tag{7.268}$$

then

$$\frac{\partial}{\partial x} \int_{-\infty}^{\infty} \rho u^2 dy = 0 \tag{7.269}$$

or

$$\int_{-\infty}^{\infty} \rho u^2 dy = M \tag{7.270}$$

Similarly, if the energy equation is written in conservative form and integrated in the y direction, then

$$\frac{\partial}{\partial x} \int_{-\infty}^{\infty} \rho c_p u T dy + \rho c_p v T \Big|_{\infty} - \rho c_p v T \Big|_{-\infty} = k \left[\frac{\partial T}{\partial y} \Big|_{\infty} - \frac{\partial T}{\partial y} \Big|_{-\infty} \right] \tag{7.271}$$

Since

$$v \text{ and } \frac{\partial T}{\partial y} \to 0 \text{ as } y \to \pm\infty \tag{7.272}$$

then

$$\frac{\partial}{\partial x} \int_{-\infty}^{\infty} \rho c_p u T dy = 0 \tag{7.273}$$

or

$$\int_{-\infty}^{\infty} \rho c_p u T dy = H \tag{7.274}$$

where H is the jet's total enthalpy which remains constant as the jet expands. If the jet exits the nozzle with velocity U_j, and temperature T_j, and the jet width is $2b$, then

$$M = 2\int_{0}^{b} \rho U_j^2 dy = 2b\rho U_j^2 \tag{7.275}$$

$$H = 2\int_{0}^{b} \rho c_p U_j T_j dy = 2b\rho c_p U_j T_j \tag{7.276}$$

To obtain the solution, consider two axial locations; at each location, the velocity is maximum at the center, U_0, and drops to zero some distance away from the centerline, δ. The centerline velocity U_0 is a decreasing function of x, and the situation has some similarity to the Falkner–Skan flow over a negative wedge (sloping down surface) where the free stream velocity decreases according to some power of x. Therefore, it may be reasonable to expect the velocity and temperature profiles to be similar. If we define the similarity variables,

$$\eta = \frac{y}{\delta} \tag{7.277}$$

As the case for flow over a wedge, the boundary layer thickness is expected to be

$$\delta = \sqrt{C\frac{\nu x}{U_0}} \tag{7.278}$$

and assume the center velocity to drop with some yet unknown power $(-1/n)$ of x,

$$U_0 = Bx^{-1/n} \tag{7.279}$$

$$\delta = \sqrt{\frac{C}{B}\nu x^{\frac{n+1}{n}}} \tag{7.280}$$

where C, B, and n are constants. Therefore, the similarity variable becomes

$$\eta = \sqrt{\frac{B}{C}} \frac{y}{\nu^{1/2} x^{\frac{n+1}{2n}}} \tag{7.281}$$

for the similarity solution to exist, there must exist a function f, such that

$$f'(\eta) = \frac{u}{U_0} = \frac{u}{Bx^{-1/n}}$$ (7.282)

The momentum flux is

$$M = 2\rho \int_0^\infty u^2 dy = 2\rho \int_0^\infty \left[Bx^{-1/n} f' \right]^2 d\eta \frac{dy}{d\eta}$$ (7.283)

$$M = 2\rho \sqrt{B^3 C} v^{1/2} x^{\frac{n-3}{2n}} \int_0^\infty [f']^2 d\eta$$ (7.284)

and for M to be constant, independent of x, $n = 3$
   Therefore,

$$U_0 = Bx^{-1/3}$$ (7.285)

$$\delta = \sqrt{\frac{C}{B}} v^{1/2} x^{2/3}$$ (7.286)

$$\eta = \sqrt{\frac{B}{C}} \frac{y}{v^{\frac{1}{2}}} x^{-2/3}$$ (7.287)

For the similarity solution to exist, there must exist a function f, such that

$$f(\eta) = \frac{\psi}{U_0 \delta} = \frac{\psi}{\sqrt{BC} v^{1/2} x^{1/3}}$$ (7.288)

and

$$\psi = \sqrt{BC} v^{1/2} x^{1/3} f$$

Then

$$f'(\eta) = \frac{\partial}{\partial y} \left(\frac{\psi}{\sqrt{BC} v^{1/2} x^{1/3}} \right) \frac{\partial y}{\partial \eta} = \frac{u}{\sqrt{BC} v^{1/2} x^{1/3}} \frac{1}{\sqrt{\frac{B}{C}} \frac{1}{v^{1/2} x^{2/3}}} = \frac{u}{Bx^{-1/3}} = \frac{u}{U_0}$$ (7.289)

$$f'(\eta) = \frac{u}{Bx^{-1/3}}$$ (7.290)

Expressing the momentum and energy equations in terms of the stream function, and following a process similar to that flow flat plate, the momentum equation becomes

$$f''' + \frac{C}{3}f'^2 + \frac{C}{3}ff'' = 0 \tag{7.291}$$

Since C is an arbitrary constant, we assume $C = 6$

$$f''' + 2f'^2 + 2ff'' = 0 \tag{7.292}$$

$$\eta = 0, \quad f = 0, \quad f'' = 0$$
$$\eta \to \infty, \quad f' \to 0 \tag{7.293}$$

Schlichting [1] showed that this equation can be solved analytically. Integrating it once,

$$f'' + 2ff' = c_1 \tag{7.294}$$

From the boundary conditions for $\eta = 0$, $c_1 = 0$. Integrating Equation 7.294 once more,

$$f' + f^2 = c_2 \tag{7.295}$$

The solution to this equation is hyperbolic tangent. Assume

$$f = \tan h(b\eta) = \frac{\sin h(b\eta)}{\cos h(b\eta)} \tag{7.296}$$

$$f' = b\left[1 - \tan h^2(b\eta)\right] \tag{7.297}$$

substituting in Equation 7.295

$$b - b\tan h^2(b\eta) + \tan h^2(b\eta)^2 = c_2 \tag{7.298}$$

for the solution to exist

$$b = c_2 = 1 \tag{7.299}$$

and the solution becomes

$$f = \tan h(\eta) \tag{7.300}$$

$$f' = [1 - \tan h^2(\eta)] \tag{7.301}$$

We now have to determine the constant B, by relating it to the jet's momentum. From Equation 7.284,

$$M = 2\rho\sqrt{B^3 C}v^{1/2}\int_0^\infty \left[1 - \tan h^2(\eta)\right]^2 d\eta \qquad (7.302)$$

since

$$\int_0^\infty \left[1 - \tan h^2[ay]\right]^2 dy = \frac{2}{3a} \qquad (7.303)$$

Then

$$B = \left(\frac{3}{32}\frac{M^2}{\rho^2 v}\right)^{1/3} \qquad (7.304)$$

and

$$\eta = \left(\frac{1}{48}\frac{M}{\rho v^2 x^2}\right)^{1/3} y \qquad (7.305)$$

Substituting for B and C in Equation 7.301, and simplifying, results in

$$u = 0.4543\left(\frac{M^2}{\rho^2 v x}\right)^{1/3}\left[1 - \tan h^2\left(0.2752\left(\frac{M}{\rho v^2 x^2}\right)^{1/3} y\right)\right] \qquad (7.306)$$

The centerline velocity is

$$U_0 = 0.4543\left(\frac{M^2}{\rho^2 v x}\right)^{1/3} \qquad (7.307)$$

This equation shows that the centerline velocity decreases with the power of $x^{-1/3}$. The solution also indicates that at the origin, the centerline velocity approaches infinity. The boundary layer approximation and therefore the resulting similarity solution are not expected to be valid at small values of x or close to the jet exit plane. The similarity solution corresponds to the case of a jet having zero width at the origin with infinite velocity. This point is called the virtual origin [25].

To determine the thickness of the jet boundary layer, the boundary layer thickness is defined as the location where velocity is 0.01 of the centerline velocity:

$$\frac{u}{U_0} = \left[1 - \tan h^2\left(0.2752\left(\frac{M}{\rho v^2 x^2}\right)^{1/3} y\right)\right] = 0.01 \qquad (7.308)$$

$$\delta = \frac{10.876}{\left(\dfrac{M}{\rho v^2 x^2}\right)^{1/3}} \qquad (7.309)$$

and the jet thickness is twice the boundary layer thickness

$$d = 2\delta = 21.8 \left(\frac{\rho v^2}{M} \right)^{1/3} x^{2/3} \tag{7.310}$$

Equation 7.310 shows that the jet width is proportional to $x^{2/3}$. As the jet expands, it entrains the surrounding fluid. The mass flow rate can be determined by

$$\dot{m} = 2\rho \int_0^\infty u\,dy = 2\rho U_0 \int_0^\infty \frac{1}{\cos h^2 ay}\,dy = 3.3021(\rho \mu x M)^{1/3} \tag{7.311}$$

The ratio of the mass flow rate of the fluid to that exiting the nozzle is

$$\frac{\dot{m}}{2\rho U_j b} = 3.3021 \left(\frac{x/b}{Re} \right)^{1/3} \tag{7.312}$$

where

$$Re = \frac{\rho U_j 4b}{\mu} \tag{7.313}$$

is the jet's Reynolds number, based on the nozzle hydraulic diameter and average exit velocity. For example, for $Re = 30$, and 20 jet width from the nozzle, the jet flow rate is 3.63 times of what exited the nozzle.

The temperature distribution in the jet can be obtained in a similar fashion. If we define a nondimensional temperature

$$T^* = \frac{T - T_\infty}{T_0 - T_\infty} \tag{7.314}$$

where T_0 is the centerline temperature, then $T_0 - T_\infty$ must also change according to $x^{-1/3}$ for the similarity solution to exist. Therefore, there exists a function

$$T^*(\eta) = \frac{T - T_\infty}{Bx^{-1/3}} \tag{7.315}$$

therefore

$$T = T_\infty + Bx^{-1/3} T^*(\eta) \tag{7.316}$$

Substituting in the energy equation and changing variables result in

$$\frac{1}{Pr} \frac{\partial^2 T^*}{\partial \eta^2} + 2(fT^*)' = 0 \tag{7.317}$$

Integrating Equation 7.317 once,

$$\frac{1}{Pr} \frac{\partial T^*}{\partial \eta} + 2fT^* = d_1 \tag{7.318}$$

since at $\eta = 0$, $\dfrac{\partial T^*}{\partial \eta} = 0$

$$\frac{1}{Pr}\frac{\partial T^*}{\partial \eta} + 2fT^* = 0 \tag{7.319}$$

$$\frac{\dfrac{\partial T^*}{\partial \eta}}{T^*} = -2Prf \tag{7.320}$$

$$\ln T^* = -2Pr\int f d\eta = -2Pr\int \tan h(\eta)d\eta = -2Pr\ln\left[\cos h(\eta)\right] + d_2 \tag{7.321}$$

resulting in

$$T^* = \frac{1}{\left[\cos h\left(\left(\dfrac{1}{48}\dfrac{M}{\rho v^2 x^2}\right)^{1/3}y\right)\right]^{2Pr}} \tag{7.322}$$

The centerline temperature can be determined from

$$\int_{-\infty}^{\infty} \rho c_p u T dy = H \tag{7.323}$$

For $Pr = 1$, the integration results in

$$T_0 - T_\infty = \frac{3^{2/3}}{4\sqrt{2}}\frac{H}{\rho c_p}(vxM)^{-1/3} \tag{7.324}$$

The nondimensional velocity and temperature distribution for different Prandtl numbers are plotted in Figure 7.19. Again, note that for $Pr = 1$, $f' = T^*$.

7.6.2 Axisymmetric Jet

The solution for axisymmetric jet was also solved by Schlichting [1] first. Starting with the governing equations

$$\frac{\partial u}{\partial x} + \frac{1}{r}\frac{\partial(rv)}{\partial r} = 0 \tag{7.325}$$

$$\rho\left(u\frac{\partial u}{\partial x} + v\frac{\partial u}{\partial r}\right) = \mu\frac{1}{r}\frac{\partial}{\partial r}\left(r\frac{\partial u}{\partial r}\right) \tag{7.326}$$

$$\rho c_p\left(u\frac{\partial T}{\partial x} + v\frac{\partial T}{\partial r}\right) = k\frac{1}{r}\frac{\partial}{\partial r}\left(r\frac{\partial T}{\partial r}\right) \tag{7.327}$$

$$r = 0, \quad v = 0, \quad \frac{\partial u}{\partial r} = 0, \quad \frac{\partial T}{\partial r} = 0$$
$$r \to \infty, \quad u = 0, \quad T = T_\infty \tag{7.328}$$

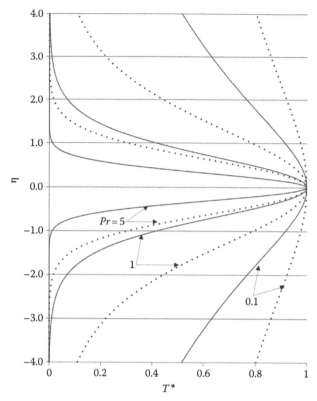

FIGURE 7.19
Temperature distribution for planar (solid) and axisymmetric (dashed) jets for different *Pr*.

The stream function for axisymmetric flow is given by

$$u = \frac{1}{r}\frac{\partial \psi}{\partial r}$$

$$v = -\frac{1}{r}\frac{\partial \psi}{\partial x}$$

(7.329)

For the similarity solution to exist, U_0 and $(T_0 - T_\infty)$ must be inversely proportional to x:

$$u^* = \frac{u}{U_0} = \frac{u}{Bx^{-1}} = \frac{f'}{\eta}$$

$$T^* = \frac{T - T_\infty}{T_0 - T_\infty}$$

(7.330)

The transformed momentum and energy equations become

$$\frac{\partial}{\partial \eta}\left(f'' - \frac{f'}{\eta}\right) = \frac{1}{\eta^2}(ff' - \eta f'^2 - \eta ff'')$$

(7.331)

$$\frac{\partial}{\partial \eta}\left(\frac{\eta}{Pr}T^{*\prime} + fT^*\right) = 0 \tag{7.332}$$

The solution is

$$f(\eta) = \frac{\dfrac{1}{2}\eta^2}{1 + \dfrac{1}{8}\eta^2} \tag{7.333}$$

$$\eta = \left(\frac{3M}{16\pi\rho\nu^2}\right)^{1/2}\frac{r}{x} \tag{7.334}$$

$$\frac{f'}{\eta} = \frac{1}{\left(1 + \dfrac{1}{8}\eta^2\right)^2} \tag{7.335}$$

and

$$T^* = \frac{1}{\left(1 + \dfrac{1}{8}\eta^2\right)^{2Pr}} \tag{7.336}$$

The nondimensional temperature distribution for axi-symmetric jets are also plotted in Figure 7.19.

7.6.3 Jet Impingent

Jet impingent on a surface is an effective method for removing a large amount of heat from the striking surface and is generally accomplished by using an array of round or rectangular jets. The fluid exits the nozzles, located a short distance from the surface, impinges on the surface, and travels along the surface. There are a number of parameters that impact the heat removal rate, including the jet distance to the surface, the flow rate, the jet diameter or aspect ratio, the jet and target temperatures, the ambient temperature, and the jet angle. A clear understanding of the impact of the different parameters is needed for the effective design of such systems.

There is a large body of literature on the study of impinging jet. Gardon and Akfirat [17], Martin [14], and Craft et al. [22] used computational fluid dynamic simulations. As indicated by Martin [14], the flow pattern in an impingent jet can be divided into three regions. Close to the nozzle exit, there is a free jet region, followed by the stagnation region, and finally a region of lateral flow, also called the wall jet.

Miyazaki and Silberman [19] used boundary layer equations to numerically solve for velocity and temperature distributions. The velocity outside of the boundary layer was determined from the potential flow solutions. The Nusselt number [26] variation along the plate is shown in Figure 7.18, for nozzle placed 0.5, 1, and 1.5 jet width (D) from the plate. There is a peak in heat transfer at the point near the nozzle rim ($x/D = 0.5$) projection on the plate. The Nusselt number and velocity variation along the striking surface are shown in Figures 7.20 and 7.21.

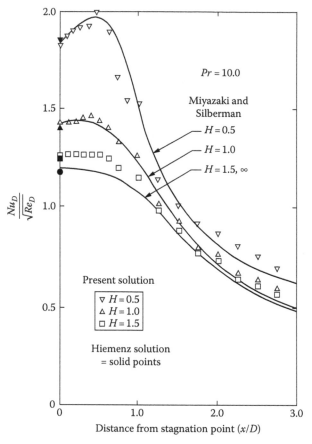

FIGURE 7.20
Nusselt number along the plate for different nozzle distances from the plate. (From Lipsett, A.W. and Gilpin, R.R., *Int. J. Heat Mass Trans.*, 21(1), 25, 1978).

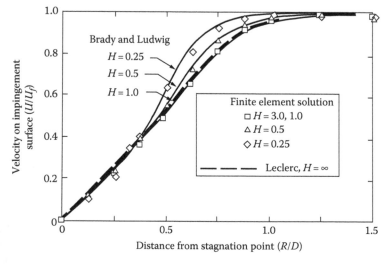

FIGURE 7.21
Velocity variation along the plate for different nozzle distances from the plate. (From Lipsett, A.W. and Gilpin, R.R., *Int. J. Heat Mass Trans.*, 21(1), 25, 1978).

7.7 Flow over Rotating Surfaces

Understanding flow and heat transfer involving rotating surfaces is important in the design of turbomachinery, like pumps and turbines. To gain a better understanding of the fundamental behavior of these problems, consider a circular disk of radius R rotating with the angular velocity ω in a fluid. For a rotating cylinder, the fluid is spun out of the disk which requires a downflow; therefore, we are dealing with a 3D flow. Consider a point on the disk a distance r away from the center as shown in Figure 7.22. The r, θ, z components of velocity are axially symmetric and will not change with θ. At a given r location, as one moves away from the surface, the r component starts from zero, reaches a maximum value, and drops to zero in the z direction. On a z plane, the r component of velocity starts from zero value at the center and reaches a maximum at R. The tangential component of velocity is related to the rotational velocity and starts from zero at the center and reaches maximum at R. The axial component of velocity starts from a small value far from the disk and increases to a given value and then goes to zero on the plate.

For axially symmetric condition, $\dfrac{\partial}{\partial \phi} = 0$, and steady state, $\dfrac{\partial}{\partial t} = 0$, conditions, the conservation equation reduces to

$$\frac{\partial u_z}{\partial z} + \frac{1}{r}\frac{\partial(r u_r)}{\partial r} = 0 \tag{7.337}$$

$$\rho\left(u_r \frac{\partial u_r}{\partial r} + u_z \frac{\partial u_r}{\partial z} - \frac{u_\phi^2}{r} \right) = -\frac{\partial P}{\partial r} + \mu\left[\frac{\partial^2 u_r}{\partial z^2} + \frac{1}{r}\frac{\partial}{\partial r}\left(r\frac{\partial u_r}{\partial r} \right) - \frac{u_r}{r^2} \right] \tag{7.338}$$

$$\rho\left(u_r \frac{\partial u_z}{\partial r} + u_z \frac{\partial u_z}{\partial z} \right) = -\frac{\partial P}{\partial z} + \mu\left[\frac{\partial^2 u_z}{\partial z^2} + \frac{1}{r}\frac{\partial}{\partial r}\left(r\frac{\partial u_z}{\partial r} \right) \right] \tag{7.339}$$

$$\rho\left(u_r \frac{\partial u_\phi}{\partial r} + u_z \frac{\partial u_\phi}{\partial z} + \frac{u_r u_\phi}{r} \right) = \mu\left[\frac{\partial^2 u_\phi}{\partial z^2} + \frac{1}{r}\frac{\partial}{\partial r}\left(r\frac{\partial u_\phi}{\partial r} \right) - \frac{u_\phi}{r^2} \right] \tag{7.340}$$

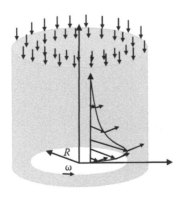

FIGURE 7.22
Disk of radius R rotating with an angular velocity ω.

$$\rho c_p \left(u_r \frac{\partial T}{\partial r} + u_z \frac{\partial T}{\partial z} \right) = k \left[\frac{1}{r} \frac{\partial}{\partial r} \left(r \frac{\partial T}{\partial r} \right) + \frac{\partial^2 T}{\partial z^2} \right] \tag{7.341}$$

$$z = 0, \quad u = 0, \quad v = 0, \quad w = r\omega, \quad T = T_w \tag{7.342}$$

$$z \to \infty, \quad v = 0, \quad w = 0, \quad T = T_\infty \tag{7.343}$$

The velocity components, pressure, and temperature are a function of r and z. We define a nondimensional distance z

$$\eta = \frac{z}{\sqrt{\dfrac{\nu}{\omega}}} \tag{7.344}$$

and define the following nondimensional variables [1]

$$F(\eta) = \frac{u_r}{r\omega}, \quad G(\eta) = \frac{u_\phi}{r\omega}, \quad H(\eta) = \frac{u_z}{\sqrt{\nu\omega}}, \quad P^*(\eta) = \frac{P - P_{ref}}{\rho\nu\omega}, \quad T^*(\eta) = \frac{T - T_\infty}{T_w - T_\infty} \tag{7.345}$$

where P_{ref} is an appropriately defined reference pressure. Assuming similarity, i.e., the non-dimensional variables above to be only a function of η and independent of r, then the conservation equations reduce to the following set of ordinary differential equations [1,27,28]:

Conservation of mass	$2F + H' = 0$	(7.346)
Conservation of r momentum	$F'' - HF' = F^2 - G^2$	(7.347)
Conservation of ϕ momentum	$2FG + HG' - G'' = 0$	(7.348)
Conservation of z momentum	$P^{*'} = H'' - HH'$	(7.349)
Conservation of energy	$T^{*''} - PrHT^{*'} = 0$	(7.350)

where prime indicates differentiation with respect to η. The boundary conditions are

$$\eta = 0, \quad F = 0, \quad G = 1, \quad H = 0, \quad P = 0, \quad T^* = 1$$
$$\eta = \infty, \quad F = 0, \quad G = 0, \quad T^* = 0 \tag{7.351}$$

Equation 7.349 can be integrated directly,

$$P^* = -\frac{1}{2} H^2 + H' + C \tag{7.352}$$

$$\text{at } z \to \infty, \ P = P_\infty \tag{7.353}$$

$$P_\infty^* = -\frac{1}{2}H_\infty^2 + C \tag{7.354}$$

$$P^* = -\frac{1}{2}H^2 + H' + P_\infty^* + \frac{1}{2}H_\infty^2 \tag{7.355}$$

Note that H_∞ is a constant, yet to be determined. Therefore, assuming the nondimensional reference pressure as

$$P_{ref}^* = \frac{P_{ref}}{\rho\nu\omega} = P_\infty^* + \frac{1}{2}H_\infty^2$$

$$P^* = -\frac{1}{2}H^2 + H' \tag{7.356}$$

The finite difference form of the equations results in

$$\left(\frac{1}{2d^2}\right)H_{i+1} - \left(\frac{1}{2d}\right)H_{i-1} = -2F_i \tag{7.357}$$

$$\left(\frac{1}{d^2} + \frac{H_i}{2d}\right)F_{i-1} - \left(\frac{2}{d^2} + F_i\right)F_i + \left(\frac{1}{d^2} - \frac{H_i}{2d}\right)F_{i+1} = -G_i^2 \tag{7.358}$$

$$\left(\frac{1}{d^2} + \frac{H_i}{2d}\right)G_{i-1} - \left(\frac{2}{d^2} + 2F_i\right)G_i + \left(\frac{1}{d^2} - \frac{H_i}{2d}\right)G_{i+1} = 0 \tag{7.359}$$

$$P_i = -\frac{H_i^2}{2} + \frac{H_{i+1} - H_{i-1}}{2d} \tag{7.360}$$

$$\left(\frac{1}{d^2} + \frac{\Pr H_i}{2d}\right)T_{i-1} - \frac{2}{d^2}T_i + \left(\frac{1}{d^2} - \frac{\Pr H_i}{2d}\right)T_{i+1} = 0 \tag{7.361}$$

Using the subroutine TRIDI, these equations can be solved using spreadsheets, and the results are given in Figure 7.23.

The results are very close to Schlichting [1, p. 95], or White [9, Table 3.6], except for P which has the opposite sign, due to the reference pressure used. The results show that far above the disk, the fluid has a downward velocity component only. As the fluid gets closer to the disk, the z component of velocity decreases and the r and ϕ components increase. The r component reaches a peak and then goes to zero to satisfy the no-slip condition, and the ϕ component increases to its maximum value of $r\omega$ on the surface of the disk. The r and ϕ components of velocity reach a value of 0.01 at $\eta = 5$, which can be considered to be the thickness of the boundary layer formed. Outside of this region, the fluid is moving with a uniform velocity along the z direction toward the plate. In terms of the dimensional variable,

$$\delta = 5\sqrt{\frac{\nu}{\omega}} \tag{7.362}$$

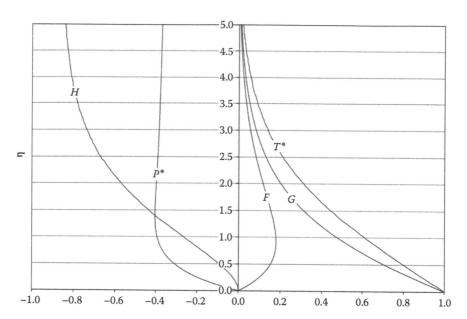

FIGURE 7.23
Solution results for a rotating disk.

It is important to note that the boundary layer thickness is very small and independent of r. For a disk rotating at 100 rpm, in water, the boundary layer thickness is about 1.5 mm.

The net effect of the rotation of the disk is fluid moving along the z direction and then making a 90° turn and thrown radially and tangentially in a very thin region near the disk:

$$\dot{m} = \int_0^z 2\pi r \rho u_r dz = 2\pi r \rho r \omega \int_0^z \frac{u}{r\omega} \sqrt{\frac{\nu}{\omega}} d\eta = 2\pi \rho r^2 \omega \sqrt{\frac{\nu}{\omega}} \int_0^\eta F d\eta \qquad (7.363)$$

$$\int_0^\eta F d\eta = -\frac{1}{2} H(\eta) \qquad (7.364)$$

$$\dot{m} = -H\pi r^2 \rho \sqrt{\omega \nu} \qquad (7.365)$$

The mass flow rate must also be equal to

$$\dot{m} = -\int_0^r 2\pi r \rho u_z dr = -\int_0^r 2\pi r \rho \sqrt{\nu \omega} H dr = -H\pi r^2 \rho \sqrt{\nu \omega} \qquad (7.366)$$

H approaches a value of 0.85; therefore

$$\dot{m} = 0.85\pi r^2 \rho \sqrt{\nu \omega} \qquad (7.367)$$

The average radial velocity is

$$\bar{u}_r = \frac{\dot{m}}{\rho 2\pi r\delta} = \frac{0.85\pi r^2 \rho\sqrt{\nu\omega}}{\rho 2\pi r 5\sqrt{\dfrac{\nu}{\omega}}} = 0.085 r\omega \tag{7.368}$$

A 20 cm diameter disk rotating at 1000 rpm will produce a mass flow rate of 0.11 kg/s with water flowing radially at the tip with an average velocity of 0.14 m/s. The tangential component of velocity is 1.67 m/s.

The drag coefficient is

$$c_f = \frac{\mu \dfrac{\partial u_\phi}{\partial z}}{\dfrac{1}{2\rho(R\omega)^2}} = \frac{2\sqrt{\dfrac{\omega}{\nu}}rG'}{Re} \tag{7.369}$$

$$\bar{c}_f = \frac{1}{R}2\sqrt{\frac{\omega}{\nu}}\frac{G'}{Re}\int_0^R r\,dr = \frac{G'}{Re^{1/2}} = \frac{0.616}{Re^{1/2}} \tag{7.370}$$

where

$$Re = \frac{\rho R(R\omega)}{\mu} \tag{7.371}$$

is based on the angular velocity and radius. The local value of the Nusselt number is

$$Nu_r = \frac{-k\dfrac{\partial T}{\partial z}r}{k(T_w - T_\infty)} = \frac{-k(T_w - T_\infty)\dfrac{1}{\sqrt{\dfrac{\nu}{\omega}}}\dfrac{\partial T^*}{\partial \eta}r}{(T_w - T_\infty)k} = -\sqrt{\frac{rr\omega}{\nu}}\frac{\partial T^*}{\partial \eta} \tag{7.372}$$

$$Nu_r = -Re^{1/2}\frac{\partial T^*}{\partial \eta} = 0.4\,Re_r^{1/2}$$

for Prandtl number 1 and by regression analysis can be approximated to within 5% by

$$Nu_r = 0.417 Pr^{0.425} Re_r^{1/2} \tag{7.373}$$

The local value of the heat transfer coefficient is independent of the radial position, and therefore local and average values are the same.

7.8 Summary

This chapter covered the basic features of a broad range of external flow problems, presenting the analytical and numerical solutions to many classical problems in forced convection. Many external flow problems can be analyzed using the boundary layer approximation, greatly simplifying the equations, with some having similarity solution. The chapter also covered the concept of flow separation, that accompanies flow in many practical situations, and the complications that flow separation creates in analyzing such problems.

Problems

7.1 Starting with the NS equations, come up with the governing, the computational domain, the boundary conditions and nondimensionalize the equations and boundary conditions for flow over rectangular fin array.

7.2 Starting with the NS equations, come up with the governing, the computational domain, the boundary conditions and nondimensionalize the equations and boundary conditions for flow over an infinitely long circular cylinder.

7.3 Simplify the NS equations for flow over the step shown in the figure. Define the solution domain and list the boundary conditions. Nondimensionalize the governing equations and boundary conditions.

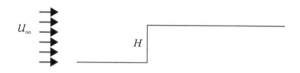

7.4 Starting with the NS equations, come up with the governing, the computational domain, the boundary conditions and nondimensionalize the equations and boundary conditions for flow exiting a rectangular orifice.

7.5 Starting with the NS equations, come up with the governing, the computational domain, the boundary conditions and nondimensionalize the equations and boundary conditions for flow exiting a circular orifice.

7.6 Numerically determine the thickness of thermal boundary layer, and Nusselt number for flow over a flat plate subject to a linearly varying temperature distribution, from T_∞ at the leading edge to T_w at $x = L$.

7.7 Repeat Problem 7.5, for a plate that has an unheated length and then maintained at a constant temperature.

7.8 Reproduce Figure 7.8, by numerically solving the boundary layer equations.

7.9 Repeat Problem 7.6 for unheated length after which it is subjected to a uniform heat flux.

7.10 Determine the thermal boundary layer thickness and Nusselt number for flow over an isothermal wedge.

7.11 Determine the thermal boundary layer thickness and Nusselt number for flow over a wedge subjected to a uniform heat flux.

7.12 Reproduce Figure 7.16 by numerically solving the governing equations.

7.13 Reproduce Figure 7.17 by numerically solving the governing equations.

7.14 Numerically solve Equation 7.292 and compare with the exact solution.

7.15 Numerically solve Equation 7.317 and compare with the exact solution.

7.16 Numerically solve Equation 7.331 and compare with the exact solution.

7.17 Numerically solve Equation 7.332 and compare with the exact solution.

7.18 Reproduce Figure 7.23 by numerically solving the governing equations.

7.19 Numerically verify Equation 7.373.

7.20 Air jet velocity coming out of a nozzle is 15 m/s. What is the heaviest ball having a diameter of 6 cm and a surface roughness of 0.075 cm that can be suspended?

7.21 An infinitely large flat plate is initially at rest in a fluid. It suddenly starts moving in the x direction with a constant velocity V. Show that by defining $\eta = y/\sqrt{4\upsilon t}$ and $u^* = u/V$, the momentum equation simplifies to

$$\frac{d^2 u^*}{d\eta^2} + 2\eta \frac{du^*}{d\eta} = 0$$

7.22 Obtain the velocity distribution for Problem 7.21 numerically.

7.23 A spherically shaped hailstone with a diameter of 12 mm at −5°C falls through the air at 10°C and 100 kPa. It reaches its terminal velocity when the drag force acting on it equals its weight. Neglecting buoyancy and evaporation,
   a. Determine its terminal velocity.
   b. Determine the rate of heat transfer from it.

7.24 A nuclear reactor fuel rod is a circular cylinder, 1 cm in diameter. The rod is to be tested by cooling it with liquid sodium at 232°C, flowing perpendicular to the fuel rod axis with a velocity of 6 cm/s. If the rod surface temperature is not to exceed 300°C, estimate the maximum allowable power dissipation in the rod.

References

1. Schlichting, H. (1968) *The Boundary-Layer Theory*, 6th edn. New York: McGraw-Hill.
2. Bejan, A. (1984) *Convection Heat Transfer*. New York: Wiley.
3. Oosthuizen, P. H. and Naylor, D. (1999) *An Introduction to Convective Heat Transfer Analysis*. New York: WCB/McGraw-Hill.

4. Lin, F. N. and Chern, S. Y. (1979) Laminar boundary-layer flow of non-Newtonian fluid. *International Journal of Heat and Mass Transfer* 22(7-10), 1323–1331, 1979.

5. Kays, W. M. and Crawford, M. E. (1993) *Convective Heat and Mass Transfer*, 3rd edn. New York: McGraw-Hill.

6. Burmeister, L. C. (1993) *Convective Heat Transfer*, 2nd edition. New York: Wiley.

7. Fakheri, A. (2009) Flow separation. *ASME International Mechanical Engineering Conference and Exposition*, November 13–19. Orlando, FL.

8. Roshko, A. (1954) *On the Development of Turbulent Wakes from Vortex Streets*. National Advisory Committee for Aeronautics. NACA Tech Report 1191.

9. White, F. M. (1974) *Viscous Fluid Flow*. New York: McGraw-Hill.

10. Smith, A. M. O. and Clutter, D. W. (1963) Machine calculation of incompressible boundary layer equations. *AIAA Journal*, 1, 2062–2071.

11. Schwarz, W. H. (1963) The radial free jet. *Chemical Engineering Science* 18(12), 779–786.

12. Schwarz, W. H. and O'Nan, M. (1965) The laminar cylindrical wall-jet in the early part of the flow. *Chemical Engineering Science* 20(5), 365–372.

13. Schneider, W. (1981) Flow induced by jets and plumes. *Journal of Fluid Mechanics* 108, 55–65.

14. Martin, H. (1977) Heat and mass transfer between impinging gas jets and solid surfaces. *Advances in Heat Transfer* 13, 1.

15. Kalteh, M. and Abbassi, A. (2006) Similarity solution of laminar axisymmetric jets with effect of viscous dissipation. *Journal of Heat Transfer* 128, 1099.

16. Goldstein, R. J. and Franchett, M. E. (1988) Heat transfer from a flat surface to an oblique jet. *ASME Journal of Heat Transfer* 110, 87–93.

17. Gardon, R. and Akfirat, J. C. (1965) The role of turbulence in determining the heat-transfer characteristics of impinging jets. *International Journal of Heat Mass Transfer* 8s, 1261–1272.

18. Striegl, S. A. and Diller, T. E. (1984) The effect of entrainment on jet impingement heat transfer. *ASME Journal of Heat Transfer* 106, 27–33.

19. Miyazaki, H. and Silberman, E. (1972) Flow and heat transfer on a flat plate normal to a two-dimensional laminar jet issuing from a nozzle of finite height. *International Journal of Heat and Mass Transfer* 15(7-11), 2097–2107.

20. Parneix, S., Behnia, M., and Durbin. P. (1998) Prediction of heat transfer in a jet impinging on a heated pedestal. *ASME Journal of Heat Transfer* 121, 43–49.

21. Looney, M. K. and Walsh, J. J. (1984) Mean-flow and turbulent characteristics of free and impinging jet-flows. *Journal of Fluid Mechanics* 147, 397–429.

22. Craft, T. J., Graham, L. J. W., and Launder, B. E. (1993) Impinging jet studies for turbulence model assessment—II. An examination of the performance of four turbulence models. *International Journal of Heat and Mass Transfer* 36, 2685–2697.

23. Steven, J., Pan, Y., and Webb, N. W. (1992) Effect of nozzle configuration on transport in the stagnation zone of axisymmetric, impinging free surface liquid jets—I—Turbulence flow structure. *Journal of Heat Transfer* 114, 874–879.

24. Pan, Y., Steven, J., and Webb, N. W. (1992) Effect of nozzle configuration on transport in the stagnation zone of axisymmetric, impinging free-surface liquid jets—II—Local heat transfer. *Journal of Heat Transfer* 114, 880–886.

25. Rankin, G. W. and Sridhar, K. (1978) *Journal of Fluids Engineering/Transactions of ASME* 100, 55–59.

26. Lipsett, A. W. and Gilpin, R. R. (1978) Laminar jet impingement heat transfer including the effects of melting. *International Journal of Heat and Mass Transfer* 21(1), 25–33.

27. Kreith, F. (1968) Convection heat transfer in rotating systems. *Advances in Heat Transfer* 5, 129.

28. Barbee, D. and Shih, T. S. (1968) Incompressible flow induced by an infinite isothermal disk rotating in a rarefied gas. *Journal of Heat Transfer* 359–361.

8

Internal Flow

8.1 Introduction

Internal or duct flows are of great practical significance, in many industrial applications, including the piping systems used to transport fluids, heat exchangers, and combustion chambers. Internal flows refer to situations where the fluid is encased by solid boundaries, except for the inlet and exit regions. For external flows, the fluid is not constrained by a solid boundary in at least one direction and the free stream velocity and temperature serve as the appropriate references. For internal flows, such explicit constant scales are not available and instead mean velocity and mean temperature are defined as the appropriate reference values.

The mean or average velocity is defined as the velocity that when multiplied by the density and area provides the mass flow rate, or

$$\bar{u} = \frac{1}{\rho A} \int_A \rho u \, dA \tag{8.1}$$

Note that mean velocity in general is not constant. From conservation of mass, for an incompressible flow, in a constant area pipe, the average velocity at each section is constant. Furthermore, if the fluid enters the pipe with a uniform velocity U, then the average (mean) velocity is the same as the uniform inlet velocity

$$\bar{u} = U \tag{8.2}$$

The Reynolds number for duct flow is defined as

$$Re = \frac{\rho \bar{u} D_h}{\mu} \tag{8.3}$$

where the characteristic length for internal flow, the hydraulic diameter, is defined as four times the cross-sectional area divided by the wetted circumference, which is the perimeter of the duct that is in contact with the fluid

$$D_h = \frac{4A}{p} \tag{8.4}$$

The Reynolds number is related to mass flow rate through

$$Re = \frac{4\dot{m}}{\mu p} \tag{8.5}$$

For duct flow, the critical Reynolds number, where flow makes the transition from laminar to turbulent flow, is generally assumed to be

$$Re_c = \frac{\rho U D_h}{\mu} \approx 2300 \tag{8.6}$$

For a pipe with constant cross section, if viscosity can be assumed to be constant, isothermal flow, Reynolds number would remain constant, or if the flow enters the duct as laminar, it would remain laminar throughout the pipe. That is not necessarily the case if the fluid temperature changes along the pipe, and fluid properties are a strong function of temperature, like oils.

The average temperature $\bar{T}$, mean temperature T_m, or bulk temperature T_b of the fluid at the given cross section is the temperature that when multiplied by the mass flow rate and specific heat provides the amount of enthalpy (energy) that crosses a section of the duct

$$\bar{T} = \frac{1}{\dot{m}c_p} \int \rho u c_p T dA \tag{8.7}$$

The mean temperature is also sometimes called the mixing cup temperature, the temperature that the fluid will reach if it is placed in a cup and stirred adiabatically.

For a circular pipe, assuming fluid properties to be constant, Equation 8.7 simplifies to

$$\bar{T} = \frac{2}{R^2 \bar{u}} \int_0^R r u T dr \tag{8.8}$$

Mean temperature, unlike free stream temperature, is not constant along the pipe. The local heat transfer coefficient is based on the difference between the local surface temperature of the duct (T_w) and the mean fluid temperature

$$h_x = \frac{q''}{T_w - \bar{T}} = \frac{-k \left. \frac{\partial T}{\partial n} \right|_{n=0}}{T_w - \bar{T}} \tag{8.9}$$

where n is coordinate normal to the surface. The local Nusselt number is defined as

$$Nu_x = \frac{h_x D_h}{k} \tag{8.10}$$

To determine the heat transfer coefficient, we start with nondimensionalizing Equation 8.8. If y is the coordinate measured away from the pipe wall, and defining the following nondimensional variables

$$y^* = \frac{y}{D_h} \tag{8.11}$$

$$u^* = \frac{u}{\bar{u}} \tag{8.12}$$

$$T^* = \frac{T - T_r}{T_s} \tag{8.13}$$

where T_r and T_s are appropriately defined reference and scale temperatures, then the non-dimensional form of Equation 8.1 then becomes

$$\frac{1}{\rho A}\int_A \rho u^* dA = 1 \tag{8.14}$$

which for incompressible flow becomes

$$\frac{1}{A}\int_A u^* dA = 1 \tag{8.15}$$

and for a circular pipe $dA = 2\pi r dr$ and Equation 8.14 simplifies to

$$8\int_A u^* r^* dr^* = 1 \tag{8.16}$$

where $r^* = \dfrac{r}{D}$.

Nondimensionalizing Equation 8.7 results in the dimensionless form of the mean temperature

$$\bar{T} = \frac{1}{\dot{m}c_p}\int_A \rho u^* u_s c_p (T_r + T^* T_s) dA = \frac{1}{\dot{m}c_p}\left[c_p T_r \bar{u} \int_A \rho u^* dA + c_p T_s \bar{u} \int_A \rho u^* T^* dA \right] \tag{8.17}$$

Using Equation 8.14

$$\bar{T} = \frac{1}{\dot{m}c_p}\left[c_p T_r \bar{u}\rho A + c_p T_s \bar{u}\rho A \frac{1}{\rho A}\int_A \rho u^* T^* dA \right] \tag{8.18}$$

then

$$\bar{T}^* = \frac{\bar{T} - T_r}{T_s} = \frac{1}{\rho A}\int_A \rho u^* T^* dA \tag{8.19}$$

Again, for an incompressible flow in a circular pipe,

$$\bar{T}^* = 8\int_A u^* T^* r^* dr^* \tag{8.20}$$

To determine the Nusselt number (heat transfer coefficient), substitute for h from Equation 8.9 into Equation 8.10 and nondimensionalize the RHS

$$Nu_x = \frac{\dfrac{-k\dfrac{\partial T}{\partial y}\bigg|_w}{T_w - \bar{T}}D_h}{k} = \frac{-\dfrac{T_s}{D_h}\dfrac{\partial T^*}{\partial y^*}\bigg|_w}{(T_w - T_r) - (\bar{T} - T_r)}D_h \tag{8.21}$$

which simplifies to

$$Nu_x = \frac{-\left.\frac{\partial T^*}{\partial y^*}\right|_w}{T_w^* - \bar{T}^*} \tag{8.22}$$

Determination of Nusselt number requires the knowledge of wall temperature, mean temperature, and the temperature gradient at the wall, or simply the temperature distribution in the fluid which can then be used to calculate them.

The actual thermal boundary conditions that a duct is subjected to, could be difficult to model, however they generally fall between the two limits of the duct receiving a uniform heat flux or having an isothermal surface, although either of these is rarely satisfied. In many instances, the peripheral variations must be taken into account, since the surface temperature and the temperature gradient at the wall are not constant around the circumference of the duct. The value of the peripherally averaged heat transfer coefficient at a given axial location is given by

$$\bar{h}_x = \frac{1}{p} \int_p \frac{-k \left.\frac{\partial T}{\partial n}\right|_{n=0}}{T_w - T} dp \tag{8.23}$$

and the peripherally averaged local Nusselt number becomes

$$Nu_x = \frac{\bar{h}_x D_h}{k} = \frac{D_h}{k} \frac{1}{p} \int_p \frac{-k \left.\frac{\partial T}{\partial n}\right|_{n=0}}{T_w - T} dp \tag{8.24}$$

For example, consider a circular pipe, insulated in the bottom half and receiving a uniform heat flux for $0 < \phi < 180$ then

$$Nu_x = \frac{\bar{h}_x D_h}{k} = \frac{D_h}{k} \frac{q''}{\pi} \int_0^\pi \frac{-d\phi}{T_w - \bar{T}} \tag{8.25}$$

The total amount of heat transferred to the fluid from the inlet of the pipe to any location x is obtained from

$$q = \int_A q'' dA = \int_A -k \left.\frac{\partial T}{\partial n}\right|_{n=0} dA \tag{8.26}$$

The axially averaged heat transfer coefficient is determined by averaging local peripherally averaged heat transfer coefficient over the length of the pipe, or

$$\bar{h} = \frac{1}{A} \int_A \bar{h}_x dA \tag{8.27}$$

where h_x is given by Equation 8.23. The axially averaged Nusselt number is then

$$\overline{Nu} = \frac{\bar{h}D_h}{k} \tag{8.28}$$

As will be shown later, the Nusselt number changes over a finite length from the inlet of the pipe, called the entrance length (x_{fd}), and then it becomes constant,

$$q = \int_0^{x_{fd}} h_x \left(T_w - \overline{T}\right) dA + A\bar{h}\frac{1}{A} \int_{x_{fd}}^{x} \left(T_w - \overline{T}\right) dA \tag{8.29}$$

The term $\dfrac{1}{A}\displaystyle\int_{x_{fd}}^{x} \left(T_w - \overline{T}\right) dA$ on the right hand side is the appropriate temperature difference in the fully developed region. As will be discussed in more details later, the average temperature difference takes different forms, depending on the wall boundary condition.

8.2 Couette Flow

To study the basic features of internal flows, we consider one of the classical internal flow problems, the Couette flow, between two infinitely long parallel plates that are a distance D apart in the y direction and extending to infinity in x and z directions, as shown in Figure 8.1. The lower plate is stationary and maintained at a constant temperature T_0, and the upper plate is moving with a constant velocity V and is at a constant temperature T_D. Starting with the 2D NS equations for steady 2D incompressible flow,

$$\frac{\partial u}{\partial x} + \frac{\partial v}{\partial y} = 0 \tag{8.30}$$

$$\rho\left(u\frac{\partial u}{\partial x} + v\frac{\partial u}{\partial y}\right) = -\frac{\partial P}{\partial x} + \mu\left[\frac{\partial^2 u}{\partial x^2} + \frac{\partial^2 u}{\partial y^2}\right] + \rho g_x \tag{8.31}$$

$$\rho\left(u\frac{\partial v}{\partial x} + v\frac{\partial v}{\partial y}\right) = -\frac{\partial P}{\partial y} + \mu\left[\frac{\partial^2 v}{\partial x^2} + \frac{\partial^2 v}{\partial y^2}\right] + \rho g_y \tag{8.32}$$

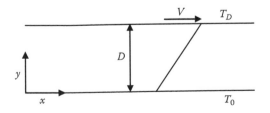

FIGURE 8.1
Geometry for Couette flow.

$$\rho c_p \left(u \frac{\partial T}{\partial x} + v \frac{\partial T}{\partial y} \right) = k \left[\frac{\partial^2 T}{\partial x^2} + \frac{\partial^2 T}{\partial y^2} \right] \tag{8.33}$$

The conservation equations simplify greatly for this idealized case, by noting that the gradient of all dependent variables must be zero in the x direction; otherwise, that variable would be unbounded. For example, $\frac{\partial P}{\partial x}$ cannot be positive or negative; otherwise, the pressure will eventually approach infinity which is not possible. Similarly $\frac{\partial u}{\partial x} = 0$; then from continuity equation,

$$\frac{\partial v}{\partial y} = 0 \tag{8.34}$$

or the y component of velocity must be constant in the y direction (as well as x):

$$v = \text{constant} \tag{8.35}$$

Since at the plates, the y component of velocity is zero, v must be zero everywhere ($v = 0$) and the x momentum equation simplifies to

$$\rho \left(u0 + 0 \frac{\partial u}{\partial y} \right) = -0 + \mu \left[0 + \frac{\partial^2 u}{\partial y^2} \right] + \rho 0 \tag{8.36}$$

$$\frac{\partial^2 u}{\partial y^2} = 0 \tag{8.37}$$

which indicates that the x component of velocity is independent of x and changes linearly in the y direction. The y momentum equation simplifies to

$$\rho \left(u0 + 0 * 0 \right) = -\frac{\partial P}{\partial y} + \mu \left[0 + 0 \right] + \rho g_y \tag{8.38}$$

$$P = P_0 - \rho g y \tag{8.39}$$

If D is small, then the pressure variation in the y direction would be negligible. The boundary conditions for Equation 8.37 are

$$y = 0, \quad u = 0$$
$$y = D, \quad u = V \tag{8.40}$$

resulting in

$$u = \frac{V}{D} y \tag{8.41}$$

Once the velocity distribution is determined, the average velocity can be calculated from Equation 8.1

$$\bar{u} = \frac{1}{\rho DW} \int_0^D \rho V \frac{y}{D} W dy = \frac{V}{D^2} \int_0^D (y) dy = \frac{V}{D^2} \frac{D^2}{2} = \frac{V}{2} \tag{8.42}$$

where W is the plate depth, which approaches infinity. The hydraulic diameter for flow between two parallel plates is

$$D_h = \frac{4A}{p} = \lim_{W \to \infty} \frac{4(D \times W)}{2(D+W)} = 2D \tag{8.43}$$

or the hydraulic diameter is twice the plate spacing. Defining the nondimensional velocity as

$$u^* = \frac{u}{\bar{u}} = \frac{u}{\frac{1}{2}V} = \frac{\frac{V}{D} y}{1/2} = \frac{2y}{D} = \frac{4y}{2D} = 4y^* \tag{8.44}$$

where $y^* = \frac{y}{D_h} = \frac{y}{2D}$ and varies between 0 and 0.5; therefore, the dimensionless velocity varies between 0, at the lower plate, and 2, at the upper plate.

The shear stress in the flow is

$$\tau = \mu \frac{\partial u}{\partial y} = \frac{V}{D} \tag{8.45}$$

which is a constant, proportional to the upper plate velocity and inversely proportional to the distance between the two plates. The drag coefficient is dimensionless form of the shear stress, which in this case becomes

$$c_f = \frac{\tau}{\frac{1}{2}\rho \bar{u}^2} = \frac{\mu \frac{V}{D}}{\frac{1}{2}\rho \bar{u}^2} = \frac{\mu \frac{2\bar{u}}{2D} 2}{\frac{1}{2}\rho \bar{u}^2} = \frac{8\mu}{\rho \bar{u} D_h} \tag{8.46}$$

or

$$c_f = \frac{8}{Re} \tag{8.47}$$

The energy equation simplifies to

$$\frac{\partial^2 T}{\partial y^2} = 0 \tag{8.48}$$

solving the energy equation

$$T = T_0 + \frac{T_D - T_0}{D} y \tag{8.49}$$

and nondimensionalizing

$$T^* = 2y^* \tag{8.50}$$

where

$$T^* = \frac{T - T_0}{T_D - T_0} \tag{8.51}$$

To find the mean temperature

$$\bar{T}^* = \frac{1}{A} \int_A u^* T^* dA = \frac{1}{A} \int_A 2y^* \, 2y^* dA = \frac{1}{DW} \int_0^D 4y^{*2} W d(y) = \int_0^{0.5} 8y^{*2} dy^* = \frac{8}{3}(0.5)^3 = \frac{1}{3} \tag{8.52}$$

which shows that the mean temperature of the fluid is constant, implying that the net heat transfer to the fluid must be zero. In terms of the dimensional variables,

$$\bar{T} = \frac{2}{3} T_0 + \frac{1}{3} T_D \tag{8.53}$$

The heat transfer coefficient can be determined from Equation 8.22

$$Nu_x = \frac{-\dfrac{\partial T^*}{\partial y^*}\bigg|_w}{T_w^* - \bar{T}^*} \tag{8.54}$$

where nondimensional temperature is defined by Equation 8.51 and will be different for the two plates, because they are at different temperatures. The Nusselt number for the lower plate becomes

$$Nu_{x,0} = \frac{-2}{0 - \dfrac{1}{3}} = 6 \tag{8.55}$$

Note that to determine the heat transfer coefficient for the upper plate, Equation 8.54 must be multiplied by a negative sign, since $-k\left(\dfrac{\partial T}{\partial y}\right)$ is heat conducted in the positive y direction which will be heat transferred from the fluid to the wall and not to the fluid from the upper plate:

$$Nu_{x,D} = -\frac{-2}{1 - \dfrac{1}{3}} = 3 \tag{8.56}$$

Since the local values of the Nusselt number are constant, they will be equal to the average value. Therefore, the local and average values of the heat transfer coefficient are given by

$$h_0 = \overline{h_0} = 6\frac{k}{2D} = 3\frac{k}{D} \tag{8.57}$$

$$h_D = \overline{h_D} = 3\frac{k}{2D} = 1.5\frac{k}{D} \tag{8.58}$$

The net heat transfer to the fluid is

$$q = h_0 A_0 (T_0 - \bar{T}) + h_D A_D (T_D - \bar{T}) \tag{8.59}$$

which can be rearranged as

$$q = A(T_D - T_0) \left[h_0 \frac{T_0 - \bar{T}}{T_D - T_0} + h_D \frac{T_D - \bar{T}}{T_D - T_0} \right] = A(T_D - T_0)[-h_0 \bar{T}^* + h_D(1 - \bar{T}^*)] \tag{8.60}$$

$$q = A(T_L - T_0) \left[-3\frac{k}{D}\frac{1}{3} + 1.5\frac{k}{D}\left(1 - \frac{1}{3}\right) \right] = 0 \tag{8.61}$$

which confirms the expectation that since the average temperature of the fluid is constant, the net heat transfer to it must be zero; therefore, all the energy transferred to the fluid from the upper plate must be removed from the fluid at the lower plate, or heat is transferred from the upper plate to the lower plate. This heat transfer can be calculated by either equation for upper or lower plate, for example, using the heat transfer from the lower plate and expressing the mean temperature in terms of upper and lower plate temperatures

$$q = 3\frac{k}{D} A_0 \left[T_0 - \left(\frac{2}{3}T_0 + \frac{1}{3}T_D\right) \right] = kA_0 \frac{T_0 - T_D}{D} \tag{8.62}$$

This shows that the heat transfer in this case is by conduction only through the fluid in the y direction. This is not surprising, since heat transfer is in the y direction and fluid is moving only in the x direction, leaving the only mechanism for transfer of heat to be molecular diffusion in the y direction.

8.3 General Considerations in Duct Flow

The Couette flow serves to demonstrate some of the more important features of internal flows. In Couette flow, the velocity profile is independent of x. For all internal flows, far from the duct inlet, the flow eventually reaches a state called hydrodynamically fully developed where velocity profile no longer changes along the pipe. There is also a region in all duct flows, where Nusselt number becomes independent of axial the position. We can define a new nondimensional temperature

$$T^+ = \frac{T(x,y) - T_w(x)}{\bar{T}(x) - T_w(x)} \tag{8.63}$$

For Couette flow using the temperature of the lower plate

$$T^+ = \frac{T_0 + \dfrac{T_D - T_0}{D} y - T_0}{\dfrac{2}{3}T_0 + \dfrac{1}{3}T_D - T_0} = \frac{\dfrac{T_D - T_0}{D} y}{\dfrac{T_D - T_0}{3}} = 6y^* \tag{8.64}$$

This nondimensional temperature becomes independent of x for Couette flow everywhere. This condition is known as thermally fully developed, i.e., the flow is said to be thermally fully developed if

$$\frac{\partial T^+}{\partial x} = 0 \tag{8.65}$$

As the derivation shows, in Couette flow, the flow is thermally fully developed everywhere, and the Nusselt number is constant. This same behavior is observed in other internal flows, reaching a region where the flow becomes thermally fully developed and the dimensionless temperature defined by Equation 8.63 becomes independent of axial location. As will be shown later, in the thermally fully developed region, the heat transfer coefficient and therefore Nusselt number will also become constant and independent of axial location. The fully developed region is preceded by a region where the velocity and temperature fields develop, called the entrance length.

8.3.1 Entrance Length

Consider flow between two parallel plates a distance D apart that are assumed to be very large in the z direction as shown in Figure 8.2. As the fluid enters the duct, two boundary layers will form on the top and bottom plates, with the flow between the two boundary layers being inviscid. The boundary layers grow with the distance from the inlet, and the inviscid core decreases until they become half the spacing between the plates and merge together, with the inviscid core disappearing. This distance from the inlet where this happens is sometimes referred to as the inviscid core length. Past this point, the velocity profiles continue to develop, and eventually no longer change in the x direction. The distance from the channel entrance to the point where the velocity no longer changes axially is called the entrance length and the region is called hydrodynamically developing region. In this region, the velocity field is two dimensional (2D) i.e., both u and v are nonzero. The inviscid core length is around one-third of the entrance length.

In the fully developed region, the x component of velocity, u, is no longer a function of x and the y component of velocity becomes zero, i.e., the velocity field becomes one dimensional (1D). The comparison of the entrance length to the total duct length establishes if the entire flow in the duct can be assumed to be developing, fully developed, or mixed.

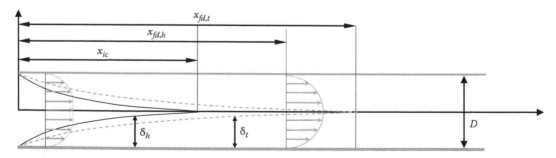

FIGURE 8.2
Different flow regions for flow between two parallel plates.

To estimate the entrance length for laminar flow, we go back to the results that we obtained for flow over a flat plate where the boundary layer thickness is given by

$$\frac{\delta}{x} = \frac{5}{\sqrt{Re_x}} \tag{8.66}$$

The inviscid core length, x_{ic}, is the location where the boundary layer thickness becomes $D/2$

$$\frac{D}{2x_{ic}} = \frac{5}{\sqrt{\dfrac{\rho U x_{ic}}{\mu}}} \tag{8.67}$$

Rearranging Equation 8.67 and noting that the hydraulic diameter is $2D$ result in

$$x_{ic} = 0.0025 D_h Re_{D_h} \tag{8.68}$$

The entrance length is expected to be longer than this, perhaps by a factor of two. However, the actual entrance length is one order of magnitude longer. This can be partly explained by realizing that for boundary layer flow over a flat plate, the free stream velocity is constant. For the flow between two plates, as the thickness of the boundary layer increases, more and more of the fluid slows down. Since the mass flow rate at every section is the same, to ensure that the conservation of mass is satisfied, the velocity outside of the boundary layer must increase to balance the slow down inside the boundary layer. This means that the velocity outside the boundary layer, "free stream velocity," increases with distance from the entrance. The boundary layer thickness is inversely proportional to the square root of the free stream velocity (see Equation 8.66). Therefore, the boundary layer thickness predicted by Equation 8.66 for flow over a flat plate overestimates the actual boundary layer thickness, which means that the boundary layers grow at a lower rate compared to that predicted by Equation 8.66 or the two boundary layers will merge at a distance longer than that based on Equation 8.66, and therefore the entrance length would also be longer than twice the entrance length predicted by the results based on Equation 8.66. The accepted expression for entrance length in laminar flow is

$$\frac{x_{fd,h}}{D_h Re_{D_h}} = 0.06 \tag{8.69}$$

For turbulent flow, the generally accepted expression for the entrance length is

$$\frac{x_{fd}}{D_h Re_{D_h}^{1/6}} = 0.06 \tag{8.70}$$

or for both flows the entrance length is given by

$$\frac{x_{fd}}{D_h Re_{D_h}^{n}} = 0.06 \tag{8.71}$$

with $n = 1$ for laminar flow and $n = 1/6$ for turbulent flow. At the critical $Re = 2300$, the entrance length is around 16 times the hydraulic diameter for turbulent flow and 138 for laminar flow, or if we increase the Re beyond 2300, the flow makes a transition to turbulent and becomes fully developed only after 16 hydraulic diameters, as opposed to 138.

8.3.2 Heat Transfer in Duct Flow

Similar to the hydrodynamic boundary layer, if the duct wall temperature is different than the fluid inlet temperature, a thermal boundary layer will be established on each plate and two boundary layers will merge together at a distance called thermal entrance length from the inlet of the pipe. For both laminar and turbulent flows, the thermal entrance length is proportional to the hydrodynamic entrance length with the proportionality constant being the Prandtl number, or

$$\frac{x_{fd,t}}{D_h Re_{D_h}^n Pr} = 0.06 \tag{8.72}$$

For $Pr > 1$, the thermal entrance length is longer than the hydrodynamic one, and we can identify three regions, starting from the pipe inlet:

  a. Thermally and hydrodynamically developing flow
  b. Hydrodynamically developed and thermally developing flow
  c. Thermally and hydrodynamically fully developed flow

If Prandtl number is equal to 1, then the two boundary layers merge together at the same point and the thermal and hydrodynamic length will be the same and there will be two regions. For $Pr < 1$, as in the case of liquid metals, the flow will become thermally fully developed before it becomes hydrodynamically developed. Therefore, region 2 would be thermally developed and hyrodynamically developing.

The dimensionless axial distance

$$x^+ = \frac{x}{D_h RePr} \tag{8.73}$$

is an important variable in internal flow and its inverse is known as the Graetz number. Therefore, laminar flow becomes thermally fully developed for $Gz = 1/0.06 = 17$.

As a cold fluid flows in a hot duct, it receives thermal energy from the duct and its average temperature at every cross section constantly increases. Therefore, inside the tube, the temperature will be both a function of x and y. As mentioned earlier, the concept of thermally fully developed refers to the condition where the nondimensional temperature defined by Equation 8.63

$$T^+ = \frac{T(x,y) - T_w(x)}{\overline{T}(x) - T_w(x)} \tag{8.74}$$

becomes independent of axial direction,

$$\frac{\partial T^+}{\partial x} = 0 \tag{8.75}$$

even though every temperature on the RHS of Equation 8.74 can be a function of x, in the thermally fully developed region. Differentiating Equation 8.74 and setting it to zero:

$$\left[\frac{\partial T(x,y)}{\partial x} - \frac{\partial T_w(x)}{\partial x}\right][\overline{T}(x) - T_w(x)] - \left[\frac{\partial \overline{T}(x)}{\partial x} - \frac{\partial T_w(x)}{\partial x}\right][T(x,y) - T_w(x)] = 0 \tag{8.76}$$

In addition, since T^+ is independent of x in the fully developed region, its derivative in the lateral direction would also be independent of x:

$$\frac{\partial T^+}{\partial y} = \frac{\partial}{\partial y}\left[\frac{T(x,y)-T_w(x)}{\overline{T}(x)-T_w(x)}\right] = \frac{\dfrac{\partial T}{\partial y}}{\overline{T}-T_w} \neq f(x) \tag{8.77}$$

From Equation 8.9

$$\frac{\dfrac{\partial T}{\partial y}}{\overline{T}-T_w} = \frac{\dfrac{q''}{k}}{T_w-\overline{T}} = \frac{h}{k} \neq f(x) \tag{8.78}$$

Since k is not a function of x, then h is not a function of x. This means that in the thermally fully developed region, the heat transfer coefficient and therefore Nusselt number are constant. This was also shown to be the case for Couette flow.

8.4 Mean Temperature Analysis

The mean temperature plays an important role in internal flows; therefore, we need to look at how it can be determined. Consider the energy analysis on a differential control volume in the duct as shown in Figure 8.3:

$$(\dot{m}c_p\overline{T})_x + q''p\Delta x = (\dot{m}c_p\overline{T})_{x+\Delta x} = (\dot{m}c_p\overline{T})_x + \frac{d(\dot{m}c_p\overline{T})}{dx}\cdot\frac{\Delta x}{1!} + \cdots \tag{8.79}$$

which simplifies to as the size of the control volume approaches zero,

$$\frac{d\overline{T}}{dx} = \frac{q''p}{\dot{m}c_p} \tag{8.80}$$

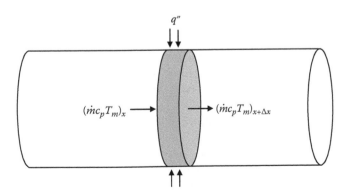

FIGURE 8.3
Energy balance in duct flow.

Substituting for heat flux in terms of the heat transfer coefficient from Equation 8.9 results in an equivalent form

$$\frac{d\bar{T}}{dx} = \frac{q''p}{\dot{m}c_p} = \frac{h(T_w - \bar{T})p}{\dot{m}c_p} \tag{8.81}$$

This equation applies to laminar or turbulent flow and developing or fully developed flow. We consider two wall boundary conditions of uniform heat flux and uniform surface temperature.

8.4.1 Uniform Heat Flux Pipe

The practical examples of a duct subjected to a uniform heat flux include when the duct wall is heated electrically or subjected to radiant flux like exposed to sun, or when a second fluid flows on the outside of the tube, like in heat exchangers. For the uniform heat flux boundary condition, the RHS of Equation 8.81 is constant; therefore, the mean temperature will change linearly for a uniform heat flux pipe resulting in

$$\bar{T} = \frac{q''p}{\dot{m}c_p}x + T_i \tag{8.82}$$

Substituting for mass flow rate from Equation 8.5, and manipulating the terms in Equation 8.82

$$\bar{T} = \frac{q''p}{Re\frac{\mu p}{4}\frac{c_p}{k}xk}\frac{x}{D_h}D_h + T_i \tag{8.83}$$

which is then rearranged to arrive at an expression for nondimensional mean temperature in a pipe subjected to a uniform heat flux:

$$\bar{T}^* = \frac{\bar{T} - T_{in}}{\frac{q''}{k}D_h} = 4x^+ \tag{8.84}$$

where x^+ is given by Equation 8.73 and is inverse of the Graetz number. This equation shows that the dimensionless mean temperature is linearly proportional to the dimensionless distance along the duct, with the slope of 4. Again, this result is valid for laminar or turbulent flow, in the developing or fully developed region, regardless of the duct cross section, as long as the cross section and the properties are independent of axial position.

Far from the inlet, the flow becomes thermally fully developed, and as was shown before the heat transfer coefficient also becomes constant. Therefore, since

$$q'' = h(T_w - \bar{T}) \tag{8.85}$$

for uniform heat flux wall, the difference between the wall and mean temperature will also be constant, and since the mean temperature changes linearly in the duct, in the thermally fully developed region, the wall temperatures will also change linearly so that the difference between wall and mean temperatures will be constant, and the axial variations of mean and wall temperatures are shown in Figure 8.4.

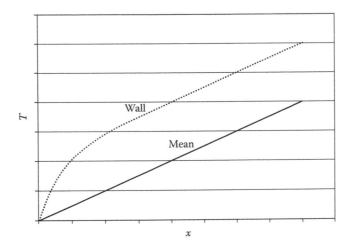

FIGURE 8.4
Mean and wall temperature variation for a pipe subjected to a uniform heat flux.

8.4.2 Isothermal Duct

The second case we consider is that of a pipe subject to a constant wall temperature (Equation 8.81)

$$\frac{d\bar{T}}{dx} = \frac{h(T_w - \bar{T})p}{\dot{m}c_p} \tag{8.86}$$

which can be integrated by separation of variables to yield

$$\ln\frac{\bar{T} - T_w}{T_{in} - T_w} = -\int_0^x \frac{hp}{\dot{m}c_p}\,dx = \frac{\bar{h}px}{\dot{m}c_p} \tag{8.87}$$

where $\bar{h} = -\dfrac{1}{x}\displaystyle\int_0^x h\,dx$ is the axially averaged heat transfer coefficient. Equation 8.87 can be rearranged into

$$\frac{\bar{T} - T_w}{T_i - T_w} = e^{-\frac{\bar{h}px}{\dot{m}c_p}} \tag{8.88}$$

The fluid mean temperature increases exponentially and asymptotes to the wall temperature, which is plotted in Figure 8.5. In the fully developed region, h is constant, and local and average values will be equal in that region; however, at large values of Graetz number, near the inlet of the pipe, h is not constant. This will be shown later, when we obtain the numerical solution for developing flow.

Between the inlet and exit of the pipe, from the first law for a control volume, the amount of heat transferred to the fluid is

$$q = \dot{m}c_p(T_o - T_i) \tag{8.89}$$

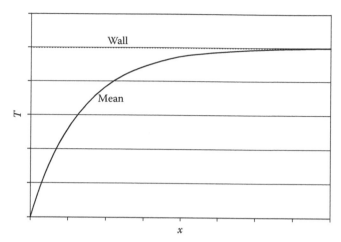

FIGURE 8.5
Mean and wall temperature variation for a pipe with constant wall temperature.

Solving for $\dot{m}c_p$ from Equation 8.89 and substituting in Equation 8.87, noting that $A = px$, and solving for q, result in

$$q = -\bar{h}A \frac{(T_o - T_i)}{\ln \dfrac{T_o - T_w}{T_i - T_w}} \tag{8.90}$$

If we add and subtract the surface temperature to the numerator, then

$$q = \bar{h}A \frac{(T_w - T_i) - (T_w - T_o)}{\ln \dfrac{T_w - T_i}{T_w - T_o}} \tag{8.91}$$

In this equation, the total rate of heat transfer is equal to the product of heat transfer coefficient, area, and what must be some type of temperature difference. This temperature difference is called the logarithmic mean temperature difference (LMTD) and appears frequently in the heat transfer calculation. LMTD is defined as the temperature difference at one end minus temperature difference at the other end divided by the natural log of the ratio of the temperature differences:

$$q = \bar{h}A \; LMTD \tag{8.92}$$

where

$$LMTD = \frac{\Delta T_i - \Delta T_o}{\ln \dfrac{\Delta T_i}{\Delta T_o}} \tag{8.93}$$

Note that ΔT_i and ΔT_o must be positive. To calculate the heat transfer rate from 8.92, the heat transfer coefficient is needed. We have learned much about the behavior of Nusselt number in internal flow; however, the determination of the actual values requires the knowledge of the temperature distribution in the duct, as shown later.

8.5 Nusselt Number Calculation

The local value of Nusselt number is given by Equation 8.24 and since it depends on the temperature gradient at the wall, determination of the Nusselt number requires the knowledge of temperature distribution in the fluid. Nusselt number can also be directly related to the mean fluid temperature. Nondimensionalizing Equation 8.81, using the inlet temperature as the reference and an arbitrary temperature for scale, and inverse Graetz number as the dimensionless distance,

$$\frac{T_s}{D_h RePr}\frac{d\bar{T}^*}{dx^+} = -\frac{4h[(T_w - T_{in}) - (\bar{T} - T_{in})]p}{\mu pRec_p} \tag{8.94}$$

which simplifies to

$$Nu_x = \frac{1}{4}\frac{1}{T_w^* - \bar{T}^*}\frac{d\bar{T}^*}{dx^+} \tag{8.95}$$

This equation shows that for internal flow the mean temperature and Nusselt number are not independent concepts, i.e., if one is known, so is the other. To find Nu from Equation 8.95, the mean temperature is needed which requires temperature distributions to be obtained from the solution of NS equations. For the uniform heat flux boundary condition, using Equation 8.84, Equation 8.95 simplifies to

$$Nu_x = \frac{1}{T_w^* - \bar{T}^*} \tag{8.96}$$

For the isothermal case, taking the temperature scale to be

$$T_s = T_w - T_{in} \tag{8.97}$$

Equation 8.95 can be written as

$$Nu_x = \frac{1}{4}\frac{1}{1-\bar{T}^*}\frac{d\bar{T}^*}{dx^+} \tag{8.98}$$

which simplifies to

$$Nu_x = -\frac{1}{4}\frac{d\ln(1-\bar{T}^*)}{dx^+} \tag{8.99}$$

8.5.1 Average Nusselt Number

If the conditions are uniform peripherally, to calculate the total amount of heat transfer from the inlet to any location x, we need the axially averaged value of h:

$$\overline{Nu} = \frac{\bar{h}D}{k} \tag{8.100}$$

where

$$\bar{h} = \frac{1}{x} \int_0^x h_x dx = \frac{1}{x^+} \int_0^{x^+} h_x dx^+ \tag{8.101}$$

or

$$\overline{Nu} = \frac{1}{x^+} \int_0^{x^+} Nu_x dx^+ \tag{8.102}$$

From Equation 8.22

$$\overline{Nu} = \frac{1}{x^+} \int_0^{x^+} \frac{-\dfrac{\partial T^*}{\partial y^*}\bigg|_w}{T_w^* - \overline{T}^*} dx^+ \tag{8.103}$$

and from Equation 8.95

$$\overline{Nu} = \frac{1}{4x^+} \int_0^{\overline{T}^*} \frac{1}{T_w^* - \overline{T}^*} d\overline{T}^* \tag{8.104}$$

Both of these equations can be used to determine the mean Nusselt number. For a uniform heat flux wall, substituting from Equation 8.84 into 8.104

$$\overline{Nu} = \frac{1}{x^+} \int_0^{x^+} \frac{1}{T_w^* - \overline{T}^*} dx^+ \tag{8.105}$$

This integral can be written in terms of the sum of two integrals:

$$\overline{Nu} = \frac{1}{x^+} \left[\int_0^{x_{fd}^+} \frac{1}{T_w^* - \overline{T}^*} dx^+ + \int_{x_{fd}^+}^{x^+} \frac{1}{T_w^* - \overline{T}^*} dx^+ \right] \tag{8.106}$$

As was shown earlier, for a uniform heat flux pipe, in the fully developed region, the difference between wall and mean temperatures is constant and therefore

$$\overline{Nu} = \frac{1}{x^+} \int_0^{x_{fd}^+} \frac{1}{T_w^* - \overline{T}^*} dx^+ + \frac{1}{T_w^* - \overline{T}^*} \frac{x^+ - x_{fd}^+}{x^+} \tag{8.107}$$

Since x_{fd}^+ is small, then, for a uniform heat flux pipe,

$$\overline{Nu} \approx \frac{1}{T_w^* - \overline{T}^*} \tag{8.108}$$

where $\overline{Nu}$ is the Nusselt number in the fully developed region. It was also shown that for uniform heat flux pipe, the mean fluid temperature is given by

$$\overline{T}^* = 4x^+ \tag{8.109}$$

and therefore in the fully developed region of a uniform heat flux pipe

$$T_w^* = 4x^+ + \frac{1}{\overline{Nu}} \tag{8.110}$$

For an isothermal wall from Equation 8.98

$$\overline{Nu} = \frac{1}{4x^+} \int_0^{\overline{T}^*} \frac{1}{1 - \overline{T}^*} d\overline{T}^* \tag{8.111}$$

which can be integrated to provide

$$\overline{Nu} = -\frac{\ln(1 - \overline{T}^*)}{4x^+} \tag{8.112}$$

This equation is very useful, in that it directly relates the average Nusselt number from the inlet of the pipe to a particular location x, to the mean fluid temperature at that particular location for an isothermal pipe. Solving for the mean fluid temperature

$$\overline{T}^* = 1 - e^{-4\overline{Nu}\,x^+} \tag{8.113}$$

In the fully developed region, Nusselt number is constant, therefore, and if $x_{fd}^+ = 0.06Pr$ is small (liquids and gases), the mean temperature in the fully developed region can be reasonably approximated by Equation 8.113, where $\overline{Nu}$ is replaced with the Nusselt number in the fully developed region.

8.6 Laminar Fully Developed Flow between Two Parallel Plates (Hagen–Poiseuille Flow)

Consider a fluid flowing between two parallel plates, which also approximates high aspect ratio rectangular ducts as shown in Figure 8.6. The conservation of mass, momentum, and energy equations result in

$$\frac{\partial u}{\partial x} + \frac{\partial v}{\partial y} = 0 \tag{8.114}$$

$$\rho\left(u\frac{\partial u}{\partial x} + v\frac{\partial u}{\partial y}\right) = -\frac{\partial P}{\partial x} + \mu\left[\frac{\partial^2 u}{\partial x^2} + \frac{\partial^2 u}{\partial y^2}\right] + \rho g_x \tag{8.115}$$

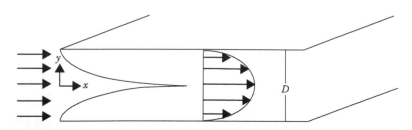

FIGURE 8.6
Laminar fully developed flow between two parallel plates.

$$\rho\left(u\frac{\partial v}{\partial x}+v\frac{\partial v}{\partial y}\right)=-\frac{\partial P}{\partial y}+\mu\left[\frac{\partial^2 v}{\partial x^2}+\frac{\partial^2 v}{\partial y^2}\right]+\rho g_y \tag{8.116}$$

$$\rho c_p\left(u\frac{\partial T}{\partial x}+v\frac{\partial T}{\partial y}\right)=k\left[\frac{\partial^2 T}{\partial x^2}+\frac{\partial^2 T}{\partial y^2}\right] \tag{8.117}$$

In the hydrodynamically fully developed region, $\frac{\partial u}{\partial x}=0$ or $u\neq u(x)$ and thus $u=u(y)$. Therefore, from the continuity equation, $\frac{\partial v}{\partial y}=0$ or v is not a function of y and must be constant in the y direction. Since v is a constant in the y direction and it is equal to zero at the wall, it must be equal to zero ($v=0$) everywhere in the fully developed region. Therefore from the y momentum equation, assuming the spacing between the plates to be small, then hydrodynamic pressure is negligible, resulting in

$$\frac{\partial P}{\partial y}=0 \tag{8.118}$$

or in the fully developed region, pressure is only a function of x. Therefore, the governing equations reduce to

$$\frac{1}{\mu}\frac{dP}{dx}=\frac{d^2 u}{dy^2} \tag{8.119}$$

$$\rho c_p u\frac{\partial T}{\partial x}=k\left[\frac{\partial^2 T}{\partial x^2}+\frac{\partial^2 T}{\partial y^2}\right] \tag{8.120}$$

8.6.1 Velocity Distribution for Fully Developed Flow

The LHS of the x momentum equation (Equation 8.119) is only a function of x and the RHS is only a function of y, and x and y are independent of each other. The equality can only hold if the two are equal to a constant:

$$\frac{1}{\mu}\frac{dP}{dx}=\frac{d^2 u}{dy^2}=C \tag{8.121}$$

Integrating for the velocity distribution

$$u = \frac{C}{2}y^2 + C_1 y + C_2 \tag{8.122}$$

and using the boundary conditions

$$y = 0, \quad \frac{du}{dy} = 0 \quad \Rightarrow C_1 = 0 \tag{8.123}$$

$$y = \frac{D}{2}, \quad u = 0 \quad \Rightarrow C_2 = -\frac{CD^2}{8} \tag{8.124}$$

result in

$$u = C\left(\frac{y^2}{2} - \frac{D^2}{8}\right) \tag{8.125}$$

The constant C is next expressed in terms of the mean velocity, U, using the conservation of mass

$$\dot{m} = \rho U W \frac{D}{2} = \int_A \rho u \, dA = \int_0^{D/2} \rho C\left(\frac{y^2}{2} - \frac{D^2}{8}\right)W dy = \rho C W \left[\frac{y^3}{6} - \frac{D^2}{8}y\right]_0^{D/2} \tag{8.126}$$

that results in

$$C = -\frac{12U}{D^2} \tag{8.127}$$

Substituting for C in Equations 8.125 and 8.121, the expressions for velocity distribution and pressure gradient become

$$\frac{u}{U} = \frac{3}{2}\left[1 - \left(\frac{y}{D/2}\right)^2\right] \tag{8.128}$$

$$\frac{dP}{dx} = -12\frac{\mu U}{D^2} \tag{8.129}$$

8.6.2 Friction Factor

The internal flows are generally caused by a pressure gradient, and therefore the determination of pressure drop is an important consideration in duct flows. The dimensionless pressure drop is expressed in terms of Darcy (or Moody) friction factor defined by

$$f = \frac{-\dfrac{dP}{dx}D_h}{\dfrac{1}{2}\rho U^2} \tag{8.130}$$

For flow between two parallel plates from Equation 8.113

$$f = \frac{12\frac{\mu U}{D^2}D_h}{\frac{1}{2}\rho U^2} = \frac{24\mu D_h}{\rho U D^2} \tag{8.131}$$

with $D_h = 2D$, it simplifies to

$$f = \frac{96\mu}{\rho U D_h} = \frac{96}{Re} \tag{8.132}$$

The friction factor defined by Equation 8.130 is the dimensionless pressure gradient in the pipe. The nondimensional wall shear stress, friction, or drag coefficient,

$$c_f = \frac{\tau}{\frac{1}{2}\rho \bar{U}^2} \tag{8.133}$$

that was defined for external flows, can also be calculated for internal flows and is known as Fanning friction factor. For flow between two parallel plates

$$c_f = \frac{-\mu\frac{\partial u}{\partial y}\Big|_{y=D/2}}{\frac{1}{2}\rho \bar{U}^2} = \frac{-\mu U \frac{3}{2}\left[-2\left(\frac{1}{D/2}\right)\right]}{\frac{1}{2}\rho \bar{U}^2} = \frac{24\mu}{\rho \bar{U} D_h} \tag{8.134}$$

$$c_f = \frac{24}{Re} \tag{8.135}$$

Comparing Equations 8.132 and 8.134, it can be seen that Darcy friction factor is four times the Fanning friction factor (friction coefficient), or

$$\frac{f}{8} = \frac{c_f}{2} \tag{8.136}$$

This is a useful relationship between friction factor and drag coefficient which is valid for both laminar and turbulent fully developed flows. In fully developed flows, the velocity profile does not change axially, implying that the change in momentum is zero; therefore, sum of the forces must add up to zero, or pressure forces must balance frictional forces. Consider a section of fluid of length L in the fully developed region as shown in Figure 8.7. The net force must be zero, or the pressure force must balance the friction force:

$$[P(x) - P(x+L)]D = \tau_w 2L \tag{8.137}$$

$$-\frac{P(x+L) - P(x)}{L}2D = 4\tau_w \tag{8.138}$$

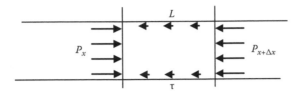

FIGURE 8.7
Force balance for fully developed flow.

$$-\frac{\frac{\Delta P}{L} D_h}{\frac{1}{2}\rho U^2} = 4\frac{\tau_w}{\frac{1}{2}\rho U^2} \tag{8.139}$$

$$f = 4c_f \tag{8.140}$$

which again confirms Equation 8.136. Note that the equation does not require the flow to be laminar and is valid for laminar or turbulent flow.

8.6.3 Temperature Distribution for Fully Developed Flow

As stated by Equation 8.120, the energy equation in the hydrodynamically developed and thermally developing region is

$$\rho c_p u \frac{\partial T}{\partial x} = k\left[\frac{\partial^2 T}{\partial x^2} + \frac{\partial^2 T}{\partial y^2}\right] \tag{8.141}$$

and the boundary conditions are

$$x = 0, \quad T = T_{in} \tag{8.142}$$

$$y = 0, \quad \frac{\partial T}{\partial y} = 0 \tag{8.143}$$

$$y = \frac{D}{2}, \quad T = T_w, \quad \text{or} \quad k\frac{\partial T}{\partial y} = q'' \tag{8.144}$$

Equation 8.142 is not the exact initial condition, because, at the inlet of the pipe, flow is not fully developed and Equation 8.117 rather than Equation 8.141 is valid; however, the errors quickly disappear and accurate results are obtained by using Equation 8.141. Defining the nondimensional variables

$$y^* = \frac{y}{D_h}, \quad T^* = \frac{T - T_{in}}{T_s}, \quad x^* = \frac{x}{D_h}, \quad u^* = \frac{u}{U} \tag{8.145}$$

the nondimensional energy equation becomes

$$u^* \frac{\partial T^*}{\partial x^*} = \frac{1}{RePr}\left[\frac{\partial^2 T^*}{\partial x^{*2}} + \frac{\partial^2 T^*}{\partial y^{*2}}\right] \tag{8.146}$$

Performing an order of magnitude analysis

$$u^* \frac{\partial T^*}{\partial x^*} = \frac{1}{RePr} \left[\frac{1}{x^{*2}} \quad \frac{1}{1} \right]$$
(8.147)

As can be seen, a few hydraulic diameters from the inlet of the pipe, the first term in the bracket on the RHS becomes much smaller than the second term, which is of the order of 1, and therefore can be neglected, simplifying the energy equation to

$$u^* \frac{\partial T^*}{\partial x^*} = \frac{1}{RePr} \frac{\partial^2 T^*}{\partial y^{*2}}$$
(8.148)

For fully developed flow, we can define a new dimensionless axial distance

$$x^+ = \frac{x^*}{RePr}$$
(8.149)

which as mentioned earlier is the inverse of Graetz number. The use of this dimensionless variable causes the Reynolds and Prandtl numbers to be scaled out of the energy equation, simplifying the energy equation to

$$u^* \frac{\partial T^*}{\partial x^+} = \frac{\partial^2 T^*}{\partial y^{*2}}$$
(8.150)

Rearranging the energy equation,

$$u^* = \frac{\dfrac{\partial^2 T^*}{\partial y^{*2}}}{\dfrac{\partial T^*}{\partial x^+}}$$
(8.151)

Note that in the fully developed region, the LHS is only a function of y and so should the RHS. Temperature is a function of x and y; however, the dependence on x must be such that the ratio on the RHS is independent of x. A possible solution for temperature distribution that satisfies Equation 8.151 is

$$T^* = f(y^*) + ax^+$$
(8.152)

where a is a constant.

8.6.3.1 Uniform Heat Flux Duct

The boundary conditions for the uniform heat flux case are

$$x^* = 0, \quad T^* = 0$$

$$y^* = 0, \quad \frac{\partial T^*}{\partial y^*} = 0$$
(8.153)

$$y^* = \frac{1}{4}, \quad \frac{\partial T^*}{\partial y^*} = 1 \quad \text{where } T_s = \frac{q'' D_h}{k}$$

Note that for a uniform heat flux pipe, the surface temperature is not constant, and the scale used eliminates several parameters. Substituting for the fully developed velocity distribution, the energy equation becomes

$$\frac{3}{2}(1-16y^{*2})\frac{\partial T^*}{\partial x^+} = \frac{\partial^2 T^*}{\partial y^{*2}} \tag{8.154}$$

As discussed in Section 8.4.1, for thermally fully developed flow in a pipe subjected to a uniform heat flux, the difference between surface and mean temperatures is constant; therefore,

$$\frac{dT_w}{dx} = \frac{d\overline{T}}{dx} \tag{8.155}$$

and from Equation 8.76, since $\overline{T}(x) - T_w(x) \neq 0$,

$$\frac{d\overline{T}}{dx} = \frac{dT_w}{dx} \tag{8.156}$$

or for a pipe subjected to a uniform heat flux in the fully developed flow region, all temperatures change linearly in the x direction and they all have the same slope. From Equation 8.84, the slope is

$$\frac{d\overline{T}^*}{dx^+} = 4 \tag{8.157}$$

and the energy equation becomes

$$6(1-16y^{*2}) = \frac{\partial^2 T^*}{\partial y^{*2}} \tag{8.158}$$

Integrating Equation 8.158 once

$$\frac{\partial T^*}{\partial y^*} = 6\left(y^* - \frac{16}{3}y^{*3}\right) + C_1 \tag{8.159}$$

C_1 is zero from the first boundary condition. This equation also satisfies the wall boundary condition. Integrating the equation one more time

$$T^* = (3y^{*2} - 8y^{*4}) + f(x^+) \tag{8.160}$$

Since $T^*(x^+, y^* = 0) = f(x^+)$, then the physical interpretation of the function is that it is the centerline temperature, $f(x^+) = T_{CL}^*$, which varies along the pipe

$$T^*(x^+, y^*) = 3y^{*2} - 8y^{*4} + T_{CL}^*(x^+) \tag{8.161}$$

This equation provides the temperature distribution in the fully developed region. As can be seen, relative to the centerline temperature, the temperature variation has the same y

dependence and is independent of the x location. The wall temperature, for example, can be determined by evaluating Equation 8.161 at $y^* = \dfrac{1}{4}$, resulting in

$$T_w^* = \frac{5}{32} + T_{CL}^* \tag{8.162}$$

We now have to determine the x dependence of the temperature distribution. Using Equation 8.19 to determine the mean temperature in the pipe

$$\bar{T}^* = 4 \int\limits_0^{0.25} u^* T^* dy^* = 4 \int\limits_0^{0.25} \frac{3}{2}(1-16y^{*2})\left[(3y^{*2} - 8y^{*4}) + T_{CL}^*(x^+)\right] dy^* \tag{8.163}$$

Evaluating the integral results in

$$\bar{T}^* = \frac{39}{1120} + T_{CL}^*(x^+) \tag{8.164}$$

Equating this equation with Equation 8.84, the expression for the $T_{CL}^*(x^+)$ is

$$T_{CL}^*(x^+) = 4x^+ - \frac{39}{1120} \tag{8.165}$$

Note that this equation is only valid in the fully developed region, and will fail near the pipe inlet (negative temperature!).

The Nusselt number can be determined from Equation 8.96:

$$Nu_x = \frac{1}{T_w^* - \bar{T}} = \frac{1}{\dfrac{5}{32} + T_{CL}^*(x^+) - \dfrac{39}{1120} - T_{CL}^*(x^+)} = \frac{1}{\dfrac{5}{32} - \dfrac{39}{1120}} = 8.2353 \tag{8.166}$$

The fluid and wall temperatures in the fully developed region ($x^+ > 0.06$) for the uniform heat flux case are then given by

$$T^* = (3y^{*2} - 8y^{*4}) + 4x^+ - \frac{39}{1120} \tag{8.167}$$

$$T_w^* = 4x^+ + \frac{136}{1120} \tag{8.168}$$

8.6.3.2 Isothermal Duct

The isothermal boundary condition is closely approximated in condensers and evaporators. For the case when the two plates are at a constant temperature, the boundary conditions for Equation 8.141 are

$$x^* = 0, \quad T^* = \frac{T - T_{in}}{T_w - T_{in}} = 0$$

$$y^* = 0, \quad \frac{\partial T^*}{\partial y^*} = 0 \tag{8.169}$$

$$y^* = \frac{D/2}{2D} = \frac{1}{4} \quad T^* = 1 \quad \text{for isothermal pipe, } T_s = T_w - T_{in}$$

Unlike the uniform heat flux case, the axial temperature gradient is not constant and Equation 8.76 simplifies to

$$\frac{\partial T}{\partial x}[\bar{T} - T_w] - \frac{d\bar{T}}{dx}[T - T_w] = 0 \tag{8.170}$$

which when nondimensionalized becomes

$$\frac{\partial T^*}{\partial x^+} = \frac{1 - T^*}{1 - \bar{T}^*}\frac{d\bar{T}^*}{dx^+} \tag{8.171}$$

Therefore, the energy equation reduces to

$$u\frac{1 - T^*}{1 - \bar{T}^*}\frac{d\bar{T}^*}{dx^+} = \frac{\partial^2 T^*}{\partial y^{*2}} \tag{8.172}$$

Substituting from Equations 8.98 and 8.128 reduces the energy equation to

$$\frac{3}{2}(1 - 16y^{*2})4Nu_x(1 - T^*) = \frac{\partial^2 T^*}{\partial y^{*2}} \tag{8.173}$$

In the fully developed region, Nusselt number is constant, and therefore the solution can be obtained iteratively, by guessing a value for Nu, solving the energy equation (Equation 8.173), calculating the mean temperature, and finding the Nu number from Equation 8.22:

$$Nu_x = \frac{\left.\frac{\partial T^*}{\partial y^*}\right|_w}{1 - \bar{T}^*} \tag{8.174}$$

If this new value is not close to the old value, then repeat the process. The solution may be obtained analytically using series solution and will result in

$$Nu = 7.54 \tag{8.175}$$

The solution can also be obtained numerically. The finite difference form of Equation 8.173 is

$$T_{i-1} - [2 - 4Nu\Delta y^2 u_i]T_i + T_{i+1} = 4Nu\Delta y^2 u_i \tag{8.176}$$

Using TRIDI function in Excel®, the temperature distribution is obtained, which is used to calculate mean temperature, and then the new value of Nu. The results of the iterative numerical solution are shown in Table 8.1 and as can be seen, the solution quickly converges to the analytical solution.

TABLE 8.1

Iterative Numerical Solution for Isothermal Parallel Plates

Iteration	1	2	3	4	5	6	7
Nu	1.00	6.49	7.32	7.50	7.53	7.54	7.54

8.7 Laminar Fully Developed Flow in Circular Duct

Circular ducts are widely used to confine and transport fluids. For flow in a circular duct, the conservation equations are

$$\frac{\partial u}{\partial x} + \frac{1}{r}\frac{\partial (rv)}{\partial r} = 0 \tag{8.177}$$

$$\rho\left(u\frac{\partial u}{\partial x} + v\frac{\partial u}{\partial r}\right) = -\frac{\partial P}{\partial x} + \mu\left[\frac{\partial^2 u}{\partial x^2} + \frac{1}{r}\frac{\partial}{\partial r}\left(r\frac{\partial u}{\partial r}\right)\right] + \rho g_x \tag{8.178}$$

$$\rho\left(u\frac{\partial v}{\partial x} + v\frac{\partial v}{\partial r}\right) = -\frac{\partial P}{\partial r} + \mu\left[\frac{\partial^2 v}{\partial x^2} + \frac{1}{r}\frac{\partial}{\partial r}\left(r\frac{\partial v}{\partial r}\right) - \frac{v}{r^2}\right] + \rho g_r \tag{8.179}$$

$$\rho c_p\left(u\frac{\partial T}{\partial x} + v\frac{\partial T}{\partial r}\right) = k\left[\frac{\partial^2 T}{\partial x^2} + \frac{1}{r}\frac{\partial}{\partial r}\left(r\frac{\partial T}{\partial r}\right)\right] \tag{8.180}$$

For fully developed flow, $v = 0$. Therefore,

$$\frac{1}{\mu}\frac{dP}{dx} = \frac{1}{r}\frac{d}{dr}\left(r\frac{du}{dr}\right) = C \tag{8.181}$$

Integrating twice to obtain the velocity distribution

$$u = C\frac{r^2}{4} + C_1\ln(r) + C_2 \tag{8.182}$$

The boundary conditions are no slip at the wall and because of the symmetry, maximum velocity at the centerline

$$r = R, \quad u = 0 \tag{8.183}$$

$$r = 0, \quad \frac{du}{dr} = 0 \tag{8.184}$$

$$u = -\frac{CR^2}{4}\left(1 - \frac{r^2}{R^2}\right) \tag{8.185}$$

The constant C is related to the average velocity by using the conservation of mass

$$\dot{m} = \rho\pi R^2 U = \int_A \rho u\, dA = \int_0^R \rho\frac{-CR^2}{4}\left(1 - \frac{r^2}{R^2}\right)2\pi r\, dr \tag{8.186}$$

Performing the integration and solving for C,

$$C = -32\frac{U}{D^2}$$
(8.187)

Therefore,

$$u = 2U\left(1 - \frac{r^2}{R^2}\right)$$
(8.188)

and in terms of nondimensional variables

$$u^* = 2(1 - 4r^{*2})$$
(8.189)

where

$$r^* = \frac{r}{D_h}$$

and

$$D_h = \frac{4A}{p} = \frac{4\frac{\pi}{4}D^2}{\pi D} = D$$
(8.190)

From Equation 8.181, pressure gradient can be calculated as

$$\frac{dP}{dx} = -32\frac{\mu U}{D^2}$$
(8.191)

and the friction factor is

$$f = \frac{32\frac{\mu U}{D_h^2}D_h}{\frac{1}{2}\rho U^2} = \frac{64\mu}{\rho U D_h} = \frac{64}{Re}$$
(8.192)

8.7.1 Uniform Heat Flux Pipe

Similar to flow between two parallel plates, the energy equation becomes

$$2(1 - 4r^{*2})\frac{\partial T^*}{\partial x^+} = \frac{1}{r^*}\frac{\partial}{\partial r^*}\left(r^*\frac{\partial T^*}{\partial r^*}\right)$$
(8.193)

This equation has to be solved subject to the following

$$x^* = 0, \quad T^* = \frac{T - T_{in}}{T_s} = 0 \quad \text{where } T_s = \frac{q''D}{k}$$

$$r^* = 0, \quad \frac{\partial T^*}{\partial r^*} = 0 \tag{8.194}$$

$$r^* = \frac{1}{2}, \quad \frac{\partial T^*}{\partial r^*} = 1$$

Again the mean temperature is $\bar{T}^* = 4x^+$ and the energy equation becomes

$$8(r^* - 4r^{*3}) = \frac{\partial}{\partial r^*}\left(r^* \frac{\partial T^*}{\partial r^*}\right) \tag{8.195}$$

Integrating and evaluating the constants, the fluid temperature becomes

$$T^* = 2(r^{*2} - r^{*4}) + 4x^+ - \frac{7}{48} \tag{8.196}$$

This equation provides the fluid temperature in the fully developed region. The centerline temperature is

$$T_{cl}^* = 4x^+ - \frac{7}{48} \tag{8.197}$$

and the wall temperature

$$T_w^* = 4x^+ + \frac{11}{48} \tag{8.198}$$

$$T_w^* - \bar{T}^* = \frac{11}{48} \tag{8.199}$$

$$Nu = \frac{hD}{k} = \frac{1}{[T_w^* - \bar{T}^*]} = \frac{48}{11} = 4.36 \tag{8.200}$$

8.7.2 Isothermal Pipe

Similar to flow between two parallel plates, the energy equation becomes

$$2(1 - 4r^{*2})(1 - T^*)4Nu = \frac{1}{r^*}\frac{\partial}{\partial r^*}\left(r^* \frac{\partial T^*}{\partial r^*}\right) \tag{8.201}$$

This equation has to be solved subject to the following

$$r^* = 0, \quad \frac{\partial T^*}{\partial r^*} = 0$$

(8.202)

$$r^* = \frac{1}{2}, \quad T^* = 1$$

Shah et al. [1] provide a solution developed by Bhatti and cited by Kays and Crawford [2] of the form

$$\frac{T^* - 1}{T_{cl}^* - 1} = \sum_{i=0}^{\infty} C_{2i}(2r^*)^{2i}$$

(8.203)

where

$$C_0 = 1$$

$$C_2 = -\frac{1}{4}\lambda_0^2$$

(8.204)

$$C_{2i} = \frac{\lambda_0^2}{(2i)^2}(C_{2i-4} - C_{2i-2}) \quad i > 1$$

$$\lambda_0 = 2.704364$$

Substituting for the nondimensional temperature from Equation 8.203 into 8.201 and solving for *Nu* result in

$$Nu = -\frac{\sum_{i=1}^{\infty} C_{2i}16i^2(2r^*)^{2i-2}}{8(1 - 4r^{*2})\left(\sum_{i=0}^{\infty} C_{2i}(2r^*)^{2i}\right)}$$

(8.205)

which is approximately independent of the radial position. Evaluating the RHS for a small value of r^* results in

$$Nu = 3.657$$

(8.206)

which is the Nusselt number for laminar fully developed flow in an isothermal circular pipe. The finite difference approximation of the energy equation is

$$u_i(1 - T_i)4Nu = \frac{1}{r_i}\frac{T_{i+1} - T_{i-1}}{2\Delta r} + \frac{T_{i+1} - 2T_i + T_{i-1}}{\Delta r^2}$$

(8.207)

which results in

$$A_i = \left(1 - \frac{1}{r_i}\frac{\Delta r}{2}\right)$$

$$B_i = -(2 - 4uNu\Delta r^2)$$

(8.208)

$$C_i = \left(1 + \frac{1}{r_i}\frac{\Delta r}{2}\right)$$

$$R_i = 4u_iNu\Delta r^2$$

The finite difference form of the boundary condition at center can be obtained by applying Equation 8.207 to $i = 1$ and noting that $T_0 = T_2$ resulting in

$$-\left(\frac{2}{\Delta r^2} - 4u_1 Nu\right)T_1 + \left(\frac{2}{\Delta r^2}\right)T_2 = 4u_1 Nu \tag{8.209}$$

The solution proceeds by guessing a value for Nu, solving the energy equation, calculating the mean temperature from Equation 8.20, and finding the Nu number from

$$Nu_x = \frac{\frac{\partial T^*}{\partial r^*}}{(1 - \bar{\bar{T}}^*)} \tag{8.210}$$

The solution is obtained using 101 equally spaced nodes in the r direction. It is also interesting to note that $T^* = 1$ is a solution to Equation 8.201 and therefore rather than using the temperature symmetry boundary condition, the solution is obtained for a location which has an arbitrary assigned centerline temperature. The results of the iterative numerical solution are shown in Table 8.2 and as can be seen, the solution quickly converges to the analytical solution.

Table 8.3 provides a summary of the fully developed results for different duct geometries.

TABLE 8.2

Iterative Numerical Solution for Isothermal Pipe

Iteration	1	2	3	4	5	6	7	8
Nu	1.00	2.91	3.44	3.60	3.66	3.67	3.68	3.68

TABLE 8.3

Laminar Fully Developed Flow Results

	b/a	D_h	fRe	Isothermal	q'
D	—	D	64	3.66	4.36
a, b	1	a	57	2.98	3.61
	0.7	$\frac{14}{17}a$	59	3.08	3.73
	1/2	$\frac{2}{3}a$	62	3.39	4.12
	1/3	$\frac{1}{2}a$	69	3.96	4.79
	1/4	$\frac{2}{5}a$	73	4.44	5.33
	1/8	$\frac{2}{9}a$	82	5.6	6.49
	0	$2a$	96	7.54	8.23
	0	$2a$	96	4.86	5.39
a	—	$\frac{a}{\sqrt{3}}$	53	2.47	3.11

8.8 Hydrodynamically Developed and Thermally Developing Flow

For fluids that have high Prandtl number, the velocity profile develops much faster than the temperature profile; therefore over a significant portion of the duct, the flow is hydrodynamically developed and thermally developing. The numerical solution to this problem is presented next. For $Pe \gg 1$, the energy equation is

$$u^* \frac{\partial T^*}{\partial x^+} = \frac{1}{r^*} \frac{\partial T^*}{\partial r^*} + \frac{\partial^2 T^*}{\partial r^{*2}} \tag{8.211}$$

$$x^+ = 0, \quad T^* = 0$$

$$r^* = 0, \quad \frac{\partial T^*}{\partial r^*} = 0 \tag{8.212}$$

$$r^* = \frac{1}{2}, \quad \frac{\partial T^*}{\partial r^*} = 1$$

$$T^* = 1$$

Using central difference approximation in the r direction and backward difference approximation in the x direction for the first derivative, the discretized equation again becomes

$$A_i T_{i-1} + B_i T_i + C_i T_{i+1} = R_i \tag{8.213}$$

where the coefficients for the finite difference approximation are

$$A_i = \left(\frac{1}{\Delta r^2} - \frac{1}{2r_i \Delta r} \right)$$

$$B_i = -\left(\frac{2}{\Delta r^2} + u_i \frac{1}{\Delta x} \right)$$

$$C_i = \left(\frac{1}{\Delta r^2} + \frac{1}{2r_i \Delta r} \right) \tag{8.214}$$

$$R_i = -u_i \frac{T_{i-1,j}}{\Delta x}$$

The finite difference form of the boundary condition at center can be obtained by applying Equation 8.213 to $j = 1$ and noting that $T_{i,0} = T_{i,2}$ resulting in

$$-T_{i,1} \left(\frac{2}{\Delta r^2} + 2(1 - 4r^2) \frac{1}{\Delta x} \right) + T_{i,2} \left(\frac{2}{\Delta r^2} \right) = -2(1 - 4r^2) \frac{T_{i-1,1}}{\Delta x} \tag{8.215}$$

the wall boundary condition being

$$-T_{i,N-1} + T_{i,N} = \Delta r$$

or

$$T_{i,N} = 1 \tag{8.216}$$

depending on whether the pipe is subjected to a uniform heat flux or is isothermal.

The solution was obtained using 51 equally spaced nodes in the r direction. Since Equation 8.211 is parabolic (first order) with respect to x, we can use variable spacing in the x direction. The axial distance can be expressed as

$$x_i = x_{i-1} + \Delta x_i \tag{8.217}$$

The increment in the x direction starts with a small value and increases geometrically

$$\Delta x_i = \Delta x_{i-1} d \tag{8.218}$$

In the results presented later, we assumed

$$\Delta x_1 = 8 \times 10^{-6}$$
$$d = 1.3 \tag{8.219}$$

A total of 51 nodes were also used in the x direction. Equation 8.218 was used for the first 27 nodes and thereafter constant $\Delta x = \Delta x_{27}$ was used. Once the temperature distribution is determined, the mean temperature is calculated by numerically integrating Equation 8.20 using trapezoidal rule and the Nu number from

$$Nu_x = \frac{\dfrac{\partial T^*}{\partial r^*}}{(1 - \bar{T}^*)} \tag{8.220}$$

8.8.1 Uniform Heat Flux

Figure 8.8 shows the variation of the wall, mean, and centerline temperatures along the pipe. The solid lines are from the numerical solution of hydrodynamically developed, thermally developing flow and the points are the same quantities obtained analytically for the fully developed case. The mean temperature is calculated from Equation 8.84, and the centerline and wall temperatures from Equations 8.179 and 8.198, respectively. As can be seen, there is excellent agreement between the numerical and analytical solutions in the fully developed region. It is also important to note that the mean temperature is accurately represented by Equation 8.84 even in the developing region.

The numerical solution also confirms the expectation that the temperatures develop quickly and they all increase linearly with the same slope in the fully developed region. Once the temperature distribution is obtained, the other quantities can be calculated. For example, we can calculate T^+ at different radial positions. On the same figure, the two dashed curves are T^+ along the pipe, at the center, and half distance from the center. In both locations, the nondimensional temperature T^+ increases and asymptotes to a constant value, establishing the fully developed condition, which is generally assumed to occur around $x^+ = 0.06$, which is not exact.

Once the mean temperature is calculated, then the local value of the Nusselt number can be calculated from Equation 8.22 or 8.98. The derivative in Equation 8.22 was approximated using second-order backward approximation:

$$\left. \frac{\partial T^*}{\partial r^*} \right|_{r^*=1/2} = \frac{3T_N^* - 4T_{N-1}^* + T_{N-2}^*}{2\Delta r^*} \tag{8.221}$$

Figure 8.9 shows the variation of the local Nusselt number along the pipe, using both approaches and as can be seen the two approaches provide close results. As it can also be seen, the Nusselt number asymptotes to the constant value of 4.36.

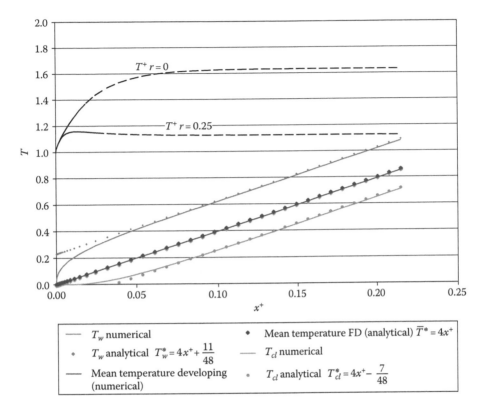

FIGURE 8.8
Temperature variation along a uniform heat flux pipe.

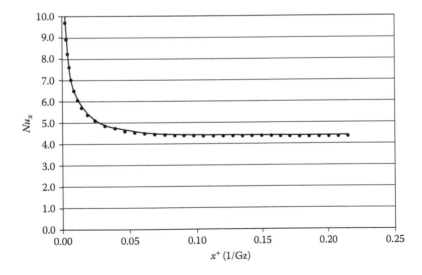

FIGURE 8.9
The variation of the local Nusselt number along a uniform heat flux pipe.

8.8.2 Isothermal Pipe

For fluid flowing in a circular pipe whose surface is maintained at a constant temperature, the energy equation and boundary conditions are the same as those given for the uniform heat flux case, except the wall condition, which is

$$r^* = \frac{1}{2}, \quad T^* = 1 \tag{8.222}$$

Since in this case, $\dfrac{\partial T^*}{\partial x^+}$ is not constant in the developing or fully developed region, the developing region must be solved numerically. We have already seen the analytical solution for the fully developed region which is somewhat more involved, and is obtained by an iterative approach or by series solution [2]. The approach taken here is to obtain the solution numerically. The procedure is identical to that used for the previous case of uniform heat flux, with the only difference being changing the wall temperature to 1.

Figure 8.10 shows the variation of the mean temperature and centerline temperature a long the pipe. As can be seen, they both increase axially. On the same figure, we have also plotted T^+ at two radial positions and again, as can be seen, they both quickly asymptote to different constant values that indicate the fully developed condition. Figure 8.11 shows the variation of the local Nusselt number along the pipe, calculated from Equations 8.85 and 8.87, and as can be seen the two approaches provide close results. As it can also be seen, the Nusselt number asymptotes to the constant value of 3.66, which is the fully developed limit.

Shah and London [3] provide the following expression for Nu in developing flow

$$Nu = \begin{cases} 1.077(x^+)^{-1/3} & x^+ \leq 0.01 \\ 3.657 + 6.874(10^3 x^+)^{-0.488} e^{-57.2 x^+} & x^+ > 0.01 \end{cases} \tag{8.223}$$

Equation 8.223 is also plotted in Figure 8.11. As can be seen, there is in general close agreement between the two. The difference around $x^+ = 0.01$ may be due to the hydrodynamically

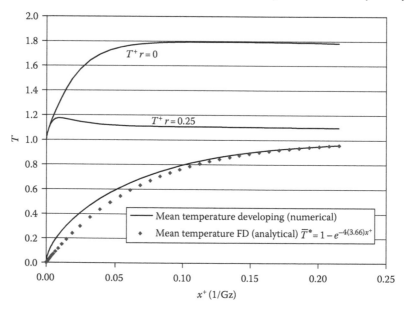

FIGURE 8.10
Temperature variation along an isothermal pipe.

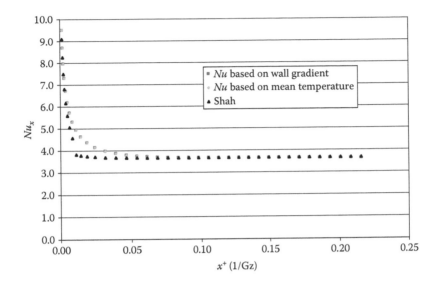

FIGURE 8.11
The variation of the local Nusselt number along an isothermal pipe.

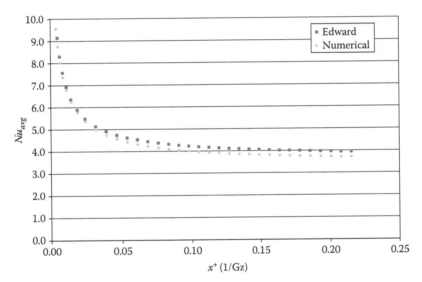

FIGURE 8.12
The axial variation of the mean Nusselt number for an isothermal pipe.

fully developed assumption and neglecting of the axial diffusion in the energy equation made in the numerical solution.

The mean Nusselt number can be determined from Equation 8.112 and is plotted in Figure 8.12. Edwards et al. [4] provide the following expression for average Nusselt number in the entrance region:

$$\overline{Nu} = 3.66 + \frac{0.065}{x^+ + 0.04 x^{+1/3}} \tag{8.224}$$

This equation is also plotted on Figure 8.12, and there is reasonable agreement between the two.

8.8.3 Concentric Pipes

The flow in concentric pipes, shown in Figure 8.13, is important in a number of applications, including in double-pipe parallel and counter flow heat exchangers. For fully developed flow, defining

$$r^* = \frac{r}{r_o}, \quad u^* = \frac{u}{U} \tag{8.225}$$

the dimensionless momentum equation becomes

$$\frac{1}{r^*}\frac{d}{dr^*}\left(r^*\frac{du^*}{dr^*}\right) = \frac{(r_o)^2}{\mu U}\frac{dP}{dx} = C \tag{8.226}$$

with boundary conditions

$$r_i^* = \frac{r_i}{r_o}, \quad u^* = 0$$
$$r^* = 1, \quad u^* = 0 \tag{8.227}$$

Integrating and evaluating the constants,

$$u^* = \frac{C}{4}\left[\left(1 - r_i^{*2}\right)\frac{\ln r^*}{\ln r_i^*} - \left(1 - r^{*2}\right)\right] \tag{8.228}$$

The constant C is related to the average velocity, by using the conservation of mass

$$\dot{m} = \rho\pi\left(r_o^2 - r_i^2\right)U = \int_{r_i}^{r_o}\rho 2\pi r u\, dr \tag{8.229}$$

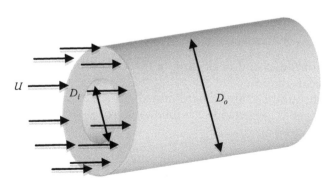

FIGURE 8.13
Flow between two concentric pipes.

which simplifies to

$$\frac{2}{1-r_i^{*2}}\int_{r_i^*}^{1} r^* u^* dr^* = 1 \tag{8.230}$$

Substituting for the velocity and performing the integration, the constant becomes

$$C = -\frac{8}{\left(1+r_i^{*2}\right)+\dfrac{1-r_i^{*2}}{\ln r_i^*}} \tag{8.231}$$

Therefore, the velocity distribution is

$$u^* = \frac{2\left[1-r^{*2}-\left(1-r_i^{*2}\right)\dfrac{\ln r^*}{\ln r_i^*}\right]}{\left(1+r_i^{*2}\right)+\dfrac{1-r_i^{*2}}{\ln r_i^*}} \tag{8.232}$$

The location of maximum velocity is obtained by differentiating Equation 8.232, and setting it to zero, resulting in

$$r^* = \sqrt{\frac{1-r_i^{*2}}{-2\ln\left(r_i^*\right)}} \tag{8.233}$$

which is closer to the inner cylinder. The friction factor becomes

$$f = \frac{-\dfrac{\partial P}{\partial x}D_h}{\dfrac{1}{2}\rho U^2} = \frac{-8C\left(1-r_i^*\right)^2}{Re} \tag{8.234}$$

where

$$D_h = \frac{4\dfrac{\pi}{4}\left(D_o^2-D_i^2\right)}{\pi\left(D_o+D_i\right)} = D_o - D_i \tag{8.235}$$

which simplifies to

$$f = \frac{\left(1-r_i^*\right)^2}{\left(1+r_i^{*2}\right)+\dfrac{1-r_i^{*2}}{\ln r_i^*}}\frac{64}{Re} \tag{8.236}$$

Note that $r_i^* = 0$ corresponds to flow in a circular cylinder and the first fraction on the RHS is equal to one, reducing Equation 8.236 to $f = \dfrac{64}{Re}$, which is for the circular pipe. The limit of $r_i^* \to 1$ represents the case of the radius of curvature approaching infinity or of flow between parallel plate, then

$$\frac{\left(1-r_i^*\right)^2}{\left(1+r_i^{*2}\right)+\dfrac{1-r_i^{*2}}{\ln r_i^*}} \to 1.5 \tag{8.237}$$

or $f = \dfrac{96}{Re}$, which is for flow between parallel plates.

8.8.4 Numerical Solution of Thermally Developing and Hydrodynamically Fully Developed Flow

The dimensionless form of the energy equation and boundary conditions for hydrodynamically fully developed and thermally developing developing flow in concentric pipes is

$$u^* \frac{\partial T^*}{\partial x^-} = \frac{1}{r^*} \frac{\partial}{\partial r^*}\left(r^* \frac{\partial T^*}{\partial r^*}\right) \tag{8.238}$$

The boundary conditions are

$$r^* = r_i^*, \quad u^* = 0, \quad \frac{\partial T^*}{\partial r^*} = 0$$

$$r^* = 1, \quad u^* = 0, \quad T^* = 1 \tag{8.239}$$

where

$$r^* = \frac{r}{r_o}, \quad u^* = \frac{u}{U}, \quad x^- = \frac{x}{r_o RePr}, \quad Re = \frac{Ur_o}{\nu} \tag{8.240}$$

$$Re = \frac{Ur_o}{\nu} = Re_{D_h} \frac{r_o}{D_o - D_i} = \frac{Re_{D_h}}{2\left(1-r_i^*\right)} \tag{8.241}$$

$$x^- = \frac{x}{r_o RePr} = x^+ \left(\frac{D_h}{r_o}\right)^2 = x^+ 4\left(1-r_i^*\right)^2 \tag{8.242}$$

In the developing part of the pipe $\dfrac{\partial T^*}{\partial x^-}$ is not constant, and the solution is to be obtained numerically. The coefficients for the finite difference approximation are the same as those for a circular pipe, given by Equation 8.214. The same spreadsheet used to obtain solution in the circular pipe can be used; the difference being that $r_i \leq r \leq r_o$ rather than starting at zero; and the velocity distribution is different. The numerical solution using spreadsheet and subroutine TRIDI is shown in Figure 8.14 for the velocity distribution and temperature profiles at three axial locations.

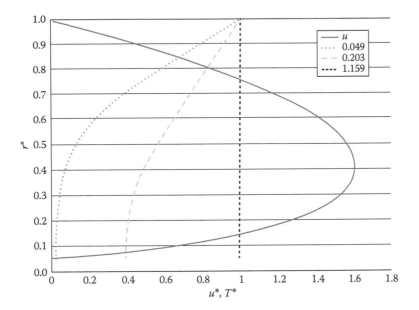

FIGURE 8.14
Velocity distribution and temperature profiles at three axial locations (dashed lines).

To calculate the Nusselt number, the mean temperature needs to be calculated. The mean temperature can be determined from 8.19

$$\bar{T}^* = \frac{2}{1-r_i^{*2}} \int_{r_i^*}^{1} u^* T^* r^* dr^* \tag{8.243}$$

which can be integrated numerically, and used to calculate the Nusselt number on each surface from

$$Nu = \frac{hD_h}{k} = \frac{-\dfrac{\partial T}{\partial r}}{T_w - \bar{T}} D_h \tag{8.244}$$

An energy balance on the fluid results in

$$\dot{m}c_p \frac{d\bar{T}}{dx} = p_i q_i'' + p_o q_o'' = h_i p_i (T_i - \bar{T}) + h_o p_o (T_o - \bar{T}) \tag{8.245}$$

which in terms of the dimensionless variables reduces to

$$\left(1 - r_i^{*2}\right) \frac{d\bar{T}^*}{dx^-} = Nu_i \frac{r_i^*}{1 - r_i^*} \left(T_i^* - \bar{T}^*\right) + Nu_o \frac{1}{1 - r_i^*} \left(T_o^* - \bar{T}^*\right) \tag{8.246}$$

This equation can be used as a check for the accuracy of the results obtained from Equation 8.244.

8.8.5 Parallel Flow Heat Exchanger

A practically important problem is a double-pipe heat exchanger, where hot and cold fluids flow in the inner and outer pipes, respectively. Consider a fluid entering the inner cylinder at temperature T_{ci} and the other fluid entering the annulus at temperature T_{ai} and the outer shell is insulated. If $T_c(x, r)$ and $T_a(x, r)$ are the temperatures of fluids in the cylinder and annulus, respectively, then for hydrodynamically fully developed, thermally developing flow

$$u_c \frac{\partial T_c}{\partial x} = \alpha_c \left(\frac{1}{r} \frac{\partial T_c}{\partial r} + \frac{\partial^2 T_c}{\partial r^2} \right) \tag{8.247}$$

$$u_a \frac{\partial T_a}{\partial x} = \alpha_a \left(\frac{1}{r} \frac{\partial T_a}{\partial r} + \frac{\partial^2 T_a}{\partial r^2} \right) \tag{8.248}$$

where u_c and u_a are given by Equations 8.189 and 8.232, respectively (Figure 8.15).
The boundary conditions are

$$r = 0, \quad \frac{\partial T_c}{\partial r} = 0 \tag{8.249}$$

$$-k_c \frac{\partial T_c}{\partial r} = -k_a \frac{\partial T_a}{\partial r}$$
$$r = r_i \tag{8.250}$$
$$T_c = T_a$$

$$r = r_o, \quad \frac{\partial T_a}{\partial r} = 0 \tag{8.251}$$

Defining the nondimensional variables

$$T_c^* = \frac{T_c - T_{ai}}{T_{ci} - T_{ai}}, \quad T_a^* = \frac{T_a - T_{ai}}{T_{ci} - T_{ai}}, \quad u_c^* = \frac{u_c}{U_c}, \quad u_a^* = \frac{u_a}{U_a}, \quad r^* = \frac{r}{r_o}, \quad x^- = \frac{x}{U_a \frac{r_o^2}{\alpha_a}} \tag{8.252}$$

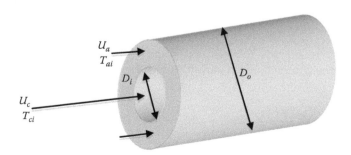

FIGURE 8.15
Flow in a double pipe parallel flow heat exchanger.

the energy equations reduce to

$$Hu_c^* \frac{\partial T_c^*}{\partial x^-} = \left(\frac{1}{r^*} \frac{\partial T_c^*}{\partial r^*} + \frac{\partial^2 T_c^*}{\partial r^{*2}} \right) \tag{8.253}$$

$$u_a^* \frac{\partial T_a^*}{\partial x^-} = \left(\frac{1}{r^*} \frac{\partial T_a^*}{\partial r^*} + \frac{\partial^2 T_a^*}{\partial r^{*2}} \right) \tag{8.254}$$

where the parameter H is

$$H = \frac{\dot{m}_c c_{pc} k_a}{\dot{m}_a c_{pa} k_c} \left(\frac{r_o^2}{r_i^2} - 1 \right) = \frac{Re_{c,D_h} Pr_c}{Re_{a,D_h} Pr_a} \left(1 + \frac{1}{r_i^*} \right) \tag{8.255}$$

The boundary conditions are

$$r^* = 0, \quad \frac{\partial T_c^*}{\partial r^*} = 0 \tag{8.256}$$

$$r^* = r_i^*, \quad \begin{aligned} \frac{\partial T_c^*}{\partial r^*} &= \frac{k_a}{k_c} \frac{\partial T_a^*}{\partial r^*} \\[6pt] T_c^* &= T_a^* \end{aligned} \tag{8.257}$$

$$r^* = 1, \quad \frac{\partial T_a^*}{\partial r^*} = 0 \tag{8.258}$$

Assume there are a total of N nodes in the cylinder and annulus, with nodes 1 to M in the cylinder (M on the cylinder wall) and $(M + 1)$ to N in the annulus. The finite difference form of the energy equations are given by

$$A_i X_{i-1} + B_i X_i + C_i X_{i+1} = R_i \tag{8.259}$$

where $X_i = T_c$ for $1 \leq i \leq M$ and $X_i = T_a$ for $M \leq i \leq N$ and

$$\begin{aligned} A_i &= 1 - \frac{\Delta r}{2r_i} \\[8pt] B_i &= -2 - Hu_i \frac{\Delta r^2}{\Delta x} \\[8pt] C_i &= 1 + \frac{\Delta r}{2r_i} \\[8pt] R_i &= -\frac{\Delta r^2}{\Delta x} Hu_i T_{i-1,j} \end{aligned} \tag{8.260}$$

where $H = 1$ for the annulus and given by Equation 8.255 for cylinder. The finite difference form of the boundary condition at the interface becomes

$$\frac{T_{c,M}^* - T_{c,M-1}^*}{r_M^* - r_{M-1}^*} = \frac{k_a}{k_c} \frac{T_{a,M+1}^* - T_{a,M}^*}{r_{M+1}^* - r_M^*} \tag{8.261}$$

From continuity of temperatures $T_{c,M}^* = T_{a,M}^*$, and Equation 8.261 becomes

$$T_{c,M-1}^* - \left(\frac{k_a}{k_c}\frac{r_M^* - r_{M-1}^*}{r_{M+1}^* - r_M^*} + 1\right)T_{a,M}^* + \frac{k_a}{k_c}\frac{r_M^* - r_{M-1}^*}{r_{M+1}^* - r_M^*}T_{a,M+1}^* = 0 \qquad (8.262)$$

Using subroutine TRIDI, the two equations are solved simultaneously for laminar flow of water in both the cylinder and annulus for the parameters listed in Table 8.4. From the first law analysis, the equilibrium exit temperature, the exit temperature of both fluids in an infinitely long heat exchanger, is

$$\bar{T} = \frac{C_a T_{ai} + C_c T_{ci}}{C_a + C_c} \qquad (8.263)$$

TABLE 8.4

Parameters Used for the Double Pipe Heat Exchanger Simulation

Parameter	Value	Unit
r_i	0.015	m
r_o	0.05	m
m_c	0.022	kg/s
m_a	0.12	kg/s
U_c	0.030	m/s
U_a	0.02	m/s
ρ	1000	kg/m^3
c_p	4180	J/kg K
μ	0.001	kg/m s
k_c	0.5	W/m K
k_a	0.7	W/m K
H	2.548	
Re_{c,D_h}	913.8	
Re_{a,D_h}	2175.7	
r_i^*	0.3	
N	101	
M	51	
Δx_0^-	1.54E−03	
Δr_c	0.006	
Δr_a	0.01	
d	1.08	
Δx_i^-	$d * \Delta x_{i-1}^-$	
$\bar{T}_C$	0.157	
$\bar{T}_a$	0.157	
$\bar{T}$ Equilibrium	0.153	
Error	2.8%	

The dimensionless mean temperatures are given by

$$\bar{T}_a^* = \frac{2}{1-r_i^{*2}} \int_{r_i^*}^{1} u^* T^* r^* dr^* \tag{8.264}$$

$$\bar{T}_c^* = \frac{2}{r_i^{*2}} \int_{0}^{r_i^*} u_c^* T_c^* r^* dr^* \tag{8.265}$$

and if the heat exchanger is long enough, the two fluids exit at the same temperature given by Equation 8.263.

The numerical solutions are plotted in Figure 8.16, and as can be seen the two fluids reach an exit temperature of 0.157 which is within 2.8% of the exact solution i.e., 0.153. The development of the wall temperature is also plotted in Figure 8.16, and is compared with the average of the mean temperatures of the two fluids, and as seen, the wall separating the two fluids is at the average of the mean temperature of the two fluids at each axial location.

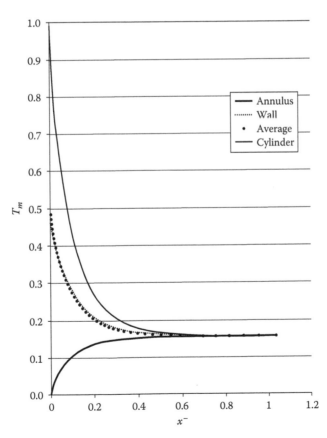

FIGURE 8.16
Temperature variation in a double pipe heat exchanger.

8.9 Hydrodynamically and Thermally Developed Flow in Rectangular Duct

For fully developed flow in a rectangular duct, the y and z components of velocity will be zero, and the x component of velocity will not change in the x direction, Figure 8.17. Therefore, the x momentum equation becomes

$$\frac{\partial^2 u}{\partial y^2} + \frac{\partial^2 u}{\partial z^2} = \frac{1}{\mu}\frac{\partial P}{\partial x} \tag{8.266}$$

$$\rho c_p u \frac{\partial T}{\partial x} = k\left[\frac{\partial^2 T}{\partial y^2} + \frac{\partial^2 T}{\partial z^2}\right] \tag{8.267}$$

Note that both sides of Equation 8.266 must be equal to a constant, because the LHS is a function of y and z and the RHS is a function of x. If the duct dimensions are a and b where a is the smaller dimension, then the aspect ratio is defined as

$$\alpha = \frac{a}{b} \tag{8.268}$$

and the hydraulic diameter is

$$D_h = \frac{4ab}{2(a+b)} = \frac{2a}{\alpha+1} \tag{8.269}$$

nondimensionalizing the momentum and energy equations

$$x^* = \frac{x}{D_h}, \quad y^* = \frac{y}{D_h}, \quad z^* = \frac{x}{D_h}, \quad u^* = \frac{u}{u_s}, \quad T^* = \frac{T-T_{in}}{T_s} \tag{8.270}$$

$$\frac{\partial^2 u^*}{\partial y^{*2}} + \frac{\partial^2 u^*}{\partial z^{*2}} = \frac{D_h^2}{u_s}\frac{1}{\mu}\frac{\partial P}{\partial x} \tag{8.271}$$

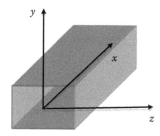

FIGURE 8.17
Rectangular duct.

The RHS of Equation 8.271 is equal to a negative constant and by setting

$$u_s = -\frac{D_h^2}{\mu}\frac{\partial P}{\partial x} \tag{8.272}$$

$$\frac{\partial^2 u^*}{\partial y^{*2}} + \frac{\partial^2 u^*}{\partial z^{*2}} = -1 \tag{8.273}$$

The boundary conditions are

$$
\begin{aligned}
y^* &= 0, & \frac{\partial u^*}{\partial z^*} &= 0 \\
y^* &= \frac{a^*}{2} = \frac{\alpha+1}{4}, & u^* &= 0 \\
z^* &= 0, & \frac{\partial u^*}{\partial y^*} &= 0 \\
z^* &= \frac{b^*}{2} = \frac{\alpha+1}{4\alpha}, & u^* &= 0
\end{aligned}
\tag{8.274}
$$

It is easy to show that

$$f = \frac{2}{\bar{u}^*}\frac{1}{Re} \tag{8.275}$$

where the average dimensionless velocity can be calculated using Equation 8.1. As can be seen, the momentum equation is only dependent on the aspect ratio. The solution can be obtained by the separation of variables or numerically. The finite difference solution results in

$$u_{i,j}^* = \frac{\dfrac{u_{i+1,j}^* + u_{i-1,j}^*}{\Delta z^{*2}} + \dfrac{u_{i,j+1}^* + u_{i,j-1}^*}{\Delta y^{*2}} + 1}{\dfrac{2}{\Delta y^{*2}} + \dfrac{2}{\Delta z^{*2}}} \tag{8.276}$$

The numerical solution is obtained iteratively for a square duct and plotted in Figure 8.18. The mean velocity is obtained by numerical integration and is given by

$$\bar{u}^* = \frac{4}{a^* b^*}\left[2\sum_{i=1}^{N} s_i - (s_1 + s_N)\right]\frac{\Delta y^*}{2}\frac{\Delta z^*}{2} \tag{8.277}$$

where

$$s_i = 2\sum_{j=1}^{M} u_{i,j}^* - \left(u_{i,1}^* + u_{i,M}^*\right) \tag{8.278}$$

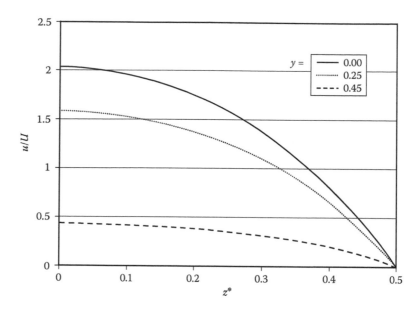

FIGURE 8.18
Velocity distribution for fully developed flow in a square duct.

The maximum velocity is at the center and is 2.036 times the average velocity. The calculated mean velocity for a square duct is $\bar{u}^* = 0.0351$ and from Equation 8.275

$$fRe = 56.98 \tag{8.279}$$

which is close to $fRe = 57$, which is the value given in Table 8.3.

8.9.1 Uniform Heat Flux

The dimensionless form of the energy equation for a rectangular duct is

$$\frac{u^*}{\bar{u}^*}\frac{\partial T^*}{\partial x^+} = \left[\frac{\partial^2 T^*}{\partial y^{*2}} + \frac{\partial^2 T^*}{\partial z^{*2}}\right] \tag{8.280}$$

and the boundary conditions for the uniform heat flux wall are

$$
\begin{aligned}
y^* &= 0, & \frac{\partial T^*}{\partial z^*} &= 0 \\
y^* &= \frac{a^*}{2} = \frac{\alpha+1}{4}, & \frac{\partial T^*}{\partial y^*} &= 1 \\
z^* &= 0, & \frac{\partial T^*}{\partial y^*} &= 0 \\
z^* &= \frac{b^*}{2} = \frac{\alpha+1}{4\alpha}, & \frac{\partial T^*}{\partial z^*} &= 1
\end{aligned} \tag{8.281}
$$

where

$$T_s = \frac{q''D_h}{k} \tag{8.282}$$

For a duct subjected to a uniform heat flux,

$$\frac{\partial T^*}{\partial x^+} = \frac{d\bar{T}^*}{dx^+} = 4 \tag{8.283}$$

and the finite difference form of the energy equation becomes

$$\frac{T^*_{i,j-1}}{\Delta y^{*2}} - \left(\frac{2}{\Delta y^{*2}} + \frac{2}{\Delta z^{*2}}\right)T^*_{i,j} + \frac{T^*_{i,j+1}}{\Delta y^{*2}} = 4\frac{u^*_{i,j}}{\bar{u}^*} - \frac{T^*_{i+1,j} + T^*_{i-1,j}}{\Delta z^{*2}} \tag{8.284}$$

This equation converges faster using the function TRIDI as opposed to directly iterating. For a uniform heat flux case, the wall temperature is not constant around the duct, and therefore the heat transfer coefficient changes around the periphery of the duct. The local value of Nusselt number is given by

$$Nu_x = \frac{h_x D_h}{k} = \frac{1}{T^*_w - \bar{T}^*} \tag{8.285}$$

The average temperature can be determined in a similar manner as the average velocity:

$$\bar{T}^* = \frac{1}{\bar{u}^* A}\int_A u^* T^* dA \tag{8.286}$$

The peripherally averaged value of the Nusselt number is

$$\overline{Nu_x} = \frac{\bar{h}_x D_h}{k} \tag{8.287}$$

where $\bar{h}_x$ is the local value of the peripherally averaged heat transfer coefficient, determined from

$$\bar{h}_x = \frac{2}{a+b}\left[\int_0^{b/2} h\,dz + \int_0^{a/2} h\,dy\right] \tag{8.288}$$

Substituting in Equation 8.287 results in

$$\overline{Nu_x} = \frac{4\alpha}{(\alpha+1)^2}\left[\int_0^{(\alpha+1)/4\alpha} \frac{1}{T^*_w - \bar{T}^*}dz^* + \int_0^{(\alpha+1)/4} \frac{1}{T^*_w - \bar{T}^*}dy^*\right] \tag{8.289}$$

For a square duct, $\alpha = 1.0$. The numerical results using spreadsheet converge to $Nu = 3.43$ which is within 5% of the value given in Table 8.3. Shah and London [3] proposed the following equation to determine friction factor and Nusselt number in a straight pipe of rectangular cross section:

$$fRe = 96(1 - 1.3553\alpha + 1.9467\alpha^2 - 1.7012\alpha^3 + 0.9564\alpha^4 - 0.2537\alpha^5)$$ (8.290)

$$Nu_T = 7.541(1 - 2.610\alpha + 4.970\alpha^2 - 5.119\alpha^3 + 2.702\alpha^4 - 0.548\alpha^5)$$ (8.291)

Note that α must be <1 regardless of the orientation of the cross section.

8.10 Helicoidal Pipes

Curved pipes are commonly used in industrial operations, and helicoidal tube configurations are used extensively in heat exchanger applications. They exhibit interesting characteristics including enhanced heat transfer rates, compactness, and relatively low pressure drop. Extensive research has been conducted to quantify the heat transfer and pressure drop in helicoidal pipes made from round tubes shown in Figure 8.19, and a review can be found in Refs. [5,6]. A distinguishing feature of curved pipes is the presence of secondary flows that develop in planes perpendicular to the main flow direction. The secondary flows result in higher heat transfer and also increase in pressure drop when compared to the same length straight tube. However, the curved pipes are far more compact and to achieve similar compactness in straight tubes, for example, in heat exchangers, 180° tube bends or "hairpins" must be used that significantly increase the pressure drop, and in fact may be responsible for over 70% of the pressure drop in compact heat exchangers.

In 1927, Dean [7] pioneered the study of the flow characteristics in curved tubes, and proposed a nondimensional parameter, now referred to as Dean number:

$$De = Re_D \sqrt{\frac{d}{D}}$$ (8.292)

that characterizes the secondary flows in curved tubes. For flow in circular helicoidal pipes, Dean developed velocity profiles using perturbation analysis for specific cases

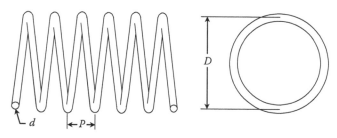

FIGURE 8.19
Helicoidal pipe.

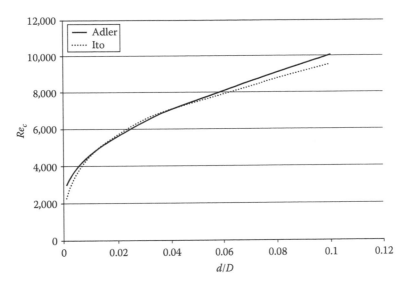

FIGURE 8.20
Critical Reynolds number for flow in helicoidal pipes.

in which the Dean number is less than 20, which generally correspond to low curvature effects, and fully developed flow exists [8]. Mori and Nakayama [9] have developed velocity profiles for fully developed flow for Dean numbers greater than 100 so long as the coil diameter (D) is much greater than the diameter of the curved tube (d).

Ito [10] experimentally studies flow in helicoidal pipes using water as the working fluid, and provides the following correlation for the critical Reynolds number for change from laminar to transitional to turbulent regimes

$$Re_{crit} = 20,000\left(\frac{d}{D}\right)^{0.32}, \quad 0.0012 < \left(\frac{d}{D}\right) < 0.0667 \tag{8.293}$$

Adler [11] provides the following criterion for the critical Reynolds number:

$$Re_{crit} = 2100 \times \left[1 + 12\left(\frac{d}{D}\right)^{0.5}\right] \tag{8.294}$$

Figure 8.20 is a comparison of the two criteria and they are reasonably close specifically at smaller d/D ratios that correspond to smaller curvature and behavior closer to the straight pipe. It is important to keep in mind that transition to turbulence happens at much larger Reynolds numbers compared to straight tubes.

8.10.1 Basic Analysis of Curved Pipes

The basic features of the flow in curved pipes is demonstrated by considering fully developed flow between two infinitely long concentric cylinders, with a fluid flowing in the angular direction in between the two. This is similar to Hagen–Poiseuille flow except the flow is in the θ direction. It is conceivable that the flow may reach a fully developed condition where $\dfrac{\partial}{\partial \theta} = 0$ for all variables except for P. The z component of velocity is zero; let us further assume that the radial component of velocity is also zero, leaving u_θ as the only nonzero component of velocity. The conservation of momentum in r direction becomes

$$-\frac{u_\theta^2}{r} = -\frac{\partial P}{\partial r} \tag{8.295}$$

The term $\dfrac{u_\theta^2}{r}$ is the centrifugal force, which results from the fluid motion in the θ direction. As seen from Equation 8.295, this force then results in a pressure gradient in the r direction, which could result in the flow in the same direction which is perpendicular to the main flow direction θ. Therefore, the assumption of no radial velocity may not be valid, particularly for pipes with small radius of curvature. As the radial flow encounters the cylinder walls, vortices could form, creating the secondary flow (Figure 8.21).

We can arrive at an approximate solution for the angular velocity, by neglecting the r velocity component, which simplifies the momentum in the θ direction significantly. Since the pressure variation is assumed to be only a function of θ and velocity a function of r, the equality in the resulting momentum equation holds if the two sides are equal to a constant C:

$$\frac{1}{\mu}\frac{dP}{d\theta} = \frac{d}{dr}\left(r\frac{du_\theta}{dr}\right) - \frac{u_\theta}{r} = C \tag{8.296}$$

If we define the nondimensional variables as

$$r^* = \frac{r}{r_0} \tag{8.297}$$

$$w^* = \frac{u_\theta}{-Cr_0} \tag{8.298}$$

$$-\frac{1}{C\mu}\frac{dP}{d\theta} = \frac{d}{dr^*}\left(r^*\frac{dw^*}{dr^*}\right) - \frac{w^*}{r^*} = -1 \tag{8.299}$$

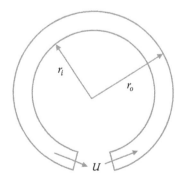

FIGURE 8.21
Angular flow between two infinitely long concentric cylinders.

which results in the following equation for the momentum equation:

$$r^{*2} \frac{d^2 w^*}{dr^{*2}} + r^* \frac{dw^*}{dr^*} - w^* + r^* = 0 \tag{8.300}$$

The boundary conditions are

$$r^* = r_i^*, \quad w^* = 0$$
$$r^* = 1, \quad w^* = 0 \tag{8.301}$$

The analytical solution to the earlier differential Equation 8.300 is

$$w^* = \frac{\left(1 - r^{*2}\right)}{\left(1 - r_i^{*2}\right)} \frac{r_i^{*2}}{2r^*} \ln r_i^* - \frac{r^*}{2} \ln r^* \tag{8.302}$$

To find C and relate it to the mean velocity

$$U = \frac{\int_{r_i}^{r_o} \rho w \, dr}{\rho(r_o - r_i)} = \frac{-C r_o \int_{r_i^*}^{1} w^* \, dr^*}{\left(1 - r_i^*\right)} \tag{8.303}$$

which results in

$$C = \frac{1}{r_o} \frac{U\left(1 - r_i^*\right)}{\frac{r_i^{*2} \ln r_i^*}{4\left(1 - r_i^{*2}\right)}(1 + 2\ln r_i^* - r_i^{*2}) - \frac{1}{8} - \frac{1}{4} r_i^{*2} \ln r_i^* + \frac{1}{8} r_i^{*2}} \tag{8.304}$$

After some algebra, the final expression for the radial variation of the angular velocity becomes

$$\frac{v}{U} = \frac{4\left(1 - r_i^*\right)}{-2r_i^* \ln r_i^{*2} + \left(1 - r_i^{*2}\right)^2} \left[r_i^{*2} \ln r_i^* \left(\frac{1}{r^*} - r^*\right) - \left(1 - r_i^{*2}\right) r^* \ln r^* \right] \tag{8.305}$$

The solution that can also be found numerically for finite difference form of Equation 8.300 is

$$\left(r_i^2 \frac{1}{\Delta r^2} - r_i \frac{1}{2\Delta r} \right) v_{i-1} - \left(r_i^2 \frac{2}{\Delta r^2} + 1 \right) v_i + \left(r_i^2 \frac{1}{\Delta r^2} + r_i \frac{1}{2\Delta r} \right) v_{i+1} = -r_i \tag{8.306}$$

Figure 8.22 is a comparison of the numerical and analytical solutions which match perfectly. The centrifugal component of the force is also plotted on the same plot and as can be seen velocity peak is pushed closer to the inner wall as a result of the centrifugal force.

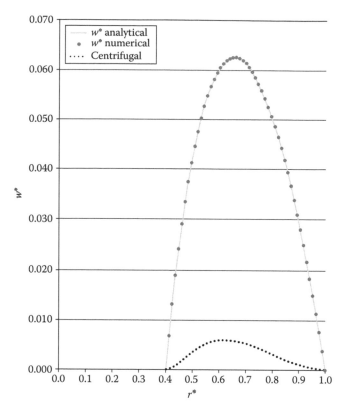

FIGURE 8.22

Comparison of the numerical and analytical solutions for angular flow between two infinitely long concentric cylinders.

8.10.2 Modified Friction Factor

The friction factor for helicoidal pipes is generally defined as

$$f = \frac{-\dfrac{\partial P}{\partial x}d}{\dfrac{1}{2\rho U^2}} \qquad (8.307)$$

Bowman and Park [12,13] compiled information about several previously developed correlations and experimental data for both friction factor and heat transfer coefficient. They gave the following correlation for the ratio of the friction factor for a coil and that of the corresponding straight tube [11]:

$$\frac{f_c}{f_s} = 1.07 + 0.0726 Re^{0.278}\left(\frac{d}{D}\right)^{0.59} \qquad (8.308)$$

which shows that for typical applications, the friction factor in a helical pipe of length L is 20% to 30% higher than that in a straight pipe of the same diameter, length, and flow rate, so long as this correlation is utilized within the proper Reynolds number bounds of $10^4 < Re < 10^6$. A compilation of the available correlations for friction factor and heat transfer is given in Katinas [5].

In an attempt to better correlate the available data, a modified friction factor has been proposed by Katinas and Fakheri [6] by using the geometric average of the coil and tube diameters as the relevant characteristic length, resulting in

$$f_m = \frac{\frac{\Delta P}{L}\sqrt{dD}}{\frac{1}{2}\rho U^2} \tag{8.309}$$

The goal of the modified friction factor is to normalize the friction factor results in an attempt to create an all-encompassing correlation for various coil diameters, cross-sectional aspect ratios, and Reynolds numbers. As indicated earlier, a correlation for the friction factor requires both Reynolds number and curvature effects, but Dean number alone does not fully describe friction factors observed at low Reynolds number and high curvature. Since most correlations have been developed for higher Dean numbers with moderate curvature effects, this will be the focus of the correlation. Note that

$$f_m = \frac{f}{\sqrt{d/D}} \tag{8.310}$$

Figure 8.23 is a plot of modified friction factor calculated from Equation 8.308 for three d/D ratios, as a function of the Dean number and shows the usefulness of the modified friction factor. Katinas and Fakheri [6] proposed the following correlation for modified friction factor for helicoidal pipes of rectangular cross section:

$$f_m = 0.11662 + \frac{17.499\ln(De)}{De} + \frac{208.742}{De^2} \qquad 10 \le De \le 565 \tag{8.311}$$

Unlike the friction factor versus Reynolds number, the modified friction factor appears to be nearly independent of coil diameter ratio. In fact, in the low Dean number region, the modified friction factor also seems to be nearly independent of aspect ratio as well.

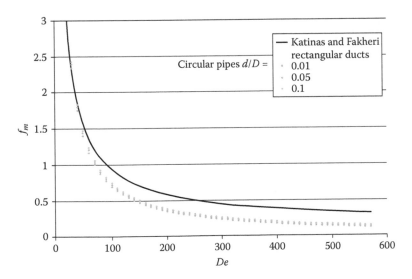

FIGURE 8.23
Modified friction factor for rectangular and circular ducts.

Note that Equation 8.311 can be utilized to capture the friction factor of a curved square or rectangular duct, regardless of the coil diameter or the aspect ratio of the pipe, since modified friction factor was utilized to remove the effect of coil diameter.

Manlapaz and Churchill [14] provide the following two correlations for Nusselt number for a helicoidal pipe of circular cross section or isothermal and uniform heat flux cases:

$$Nu_T = \left[\left(3.657 + \frac{4.343}{\left(1 + \frac{957}{De^2 Pr} \right)^2} \right)^3 + 1.158 \left(\frac{De}{1.0 + \frac{0.477}{Pr}} \right)^{1.5} \right]^{1/3} \tag{8.312}$$

$$Nu_H = \left[\left(4.364 + \frac{4.636}{\left(1 + \frac{1342}{De^2 Pr} \right)^2} \right)^3 + 1.816 \left(\frac{De}{1.0 + \frac{1.15}{Pr}} \right)^{1.5} \right]^{1/3} \tag{8.313}$$

In addition, several correlations for thermally developed heat transfer coefficients in circular cross-sectional tubes compare Nusselt numbers for a curved pipe to those of a straight pipe. Cheng et al. [15] provide a correlation for use in square ducts for Dean numbers from 20 to 705:

$$Nu_T = 0.152 + 0.7457 De^{1/2} Pr^{1/4} \tag{8.314}$$

8.11 Moving Solids

In applications like continuous drying or extrusion, the solid moves relative to a stationary coordinate system. Consider a plate being extruded as shown in Figure 8.24, it leaves the furnace at some temperature T_0, with a velocity U and we are interested in determining the cooling rate. The solid can be assumed to be a fluid moving with a uniform velocity, and therefore the dimensionless energy equation becomes

$$Pe \frac{\partial T^*}{\partial x^*} = \frac{\partial^2 T^*}{\partial y^{*2}} \tag{8.315}$$

$$x^* = 0, \quad T^* = 1$$

$$x^* \to \infty, \quad T^* = 0$$

$$y^* = 0, \quad \frac{\partial T^*}{\partial y^*} = 0 \tag{8.316}$$

$$y^* = 1, \quad \frac{\partial T^*}{\partial y^*} = -Bi T^*$$

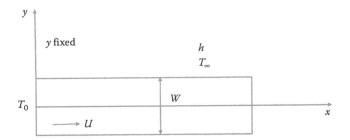

FIGURE 8.24
Moving plate.

where

$$x^* = \frac{x}{W/2}$$

$$y^* = \frac{y}{W/2}$$

$$T^* = \frac{T - T_\infty}{T_0 - T_\infty}$$

$$Bi = \frac{hW}{2k}$$

$$Pe = \frac{UW}{2\alpha}$$

Assume the solution is separable

$$T^* = XY \tag{8.317}$$

$$Pe\,\frac{1}{X}\frac{\partial X}{\partial x^*} = \frac{1}{Y}\frac{\partial^2 Y}{\partial y^{*2}} = -\lambda^2 \tag{8.318}$$

$$T^* = e^{-\frac{\lambda^2}{Pe}x^*}\left(C_1 \sin(\lambda y^*) + C_2 \cos(\lambda y^*)\right) \tag{8.319}$$

evaluating C_1 and C_2 using the boundary conditions in the y direction

$$T^* = \sum_0^\infty C_n e^{-\lambda_n^2 \frac{x^*}{Pe}} \cos(\lambda_n y^*) \tag{8.320}$$

where the eignevalues are determined from

$$\lambda_n \tan(\lambda_n) = Bi \tag{8.321}$$

For *Bi* approaching infinity (impingent jet cooling), the top surface will be at the ambient temperature, and eigenvalues become

$$\lambda_n = \frac{2n+1}{2}\pi \tag{8.322}$$

The boundary condition at $x = 0$ can be used to determine the remaining constants, using orthogonal functions, and the solution is given by Arpaci [16, equation 4.85]:

$$T^* = 2\sum_0^\infty \frac{(-1)^n}{\frac{2n+1}{2}\pi}e^{-\left(\frac{2n+1}{2}\pi\right)^2\frac{x^*}{Pe}}\cos\left(\frac{2n+1}{2}\pi y^*\right) \tag{8.323}$$

The solution can also be easily obtained numerically. The finite difference approximation of Equation 8.289 is

$$T_{i,j} = \frac{T_{i,j+1} + T_{i,j-1} + Pe\frac{\Delta y^2}{\Delta x}T_{i-1,j}}{2 + Pe\frac{\Delta y^2}{\Delta x}} \tag{8.324}$$

and can be solved iteratively. Figure 8.25 is a comparison of the numerical solution using 51 and 101 equally spaced nodes in the x and y directions, respectively, and again the two agree closely. The numerical solution can easily be modified to study the impact of Biot number on the temperature distribution.

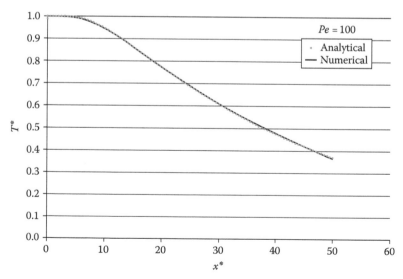

FIGURE 8.25
Comparison of the analytical and numerical solutions for a moving plate.

Problems

8.1 Solve the developing flow in a pipe subjected to a uniform heat flux using Excel and compare the results to those of Figure 8.8.

8.2 Solve the developing flow between two isothermal plates numerically.

8.3 Solve the developing flow between an isothermal and an insulated plate numerically, and compare the results with the analytical solution.

8.4 Solve the developing flow in an isothermal circular pipe numerically, and compare the results with the analytical solution.

8.5 Analytically determine the velocity distribution for flow between two rotating cylinders.

8.6 A viscous fluid is falling between two parallel plates at distance h apart as a result of gravity. There is no applied pressure. Calculate the fully developed velocity field.

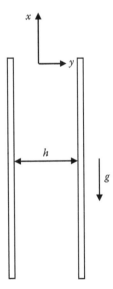

8.7 The solar flux at a location is 300 W/m². Determine the diameter of a solar collector to heat water at a rate of 0.2 GPM in a 1 cm diameter copper pipe from 15°C to 85°C.

8.8 A fluid is flowing between two concentric pipes:
  a. Simplify the governing equation and boundary conditions for fully developed flow.
  b. Nondimensionalize the governing equation and the boundary conditions.
  c. Analytically solve for the fully developed velocity distribution between two concentric pipes.

8.9 Air with an average velocity of 0.5 m/s is flowing between two concentric cylinders, having radii 3 and 5 cm:
  a. Solve the problem using spreadsheet.
  b. Compare the results with the analytical solution of the previous problem.
  c. Determine the shear stress on each pipe surface.

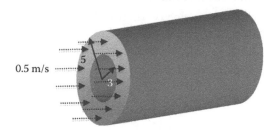

0.5 m/s

8.10 Numerically determine the Nusselt number for hydrodynamically developed, thermally developing flow between two concentric pipes, if they are at different temperatures.

8.11 Oil with viscosity of 0.18 N s/m² and a density of 915 kg/m³ enters a 3 m long 0.5 cm pipe. What is the flow rate if pressure drop is 80 kPa (20 pts)?

8.12 Air at 20°C enters a 1 cm pipe with a velocity of 0.5 m/s. The pipe surface is constant at 90°C. Determine the exit temperature if the pipe is 1 m long.

8.13 Air at 20°C enters a 1 cm pipe with a velocity of 0.5 m/s. The pipe surface is constant at 90°C. Determine the pipe length for the air to leave at 75°C.

8.14 Water enters a 3 m long equilateral triangular pipe having a 1.5 cm side with an average velocity of 0.25 m/s. Determine the pressure drop along the pipe.

8.15 Steam is condensing inside the inner pipe of a double-pipe heat exchanger using water, flowing in the outer pipe. If the outside surface of the outer pipe is insulated, numerically determine the Nusselt number for hydrodynamically developed, thermally developing flow.

8.16 Repeat the previous problem, if the inner pipe provides a uniform heat flux.

8.17 Determine the Nusselt number for hydrodynamically developed, thermally developing flow in a circular pipe, if the pipe wall temperature increases linearly.

8.18 Determine the Nusselt number for hydrodynamically developed, thermally developing flow in a circular pipe, if the pipe wall is subjected to a linearly increasing heat flux.

8.19 Determine the pipe surface temperature for a pipe receiving a heat flux that varies sinusoidally between a maximum and minimum value along the tube length as

$$q'' = \frac{q''_{max} + q''_{min}}{2} + \frac{q''_{max} - q''_{min}}{2} \sin\left(\frac{2\pi x}{L}\right)$$

where L is the period. Assume the flow in the pipe to be laminar and fully developed. Compare the results with those of a pipe subjected to a uniform heat flux equal to the average heat flux.

8.20 Consider a chemically reactive flow in an insulated circular pipe where there is constant energy generation $\dot{q}$ per unit volume in the gas phase due to chemical reactions. Assume hydrodynamically fully developed, thermally developing flow, nondimensionalize the energy equation and the boundary conditions.

8.21 Determine the temperature distribution in a chemically reactive flow in an insulated circular pipe where there is constant energy generation $\dot{q}$ per unit volume in the gas phase due to chemical reactions. Assume hydrodynamically fully developed, thermally developing flow.

8.22 In a catalytic converter, there is catalytic reaction on the surface of the pipe only (heterogeneous combustion). The energy generated on the surface results in a heat flux that can be approximated by

$$q'' = A e^{-\frac{E}{RT}}$$

where
  A is a constant
  E is the activation energy
  R is the gas constant

The outer surface of the pipe is insulated. Neglecting the gas phase reactions, assume hydrodynamically fully developed, thermally developing flow, nondimensionalize the energy equation.

8.23 Numerically solve the energy equation for Problem 8.22 to determine the pipe surface temperature.

8.24 The energy equation for fully developed flow between two parallel plates that are $2L$ apart is given by distribution in

$$1.5 U_{in}\left[1-\left(\frac{y}{L}\right)^2\right]\frac{\partial T}{\partial x} = \alpha\frac{\partial^2 T}{\partial y^2}$$

a. What are the boundary conditions if the plates are subjected to a uniform heat flux? Assume at $x = 0$, $u = U_{in}$, $T = T_{in}$.
b. Nondimensionalize the governing equation and the boundary conditions.
c. Solve for the temperature distribution using finite difference method.

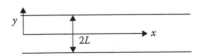

8.25 Water at 90°C and the rate of 0.5 kg/s is cooled to 45°C in a copper tube of 2 cm outer diameter, 2 mm thick by a vapor that is condensing on its outer surface at 30°C:
a. Determine the tube length.
b. Determine the pressure drop.

8.26 Air with a uniform inlet velocity and temperature of U_{in}, and T_{in} flows between two infinitely long parallel plates "a" meters apart. The top plate receives a uniform heat flux "q" while the bottom plate is insulated. Assuming laminar flow,
   a.   Write the governing equations and the boundary conditions.
   b.   Nondimensionalize the governing equations and the boundary conditions.
   c.   Obtain the solution for the fully developed flow.

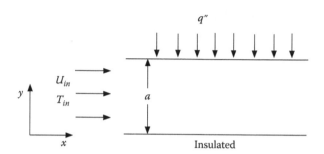

8.27 Air with a uniform velocity and temperature of U_{in}, and T_{in} enters a square duct "a" meters wide. The top surface receives a uniform heat flux "q" while the other three surfaces are insulated. Assuming laminar flow,
   a.   Write the governing equations and the boundary conditions.
   b.   Nondimensionalize the governing equations and the boundary conditions.

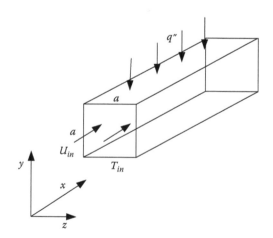

8.28 Water at 90°C and the rate of 0.5 kg/s is cooled to 45°C in a copper tube of 2 cm outer diameter, 2 mm thick by a vapor that is condensing on its outer surface at 30°C:
   a.   Determine the tube length.
   b.   Determine the pressure drop.

8.29 Water at 90°C and the rate of 0.5 kg/s is cooled in a stainless steel (AISI 302) tube of 2 cm outer diameter, 2 mm thick which is fitted with a steel spiral annular fin of 5 cm outer diameter, 0.2 mm thick wound at a pitch of 3 mm:
   a.   Estimate the fin efficiency, if the heat transfer coefficient is 15 W/m² K.
   b.   Determine the surface effectiveness.

c. Determine the water heat transfer coefficient.
d. Determine the length required for water to exit at 45°C.

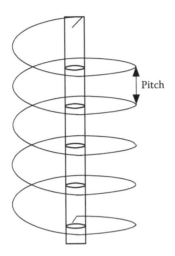

Pitch

References

1. Shah, R. K., Aung, W., and Kakaç, S., eds. (1987) *Handbook of Single-Phase Convective Heat Transfer*. New York: Wiley.
2. Kays, W. M. and Crawford, M. E. (1993) *Convective Heat and Mass Transfer*. New York: McGraw-Hill.
3. Shah, R. K. and London, A. L. (1978) *Advances in Heat Transfer Supplement*. New York: Academic Press.
4. Edwards, D. K., Denny, V. E., and Mills, A. F. (1979) *Transfer Processes*, 2nd edn. Washington, DC: Hemisphere.
5. Katinas, C. M. Numerical solution of flow and heat transfer in rectangular helicoidal pipes. MS dissertations, Bradley University, Peoria, IL.
6. Katinas, C. and Fakheri, A. (2010) Heat transfer and pressure drop correlation in helicoidal pipes of rectangular cross section. *ASME International Mechanical Engineering Congress and Exposition*. Vancouver, CA.
7. Dean, W. R. (1927) Note on the motion of fluid in a curved pipe. *Philosophical Magazine Series* 7(4), 208–223.
8. Dean, W. R. (1928) The stream-line motion of fluid in a curved pipe. *Philosophical Magazine Series* 7(5), 673–695.
9. Mori, Y. and Nakayama, W. (1965) Study on forced convective heat transfer in coiled circular pipes. *International Journal of Heat and Mass Transfer* 8, 67–82.
10. Ito, H. (1959) Friction factors for turbulent flow in curved pipes. *Journal of Basic Engineering, Transactions of the ASME Series D* 81, 123–134.
11. Adler, M. (1934) Flow in curved pipes. *Zeitschrift für Angewandte Mathematik und Mechanic* 14, 257–275.
12. Bowman, A. J. and Park, H. (2004) CFD study on laminar flow pressure drop and heat transfer characteristics in toroidal and spiral coil systems. *ASME International Mechanical Engineering Congress*. Anaheim, CA.

13. Bowman, A. J. and Park, H. (2003) Numerical investigation of generalized correlation for turbulent flow pressure drop and heat transfer applied in helically coiled tube system. *ASME International Mechanical Engineering Congress*. Washington, DC.
14. Manlapaz, R. L. and Churchill, S. W. (1981) Fully developed laminar convection from a helical coil. *Chemical Engineering Communications* 9, 185–200.
15. Cheng, K. C., Lin, R. C., and Ou, J. W. (1975) Graetz problem in curved square channels. *Journal of Heat Transfer* 97, 244–248.
16. Arpaci, V. S. (1966) *Conduction Heat Transfer*. Reading, MA: Addison-Wesley Pub. Co.

9

Turbulent Flow

9.1 Introduction

The motion of the fluid along regular predictable paths characterizes laminar flows. This predictability and relative ease of obtaining analytical solutions are used to study the physics of different problems and develop insights into the factors that impact flow and heat transfer, and have been our primary focus so far. However, most problems of practical interest are turbulent. Turbulent flows are unsteady where fluid properties, such as velocity, pressure, and temperature, continuously change with time, with the behavior appearing random and chaotic.

Consider flow over a flat plate, where we use a velocity probe to measure the velocity at the same y location at two different x locations. At location x_1, and y, for x_1 close to the leading edge, the flow will be laminar and the probe will read a constant velocity, for example, equal to 0.53 m/s (Figure 9.1).

If we move the probe parallel to the plate to a new location x_2, and if x_2 is far enough from the leading edge, the output of the probe would become unstable similar to what is shown in Figure 9.1. In the turbulent flow region, the eddies produce fluctuations in the flow properties like velocity components and pressure, i.e., they change with time, even though they fluctuate around an average value, in this case 0.5. The period of fluctuations is of the order of milliseconds, or the velocity fluctuates around its mean value with the amplitude changing more than 1000 times every second.

A basic question is why does the flow become turbulent? Even though the external conditions remain the same, why does the flow make a transition from a steady orderly behavior near the leading edge of the plate to an unsteady seemingly random behavior further down the plate? A clear answer still alludes us; however, an explanation can be that the NS equations that describe the motion of a fluid are nonlinear partial differential equations. Many nonlinear phenomena become unstable for the right set of parameters. For a given geometry, the flow of a fluid primarily depends on a single nondimensional group, called Reynolds number. Reynolds number is the ratio of inertial to viscous forces, and viscous forces act to dampen the flow instabilities. In laminar flow, Reynolds number is small and therefore any fluctuation in the velocity field will die down due to viscous dissipation. As Reynolds number increases, the relative significance of viscous forces decreases and therefore for the large enough values of Re, the instabilities in the flow not only do not die but also in fact will amplify and the flow becomes unstable or turbulent. Turbulent flows are unsteady and 3D, with fluid velocity and pressure changing with respect to time in all three directions at every point in the flow field. For external

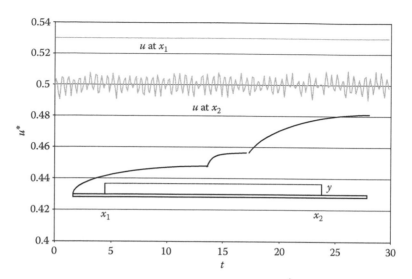

FIGURE 9.1
Laminar and turbulent velocity change with time.

flows, the transition to turbulence happens at Reynolds number of around 10^5, based on the distance along the free stream velocity, and for internal flows, it is around 10^3 based on the mean velocity and hydraulic diameter (lateral dimension). For example, for flow over a relatively smooth flat plate, the critical Reynolds number $Re_c = \dfrac{\rho U_\infty x}{\mu} \approx 5 \times 10^5$, and for ducts, the transition to turbulence occurs at $Re_c = \dfrac{\rho U D_h}{\mu} \approx 2300$. For smooth surfaces, and low fluctuations in the free stream or at the inlet, these limits can raise significantly.

One characteristic of turbulent flows is the formation of unsteady fluid patches, or eddies, or vortices on many scales that move randomly in the three directions, often swirling, colliding with one another forming additional eddies. Large eddies contain most of the kinetic energy of the turbulence, but are unstable and break down into smaller ones, which inherit the kinetic energy of the initial larger eddies. These smaller eddies in turn form the next generation of even smaller eddies, further breaking down the kinetic energy and so on. The net effect is that the energy is passed down from the large scales of the motion to smaller scales creating an energy spectrum (cascade). Eventually the eddies become small enough such that molecular diffusion becomes important and viscous dissipation of energy finally takes place [1,2].

The size of the largest scales of fluid motion (eddies) is of the order of the physical characteristic length of the flow geometry, and the size of the smallest eddies are set by the Reynolds number. The higher the Reynolds number, the smaller the size of the smaller eddies; therefore, Reynolds number in turbulent flow can be considered to indicate the ratio of the size of the largest to smallest eddies.

Kolmogorov hypothesized that at high Reynolds numbers the behavior at small scales is determined by the viscosity (ν) and the rate of energy dissipation (ε). These two quantities also define a length given by [1,2]

$$\eta = \left(\frac{\nu^3}{\varepsilon} \right)^{1/4}$$

$$(9.1)$$

which is known as the Kolmogorov length scale, and is also equal to

$$\eta = \frac{L}{Re^{3/4}} \tag{9.2}$$

where
   L is the physical length scale
   Re is the Reynolds number

As eddy sizes approach this scale, viscous effects become important, whereas at scales much larger than η, kinetic energy is essentially conserved, since at these scales Reynolds number is large, implying that viscous forces are not important.

The motion of eddies is an effective mixing mechanism that in turn enhances the rate of momentum and energy transfer. The transverse motions also smooth out property gradients more effectively than in a laminar flow, causing the velocity and temperature profiles to be more uniform than in a laminar flow. The more uniform velocity and temperature profiles also mean that the change in velocity and temperature from their wall value to the main flow value takes place over a shorter distance making the gradients of velocity and temperature at the wall much higher than those in a laminar flow at the same Reynolds number, so that the wall shear stress and heat coefficient are larger.

9.2 Solution of Turbulent Flows

The NS equations describe the behavior of laminar as well as turbulent flows, and complete knowledge of turbulent flows requires the solution of full NS equations.

The computational fluid dynamics (CFD) approaches have led to much better insight into turbulent flows, and later we review some of the solution methodologies available for turbulent flows. As will be seen, all available solution approaches involve computational powers, currently unavailable, or modeling and approximations, whose validity can only be verified through experimental verifications. Therefore, experimental analysis is still an integral part of studying turbulent flows.

9.2.1 Direct Solution of Turbulent Flows

The solution of transient 3D full NS equations is called direct numerical simulation (DNS), and provides detailed description of the turbulent velocity, pressure, and temperature fields. DNS is the most desirable solution technique for turbulent flow problems; however, they are not yet feasible for most problems of practical interest. As mentioned earlier, turbulent flows (eddies) involve a wide range of scales, from the order of Kolmogorov to the physical dimensions of the system. With increased Reynolds number, the ratio of the largest to the smallest scales also increases. Generally, it is not possible to neglect any of the scales; therefore, turbulent flow simulations require very fine grid sizes (of the order of Kolmogorov scale) which necessitate very fast computers, with very large memory, and highly specialized numerical techniques.

The available computational power is expanding rapidly, yet currently, only DNS of relatively simple geometries at moderate Reynolds numbers are possible. Practical turbulent flow problems require computational powers that are several orders of magnitude higher than the currently available computers.

9.2.2 Large Eddy Simulation

Large eddy simulation (LES) is a numerical solution technique in which the eddies are divided into two categories of large and small. The numerical grid chosen is small enough to capture the details of the flow in the large eddies. For eddies that are smaller than the grid size, approximate models are used that are similar to the turbulence models discussed later for Reynolds-averaged NS (RANS) calculations.

LES was initially proposed in 1963 by Joseph Smagorinsky [3–5] to simulate atmospheric air currents and is currently applied to a wide variety of engineering applications. This approach reduces the computational cost of the simulation, although it is still fairly computationally intensive.

9.2.3 RANS Calculations

Osborne Reynolds in 1895 assumed that the instantaneous values of the velocity components in turbulent flow, as well as other quantities of interest like temperature and pressure, can be decomposed into two components, the mean or time-averaged value and the magnitude of the fluctuations. For example, as shown in Figure 9.2, the instantaneous x component of velocity u can be written as

$$u = \bar{u} + u' \tag{9.3}$$

where

$$\bar{u} = \lim_{t \to \infty} \frac{1}{t} \int_0^t u \, dt \tag{9.4}$$

$$\bar{u} = \lim_{t \to \infty} \frac{1}{t} \int_0^t (\bar{u} + u') \, dt = \lim_{t \to \infty} \frac{1}{t} \int_0^t \bar{u} \, dt + \lim_{t \to \infty} \frac{1}{t} \int_0^t u' \, dt = \bar{u} + \overline{u'} \tag{9.5}$$

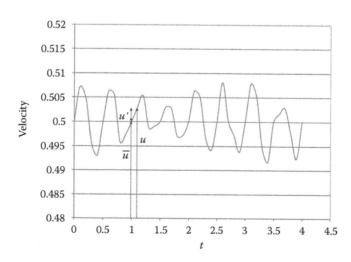

FIGURE 9.2
Turbulent velocity components.

therefore,

$$\overline{u'} = 0 \tag{9.6}$$

or the average value of the fluctuations is zero or the fluctuations are as likely to be positive as they are negative.

RANS or time-averaged equations are the oldest approach to turbulence modeling. The NS equations are averaged over time to arrive at equations for the determination of the average properties of interest, like velocity components, pressure, temperature, etc. This approach results in additional unknowns in all the conservation equations. For example, in addition to laminar shear stresses, new apparent stresses known as Reynolds stresses will also appear in the momentum equations. The determination of the mean values requires the knowledge of the turbulent fluctuations that are considered to be stochastic variables.

The derivation of the RANS is demonstrated by considering the case of 2D steady incompressible flow. The continuity is

$$\frac{\partial u}{\partial x} + \frac{\partial v}{\partial y} = 0 \tag{9.7}$$

If we substitute for instantaneous velocity in terms of average and fluctuations from Equation 9.3

$$\frac{\partial(\overline{u} + u')}{\partial x} + \frac{\partial(\overline{v} + v')}{\partial y} = 0 = \frac{\partial(\overline{u})}{\partial x} + \frac{\partial(u')}{\partial x} + \frac{\partial(\overline{v})}{\partial y} + \frac{\partial(v')}{\partial y} \tag{9.8}$$

and average this equation

$$\overline{\frac{\partial(\overline{u} + u')}{\partial x}} + \overline{\frac{\partial(\overline{v} + v')}{\partial y}} = 0 \tag{9.9}$$

$$\overline{\frac{\partial(\overline{u})}{\partial x}} + \overline{\frac{\partial(u')}{\partial x}} + \overline{\frac{\partial(\overline{v})}{\partial y}} + \overline{\frac{\partial(v')}{\partial y}} = 0 \tag{9.10}$$

$$\frac{\partial(\overline{u})}{\partial x} + \frac{\partial(\overline{u'})}{\partial x} + \frac{\partial(\overline{v})}{\partial y} + \frac{\partial(\overline{v'})}{\partial y} = 0 \tag{9.11}$$

since average fluctuations is zero, then

$$\frac{\partial(\overline{u})}{\partial x} + \frac{\partial(\overline{v})}{\partial y} = 0 \tag{9.12}$$

or the mean values satisfy the original continuity equation. Subtracting Equation 9.12 from Equation 9.8

$$\frac{\partial(u')}{\partial x} + \frac{\partial(v')}{\partial y} = 0 \tag{9.13}$$

Thus, the instantaneous, average, and fluctuations satisfy the same equation. The x momentum equation

$$\frac{\partial \rho u u}{\partial x} + \frac{\partial \rho u v}{\partial y} = -\frac{\partial P}{\partial x} + \frac{\partial}{\partial x}\left(\mu \frac{\partial u}{\partial x}\right) + \frac{\partial}{\partial y}\left(\mu \frac{\partial u}{\partial y}\right) \tag{9.14}$$

can also be averaged

$$\frac{\partial \overline{\rho uu}}{\partial x} + \frac{\partial \overline{\rho uv}}{\partial y} = -\frac{\partial \overline{P}}{\partial x} + \frac{\partial}{\partial x}\left(\mu \frac{\partial \overline{u}}{\partial x}\right) + \frac{\partial}{\partial y}\left(\mu \frac{\partial \overline{u}}{\partial y}\right) \tag{9.15}$$

The average of the product in the first term becomes

$$\overline{\rho uu} = \overline{\rho(\overline{u}+u')(\overline{u}+u')} = \overline{\rho(\overline{u}u' + u'u' + \overline{u}\,\overline{u} + \overline{u}u')} = \rho(\overline{\overline{u}u'} + \overline{u'u'} + \overline{\overline{u}\,\overline{u}} + \overline{\overline{u}u'}) \tag{9.16}$$

which simplifies to

$$\overline{\rho uu} = \rho \overline{u}\,\overline{u} + \rho \overline{u'u'} \tag{9.17}$$

similarly

$$\overline{\rho uv} = \rho \overline{u}\,\overline{v} + \rho \overline{u'v'} \tag{9.18}$$

substituting these in the momentum equation

$$\frac{\partial \rho \overline{u}\,\overline{u} + \rho \overline{u'u'}}{\partial x} + \frac{\partial \rho \overline{u}\,\overline{v} + \rho \overline{u'v'}}{\partial y} = -\frac{\partial \overline{P}}{\partial x} + \frac{\partial}{\partial x}\left(\mu \frac{\partial \overline{u}}{\partial x}\right) + \frac{\partial}{\partial y}\left(\mu \frac{\partial \overline{u}}{\partial y}\right) \tag{9.19}$$

and rearranging the equation by moving the terms involving fluctuations to the RHS

$$\frac{\partial \rho \overline{u}\,\overline{u}}{\partial x} + \frac{\partial \rho \overline{u}\,\overline{v}}{\partial y} = -\frac{\partial \overline{P}}{\partial x} + \frac{\partial}{\partial x}\left(\mu \frac{\partial \overline{u}}{\partial x} - \rho \overline{u'u'}\right) + \frac{\partial}{\partial y}\left(\mu \frac{\partial \overline{u}}{\partial y} - \rho \overline{u'v'}\right) \tag{9.20}$$

and a similar expression can be obtained for the *y* momentum equation, and energy equation, resulting in the following set for steady 2D turbulent flow:

$$\frac{\partial \overline{u}}{\partial x} + \frac{\partial \overline{v}}{\partial y} = 0 \tag{9.21}$$

$$\rho \overline{u} \frac{\partial \overline{u}}{\partial x} + \rho \overline{v} \frac{\partial \overline{u}}{\partial y} = -\frac{\partial \overline{P}}{\partial x} + \frac{\partial}{\partial x}\left(\mu \frac{\partial \overline{u}}{\partial x} - \rho \overline{u'u'}\right) + \frac{\partial}{\partial y}\left(\mu \frac{\partial \overline{u}}{\partial y} - \rho \overline{u'v'}\right) \tag{9.22}$$

$$\rho \overline{u} \frac{\partial \overline{v}}{\partial x} + \rho \overline{v} \frac{\partial \overline{v}}{\partial y} = -\frac{\partial \overline{P}}{\partial y} + \frac{\partial}{\partial x}\left(\mu \frac{\partial \overline{v}}{\partial x} - \rho \overline{u'v'}\right) + \frac{\partial}{\partial y}\left(\mu \frac{\partial \overline{v}}{\partial y} - \rho \overline{v'v'}\right) \tag{9.23}$$

$$\rho c_p \overline{u} \frac{\partial \overline{T}}{\partial x} + \rho c_p \overline{v} \frac{\partial \overline{T}}{\partial y} = +\frac{\partial}{\partial x}\left(k \frac{\partial \overline{T}}{\partial x} - \rho c_p \overline{u'T'}\right) + \frac{\partial}{\partial y}\left(k \frac{\partial \overline{T}}{\partial y} - \rho c_p \overline{v'T'}\right) \tag{9.24}$$

The RANS and energy equation have the same form as their laminar flow version, except for the additional terms like $-\rho\overline{u'v'}$. These terms have the same dimension as $\mu\dfrac{\partial\overline{u}}{\partial y}$, which is the laminar shear stress, and are called turbulent, apparent, or Reynold stresses.

The turbulent stress tensor becomes

$$
\begin{pmatrix}
-\rho\overline{u'u'} & -\rho\overline{u'v'} & -\rho\overline{u'w'} \\
-\rho\overline{u'v'} & -\rho\overline{v'v'} & -\rho\overline{v'w'} \\
-\rho\overline{u'w'} & -\rho\overline{v'w'} & -\rho\overline{w'w'}
\end{pmatrix}
\tag{9.25}
$$

similarly turbulent heat fluxes become

$$
-\rho c_p\overline{u'T'}, \quad -\rho c_p\overline{v'T'}, \quad -\rho c_p\overline{w'T'}
\tag{9.26}
$$

Therefore by averaging the NS equations, for the general 3D case, we have six additional unknown Reynolds stresses and three additional turbulent heat fluxes that we need to somehow find expressions for to be able to close the set of equations.

This situation is known as the closure problem in turbulence, that is, in RANS one always ends up with more unknowns than there are equations. To close the system and arrive at the same number of equations as unknowns, the turbulent shear stresses need to be related to other flow properties or come up with ways to model turbulence (turbulence modeling).

The apparent stresses can be written to resemble laminar shear stress form, for example, for boundary layer flow, as

$$
\tau_t = \mu_t\frac{\partial\overline{u}}{\partial y}
\tag{9.27}
$$

where μ_t is the turbulent viscosity or eddy viscosity and is an unknown to be determined. Similar to molecular viscosity, eddy viscosity can be written as

$$
\mu_t = \rho\nu_t = \rho\varepsilon_m
\tag{9.28}
$$

where ν_t or ε_m is known as the turbulent kinematic viscosity, generally referred to as eddy diffusivity.

Since u' and v' are of the same order of magnitude, the conservation equations can be written as

$$
\frac{\partial\overline{u}}{\partial x}+\frac{\partial\overline{v}}{\partial y}=0
\tag{9.29}
$$

$$
\overline{u}\frac{\partial\overline{u}}{\partial x}+\overline{v}\frac{\partial\overline{u}}{\partial y}=-\frac{1}{\rho}\frac{\partial\overline{P}}{\partial x}+\frac{\partial}{\partial x}\left[(\nu+\nu_t)\frac{\partial\overline{u}}{\partial x}\right]+\frac{\partial}{\partial y}\left[(\nu+\nu_t)\frac{\partial\overline{u}}{\partial y}\right]
\tag{9.30}
$$

$$
\overline{u}\frac{\partial\overline{v}}{\partial x}+\overline{v}\frac{\partial\overline{v}}{\partial y}=-\frac{1}{\rho}\frac{\partial\overline{P}}{\partial y}+\frac{\partial}{\partial x}\left[(\nu+\nu_t)\frac{\partial\overline{v}}{\partial x}\right]+\frac{\partial}{\partial y}\left[(\nu+\nu_t)\frac{\partial\overline{v}}{\partial y}\right]
\tag{9.31}
$$

$$\bar{u}\frac{\partial \bar{T}}{\partial x} + \bar{v}\frac{\partial \bar{T}}{\partial y} = \frac{\partial}{\partial x}\left[\left(\frac{\nu}{Pr} + \frac{\nu_t}{Pr_t}\right)\frac{\partial \bar{T}}{\partial x}\right] + \frac{\partial}{\partial y}\left[\left(\frac{\nu}{Pr} + \frac{\nu_t}{Pr_t}\right)\frac{\partial \bar{T}}{\partial y}\right] \qquad (9.32)$$

where Pr_t is the turbulent Prandtl number which is generally assumed to be constant and around 1, since the mechanism responsible for the transport of momentum and energy is the same in turbulent flow. The challenge in solving the earlier equations is the determination of eddy diffusivity, which is not a constant, and changes with position, and often with fluid and flow. The determination of $\nu_t(\varepsilon_m)$ is the subject of turbulence modeling.

9.3 Turbulence Models

The turbulence models broadly fall under three categories of algebraic, one-equation, and two-equation models. The simplest turbulence models are the algebraic ones, where the turbulent viscosity (eddy diffusivity) is related to flow parameters through an algebraic equation. In one-equation model, a transport equation similar to the momentum equation is derived for the determination of the turbulent kinetic energy (k). The eddy diffusivity is then related to the turbulent kinetic energy. The two-equation models, like the k–ε, use a second transport equation for determining the rate by which turbulent energy is dissipated (ε) and the eddy viscosity is determined at each location from the known local values of turbulence kinetic energy (k) and its rate of dissipation (ε).

9.3.1 Algebraic Models

Algebraic models are based on the analogy between collision of molecules that are the cause of fluid viscosity, and the collision of packets of fluid (eddies) that characterize turbulence. Similar to particle mean free path, Prandtl's mixing length theory is based on the assumption that eddies travel an average distance (mixing length) before they collide and the turbulent viscosity is related to the mixing length.

Consider a small packet of fluid located a distance y from a surface (Figure 9.3). This packet fluctuates around y, by an amount l:

$$u' \approx \frac{\partial u}{\partial t}\Delta t \approx u(y+l) - u(y) \approx l\frac{\partial \bar{u}}{\partial y} \qquad (9.33)$$

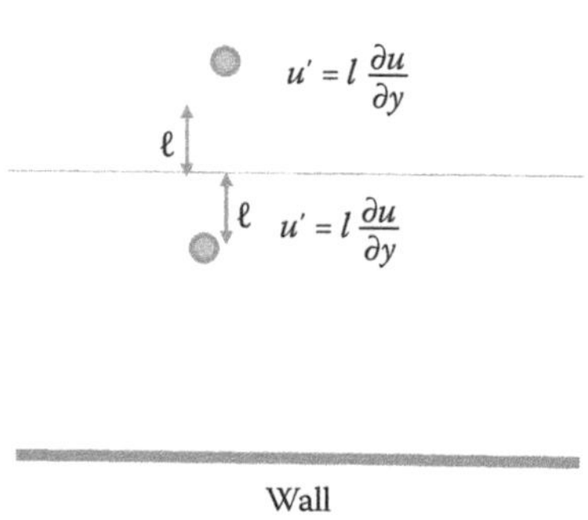

FIGURE 9.3
Mixing length.

Since u' and v' are of the same order of magnitude, then

$$\tau_t = -\rho\overline{u'v'} = \rho l^2 \left(\frac{\partial \overline{u}}{\partial y}\right)^2 \tag{9.34}$$

This equation can be rearranged as

$$\tau_t = \rho l^2 \left|\frac{\partial \overline{u}}{\partial y}\right| \left(\frac{\partial \overline{u}}{\partial y}\right) = \mu_t \left(\frac{\partial \overline{u}}{\partial y}\right) \tag{9.35}$$

Prandtl's analogy results in the turbulent viscosity to be given by

$$\mu_T = \rho l^2 \frac{\partial \overline{u}}{\partial y} \tag{9.36}$$

where l is the mixing length and Prandtl assumed it to be proportional to the packets' distance from the solid boundary

$$l = \kappa y \tag{9.37}$$

where y is the distance from the solid boundary and κ is an experimentally determined constant, known as Prandtl's constant, approximately equal to 0.4, as we will later verify.

9.4 Turbulent Fully Developed Duct Flow

In this section, we examine the solution to fully developed turbulent flow between two parallel plates that are a distance $2h$ apart, as shown in Figure 9.4. The entrance length in turbulent flows is relatively short and the hydrodynamic entrance length for turbulent flow is given by a number of different correlations, including 10 to 100 hydraulic diameters or by the following expression [6]

$$x_{h,fd,t} = 1.359 D Re_D^{1/4} \tag{9.38}$$

or a similar one due to Hinze [7] where the constant is 0.693.

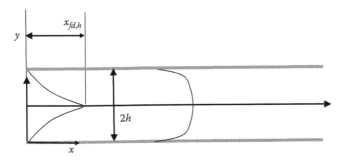

FIGURE 9.4
Turbulent flow between two parallel plates $2h$ apart.

The fully developed turbulent flow solution is used to examine the basic features of turbulent flows. The close relationship between theory and experiments is also demonstrated by comparing theoretical predictions with experimental results, to determine Prandtl's constant.

In the fully developed region, the momentum equation (Equation 9.30) reduces to

$$0 = -\frac{\partial \bar{P}}{\partial x} + \frac{\partial}{\partial y}\left[(\mu + \mu_t)\frac{\partial \bar{u}}{\partial y}\right]$$
(9.39)

Note that the term in the square bracket is the total shear stress, which is made up of the laminar and turbulent components.

Again, as in fully developed laminar duct flow, the pressure is only a function of x and velocity and turbulent viscosity are only a function of y (y measured from the wall). Since x and y are independent variables, the two terms must be equal to a constant. Factoring out viscosity and substituting for the turbulent viscosity from Prandtl's mixing length theory result in

$$\frac{\partial \tau}{\partial y} = \frac{\partial}{\partial y}\left[\mu\left(1 + \frac{\mu_t}{\mu}\right)\frac{\partial \bar{u}}{\partial y}\right] = \frac{\partial}{\partial y}\left[\rho\nu\left(1 + \frac{l^2}{\nu}\left|\frac{\partial \bar{u}}{\partial y}\right|\right)\frac{\partial \bar{u}}{\partial y}\right] = C$$
(9.40)

This equation states that the total shear stress varies linearly. If τ_o is the wall shear stress, Equation 9.40 is nondimensionalized by selecting the following velocity and length scales:

$$u_s = \left(\frac{\tau_o}{\rho}\right)^{1/2}$$
(9.41)

$$L_s = \frac{\nu}{u_s} = \nu\left(\frac{\rho}{\tau_o}\right)^{1/2}$$
(9.42)

which are called friction velocity and length scales. The dimensionless velocity and y coordinate then become

$$u^+ = \frac{\bar{u}}{u_s}$$
(9.43)

$$y^+ = \frac{y}{L_s}$$
(9.44)

Substituting for the nondimensional variables in Equation 9.40 results in

$$\frac{1}{L_s}\frac{\partial \tau}{\partial y} = \frac{u_s}{L_s^2}\frac{\partial}{\partial y^+}\left[\rho\nu\left(1 + \frac{l^2}{\nu}\frac{u_s}{L_s}\left|\frac{\partial u^+}{\partial y^+}\right|\right)\frac{\partial u^+}{\partial y^+}\right] = C$$
(9.45)

which simplifies to

$$\frac{\partial}{\partial y^+}\left(\frac{\tau}{\tau_0}\right) = \frac{\partial}{\partial y^+}\left[\left(1 + l^{+2}\left|\frac{\partial u^+}{\partial y^+}\right|\right)\frac{\partial u^+}{\partial y^+}\right] = C^+$$
(9.46)

where

$$C^+ = C\frac{\rho^{1/2}\nu}{\tau_o^{3/2}} \tag{9.47}$$

and

$$l^+ = \frac{l}{L_s} \tag{9.48}$$

From Equation 9.36,

$$\frac{\nu_t}{\nu} = \frac{l^2}{\nu}\frac{\partial\bar{u}}{\partial y} \tag{9.49}$$

This ratio often shows up in the energy and momentum equations, and in terms of the nondimensional variables, the ratio of turbulent to laminar viscosity becomes

$$\frac{\nu_t}{\nu} = \frac{l^{+2}}{\nu}\frac{\partial u^+}{\partial y^+} \tag{9.50}$$

From Equation 9.46, the dimensionless shear stress is given by

$$\frac{\tau}{\tau_0} = \left[\left(1+l^{+2}\left|\frac{\partial u^+}{\partial y^+}\right|\right)\frac{\partial u^+}{\partial y^+}\right] \tag{9.51}$$

which at the wall is one. At the wall, the mixing length is zero, and therefore

$$\frac{\partial u^+}{\partial y^+} = 1 \tag{9.52}$$

Integrating Equation 9.46,

$$\frac{\tau}{\tau_0} = C^+ y^+ + C_1 \tag{9.53}$$

The boundary conditions are

$$y^+ = 0, \quad \tau = \tau_0 \tag{9.54}$$

$$y^+ = h^+, \quad \tau = 0 \tag{9.55}$$

which result in

$$C_1 = 1 \quad \text{and} \quad C^+ = -\frac{1}{h^+} \tag{9.56}$$

and

$$\frac{\tau}{\tau_0} = 1 - \frac{y^+}{h^+} \tag{9.57}$$

Substituting from Equation 9.51 into Equation 9.57, an equation for velocity distribution is obtained:

$$\left[1+l^{+2}\left|\frac{\partial u^+}{\partial y^+}\right|\right]\frac{\partial u^+}{\partial y^+}=1-\frac{y^+}{h^+} \tag{9.58}$$

This is a nonlinear first-order differential equation that has to be solved subject to

$$y^+=0, \quad u^+=0 \tag{9.59}$$

Equation 9.58 is a quadratic equation that can be rearranged as

$$l^{+2}\left(\frac{\partial u^+}{\partial y^+}\right)^2+\frac{\partial u^+}{\partial y^+}-\left[1-\frac{y^+}{h^+}\right]=0 \tag{9.60}$$

Solving for the positive root of Equation 9.60,

$$\frac{du^+}{dy^+}=\frac{\sqrt{1+4l^{+2}\left[1-\dfrac{y^+}{h^+}\right]}-1}{2l^{+2}} \tag{9.61}$$

This is the same as equation 10.2.20 of Arpaci and Larsen [8]. It is unlikely that this first-order differential equation will have a closed form solution for the mixing length expressions given by Equation 9.37. Therefore, the solution is obtained numerically. Using backward difference approximation, the finite difference form of Equation 9.61 becomes

$$u_i^+=u_{i-1}^++\Delta y^+\ \frac{2\left[1-\dfrac{y^+}{h^+}\right]}{1+\sqrt{1+4l^{+2}\left[1-\dfrac{y^+}{h^+}\right]}} \tag{9.62}$$

Since Equation 9.61 is of first order, the Δy^+ does not have to be constant, and one can use small values of Δy^+ near the wall where the gradients are large and increase as the distance increases.

9.4.1 Friction Factor

To solve Equation 9.62, the value of h^+ must also be known. From the continuity equation

$$\rho Uh=\int_0^h \rho u\,dy=\int_0^{h^+}\rho u_s u^+ L_s dy^+=\rho u_s L_s\int_0^{h^+}u^+dy^+ \tag{9.63}$$

which simplifies to

$$\frac{Uh}{\nu}\frac{\nu}{u_s L_s}=\frac{Uh}{\nu}=\int_0^{h^+}u^+dy^+ \tag{9.64}$$

Also for two parallel plates that are $2h$ apart, the hydraulic diameter is $4h$, thus the LHS of Equation 9.64 becomes $\dfrac{Re}{4}$ and it can be rearranged into

$$Re = 4 \int_0^{h^+} u^+ dy^+ \tag{9.65}$$

Therefore, for an assumed value of h^+, Equation 9.62 can be used to determine the velocity distribution, which can then be substituted in Equation 9.65 to find the corresponding Reynolds number, for the assumed value of h^+.

Friction factor is defined as

$$f = \frac{-\dfrac{\partial \bar{P}}{\partial x} D_h}{\dfrac{1}{2} \rho U^2} \tag{9.66}$$

By using Equations 9.40, 9.47, and 9.66, friction factor can be written as

$$f = \frac{-C(4h)}{\dfrac{1}{2}\rho U^2} = \frac{-\dfrac{\tau_0^{3/2} C^+}{\rho^{1/2}\nu} 4h}{\dfrac{1}{2}\rho U^2} = \frac{-2u_s^3 C^+ 4h}{\nu U^2} = \frac{2}{h^+}\left(\frac{4h}{\dfrac{\nu}{u_s}}\right)^3 \frac{1}{\dfrac{U^2}{\nu^2}} D^2 \tag{9.67}$$

which simplifies to

$$f = 128 \frac{h^{+2}}{Re^2} \tag{9.68}$$

To obtain the solution, a value for h^+ is assumed and the solution for u^+ is obtained from Equation 9.62, which shows the solution at a node depends on the solution at the previous node. Once the values of u^+ are known, they can be integrated to get Reynolds number from Equation 9.65.

The numerical integration of Equation 9.65 using trapezoidal rule becomes

$$Re = 4 \sum_{i=2}^{N} \frac{1}{2}\left(u_{i-1}^+ + u_i^+\right)\left(y_i^+ - y_{i-1}^+\right) \tag{9.69}$$

where N is the number of nodes selected to discretize the domain from wall to the center of the duct. Knowing Re and the assumed h^+, f is determined from Equation 9.68 for the calculated Reynolds number that corresponds to the value of the assumed h^+.

9.4.2 Determining Prandtl's Constant

The first step in examining the turbulent flow is the determination of Prandtl's constant κ. Based on analytical modeling, Spalding [9] proposed the following correlation

$$y^+ = u^+ + e^{-\kappa B}\left[e^{\kappa u^+} - 1 - \kappa u^+ - \frac{(\kappa u^+)^2}{2} - \frac{(\kappa u^+)^3}{6}\right] \tag{9.70}$$

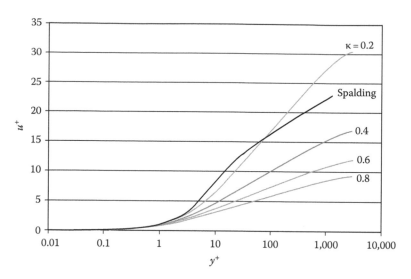

FIGURE 9.5
Numerical solution of fully developed turbulent flow for four different values κ.

for velocity profile. In this equation, κ = 0.4 and B = 5. This is sometimes referred to as Spalding's law of the wall. The accuracy of Equation 9.70 in predicting the velocity profile has been supported by many experimental data, and therefore can be used to represent the experimental results. To determine the value of κ which provides best agreement with experimental results, the solution is obtained for different values of κ, and compared with Equation 9.70.

Figure 9.5 is the numerical solution to Equation 9.62 for four different values of the constant κ, using Prandtl's mixing length model (Equation 9.37). As can be seen, there is good agreement between the experimental results and those of the mixing length model up to about $y^+ = 5$, regardless of the value of the κ. At higher values of κ, there is poor agreement between the experimental data, represented by Spalding's law of the wall, and the mixing length model regardless of κ used.

This appears discouraging. However, a close examination of the figure reveals that for κ = 0.4, the mixing length results and experimental results (Spalding's law of the wall) match closely until $y^+ = 5$, and are almost parallel for y^+ values greater than around 30. The discrepancy between the model and the experiment is primarily due to the poor performance of the model between 5 and 26, which then impacts the solution past 26. It appears that if the Prandtl's solution is shifted up at $y^+ = 26$, it would accurately match the experimental results past that point. Therefore correcting for the intermediate region, the mixing length model with κ = 0.4 provides accurate solution for $y^+ < 5$ or $y^+ > 26$.

The region between 5 and 26 is called the transition zone or buffer zone, where both laminar and turbulent shear stresses are important. Prandtl's mixing length model predicts the behavior in the laminar sublayer and the fully turbulent region, but not in the transition region. The experimental results in this region can be curve fit using the correlation

$$u^+ = 5\ln\left(\frac{y^+}{5}\right) + 5 = 5\ln(y^+) - 3.05 \tag{9.71}$$

The result of using this correlation for $5 < y^+ < 26$ instead of the mixing length model is shown in Figure 9.6. In this case (Prandtl modified), the solution is obtained from

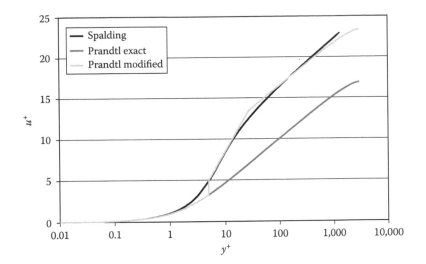

FIGURE 9.6
Velocity profile for fully developed turbulent flow between two parallel plates, $\kappa = 0.4$.

Equation 9.62 up to $y^+ = 5$, and then Equation 9.71 is used up to 26, beyond which the solution switches back to Equation 9.62. As can be seen, adjusting for the transition or buffer zone, there is close agreement between the mixing length model and the experimental data.

9.4.3 Improved Algebraic Models

To improve on Prandtl's mixing length model in the transition zone, Van Driest [10] proposed an exponential damping factor to Prandtl's mixing length model. In terms of the nondimensional variables, the modified mixing length model becomes

$$l^+ = ky^+\left[1-\exp\left(-\frac{y^+}{A^+}\right)\right] \quad \text{where } A^+ = 26 \tag{9.72}$$

Note that far from the wall, Van Driest's model approaches Prandtl's mixing length model. The solution for fully developed turbulent flow, using Van Driest's model for the mixing length, is shown in Figure 9.7 and as can be seen, the model accurately predicts the experimental results up to $y^+ = 1000$.

The experimental or analytical results for $y^+ > 26$ [11] can be accurately approximated by the algebraic equation

$$u^+ = 2.5\ln(y^+)+5.5 \tag{9.73}$$

The correlation is known as the "law of the wall." The turbulent flow velocity profile can therefore be represented by three equations:

$$u^+ = y^+, \quad y^+ < 5 \tag{9.74}$$

$$u^+ = 5\ln(y^+)-3.05, \quad 5 \le y^+ \le 26 \tag{9.75}$$

$$u^+ = 2.5\ln(y^+)+5.5, \quad y^+ > 26 \tag{9.76}$$

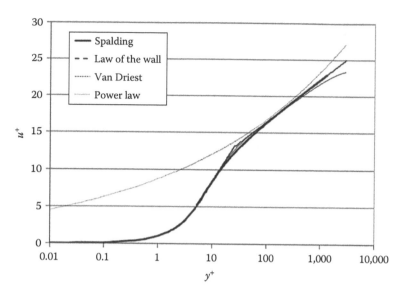

FIGURE 9.7
Velocity profile for fully developed turbulent using different models.

Figure 9.7 also compares the law of the wall representation of the velocity profile. Away from the wall, a simple power law can approximate the velocity profile

$$u^+ = 8.7 y^{+^{1/7}} \tag{9.77}$$

This is known as the 1/7th power law, which is also shown in Figure 9.7.

Figure 9.8 is a plot of the shear stress across the pipe. To show the details near the wall, a log scale is used in the y direction. The shear stress for fully developed pipe flow varies linearly across the pipe and reaches zero at the center of the pipe. The slope is small, which means that, near the wall, the shear stress is very close to the wall shear stress.

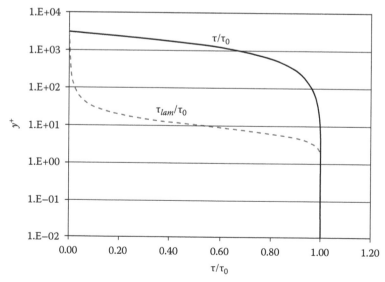

FIGURE 9.8
Laminar and turbulent contributions of the shear stress.

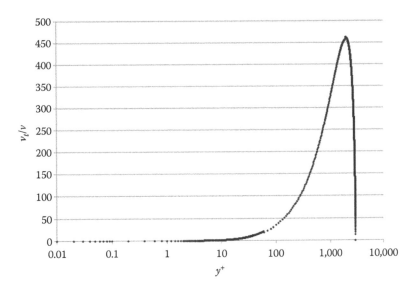

FIGURE 9.9
Ratio of the turbulent to laminar viscosity across the duct.

From Equation 9.51, the ratio of the laminar component of the total shear stress ($l = 0$) to total shear stress is

$$\frac{\tau_{lam}}{\tau_0} = \frac{du^+}{dy^+} \tag{9.78}$$

As can be seen from Figure 9.8, near the wall the total shear stress is due to the laminar component. The laminar shear stress rapidly drops to zero, as the distance from the wall increases. This can also be seen in Figure 9.9 which is a plot of the ratio of the turbulent to laminar viscosity across the duct. The turbulent viscosity is zero at the wall and rapidly increases away from the wall, and drops to zero at the center, due to symmetry.

9.4.4 Turbulent Fully Developed Flow in a Circular Pipe

The numerical solution for turbulent flow in a circular pipe using a more advanced turbulence model, k-ε, discussed in Section 9.6.3, is presented in Chapter 16. Here, for fully developed flow in a circular pipe an analysis similar to the above for flow between parallel plates is performed. The x momentum equation simplifies to

$$\frac{\partial \bar{P}}{\partial x} = \frac{1}{r}\frac{\partial}{\partial r}\left[r\rho v\left(1+\frac{l^2}{v}\left|\frac{\partial \bar{u}}{\partial r}\right|\right)\frac{\partial \bar{u}}{\partial r}\right] = C \tag{9.79}$$

Here we are using absolute value, since turbulent viscosity has to be positive and $\frac{\partial \bar{u}}{\partial r} < 0$. Equation 9.79 is nondimensionalized using the friction velocity and length scales and simplifies to

$$\frac{1}{r^+}\frac{\partial}{\partial r^+}\left[r^+\left(1+l^{+2}\left|\frac{\partial u^+}{\partial r^+}\right|\right)\frac{\partial u^+}{\partial r^+}\right] = C^+ \tag{9.80}$$

where

$$C^+ = C \frac{\rho^{1/2} v}{\tau_o^{3/2}} \tag{9.81}$$

Integrating Equation 9.80

$$\left(1 + l^{+2} \left|\frac{\partial u^+}{\partial r^+}\right|\right) \frac{\partial u^+}{\partial r^+} = C^+ \frac{r^+}{2} + \frac{C_1}{r^+} \tag{9.82}$$

and using the boundary condition

$$r^+ = 0, \quad \frac{du^+}{dr^+} = 0 \tag{9.83}$$

which requires $C_1 = 0$, simplifying the momentum equation to

$$\left(1 + l^{+2} \left|\frac{\partial u^+}{\partial r^+}\right|\right) \frac{\partial u^+}{\partial r^+} = C^+ \frac{r^+}{2} \tag{9.84}$$

Note that $y^+ = R^+ - r^+$, and then from Equation 9.51 $\frac{du^+}{dr^+} = -1$, which result in $C^+ = \frac{-2}{R^+}$, simplifying Equation 9.84 to

$$\left[1 + l^{+2} \left|\frac{\partial u^+}{\partial r^+}\right|\right] \frac{\partial u^+}{\partial r^+} = -\frac{r^+}{R^+} \tag{9.85}$$

This is a nonlinear first-order differential equation that has to be solved subject to

$$r^+ = R^+, \quad u^+ = 0 \tag{9.86}$$

Equation 9.85 is also a quadratic equation. Solving and simplifying the negative root of Equation 9.85 result in

$$\frac{du^+}{dr^+} = \frac{-2 \dfrac{r^+}{R^+}}{\sqrt{1 + 4l^{+2} \dfrac{r^+}{R^+} + 1}} \tag{9.87}$$

The solution is obtained numerically using spreadsheets, similar to that for the flow between parallel plates. To determine Reynolds number, consider the mass flow rate

$$\rho U \pi R^2 = \int_0^R \rho u 2\pi r dr \tag{9.88}$$

substituting in terms of the dimensionless quantities and simplifying result in

$$Re = \frac{4}{R^+} \int_0^{R^+} u^+ r^+ dr^+ \tag{9.89}$$

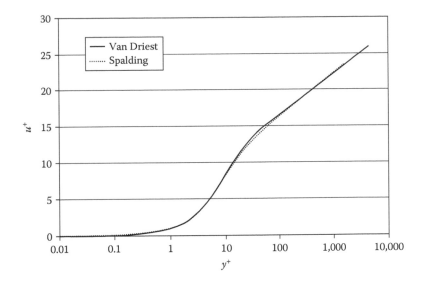

FIGURE 9.10
Velocity profile for fully developed turbulent flow in a circular pipe.

Similar to flow between two parallel plates, the friction factor becomes

$$f = 32\frac{R^{+2}}{Re^2} \tag{9.90}$$

Therefore, for an assumed value of R^+, velocity distribution is obtained (Equation 9.87), using Prandtl's or Van Driest's mixing length model which can then be used to find Reynolds number from Equation 9.89. Figure 9.10 is a plot of velocity distribution from the numerical solution of Equation 9.87 using Van Driest's model and Spalding's correlation.

There are a number of available correlations for turbulent flow in ducts, and several of the more commonly used ones for smooth pipes are listed as follows:

$$f = 0.184Re^{-0.2}, \quad Re > 10^5 \tag{9.91}$$

$$f = 0.316Re_d^{-1/4}, \quad 4000 < Re < 10^5 \tag{9.92}$$

$$f = \left(1.8\log\frac{Re_d}{6.9}\right)^{-2} \tag{9.93}$$

$$f = (0.79\ln Re - 1.64)^{-2} \tag{9.94}$$

$$\frac{1}{f^{1/2}} = 2.0\log(Re_d f^{1/2}) - 0.8 \tag{9.95}$$

The earlier correlations are plotted in Figure 9.11, and again as can be seen, the first two are accurate over their range of applicability, and the last three over the entire range.

Figure 9.12 is a plot of friction factor obtained from the numerical solution using Van Driest's model and as a function of Reynolds number. Equation 9.96 is also plotted in Figure 9.12. As can be seen, there is good agreement between the turbulence

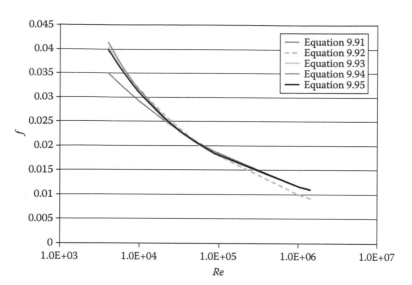

FIGURE 9.11
Comparison of the different f correlations for turbulent flow in smooth pipe.

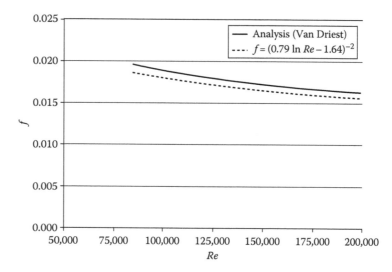

FIGURE 9.12
Comparison of the numerical and experimental values of friction factor.

modeling using Van Driest's mixing length model and the experimental results, with the difference being <5%.

The surface roughness plays an important role in turbulence, increasing both the heat transfer and pressure drop. For rough pipes, a general correlation for determining the friction factor is

$$\frac{1}{f^{1/2}} = -2.0\log\left(\frac{\frac{\varepsilon}{d}}{3.7} + \frac{2.51}{Re_d f^{1/2}}\right) \tag{9.96}$$

This is an implicit equation. An explicit equation for friction factor is

$$\frac{1}{f^{1/2}} = -1.8\log\left(\frac{\frac{\varepsilon}{d}}{3.7} + \frac{6.9}{Re_d}\right) \qquad (9.97)$$

There are two factors contributing to pressure drop in a pipe, viscosity, and the surface roughness. As Reynolds number increases, the contribution of viscosity decreases (the second term) and eventually further increase in Reynolds number will have insignificant impact on pressure drop, and the pipe is said to have reached the fully rough limit. For the fully rough limit, taking the limit in Equation 9.96 results in

$$\frac{1}{f^{1/2}} = -2.0\log\left(\frac{\frac{\varepsilon}{d}}{3.7}\right) \qquad (9.98)$$

9.4.5 Thermally Fully Developed Turbulent Flow

We now consider heat transfer in thermally fully developed turbulent flow in a circular pipe. The energy equation (Equation 9.32) in cylindrical coordinates simplifies to

$$\bar{u}\frac{\partial \bar{T}}{\partial x} = \frac{1}{r}\frac{\partial}{\partial r}\left[\left(\frac{\nu}{Pr} + \frac{\nu_t}{Pr_t}\right)r\frac{\partial \bar{T}}{\partial r}\right] \qquad (9.99)$$

where $\bar{T}$ is the time-averaged temperature, not to be confused with the radially averaged mean temperature. For a pipe subjected to a uniform heat flux, if T_b is the mean, bulk, or radially averaged temperature of the fluid, then

$$\frac{\partial \bar{T}}{\partial x} = \frac{dT_b}{dx} = \frac{q''p}{\dot{m}c_p} \qquad (9.100)$$

which is a constant, simplifying Equation 9.99 to

$$\bar{u}\frac{q''p}{\dot{m}c_p} = \frac{1}{r}\frac{\partial}{\partial r}\left[\left(\frac{\nu}{Pr} + \frac{\nu_t}{Pr_t}\right)r\frac{\partial \bar{T}}{\partial r}\right] \qquad (9.101)$$

The boundary conditions are

$$r = 0, \quad \frac{d\bar{T}}{dr} = 0 \qquad (9.102)$$

$$r = R, \quad k\frac{d\bar{T}}{dr} = q'' \qquad (9.103)$$

9.4.6 Nondimensionalization

Defining $u^+ = \dfrac{\bar{u}}{u_s}$, $r^+ = \dfrac{r}{L_s}$, and $T^+ = \dfrac{\bar{T}-T_w}{T_s}$, where T_s is the temperature scale to be determined, then

$$u^+u_s\frac{q''p}{\dot{m}c_p} = \frac{1}{L_s^2}\frac{1}{r^+}\frac{\partial}{\partial r^+}\left[\left(\frac{\nu}{Pr} + \frac{\nu_t}{Pr_t}\right)r^+L_s\frac{T_s}{L_s}\frac{\partial T^+}{\partial r^+}\right] \qquad (9.104)$$

$$r^+ = R^+, \quad -k\frac{T_s}{L_s}\frac{dT^+}{dr^+} = q'' \tag{9.105}$$

From the boundary condition

$$T_s = \frac{q'' L_s}{k} \tag{9.106}$$

$$u^+ \frac{k u_s p}{v \dot{m} c_p} L_s = \frac{1}{r^+}\frac{\partial}{\partial r^+}\left[\left(\frac{1}{Pr} + \frac{v_t}{v Pr_t}\right)r^+ \frac{\partial T^+}{\partial r^+}\right] \tag{9.107}$$

which simplifies to

$$\frac{4}{RePr}u^+ = \frac{1}{r^+}\frac{\partial}{\partial r^+}\left[\left(\frac{1}{Pr} + \frac{1}{Pr_t}l^{+2}\left|\frac{\partial u^+}{\partial y^+}\right|\right)r^+ \frac{\partial T^+}{\partial r^+}\right] \tag{9.108}$$

The boundary conditions are

$$r^+ = 0, \quad \frac{dT^+}{dr^+} = 0 \tag{9.109}$$

$$r^+ = R^+, \quad \frac{dT^+}{dr^+} = 1 \tag{9.110}$$

Multiplying both sides by r^+ and integrating Equation 9.108 once, and using the boundary condition at zero

$$\left(\frac{1}{Pr} + \frac{1}{Pr_t}l^{+2}\left|\frac{\partial u^+}{\partial y^+}\right|\right)\frac{\partial T^+}{\partial r^+} = \frac{1}{r^+}\frac{4}{RePr}\int_0^{r^+} r^+ u^+ dr^+ \tag{9.111}$$

The solution to this first-order differential equation provides the temperature distribution in the thermally fully developed region of a circular pipe subjected to a uniform heat flux.

9.4.7 Solution

Approximate solutions to Equation 9.111 can be obtained by integrating it one more time and breaking the integrals into three parts, laminar sublayer, the buffer zone, and the fully turbulent region, and using the available approximations for the velocity and turbulent viscosity. Using this approach, Petukhov and Popov [12,13] as reported by Burmeister [14, p. 335] provide the following expression for Nusselt number for fully developed flow in a circular pipe subjected to a uniform heat flux

$$\frac{Nu_D}{RePr} = \frac{1}{1.07 + 12.7(Pr^{2/3} - 1)\sqrt{\frac{f}{8}}}\frac{f}{8} \tag{9.112}$$

where f is found from the appropriate correlation given by Equations 9.91 through 9.98. Gnielinski [15] provides the following correlation for determining Nusselt number which is more accurate for lower Reynolds numbers:

$$\frac{Nu}{RePr} = \frac{\left(1-\dfrac{1000}{Re}\right)}{1+12.7(Pr^{2/3}-1)\sqrt{\dfrac{f}{8}}}\frac{f}{8} \qquad \begin{array}{l} 0.5 \le Pr \le 2000 \\ 3\times10^3 \le Re \le 5\times10^6 \end{array} \qquad (9.113)$$

For large values of Reynolds number, the second term in the numerator of Equation 9.113 becomes small, making the last two correlations very close. Rearranging Equation 9.113 results in

$$\frac{StPr^{2/3}}{\dfrac{f}{8}} = \frac{Pr^{2/3}\left(1-\dfrac{1000}{Re}\right)}{1+12.7(Pr^{2/3}-1)\sqrt{\dfrac{f}{8}}} \qquad \begin{array}{l} 0.5 \le Pr \le 2000 \\ 3\times10^3 \le Re \le 5\times10^6 \end{array} \qquad (9.114)$$

Figure 9.13 plots Equation 9.114 as a function of Pr for the Reynolds numbers specified. In the limit of $Pr \to \infty$, the RHS approaches $\dfrac{\left(1-\dfrac{1000}{Re}\right)}{12.7\sqrt{\dfrac{f}{8}}}$ which ranges between 0.25 and 0.83, even-though based on Reynolds–Colburn analogy, the RHS of the equation is expected be close to 1.

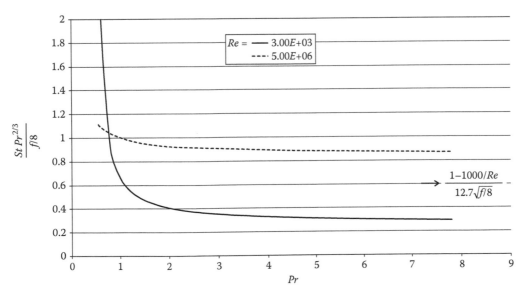

FIGURE 9.13
Impact of Pr on the heat transfer in turbulent flow.

Equation 9.111 can also be solved numerically. Multiplying both sides of the equation by the Prandtl number, and using the backward difference approximation, the finite difference form becomes

$$T^+{}_i = T^+{}_{i-1} + \left(r_i^+ - r_{i-1}^+ \right) \frac{\dfrac{1}{r^+} \dfrac{4}{Re} \displaystyle\int_0^{r^+} r^+ u^+ dr^+}{1 + \dfrac{Pr}{Pr_t} l^{+2} \left| \dfrac{\partial u^+}{\partial y^+} \right|} \tag{9.115}$$

where l is Prandtl's mixing length. The wall boundary condition becomes

$$T_2^+ = T_1^+ - \left(r_2^+ - r_1^+ \right) \tag{9.116}$$

The integral on the RHS of the equation only depends on the velocity distribution and is available from its solution. This equation can again be solved using spreadsheet. The solution can be obtained by assuming an arbitrary temperature for the wall, calculating temperature at node 2 from Equation 9.116, and at the other nodes from Equation 9.115.

9.5 Turbulent Boundary Layer

As was discussed earlier, and shown in Figure 9.1, for boundary layer flows, as the local Reynolds number increases, the boundary layer makes a transition from laminar to turbulent. To obtain the solution to turbulent boundary layer flow, we use the integral method, which can be used for laminar flow as well. Consider laminar boundary layer flow over a flat plate. In conservative form, the continuity and momentum equations are

$$\frac{\partial u}{\partial x} + \frac{\partial v}{\partial y} = 0 \tag{9.117}$$

$$\rho \left(\frac{\partial uu}{\partial x} + \frac{\partial vu}{\partial y} \right) = \frac{\partial}{\partial y} \left(\mu \frac{\partial u}{\partial y} \right) \tag{9.118}$$

9.5.1 Integral Solution

Integrating the continuity equation across the boundary layer, and changing the order of integration and differentiation and evaluating the second integral

$$\frac{\partial}{\partial x} \int_0^\delta u\, dy + v_\delta - v_0 = 0 \tag{9.119}$$

Equation 9.119 results in an expression for the y component of velocity at the edge of the boundary layer:

$$v_\delta = -\frac{\partial}{\partial x} \int_0^\delta u\, dy \tag{9.120}$$

Integrating the momentum equation

$$\int_0^\delta \frac{\partial uu}{\partial x} dy + uv|_\delta - uv|_0 = v\frac{\partial u}{\partial y}\bigg|_\delta - v\frac{\partial u}{\partial y}\bigg|_0 \tag{9.121}$$

and substituting for the y component of velocity from Equation 9.120 results in

$$\frac{\partial}{\partial x}\int_0^\delta u^2 dy - u_\infty \frac{\partial}{\partial x}\int_0^\delta u\, dy = -v\frac{\partial u}{\partial y}\bigg|_0 = -\frac{\tau_w}{\rho} \tag{9.122}$$

Defining $u^* = \dfrac{u}{u_\infty}$, $\eta = \dfrac{y}{\delta}$, the nondimensional form of Equation 9.122 results in

$$\left[\int_0^1 u^*(1-u^*)d\eta\right]\frac{d\delta}{dx} = \frac{1}{\delta}\frac{\mu}{\rho u_\infty}\frac{du^*}{d\eta} = \frac{c_f}{2} \tag{9.123}$$

Solving and rearranging Equation 9.123

$$\frac{\delta}{x} = \left[\frac{2\dfrac{du^*}{d\eta}\bigg|_0}{\displaystyle\int_0^1 u^*(1-u^*)d\eta}\right]^{1/2} Re_x^{-1/2} \tag{9.124}$$

This equation is exact and is valid for both laminar and turbulent flows. An approximate solution to Equation 9.124 can be found by assuming a velocity profile. Reasonable estimates can be obtained by noting that from the conservation of momentum the velocity profile satisfies the following conditions at the wall

$$\eta = 0, \quad u^* = 0, \quad \text{and} \quad \frac{d^n u^*}{d\eta^n} \quad \text{for } n > 1 \tag{9.125}$$

and from the definition of the boundary layer flow, at the edge of the boundary layer

$$\eta = 1, \quad u^* = 1, \quad \text{and} \quad \frac{d^n u^*}{d\eta^n} \quad \text{for } n > 0 \tag{9.126}$$

For example, assuming a linear profile, for laminar flow

$$u^* = \eta \tag{9.127}$$

the solution becomes

$$\frac{\delta}{x} = \left[\frac{2\times 1}{\displaystyle\int_0^1 \eta(1-\eta)d\eta}\right]^{1/2} Re_x^{-1/2} = [12]^{1/2} Re_x^{-1/2} = 3.46 Re_x^{-1/2} \tag{9.128}$$

which is very close to the exact solution of $5Re_x^{-1/2}$.

For turbulent flow, the 1/7th power law can be used to evaluate the integral in Equation 9.124:

$$u^+ = 8.7 y^{+^{1/7}}$$ (9.129)

Evaluating this equation at the edge of the boundary layer

$$u_\infty^+ = 8.7 \delta^{+^{1/7}}$$ (9.130)

and dividing the two equations results in

$$\frac{u}{u_\infty} = \left(\frac{y}{\delta}\right)^{1/7} = \eta^{1/7}$$ (9.131)

Evaluating the integral

$$\int_0^1 u^*(1-u^*)d\eta = \int_0^1 (\eta^{1/7} - \eta^{2/7})d\eta = \frac{7}{8} - \frac{7}{9} = \frac{7}{72}$$ (9.132)

Evaluating the numerator of Equation 9.124 using the 1/7th power law is not appropriate, since the power law is not accurate at the wall, instead we substitute in Equation 9.123, and it simplifies to

$$\frac{d\delta}{dx} = \frac{72}{7}\frac{c_f}{2}$$ (9.133)

We need to arrive at an expression, relating the drag coefficient to the boundary layer thickness, which is given by Equation 9.130. Substituting for the scales

$$\frac{u_\infty}{u_\infty\sqrt{\frac{\tau_w}{\rho u_\infty^2}}} = 8.7\left(\frac{\delta}{\nu}u_\infty\sqrt{\frac{\tau_w}{\rho u_\infty^2}}\right)^{1/7}$$ (9.134)

it simplifies to

$$\frac{1}{\sqrt{\frac{c_f}{2}}} = 8.7\left(\frac{\delta}{\nu}u_\infty\sqrt{\frac{c_f}{2}}\right)^{1/7}$$ (9.135)

solving for the drag coefficient

$$\frac{c_f}{2} = \frac{1}{8.7^{7/4}\left(\frac{\delta}{x}Re\right)^{1/4}}$$ (9.136)

Then this equation can be used to eliminate drag coefficient in Equation 9.133, resulting in

$$\frac{\partial \delta^{5/4}}{\partial x} = \frac{5}{4}\frac{72}{7}\frac{1}{8.7^{7/4}}\left(\frac{\nu}{u_\infty}\right)^{1/4}$$ (9.137)

solving for the boundary layer thickness

$$\frac{\delta}{x} = 0.3732Re^{-1/5} \tag{9.138}$$

which can then be substituted in Equation 9.136 to yield an expression for the drag coefficient

$$\frac{c_f}{2} = \frac{1}{8.7^{7/4}(0.373Re^{4/5})^{1/4}} \tag{9.139}$$

simplifying to

$$c_f = 0.0581Re^{-1/5} \tag{9.140}$$

This equation is accurate for the range $5 \times 10^5 < Re_x < 10^7$. At higher Reynolds numbers, the drag coefficient is obtained from

$$c_{f,x} = 0.37\left[\log_{10}^{Re_x}\right]^{-2.584} \tag{9.141}$$

To determine the thickness of the sublayers, from the definition of y^+

$$y^+ = \frac{y}{L_s} = \frac{yu_\infty}{\nu}\left[\frac{1}{2}c_{f,x}\right]^{1/2} \tag{9.142}$$

The viscous sublayer is at $y^+ = 5$ and the buffer zone ends at around $y^+ = 27$, and thus

$$\frac{y_{vsl}}{x} = 29.36Re_x^{-\frac{9}{10}} \tag{9.143}$$

and

$$\frac{y_b}{x} = 158.5Re_x^{-\frac{9}{10}} \tag{9.144}$$

as can be seen, both change very slowly with respect to x ($x^{0.1}$) whereas the boundary layer thickness is proportional to $x^{0.8}$, changing almost linearly with position.

Using the Reynolds–Colburn analogy, the Nusselt number is

$$Nu_x = 0.029Re_x^{4/5}Pr^{1/3} \tag{9.145}$$

9.5.2 Numerical Solution of the Turbulent Boundary Layer Flow

Cebeci et al. [16] have conducted an extensive analysis of the turbulent boundary layer flows. They transformed the general boundary layer equations using similarity type variables, and obtained the solution to the transformed equations numerically. The boundary layer equations are given by

$$\frac{\partial r^k \bar{u}}{\partial x} + \frac{\partial r^k \bar{v}}{\partial y} = 0 \tag{9.146}$$

$$\rho \bar{u} \frac{\partial \bar{u}}{\partial x} + \rho \bar{v} \frac{\partial \bar{u}}{\partial y} = \rho u_\infty \frac{du_\infty}{dx} + \frac{1}{r^k} \frac{\partial}{\partial y} \left[r^k \left(\mu \frac{\partial \bar{u}}{\partial y} - \rho \overline{u'v'} \right) \right]$$ (9.147)

$$\rho \bar{u} \frac{\partial \bar{H}}{\partial x} + \rho \bar{v} \frac{\partial \bar{H}}{\partial y} = \frac{1}{r^k} \frac{\partial}{\partial y} \left[r^k \frac{\mu}{Pr} \frac{\partial \bar{H}}{\partial y} - \rho \overline{v'H'} + \mu \left(1 - \frac{1}{Pr} \right) \bar{u} \frac{\partial \bar{u}}{\partial y} \right]$$ (9.148)

where

$H = h + \dfrac{u^2}{2}$ is the total enthalpy

k is the flow index (zero for 2D flow and one for axisymmetric flow)
r is the radial distance from the axis of revolution

Cebeci et al. [16] transformed the earlier equations by defining two new variables

$$\xi = \rho \mu u_\infty \left(\frac{R}{L} \right)^{2k} x$$ (9.149)

$$\eta = \frac{\rho u_\infty}{\sqrt{2\xi}} \left(\frac{r}{L} \right)^k y$$ (9.150)

$$\eta = \sqrt{\frac{u_\infty}{2vx}} y$$

and the dimensionless stream function by

$$f(\xi, \eta) = \frac{\psi(x, y)}{\sqrt{2\xi}}$$ (9.151)

The transformed momentum and energy equations become

$$\left[(1+\tau)^{2k} \left(1 + \frac{v_t}{v} \right) f'' \right]' + ff'' + \beta[1 - f'^2] = 2\xi \left[f' \frac{\partial f'}{\partial \xi} - f'' \frac{\partial f}{\partial \xi} \right]$$ (9.152)

$$\left\{ (1+\tau)^{2k} \left[\left(1 + \frac{v_t}{v} \frac{Pr}{Pr_t} \right) \frac{g'}{Pr} + \frac{u_\infty^2}{H_\infty} \left(1 - \frac{1}{Pr} \right) ff'' \right] \right\}' + fg' = 2\xi \left[f' \frac{\partial g}{\partial \xi} - g' \frac{\partial f}{\partial \xi} \right]$$ (9.153)

In Equation 9.152, τ is the transverse curvature and is given by

$$\tau = -1 + \left[\left(1 + \frac{2L\cos\alpha}{R^2} \frac{\sqrt{2\xi}}{\rho_\infty u_\infty} \right) \eta \right]^{1/2}$$ (9.154)

and $\beta = \dfrac{2\xi}{u_\infty} \dfrac{du_\infty}{d\xi}$ which is the pressure gradient, α is the angle between y and r, and $g = \dfrac{H}{H_\infty}$.
For 2D flow, $k = 0$, and for axisymmetric flow without curvature effects, $k = 1$, and $\tau = 0$.

For boundary layer flow over a flat plate, as long as the free stream velocity is not comparable to the speed of sound, then the total enthalpy will be close to the enthalpy ($H_\infty \gg u_\infty$) and therefore the boundary layer equations become

$$\left[\left(1+\frac{v_t}{v}\right)f''\right]' + ff'' = 2\xi\left[f'\frac{\partial f'}{\partial \xi} - f''\frac{\partial f}{\partial \xi}\right] \tag{9.155}$$

$$\left[\left(1+\frac{v_t}{v}\frac{Pr}{Pr_t}\right)\frac{T^{*\prime}}{Pr}\right]' + fg' = 2\xi\left[f'\frac{\partial T^{*\prime}}{\partial \xi} - T^{*\prime}\frac{\partial f}{\partial \xi}\right] \tag{9.156}$$

Close to the wall up to a distance Y_c, Cebeci et al. [16] assumed turbulent viscosity $v_{t,i}$ to be given by the Van Driest's model. Further from the wall, the turbulent viscosity was given by

$$v_{t,o} = k_2 u_\infty \delta^* \gamma \tag{9.157}$$

where

$$k_2 = 0.0168$$

$$\delta^* = \int_0^\infty (1-u^*)dy \tag{9.158}$$

is the displacement thickness and γ is intermittency factor given by

$$\gamma = \left[1+5.5\left(\frac{y}{\delta}\right)^6\right]^{-1} \tag{9.159}$$

The distance Y_c is the position where $v_{t,i} = v_{t,o}$. The local Stanton number in terms of the transformed variables is

$$St = \frac{h}{\rho c_p u_\infty} = \frac{-\left.\dfrac{\partial T^*}{\partial \eta}\right|_0}{Pr\sqrt{2Re_x}} \tag{9.160}$$

The numerical simulations of Cebeci et al. [16] for the velocity profiles and Stanton numbers, which are based on the mixing length model, closely match the experimental measurements of Seban and Doughty [17] and Moretti and Kays [18].

9.6 Other Turbulence Models

The mixing length model provides reasonable results for constant pressure flows or those subject to favorable pressure gradient, i.e., decreasing pressure in the direction of flow. In situations where the fluid is flowing against an increasing pressure, for example, behind a cylinder, the mixing length predictions are not very accurate. Also, the mixing length model predicts zero turbulent diffusion where the velocity gradient is zero, for example, no turbulent heat diffusion across the centerline between two parallel plates at different temperature, which obviously is not correct.

To better predict different turbulent flows, more advanced turbulence models need to be used. Unfortunately, no turbulence model works well in all cases, and the choice depends on a number of factors, including the level of accuracy needed and the type of problem being solved. Many of the more advanced turbulence models are based on solving one or more transport equations for determining flow properties that lead to the calculation of the eddy diffusivity. The general form of the transport equations is

$$\frac{D\rho\phi}{Dt} = \nabla \cdot (\mu_{eff}\nabla\phi) + P - D \tag{9.161}$$

where ϕ is the quantity of interest, or the conserved scalar. The terms on the LHS are the convection terms. The first term on the RHS is the diffusion term, followed by "P" which represents the terms that account for the production of ϕ and "D" that describe the dissipation or destruction of ϕ. This equation is similar in form to momentum, energy, or species equations, for which efficient solution techniques have been developed.

9.6.1 One-Equation Model: Spalart–Allmaras

In one-equation models, a single transport equation is arrived at whose solution directly or indirectly results in the calculation of eddy viscosity. In the Spalart–Allmaras model [19], the transport equation solved effectively determines the eddy viscosity. They introduce two variables:

$$\tilde{v} = \frac{v_t}{f_{v1}} \tag{9.162}$$

and

$$X = \frac{\tilde{v}}{v} \tag{9.163}$$

where

$$f_{v1} = \frac{X^3}{X^3 + C_{v1}^3} \tag{9.164}$$

is a damping function used primarily in the viscous region near the wall [20], and $C_{v1} = 7.1$ is an empirically determined constant. Note that $\tilde{v}$ is equal to v_t except near the wall. The transport equation for the Spalart–Allmaras model is

$$\frac{D\rho\tilde{v}}{Dt} = \frac{1}{\sigma}\nabla\cdot((\mu+\rho\tilde{v})\nabla\tilde{v}) + \frac{c_{b2}}{\sigma}\rho(\nabla\tilde{v})^2 + c_{b1}\rho\tilde{S}\tilde{v} - c_{w1}\rho f_w\left(\frac{\tilde{v}}{2}\right)^2 \qquad (9.165)$$

where the second and third terms on the RHS account for the production of eddy viscosity and the last term is its rate of destruction. The functions $\tilde{S}$ and f_w are related to other flow properties. In addition to C_{v1}, there are eight other empirically determined constants in the SA model, including $\sigma = \dfrac{2}{3}$, $c_{b1} = 0.1355$, $c_{b2} = 0.622$, $c_{w1} = \dfrac{c_{b1}}{\kappa^2} + \dfrac{1+c_{b2}}{\sigma}$, $\kappa = 0.41$, $c_{w2} = 0.3$, $c_{w3} = 2$, and $c_{t4} = 0.5$ [19].

The boundary conditions for this equation are $v_t = 0$ at the wall and specified turbulent eddy viscosity at the free stream boundary condition which often is taken to be a multiple of the molecular viscosity.

9.6.2 One-Equation Model: *k* Equation

Drawing an analogy between molecular viscosity and turbulent viscosity, as was done by Prandtl in arriving at the mixing length model, one can reasonably argue that the turbulent viscosity must be proportional to

$$v_t \propto \text{Characteristic velocity} \times \text{Characteristic length} \qquad (9.166)$$

An important quantity in turbulent flows is the turbulent kinetic energy defined as

$$k = \frac{1}{2}(u'^2 + v'^2 + w'^2) \qquad (9.167)$$

which is a measure of intensity of the turbulence. As with other turbulent flow properties,

$$k = \overline{k} + k' \qquad (9.168)$$

The time-averaged turbulent kinetic energy is then

$$\overline{k} = \frac{1}{2}\overline{(u'^2 + v'^2 + w'^2)} \qquad (9.169)$$

and its square root represents an appropriate velocity scale. Therefore, it is reasonable to assume

$$\mu_t = c_\mu \rho k^{1/2} l \qquad (9.170)$$

where
  c_μ is an experimentally determined constant
  l is the characteristic length which is similar to the mixing length

The conservation equation for mean turbulent kinetic energy takes the form

$$\frac{D\rho k}{Dt} = \nabla \cdot \left(\left(\mu + \frac{\mu_t}{\sigma_k} \right) \nabla k \right) + G_k - \rho \varepsilon \tag{9.171}$$

The first term on the RHS accounts for the diffusion of the kinetic energy, G_k is the production rate and includes terms like $\mu_t \left(\frac{\partial \bar{u}}{\partial y} \right)^2$, and the dissipation rate is the last term on the RHS, and is given by

$$\varepsilon = C_D \frac{k^{3/2}}{l} \tag{9.172}$$

where C_D is a constant.

The one-equation model still requires some algebraic equation for the determination of the mixing length, which is problem-dependent. This limits the application of the one-equation model, essentially to the same types of problems as those for mixing length.

9.6.3 Two-Equation Model: *k*–ε Model

In two-equation models, in addition to the turbulent kinetic energy, another equation is arrived at that is used to determine the turbulent length scale and/or turbulence dissipation rate. In the *k*–ε model, the second transport equation determines the rate of turbulence dissipation or rate of destruction of turbulent kinetic energy. The equation is of the form

$$\frac{D\rho \varepsilon}{Dt} = \nabla \cdot \left(\left(\mu + \frac{\mu_t}{\sigma_\varepsilon} \right) \nabla \varepsilon \right) + C_{1\varepsilon} \frac{\varepsilon}{k} G_k - C_{2\varepsilon} \rho \frac{\varepsilon^2}{k} \tag{9.173}$$

with

$$\mu_t = \rho C_\mu \frac{k^2}{\varepsilon} \tag{9.174}$$

This equation, along with Equation 9.171, and the NS equations are solved to determine the flow field. The constants in this model are $C_\mu = 0.09$, $C_{1\varepsilon} = 1.44$, $C_{2\varepsilon} = 1.92$, $\sigma_k = 1$, and $\sigma_\varepsilon = 1.3$[19].

The *k*–ε model is primarily valid far from the wall in the fully turbulent region, and it breakdowns near the wall for wall-bounded flows. As was shown earlier, very close to the wall there is a laminar or viscous sublayer, followed by the buffer zone, where laminar and turbulent stresses are important. There are two main approaches for modeling the flow details near the wall. In one approach, the grid near the wall is made fine enough to capture the flow details. In another approach, the grid size is large enough that it will not be able to capture the flow details outside of the fully turbulent region and the flow details near the wall are obtained from experimentally based formulas called "wall functions" like logarithmic law of the wall, discussed earlier.

9.6.4 Two-Equation Model: *k*–ω Model

Another widely used two-equation model is the *k*–ω model. In this model, ω is the specific dissipation rate, which can be considered to be proportional to the ratio of $\frac{\varepsilon}{k}$. The model has evolved greatly over time and additional terms have been added to both the k and ω equations [19]. Again, the transport equation for ω is

$$\frac{D\rho\omega}{Dt} = \nabla \cdot \left(\left(\mu + \frac{\mu_t}{\sigma_\omega} \right) \nabla\omega \right) + G_\omega - D_\omega \qquad (9.175)$$

and

$$\mu_t = \rho \frac{k}{\omega} \qquad (9.176)$$

Generally the k–ε model is more accurate for free shear flows, flows that are not bounded by solid walls, and k–ω performs better for wall-bounded flows. Examples of free shear flows include wakes that are formed behind objects placed in a flow field, jets that are formed when a fluid is injected into a second quiescent fluid, and mixing layers that are formed when two streams move parallel to each other at different speeds.

9.6.5 Detached Eddy Simulation

Flow around airplanes and automobiles are high Reynolds number turbulent flow and typically involve flow separation. Traditionally, high Reynolds number separated flows have been simulated by steady or unsteady RANS equations; however, the empirical constants used in RANS models are generally arrived at from simpler flows, and typically not applicable to separated flows. On the other hand, large eddy simulation of these types of problems is prohibitively expensive, and not very practical. Detached eddy simulation (DES) uses unsteady three dimensional model and can be considered as a hybrid that combines LES with RANS modeling [21]. The RANS modeling is used in the boundary layer region where the flow is attached, and LES in the separated regions, to simulate time-dependent, 3D large eddies.

The model generally uses the Spalart–Allmaras one-equation model, with the modification that the length scale used is the minimum of the distance to the solid wall or a constant fraction (0.65) of the largest dimension of the computational grid [21]. This approach can be viewed as LES with a wall model. Other RANS models can also be used with LES, although the length scale must be modified.

Problems

9.1 Water is to be heated from 20°C to 50°C at a rate of 15 kg/s in 10 parallel circular tubes each having a diameter of 3 cm. If the surface temperature of the pipes is 80.0°C,
a. Determine the required tube lengths.
b. Determine the required tube lengths, if the tube diameter is reduced in half.
c. Determine the pressure drop for a and b.

9.2 Reichardt has proposed the following expression for the turbulent viscosity:

$$\frac{\mu_T}{\mu} = \kappa \left[y^+ - 5 \tanh \frac{y^+}{5} \right]$$

Using this equation, determine the velocity at $y^+ = 1$ for fully developed turbulent flow between parallel plates.

9.3 Reichardt has proposed the following expression for the turbulent viscosity:

$$\frac{\mu_T}{\mu} = \kappa \left[y^+ - 5 \tan h \frac{y^+}{5} \right]$$

where $\kappa = 0.4$

a. Show that for fully developed flow between two parallel plates, the momentum equation becomes

$$\frac{du^+}{dy^+} = \frac{1 - \dfrac{y^+}{h^+}}{1 + \kappa \left(y^+ - A \tan h \dfrac{y^+}{A} \right)}$$

b. Using this equation, determine the velocity distribution for water ($\rho = 997$ kg/m^3, $\mu = 8.94 \times 10^{-4}$) flowing between two parallel plates 5 cm apart, and complete the table as follows for $h^+ = 3000$, taking 3000 nodes.

c. Plot velocity distribution.

d. Reynolds number.

e. Friction factor.

f. Velocity at the centerline.

9.4 Using Reichardt's model, determine the velocity at $y^+ = 150$ for fully developed turbulent flow between parallel plates, if $Re = 80,000$.

9.5 Water is flowing between two parallel plates 5 cm apart. For $h^+ = 3270$ and taking 2000 nodes and using Van Driest's model, determine the

a. Reynolds number

b. Friction factor

c. Average velocity

d. Velocity at the centerline

e. Velocity at 0.5 cm from the wall

9.6 Oil with viscosity of 0.18 N s/m^2 and a density of 915 kg/m^3 enters a 2.5 cm pipe at a rate of 750 cm^3/s and a pressure of 180 kPa. Determine the length, if the exit pressure is 100 kPa.

9.7 Using Van Driest's model, $l = \kappa y \left[1 - \exp\left(-\dfrac{y}{A} \right) \right]$ where $\kappa = 0.4$, $A = 26$

a. Determine the velocity distribution for fully developed turbulent flow in a circular pipe, for Re = 200,000.

b. Compare the results with that of Spalding's law of the wall.

c. Compare the friction factor to the value obtained from available correlations.

d. Determine the ratio of centerline velocity to mean velocity.

9.8 Water ($\rho = 997$ kg/m^3, $\mu = 8.94 \times 10^{-4}$) is flowing in a circular pipe having a 5 cm diameter. Numerically solve for fully developed flow assuming $h^+ = 3270$, taking 2000 nodes, and using Van Driest's model:

a. Reynolds number

b. Friction factor

c. Velocity scale

d. Length scale

     e. Average velocity

     f. Velocity at the centerline

     g. Velocity at 0.5 cm from the wall

9.9 Derive the integral boundary layer equation for flows where the free stream velocity is given by $U_\infty = Cx^m$. Determine the differential equation for the boundary layer thickness using linear velocity profile.

9.10 Numerically solve the differential equation in Problem 9.7 to determine the thickness of the boundary layer.

9.11 Water at 25°C is flowing between two parallel 2-cm plates, if and the velocity distribution is given by $u^+ = 8.7y^{+1/7}$

     a. Determine the friction factor and compare with value from Moody diagram.

     b. Determine the mean velocity.

9.12 For turbulent boundary-layer flow over a flat plate, using the integral method and law of the wall for velocity distribution,

     a. Set up the equations for determining the thickness of the boundary layer.

     b. Numerically solve for the thickness of the boundary layer.

References

1. Schlichting, H. and Gersten, K. (2003) *The Boundary-Layer Theory*, 8th revised and English edition. New York: Springer.
2. *Turbulence—Wikipedia, The Free Encyclopedia* (n.d.). http://en.wikipedia.org/wiki/Turbulence (retrieved January 2, 2012).
3. *Large Eddy Simulation—Wikipedia, The Free Encyclopedia* (n.d.). http://en.wikipedia.org/wiki/Large_eddy_simulation (retrieved January 2, 2012).
4. Piomelli, U., Cabot, W., Moin, P., and Lee, S. (1991) Subgrid-scale backscatter in turbulent and transitional flows. *Physics of Fluids A* 3(7), 1766–1771.
5. Ghosal, S. (1996) An analysis of numerical errors in large-eddy simulations of turbulence. *Journal of Computational Physics* 125(1), 187–206.
6. Bhatti M. S. and Shah, R. K. (1987) Turbulent and transition convective heat transfer in ducts. In *Handbook of Single Phase Convective Heat Transfer*, S. Kakac, R. K. Shah, and W. Aung, eds. Chapter 4, 166 pp. John Wiley: New York.
7. Hinze, J. O. (1975) *Turbulence*. New York: McGraw-Hill.
8. Arpaci, V. S. and Larsen, P. S. (1984) *Convection Heat Transfer*. Englewood Cliffs, NJ: Prentice Hall, 1984.
9. Spalding, D. B. (1961) A single formula for the law of the wall. *Transactions of the ASME, Series E: Journal of Applied Mechanics* 28, 455–458.
10. Van Driest, E. R. (1956) On turbulent flow near a wall. *Journal of Aeronautical Science* 23, 1007–1011.
11. Martinelli, R. (1947) *Transactions of ASME* 69, 947–959.
12. Petukhov, B. S. and Popov, V. N. (1963) *Teplofiz. Vysok. Temperatur (High Temperature Heat Physics)* 1(1), 69–83.
13. Petukhov, B. S. (1970) Heat transfer and friction in turbulent in pipe flow with variable physical properties. In *Advances Heat Transfer*, Vol. 6, T. F. Irvine and J. P. Hartnett, eds. New York: Academic Press, pp. 503–565.
14. Burmeister, L. C. (1993) *Convective Heat Transfer*, 2nd edn. New York: Wiley.

15. Gnielinski, V. (1976) New equations for heat and mass transfer in turbulent pipe and channel flow. *International Chemical Engineering* 16, 359–368.
16. Cebeci, T., Smith, A. M. O., and Mosinskis, G. (1970) Solution of the incompressible turbulent boundary-layer equations with heat transfer. *Journal of Heat Transfer* 92, 133.
17. Seban, R. A. and Doughty, D. L. (1956) Heat transfer to turbulent boundary layers with variable free stream velocity. *Journal of Heat Transfer* 78, 217.
18. Moretti, P. M. and Kays, W. M. (1965) Heat transfer to a turbulent boundary layer with varying free-stream velocity and varying surface temperature—An experimental study. *International Journal of Heat Mass Transfer* 8, 1187–1202.
19. FLUENT Theory Guide, ANSYS Corp., https://support.ansys.com/AnsysCustomerPortal/en_us/Products/All+Products/Fluid+Dynamics/ANSYS+Fluent?prodid=P22 (retrieved March 18, 2013).
20. Roy, C. J. and Blottner, F. G. (2001) Assessment of one- and two-equation turbulence models for hypersonic transitional flows. *Journal of Spacecraft and Rockets* 38(5), 699–710.
21. Travin, A., Shur, M., Strelets, M., and Spalart, P. (2000) Detached-eddy simulations past a circular cylinder. *Flow, Turbulence and Combustion*, 63(1), 293–313.

10

Heat Transfer by Natural Convection

10.1 Introduction

In forced convection, the fluid motion is caused by an external mean, like a fan, a pump, or the motion of the solid boundary. When the fluid motion is caused by naturally occurring forces, for example, gravity, the mode of heat transfer is called natural or free convection. Natural convection is the dominant mode of heat transfer in hot water radiators for space heating, refrigerator coils, and the primary cause of thermosiphon (fluid circulation due to heating), fire spread, winds, tornadoes, etc. Although for comparable size systems, the fluid velocity and heat transfer coefficients are typically smaller than those in forced convection, natural convection is still very important in many situations since it is "naturally occurring" (self-sustaining). Additionally, in tornadoes, and hurricanes, the length scales are very large, resulting in high velocities.

Natural or free convection occurs when a force acts on the mass of the fluid (body force). The most common situation is the action of gravitational field on the density gradient within a fluid that results in the creation of the buoyancy force in the fluid, causing its motion. The discovery of this effect is attributed to Archimedes of Syracuse (287–212 BC), a Greek mathematician, generally considered to be the greatest mathematician of the antiquity and one of the greatest of all time [1].

As the story goes, upon stepping into a bath, Archimedes noticed that the water level rises the further he sinks in the bath, and thus found a way of determining the volume of irregular objects. In addition, he is credited for discovering the existence of the buoyancy force.

The Archimedes' principle [2] states that a body immersed in a fluid experiences a buoyant force equal to the weight of the fluid it displaces. He is said to have been so excited at potentially finding a way of assessing the purity of King's gold crown, that he jumped out of the bathtub and ran through the streets naked, shouting Eureka, Eureka (*I found it*) and hence also the origin of an "Eureka moment." Using Archimedes' principle, a simple method of determining if the King's crown is made of pure gold is to balance the crown with pure gold on a scale, and then submerge the scale, crown, and gold in water. If the balance holds, then it is pure gold [3].

This phenomenon explains why ships or hot air balloons float, as their average density is less than the density of the fluid that they displace, resulting in a buoyancy force acting on them in the opposite direction to gravity. As shown in Figure 10.1, the net force per unit volume acting on a submerged object in the positive z direction, which is opposite to the direction of the gravitational acceleration, is

$$\frac{F_{net}}{V} = (\rho_\infty - \rho)g \qquad (10.1)$$

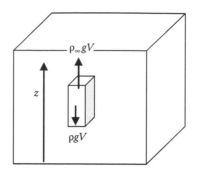

FIGURE 10.1
Forces acting on a fluid element.

This equation is also valid in the limit when the size of the object approaches zero, or if there is no foreign object in the fluid but there is density gradient in the fluid, at every point within the fluid, two body forces act on fluid particles at every point, one, ρg, due to gravity in the direction of gravitational acceleration, and the second buoyancy, $\rho_\infty g$, in the opposite direction. Therefore, temperature or concentration variation in the flow causes density gradients, resulting in a force imbalance, and fluid motion. Once the motion commences, viscous forces also develop that oppose the motion.

If temperature variation is the cause of density gradient, and the fluid is an ideal gas, then

$$\frac{F_{net}}{V} = \frac{P}{R} g \left(\frac{T - T_\infty}{TT_\infty} \right) \tag{10.2}$$

Or the force per unit volume exerted at every point in a non-isothermal ideal gas is proportional to the temperature difference and inversely proportional to the product of local and free stream temperatures. In the case of a hot air balloon, Figure 10.2, assuming the ambient to be at 20°C and the air inside the balloon at 120°C, then

$$\frac{F_{net}}{V} = \frac{101,325}{287} 9.81 \left(\frac{120 - 20}{293 \times 393} \right) = 3 \frac{N}{m^3} \tag{10.3}$$

Most balloons are roughly 100,000 ft³ (2,832 m³) and therefore the net force will be around 8,500 N (870 kg) enough to carry three to five people (350 kg) and leaving about 500 kg for the weight of the balloon, the gondola, the propane tanks, etc.

Natural convection is also induced in a fluid mixture, if the concentrations of the different species making up the fluid are not spatially uniform. Natural convection can also be caused by both effects, i.e., concentration gradients and temperature gradients.

The deep ocean currents such as the Gulf Stream [4] are caused by a combination of temperature and salinity change. As shown in Figure 10.3, the Gulf Stream is an Atlantic Ocean current that starts in the west coast of Africa, crosses the Atlantic Ocean, follows the eastern coastlines of south America and the United States before moving east across the Atlantic Ocean again. In the middle of the Atlantic Ocean, it splits in two, with the northern stream crossing to northern Europe and the southern stream recirculating off West Africa.

As it travels north, the wind moving over the Gulf Stream's warm water causes its evaporation and increases its salinity (leaving a saltier brine), and also cools it down, all of which cause its density to increase. In the North Atlantic Ocean, freezing removes additional mass

FIGURE 10.2
Hot air balloon.

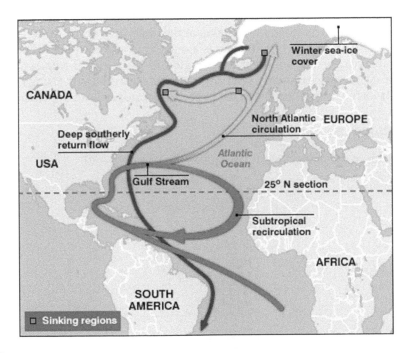

FIGURE 10.3
Gulf Stream. (From Thermohaline circulation Gulf Stream shutdown, http://thewe.cc/weplanet/poles/ thermohaline_circulation.htm, retrieved on December 22, 2011.)

of fresh water, leaving the salt out, further increasing the density of the remaining water, causing it to sink down. This downdraft of heavy, cold, salty, and dense water becomes a part of the North Atlantic Deep Water, a south flowing stream, completing the cycle.

Gulf Stream is 80–150 km wide, 800–1200 m deep, with a flow rate of 30–80 million m^3/s. In comparison, the flow rate of Mississippi River at New Orleans is 17,000 m^3/s. Gulf Stream carries about 1.4 petawatt (10^{15}) of thermal energy, 100 times the world's energy demand [4,5]. The impact of Gulf Stream on Western Europe climate can be better appreciated by noting that there are few major population centers in the United States or Canada above 50° latitude; for example, Montreal, Canada is at 45° latitude, whereas the whole of United Kingdom and Ireland are above 50°.

10.2 Natural Convection Boundary Layer Flow

To study the basic aspects of natural convection, we again consider laminar boundary layer flow. Consider a hot plate placed vertically in a room, with x measured along the plate and y measured perpendicular to the plate, as shown in Figure 10.4. The hot gases closer to the plate have a lower density and therefore experience a buoyancy force and move up, pulling and dragging with them the colder fluid, and are replaced by colder heavier fluid and therefore a fluid motion is induced as a result of the density gradient. On the plate, the fluid velocity is zero, and far away from the plate, the fluid temperature is the same as the ambient temperature and therefore there is no buoyancy force and the fluid velocity must reach zero. Therefore, velocity profile must reach a peak in the region where it changes (the boundary layer region). The fluid temperature starts from a high value at the plate, and drops to the ambient value, and therefore, as in forced convection, there is a region close to the plate where the velocity and temperature change—the boundary layer region. If this region is small compared to the length of the plate, then the boundary layer approximation can be used. Using similar reasoning as those for forced convection over a flat plate, the boundary layer approximation means

1. Axial diffusion is negligible, $\dfrac{\partial^2}{\partial x^2} \approx 0$

2. Momentum in the y direction is negligible; therefore, $\dfrac{\partial P}{\partial y} \approx 0$

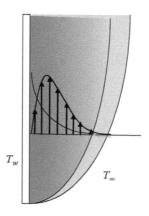

FIGURE 10.4
Natural convection from a vertical plate.

Since the fluid motion is caused by density change and gravity, the compressibility effect and body force terms must be retained. Therefore, the conservation equations become

$$\frac{\partial \rho u}{\partial x} + \frac{\partial \rho v}{\partial y} = 0 \tag{10.4}$$

$$\rho u \frac{\partial u}{\partial x} + \rho v \frac{\partial u}{\partial y} = -\frac{\partial P}{\partial x} + \mu \frac{\partial^2 u}{\partial y^2} - \rho g \tag{10.5}$$

$$\rho c_p \left(u \frac{\partial T}{\partial x} + v \frac{\partial T}{\partial y} \right) = k \frac{\partial^2 T}{\partial y^2} \tag{10.6}$$

At the edge of the boundary layer, the momentum equation becomes

$$\rho_\infty \left(0 \times \frac{\partial u}{\partial x} + 0 \times 0 \right) = -\frac{\partial P_\infty}{\partial x} + \mu \times 0 - \rho_\infty g \tag{10.7}$$

and since pressure does not change across the boundary layer, then

$$-\frac{\partial P_\infty}{\partial x} = \rho_\infty g = -\frac{\partial P}{\partial x} \tag{10.8}$$

where P is the pressure inside the boundary layer. The momentum equation becomes

$$\rho u \frac{\partial u}{\partial x} + \rho v \frac{\partial u}{\partial y} = \mu \frac{\partial^2 u}{\partial y^2} + (\rho_\infty - \rho)g \tag{10.9}$$

The thermodynamic property, coefficient of thermal expansion, is defined as

$$\beta = -\frac{1}{\rho} \left(\frac{\partial \rho}{\partial T} \right)_p \tag{10.10}$$

It represents percentage change in the density of the substance as a result of change in temperature, when pressure is held constant. For an ideal gas, $\rho = \frac{P}{RT}$, therefore

$$\beta = -\frac{1}{\rho} \left(-\frac{\rho}{T} \right) = \frac{1}{T} \approx \frac{1}{T_f} \tag{10.11}$$

where the film temperature $T_f = \frac{T_w + T_\infty}{2}$ must be expressed in kelvin or Rankine. For other materials, β must be obtained from tables of thermodynamic properties. The coefficient of thermal expansion can be approximated by

$$\beta \approx \frac{1}{\rho} \frac{\rho_\infty - \rho}{T - T_\infty} \tag{10.12}$$

The governing equations are further simplified by invoking the Boussinesq approximation, named after the French mathematician and physicist Joseph Valentin Boussinesq (1842–1929).

He proposed that density change can be neglected everywhere in the three equations, except where it is multiplied by gravitational acceleration, simplifying the governing equations to

$$\frac{\partial u}{\partial x} + \frac{\partial v}{\partial y} = 0 \tag{10.13}$$

$$u\frac{\partial u}{\partial x} + v\frac{\partial u}{\partial y} = \nu\frac{\partial^2 u}{\partial y^2} + g\beta(T - T_\infty) \tag{10.14}$$

$$u\frac{\partial T}{\partial x} + v\frac{\partial T}{\partial y} = \alpha\frac{\partial^2 T}{\partial y^2} \tag{10.15}$$

Here, we have three coupled equations for the two velocity components and temperature. The two terms on the LHS of the momentum equation are the inertial force per unit volume, the first term on the RHS is shear force and the second is buoyancy force per unit volume, which for a heated gas points in the positive x direction.

10.2.1 Nondimensional Boundary Layer Equations

Again, before proceeding to obtain the solution, the governing equations and boundary conditions are nondimensionalized. For natural convection, there is no explicit velocity scale in the governing equations or boundary conditions, and therefore, the appropriate velocity scale needs to be determined.

Defining $x^* = \dfrac{x}{L}$, $y^* = \dfrac{y}{L}$, $u^* = \dfrac{u}{u_s}$, and $T^* = \dfrac{(T - T_\infty)}{T_s}$, then the nondimensional continuity equation becomes

$$\frac{\partial u^*}{\partial x^*} + \frac{\partial u^*}{\partial y^*} = 0 \tag{10.16}$$

and the nondimensional momentum equation reduces to

$$u^*\frac{\partial u^*}{\partial x^*} + v^*\frac{\partial u^*}{\partial y^*} = \frac{\nu}{u_s L}\frac{\partial^2 u^*}{\partial y^{*2}} + \frac{g\beta L T_s}{u_s^2}T^* \tag{10.17}$$

To determine the relevant velocity scale, we could assume

$$u_s = \frac{\nu}{L} \tag{10.18}$$

This velocity scale is too small compared with those encountered in natural convection and is not the relevant scale. This choice also eliminates the equivalent of Reynolds number. A more appropriate choice is to assume the velocity scale to be

$$u_s = \sqrt{g\beta L T_s} \tag{10.19}$$

which eliminates four parameters from the momentum equation, while retaining the equivalent of Reynolds number. For an isothermal plate, the relevant temperature scale is the difference between surface and ambient temperatures, and thus

$$T^* = \frac{T - T_\infty}{T_w - T_\infty} \tag{10.20}$$

and

$$u_s = \sqrt{g\beta L(T_W - T_\infty)} \tag{10.21}$$

This intrinsic velocity scale is proportional to the peak velocity inside the boundary layer, and as will be shown later, for $Pr = 1$, it is about twice the maximum velocity at $x = L$. Once the velocity scale is established, the nondimensional group $\frac{u_s L}{\nu}$, which is the equivalent Reynolds number for natural convection, becomes

$$\frac{u_s L}{\nu} = \sqrt{\frac{g\beta L^3 (T_W - T_\infty)}{\nu^2}} \tag{10.22}$$

The nondimensional group under the square root is called Grashof number after Franz Grashof (1826–1893), a German engineer and professor:

$$Gr = \frac{g\beta L^3 (T_W - T_\infty)}{\nu^2} \tag{10.23}$$

The square root of Grashof number in natural convection is almost equivalent of Reynolds number in forced convection. Thus the boundary layer thickness for natural convection should be inversely proportional to the local value of the Grashof number to the one-fourth power

$$\frac{\delta}{x} \propto \frac{1}{Gr_x^{1/4}} \tag{10.24}$$

As shown in the order of magnitude analysis as follows and later verified by the actual solution, Rayleigh number which is the product of Grashof and Prandtl numbers

$$Ra = GrPr = \frac{g\beta L^3 (T_W - T_\infty)}{\nu\alpha} \tag{10.25}$$

is a more appropriate nondimensional parameter in natural convection, and the solutions are generally expressed in terms of Rayleigh number.

The nondimensional governing equations become

$$\frac{\partial u^*}{\partial x^*} + \frac{\partial v^*}{\partial y^*} = 0 \tag{10.26}$$

$$u^* \frac{\partial u^*}{\partial x^*} + v^* \frac{\partial u^*}{\partial y^*} = \frac{\sqrt{Pr}}{\sqrt{Ra}} \frac{\partial^2 u^*}{\partial y^{*2}} + T^*$$ (10.27)

$$u^* \frac{\partial T^*}{\partial x^*} + v^* \frac{\partial T^*}{\partial y^*} = \frac{1}{\sqrt{Ra}\sqrt{Pr}} \frac{\partial^2 T^*}{\partial y^{*2}}$$ (10.28)

Also, since T^* appears in the momentum equation, the momentum and energy equations are now directly linked. The boundary conditions are

$$x^* = 0, \quad u^* = v^* = T^* = 0$$ (10.29)

$$y^* = 0, \quad u^* = v^* = 0, \quad T^* = 1$$ (10.30)

$$y^* \to \infty, \quad u^* = T^* = 0$$ (10.31)

The nondimensional temperature distribution is therefore a function of $T^*(x^*, y^*, Ra, Pr)$. Therefore, the Nusselt number is a function of Rayleigh and Prandtl numbers.

10.2.2 Order of Magnitude Analysis

For $Pr \gg 1$, the thermal boundary layer is much smaller than the hydrodynamic boundary layer (see Figure 10.7). This is interesting, indicating that the fluid continues its motion long after buoyancy effects have dissipated. The order of magnitude analysis of the energy equation results in

$$u^* \frac{1}{1} + v^* \frac{1}{\delta_T^*} \sim \frac{1}{\sqrt{Ra}\sqrt{Pr}} \frac{1}{\delta_T^{*2}}$$ (10.32)

or

$$u^* \sim \frac{1}{\sqrt{Ra}\sqrt{Pr}} \frac{1}{\delta_T^{*2}}$$ (10.33)

Performing the same on the momentum equation

$$u^* \frac{u^*}{1} \quad v^* \frac{u^*}{\delta_T^*} \sim \frac{\sqrt{Pr}}{\sqrt{Ra}} \frac{u^*}{\delta_T^{*2}} + 1$$ (10.34)

Substituting for velocity from Equation 10.33 and simplifying

$$\frac{1}{Pr} \frac{1}{Ra\delta_T^{*4}} \sim \frac{1}{Ra\delta_T^{*4}} \quad 1$$ (10.35)

For $Pr \gg 1$, the inertial terms are of the order of $\frac{1}{Pr}$ which is much smaller than one or the inertial forces are small everywhere. Inside the thermal boundary layer, the force balance is between friction and buoyancy. As one moves away from the plate, they both decrease, while being of the same order of magnitude and essentially balancing each other. Therefore, the first term on the RHS of Equation 10.35 must be of the order of one as well

$$\frac{\frac{1}{\sqrt{Pr}}\frac{1}{\sqrt{Ra}}\sqrt{Pr}\frac{1}{\delta_T^{*2}}}{\sqrt{Ra}\frac{1}{\delta_T^{*2}}} \sim 1 \tag{10.36}$$

and simplifying results in

$$\delta_T^* \sim Ra^{-1/4} \tag{10.37}$$

or the thickness of thermal boundary layer is dependent on the Rayleigh number. Nusselt number is inversely proportional to the thermal boundary layer thickness, and therefore is expected to be a function of Rayleigh number to power $\frac{1}{4}$.

For $Pr \sim 1$, all three forces are of the same order of magnitude and so are the thermal and hydrodynamic boundary layers.

10.2.3 Solution for Laminar Free Convection on a Vertical Isothermal Plate

Similar to forced convection, the solution can be obtained by similarity method, where the similarity variable is

$$\eta = \frac{y}{\delta} = \frac{y}{x}(Ra_x)^{1/4} \tag{10.38}$$

Using the stream function definition $u = \frac{\partial \psi}{\partial y}$ and $v = \frac{-\partial \psi}{\partial x}$ and defining the nondimensional stream function f as

$$f(\eta) = \frac{\psi}{\alpha Ra_x^{1/4}} \tag{10.39}$$

then

$$u = \frac{\partial \psi}{\partial y} = \frac{\alpha Ra_x^{1/2}}{x}f' \tag{10.40}$$

which can be rearranged into

$$u^* = \frac{u}{u_s} = \sqrt{\frac{x/L}{Pr}}f' \tag{10.41}$$

Similarly,

$$v = -\frac{\partial \psi}{\partial x} = -\frac{\partial\left(\alpha Ra_x^{1/4}f\right)}{\partial x} = -\alpha f \frac{\partial\left(Ra_x^{1/4}\right)}{\partial x} - \alpha Ra_x^{1/4}\frac{\partial f}{\partial \eta}\frac{\partial \eta}{\partial x} \tag{10.42}$$

which simplifies to

$$v^* = \frac{v}{\sqrt{g\beta L(T_W - T_\infty)}} = \sqrt{\frac{x/L}{Pr}} Ra_x^{-1/4} \left[\frac{1}{4}\eta f' - \frac{3}{4} f \right] \tag{10.43}$$

Dividing Equation 10.41 by 10.43, the ratio of x and y velocity components becomes

$$\frac{u}{v} = \frac{f'}{Ra_x^{-1/4} \left[\frac{1}{4}\eta f' - \frac{3}{4} f \right]} \approx Ra_x^{1/4} \gg 1 \tag{10.44}$$

or the boundary layer approximation will be valid, as long as fourth root of Rayleigh number is much larger than 1.

Using the variables defined, the momentum and energy equations become

$$\frac{1}{2Pr} f'^2 - \frac{3}{4Pr} ff'' = f''' + T^* \tag{10.45}$$

$$-\frac{3}{4} fT^{*'} = T^{*''} \tag{10.46}$$

Many authors define similarity variable in terms of the Grashof number and Table 10.1 is a comparison between the two approaches and the final forms of the similarity solutions.

The boundary conditions are

$$\eta = 0, \quad f = f' = 0, \quad T^* = 1$$
$$\eta \to \infty, \quad f' = 0, \quad T^* = 0 \tag{10.47}$$

Note, using the similarity variable defined in terms of the Rayleigh number eliminates Prandtl number from the energy equation or the temperature distribution is not directly a function of Pr. The Prandtl number effects are conveyed through the nondimensional stream function, f, that is obtained from the solution of the momentum equation which does directly depend on Prandtl number. Therefore, the thermal boundary layer, and thus Nusselt number, must primarily depend on the Rayleigh number and have weak dependence on the Prandtl number, which also verifies the order magnitude analysis performed earlier. The hydrodynamic boundary layer is expected to be a strong function of both.

The two terms on the LHS of the momentum equation are the dimensionless mass times acceleration per unit volume, often referred to as inertial force. The first term on the RHS of the momentum equation is the dimensionless viscous force and the second term is buoyancy force per unit volume.

The solution to Equations 10.45 and 10.46 for $Pr = 1$ is shown in Figure 10.5. The nondimensional velocity increases to a maximum value of about 0.49 and drops back to zero and

TABLE 10.1

Two Choices of Similarity Variables and the Resulting Equations

$$\eta = \frac{y}{x}(Ra_x)^{1/4} \qquad\qquad \varsigma = \frac{y}{x}\left(\frac{Gr_x}{4}\right)^{1/4}$$

$$f(\eta) = \frac{\psi}{\alpha\, Ra_x^{1/4}} \qquad\qquad F(\varsigma) = \frac{\psi}{4\nu\left(\dfrac{Gr}{4}\right)^{1/4}}$$

$$T^* = \frac{T - T_\infty}{T_w - T_\infty} \qquad\qquad \theta = \frac{T - T_\infty}{T_w - T_\infty}$$

$$f' = \frac{\partial f}{\partial \eta} \qquad\qquad F' = \frac{\partial f}{\partial \varsigma}$$

$$T^{*'} = \frac{\partial T^*}{\partial \eta} \qquad\qquad T^{*'} = \frac{\partial T^*}{\partial \varsigma}$$

$$u = \frac{\alpha Ra_x^{1/2}}{x} f' \qquad\qquad u = \frac{2\nu Gr_x^{1/2}}{x} F'$$

$$u^* = \sqrt{\frac{x/L}{Pr}} f' \qquad\qquad u^* = 2\sqrt{\frac{x}{L}} F'$$

$$f''' + \frac{3}{4Pr} ff'' - \frac{1}{2Pr} f'^2 + T^* = 0 \qquad F''' + 3FF'' - 2F'^2 + \theta = 0$$

$$T^{*''} + \frac{3}{4} fT^{*'} = 0 \qquad\qquad \theta'' + 3PrF\theta' = 0$$

$$\eta = \varsigma(4Pr)^{1/4}, \quad f = F(4Pr)^{3/4}, \quad f' = (4Pr)^{1/2}F',$$
$$f'' = (4Pr)^{1/4}F'', \quad f''' = F''', \quad T^* = \theta$$

the temperature starts from the wall temperature and reaches the ambient temperature far from the wall. For this case, $Pr = 1$, the thermal and hydrodynamic boundary layers essentially have the same thickness.

The forces and velocity distribution for $Pr = 1$ are shown in Figure 10.6, and as can be seen, near the wall, the inertial forces are very small and the force balance is between buoyancy and friction. Close to the plate the buoyancy, acting in the positive x direction, is opposed by the friction. The fluid is accelerating inside the boundary layer, and therefore the inertial forces will always be positive even though the velocity and therefore the inertial forces are small close to the wall. As the distance from the wall increases, the drag forces drop faster than the buoyancy force increasing the inertial force which peaks and drops and near the edge of the boundary layer the buoyancy force is negligible and the balance is between inertial and viscous forces.

Figure 10.7 is a comparison of velocity and temperature profiles for different Prandtl numbers. The temperature distribution is not a strong function of Prandtl number. The thickness of thermal boundary layer decreases slightly with increasing Pr. This weak dependence also means that the Nusselt number is primarily a function of Rayleigh number and not a strong function of Prandtl number.

The hydrodynamic boundary layer, on the other hand, is a strong function of Prandtl number. As can be seen, for $\eta > 6$, the dimensionless temperature is essentially zero,

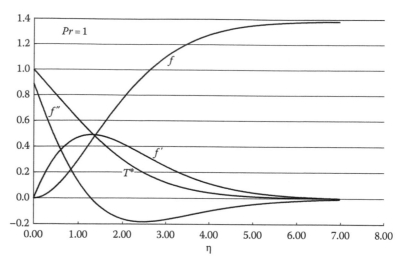

FIGURE 10.5
Velocity and temperature profiles for isothermal vertical plate.

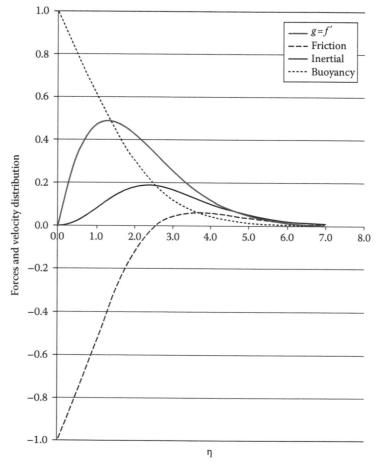

FIGURE 10.6
Forces and velocity distribution for $Pr = 1$.

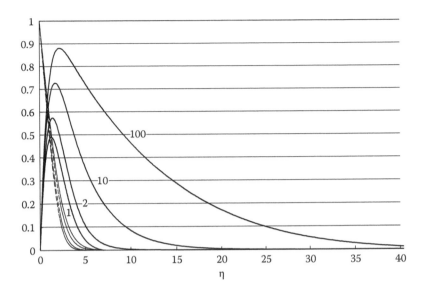

FIGURE 10.7
Velocity and temperature profiles for isothermal vertical plate for different Prandtl numbers.

meaning that the buoyancy effects are negligible, and the inertial forces balance the viscous forces and drag the fluid up, even though there is no buoyancy.

As Prandtl number increases, the impact of inertial forces that are inversely proportional to Prandtl number, as argued by Bejan [7] and seen from Equation 10.45, decreases and the force balance is between the buoyancy and friction, with negligible inertial force inside the thermal boundary layer. The thermal boundary layer is much thinner than the hydrodynamic one, and thus between the edges of thermal and hydrodynamic boundary layers, the buoyancy is negligible, and the fluid is moving by being dragged, and the force balance is between inertial and friction forces, both of which are small but not negligible.

10.2.4 Nusselt Number

Nusselt number is obtained from the solution of the energy equation, and is given by

$$Nu_x = \frac{hx}{k} = \frac{-k\dfrac{\partial T}{\partial y}\Big|_{y=0}}{T_W - T_\infty}\frac{x}{k} = -\frac{\partial T^*}{\partial \eta}\Big|_{\eta=0}(Ra_x)^{1/4} \tag{10.48}$$

The first term on the RHS is obtained from the numerical solution of the governing equations and is only a function of Prandtl number and from the numerical solution for $Pr = 1$,

$$\frac{\partial T^*}{\partial \eta} = -0.4014 \tag{10.49}$$

Based on the results of Lefevre [8], the following interpolation formula can be obtained for the temperature gradient

$$\frac{\partial T^*}{\partial \eta} = -\left(\frac{0.13Pr}{1+2Pr^{1/2}+2.03Pr}\right)^{1/4} \tag{10.50}$$

which when substituted in Equation 10.48 results in

$$Nu_x = \left(\frac{0.13Pr}{1+2Pr^{1/2}+2.03Pr} \right)^{1/4} Ra_x^{1/4} \tag{10.51}$$

This is not a strong function of Prandtl number, as the coefficient ranges between 0.4 and 0.5 for Pr ranging between 1 and 1000. LeFevre [8] has shown that in the limit of $Pr \to \infty$, the constant should be 0.5027. Using an average value of 0.45, then

$$Nu_x = 0.45Ra_x^{1/4} \tag{10.52}$$

The fact that Nusselt number is not a strong function of Prandtl number also shows the appropriateness of using the Rayleigh number as the relevant nondimensional parameter.

Churchill and Usagi [9] used the available results to provide the following expression for the local Nusselt number for isothermal flat plate

$$Nu_x = \frac{0.503Ra_x^{1/4}}{\left[1+\left(\frac{0.492}{Pr} \right)^{9/16} \right]^{4/9}} \tag{10.53}$$

The difference between Equations 10.51 and 10.53 is <0.3% in the range $1 \le Pr \le 1000$.

The average Nusselt number

$$\overline{Nu} = \frac{\bar{h}L}{k} = \frac{L}{k}\left[\frac{1}{L}\int_0^L h\,dx \right] = \frac{L}{k}\left[\frac{1}{L}\int_0^L -k\frac{\partial T^*}{\partial \eta}\frac{1}{x}(Ra_x)^{1/4}dx \right] = -\frac{\partial T^*}{\partial \eta}\int_0^L \frac{1}{x}(Ra_x)^{1/4}dx \tag{10.54}$$

$$\overline{Nu} = -\frac{\partial T^*}{\partial \eta}\left(\frac{g\beta(T_W-T_\infty)}{\alpha v} \right)^{1/4}\int_0^L x^{-1/4}dx = -\frac{4}{3}\frac{\partial T^*}{\partial \eta}Ra^{1/4} \tag{10.55}$$

$$\overline{Nu} = \frac{4}{3}Nu_L \tag{10.56}$$

Using Equation 10.52, a simple equation for laminar flow is

$$\overline{Nu} = 0.6Ra_L^{1/4} \quad \text{for } 10^4 < Ra < 10^9 \tag{10.57}$$

which is close to the expression developed by McAdams [10, p. 523] which has a coefficient of 0.555 instead of 0.6. Over the entire range for laminar flow,

$$\overline{Nu} = 0.68 + \frac{0.67Ra_L^{1/4}}{\left[1+\left(\frac{0.492}{Pr} \right)^{9/16} \right]^{4/9}} \quad (0 < Ra_L < 10^9) \tag{10.58}$$

Figure 10.8 is a comparison of the Equations 10.57 and 10.58 over a broad range of Rayleigh and Prandtl numbers. For combined laminar and turbulent flow, the Nusselt number is

$$\overline{Nu} = 0.825 + \frac{0.387Ra_L^{1/6}}{\left[1+\left(\frac{0.492}{Pr} \right)^{9/16} \right]^{8/27}} \tag{10.59}$$

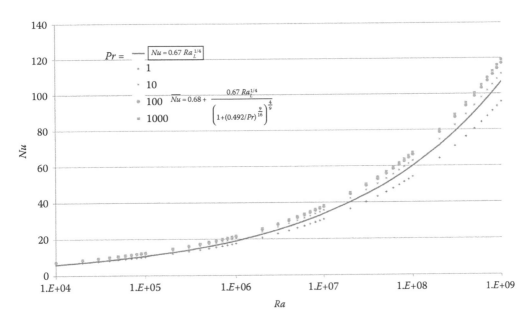

FIGURE 10.8
Impact of Prandtl number on Nusselt number.

10.2.5 Numerical Solution for Laminar Free Convection on a Vertical Isothermal Plate

Assuming $g = f'$, the solution to these equations can be obtained numerically and the finite difference form of the two equations is

$$\left(\frac{1}{\Delta\eta^2} - \frac{3f_i}{4Pr} \frac{1}{2\Delta\eta} \right) g_{i-1} - \left(\frac{1}{2Pr} g_i + \frac{2}{\Delta\eta^2} \right) g_i + \left(\frac{1}{\Delta\eta^2} + \frac{3f_i}{4Pr} \frac{1}{2\Delta\eta} \right) g_{i+1} = -T_i^* \qquad (10.60)$$

$$\left(\frac{1}{\Delta\eta^2} - \frac{3}{4} f_i \frac{1}{2\Delta\eta} \right) T_{i-1}^* - \left(\frac{2}{\Delta\eta^2} \right) T_i^* + \left(\frac{1}{\Delta\eta^2} + \frac{3}{4} f_i \frac{1}{2\Delta\eta} \right) T_{i+1}^* = 0 \qquad (10.61)$$

The coupled set can be solved using spreadsheet's iterative functionality. Using subroutine TRIDI also requires iteration, since the two equations are coupled together. The results shown in Figures 10.5 through 10.8 were obtained using iterative functionality of the spreadsheet, by solving for g_i and T_i^* from Equations 10.60 and 10.61. As in forced convection, $f_i = f_{i-1} + (g_i + g_{i-1}) \dfrac{\Delta\eta}{2}$.

10.2.6 Uniform Heat Flux Vertical Plate

Many practical cooling situations, like natural convection cooling of printed circuit boards, can be approximated by a vertical flat plate subject to a uniform heat flux. For this case, the governing equations and boundary conditions are all the same, except the condition for the energy equation at the wall, where the boundary condition is

$$q'' = -k \frac{\partial T}{\partial y} \qquad (10.62)$$

which when nondimensionalized becomes

$$\frac{\partial T^*}{\partial y^*} = -1 \tag{10.63}$$

where

$$T^* = \frac{T - T_\infty}{T_s} \tag{10.64}$$

and

$$T_s = q'' \frac{L}{k} \tag{10.65}$$

and from Equation 10.19, the relevant velocity scale becomes

$$u_s = \sqrt{\frac{g\beta q'' L^2}{k}} \tag{10.66}$$

which can then be used to define a modified Grashof number

$$Gr^* = \frac{g\beta q'' L^4}{v^2 k} \tag{10.67}$$

Sparrow and Gregg [11] have shown that the uniform heat flux case also has similarity solution by defining

$$\eta = \frac{y}{x} Ra_x^{*1/5} \tag{10.68}$$

$$f = \frac{\psi}{\alpha \left(Ra_x^*\right)^{1/5}} \tag{10.69}$$

and define a new dimensionless temperature

$$T^* = \frac{T - T_\infty}{\dfrac{q''}{k} x} Ra_x^{*1/5} \tag{10.70}$$

where

$$Ra_x^* = \frac{g\beta q'' x^4}{v\alpha k} \tag{10.71}$$

Then the momentum and energy equations become

$$f''' + \frac{4}{5Pr} ff'' - \frac{3}{5Pr} f'^2 + T^* = 0 \tag{10.72}$$

$$\mathbf{T}^{*''} + \frac{1}{5}(4f\mathbf{T}^{*'} - \mathbf{T}^* f') = 0 \tag{10.73}$$

Again, the energy equation is not directly dependent on Rayleigh number. The finite difference approximation to momentum and energy equations becomes

$$g_i = \frac{\left(\dfrac{1}{\Delta\eta^2} - \dfrac{4f}{5Pr}\dfrac{1}{2\Delta\eta}\right)g_{i-1} + \left(\dfrac{1}{\Delta\eta^2} + \dfrac{4f}{5Pr}\dfrac{1}{2\Delta\eta}\right)g_{i+1} + T_i^*}{\dfrac{3}{5Pr}g_i + \dfrac{2}{\Delta\eta^2}} \tag{10.74}$$

$$T_i^* = \frac{\left(\dfrac{1}{\Delta\eta^2} + \dfrac{4}{5}f\dfrac{1}{2\Delta\eta}\right)T_{i+1}^* + \left(\dfrac{1}{\Delta\eta^2} - \dfrac{4}{5}f\dfrac{1}{2\Delta\eta}\right)T_{i-1}^*}{2\dfrac{1}{\Delta\eta^2} + \dfrac{1}{5}f'} \tag{10.75}$$

the boundary conditions become

$$f_1 = g_1 = 0 \tag{10.76}$$

$$T_2^* + \Delta\eta = T_1^* \tag{10.77}$$

$$g_N = T_N^* = 0 \tag{10.78}$$

The solution can be obtained numerically using spreadsheets and is shown in Figure 10.9. From the numerical solutions, $T^*(0) = 1.873$ which can be substituted in Equation 10.70, resulting in

$$T_w - T_\infty = 1.873 \left(\frac{q''^4}{k^4}\frac{v^2}{g\beta}\right)^{1/5} x^{1/5} \tag{10.79}$$

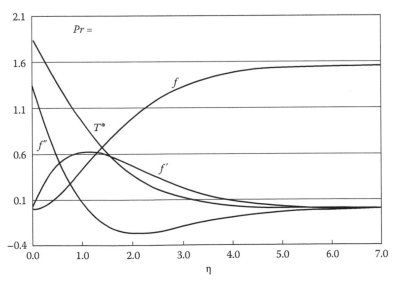

FIGURE 10.9
Velocity and temperature profiles for vertical plate subject to uniform heat flux.

which shows that the wall temperature increases proportional to the one-fifth power of the distance from the leading edge. The average wall temperature is

$$\bar{T}_w = \frac{1}{L}\int_0^L T_w dx = T_\infty + \frac{1}{L}\int_0^L \mathbf{T}^*(0)\frac{q''}{k}xRa_x^{*-1/5}dx = T_\infty + 0.833\mathbf{T}^*(0)\frac{q''L}{k}Ra_L^{*-1/5} \tag{10.80}$$

and the temperature at the middle of the plate is

$$T_w(L/2) = T_\infty + \frac{16^{1/5}}{2}\mathbf{T}^*(0)\frac{q''L}{k}Ra_L^{*-1/5} = T_\infty + 0.871\mathbf{T}^*(0)\frac{q''L}{k}Ra_L^{*-1/5} \tag{10.81}$$

As can be seen, the nondimensional average temperature and the temperature in the middle of the plate are very close. The local value of the heat transfer coefficient is determined from

$$h = \frac{q''}{T_W - T_\infty} \tag{10.82}$$

which drops along the plate, since the wall temperature increases. Local value of the Nusselt number becomes

$$Nu = \frac{hx}{k} = \frac{q''}{T_W - T_\infty}\frac{x}{k} = \frac{q''}{\mathbf{T}^*(0)\dfrac{q''}{k}xRa_x^{*-1/5}}\frac{x}{k} \tag{10.83}$$

which simplifies to

$$Nu = \frac{1}{\mathbf{T}^*(0)}Ra_x^{*1/5} \tag{10.84}$$

For $Pr = 1$, from the numerical solution, $\mathbf{T}^*(0) = 1.873$, thus

$$Nu = 0.534 Ra_x^{*1/5} \tag{10.85}$$

Sparrow [12] provided the following expression for the local Nusselt number

$$Nu_x = \frac{2}{360^{1/5}}\left(\frac{Pr}{\dfrac{4}{5}+Pr}\right)^{1/5}Ra_x^{*1/5} \tag{10.86}$$

which for $Pr = 1$ becomes

$$Nu = 0.548 Ra_x^{*1/5} \tag{10.87}$$

which is close to Equation 10.85. It is easy to show that the average Nusselt number is

$$\overline{Nu} = \frac{\bar{h}L}{k} = \frac{5}{4}Nu_L \tag{10.88}$$

TABLE 10.2

Comparison of the Nusselt Number for
Uniform Heat Flux and Isothermal Plate
Using Mean Plate Temperature and
Halfway Temperature

Pr	$\dfrac{\overline{Nu_{q''}}}{Nu_T}$ Mean Temperature	$\dfrac{\overline{Nu_{q''}}}{Nu_T}$ Halfway Temperature
0.1	1.08	1.02
1	1.07	1.013
10	1.08	1.01
100	1.05	1.0

Sparrow and Gregg [11] have shown that the average Nusselt number for uniform heat flux case is very similar to isothermal case, if the temperature of the plate in the middle is used.

Therefore, the correlations available for an isothermal plate can be used for a uniform heat flux plate if the plate is assumed to be at a uniform temperature equal to the halfway temperature. The halfway temperature can therefore be determined from

$$T_{W @ L/2} = T_\infty + \frac{q''}{h} \tag{10.89}$$

where the mean heat transfer coefficient is obtained iteratively by guessing a value for the average surface temperature (close to mid-surface temperature), calculating Rayleigh number and then h from Equation 10.51, and using Equation 10.89 to get the new surface temperature (Table 10.2).

10.2.7 Vertical Cylinder

If the diameter of the cylinder is large enough, the curvature effects become negligible, then natural convection from a vertical cylinder can be approximated by the results for a vertical flat pate. The criterion for this approximation is for the diameter to be

$$D \geq \frac{35L}{Gr^{1/4}}$$

For small curvature cylinders, starting with the boundary layer equations in cylindrical coordinates

$$\frac{\partial ru}{\partial x} + \frac{\partial rv}{\partial r} = 0 \tag{10.90}$$

$$u\frac{\partial u}{\partial x} + v\frac{\partial u}{\partial r} = \nu\frac{1}{r}\frac{\partial}{\partial r}\left(r\frac{\partial u}{\partial r}\right) + g\beta(T - T_\infty) \tag{10.91}$$

$$u\frac{\partial T}{\partial x} + v\frac{\partial T}{\partial r} = \alpha\frac{1}{r}\frac{\partial}{\partial r}\left(r\frac{\partial T}{\partial r}\right) \tag{10.92}$$

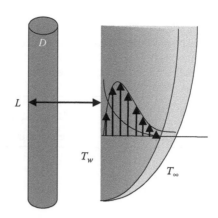

FIGURE 10.10
Vertical cylinder.

Millsaps and Pohlhausen [13] showed that these equations have similarity solution only when the wall temperature varies linearly. For other cases, including isothermal wall, Minkowycz and Sparrow [14] used local nonsimilarity method. Using the boundary layer coordinates (similarity type variables) (Figure 10.10),

$$\eta = \frac{1}{\sqrt{2}} Gr_D^{1/4} \frac{\left(\dfrac{r}{D}\right)^2 - \dfrac{1}{4}}{\left(\dfrac{x}{D}\right)^{1/4}} \tag{10.93}$$

$$\xi = \sqrt{32} Gr_D^{-1/4} \left(\frac{x}{D}\right)^{1/4} \tag{10.94}$$

then defining the nondimensional stream function and temperature as

$$f(\xi, \eta) = \frac{\psi}{\sqrt{2}\nu D \left(\dfrac{x}{D}\right)^{3/4} Gr_D^{1/4}} \tag{10.95}$$

$$T^* = \frac{T - T_\infty}{T_w - T_\infty} \tag{10.96}$$

where the stream function satisfies

$$ru = \frac{\partial \psi}{\partial r} \quad \text{and} \quad rv = -\frac{\partial \psi}{\partial x}$$

The boundary layer equations can be transformed into

$$\frac{\partial}{\partial \eta}\left[(1+\xi\eta)\frac{\partial^2 f}{\partial \eta^2}\right] + 3f\frac{\partial^2 f}{\partial \eta^2} - 2\left(\frac{\partial f}{\partial \eta}\right)^2 + T^* = \xi\left[\frac{\partial^2 f}{\partial\xi\partial\eta}\frac{\partial f}{\partial \eta} - \frac{\partial f}{\partial\xi}\frac{\partial^2 f}{\partial\eta^2}\right] \tag{10.97}$$

$$\frac{1}{Pr}\frac{\partial}{\partial \eta}\left[(1+\xi\eta)\frac{\partial T^*}{\partial \eta^2}\right] + 3f\frac{\partial T^*}{\partial \eta} = \xi\left[\frac{\partial T^*}{\partial\xi}\frac{\partial f}{\partial \eta} - \frac{\partial f}{\partial\xi}\frac{\partial T^*}{\partial \eta}\right] \tag{10.98}$$

They used local similarity and finite difference method to obtain the solution. Their results indicate that local similarity provides accurate results. The solution can also be easily determined numerically without assuming local similarity, as these equations have the same form as those for nonsimilar boundary layers, already discussed in Chapter 7.

Nagendra et al. [15,16] obtained the analytical solution for natural convection from short and long cylinders and wires assuming isothermal [15] and uniform heat flux [16] wall, and also conducted an experimental investigation [17]. Their results are summarized in Equation 10.99.

$$\overline{Nu_D} = m\left(Ra_D\frac{D}{L}\right)^n \quad \text{where} \quad \begin{array}{lll} m = 0.93 & n = 0.05 & Ra_D\dfrac{D}{L} < 0.05 \\[2mm] m = 1.37 & n = 0.16 & \text{for } 0.05 < Ra_D\dfrac{D}{L} < 10^4 \\[2mm] m = 0.6 & n = 0.25 & Ra_D\dfrac{D}{L} > 10^4 \end{array} \tag{10.99}$$

and are also shown in Figure 10.11. For vertical cylinders, the curvature effects can be neglected and the correlations available for vertical plates, like Equation 10.57 can be used, if $Ra_D\dfrac{D}{L} > 10^4$. The values of m and n in the range between 0.05 and 10,000 (long cylinders), and for values less than 0.05 (wires) are also given. The same correlations can be used for the uniform heat flux case, if modified Rayleigh number, Equation 10.71, is used.

10.2.8 Horizontal Cylinder

The laminar natural convection from a horizontal cylinder has been studied using a variety of techniques, and a review of the early works is given in Muntasser and Mulligan [18]. Making boundary layer approximation and again using similarity type transformation, Muntasser and Mulligan defined the variables

$$\eta = y^* \frac{(\sin x^*)^{1/3}}{(4\xi)^{1/4}} \tag{10.100}$$

$$\xi = \int_0^{x^*} (\sin z)^{1/3}\, dz \tag{10.101}$$

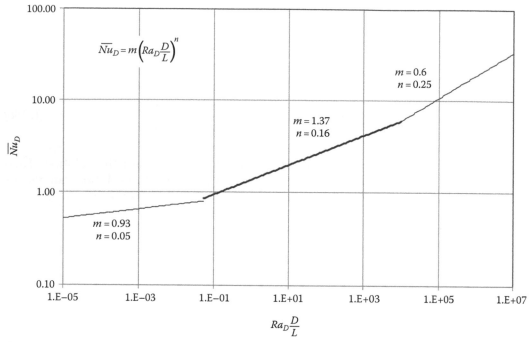

FIGURE 10.11

Nusselt number for isothermal and uniform heat flux vertical cylinder.

where x is the distance along the surface of the cylinder and y is perpendicular to the surface or in the r direction, and the nondimensional x and y are defined as

$$x^* = \frac{x}{R} \tag{10.102}$$

$$y^* = \frac{y}{R} Gr^{1/4} \tag{10.103}$$

Using the earlier and the stream function, the momentum and energy equations become

$$\frac{\partial^3 f}{\partial \eta^3} + 3f\left(\frac{\partial^2 f}{\partial \eta^2}\right) - (b+3)\left(\frac{\partial f}{\partial \eta}\right)^2 + T^* = 4\xi\left[\frac{\partial f}{\partial \eta}\frac{\partial^2 f}{\partial \xi \partial \eta} - \frac{\partial f}{\partial \xi}\frac{\partial^2 f}{\partial \eta^2}\right] \tag{10.104}$$

$$\frac{1}{Pr}\frac{\partial^2 T^*}{\partial \eta^2} + 3f\frac{\partial T^*}{\partial \eta} = 4\xi\left[\frac{\partial f}{\partial \eta}\frac{\partial T^*}{\partial \xi} - \frac{\partial f}{\partial \xi}\frac{\partial T^*}{\partial \eta}\right] \tag{10.105}$$

with

$$b = \frac{4}{3}(\sin x^*)^{-4/3}\cos x^* \int_0^{x^*}(\sin z)^{1/3}dz - 1 \tag{10.106}$$

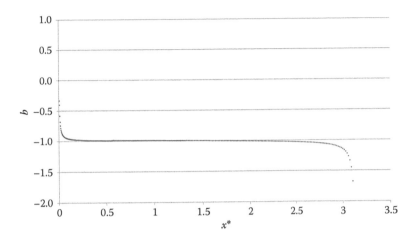

FIGURE 10.12
Variation of the parameter b for isothermal horizontal cylinder.

Hermann [19] assumed $b = -1$ and local similarity which simplify the momentum and energy equations to

$$f''' + 3ff'' - 2f'^2 + T^* = 0 \tag{10.107}$$

$$\frac{1}{Pr}T^{*''} + 3fT^{*'} = 0 \tag{10.108}$$

which can readily be solved numerically. Figure 10.12 shows the variation of b along the cylinder, and as can be seen it is very close to -1, except near the top and bottom of the cylinder.

Muntasser and Mulligan [18] solved Equations 10.104 and 10.105 using local nonsimilarity. Based on their results, the local value of Nusselt number can be expressed as

$$Nu_\theta = CRa^{1/4} \tag{10.109}$$

where the values of C are given in Table 10.3 for different Prandtl numbers at different angles around the cylinder. Churchill and Chu [20] provide the following correlation for the average Nusselt number for natural convection around a cylinder:

TABLE 10.3
Constant C in Equation 10.109

θ\Pr	0.72	0.733	1	5	10
0 (0)	0.4066	0.4074	0.4214	0.4776	0.4961
$\pi/6$ (30)	0.4017	0.4025	0.4162	0.4715	0.4892
$\pi/3$ (60)	0.3872	0.3880	0.4010	0.4541	0.4696
$\pi/2$ (90)	0.3630	0.3637	0.3753	0.4230	0.4360
$4\pi/6$ (120)	0.3278	0.3284	0.3379	0.3756	0.3867
$5\pi/6$ (150)	0.2820	0.2822	0.2853	0.2999	0.3056
Arithmetic mean	0.3614	0.3620	0.3729	0.4170	0.4305

$$Nu = \left[0.67 + \frac{0.387 Ra_D^{1/6}}{\left[1 + \left(\dfrac{0.559}{Pr} \right)^{9/16} \right]^{8/27}} \right]^2 \quad (10^5 < Ra_D < 10^{12}) \qquad (10.110)$$

10.2.9 Sphere

Natural convection around sphere has been the subject of a number of analytical and experimental investigations. Acrivos [21] defined

$$\eta = \left[\frac{3 Ra (\sin \theta)^{8/3}}{32 \displaystyle\int_0^\theta (\sin \theta)^{5/3} d\theta} \right]^{1/4} \frac{y}{R} \qquad (10.111)$$

In the limit of $Pr \to \infty$, the momentum and energy equations reduce to

$$f''' + T^* = 0 \qquad (10.112)$$

$$T^{*''} + f T^{*'} = 0 \qquad (10.113)$$

$$\eta = 0, \quad f = f' = 0, \quad T^* = 1$$
$$\eta = \infty, \quad f'' = T^* = 0 \qquad (10.114)$$

Outside the thermal boundary layer, the balance is between drag and inertial forces, and the drag forces must eventually approach zero. The finite difference solution of the earlier equations is shown in Figure 10.13. From the results, $f''(0) = 1.08$ and $T^{*'}(0) = 0.54$, which match the values reported by Acrivos [21]. Again, as can be seen, the velocity asymptotes to the value of 0.884, which is incorrect (should go to zero). Equations 10.112 and 10.113 are valid near the wall only, where the force balance is between buoyancy and drag.

Experimental measurements of Amato and Tien [22] are in good agreement with the numerical solution shown in Figure 10.13. The average Nusselt number is given to be

$$Nu = 0.583 Ra_D^{1/4} \qquad (10.115)$$

which is based on limit of large Prandtl number. A more general expression is

$$Nu = 2 + \frac{0.589 Ra_D^{1/4}}{\left[1 + \left(\dfrac{0.469}{Pr} \right)^{9/16} \right]^{4/9}} \quad Ra < 10^{11} \qquad (10.116)$$

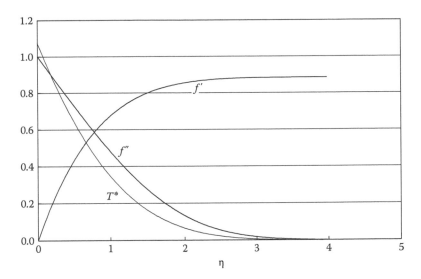

FIGURE 10.13
Velocity and temperature profiles for an isothermal sphere.

10.2.10 Horizontal and Inclined Plates

The physics of natural convection from a horizontal or inclined flat plate is somewhat different than that of a vertical plate. For a horizontal flat plate, the buoyancy force acts perpendicular to the plate, and results in a pressure gradient along the plate. This induced pressure gradient leads to the flow of the fluid along the plate. For a finite sized heated plate, the flow can proceed from the leading and trailing edges of the plate toward the center of the plate where the flow will turn to form a "thermal jet" above the plate [23]. If the plate is sufficiently long, it has been observed [24] that the gravitationally unstable layer will separate from the heated boundary and give rise to the formation of eddies along the plate. The boundary layer flow is still applicable and is also experimentally observed near the leading edge of the heated horizontal plate [25].

If the acute angle θ is the angle the plate makes from the vertical, as shown in Figure 10.14, the boundary layer equations for natural convection along a horizontal or inclined flat plate are

$$\frac{\partial u}{\partial x} + \frac{\partial v}{\partial y} = 0 \tag{10.117}$$

$$u\frac{\partial u}{\partial x} + v\frac{\partial u}{\partial y} = -\frac{1}{\rho}\frac{\partial P}{\partial x} + g\beta\cos\theta(T - T_\infty) + \nu\frac{\partial^2 u}{\partial y^2} \tag{10.118}$$

$$0 = -\frac{1}{\rho}\frac{\partial P}{\partial y} + g\beta\sin\theta(T - T_\infty) \tag{10.119}$$

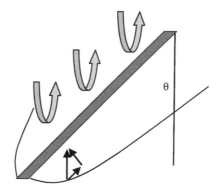

FIGURE 10.14
Inclined plate.

$$u\frac{\partial T}{\partial x} + v\frac{\partial T}{\partial y} = \alpha\frac{\partial^2 T}{\partial y^2} \tag{10.120}$$

Integrating Equation 10.119 with respect to y and differentiating it with respect to x provided an expression for axial variation of pressure, due to buoyancy

$$-\frac{1}{\rho}\frac{\partial P}{\partial x} = g\beta\sin\theta\frac{\partial}{\partial x}\int_y^\infty (T - T_\infty)dy \tag{10.121}$$

which can be substituted in the x momentum equation

$$u\frac{\partial u}{\partial x} + v\frac{\partial u}{\partial y} = g\beta\sin\theta\frac{\partial}{\partial x}\int_y^\infty (T - T_\infty)dy + g\beta\cos\theta(T - T_\infty) + v\frac{\partial^2 u}{\partial y^2} \tag{10.122}$$

Chen et al. [26] examined laminar free convection along horizontal, inclined, and vertical flat plates with subject to nonuniform wall temperature and heat flux. The boundary layer equations in general do not have similarity solution for these cases; however, the numerical solution is greatly facilitated by transforming the governing equations using the boundary layer coordinates, as was also done for forced convection. Table 10.4 summarized the variables chosen by Chen et al. [26] and the final form of the transformed equations. Note that the equations given in Table 10.4 are valid for $0 < \theta \le 90$.

Numerical results for both cases of power law variation of wall temperature and surface heat flux are presented for different values of the exponents and Prandtl numbers are presented. Table 10.5 is a comparison between the numerical results of Chen and the experimental results of Vliet and Ross [27,28] for free convection from inclined plates subjected to a uniform heat flux, and the numerical and experimental results are in reasonable agreement.

TABLE 10.4

Transformed Equations ($\theta > 0$)

Power Law Variation of Wall Temperature		Power Law Variation of Surface Heat Flux	
$T_w(x) - T_\infty = ax^n$	(10.123a)	$q_w''(x) = bx^m$	(10.123b)
$\xi = \dfrac{\left(\dfrac{Gr_x}{5}\right)^{1/5}}{\tan\theta}$	(10.124a)	$\xi = \dfrac{\left(\dfrac{Gr_x^*}{6}\right)^{1/6}}{\tan\theta}$	(10.124b)
$\eta = \dfrac{y}{x}\left(\dfrac{Gr_x}{5}\right)^{1/5}$	(10.125a)	$\eta = \dfrac{y}{x}\left(\dfrac{Gr_x^*}{6}\right)^{1/6}$	(10.125b)
$Gr_x = \dfrac{g\beta\Delta T x^3}{\nu^2}\sin\theta$	(10.126a)	$Gr_x^* = \dfrac{g\beta q'' x^4}{k\nu^2}\sin\theta$	(10.126b)
$Ra_x = PrGr_x = \dfrac{g\beta\Delta T x^3}{\nu\alpha}\sin\theta$	(10.127a)	$Ra_x^* = PrGr_x^* = \dfrac{g\beta q'' x^4}{k\nu\alpha}\sin\theta$	(10.127b)
$f = \dfrac{\psi(x,y)}{5\nu\left(\dfrac{Gr_x}{5}\right)^{1/5}}$	(10.128a)	$f = \dfrac{\psi(x,y)}{6\nu\left(\dfrac{Gr_x^*}{6}\right)^{1/6}}$	(10.128b)
$T^* = \dfrac{T - T_\infty}{T_w(x) - T_\infty}$	(10.129a)	$T^* = \dfrac{(T - T_\infty)\left(\dfrac{Gr_x^*}{6}\right)^{1/6}}{q'' x / k}$	(10.129b)

$$f''' + (n+3)ff'' - (2n+1)f'^2 + \xi T^* +$$
$$\frac{1}{5}\left[(2-n)\eta T^* + (4n+2)\int_\eta^\infty T^* d\eta + (n+3)\xi\int_\eta^\infty \frac{\partial T^*}{\partial\xi}d\eta\right]$$
$$= (n+3)\xi\left(f'\frac{\partial f'}{\partial\xi} - f''\frac{\partial f}{\partial\xi}\right) \qquad (10.130a)$$

$$f''' + (m+4)ff'' - 2(m+1)f'^2 + \eta T^* +$$
$$\frac{1}{6}\left[(2-m)\eta T^* + 4(m+1)\int_\eta^\infty T^* d\eta + (m+4)\xi\int_\eta^\infty \frac{\partial T^*}{\partial\xi}d\eta\right]$$
$$= (m+4)\xi\left(f'\frac{\partial f'}{\partial\xi} - f''\frac{\partial f}{\partial\xi}\right) \qquad (10.130b)$$

$$\frac{1}{Pr}T^{*''} + (n+3)fT^{*'} - 5nf'T^*$$
$$= (n+3)\xi\left(f'\frac{\partial T^*}{\partial\xi} - T^{*'}\frac{\partial f}{\partial\xi}\right) \qquad (10.131a)$$

$$\frac{1}{Pr}T^{*''} + (m+4)fT^{*'} - (5m+2)f'T^*$$
$$= (m+4)\xi\left(f'\frac{\partial T^*}{\partial\xi} - T^{*'}\frac{\partial f}{\partial\xi}\right) \qquad (10.131b^1)$$

| $\eta = 0, \quad f = f' = 0, \quad T^* = 1$ | | $\eta = 0, \quad f = f' = 0, \quad T^{*'} = -1$ | |
| $\eta \to \infty, \quad f' = 0, \quad T^* = 0$ | (10.132a) | $\eta \to \infty, \quad f' = 0, \quad T^* = 0$ | (10.132b) |

[1] The authors use a plus sign in the parenthesis on the RHS of Equation 10.131b.

Based on their numerical solutions, Chen et al. [26] provide the following expression for the local value of the Nusselt number, for a uniform wall temperature

$$Nu_x Ra_x^{-1/4} = \frac{3}{4}\left[\frac{2Pr}{5(1+2Pr^{1/2}+2Pr)}\right]^{1/4} \qquad 0 \le \theta \le 75 \quad \text{and} \quad 5\times10^3 \le Ra \le 5\times10^9 \qquad (10.133)$$

$$Nu_x\left(\frac{Ra_x}{5}\right)^{-(1/5+0.07\sqrt{\cos\theta})} = \frac{\sqrt{Pr}}{0.25+1.6\sqrt{Pr}} \qquad 75 \le \theta \le 90 \quad \text{and} \quad 10^3 \le Ra \le 10^9 \qquad (10.134)$$

TABLE 10.5

Local Value of Nusselt Number

Gr_x^*	Pr = 0.7		Pr = 7	
	$\theta = 60$		$\theta = 60$	
	Chen	Ref. [27]	Chen	Ref. [26]
1.1972×10^4	2.94	2.92	5.09	5.04
1.3637×10^5	4.67	4.74	8.20	8.20
7.6620×10^5	6.51	6.70		
2.9228×10^6	8.46	8.76	15.02	15.14
2.2008×10^7	12.57	13.11	22.42	22.67
9.9412×10^7	16.93	17.73	30.27	30.65
5.5856×10^8	23.82	25.04	42.68	43.29
3.1384×10^9	33.57	35.36	60.22	61.14
	$\theta = 15$		$\theta = 15$	
2.5619×10^4	3.69	3.87	6.61	6.70
6.4600×10^4	4.43	4.66	7.94	8.06
5.4910×10^5	6.78	7.15	12.17	12.36
4.1345×10^6	10.14	10.71	18.21	18.51
3.5142×10^7	15.54	16.43	27.93	28.40
2.3749×10^8	22.75	24.07	40.93	41.62
1.0094×10^9	30.38	32.15		
3.2358×10^9	38.34	40.59	69.00	70.17

and for a uniform wall heat flux

$$Nu_x Ra_x^{*-1/5} = \left[\frac{Pr}{4 + 9Pr^{1/2} + 10Pr} \right]^{1/5} \quad 0 \le \theta \le 75 \quad \text{and} \quad 5 \times 10^4 \le Ra \le 5 \times 10^{10} \tag{10.135}$$

$$Nu_x \left(\frac{Ra_x^*}{6} \right)^{-(1/6 + 0.038\sqrt{\cos\theta})} = \frac{\sqrt{Pr}}{0.12 + 1.2\sqrt{Pr}} \quad 75 \le \theta \le 90 \quad \text{and} \quad 10^4 \le Ra \le 10^{10} \tag{10.136}$$

The average Nusselt numbers for uniform wall temperature are

$$\overline{Nu} Ra_L^{-1/4} = \left[\frac{2Pr}{5(1 + 2Pr^{1/2} + 2Pr)} \right]^{1/4} \quad 0 \le \theta \le 75 \quad \text{and} \quad 5 \times 10^3 \le Ra \le 5 \times 10^9 \tag{10.137}$$

$$\overline{Nu} \left(\frac{Ra_L}{5} \right)^{-(1/5 + 0.07\sqrt{\cos\theta})} = \frac{\dfrac{\sqrt{Pr}}{(0.25 + 1.6\sqrt{Pr})}}{0.6 + 0.21\sqrt{\cos\theta}} \quad 75 \le \theta \le 90 \quad \text{and} \quad 10^3 \le Ra \le 10^9 \tag{10.138}$$

and for a uniform wall heat flux

$$\overline{Nu} Ra_L^{*-1/5} = \frac{5}{4} \left[\frac{Pr}{4+9Pr^{1/2}+10Pr} \right]^{1/5} \quad 0 \leq \theta \leq 75 \quad \text{and} \quad 5 \times 10^4 \leq Ra_L^* \leq 5 \times 10^{10} \quad (10.139)$$

$$\overline{Nu} \left(\frac{Ra_L^*}{6} \right)^{-(1/6+0.038\sqrt{\cos\theta})} = \frac{\dfrac{\sqrt{Pr}}{(0.12+1.2\sqrt{Pr})}}{0.667+0.152\sqrt{\cos\theta}} \quad 75 \leq \theta \leq 90 \quad \text{and} \quad 10^4 \leq Ra_L^* \leq 10^{10} \quad (10.140)$$

For finite size plates, the following correlations can be used. For the upper surface of a heated plate or the lower surface of a cooled plate

$$Nu = 0.54 Ra^{1/4}, \quad Ra = 10^4-10^7 \tag{10.141}$$

$$Nu = 0.15 Ra^{1/3}, \quad Ra = 10^7-10^{11} \tag{10.142}$$

and for lower surface of a heated plate or the upper surface of a cooled plate

$$Nu = 0.27 Ra^{1/4}, \quad Ra = 10^5-10^{11} \tag{10.143}$$

For a horizontal plate, the characteristic length used is $L_c = \dfrac{A_s}{p}$, where A_s and p are plate's surface area and perimeter.

10.3 Combined Natural and Forced Convection

Combined forced and free convection heat transfer is important in low speed flows where buoyancy plays an important role. The buoyancy force effectively produces a stream-wise pressure gradient in the fluid close to the plate surface. The pressure gradient is favorable when the fluid flows over a hot plate, and adverse when the plate is cold. Therefore, the flow is either accelerated or decelerated compared to the corresponding forced convection [29]. The interaction between the two heat transfer mechanisms can result in interesting flow features that are of interest from theoretical and practical perspectives. For example, if the plate is sufficiently cold, the flow can actually separate.

10.3.1 Horizontal Flat Plate

Laminar mixed convection over a horizontal flat plate with uniform wall temperature has been investigated both analytically [29,30] and experimentally [31] and a review of the literature is given in the earlier references. The relative significance of free to forced convection is generally assessed by the buoyancy parameter, $\dfrac{Gr}{Re^{5/2}}$ for isothermal and $\dfrac{Gr^*}{Re^3}$

for uniform heat flux walls. The positive values of the buoyancy parameter is for flow above a heated plate where buoyancy assists mixed convection and the negative values correspond to flow above a cooled plate, with buoyancy opposing mixed convection. Lin et al. [29] proposed new buoyancy parameters

$$\xi = \frac{\left(\dfrac{Pr}{Pr+1}Ra_x\right)^{1/5}}{\left(\dfrac{Pr}{Pr+1}Ra_x\right)^{1/5} + \left[\dfrac{Pr}{(Pr+1)^{1/3}}Re_x\right]^{1/2}} \tag{10.144}$$

$$\xi = \frac{\left(\dfrac{Pr}{Pr+1}Ra_x^*\right)^{1/6}}{\left(\left(\dfrac{Pr}{Pr+1}Ra_x^*\right)^{1/6}\right) + \left[\dfrac{Pr}{(Pr+1)^{1/3}}Re_x\right]^{1/2}} \tag{10.145}$$

for isothermal and uniform heat flux cases, respectively. The parameter ξ varies between $\xi = 0$ corresponding to forced convection and $\xi = 1$ corresponding to free convection alone. The numerical results show that the velocity could go beyond the free stream velocity, and that the velocity overshoot increases with increasing buoyancy force. For flow over a hot plate, the values of $\xi < 0.5$ correspond to negligible free convection and the friction factor results are very close to those for boundary layer flow over a flat plate, i.e., $c_f Re^{1/2} = 0.664$.

For the case of the flow of a hot fluid over a cold plate, adverse pressure gradient, as ξ increases from zero, the friction factor initially stays the same as that for flat plate, and then for some critical value, the wall friction decreases dramatically and the solution diverges. This critical point corresponds to the condition, under which boundary layer approximation is no longer valid.

The Nusselt number is given by

$$Nu = -\left(\frac{Pr}{Pr+1}Ra_x\right)^{1/5}\frac{T^{*\prime}(\xi,0)}{\xi} \tag{10.146}$$

$$Nu = \left(\frac{Pr}{Pr+1}Ra_x^*\right)^{1/6}\frac{1}{\xi T^*(\xi,0)} \tag{10.147}$$

for isothermal and uniform heat flux wall, respectively. The values of $T^{*\prime}(\xi,0)$ and $T^*(\xi,0)$ from the numerical solution of Ref. [29] for both favorable (heating) and unfavorable (cooling) pressure gradients are given in Table 10.6.

Their numerical results show that the values of $\xi < 0.3$ correspond to negligible free convection, with Nusselt number being very close to that for boundary layer flow over a flat plate. Similarly, for $\xi > 0.67$, Nusselt number approaches the values for pure free convection [29].

TABLE 10.6

Results for Isothermal and Uniform Heat Flux Heating and Cooling the Stream

$\frac{\xi}{Pr}$	0.001	0.01	0.1	0.7	1	7	10	100	1000	10,000
$-T^{*\prime}\ (\xi, 0)$ Heating, isothermal										
0	0.548	0.517	0.450	0.382	0.373	0.345	0.344	0.339	0.338	0.338
0.1	0.493	0.465	0.405	0.344	0.335	0.311	0.310	0.305	0.305	0.305
0.2	0.438	0.414	0.360	0.306	0.299	0.277	0.277	0.272	0.271	0.271
0.3	0.386	0.365	0.319	0.272	0.265	0.248	0.249	0.243	0.243	0.243
0.4	0.347	0.334	0.293	0.253	0.247	0.235	0.237	0.232	0.232	0.232
0.5	0.340	0.327	0.296	0.261	0.256	0.247	0.248	0.247	0.248	0.249
0.6	0.365	0.352	0.322	0.289	0.284	0.275	0.275	0.279	0.280	0.281
0.7	0.406	0.393	0.361	0.325	0.320	0.311	0.311	0.318	0.320	0.322
0.8	0.456	0.441	0.405	0.366	0.361	0.352	0.352	0.360	0.364	0.365
0.9	0.510	0.493	0.453	0.410	0.404	0.394	0.395	0.404	0.408	0.410
1	0.566	0.547	0.503	0.455	0.449	0.438	0.439	0.448	0.453	0.455
$-T^{*\prime}\ (\xi, 0)$ Cooling, isothermal										
0		0.517	0.450	0.382	0.373	0.345	0.344	0.339		
0.05		0.491	0.427	0.363	0.354	0.328	0.327	0.322		
0.1		0.465	0.405	0.344	0.335	0.311	0.311	0.305		
0.15		0.439	0.382	0.325	0.317	0.294	0.295	0.288		
0.2		0.413	0.360	0.305	0.298	0.277	0.280	0.271		
0.25		0.387	0.336	0.285	0.278	0.259	0.264	0.253		
0.3		0.358	0.311	0.263	0.256	0.239	0.246	0.231		
0.35		0.325	0.279	0.232	0.226	0.209	0.223	0.216		
$T^{*}\ (\xi, 0)$ Heating, uniform heat flux										
0	1.182	1.287	1.551	1.887	1.941	2.t080	2.111	2.154	2.155	2.153
0.1	1.313	1.430	1.723	2.096	2.157	2.336	2.332	2.393	2.392	2.388
0.2	1.477	1.609	1.938	2.357	2.425	2.611	2.594	2.690	2.682	2.671
0.3	1.684	1.832	2.202	2.670	2.745	2.919	2.884	3.030	3.001	2.981
0.4	1.915	2.062	2.428	2.882	2.952	3.087	3.052	3.182	3.135	3.112
0.5	2.018	2.126	2.408	2.769	2.824	2.937	2.922	2.969	2.931	2.988
0.6	1.906	1.987	2.021	2.483	2.526	2.628	2.624	2.621	2.596	2.633
0.7	1.713	1.780	1.957	2.191	2.225	2.312	2.315	2.295	2.278	2.270
0.8	1.524	1.582	1.735	1.938	1.968	2.046	2.045	2.023	2.010	2.002
0.9	1.362	1.413	1.549	1.730	1.755	1.824	1.821	1.803	1.971	1.787
1	1.227	1.273	1.395	1.558	1.580	1.642	1.638	1.623	1.613	1.609
$T^{*}\ (\xi, 0)$ Cooling, uniform heat flux										
0		1.287	1.551	1.887	1.941	2.108	2.111	2.153		
0.05		1.355	1.633	1.986	2.043	2.217	2.216	2.267		
0.1		1.430	1.723	2.096	2.157	2.336	2.332	2.393		
0.15		1.514	1.835	2.220	2.284	2.468	2.458	2.533		
0.2		1.609	1.939	2.359	2.428	2.615	2.597	2.693		
0.25		1.718	2.071	2.521	2.594	2.782	2.754	2.882		
0.3		1.846	2.231	2.724	2.805	2.990	2.997	3.133		
0.35		2.012	2.460	3.068	3.179	3.369	3.266			

Source: Lin, I.T. et al., *Warme Stoffubertragung*, 24, 225, 1989.

10.3.2 Inclined Plate

The combined forced and natural convection has also been studied for a variety of other geometries. As expected, the rate of heat transfer will increase or decrease depending on whether the buoyancy assists or hinders the forced flow. Mucoglu and Chen [32] considered mixed convection on an inclined plate. The fluid is forced to flow along the plate with buoyancy assisting the flow, when the fluid flows above the plate and hindering the flow, when fluid flows under the plate, as shown in Figure 10.15. Using the boundary layer approximation, invoking Boussinesq approximation results in

$$\frac{\partial u}{\partial x} + \frac{\partial v}{\partial y} = 0 \tag{10.148}$$

$$u\frac{\partial u}{\partial x} + v\frac{\partial u}{\partial y} = -\frac{1}{\rho}\frac{\partial P}{\partial x} + \nu\frac{\partial^2 u}{\partial y^2} \pm g\beta(T - T_\infty)\cos\theta \tag{10.149}$$

$$0 = -\frac{1}{\rho}\frac{\partial P}{\partial y} \pm g\beta(T - T_\infty)\sin\theta \tag{10.150}$$

$$u\frac{\partial T}{\partial x} + v\frac{\partial T}{\partial y} = \alpha\frac{\partial^2 T}{\partial y^2} \tag{10.151}$$

where the acute angle θ is the angle the plate makes from the vertical, as shown in Figure 10.15 and the plus and minus in Equation 10.149 signs are for upward and downward forced flow. The plus sign in Equation 10.150 is for $T_w > T_\infty$ if the fluid flows above the plate or for $T_w < T_\infty$ for fluid flowing below the plate.

The negative sign is for the reverse, $T_w < T_\infty$ for flow above the plate and $T_w > T_\infty$ for flow below the plate. Integrating Equation 10.150 with respect to y and differentiating it with respect to x provide an expression for axial variation of pressure, due to buoyancy

$$-\frac{1}{\rho}\frac{\partial P}{\partial x} = \pm g\beta\sin\theta\frac{\partial}{\partial x}\int_y^\infty (T - T_\infty)dy \tag{10.152}$$

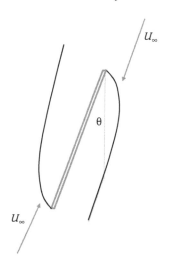

FIGURE 10.15
Mixed convection, with gravity assisting above the plate and opposing below the plate.

Substituting for the stream-wise pressure gradient in the x momentum equation

$$u\frac{\partial u}{\partial x}+v\frac{\partial u}{\partial y}=+\nu\frac{\partial^2 u}{\partial y^2}\pm g\beta(T-T_\infty)\cos\theta\left[1+\tan\theta\frac{\frac{\partial}{\partial x}\int_y^\infty(T-T_\infty)dy}{(T-T_\infty)}\right] \qquad (10.153)$$

The boundary conditions are

$$x=0,\quad u=U_\infty,\quad v=0,\quad T=T_\infty$$

$$y=0,\quad u=0,\quad v=0,\quad T=T_w,\ \text{or}\quad -k\frac{\partial T}{\partial y}=q'' \qquad (10.154)$$

$$y\to\infty,\quad u\to U_\infty,\quad T\to T_\infty$$

Consider an order of magnitude analysis of the fraction in the bracket. The integral is non-zero, only up until the edge of the boundary layer

$$\frac{\frac{\partial}{\partial x}\int_y^\infty(T-T_\infty)dy}{(T-T_\infty)}\sim\frac{\frac{\partial}{\partial x}\int_y^\delta(T-T_\infty)dy}{(T-T_\infty)}\sim\frac{\frac{\partial}{\partial x}\int_0^\delta(T_w-T_\infty)dy}{(T_w-T_\infty)}\sim\frac{\partial}{\partial x}\int_0^\delta dy\sim\frac{\Delta y}{\Delta x}\sim\frac{\delta}{x} \qquad (10.155)$$

The buoyancy-induced pressure gradient will be negligible if

$$\frac{\delta}{x}\tan\theta\ll1 \qquad (10.156)$$

If natural convection is not important, then the boundary layer thickness is $\frac{\delta}{x}=5Re_x^{-1/2}$, then the second term can be neglected if

$$\theta\ll\tan^{-1}\left(\frac{Re_x^{1/2}}{5}\right) \qquad (10.157)$$

The higher the Reynolds number, the larger the angle from vertical for which the second term is negligible. For example, for $Re=1000$, as long as $\theta\ll81°$ the second term in the bracket becomes negligible, and the x momentum equation becomes

$$u\frac{\partial u}{\partial x}+v\frac{\partial u}{\partial y}=+\nu\frac{\partial^2 u}{\partial y^2}\pm g\beta(T-T_\infty)\cos\theta \qquad (10.158)$$

Using similarity type variables, the transformed momentum and energy equations for isothermal and uniform heat flux walls are given in Table 10.7, assuming the buoyancy-induced pressure gradient term is negligible, and the momentum equation is given by Equation 10.158. If these terms cannot be neglected, then the transformed form of Equations 10.151 and 10.153 for an isothermal plate becomes

$$f'''+\frac{1}{2}ff''\pm\frac{1}{2}\Omega\sin\theta\left[\eta T^*+\int_\eta^\infty T^* d\eta+\Omega\int_\eta^\infty\frac{\partial T^*}{\partial\Omega}d\eta\right]\pm\xi T^*=\frac{1}{2}\Omega\left(f'\frac{\partial f'}{\partial\Omega}-f''\frac{\partial f}{\partial\Omega}\right) \qquad (10.159)$$

$$\frac{1}{Pr}T^{*''}+\frac{1}{2}fT^{*'}=\frac{1}{2}\Omega\left(f'\frac{\partial T^*}{\partial\Omega}-T^{*'}\frac{\partial f}{\partial\Omega}\right) \qquad (10.160)$$

where

$$\Omega = \frac{|Gr_x|}{Re_x^{5/2}} = \frac{\xi}{\cos\theta Re_x^{1/2}}$$

(10.161)

and

$$\frac{\partial}{\partial\Omega} = 2\xi\frac{\partial}{\partial\xi}$$

(10.162)

TABLE 10.7

Transformed Momentum and Energy Equations for Isothermal and Uniform Heat Flux Walls

Isothermal Plate	Uniform Heat Flux
$\xi = \dfrac{\lvert Gr_x\rvert\cos\theta}{Re_x^2}$ (10.163a)	$\xi = \dfrac{\lvert Gr_x^*\rvert\cos\theta}{Re_x^{5/2}}$ (10.163b)

$$\eta = y\sqrt{\frac{U_\infty}{vx}} \quad (10.164)$$

$Gr_x = \dfrac{g\beta\Delta Tx^3}{v^2}$ (10.165a)	$Gr_x^* = \dfrac{g\beta q''x^4}{kv^2}$ (10.165b)

$$Re = \frac{U_\infty x}{v} \quad (10.166)$$

$$f = \frac{\psi(x,y)}{\sqrt{vU_\infty x}} \quad (10.167)$$

$$f' = \frac{u}{U_\infty} \quad (10.168)$$

$T^* = \dfrac{T - T_\infty}{T_w - T_\infty}$ (10.169a)	$T^* = \dfrac{T - T_\infty}{\dfrac{q''x}{kRe_x^{1/2}}}$ (10.169b)

$$f''' + \frac{1}{2}ff'' \pm \xi T^* = C\xi\left(f'\frac{\partial f'}{\partial\xi} - f''\frac{\partial f}{\partial\xi}\right) \quad (10.170)$$

$$\frac{1}{Pr}T^{*''} + \frac{1}{2}fT^{*'} = C\xi\left(f'\frac{\partial T^*}{\partial\xi} - T^{*'}\frac{\partial f}{\partial\xi}\right) \quad (10.171)$$

$$f(\xi,0) = f'(\xi,0) = 0, \; f'(\xi,\infty) = 1$$

$$T^*(\xi,\infty) = 0$$

Isothermal wall $\quad T^*(\xi,0) = 1$

(10.172)

Uniform heat flux wall $\quad T^{*'}(\xi,0) = -1$

$C = 1$ (10.173a)	$C = \dfrac{3}{2}$ (10.173b)

$$c_f Re_x^{1/2} = 2f''(\xi,0) \quad (10.174)$$

$Nu_x Re_x^{-1/2} = T^*(\xi,0)$ (10.175a)	$Nu_x Re_x^{-1/2} = \dfrac{1}{T^*(\xi,0)}$ (10.175b)

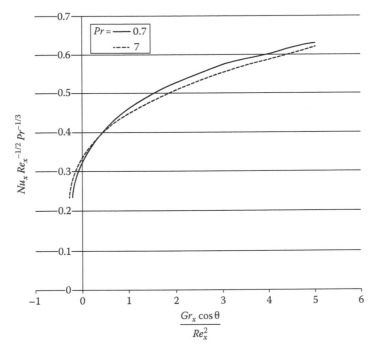

FIGURE 10.16
Impact of the buoyancy parameter on Nusselt number correlation.

The variable ξ is the buoyancy force parameter, defined by Equation 10.163. In the absence of buoyancy ($\xi = 0$), the governing equations reduce to Blasius equation for flow over a flat plate.

Mucoglu and Chen [32] obtained solutions for two Prandtl numbers of 0.7 and 7. For each Prandtl number, the solutions were obtained for two angles, vertical 0° and 60° from vertical.

Figure 10.16 is a plot of the $Nu_x Re_x^{-1/2} Pr^{-1/3}$ as a function of the buoyancy parameter for two values of Prandtl number, generated from the graphical results of Ref. [32]. As can be seen, in the range considered, Prandtl number effects can be reasonably approximated by its dependence on the one-third power. For both isothermal and uniform heat flux boundary conditions, the results show that the local Nusselt number and friction factor increase with increasing buoyancy force for favorable pressure gradient and decrease with increasing buoyancy force for the adverse pressure gradient.

Problems

10.1 Derive Equation 10.98.

10.2 Derive Equation 10.99.

10.3 Numerically obtain the solution for
   a. Velocity and temperature distribution for free convection from an isothermal vertical plate
   b. The local and average Nusselt numbers

10.4 Derive Equations 10.104 and 10.105.

10.5 Numerically obtain the solution for
   a. Velocity and temperature distribution for free convection from a vertical plate, subject to a uniform heat flux
   b. The local and average Nusselt numbers

10.6 Assuming local similarity, obtain the numerical solution to velocity and temperature distribution for a vertical cylinder, and calculate the Nusselt number and compare with the results of Nagendra.

10.7 Obtain the numerical solution to velocity and temperature distribution for a vertical cylinder.

10.8 Solve Equations 10.107 and 10.108 and determine the Nusselt number.

10.9 Determine the solution for velocity and temperature distribution for natural convection from a sphere in the limit of large Prandtl number.

10.10 Determine the average Nusselt number for mixed convection on a flat plate inclined at an angle of 45°, for Prandtl of 1 and $Re = 1000$.

10.11 Repeat problem 10 for $Pr = 7$.

10.12 Repeat problem 10 for $Re = 100$.

10.13 The momentum and energy equations for natural convection on a horizontal flat plate are

$$f''' + \frac{3}{5} ff'' - \frac{1}{5} f'^2 = \frac{2}{5}(z - \eta T^{*\prime})$$

$$z' = T^*$$

$$T^{*\prime\prime} + \frac{3}{5} Prf T^{*\prime} = 0$$

The boundary conditions are

$$\eta = 0, \quad f = f' = 0, \quad z' = T^* = 1$$
$$\eta \to \infty, \quad f' = 0, \quad T^* = 0$$

Note that z is the integral of temperature with respect to η:
   a. Write the finite difference form of the governing equations and the boundary conditions.
   b. Numerically determine $T^{*\prime}$ at $\eta = 0°$ for $Pr = 0.72$, using the subroutine TRIDI or iterative capabilities of Excel®.

10.14 An array of rectangular fins are to be cooled by natural convection. Assume that the process can be approximated by fully developed natural convection between two parallel plates, a distance *s* apart:
a. Simplify the governing equation and the boundary condition.
b. Obtain the solution numerically.

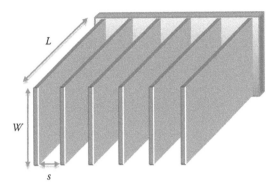

References

1. *Archimedes—Wikipedia, The Free Encyclopedia* (n.d.). http://en.wikipedia.org/wiki/Archimedes (retrieved on December 22, 2011).
2. *Buoyancy—Wikipedia, The Free Encyclopedia* (n.d.). http://en.wikipedia.org/wiki/Buoyancy (retrieved on December 22, 2011).
3. *Eureka (Word)—Wikipedia, The Free Encyclopedia* (n.d.). http://en.wikipedia.org/wiki/Eureka_ (word) ((retrieved on December 22, 2011).
4. *Gulf Stream—Wikipedia, The Free Encyclopedia* (n.d.). http://en.wikipedia.org/wiki/Gulf_Stream (retrieved on December 22, 2011).
5. *The Gulf Stream* (n.d.). http://fermi.jhuapl.edu/student/currents/gulf_stream.htm (retrieved on December 22, 2011).
6. Thermohaline circulation Gulf Stream shutdown (n.d.). http://thewe.cc/weplanet/poles/thermohaline_circulation.htm (retrieved on December 22, 2011)
7. Bejan, A. (1984) *Convection Heat Transfer*. New York: John Wiley [Print].
8. LeFevre, E. J. (1956) Laminar free convection from a vertical plane surface *Proceedings of the Ninth international Congress Applied Mechanics*, Brussels, Vol. 4, p. 168.
9. Churchill, S. W. and Usagi, R. (1972) A general expression for the correlation of rates of transfer and other phenomenon, *AIChE Journal* 18, 1121–1128.
10. Burmeister, L. C. (1993) *Convective Heat Transfer*, 2nd edn. New York: John Wiley & Sons.
11. Sparrow, E. M. and Gregg, J. L. (1956) Laminar free convection from a vertical plate with uniform surface heat flux,. *Transactions of ASME* 78, 435–440.
12. Sparrow, E. M. (1955) Laminar free convection on a vertical plate with prescribed non-uniform wall heat flux or prescribed non-uniform wall temperature, *NACA TN* 3508.
13. Millsaps, K. and Pohlhausen, K. (1958) The laminar free-convection heat transfer from the outer surface of a vertical circular cylinder, *Journal of Aeronautical Sciences* 25, 357–360.

14. Minkowycz, W. J. and Sparrow, E. M. (1974) Local nonsimilar solutions for natural convection on a vertical cylinder, *Journal of Heat Transfer, Transactions of ASME, Series C* 96, 178–183.
15. Nagendra, H. R., Tirunarayanan, M. A., and Ramachandran, A. (1971) Free convection heat transfer from vertical cylinders part I: Power law surface temperature variation, *Nuclear Engineering and Design* 16(2), 153–162.
16. Nagendra, H. R., Tirunarayanan, M. A., and Ramachandran, A. (1970) Laminar free convection from vertical cylinders with uniform heat flux, *Journal of Heat Transfer, Transactions on ASME Series C* 92(1), 191–194.
17. Nagendra, H. R., Tirunarayanan, M. A., and Ramachandran, A. (1969) Free convection heat transfer from vertical cylinders and wires, *Chemical Engineering Science* 24(9), 1491–1495.
18. Muntasser, M. A. and Mulligan, J. C. (1978) A local nonsimilarity analysis of free convection from a horizontal cylindrical surface, *Journal of Heat Transfer* 100, 165–167.
19. Hermann, R. (1954) Heat transfer by free convection from horizontal cylinders in diatomic gases., NACA Technical Member, 1366.
20. Churchill, S. W. and Chu, H. H. S. (1975) Correlating equations for laminar and turbulent free convection from a horizontal cylinder, *International Journal of Heat Mass Transfer* 18, 1049.
21. Acrivos, A. (1960) A theoretical analysis of laminar natural convection heat transfer to non-Newtonian fluids, *AIChE Journal* 6, 584–590.
22. Amato, W. S. and Tien, C. (1979) Free convection heat transfer from isothermal spheres, *International Journal of Heat and Mass Transfer* 15, 327–339.
23. Chellappa, A. K. and Singh, P. (1989) Possible similarity formulations for laminar free convection on a semi-infinite horizontal plate, *International Journal of Engineering Science* 27(2), 161–167.
24. Stewartson, K. (1958) On the free convection from a horizontal plate. *Z Angew Math Phys 9,* 276–287.
25. Rotem, Z. and Classen, L. (1969) Natural convection above unconfined horizontal surface. *Journal of Fluid Mechanics* 39, 173–192.
26. Chen, T. S., Tien, H. C., and Armaly, B. F. (1986) Natural convection on horizontal, inclined, and vertical plates with variable surface temperature or heat flux, *International Journal of Heat Mass Transfer* 29(10), 1465–1478.
27. Vliet, G. C. (1969) Natural convection local heat transfer on constant-heat-flux inclined surfaces, *Journal of Heat Transfer* 91, 511–516.
28. Vliet, G. C. and Ross, D. C. (1975) Turbulent natural convection on upward and downward facing inclined constant heat flux surfaces, *Journal of Heat Transfer* 97, 549–555.
29. Lin, I. T., Chen, C. C., and Yu, W.-S. (1989) Mixed convection from a horizontal plate to fluids of any Prandtl number, *Warme- und Stoffubertragung* 24, 225–234.
30. Schneider, W. and Wasel, M. G. (1985) Breakdown of the boundary-layer approximation for mixed convection above a horizontal plate, *International Journal of Heat Mass Transfer* 28(12), 2307–2313.
31. Wang, X. A. (1982) An experimental study of mixed, forced and free convection heat transfer from a horizontal plate to air, *Journal of Heat Transfer* 104, 139–144.
32. Mucoglu, A. and Chen, T. S. (1979) Mixed convection on inclined surfaces, *Journal of Heat Transfer* 101, 422.

11

Heat Exchangers

Heat exchangers are used to transfer heat between two or more fluids while preventing the fluids from mixing together. They are used in energy-converting devices like automobiles and air conditioners and many industries, including food processing, pharmaceutical, pulp and paper, steel, aerospace, chemical, marine, semiconductor, petrochemical, electronic, automotive, water treatment facilities, and textiles. There are a number of devices that are essentially heat exchangers, but are not necessarily known as such, for example, hot water heater, and home heating furnace.

Generally, there is a solid separating the two fluids. Heat is transferred by convection from the hot fluid to the surface of the solid separating the two, then conducted through the wall, and finally convected to the cold fluid. Because of their wide-ranging applications, heat exchangers come in many shapes and sizes. The three most widely used heat exchanger types are fin and tube, shell and tube, and plate heat exchangers as shown in Figure 11.1.

Car radiator, evaporator, or condenser in air conditioners are examples of fin and tube heat exchangers, where one fluid, typically a liquid, flows in the tubes, while the second fluid flow, typically a gas, flows over the tubes and in between the fins. They are generally used to transfer heat between a liquid and a gas and the fins are used to compensate for the lower heat transfer coefficient on the gas side, by increasing the available heat transfer area.

The shell and tube heat exchangers consist of a number of tubes where typically the process fluid or high pressure fluid flows and a shell where the other fluid flows over the outer surface of the tubes. The tubes pass through two perforated sheets, called the tubesheet, on the two ends of the shell that serve to separate the shell and the tube side fluids. Often, baffles are used to help in mixing of the shell side fluid and providing additional support for the tubes.

Plate heat exchangers are made up of a series of thin parallel plates. The hot and cold fluids flow alternately between the adjoining plates, exchanging heat through the plates. The plates are often corrugated to enhance heat transfer by directing the flow and increasing turbulence. These exchangers have high heat transfer coefficients and area; the pressure drop is also typically low; and they often provide very high effectiveness. However, they have relatively low pressure capability.

Heat exchangers are also classified according to the direction of the flow. Figure 11.2 shows parallel, counter, and cross-flow arrangements. A large class of heat exchangers is called compact heat exchangers. They are called compact because the ratio of the area of the heat exchanger to its volume is large. Generally, if the ratio of area to the volume of a heat exchanger is between 300 and 700 m^2/m^3, the heat exchanger is considered to be a compact heat exchanger.

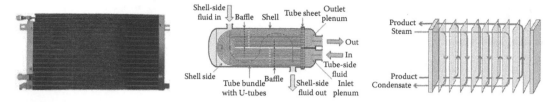

FIGURE 11.1
Finned-tube, shell and tube, and plate heat exchangers.

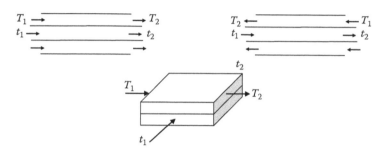

FIGURE 11.2
Parallel, counter, and cross-flow heat exchangers.

11.1 Heat Exchanger Analysis

The general heat exchanger analysis involves examination of the average behavior of fluids. Perhaps the simplest heat exchanger is a double-pipe heat exchanger where two concentric tubes carry the two fluids. The flow arrangement in a double-pipe heat exchanger can be parallel flow, where both the hot and cold fluids enter the heat exchanger at the same end and move in the same direction or counter flow, where the hot and cold fluids enter the heat exchanger at opposite ends and flow in opposite directions.

The general heat transfer analysis for heat exchangers is demonstrated first by considering a counter-flow heat exchanger shown in Figure 11.3, and then a parallel-flow heat exchanger shown in Figure 11.4. The hot fluid enters at T_1 and exits at T_2, and the cold fluid enters at t_1 and leaves at t_2. The temperature distributions for both types are shown in Figures 11.5 and 11.6. Consider a differential control volume of length Δx which provides area dA for the transfer of heat. The hot fluid temperature is T and the cold fluid is at temperature t. The heat transferred from the hot fluid to the cold fluid (δq) is first transferred by convection to the outer surface of the inner pipe, transferred through the pipe by conduction, and then transferred to the cold fluid by convection:

$$\delta q = U\,dA(T - t) \tag{11.1}$$

which is equal to the amount of heat that the hot fluid loses or the amount of heat that the cold fluid gains, if the heat loss to the ambient is zero. Applying the first law for a control volume to the hot fluid, which loses heat and whose inlet is at $x + \Delta x$ and outlet at x

$$-\delta q + (\dot{m}c_p)_h T(x + \Delta x) = (\dot{m}c_p)_h T(x) \tag{11.2}$$

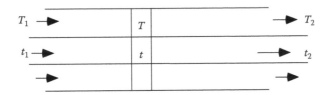

FIGURE 11.3
Parallel-flow heat exchanger.

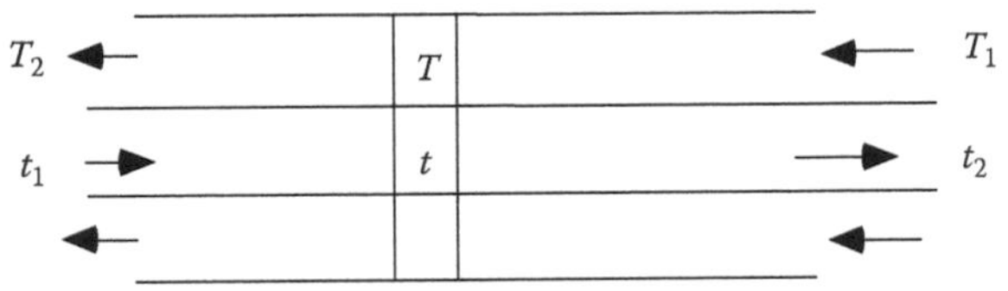

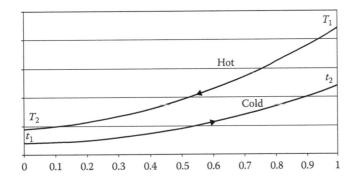

FIGURE 11.4
Counter-flow heat exchanger.

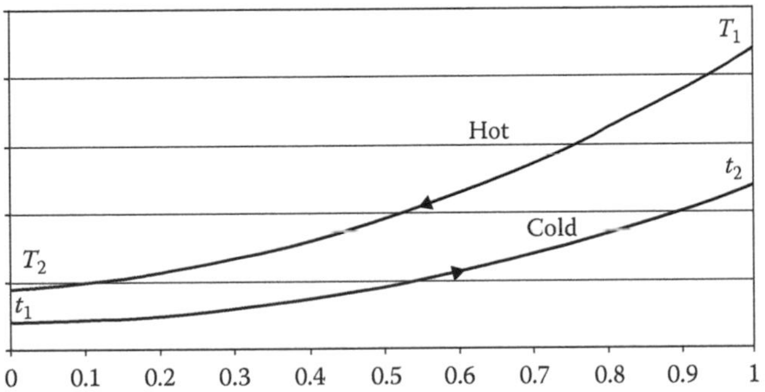

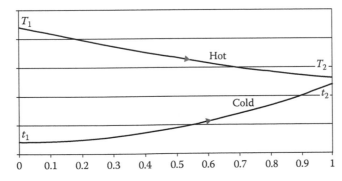

FIGURE 11.5
Temperature variation in counter-flow heat exchanger.

FIGURE 11.6
Temperature variation in parallel-flow heat exchanger.

and cold fluid

$$\delta q + (\dot{m}c_p)_c t(x) = (\dot{m}c_p)_c t(x + \Delta x) \tag{11.3}$$

Noting that

$$dT = T(x + \Delta x) - T(x) \tag{11.4}$$

$$dt = t(x + \Delta x) - t(x) \tag{11.5}$$

and defining

$$C = \dot{m}c_p \tag{11.6}$$

which is called the heat capacity of a fluid, the earlier two equations can be written as

$$\delta q = C_h dT \tag{11.7}$$

$$\delta q = C_c dt \tag{11.8}$$

Solving for the temperature differentials of the hot and cold fluids from the earlier two equations

$$dT = \frac{\delta q}{C_h} \tag{11.9}$$

$$dt = \frac{\delta q}{C_c} \tag{11.10}$$

and subtracting Equation 11.10 from Equation 11.9

$$d(T - t) = \delta q \left[\frac{1}{C_h} - \frac{1}{C_c} \right] \tag{11.11}$$

Substituting for δq from Equation 11.1 into Equation 11.11

$$d(T - t) = UdA(T - t) \left[\frac{1}{C_h} - \frac{1}{C_c} \right] \tag{11.12}$$

Integrating Equations 11.7, 11.8, 11.11, and 11.12 from the inlet ($x = 0$) to the outlet ($x = L$)

$$q = C_h(T_L - T_0) \tag{11.13}$$

$$q = C_c(t_L - t_0) \tag{11.14}$$

$$(T - t)_L - (T - t)_0 = q \left[\frac{1}{C_h} - \frac{1}{C_c} \right] \tag{11.15}$$

$$\ln \frac{(T - t)_L}{(T - t)_0} = UA \left[\frac{1}{C_h} - \frac{1}{C_c} \right] \tag{11.16}$$

A similar analysis can be performed for a parallel-flow heat exchanger shown in Figure 11.4 and the resulting equations become

$$q = -C_h(T_L - T_0) \tag{11.17}$$

$$q = C_c(t_L - t_0) \tag{11.18}$$

$$(T - t)_L - (T - t)_0 = -q\left[\frac{1}{C_h} + \frac{1}{C_c}\right] \tag{11.19}$$

$$\ln\frac{(T - t)_L}{(T - t)_0} = UA\left[\frac{1}{C_h} + \frac{1}{C_c}\right] \tag{11.20}$$

which are very similar to those for the counter-flow heat exchanger. Equations 11.13 through 11.16 or 11.17 through 11.20 can be rearranged in a number of different ways to arrive at different methods for analyzing heat exchangers. Three approaches are presented later, and although the choice of the method used is essentially dependent on what is known and what is to be determined, the use of heat exchanger efficiency provides insight into the best choice process and heat exchanger from the limitations imposed by the second law of thermodynamics.

11.2 Heat Exchanger Efficiency

A method for analyzing heat exchangers is using the concept of heat exchanger efficiency [1–10]. The heat exchanger efficiency is defined as the ratio of the actual rate of heat transfer in the heat exchanger, q, and the optimal rate of heat transfer, q_{opt},

$$\eta = \frac{q}{q_{opt}} = \frac{q}{UA\,(\overline{T} - \overline{t})} \tag{11.21}$$

The optimum (maximum) rate of the heat transfer is the product of UA of the heat exchanger under consideration and the arithmetic mean temperature difference (AMTD) in the heat exchanger

$$AMTD = \overline{T} - \overline{t} = \frac{T_1 + T_2}{2} - \frac{t_1 + t_2}{2} \tag{11.22}$$

which is the difference between the average temperatures of hot and cold fluids. The optimum heat transfer rate takes place in a balanced counter-flow heat exchanger [2]. The rate of heat transfer in any heat exchanger for the same UA and AMTD is always less than the optimum value of the heat transfer rate ($\eta \leq 1$) [1].

The number of transfer units (NTU) is defined as

$$NTU = \frac{UA}{C_{min}} \tag{11.23}$$

Therefore, the heat transfer in a heat exchanger can be calculated from

$$q = \eta NTU C_{min}\,(\overline{T} - \overline{t}) \tag{11.24}$$

Starting with Equation 11.16 for a counter-flow heat exchanger,

$$\frac{T_1 - t_2}{T_2 - t_1} = e^{UA\left[\frac{1}{C_h} - \frac{1}{C_c}\right]} \tag{11.25}$$

Rearranging this equation by subtracting denominators from the numerators and adding the numerators to the denominators results in

$$\frac{(T_1 - t_2) - (T_2 - t_1)}{(T_1 - t_2) + (T_2 - t_1)} = \frac{e^{UA\left[\frac{1}{C_h} - \frac{1}{C_c}\right]} - 1}{e^{UA\left[\frac{1}{C_h} - \frac{1}{C_c}\right]} + 1} \tag{11.26}$$

Substituting for temperature differences in the numerator from Equations 11.13 and 11.14 and in the denominator from Equation 11.22, rearranging the RHS

$$\frac{\frac{q}{C_h} - \frac{q}{C_c}}{2(\overline{T} - \overline{t})} = \frac{e^{\frac{NTU}{2}\left(\frac{C_{min}}{C_h} - \frac{C_{min}}{C_c}\right)} - e^{-\frac{NTU}{2}\left(\frac{C_{min}}{C_h} - \frac{C_{min}}{C_c}\right)}}{e^{\frac{NTU}{2}\left(\frac{C_{min}}{C_h} - \frac{C_{min}}{C_c}\right)} + e^{-\frac{NTU}{2}\left(\frac{C_{min}}{C_h} - \frac{C_{min}}{C_c}\right)}} \tag{11.27}$$

where the NTU is defined by Equation 11.23. Since

$$\tanh(x) = -\tanh(-x) \tag{11.28}$$

Regardless of which fluid has the minimum capacity, Equation 11.27 simplifies to

$$q = \frac{\tanh\left[\frac{NTU}{2}(1 - C_r)\right]}{\frac{NTU}{2}(1 - C_r)} UA(\overline{T} - \overline{t}) \tag{11.29}$$

where

$$C_r = \frac{C_{min}}{C_{max}} \tag{11.30}$$

Comparing this with Equation 11.21, we conclude that for a counter-flow heat exchanger, the efficiency is

$$\eta = \frac{\tanh\left[\frac{NTU}{2}(1 - C_r)\right]}{\frac{NTU}{2}(1 - C_r)} \tag{11.31}$$

Similar analysis for parallel-flow, shell and tube, and single stream heat exchangers results in an equation similar to Equation 11.31. Therefore, the efficiency of heat exchangers only depends on one nondimensional parameter and is given by the general expression

$$\eta = \frac{\tanh[Fa]}{Fa} \tag{11.32}$$

It is remarkable that the expression for the efficiency of these four commonly used and different heat exchangers has the same form as the efficiency of a constant area insulated tip fin. Fin analogy number, *Fa*, is a nondimensional group that characterizes the performance of different heat exchangers and is given in Table 11.1 for the four heat exchanger types.

The general form of the fin analogy number can therefore be written as

$$Fa = NTU \frac{\left(1 + mC_r{}^n\right)^{\frac{1}{n}}}{2} \tag{11.33}$$

TABLE 11.1

Fin Analogy Number of Various Heat Exchangers

Counter	Parallel	Single Stream	Single Shell
$Fa = NTU \dfrac{(1-C_r)}{2}$	$Fa = NTU \dfrac{(1+C_r)}{2}$	$Fa = \dfrac{NTU}{2}$	$Fa = \dfrac{NTU\sqrt{1+C_r^2}}{2}$

TABLE 11.2

Fin Efficiency of Various Heat Exchangers

$$\eta = \frac{\tanh\left(NTU\dfrac{(1+mC_r^n)^{\frac{1}{n}}}{2}\right)}{NTU\dfrac{(1+mC_r^n)^{\frac{1}{n}}}{2}}$$

	m	n	Maximum Error (%)
Counter flow	−1	1	0
Parallel	1	1	0
Single stream ($C_r = 0$)	1	1	0
Single shell and tube	1.	2	0
Cross flow, C_{max} unmixed, C_{min} mixed	1.2	4.4	<7.5
Cross flow, C_{max} mixed, C_{min} unmixed	1.35	4.02	<3
Cross flow, both mixed	1.2	2	<1.6
Cross flow, both unmixed	−0.1	0.37	<8.4

As shown later, the expressions for heat exchanger efficiency can be obtained from the available expressions for heat exchanger effectiveness. Fakheri [6] used regression analysis, to arrive at approximate values for m and n for different cross-flow heat exchangers, given in Table 11.2. Note that the m and n values listed for the cross-flow heat exchangers are not necessarily providing the absolute minimum errors, as minor adjustments were made to provide uniformity in the results.

The first four results are exact and thus the maximum error is zero. The maximum error reported for the cross-flow heat exchangers is for the range $0 < NTU < 3.0$ and typically occurred at large NTU values that correspond to large Fa values or low heat exchanger efficiency.

11.2.1 Rating Problem

If the size of the heat exchanger, the fluid capacities, and the inlet temperatures are known, then the objective is to determine the exit temperatures. This type of problem is called heat exchanger rating problem. To find the solution, Substituting for the mean temperature difference and then for hot and cold fluid temperature difference in terms of heat transfer and fluid capacity in Equation 11.21

$$q = \frac{UA}{2}\eta\left(2T_1 - (T_1 - T_2) - 2t_1 - (t_2 - t_1)\right) = UA\eta\left(T_1 - t_1 - \frac{q}{2}\left(\frac{1}{C_h} + \frac{1}{C_c}\right)\right) \qquad (11.34)$$

and solving for q results in

$$q = \frac{C_{min}(T_1 - t_1)}{\dfrac{1}{NTU\eta} + \dfrac{1+C_r}{2}} \tag{11.35}$$

Everything on the RHS of the equation is known, allowing the calculation of the heat transfer in the heat exchanger, which along with Equations 11.17 and 11.18 can be used to determine the exit temperatures.

11.2.2 Sizing Problem

When the fluid capacities and their inlet and exit temperatures are known or can be determined, then determining the size of the heat exchanger becomes the objective. This type of problem is called heat exchanger sizing problem. The heat transfer and the temperature difference are known, then from Equation 11.24

$$NTU\,\eta = \frac{q}{C_{min}(\bar{T} - \bar{t})} \tag{11.36}$$

From Equation 11.32

$$Fa = \tanh^{-1}\left[\eta NTU \frac{Fa}{NTU}\right] \tag{11.37}$$

Substituting from Equation 11.36 for efficiency and from Equation 11.33 for Fa results in

$$NTU = \frac{2}{\left(1+mC_r{}^n\right)^{1/n}} \tanh^{-1}\left[\frac{q}{C_{min}(\bar{T}-\bar{t})} \frac{\left(1+mC_r{}^n\right)^{1/n}}{2}\right] \tag{11.38}$$

which can also be rearranged into

$$NTU = \frac{2}{\left(1+mC_r{}^n\right)^{1/n}} \tanh^{-1}\left[\frac{1}{\dfrac{C_{min}(T_i-t_i)}{q} - \dfrac{1+C_r}{2}} \frac{\left(1+mC_r{}^n\right)^{1/n}}{2}\right] \tag{11.39}$$

and thus Equations 11.38 and 11.39 are explicit equations from which NTU can be determined.

Thus both types of heat exchanger problems can be directly and conveniently solved using the concept of heat exchanger efficiency, without any need for charts or complicated equations, using the same general expression for efficiency.

Example 11.1

Water at a rate of 1.5 kg/s is heated from 25°C to 100°C in a finned-tube cross-flow heat exchanger using exhaust gases that enter at 250°C and leave at 120°C. The exhaust gases specific heat is 1080 J/kg K, and that of water is 4200 J/kg K, and the overall heat transfer coefficient based on the gas side surface area is 150 W/m² K. Determine the gas side heat exchanger area.

Here we solve this problem using the concept of heat exchanger efficiency, using the approximate expressions of Table 11.2, $m = -0.1$, $n = 0.37$

$$C_c = 1.5 \times 4200 = 4197 \text{ W/K}$$

$$\bar{T} - \bar{t} = \frac{(250 + 120)}{2} - \frac{(100 + 25)}{2} = 122.5$$

$$q = C_c(t_2 - t_1) = 472,500 \text{ W}$$

$$C_h = \frac{q}{T_1 - T_2} = 3634.62 \text{ W/K}$$

$$C_r = 0.58$$

For cross-flow heat exchanger with both unmixed

$$NTU = \frac{2}{\left(1 + mC_r{}^n\right)^{1/n}} \tanh^{-1}\left[\frac{q}{C_{min}\left(\bar{T} - \bar{t}\right)} \frac{\left(1 + mC_r{}^n\right)^{1/n}}{2}\right] = 1.13$$

$$A = \frac{C_{min}NTU}{U} = \frac{3634.62 \times 1.13}{150} = 27.42 \text{ m}^2$$

Example 11.2

Hot exhaust gases, with a thermal capacity of 3634.62 W/K, enter a finned-tube cross-flow heat exchanger at 250°C and are used to heat water with a thermal capacity of 6300 W/K entering the heat exchanger at 20°C. The overall heat transfer coefficient based on the gas side surface area is 150 W/m² K, and the heat exchanger area is 27.42 m². Determine the rate of heat transfer.

This is the reverse problem, and an example of a heat exchanger rating problem:

$$Fa = NTU\frac{\left(1 + mC_r{}^n\right)^{1/n}}{2} = 1.13\frac{(1 - 0.1 \times 0.58^{0.37})^{1/0.37}}{2} = 0.45$$

$$\eta = \frac{\tanh(Fa)}{Fa} = 0.938$$

$$q = \frac{C_{min}(T_1 - t_1)}{\dfrac{1}{NTU\eta} + \dfrac{1 + C_r}{2}} = \frac{3634.62(250 - 25)}{\dfrac{2}{1.13 \times 0.938} + \left(\dfrac{1 + 0.58}{2}\right)} = 472,500 \text{ W}$$

which is the same as Example 11.1.

11.3 Log Mean Temperature Difference Approach

Dividing Equation 11.15 by Equation 11.16, or Equation 11.19 by Equation 11.20, and solving for the rate of heat transfer

$$q = UA\frac{(\Delta T)_L - (\Delta T)_0}{\ln\dfrac{(\Delta T)_L}{(\Delta T)_0}} = UA \ LMTD \tag{11.40}$$

where
  $(\Delta T)_L$ is the temperature difference between the hot and cold fluids at one end
  $(\Delta T)_0$ is the temperature difference at the opposite end of the heat exchanger

Therefore for counter-flow or parallel-flow heat exchangers, the rate of heat transfer in the heat exchanger is equal to the product of area, and the overall heat transfer coefficient and a temperature difference. The temperature difference is called the logarithmic mean temperature difference (LMTD). For a counter-flow heat exchanger, shown in Figure 11.4, LMTD is given by

$$LMTD = \frac{(T_2 - t_1) - (T_1 - t_2)}{\ln\dfrac{(T_2 - t_1)}{(T_1 - t_2)}} \tag{11.41}$$

and for a parallel-flow heat exchanger, shown in Figure 11.5, LMTD is

$$LMTD = \frac{(T_1 - t_1) - (T_2 - t_2)}{\ln\dfrac{(T_1 - t_1)}{(T_2 - t_2)}} \tag{11.42}$$

We also saw that the same equation applies to flow in a pipe with constant surface temperature (isothermal pipe). The simplicity of this equation and its applicability to three different cases led to its adoption for other types of heat exchangers by the introduction of a correction factor, F. Therefore, for all heat exchangers

$$q = U \ A \ F \ LMTD \tag{11.43}$$

where the LMTD used is for a counter-flow heat exchanger having the same inlet and exit fluid temperatures as the heat exchanger being considered.

An alternative view of this approach is that the relevant driving temperature potential in a heat exchanger is

$$\Delta T = F \ LMTD \tag{11.44}$$

In heat exchanger analysis, the LMTD is widely used for heat exchanger selection (sizing problems) in which the temperatures are known and the size of the heat exchanger is required. In their pioneering paper, Bowman et al. [11] compiled the available data, and presented a series of equations and charts for the determination of F for a variety of heat exchangers by choosing to express F in terms of two nondimensional variables,

$$R = \frac{T_1 - T_2}{t_2 - t_1} \tag{11.45}$$

and

$$P = \frac{t_2 - t_1}{T_1 - t_1} \tag{11.46}$$

The F correction factor for a shell and tube heat exchanger (Figure 11.7) is given by

$$F = \frac{\sqrt{R^2 + 1}}{R - 1} \frac{\ln\dfrac{1 - P}{1 - PR}}{\ln\dfrac{\left(\dfrac{2}{P}\right) - 1 - R + \sqrt{R^2 + 1}}{\left(\dfrac{2}{P}\right) - 1 - R - \sqrt{R^2 + 1}}} \tag{11.47}$$

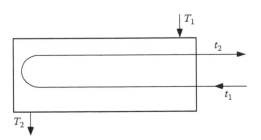

FIGURE 11.7
Shell and tube heat exchanger.

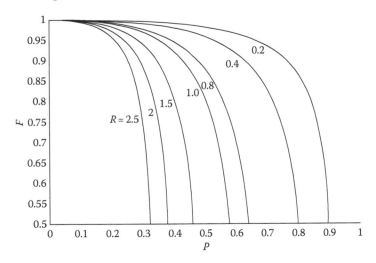

FIGURE 11.8
The F correction factor for a shell and tube heat exchanger.

and also shown in Figure 11.8. Similar charts for other heat exchanger types can be found in many heat transfer texts. Note that for a shell and tube heat exchanger, the value of F is independent of whether the hot or cold fluids are flowing in the tube or shell. The decision as to which fluid should be on the tube or shell side depends on a number of factors. In general, if the fluid is corrosive, or at very high pressure, or has significant fouling potential, it should be on the tube side [12].

For a cross-flow heat exchanger with one mixed and one unmixed

$$F = \frac{\ln\dfrac{1-P}{1-PR}}{(1-R)\ln\left[1+\dfrac{1}{R}\ln(1-PR)\right]} \tag{11.48}$$

and for a cross-flow heat exchanger with both fluids mixed, an implicit formula for the calculation of F is given by

$$F = \frac{\ln\dfrac{1-P}{1-PR}}{PR-P}\sum_{i=0}^{\infty}\sum_{j=0}^{\infty}(-1)^{(i+j)}\frac{(i+j)!}{i!(i+1)!j!(j+1)!}(F)^{-(i+j)}\left(\frac{\ln\dfrac{1-P}{1-PR}}{R-1}\right)^{(i+j)}(R)^j \tag{11.49}$$

which must be solved iteratively for given values of P and R.

Although no rationale was given in Ref. [11] for the selection of R and P, the two variables were later interpreted as the capacity ratio and a measure of heat exchanger effectiveness, respectively [13].

The LMTD method is convenient to use in heat exchanger analysis when the inlet and the outlet temperatures of the hot and cold fluids are known or can be determined from an energy balance. Once the mass flow rates and the overall heat transfer coefficient are available, the heat transfer surface area of the heat exchanger can be determined.

As can be seen from the results for shell and tube heat exchangers, the LMTD correction factor F is a strong function of P and R, which reduces the accuracy of reading the charts, particularly in the steep part of the curves ($F < 0.8$).

Example 11.3

The ship engine oil at the rate of 1.5 kg/s is to be cooled from 110°C to 45°C using sea water available at 20°C at a rate of 2 kg/s. If the overall heat transfer coefficient is 320 W/m² and the oil specific heat is 1.9 kJ/kg K, determine the area of a

  a. Double-pipe parallel-flow heat exchanger
  b. Double-pipe counter-flow heat exchanger
  c. Shell and tube heat exchanger

$$q = C_c(t_2 - t_1) = C_h(T_1 - T_2) = UA\,F\,LMTD$$

The solution is shown in the following table.

m_{oil}	1.50
c_{poil}	1900.00
m_w	2.00
c_{pw}	4180.00
T_1	110.00
T_2	45.00
t_1	20.00
t_2	42.16
U	320.00
Q	185,250.00
Counter flow	
LMTD	42.91
A	13.49
Parallel	
LMTD	25.22
A	22.95
Shell and tube	
P	0.246
R	2.933
F	0.84
A	16.01

Example 11.4

A shell and tube heat exchanger with one shell pass and eight tube passes raises the temperature of 100,000 lbm/h of water from 180°F to 300°F. The tube side fluid is air ($c_p = 0.24$ btu/lbm°F) which enters at 650°F and leaves at 350°F. If $U = 5$ btu/h ft²°F, find the heat exchanger area:

$$q = UA \, F \, LMTD$$

$$LMTD = \frac{(650 - 300) - (350 - 180)}{\ln\left[\dfrac{650 - 300}{350 - 180}\right]} = 249.3$$

$$q = \dot{m}c_p\Delta T = 100{,}000 \times 1 \times (300 - 180) = 1.2 \times 10^7$$

$$P = \frac{t_2 - t_1}{T_1 - t_1} = \frac{300 - 180}{650 - 180} = 0.255$$

$$R = \frac{T_1 - T_2}{t_2 - t_1} = \frac{650 - 350}{300 - 180} = 2.5$$

$$F = 0.89$$

$$1.2 \times 10^7 = 5 \times A \times 0.89 \times 249.3$$

$$A = 10{,}803.4 \text{ m}^2$$

11.4 Effectiveness NTU Method

The effectiveness NTU method is used for cases where fluid mass flow rates, the inlet temperatures, and the type and size of the heat exchanger are specified and the amount of heat transfer and exit temperatures are to be determined. This is called a rating problem and since the exit temperatures are unknown, the use of LMTD approach would require an iterative approach. To avoid this procedure, Kays and London [13] proposed the effectiveness NTU method, which is explained next. For any heat exchanger, absent energy loss to the surroundings,

$$C_h(T_1 - T_2) = C_c(t_2 - t_1) \tag{11.50}$$

This equation also shows that the fluid with higher capacity will experience the smaller temperature change. The maximum temperature change that a fluid may experience cannot exceed the difference between the inlet temperatures of hot and cold fluids, as required by the second law; otherwise, the cold fluid will exit the heat exchanger at a temperature higher than the inlet temperature of the hot fluid or the hot fluid will exit the heat exchanger at a temperature lower than the inlet temperature of the cold fluid, both of which violate the second law of thermodynamics. If $(T_1 - t_1)$ is experienced by a fluid in a heat exchanger, it has to be the fluid with smaller capacity; otherwise, the fluid with lower capacity will experience even larger temperature change, which is the violation of the second law. Therefore, the maximum amount of heat that can be possibly transferred in any heat exchanger is

$$q_{max} = C_{min}(T_1 - t_1) \tag{11.51}$$

Note that this maximum rate of heat transfer is different than the optimum rate of heat transfer used to define heat exchanger efficiency. The actual heat transfer rate in a heat exchanger will be less than this value; therefore, the heat exchanger effectiveness can be defined as

$$\varepsilon = \frac{q}{q_{max}} = \frac{q}{C_{min}(T_1 - t_1)} \tag{11.52}$$

substituting for heat transfer in terms of heat capacity and temperature difference, then effectiveness can be written as

$$\varepsilon = \frac{(\Delta T)_{fluid\ with\ min\ C}}{(T_1 - t_1)} = \frac{(\Delta T)_{fluid\ with\ max\ C}}{C_r(T_1 - t_1)} \tag{11.53}$$

From Equation 11.53, effectiveness can be viewed as the temperature change experienced by the fluid having the smaller capacity divided by the maximum temperature difference. It is also equal to the temperature change experienced by the fluid having the larger capacity divided by the maximum temperature difference times the capacity ratio. Therefore, if the effectiveness is known, the temperature change and thus the exit temperatures of both fluids can be quickly calculated.

The effectiveness of a heat exchanger depends on the geometry of the heat exchanger as well as the flow arrangement. Therefore, different types of heat exchangers have different effectiveness relations. In the following we derive the expression for the effectiveness of the counter-flow heat exchanger. Rearranging Equation 11.16

$$\frac{T_1 - t_2}{T_2 - t_1} = e^{UA\left[\frac{1}{C_h} - \frac{1}{C_c}\right]} \tag{11.54}$$

Adding and subtracting t_1 to the numerator and T_1 to the denominator and dividing numerator and denominator by $T_1 - t_1$

$$\frac{1 - \dfrac{t_2 - t_1}{T_1 - t_1}}{1 - \dfrac{T_1 - T_2}{T_1 - t_1}} = e^{UA\left[\frac{1}{C_h} - \frac{1}{C_c}\right]} \tag{11.55}$$

Substituting for temperature differences from Equations 11.13 and 11.14 in terms of heat transfer results in

$$\frac{1 - \dfrac{C_{min}}{C_c} \dfrac{q}{C_{min}(T_1 - t_1)}}{1 - \dfrac{C_{min}}{C_h} \dfrac{q}{C_{min}(T_1 - t_1)}} = e^{UA\left[\frac{1}{C_h} - \frac{1}{C_c}\right]} \tag{11.56}$$

which can be rearranged into

$$\varepsilon = \frac{1 - e^{\frac{UA}{C_{min}}\left[\frac{C_{min}}{C_h} - \frac{C_{min}}{C_c}\right]}}{\dfrac{C_{min}}{C_c} - \dfrac{C_{min}}{C_h} e^{\frac{UA}{C_{min}}\left[\frac{C_{min}}{C_h} - \frac{C_{min}}{C_c}\right]}} \tag{11.57}$$

Then, regardless of whether $C_c = C_{min}$ or $C_h = C_{min}$, it is easy to show that Equation 11.57 simplifies to

$$\varepsilon = \frac{1 - e^{-NTU[1-C_r]}}{1 - C_r e^{-NTU[1-C_r]}} \tag{11.58}$$

Using a similar approach, the relation between effectiveness and NTU for many other heat exchangers has been developed and some of the more widely used expressions are given in Table 11.3. Effectiveness increases rapidly with NTU for small values (up to about NTU = 1.5) but rather slowly for NTU > 3. Efficiency expressions derived from the effectiveness ones are listed in Table 11.3, while simpler expressions were provided in Table 11.2.

Example 11.5

In Example 11.4, if the mass flow rate of air drops by 30%, what is the exit temperature?

$$q = (\dot{m}c_p)_w(\Delta T)_w = (\dot{m}c_p)_a(\Delta T)_a$$

$$\dot{m}_a = \frac{1.2 \times 10^7}{0.24(650 - 350)} = 166{,}667$$

$$\dot{m}_a = 0.7 \times 166{,}667 = 116{,}667$$

$$C_H = 116{,}667 \times 0.24 = 28{,}000$$

$$C_C = 10^5 \times 1 = 100{,}000$$

$$C_r = 0.28$$

$$NTU = \frac{UA}{C_{min}} = \frac{5 \times 10{,}940}{28{,}000} = 1.954$$

$$\varepsilon = 2 \times \left[1 + C_r + \sqrt{1 + C_r^2} \, \frac{1 + e^{-NTU\sqrt{(1+C_r^2)}}}{1 - e^{-NTU\sqrt{(1+C_r^2)}}} \right] = 0.76 = \frac{q}{q_{max}}$$

$$q = 10{,}001{,}600$$

$$t_2 = 280$$

$$T_2 = 293$$

Example 11.6

The condenser of a large steam power plant is a shell and tube heat exchanger, having a single shell and 30,000 tubes, each making two passes. The tubes are thin wall, $D = 25$ mm, where steam condenses on their outside diameter, with an associated heat transfer coefficient of 11,000 W/m^2 K. The total heat removal rate is 2×10^9, which is accomplished by passing cold water through the tubes at a rate of 30,000 kg/s. The water enters at 20°C while steam condenses at 50°C. What is the temperature of cooling water exiting and what is the required tube length per pass?

The spreadsheet solution and the equations used are shown in the table. The solution is obtained by using both LMTD and ε-NTU approach.

TABLE 11.3

Heat Exchanger Effectiveness and Efficiency Expressions

Parallel flow	$\varepsilon = \dfrac{1 - \exp[-NTU(1+C_r)]}{(1+C_r)}$
Counter flow	$\varepsilon = \dfrac{1 - \exp[-NTU(1-C_r)]}{1 - C_r \exp[-NTU(1-C_r)]}$
Shell and tube	$\varepsilon = 2\left[1 + C_r + \left(1+C_r^2\right)^{1/2} \dfrac{1 + \exp\left[-NTU\left(1+C_r^2\right)^{1/2}\right]}{1 - \exp\left[-NTU\left(1+C_r^2\right)^{1/2}\right]} \right]^{-1}$
Balanced counter flow	$\varepsilon = \dfrac{NTU}{1+NTU}$
Single stream $C_r = 0$	$\varepsilon = 1 - e^{-NTU}$
C_{max} unmixed, C_{min} mixed	$\varepsilon = 1 - \exp\left[-\dfrac{1}{C_r}\left[1 - \exp[-C_r NTU]\right] \right]$
C_{max} mixed, C_{min} unmixed	$\varepsilon = \dfrac{1}{C_r}\left[1 - \exp\left[-C_r\left[1 - \exp(-NTU)\right] \right] \right]$
Both mixed	$\varepsilon = \dfrac{NTU}{\dfrac{NTU}{1 - \exp[-NTU]} + \dfrac{C_r NTU}{1 - \exp[-C_r NTU]} - 1}$
Both unmixed	$\varepsilon = 1 - \exp\left[\dfrac{1}{C_r} NTU^{.22}\left[\exp[-C_r NTU^{.78}] - 1\right] \right]$
Parallel flow	$\eta = \dfrac{\tanh\left[\dfrac{NTU}{2}(1+C_r)\right]}{\dfrac{NTU}{2}(1+C_r)}$
Counter flow	$\eta = \dfrac{\tanh\left[\dfrac{NTU}{2}(1-C_r)\right]}{\dfrac{NTU}{2}(1-C_r)}$
Shell and tube	$\eta = \dfrac{\tanh\left[\dfrac{NTU}{2}\sqrt{1+C_r^2}\right]}{\dfrac{NTU}{2}\sqrt{1+C_r^2}}$
Balanced counter flow	$\eta = 1$
Single stream $C_r = 0$	$\eta = \dfrac{\tanh\left[\dfrac{NTU}{2}\right]}{\dfrac{NTU}{2}}$
C_{max} unmixed, C_{min} mixed	$\eta = \dfrac{1}{NTU} \dfrac{1}{\dfrac{1}{1 - \exp\left[-\dfrac{1}{C_r}\left[1 - \exp[-C_r NTU]\right]\right]} - \dfrac{(1+C_r)}{2}}$
C_{max} mixed, C_{min} unmixed	$\eta = \dfrac{1}{NTU} \dfrac{1}{\dfrac{C_r}{1 - \exp\left[-C_r\left[1 - \exp[-NTU]\right]\right]} - \dfrac{(1+C_r)}{2}}$

TABLE 11.3 (continued)

Heat Exchanger Effectiveness and Efficiency Expressions

Both mixed	$\eta = \dfrac{1}{NTU} \dfrac{1}{\dfrac{1}{1-\exp[-NTU]} + \dfrac{C_r}{1-\exp[-C_r NTU]} - \dfrac{1}{NTU}} - \dfrac{(1+C_r)}{2}$
Both unmixed	$\eta = \dfrac{1}{NTU} \dfrac{1}{1-\exp\left[\dfrac{1}{C_r]NTU^{0.22}}\right]\left[\exp[-C_r NTU^{0.78}]-1\right]} - \dfrac{(1+C_r)}{2}$

Parameter	Value	
t_1	20	
t average guess	27	
c_p	4179	
μ	8.55E – 04	
k	0.613	
Pr	5.83	
Q	2.00E + 09	
m_c	30,000	
C_c	1.25E + 08	
t_2	35.95	$t_2 = t_1 + \dfrac{q}{C_c}$
t average	27.98	
N	30,000	
D	0.025	
T_s	50	
Re	59,566.76	
Nu	307.61	$Nu = 0.023 \times Re^{0.8}Pr^{0.4}$
h_i	7542.56	$h = \dfrac{kNu}{d}$
h_o	11,000.00	
U	4474.47	$U = \dfrac{1}{\dfrac{1}{h_i} + \dfrac{1}{h_o}}$
LMTD	21.02	
P	0.53	
R	0.00E + 00	
F	1.00	
A	**21,260.01**	$A = \dfrac{q}{UF\,LMTD}$
L	4.51	$L = \dfrac{A}{2N\pi D}$
Or		
ε	0.53	$\varepsilon = \dfrac{q}{C_{min}(T_1 - t_1)}$
NTU	0.76	$NTU = -\ln(1 - \varepsilon)$
UA	9.51E + 07	
A	**21,260.01**	

The LMTD correction factor, the heat exchanger effectiveness, and heat exchanger efficiency are all derived from the same basic set of equations and can therefore be related to one another. The relations are listed in Table 11.4.

TABLE 11.4

Correlations for Conversion of the Three Heat Exchanger Performance Measures

	F	ε	η
F		$F = \dfrac{1}{NTU} \dfrac{\ln\dfrac{1-\varepsilon C_r}{1-\varepsilon}}{1-C_r}$	$F = \dfrac{\tanh^{-1}\left[\eta\dfrac{NTU}{2}(1-C_r)\right]}{\dfrac{NTU(1-C_r)}{2}}$
ε	$\varepsilon = \dfrac{e^{F\,NTU(1-C_r)}-1}{e^{F\,NTU(1-C_r)}-C_r}$		$\varepsilon = \dfrac{1}{\dfrac{1}{\eta NTU}+\dfrac{(1+C_r)}{2}}$
η	$\eta = \dfrac{\tanh\left[\dfrac{F\,NTU(1-C_r)}{2}\right]}{\dfrac{NTU}{2}(1-C_r)}$	$\eta = \dfrac{1}{NTU}\dfrac{1}{\dfrac{1}{\varepsilon}-\dfrac{(1+C_r)}{2}}$	

11.5 Heat Exchanger Networks

As can be seen from the LMTD correction factor plot for shell and tube heat exchangers, for each R value, as P increases, the LMTD correction factor, F, exhibits a large decrease by a relatively small increase in P [14–17]. The charts also show that not all P and R values result in a real value for correction factor, which means it is not possible to have the fluid exit the heat exchangers at all second law acceptable values of temperature. Even if a feasible solution exits, the F may be too small for efficient operation. The general rule of thumb is that F should be kept above 0.8. Therefore, it often becomes necessary to use more than one heat exchanger to meet all the design criteria.

Consider a fluid at 0°C that is to be heated to 360°C using a hot fluid available at 410°C [14]. Table 11.5 shows the LMTD correction factor for a shell and tube heat exchanger as a function of the exit temperature of the hot fluid. As can be seen, the exit temperature of the hot fluid in a single shell and tube heat exchanger cannot go below 320°C, even though the values below this threshold do not violate the second law. If exit temperature below this threshold was a design constraint, then it needs to be accomplished in a multi-shell and tube heat exchanger. Also, for exit temperature less than 360°C, the LMTD correction factor is less than 0.8.

TABLE 11.5

F Variation with the Exit Temperature of the Hot Fluid for a Shell and Tube Heat Exchanger

t_1	0	0	0	0	0	0	0	0
t_2	360	360	360	360	360	360	360	360
T_1	410	410	410	410	410	410	410	410
T_2	310	320	330	340	350	360	370	380
P	0.24	0.22	0.20	0.17	0.15	0.12	0.10	0.07
R	3.60	4.00	4.50	5.14	6.00	7.20	9.00	12.00
F	—	—	0.59	0.70	0.77	0.82	0.87	0.91

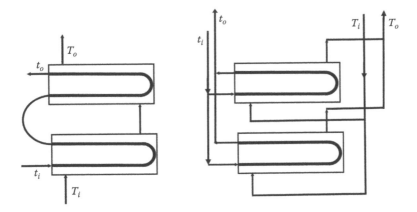

FIGURE 11.9
Heat exchangers in series and parallel.

Heat exchangers can be arranged in parallel or in series, as shown in Figure 11.9, or a combination of the two, where one of the streams is in series and the other in parallel. The parallel-flow arrangement reduces the flow rate and thus pressure drop in the individual heat exchangers and is primarily used when pressure drop is a concern. The series arrangement is used when the LMTD correction factor is too low and/or the size of the single heat exchanger becomes too large.

11.5.1 Heat Exchangers in Series

One of the challenges in designing shell and tube heat exchanger systems is the determination of the number of shells to meet the design specifications. In many applications, the value of F for a single shell heat exchanger will be infeasible or unacceptably low necessitating a multi-shell arrangement. The general rule of thumb for shell and tube heat exchangers is to have $F > 0.8$ to avoid the steep areas of the chart and also have acceptable size [14]. Several methods are available for determining the number of shells. In all these methods, the same correction factor is used for all the shells and the shells are assumed to be identical. Even with these limiting assumptions, an explicit equation for determining the number of shells has not been available. These assumptions do not necessarily lead to optimum selection, and relaxing them may indeed lead to enhancements in the overall rate of heat transfer and/or reduction in the cost (size) of the heat exchangers.

There are a number of methods for determining the number of shells. Traditionally a tedious trial and error process is used, in which the number of shells are progressively increased and the new value of F is determined from the appropriate chart, until an acceptable solution is arrived at.

The "stepping-off" method [15] is a graphical technique that results in a design in which the hot fluid temperature is never less than the cold fluid temperature or there is no "temperature cross." One difficulty associated with this method is that one has no control over the overall correction factor. Bell [15] states that following this procedure usually results in F being close to 0.8. In this method, the inlet and exit temperatures are plotted on the y axis and assumed to vary linearly as a function of an arbitrary x value, which can be considered to be the distance along an equivalent counter-flow heat exchanger, or more accurately, proportional to the amount of heat transfer. Once the two lines, referred to as the operating lines, are plotted, starting with the exit temperature of the cold fluid at $x = 0$, a horizontal line is drawn until it intersects the hot fluid operating line. At that point, a vertical line is drawn to intersect the cold fluid operating line. At the intersection point, another horizontal line is drawn and the process is continued until the vertical line intersects the cold fluid operating line at a value equal to or less than the inlet temperature of the cold fluid. The number of horizontal segments is equal to the number of shells needed.

Example 11.7

Consider a fluid at 0°C that is to be heated to 360°C using a hot fluid available at 410°C that is to exit at 110°C using shell and tube heat exchangers.

The graphical solution using "stepping off" is shown in Figure 11.10 and results in a feasible solution, requiring five shells.

An approximate solution for the determination of the number of shells [4] is

$$N \approx \frac{\sqrt{(T_1 - T_2)^2 + (t_2 - t_1)^2}}{[\bar{T} - \bar{t}]} \left[\left(\frac{80}{F} - 68.89 \right)^{1/2} - 3.33 \right]^{-1/2} \tag{11.59}$$

The value predicted from Equation 11.59, rounded up to the nearest integer, is the number of shells for the specified correction factor. Since Equation 11.59 is an approximate solution, it is recommended that other solutions near N be also checked. Assuming a value of 0.8 for the F correction factor, the number of shells predicted for Example 11.7 by Equation 11.59 becomes 3.91, which rounded up to the next digit 4. For $F = 0.85$, the number of needed shells increases to 5.

Fakheri [8] has shown that for a solution to exist the minimum number of shells in multishell and tube heat exchanger is given by

$$N > \frac{\sqrt{1 + C_r^2}}{(1 - C_r)} \tanh^{-1} \left[\frac{\Delta T_{min}}{\bar{T} - \bar{t}} \frac{(1 - C_r)}{2} \right] \tag{11.60}$$

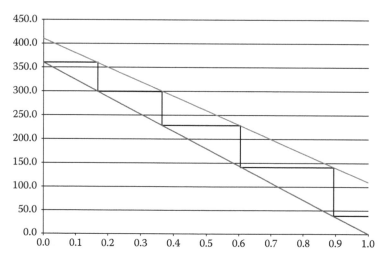

FIGURE 11.10
Stepping off method.

where ΔT_{min} is the temperature change of the fluid with minimum capacity. Since this fluid experiences the higher temperature change, then $\Delta T_{min} = max(\Delta T_h, \Delta T_c)$. From this expression, the minimum number of heat exchangers needed to have a feasible design is the next integer higher than value on the RHS of Equation 11.60. Note that a feasible solution is not necessarily the best solution, as it may have low efficiency, and therefore the number of heat exchangers must be increased to improve performance.

For the previous example, the cold fluid is experiencing larger temperature change; therefore, it must have the lower capacity, and therefore

$$C_r = \frac{410 - 110}{360 - 0} = 0.833 \tag{11.61}$$

$$N > \frac{\sqrt{1+(0.833)^2}}{1-0.833} \tanh^{-1}\left[\frac{360-0}{\frac{410+110}{2} - \frac{360+0}{2}} \frac{(1-0.833)}{2}\right] = 3.08 \tag{11.62}$$

which results in the feasible solution of four or more shells. As will be shown later, using three shells will result in inefficient heat exchangers.

11.5.2 Multishell and Tube Heat Exchangers

For N identical multi-shell and tube heat exchangers connected in series, with each shell having $2M$ tubes (M an integer), the LMTD correction factor, F, can be combined into the single general expression [3]

$$F_{N,2NM} = \frac{S \ln W}{\ln\left[\dfrac{1+W-S+SW}{1+W+S-SW}\right]} \tag{11.63}$$

where

$$W = \left[\frac{1-PR}{1-P}\right]^{1/N} \tag{11.64}$$

and

$$S = \frac{\sqrt{R^2+1}}{R-1} \tag{11.65}$$

and P and R are based on the inlet and exit temperatures of the overall system

$$P = \frac{min\left[(T_1-T_2),(t_2-t_1)\right]}{T_1-t_1} \tag{11.66}$$

and

$$R = \frac{max\left[(T_1-T_2),(t_2-t_1)\right]}{min\left[(T_1-T_2),(t_2-t_1)\right]} = \frac{1}{C_r} \tag{11.67}$$

For a balanced flow heat exchanger, $R = 1$

$$W = \frac{N - NP}{N - NP + P}$$

$$F = \sqrt{2}\,\frac{1-W}{W}\,\frac{1}{\ln\dfrac{\dfrac{W}{(1-W)} + \dfrac{1}{\sqrt{2}}}{\dfrac{W}{1-W} - \dfrac{1}{\sqrt{2}}}}$$

(11.68)

This expression combines different expressions given for multi-shell and tube heat exchangers by Bowman et al. [11] into a single general equation and is plotted in Figure 11.11.

By using four shells, each shell will have an LMTD correction factor of ~0.8. This then needs to be verified by using Equation 11.64, and the results are shown in Table 11.6 for four and five shells, and as can be seen four shells provide an LMTD correction factor of 0.76 and for five shells F improves to 0.86.

The effectiveness of N identical shell and tube heat exchangers in series is given by [8]

$$\varepsilon = \frac{1 - \left\{\dfrac{1 - \varepsilon_1 C_r}{1 - \varepsilon_1}\right\}^N}{C_r - \left\{\dfrac{1 - \varepsilon_1 C_r}{1 - \varepsilon_1}\right\}^N}$$

(11.69)

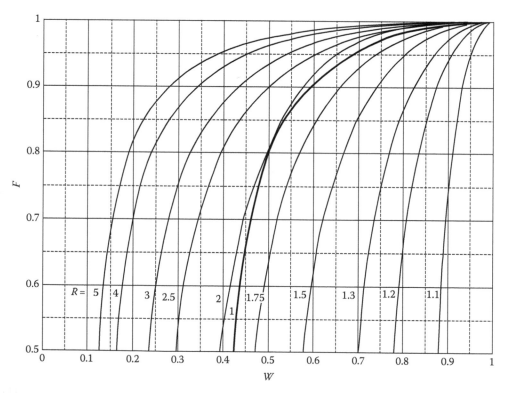

FIGURE 11.11
F correction factor for multi-shell and tube heat exchangers.

TABLE 11.6

Solution to Example 11.7
Using Figure 11.11

N	4	5
R	1.2	1.2
P	0.73	0.73
S	7.81	7.81
W	0.82	0.85
F	0.76	0.86

where the effectiveness of the individual shells of the heat exchanger is given by

$$\varepsilon_1 = \frac{2}{(1+C_r)+\dfrac{\sqrt{1+C_r^2}}{\tanh\left[\dfrac{NTU}{2N}\sqrt{1+C_r^2}\right]}} \tag{11.70}$$

The efficiency of multi-shell and tube heat exchangers is given by

$$\frac{\eta}{\eta_1} = \frac{1}{Z}\frac{1-\left\{\dfrac{1-\dfrac{Z}{N}}{1+\dfrac{Z}{N}}\right\}^N}{1+\left\{\dfrac{1-\dfrac{Z}{N}}{1+\dfrac{Z}{N}}\right\}^N} \tag{11.71}$$

where

$$Z = \frac{NTU(1-C_r)}{2}\eta_1 \tag{11.72}$$

and the efficiency of each shell is given by

$$\eta_1 = \frac{\tanh\dfrac{NTU\sqrt{C_r^2+1}}{2N}}{\dfrac{NTU\sqrt{C_r^2+1}}{2N}} \tag{11.73}$$

Figure 11.12 is a plot of Equation 11.71 as a function of Z for different values of N. As can be seen, for four or more heat exchangers connected in series, all the results fall on the same curve, which is given by

$$\frac{\eta}{\eta_1} \approx \frac{\tanh[Z]}{Z} \tag{11.74}$$

Equation 11.74 provides the overall efficiency of a system made of a large number of heat exchangers connected in series. It is the limit, as the number of heat exchangers approaches infinity. The curve labeled ∞ in Figure 11.12 is a plot of Equation 11.74 and as can be seen, it approximates Equation 11.73 for $N > 3$.

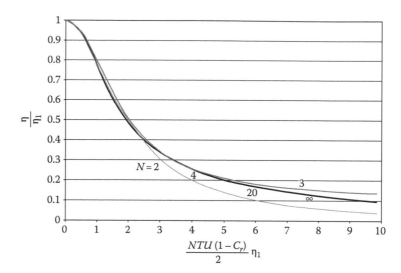

FIGURE 11.12
Efficiency of heat exchanger networks.

From Figure 11.12, the overall efficiency of the network is less than the efficiency of its components. As the number of heat exchangers is increased, the NTU of the individual heat exchangers and thus their fin analogy number, *Fa*, decrease, bringing their efficiency close to 1, making the RHS of Equation 11.74 equal to that for a counter-flow heat exchanger, or in the limit, the network behaves like that of a counter-flow heat exchanger.

For cross-flow heat exchangers connected in series where the overall behavior is counter-flow and the individual heat exchangers have the same effectiveness (ε_1) and the fluids are mixed before entering the next heat exchanger, Kays and London [13] give the following expression for the overall effectiveness of the system

$$\varepsilon = \frac{1 - \left\{ \dfrac{1 - C_r \varepsilon_1}{1 - \varepsilon_1} \right\}^N}{C_r - \left\{ \dfrac{1 - C_r \varepsilon_1}{1 - \varepsilon_1} \right\}^N} \tag{11.75}$$

Note that this equation is identical to Equation 11.70, which was derived for shell and tube heat exchangers connected in series. Therefore for any type of heat exchangers connected in series, the overall effectiveness is given by Equation 11.70, the overall efficiency by Equation 11.73 exactly, and Equation 11.75 approximately. Use of these equations is far more general compared to the traditional approaches, in that the designer has freedom to select the efficiency of the individual heat exchangers, the overall system efficiency, the number of heat exchangers, as well as the type of heat exchanger to select, by utilizing a single equation.

The solution to Example 11.7 using heat exchanger efficiency is presented next. Assume each shell to have an efficiency of 85%, then since for a shell and tube heat exchanger

$$\eta = \frac{\tanh(Fa)}{Fa} = 0.85$$

$$Fa = 0.74 = \frac{NTU\sqrt{C_r^2 + 1}}{2N}$$

For five shells

$$NTU = \frac{0.74 \times 2 \times 5}{\sqrt{C_r^2 + 1}} = 5.68$$

$$Z = \frac{NTU(1 - C_r)}{2} \eta_1 = 0.4027$$

$$\eta = \eta_1 \frac{\tanh(Z)}{Z} = 0.807$$

Or the overall network efficiency is 80.7% using five heat exchangers each having an efficiency of 85%.

11.6 Compact Heat Exchangers

Heat exchangers with large area-to-volume ratio are called compact heat exchangers and are used when large heat transfer rate per unit volume is needed, particularly when one or both fluids are gases. If α is the ratio of the total heat transfer area to the total volume of the heat exchanger, compact heat exchangers are those with $\alpha > 700$ m²/m³ then if both fluids are gas, or $\alpha > 300$ m²/m³ if one or both fluids are liquid.

Generally, the heat transfer coefficient on the gas side is significantly smaller than that on the liquid side. In single stream heat exchangers (evaporators and condensers), the difference between the resistances on the air side and the phase-change side is so large that the air side resistance is the factor that limits the transfer of heat.

To reduce the gas side resistance, fins, like those shown in Figure 11.13, are added to improve the high area-to-volume ratio. The fluid flows in small passages, and therefore the

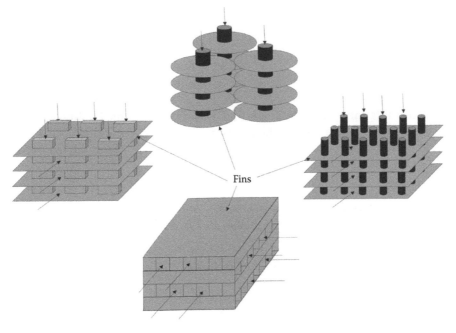

Fins

FIGURE 11.13
Compact heat exchangers.

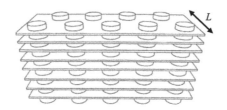

FIGURE 11.14
Tube and fin heat exchanger.

effective hydraulic diameters are small, making these heat exchangers operate at relatively low Reynolds numbers in the laminar or transition region.

The heat exchangers shown in Figure 11.13 are all cross-flow ones, and therefore the overall heat transfer can be calculated from any of the three methods for heat exchanger analysis and is given by

$$q = UA\ F\ LMTD = \varepsilon C_{min}(T_1 - t_1) = \eta UA(\bar{T} - \bar{t}) \tag{11.76}$$

where ε and η depend on UA with

$$\frac{1}{UA} = \frac{1}{(UA)_c} = \frac{1}{(UA)_h} = \sum_i R_i \tag{11.77}$$

and c and h refer to the cold and hot sides of the heat exchanger and R is the resistance. The heat exchanger shown in Figure 11.14 has at least three resistances and therefore the overall heat transfer coefficient is

$$UA = \cfrac{1}{\cfrac{1}{h_i A_i} + \cfrac{\ln\frac{r_o}{r_i}}{2\pi k L} + \cfrac{1}{h_o A\left(1 - \frac{A_f}{A}(1 - \eta)\right)}} \tag{11.78}$$

where the subscripts i and o refer to inside and outside conditions. Fouling or contact resistances may also have to be added to the terms in the denominator.

Compact heat exchangers have their own terminology that is summarized in Table 11.7.

Example 11.8

For the heat exchanger shown, the tube outside diameter $d = 0.402$, the tube spacing $s = 1$, the distance between fins (fin spacing) $b = 1/8$, and the fin thickness $t = 0.013$ in. Find the other parameters:

$$A_c = a \times c = 0.0670$$

$$A_{fr} = s \times b = 0.125$$

$$\sigma = \frac{A_c}{A_{fr}} = \frac{0.0670}{0.125} = 0.536$$

TABLE 11.7

Compact Heat Exchangers Terminology

A_{fr}	Frontal area, the area in front of the heat exchanger
A	Heat exchanger area, the total available area for heat transfer is that which includes the tube as well as the fin areas
$A_c = A_{ff}$	Minimum area (A_c), also referred to as the free flow area
L	The flow length, and V is the volume of the heat exchanger
$V = A_{fr}L$	The volume of the heat exchanger
$\alpha = \dfrac{A}{V}$	Area-to-volume ratio
$\sigma = \dfrac{A_c}{A_{fr}}$	Minimum area-to-frontal area ratio
$D_h = 4L\dfrac{A_c}{A}$	Hydraulic diameter
$\alpha = \dfrac{A}{V} = \dfrac{A}{A_{fr}L}\dfrac{A_c}{A_c} = \dfrac{4\sigma}{D_h}$	
$A_c = \sigma A_{fr} = \dfrac{AD_h}{4L} = \dfrac{A\sigma}{L\alpha}$	

The rest of the parameters are shown in the table as follows:

b	0.125
t	0.013
a	0.112
d	0.402
s	1
c	0.598
A_c	0.0670
A_{fr}	0.125
σ	0.536
L	0.866
V	0.1083
A	1.620
A	179.54
D_h	0.143
A_f/A	0.913
t_p	0.866

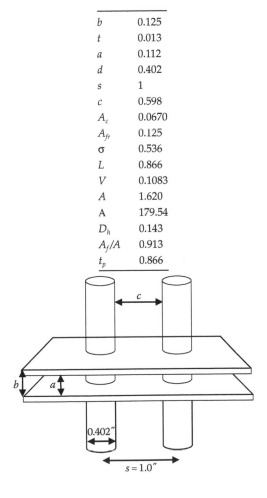

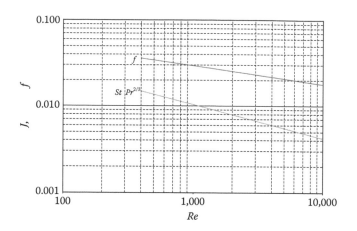

FIGURE 11.15

Heat transfer coefficient and friction factor a for compact heat exchanger where tube outside diameter = 0.402 in. (1.02 × 10⁻³ m), fin pitch = 8.0 FPI (315/m), tube spacing is 0.866 in. in the flow direction, tube spacing is 1.0 in. perpendicular to the flow direction, the hydraulic diameter, D_h = 0.143 in. (3.632 × 10⁻³ m), fin thickness = 0.013 in. (0.33 × 10⁻³ m), free flow area/frontal area (σ = 0.534), heat transfer area/total volume, α = 179 ft²/ft³ (587 m²/m³), and fin area/total area = 0.913.

To calculate the heat transfer, the values of the inside and outside heat transfer coefficient must be known. The inside heat transfer coefficient is obtained from the correlations for internal flow. The calculation of the fin efficiency was presented in Chapter 3, and the outside heat transfer coefficient is obtained from charts similar to that shown in Figure 11.15 [13]. Figure 11.10 is a typical performance chart for a compact heat exchanger. The particular chart is for a heat exchanger similar to the one in Figure 11.8.

In the chart

$$\frac{f}{4} = c_f = \frac{\rho \tau_0}{\dfrac{G^2}{2g_c}} \tag{11.79}$$

$$J = St\,Pr^{2/3} = \frac{h}{Gc_p}Pr^{2/3} \tag{11.80}$$

$$Re = \frac{\rho V_{max} D_h}{\mu} = \frac{GD_h}{\mu} \tag{11.81}$$

where $G = \rho V_{max}$ is called the mass velocity, and is based on the maximum velocity which occurs at the minimum area (A_c), also referred to as the free flow area (A_{ff}). Note that g_c is a constant that depends on the system of units, 1 in SI and 32.2 in conventional system of units. In term τ_0 used in Equation 11.79 is an equivalent shear stress based on a combination of viscous and pressure forces.

11.6.1 Pressure Drop

Figure 11.16 is the side view of the heat exchanger shown in Figure 11.14. Locations a and b are front and back of the heat exchanger and 1 and 2 represent points in the flow upstream and downstream of the heat exchanger core. As the fluid enters the heat exchanger, the velocity increases, and therefore pressure drops. As it flows in the heat exchanger, the pressure

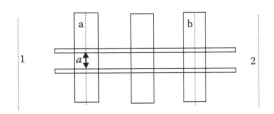

FIGURE 11.16
Side view of a compact heat exchanger.

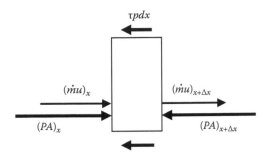

FIGURE 11.17
Force balance in a compact heat exchanger.

continues to drop, and as it exits the heat exchanger core (b–2) the fluid expands, and therefore there is some pressure recovery. If we focus in the core of the heat exchanger and consider density variation, then a differential momentum balance on a fluid element (Figure 11.17) results in

$$(\dot{m}u)_{x+\Delta x} - (\dot{m}u)_x = (PA)_x - (PA)_{x+\Delta x} - \tau p dx \tag{11.82}$$

which simplifies to

$$\dot{m}du = -AdP - \tau dA_s \tag{11.83}$$

and solving for pressure

$$dP = -\frac{\dot{m}}{A}du - \frac{\tau}{A}dA_s \tag{11.84}$$

For a compressible flow in a constant area passage, from conservation of mass

$$\frac{du}{u} + \frac{d\rho}{\rho} = 0 \tag{11.85}$$

then

$$dP = \frac{\dot{m}}{A}u\frac{d\rho}{\rho} - \frac{\tau}{A}dA_s \tag{11.86}$$

Integrating between a and b

$$P_a - P_b = G^2\left(\frac{1}{\rho_b} - \frac{1}{\rho_a}\right) + \frac{c_f}{2A}\frac{G^2}{\dfrac{\rho_b + \rho_b}{2}}A_s \tag{11.87}$$

If K_c and K_e are entrance (contraction) and exit (expansion) coefficients, then

$$\frac{P_1}{\rho_1 g} + \frac{V_1^2}{2g} = \frac{P_a}{\rho_a g} + \frac{V_a^2}{2g} + K_c \frac{V_a^2}{2g} \qquad (11.88)$$

$$\frac{P_b}{\rho_b g} + \frac{V_b^2}{2g} = \frac{P_2}{\rho_2 g} + \frac{V_2^2}{2g} + K_e \frac{V_b^2}{2g} \qquad (11.89)$$

Generally, the density changes from 1 to a and from b to 2 are small. Therefore, adding the three equations [17]

$$\frac{P_1 - P_2}{P_1} = \frac{G^2}{2P_1\rho_1}\left[(1-\sigma^2+K_c)+2\left(\frac{\rho_1}{\rho_2}-1\right)+\frac{f}{4}\frac{\rho_1}{\rho_m}\frac{A}{A_c}-[1-\sigma^2-K_e]\frac{\rho_1}{\rho_2}\right] \qquad (11.90)$$

or in terms of specific volumes

$$\frac{\Delta P}{P_1} = \frac{G^2}{2g_c}\frac{v_1}{P_1}\left[(1-\sigma^2+K_c)+2\left(\frac{v_2}{v_1}-1\right)+\frac{f}{4}\frac{A}{A_c}\frac{v_2+v_1}{2v_1}-(1-\sigma^2-K_e)\frac{v_2}{v_1}\right] \qquad (11.91)$$

In many instances, the entrance and exit effects are already incorporated in the friction factor. Furthermore, the specific volumes at 1 and 2 are generally not different, and therefore Equation 11.91 simplifies to

$$\Delta P = f \frac{G^2}{8\rho_1 g_c}\frac{A}{A_c} \qquad (11.92)$$

Example 11.9

To reduce the humidity in an indoor swimming pool, the moist air in the room is exchanged with cold dry air from outside. To save energy, a cross-flow heat exchanger with square cross section is to be used for balanced flow of air. The air flow rate is 1200 CFM and the allowable pressure drop is 400 Pa. The pool moist air at 25°C is used to heat the outside air from 0°C to 18°C. Determine suitable dimensions for the heat exchanger:

$$\varepsilon = \frac{t_2 - t_1}{T_1 - t_1} = \frac{18-0}{25-0} = 0.72$$

$$q = \varepsilon C_{min}(T_1 - t_1) = 0.72(0.696)(25-0) = 12{,}604 \text{ W}$$

If the heat exchanger is a cross-flow one with both fluids unmixed, then we can calculate the NTU, from the equation in Table 11.1, for $C_r = 1$.

The total heat transfer area is the total area of one of the streams. The area of each duct is $4pL$, the number of ducts in each row is $\dfrac{L}{p}$, and there are $\dfrac{H}{2p}$ rows; therefore, the heat transfer area becomes

$$A = \frac{L}{p}\frac{H}{2p}4pL = \frac{2HL^2}{p}$$

and the NTU

$$NTU = \frac{U}{\dot{m}c_p}\frac{2HL^2}{p}$$

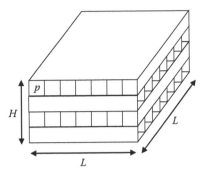

Solving for the unknowns

$$HL^2 = \frac{NTU \dot{m} c_p p}{2U}$$

Neglecting the wall resistance, the overall heat transfer coefficient can be obtained from

$$\frac{1}{U} = \frac{1}{h_c} + \frac{1}{h_h} = \frac{2}{h}$$

Since the cold and hot fluid ducts are identical. Assuming the flow to be thermally fully developed

$$U = \frac{1}{2}h = \frac{1}{2}Nu\frac{k}{D_h}$$

$$\frac{HL^2}{p^2} = \frac{NTU\,\dot{m}c_p}{kNu} \tag{11.93}$$

The pressure drop

$$\Delta P = f\frac{G^2}{8\rho_1 g_c}\frac{A}{A_c}$$

$$A_c = \frac{IIL}{2p^2}p^2 = \frac{HL}{2}$$

$$G = \frac{\dot{m}}{\dfrac{HL}{2}}$$

$$\Delta P = f\frac{\left(\dfrac{\dot{m}}{\dfrac{HL}{2}}\right)^2\left(\dfrac{2HL^2}{p}\right)}{8\rho_1 g_c\dfrac{HL}{2}} = f\frac{2(\dot{m})^2}{\rho_1 g_c pLH^2}$$

$$LH^2 = \frac{2f(\dot{m})^2}{\rho_1 g_c p\Delta P}$$

for fully developed flow in a square duct

$$f = \frac{57}{Re} = 57\frac{\mu}{\rho Vp}\frac{p}{p} = 57\frac{\mu p}{\dfrac{\dot{m}}{N}} = 57\frac{HL}{2p^2}\frac{\mu p}{\dot{m}} = \frac{\mu 57 HL}{2p\dot{m}}$$

$$\Delta P = \frac{\mu 57 HL}{2p\dot{m}}\frac{2(\dot{m})^2}{\rho_1 g_c pLH^2} = \frac{57\mu(\dot{m})}{\rho_1 g_c p^2 H}$$

$$p^2 H = \frac{57\mu(\dot{m})}{\rho_1 g_c \Delta P} \tag{11.94}$$

Here we have two equations (11.93) and (11.94) for three unknowns, p, H, and L. The following table shows the different heat exchangers obtained as a result of varying the duct size:

V dot	1200	CFM
m dot	0.696	ks/s
T_1	25	C
t_1	0	C
t_2	18	C
T_2	7	C
e	0.72	
NTU	3.89	
DP	400	Pa
C_r	1	
ε	0.720	1.0000
Tf	289	K
tf	282	K
t average	286	K
k	0.02514	W/m K
ρ	1.229057	kg/m^3
c_p	1006	J/kg K
μ	1.77E–05	kg/m s
Pr	0.71077	
q	12,604.337	W
$HL2/p2$	30,096.959	
$p2H$	1.431E–06	
NU	3.6	
Con	57	$f = Con/Re$

p (mm)	H (m)	L (m)	N	Re	xfd
1	1.43	0.15	103,774	378	0.02
1.2	0.99	0.21	72,065	454	0.03
1.4	0.73	0.28	52,946	529	0.04
1.6	0.56	0.37	40,537	605	0.05
1.8	0.44	0.47	32,029	681	0.06
2	0.36	0.58	25,944	756	0.08
2.2	0.30	0.70	21,441	832	0.09

As can be seen, for all cases the flow is laminar. If we use the design with $p = 1.8$ mm, the heat exchanger is roughly a cube, with flow Reynolds number of 681. The entrance length is 6 cm, which is much smaller than 44 and 47 cm lengths of the ducts.

11.7 Microchannel Heat Exchanger

Consider an example of a cylindrical pipe of diameter D and length L, where heat is transferred between the inside and outside fluids. For this case, the area-to-volume ratio is

$$\frac{A}{V} = \frac{\pi D L}{\dfrac{\pi}{4} D^2 L} = \frac{4}{D} \tag{11.95}$$

The tubes normally used in heat exchangers have a diameter around 1 cm ($D = 0.01$ m) and therefore the heat exchanger cannot be considered compact and the area needs to be increased by addition of the fins to the outer surface of the pipe. However, as the tube diameter is reduced, the simple heat exchanger gets closer to a compact heat exchanger, for example for $D = 2$ mm, the

$$\frac{A}{V} = 2000 \tag{11.96}$$

Note that V is the volume of a single tube and not the heat exchanger's, that also includes the volume in between the tubes. Microchannel heat exchanger channel sizes range between 0.01 and 2 mm, resulting in very large $\frac{A}{V}$ ratios. This is partly the explanation for the use of microchannels for enhanced heat transfer. Microchannels, Figure 11.18, are used in electronic cooling, automobile air conditioners, and fuel cells.

The small channel diameter means short travel distances for heat and therefore higher heat transfer coefficient. However, the drawback of using smaller tube diameter is the increase in the pressure drop, although since the heat transfer is so much higher, relatively short heat exchanger lengths are needed reducing the pressure drop. To further reduce pressure drop, fluid is distributed in parallel short-length microchannel ducts that are connected together with relatively large headers. At any given instant, the headers store a large amount of the fluid flowing in the heat exchanger. This also leads to the maldistribution of the fluid in the different channels.

A basic issue in analyzing microchannels is whether the continuum assumption is applicable in small dimensions, i.e., can the conservation equations be applied to microchannels, or are the available results based on the continuum analysis applicable to microchannels. The most widely criterion for answering that question is Knudsen number, defined as

$$Kn = \frac{\lambda}{L} \tag{11.97}$$

where
  λ is the mean free path
  L is the physical dimension of the channel, its hydraulic diameter [18]

FIGURE 11.18
Microchannel tubes.

The continuum assumption is generally considered to be valid for $Kn < 0.001$. For $0.001 < Kn < 0.1$, the continuum assumption is still applicable to the flow; however, the no slip boundary condition does not apply and the boundary conditions and needs to be modified, using a velocity slip boundary condition.

For liquids flowing in microchannels larger than 0.1 mm, the values of Kn are smaller than 0.001; therefore, the available results can be used at the microlevel. For example, for fully developed channel flow,

$$h = \frac{kNu}{d} \tag{11.98}$$

Since k and Nu are constant, h is inversely proportional to diameter, and the difference between a cm size and a mm size pipe is one order of magnitude increase in the heat transfer coefficient.

Problems

11.1 Show that regardless of which fluid has the minimum capacity, Equation 11.27 simplifies to Equation 11.29.

11.2 Water at the rate of 3 kg/s is to be heated from 35°C to 90°C by 7 kg/s of oil $c_p = 1500$ J/kg K supplied at 150°C using shell and tube heat exchanger having the overall heat transfer coefficient of 120 W/m² K. Determine:
   a. The total heat transfer area
   b. The exit temperature of water, if the oil flow rate drops by 20%

11.3 Water at the rate of 6 kg/s is to be heated from 35°C to 90°C by 14 kg/s of oil $c_p = 1500$ J/kg K supplied at 150°C using shell and tube heat exchanger. Two alternatives are being considered using a single heat exchanger or using two smaller ones connected in series on water side with oil split equally between them. If the overall heat transfer coefficient is 120 W/m² K for the heat exchangers, determine the heat transfer areas of both cases:

$$\frac{\Delta T_{min}}{\overline{T} - \overline{t}} = \frac{\Delta T_{min}}{\overline{T} - \overline{t}}$$

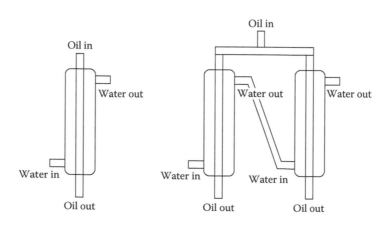

11.4 A one shell pass, two tube pass heat exchanger is used to cool vegetable oil using water. The design specifications are $U = 340$ W/m² K, $T_{c,in} = 20°C$, $T_{c,out} = 70°C$, $T_{h,in} = 150°C$, $T_{h,out} = 60°C$. After operating for 1 year, the performance is found to have deteriorated and oil is cooled only to 80°C. Fouling of the oil side is suspected; determine the apparent fouling resistance. Take $c_p = 1500$ J/kg K for the oil.

11.5 A two shell pass, four tube pass heat exchanger is available to cool 5 kg/s of liquid ammonia at 70°C against 5 kg/s of water at 15°C if the heat transfer area is 40 m² and the heat transfer coefficient is 2000 W/m² K. Determine the outlet temperatures.

11.6 To reduce the humidity of an indoor swimming pool and reduce vapor condensation on the cold windows in the winter, the warm moist indoor air at 25°C is replaced by cold dry outside air at −8°C at a rate of 0.7 m³/s. To reduce the heating requirement, the incoming air is heated by the outgoing air using a plate-type heat exchanger. The heat exchanger consists of an arrangement of 0.05 mm thick (*t*) aluminum parallel plates 0.5 m (*L*) by 0.4 m (*W*), that are separated by a distance of 3 mm (*d*). Neglect the effects of water condensation:
   a. For the counter-flow arrangement, determine the number of plates for an effectiveness of 80%.
   b. Determine the pressure drop.

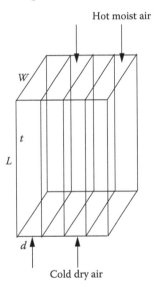
Hot moist air

W

t

L

d

Cold dry air

11.7 An alternative for the earlier problem is to take the moist air and pass it through 2.5 cm thin-walled copper pipe that is exposed to the outside air. Determine the required pipe length, for the air temperature to be reduced from 25°C to 0°C.

11.8 Repeat Example 11.9 for cold air being heated to 15°C.

11.9 Repeat Example 11.9 for a pressure drop of 1000 pa.

11.10 Water at the rate of 1 kg/s is to be heated from 35°C to 90°C by 140 kg/s of air supplied at 150°C using the heat exchanger of Figure 11.10. Determine the heat transfer size.

11.11 Water at the rate of 6 kg/s is to be heated from 35°C to 90°C by 14 kg/s of oil $c_p = 1500$ J/kg K supplied at 150°C using shell and tube heat exchanger having the overall heat transfer coefficient of 120 W/m² K:
   a. Determine the total heat transfer area.

b. Determine the exit temperature of water, if the oil flow rate drops by 20%.

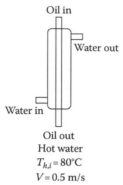

Oil in

Water out

Water in

Oil out
Hot water
$T_{h,i} = 80°C$
$V = 0.5$ m/s

11.12 A plate-type heat exchanger consists of an arrangement of 0.5 mm thick (t) aluminum parallel plates 1.0 m (L) by 0.2 m (W) that are separated by a distance of 4 mm (d):
a. For the conditions shown, determine the overall heat transfer coefficient.
b. For the counter-flow arrangement, determine the exit temperatures.

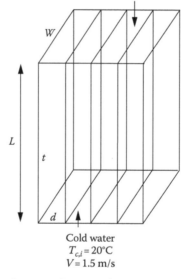

W

L

t

d

Cold water
$T_{c,i} = 20°C$
$V = 1.5$ m/s

11.13 A one shell pass, two tube pass heat exchanger is used to cool vegetable oil using water. The design specifications are $U = 340$ W/m² K, $T_{c,in} = 20°C$, $T_{c,out} = 70°C$, $T_{h,in} = 150°C$, $T_{h,out} = 60°C$. After operating for 1 year, the performance is found to have deteriorated and oil is cooled only to 80°C. Fouling of the oil side is suspected; determine the apparent fouling resistance. Take $c_p = 1500$ J/kg K for the oil.

11.14 A two shell pass, four tube pass heat exchanger is available to cool 3 kg/s of liquid ammonia at 75°C using 5 kg/s of water at 20°C if the heat transfer area is 30 m² and the heat transfer coefficient is 1000 W/m² K. Determine the outlet temperatures.

11.15 Water at 10°C and the rate of 0.5 kg/s is heated in the inner pipe of a 4 m long double-pipe coaxial stainless steel (AISI 302) counter-flow heat exchanger. The inner steel tube has a 2 cm outside diameter and 1.5 mm wall thickness. Air enters the outer jacket at 300°C. Plot the exit temperatures of water and air as a function of
a. Air mass flow rate between 0.01 and 1.0 kg/s for the outer jacket inside diameter of 4 cm.

b. Jacket diameter between 3 and 10 cm for air mass flow rate of 0.02 kg/s.

c. If the outer jacket is not insulated and has a 4 cm inner diameter and 2 mm wall thickness, estimate the heat loss from the heat exchanger to the ambient for an air flow rate of 0.02 kg/s.

11.16 The condenser of a large power plant is a shell and tube heat exchanger consisting of a single shell and 20,000 tubes, where each tube makes two passes in the shell. The tubes can be considered thin and have a diameter of 25 mm. The steam enters the shell side of the heat exchanger at a rate of 800 kg/s and condenses on the outside wall of the tubes at 60°C providing an effective heat transfer coefficient of 13,000 W/m² K on the outside of the tubes. Cooling water enters the tubes at 20°C with the total flow rate of 40,000 kg/s:

a. What is the exit temperature of the cooling water?

b. What is the required tube length per pass using LMTD method?

c. What is the required tube length using ε-NTU method?

11.17 In a double-pipe counter-flow heat exchanger, the inside tube is made of stainless steel (AISI 302) and has 8 mm ID and 10 mm OD and is surrounded by a 25 mm ID glass tube. Water at a rate of 0.3 kg/s enters the steel tube at 20°C and leaves at 85°C. Steam, condensing on the outside wall of the tube, maintains the outside wall temperature at 110°C. Determine:

a. Heat exchanger length

b. Pressure drop

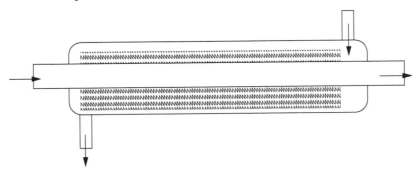

11.18 A cross-flow heat exchanger consists of 200 thin aluminum tubes that have a diameter of 1 cm and are 1 m long. Water at 25°C at a rate of 0.5 kg/s is distributed among the tubes and is heated by air flowing at 150°C across the tubes at a rate of 2 kg/s. The heat transfer coefficient on the outside is 150 W/m² K. Determine the exit temperature of both fluids.

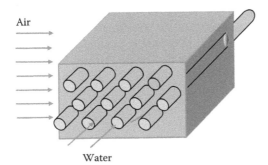

Air

Water

11.19 Determine an expression for $t(x)$, the variation of the cold fluid temperature along a parallel-flow heat exchanger.

11.20 Determine an expression for $T(x)$, the variation of the hot fluid temperature along a counter-flow heat exchanger.

References

1. Fakheri, A. (2002) Log mean temperature correction factor: An alternative representation. *Proceedings of the International Mechanical Engineering Congress and Exposition*, November 17–22, New Orleans, LA.
2. Fakheri, A. (2003) Arithmetic mean temperature difference and the concept of heat exchanger efficiency. *HT2003-47360 Proceedings of the 2003 ASME Summer Heat Transfer Conference*, July 21–23, Las Vegas, NV.
3. Fakheri, A. (2003) The exact solution for multipass shell and tube heat exchangers. *Journal of Heat Transfer* 125, 527–530.
4. Fakheri, A. (2003) An alternative approach for determining the correction factor and the number of shells in shell and tube heat exchangers. *Journal of Enhanced Heat Transfer* 10(4), 407–420.
5. Fakheri, A. (2003) The shell and tube heat exchanger efficiency and its relation to effectiveness. *Proceedings of the 2003 American Society of Mechanical Engineers (ASME)*, November 16–21, International Mechanical Engineering Congress and Exposition (IMECE), Washington, DC.
6. Fakheri, A. (2006) Thermal efficiency of the cross flow heat exchangers. *Proceedings of the 2006 ASME International Mechanical Engineering Conference and Exposition*, November 5–10, Chicago, IL.
7. Fakheri, A. (2007) Heat exchanger efficiency. *Journal of Heat Transfer* 129(9), 1268–1276.
8. Fakheri, A. (2008) Efficiency and effectiveness of heat exchanger series. *Journal of Heat Transfer* 130(8), 084502 (4 pages).
9. Fakheri, A. (2008) On application of the second law to heat exchangers. *Proceedings of the 2008 ASME International Mechanical Engineering Conference and Exposition*, October 31–November 6, Boston, MA.
10. Fakheri, A. (2010) Second law analysis of heat exchangers. *Journal of Heat Transfer* 132(11), 111802 (7 pages).
11. Bowman, R. A., Mueller, A. C., and Nagel, W. M. (1940) Mean temperature difference in design. *Transactions of ASME* 62, 283–294.
12. Bell, K. J. and Muller, A.C. (1984) *Wolverine Tube Heat Transfer Data Book*. http://www.wlv.com/products/databook/ch2_5.pdf, retrieved March 21, 2013.
13. Kays, W. M. and A. L. London, (1984) *Compact Heat Exchangers*, 3rd edn. New York: McGraw-Hill.
14. Ahmad, S., Linnhoff, B., and Smith, R. (1985) Design of multipass heat exchangers: An alternative approach. *ASME Journal of Heat Transfer* 110, 303–309.
15. Bell, K. J. (1990) *Hemisphere Handbook of Heat Exchanger Design*. New York: Hemisphere Publishing Corporation.
16. Taborek, J. (1990) *Hemisphere Handbook of Heat Exchanger Design*. New York: Hemisphere Publishing Corporation.
17. Mills, A. F. (1995) *Heat and Mass Transfer*. Burr Ridge, IL: Irwin.
18. Sobhan, C. B. and Peterson, G. P. (2008) *Microscale and Nanoscale Heat Transfer Fundamental and Engineering Applications*. Boca Raton, FL: CRC.

12

Radiation Heat Transfer

12.1 Introduction

Radiation heat transfer is the third mechanism for transfer of heat and is different than conduction and convection in a number of aspects. Radiation is internal energy leaving in the form of electromagnetic waves. All objects emit radiation due to their temperature; consequently, heat transfer by radiation occurs even if there is no temperature difference between the object and its surroundings in order to maintain thermal equilibrium. Another difference is the lack of need for a material medium for heat to be transferred, and in fact, radiation is most efficient in vacuum.

Some of the behaviors of electromagnetic radiation (EMR), for example, its travel through space, can be explained by treating it as a wave, and other behaviors can be explained by treating it as a particle, for example, it can only be emitted or absorbed as discrete quantities. This is known as the wave particle duality [1].

Consider the sinusoidal wave shown in Figure 12.1, where A is the amplitude and λ is the wavelength, which is the distance between two corresponding points on the wave. The wave will be traveling with the speed C, which is its speed of propagation. The frequency, ν, is the number of waves (fluctuations) in a second, and from the definitions of speed and wavelength, after 1 s, the number of wave fronts that have passed a point in space is $\nu = \dfrac{C}{\lambda} \text{Hz}$, which has a dimension of inverse seconds. The period of a wave, τ, is the time needed for the wave to go through one cycle, and therefore, $\tau = \dfrac{1}{\nu} = \dfrac{\lambda}{C}$. The frequency of an emitted electromagnetic wave is constant and depends on the source of the radiation and is independent of the medium through which the radiation travels. The speed of propagation and therefore the wavelength are dependent on the medium.

The propagation speed is

$$C = \frac{C_0}{n} = \frac{2.9979 \times 10^8}{n} \, \text{m/s} \tag{12.1}$$

where
C_0 is the speed of light in vacuum
n is the index of refraction of the medium

The index of refraction n is equal to 1 for vacuum and air, 1.5 for glass, and 1.33 for water. Since frequency remains constant, the wavelength of an electromagnetic wave in a medium with index of refraction n is

$$\lambda = \frac{C_0}{n\nu} \tag{12.2}$$

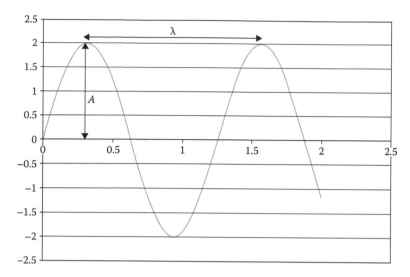

FIGURE 12.1
Sinusoidal wave traveling in the x direction with speed C.

The particle theory of radiation is based on the travel of photons that carry energy with them. The amount of energy that a photon carries is proportional to its frequency and is given by

$$e = h\nu \tag{12.3}$$

where $h = 6.6256 \times 10^{-34}\,\text{J s}$ is the Planck constant.

Generally, substances emit EMR over a broad range of frequencies (wavelengths). Figure 12.2 is a plot of the radiation spectrum in terms of wavelength in micron meters. If the wavelength is less than $10^{-7}\,\mu\text{m}$, the radiation is termed cosmic rays. At higher wavelengths, EM waves are called gamma and x-ray radiations. Between 0.01 and 0.4 μm, EMR is called ultraviolet radiation; between 0.4 and 0.7 μm, the radiation is in the visible range; and between 0.7 and 1000 μm, the radiation is called infrared (IR). Longer wavelengths correspond to radio, TV, and other communication frequencies. In terms of radiation heat transfer, the range of interest is between 10^{-2} and 1000 μm, since this is the range in which most substances absorb incident EMR.

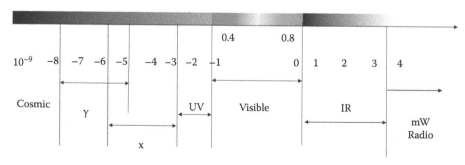

FIGURE 12.2
Radiation spectrum.

Humans are actually able to "see" electromagnetic waves that have wavelengths between 0.4 μm, which corresponds to violet, and 0.7 μm, which is for dark red light. Other colors correspond to different wavelengths, for example, green light has a wavelength of about 0.510 μm, yellow about 0.570 μm, orange about 0.590 μm, and red light about 0.650 μm. The sky appears blue, because the blue in the sun's radiation scatters much more than the other colors, including red or orange, which explains why the sun appears reddish color at sunrise and sunset, since those are the ones that scatter the least.

A frequency modulation (FM) radio receives EMR with certain frequencies, converts them to acoustic waves (audible sound waves) that have much lower frequencies between 10 and 20,000 Hz so that our ears can detect and send to our brains to further process and interpret those sound waves. In a sense we can think of our eyes as a receiver like an FM radio, they receive the EMR over the visible range and send it to the brain for further processing and interpretation. It is also no coincidence that our eyes are sensitive to the EMR range, most abundant on earth, since almost half of the sun's radiation arriving on the earth falls into visible range and the wavelength of 0.57 μm, which is almost in the middle of the visible range, is the wavelength at which the sun's intensity is maximum.

12.2 Definitions

Radiation heat transfer has its own terminology, which is discussed next. Consider a surface A at some temperature T. If we consider a differential area dA on this surface, every point of dA emits different amounts of EMR at different frequencies in different directions. Therefore in radiation heat transfer, one needs to account for dependence on temperature, direction, and wavelength, as well as incoming and outgoing radiation.

12.2.1 Solid Angle

We first define the concept of a solid angle that is the generalization of the definition of angle to three dimensions as shown in Figure 12.3. On a two-dimensional (2D) plane, an angle α is the shape formed by two rays emanating from a common point, which is the vertex of the angle. The magnitude of the angle is the "amount of rotation" that separates the two rays, defined by the length of circular arc between the two rays, L, and the distance from the arc to the vertex, r [2], or

$$\alpha = \frac{L}{r} \tag{12.4}$$

Generalizing this concept to three dimensions, a solid angle ω is the figure formed by the rays that connect an area on the surface of a sphere to the center of the sphere and its magnitude is equal to the area on the surface of the sphere divided by square of the distance from the origin

$$\omega = \frac{S}{r^2} \tag{12.5}$$

The unit for solid angle is steradian (sr) that like the radian is dimensionless. One steradian is therefore the solid angle subtended at the center of a sphere of radius r by a portion of the surface area of the sphere whose area, S, equals r^2 [3].

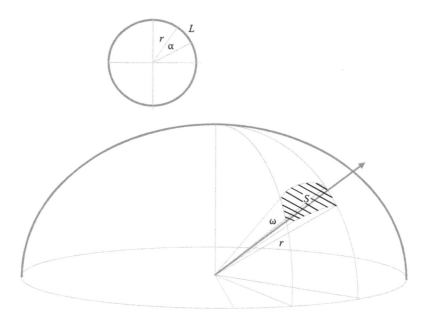

FIGURE 12.3
Two-dimensional and three-dimensional (solid) angles.

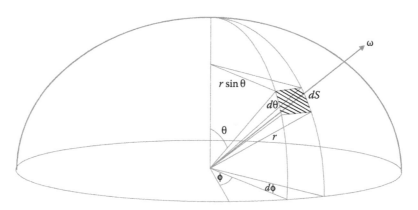

FIGURE 12.4
Solid angle in spherical coordinates.

The differential solid angle, shown in Figure 12.4, is defined as

$$d\omega = \frac{dS}{r^2} \tag{12.6}$$

where dS is the differential area on the sphere, which, using spherical coordinates, is

$$dS = (rd\theta)(r\sin\theta d\phi) \tag{12.7}$$

therefore,

$$d\omega = \sin\theta d\theta d\phi \tag{12.8}$$

where
   θ is measured from the vertical to the surface (z axis) and is called the zenith angle
   ϕ is the azimuthal angle measured from the x axis

Example 12.1

Determine the area of a spherical cap, formed by the intersection of a cone of half angle α and a sphere of radius r. Also, what value of α corresponds to 1 sr.

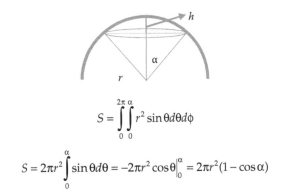

$$S = \int_0^{2\pi}\int_0^\alpha r^2 \sin\theta\, d\theta\, d\phi$$

$$S = 2\pi r^2 \int_0^\alpha \sin\theta\, d\theta = -2\pi r^2 \cos\theta\Big|_0^\alpha = 2\pi r^2(1-\cos\alpha)$$

Also, since $h = r - r\cos\alpha$, then $S = 2\pi rh$. From the definition of solid angle,

$$\omega = 2\pi(1-\cos\alpha)$$

A cone having a half angle of $1°$ subtends a solid angle of almost 0.001 sr.
For $\omega = 1$ sr, then, solving for α,

$$\alpha = \cos^{-1}\left(1 - \frac{1}{2\pi}\right) = 0.572\,\text{rad} = 32.77°$$

Distant objects may be characterized by their solid angle as shown in Figure 12.5. The solid angle subtended by an object from an origin is equal to the area of object's projection on a sphere centered at the origin having a unit radius. In Example 12.1, the ratio of the area on the sphere and the area of the cone's base is

$$\frac{S}{A} = \frac{2\pi r^2(1-\cos\alpha)}{\pi(r\sin\alpha)^2} = \frac{2(1-\cos\alpha)}{1-\cos^2\alpha} = \frac{2}{1+\cos\alpha} \tag{12.9}$$

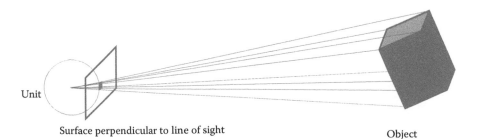

Unit

Surface perpendicular to line of sight

Object

FIGURE 12.5
Approximation of the solid angle of distance objects.

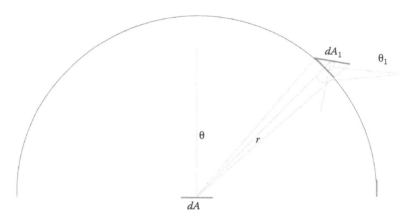

FIGURE 12.6
Solid angle of a surface at an angle θ_1 with the origin.

For small cones (small solid angles or distant objects), $\cos\alpha \approx 1$ or the area on the surface of the sphere can be approximated by the object's area projected on a plane perpendicular to the line of sight; therefore, as an approximation, rather than the area on the surface of the sphere, the area on a plane perpendicular to the line of sight can be used.

For a differential area dA_1, shown in Figure 12.6, at an angle θ_1 from the origin, the area seen from the origin is the projection of dA_1 in the direction θ, which is $dA_1 \cos\theta_1$. Therefore, the solid angle is

$$d\omega = \frac{dA_1 \cos\theta_1}{r^2} \tag{12.10}$$

12.2.2 Radiation Intensity

Consider a small disk of diameter d cm at temperature T, as shown in Figure 12.7. If q is the total amount of energy radiated by the disk, then the average heat flux can be assumed to be given by

$$q'' = \frac{q}{A} = I\cos\theta \tag{12.11}$$

where I is a constant that depends on temperature, and it will also depend on the direction defined by angles θ and ϕ. If a radiation detector is placed at $\theta = 0$, it will

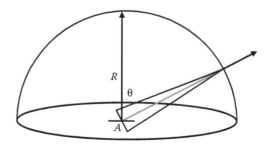

FIGURE 12.7
Radiation intensity.

detect a decreasing amount of radiation as θ is increased to 90. Dividing both sides of Equation 12.11 by cos θ

$$\frac{q}{A\cos\theta} = I \tag{12.12}$$

The term $A\cos\theta$ represents the projection of the surface area of the emitter in the direction of interest θ, and therefore, I is heat radiated per unit projected area or can be viewed as the energy that is emitted in all directions from the projection of the surface normal to the direction being considered. This quantity is called radiation intensity, which for a black-body is independent of direction (θ and φ).

The radiation emitted per unit area of the emitting surface is called the emissive power, and for a blackbody is uniformly distributed in all φ angles, but is not uniform in θ.

Now consider a differential area $dA_1 = R^2\sin\theta\,d\theta\,d\phi$ located on the surface of a sphere of radius R from the disk at an angle θ from normal to the disk. The total amount of radiation received by dA_1 is equal to the energy that is in the solid angle subtended by A from dA_1, which is

$$q = I\frac{A\cos\theta}{R^2}dA_1 \tag{12.13}$$

The term $\frac{dA_1}{R^2}$ on the right-hand side is the solid angle subtended by dA_1 when viewed from A; therefore, the amount of radiation that a surface receives from another radiating surface is equal to the intensity of radiation leaving the surface times the radiating sur-face's projected area, times the solid angle of the receiving surface. Also note that if the receiving surface, in this case dA_1, is not perpendicular to the θ direction and makes an angle θ_1, then its projection must be used.

The total amount of energy received over a sphere having a radius R becomes

$$q = \int_0^{2\pi}\int_0^{\frac{\pi}{2}} IA\cos\theta\frac{R^2\sin\theta\,d\theta\,d\phi}{R^2} = AI\int_0^{2\pi}\int_0^{\frac{\pi}{2}}\cos\theta\sin\theta\,d\theta\,d\phi \tag{12.14}$$

also

$$\int_0^{2\pi}\int_0^{\frac{\pi}{2}}\cos\theta\sin\theta\,d\theta\,d\phi = 2\pi\frac{1}{2}\sin^2\theta\Big|_0^{\frac{\pi}{2}} = \pi \tag{12.15}$$

therefore,

$$q = \pi AI \tag{12.16}$$

This is the total amount of energy that passes through a sphere located a distance R from the disk. From the first law, this is equal to the total energy emitted by the disk. The quan-tity I is the amount of energy emitted by the disk per unit projected area per unit solid angle, which is also called the total intensity. As will be seen later, for the ideal case of a blackbody, the total intensity is constant and given by

$$I = \frac{\sigma}{\pi}T^4 \tag{12.17}$$

where $\sigma = 5.67 \times 10^{-8} \text{W/m}^2\text{K}^4$ is the Stefan–Boltzmann constant. For the ideal case of a blackbody, the intensity given by Equation 12.17 is the maximum intensity at the temperature T that any surface can emit, and furthermore, as seen earlier, for a blackbody, intensity is independent of both direction and the distance from the source of radiation.

The definition of intensity can be broadened to include the wavelength effects. The spectral directional radiation intensity is defined as the energy that is leaving the surface per unit projected area in the direction of interest at a particular wavelength per unit solid angle

$$I_{\lambda,e}(\lambda,\theta,\phi) = \frac{dq}{dA\cos\theta d\omega d\lambda} \tag{12.18}$$

The directional spectral intensity is the most basic radiation property, as it accounts for area, direction, and wavelength. Solving for the amount of energy,

$$dq = I_{\lambda,e}(\lambda,\theta,\phi)dA\cos\theta d\omega d\lambda \tag{12.19}$$

Again, dq can also be viewed as the energy received by a differential surface dA_1 located at distance r and oriented at an angle θ_1 to r, making the solid angle $d\omega = \frac{dA_1\cos\theta_1}{r^2}$ when viewed from the emitting surface dA, per unit wavelength around wavelength λ, as shown in the following:

$$dq = I_{\lambda,e}(\lambda,\theta,\phi)dA\cos\theta\frac{dA_1\cos\theta_1}{r^2}d\lambda = I_{\lambda,e}(\lambda,\theta,\phi)\frac{dA\cos\theta}{r^2}dA_1\cos\theta_1 d\lambda \tag{12.20}$$

Example 12.2

Consider a surface that is a diffuse emitter having an area of 10 cm^2 that is at temperature of 500 K. Its total hemispherical emissive power is $e = 5.67 \times 10^{-8}\,T^4$. A radiation detector has a sensing area of 2 cm^2 and is placed at two different locations with different orientations as shown. Determine how much energy the detector will intercept.

The total intensity is

$$I = \frac{e}{\pi} = \frac{5.67 \times 10^{-8}(500)^4}{\pi} = 1128\,\text{W/m}^2$$

The energy intercepted by the detector is the energy in the solid angle subtended by the surface from the detector location:

$$q_1 = I\,dA\cos\theta\frac{dA_2\cos\theta_2}{r^2}d\lambda = 1128 \times (10 \times 10^{-4})\cos 20 \times \frac{(2\times10^{-4})\cos 0}{0.6^2} = 5.89 \times 10^{-4}\,\text{W at 1}$$

$$q_2 = I\,dA\cos\theta\frac{dA_2\cos\theta_2}{r^2}d\lambda = 1128 \times (10 \times 10^{-4})\cos 45 \times \frac{(2\times10^{-4})\cos 60}{0.4^2} = 4.99 \times 10^{-4}\,\text{W at 2}$$

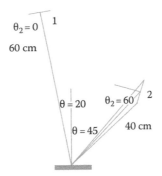

The adjective hemisphere indicates integration over all θ and φ angles or over a hemisphere above the plane. The adjective total implies integration over all wavelengths. The term diffuse refers to a quantity that is uniform over all directions, and specular means that it has directional preference or is direction dependent. For a diffuse emitter, intensity is constant and independent of direction. Another important property of intensity is that it remains constant in a given direction, along that direction, i.e., it is independent of the distance from the source, in a nonparticipating medium, such as vacuum.

If intensity is known as a function of wavelength, position on the surface, and direction, then the amount of energy emitted from a finite area A over a given solid angle and between two arbitrary wavelengths can be calculated from the following integral:

$$q = \int_A \int_{\theta_1}^{\theta_2} \int_{\phi_1}^{\phi_2} \int_{\lambda_1}^{\lambda_2} I_{\lambda,e}(\lambda,\theta,\phi)\cos\theta\sin\theta \, d\theta \, d\phi \, d\lambda \, dA \tag{12.21}$$

12.2.3 Emissive Power

An associated concept is emissive power, which is defined as the amount of energy emitted per unit area (not projected area), per unit solid angle, and per unit wavelength:

$$e_{\lambda,e}(\lambda,\theta,\phi) = \frac{dq}{dA \, d\omega \, d\lambda} \tag{12.22}$$

therefore,

$$e_{\lambda,e}(\lambda,\theta,\phi) = I_{\lambda,e}(\lambda,\theta,\phi)\cos\theta \tag{12.23}$$

Although, for a diffuse emitter, intensity is constant and independent of direction, the emissive power varies from a maximum equal to the intensity at normal to the plate, to zero at θ = 90, as shown in Figure 12.8, or for a diffuse emitter

$$e_{\lambda,e}(\lambda,0,\phi) = I_{\lambda,e}(\lambda) \tag{12.24}$$

The total directional emissive power, which is the total energy emitted over all wavelengths in a particular direction, is

$$e_e(\theta,\phi) = \int_0^\infty I_{\lambda,e}(\lambda,\theta,\phi)\cos\theta \, d\lambda \tag{12.25}$$

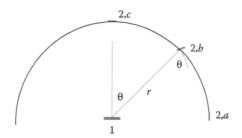

FIGURE 12.8
Emissive power variation.

And the total hemispherical emissive power, which is the total energy emitted over all wavelengths in all directions, can be determined by integrating $I_{\lambda,e}(\lambda, \theta, \phi)$ over all wavelengths and directions:

$$e = \int_0^\infty \int_0^{2\pi} \int_0^{\frac{\pi}{2}} I_{\lambda,e}(\lambda,\theta,\phi)\cos\theta\sin\theta\, d\theta\, d\phi\, d\lambda \tag{12.26}$$

For a diffuse emitter, the radiation intensity is independent of the direction, i.e., the surface emits with equal intensity in all θ and ϕ values:

$$e = I_e \int_0^{2\pi} \int_0^{\frac{\pi}{2}} \cos\theta\sin\theta\, d\theta\, d\phi = \pi I_e \tag{12.27}$$

12.3 Blackbody

A useful concept in radiation heat transfer is that of a blackbody. It is a diffuse emitter, meaning that it emits radiation with equal intensity in all directions. It also emits radiation over all wavelengths, from zero to infinity. At a given temperature and wavelength, it emits the maximum amount of radiation, i.e., no other surface can emit more. Blackbody is also a perfect absorber, i.e., it absorbs all radiations incident on it, reflects none, and hence it would appear black.

Experimentally and theoretically, it has been shown that the spectral directional intensity of a blackbody is

$$I_{\lambda,b}(\lambda) = \frac{2hC_0^2}{\lambda^5 \left[e^{\frac{hC_0}{k\lambda T}} - 1 \right]} \tag{12.28}$$

where
h is the Planck constant
$k = 1.381 \times 10^{-23}$ J/K is the Boltzmann constant

Therefore, the spectral directional emissive power of a blackbody is

$$e_{\lambda,b}(\lambda) = J_{\lambda,b}(\lambda)\cos\theta = \frac{2hC_0^2}{\lambda^5\left[e^{\frac{hC_0}{k\lambda T}}-1\right]}\cos\theta \qquad (12.29)$$

The spectral total emissive power of a blackbody is

$$e_{\lambda,e}(\lambda) = \int e_{\lambda,e}(\lambda,\theta,\phi)d\omega = \int_0^{2\pi}\int_0^{\frac{\pi}{2}} I_{\lambda,b}(\lambda,\theta,\phi)\cos\theta\sin\theta\,d\theta\,d\phi \qquad (12.30)$$

Since the intensity of the blackbody is independent of direction, from Equation 12.28,

$$e_{\lambda,b}(\lambda) = \pi I_{\lambda,b}(\lambda) = \frac{C_1}{\lambda^5\left[e^{\frac{C_2}{\lambda T}}-1\right]} \qquad (12.31)$$

where

$$C_1 = \pi 2hC_0^2 = 3.74177\times10^8\ \mathrm{W}\,\mu\mathrm{m}^4/\mathrm{m}^2$$

$$C_2 = \frac{hC_0}{k} = 1.43878\times10^4\ \mu\mathrm{m}\,\mathrm{K}$$

Figure 12.9 is a plot of the emissive power of a blackbody (Equation 12.31) as a function of the wavelength for several temperatures. It can also be viewed as π times the intensity of the blackbody. The emissive power and intensity are low at small wavelengths and reach a peak before decreasing at higher wavelengths. As the temperature increases, more and more of the energy is emitted at shorter wavelengths, and thus the wavelength at which the emissive power or intensity is maximum gets smaller.

This explains the behavior of the heating element of an electric stove. When the element is off, it appears typically as gray, meaning that it absorbs most of the incident energy that it receives from the surroundings. The human eye can detect radiation that has wavelength between 0.4 and 0.7 µm as long as the intensity of the radiation is also over a threshold. From Figure 12.9, at the ambient temperature of around 300 K, the wavelength at which most of the radiation is emitted by the element is around 10 µm, and the intensity of the radiation in the visible range is too low to be detected by the human eye. As the stove is turned on and begins to warm up, it would initially still appear gray, even though we would feel the radiation emitted from the stove. As the temperature increases, the amount of energy emitted drastically increases as does the fraction of the energy that is emitted over shorter wavelengths, and eventually there would be sufficient energy in the visible range to "see" the emitted energy, which would appear as dark red. This temperature is called Draper point [4] and is accepted to be 798 K. At this temperature, the intensity of radiation goes over 0.0145 W/m² at the wavelength of 0.7, and around this temperature the emission from the element begins to become visible. As the temperature increases, the color of the element changes, shifting to shorter and shorter values, and eventually

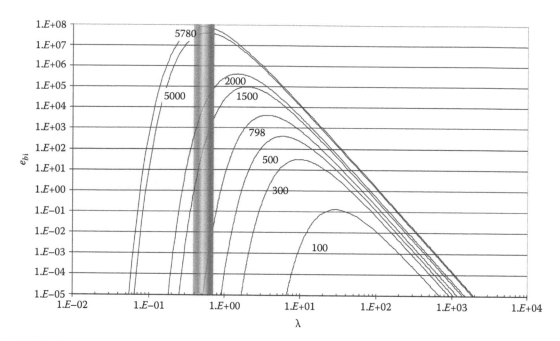

FIGURE 12.9
Spectral blackbody emissive power.

the temperature becomes high enough, and there would be enough intensity at all visible range wavelengths and the object will appear white.

The peak of the blackbody emissive power can be obtained by differentiating Equation 12.31, which results in

$$\lambda_{max}T = 2897.6 \ \mu K \tag{12.32}$$

This is known as Wiens law.

The total emissive power of a blackbody, i.e., total energy emitted over all wavelengths and angles, can be obtained by integrating Equation 12.31 over all wavelengths:

$$e_b = \int_0^\infty e_{\lambda,b}(\lambda)d\lambda = \int_0^\infty \frac{C_1}{\lambda^5 \left[e^{\frac{C_2}{\lambda T}} - 1 \right]} d\lambda = T^4 \int_0^\infty \frac{C_1}{(\lambda T)^5 \left[e^{\frac{C_2}{\lambda T}} - 1 \right]} d(\lambda T) \tag{12.33}$$

The definite integral

$$\int_0^\infty \frac{1}{x^5 \left[e^{\frac{C_2}{x}} - 1 \right]} dx = \frac{1}{15} \left(\frac{\pi}{C_2} \right)^4 \tag{12.34}$$

FIGURE 12.10
A cavity approximates a blackbody.

therefore the blackbody emissive power is

$$e_b = \sigma T^4 \tag{12.35}$$

where $\sigma = \dfrac{C_1}{15}\left(\dfrac{\pi}{C_2}\right)^4 = 5.67 \times 10^{-8}\ \text{W/m}^2\text{K}^4$ is the Stefan–Boltzmann constant. This is the maximum radiation flux that can be emitted from a surface at temperature T.

A black surface is an ideal surface, not achieved in reality; however, there are objects that come close to it. For example, an isothermal cavity (Figure 12.10) comes very close to behaving like a blackbody. Any energy incident on the opening will be absorbed and none reflected.

The radiation leaving the opening is equally likely to go in all directions, after having bounced through the cavity walls multiple times. As will be shown later, it also emits energy very close to the amount emitted by a black surface at a given temperature.

12.3.1 Solar Radiation

The sun's diameter is 1.39×10^9 m, and it is 1.5×10^{11} m from the earth. It puts out 3.85×10^{26} W of energy. Assuming the sun to be a blackbody, then its effective surface temperature can be calculated from

$$T = \left(\frac{q}{\pi D^2 \sigma}\right)^{1/4} = \left(\frac{3.85 \times 10^{26}}{\pi (1.39 \times 10^9)^2 \times 5.67 \times 10^{-8}}\right)^{1/4} = 5779.1\,\text{K}$$

or the sun can be assumed to be a blackbody at a temperature of around 5800 K. We can also calculate the intensity of the radiation emitted from the sun:

$$I = \frac{\sigma T^4}{\pi} = \frac{q}{(\pi D)^2} = 2.02 \times 10^7\ \text{W/m}^2 = 20\,\text{MW/m}^2$$

The diameter of the earth is 1.28×10^7 m (see Figure 12.11). The total amount of energy intercepted by the earth is the sun's intensity times the sun's projected area, times the earth's solid angle when viewed from the sun:

$$q = I\frac{\pi}{4}D^2 \frac{\frac{\pi}{4}d^2}{R^2} = I\left(\frac{\pi}{4}\frac{Dd}{R}\right)^2 = 2.02 \times 10^7 \left(\frac{\pi}{4}\frac{1.39 \times 10^9 \times 1.28 \times 10^7}{1.5 \times 10^{11}}\right)^2 = 1.752 \times 10^{17}\ \text{W}$$

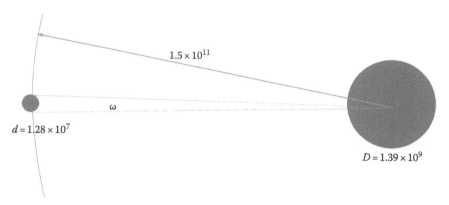

FIGURE 12.11
Solar radiation.

The total power consumption globally is around 1.5×10^7 MW, or the sun provides around 12,000 times more power than we currently consume. Dividing the energy received by the earth from the sun by the projected area of the earth

$$I = \frac{q}{\dfrac{\pi}{4}d^2} = \frac{1.752 \times 10^{17}}{\dfrac{\pi}{4}(1.28 \times 10^7)^2} = 1362 \, \text{W/m}^2$$

This is known as the solar constant, which is the amount of energy an object outside the atmosphere will receive per unit area. This is also known as total solar irradiance (TSI) and can also be calculated by dividing the total energy emitted by the sun over the area of the sphere whose radius is the distance between the sun and the earth:

$$I = \frac{q}{4\pi R^2} = \frac{3.85 \times 10^{26}}{4\pi(1.5 \times 10^{11})^2} = 1362 \, \text{W/m}^2$$

This is an approximate value, and the different satellite measurements have not been identical, but they all are close to this value. All satellite measurements show that TSI is not constant and not only varies daily due to earth's elliptical orbit around the sun, but also cyclically over what are known as solar cycles (or sunspot cycles) that happen roughly every 11 years.

The energy intercepted by the earth gets distributed over its surface; therefore, the average solar energy received per unit area on the surface of the earth is

$$I_i = \frac{q_i}{\pi D^2} = \frac{1.752 \times 10^{17}}{\pi(1.28 \times 10^7)^2} = 340 \, \text{W/m}^2$$

This is the average amount of solar power during a 24 h period. The energy received at night is zero; thus, the average daily solar flux is 680 W/m², that considers only the sunlit areas of the earth, which is half of its surface area. Also, the average energy received varies greatly depending on the latitude. At noon, at a location on the equator, the solar flux on the earth would be 1362 W/m². Of course, these numbers do not consider the portions of solar radiation that are reflected from or absorbed by the atmosphere, before reaching the surface of the earth.

12.3.2 Blackbody Fraction

The total amount of energy emitted between any two wavelengths is given by

$$e_b(\lambda_1,\lambda_2) = \int_{\lambda_1}^{\lambda_2} e_{\lambda,b}(\lambda)d\lambda = \int_{\lambda_1}^{\lambda_2} \frac{C_1}{\lambda^5\left[e^{\frac{C_2}{\lambda T}}-1\right]}d\lambda = \sigma T^4 \int_{(\lambda T)_1}^{(\lambda T)_2} \frac{C_1}{\sigma(\lambda T)^5\left[e^{\frac{C_2}{\lambda T}}-1\right]}d(\lambda T) \qquad (12.36)$$

Therefore, we can define the fraction of the total energy that is between any two wavelengths as

$$f\left[(\lambda T)_1,(\lambda T)_2\right] = \frac{e_b\left[(\lambda T)_1,(\lambda T)_2\right]}{\sigma T^4} = \int_0^{\lambda_2 T} \frac{C_1}{\sigma(\lambda T)^5\left[e^{\frac{C_2}{\lambda T}}-1\right]}d(\lambda T) - \int_0^{\lambda_1 T} \frac{C_1}{\sigma(\lambda T)^5\left[e^{\frac{C_2}{\lambda T}}-1\right]}d(\lambda T)$$

$$(12.37)$$

Each integral is only a function of λT Figure 12.12 is a plot of the fractions of blackbody radiation between zero and λT, and Table 12.1 shows the select numerical values.

Example 12.3

Determine the fractions of the sun's radiation that are in the ultraviolet, visible, and IR regions:

$$0.01 \times 5800 = 58 \qquad f_{0-58} = 0.000$$
$$0.4 \times 5800 = 2320 \qquad f_{0-2320} = 0.122 \qquad 12.4\% \text{ UV}$$
$$0.7 \times 5800 = 4060 \qquad f_{0-4060} = 0.492 \qquad 36.8\% \text{ Visible}$$
$$100 \times 5800 = 580,000 \qquad f_{0-580,000} = 1 \qquad 50.8\% \text{ IR}$$

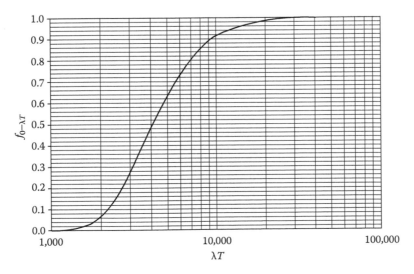

FIGURE 12.12
Blackbody fraction.

TABLE 12.1

Blackbody Fraction

λT (μK)	$f_{0-\lambda T}$	λT (μK)	$f_{0-\lambda T}$	λT (μK)	$f_{0-\lambda T}$	λT (μK)	$f_{0-\lambda T}$
0	0.000000	4300	0.532698	6700	0.789794	9500	0.903093
500	0.000000	4350	0.540820	6750	0.792995	9950	0.913173
1000	0.000328	4400	0.548798	6800	0.796135	10,000	0.914207
1100	0.000924	4450	0.556631	6850	0.799216	10,500	0.923718
1200	0.002154	4500	0.564322	6900	0.802239	11,000	0.931898
1300	0.004344	4550	0.571872	6950	0.805205	11,500	0.938968
1400	0.007824	4600	0.579283	7000	0.808116	12,000	0.945106
1500	0.012888	4650	0.586555	7050	0.810971	12,500	0.950459
1600	0.019761	4700	0.593691	7100	0.813774	13,000	0.955147
1700	0.028577	4750	0.600693	7150	0.816525	13,500	0.959269
1800	0.039385	4800	0.607562	7200	0.819224	14,000	0.962906
1900	0.052149	4850	0.614300	7250	0.821874	14,500	0.966127
2000	0.066769	4900	0.620910	7300	0.824475	15,000	0.968989
2100	0.083089	4950	0.627392	7350	0.827028	15,500	0.971541
2200	0.100922	5000	0.633750	7400	0.829534	16,000	0.973822
2300	0.120059	5050	0.639985	7450	0.831994	16,500	0.975869
2400	0.140282	5100	0.646100	7500	0.834410	17,000	0.977709
2500	0.161378	5150	0.652095	7550	0.836781	17,500	0.979368
2600	0.183140	5200	0.657974	7600	0.839110	18,000	0.980868
2700	0.205374	5250	0.663738	7650	0.841396	18,500	0.982227
2800	0.227904	5300	0.669390	7700	0.843642	19,000	0.983461
2900	0.250573	5350	0.674931	7750	0.845847	19,500	0.984585
3000	0.273241	5400	0.680364	7800	0.848012	20,000	0.985610
3050	0.284536	5450	0.685690	7850	0.850139	20,500	0.986546
3100	0.295787	5500	0.690912	7900	0.852228	21,000	0.987404
3150	0.306981	5550	0.696032	7950	0.854280	21,500	0.988191
3200	0.318107	5600	0.701051	8000	0.856296	22,000	0.988914
3250	0.329155	5650	0.705972	8050	0.858276	22,500	0.989580
3300	0.340115	5700	0.710796	8100	0.860221	23,000	0.990194
3350	0.350979	5750	0.715526	8150	0.862133	23,500	0.990761
3400	0.361739	5800	0.720164	8200	0.864011	24,000	0.991286
3450	0.372387	5850	0.724710	8250	0.865856	24,500	0.991771
3500	0.382919	5900	0.729168	8300	0.867669	25,000	0.992222
3550	0.393328	5950	0.733538	8350	0.869451	25,500	0.992641
3600	0.403609	6000	0.737823	8400	0.871203	26,000	0.993030
3650	0.413759	6050	0.742025	8450	0.872924	26,500	0.993393
3700	0.423773	6100	0.746145	8500	0.874615	27,000	0.993732
3750	0.433649	6150	0.750185	8550	0.876278	27,500	0.994047
3800	0.443384	6200	0.754146	8600	0.877913	28,000	0.994343
3850	0.452976	6250	0.758031	8650	0.879520	28,500	0.994619
3900	0.462423	6300	0.761840	8700	0.881099	29,000	0.994877
3950	0.471724	6350	0.765576	8750	0.882653	30,000	0.995348
4000	0.480878	6400	0.769240	8800	0.884180	35,000	0.997011
4050	0.489884	6450	0.772833	8850	0.885681	40,000	0.997975
4100	0.498743	6500	0.776358	8900	0.887158	45,000	0.998572
4150	0.507453	6550	0.779815	8950	0.888610	50,000	0.998961
4200	0.516015	6600	0.783205	9000	0.890037	∞	1.000000
4250	0.524430	6650	0.786531	9100	0.892823		

Example 12.4

Analyze a greenhouse whose glass transmits all of the radiations between 0.2 and 3.5 µM. Sun is at 5800 K

$$0.2 \times 5800 = 1160 \qquad f_{0-1160} = 0.001$$
$$3.5 \times 5800 = 20,300 \quad f_{0-20,300} = 0.987$$

Earth

$$0.2 \times 323 = 64.6 \qquad f_{0-64.6} = 0.000$$
$$3.5 \times 323 = 1130 \qquad f_{0-1130} = 0.001$$

which means that 98% of the sun's incident energy will pass through into the greenhouse, but less than 0.001% of radiation emitted by the greenhouse will leave the greenhouse glass by radiation. Of course, the green house looses energy by other mechanisms.

12.3.3 Irradiation

A surface may receive energy radiatively from many sources. Therefore, we can define the spectral directional intensity of incident energy as the total amount of energy per unit projected area, per unit wavelength, and per unit solid angle that is incident on the surface:

$$I_{\lambda,i}(\lambda,\theta,\phi) = \frac{dq}{dA \cos\theta \, d\omega \, d\lambda} \tag{12.38}$$

The total energy incident per unit area on the surface at a particular wavelength is obtained by integrating the incident emissive power over the hemisphere above the surface. This incident flux is termed irradiation G; thus, the spectral irradiation is

$$\frac{dq_\lambda}{dA} = q''_\lambda = G_\lambda = \int_0^{2\pi}\int_0^{\frac{\pi}{2}} I_{\lambda,i}(\lambda,\theta,\phi)\cos\theta\sin\theta \, d\theta \, d\phi \tag{12.39}$$

If the intensity of the incident radiation is known, then the incident flux can be calculated. Note that irradiation is defined in terms of the actual area and not the projected area, but the intensity is defined in terms of projected area and thus the cosine term. The total hemispherical irradiation is thus

$$G = \int_0^\infty G_\lambda \, d\lambda = \int_0^\infty\int_0^{2\pi}\int_0^{\frac{\pi}{2}} I_{\lambda,i}(\lambda,\theta,\phi)\cos\theta\sin\theta \, d\theta \, d\phi \, d\lambda \tag{12.40}$$

If the incident radiation is diffuse, i.e., it has the same intensity in all directions, then it can be pulled out of the integral and irradiation becomes

$$G = \pi \int_0^\infty I_{\lambda,i}(\lambda) \, d\lambda \tag{12.41}$$

12.3.4 Surface Properties

Irradiation is defined as the energy that is incident on a surface. This incident energy can be reflected, absorbed, or transmitted through the surface, as shown in Figure 12.13.

The energy can be reflected diffusely, meaning that it reflects with equal intensity in all directions, or it can reflect specularly, which means it reflects in a particular direction, similar to a polished surface. The directional, spectral reflectivity, $\rho_\lambda(\lambda, \theta, \phi)$, is the fraction of the incoming energy that is reflected in a particular direction at a particular wavelength:

$$\rho_\lambda(\lambda, \theta, \phi) = \frac{I_{\lambda, i, ref}(\lambda, \theta, \phi)}{I_{\lambda, i}(\lambda, \theta, \phi)} \tag{12.42}$$

The spectral hemispherical reflectivity is a number that when multiplied by the incident radiation from all directions at a particular wavelength provides the amount of energy reflected in all directions at that wavelength, which is equal to the ratio of the integrals of the numerator and denominator of Equation 12.42 over the hemisphere over the surface:

$$\rho_\lambda = \frac{G_{\lambda, ref}}{G_\lambda} = \frac{\int_0^{2\pi} \int_0^\pi \rho_\lambda(\lambda, \theta, \phi) I_{\lambda, i}(\lambda, \theta, \phi) \cos\theta \sin\theta \, d\theta \, d\phi}{\int_0^{2\pi} \int_0^\pi I_{\lambda, i}(\lambda, \theta, \phi) \cos\theta \sin\theta \, d\theta \, d\phi} \tag{12.43}$$

Note that the denominator is the spectral irradiation, given by Equation 12.39. If the incident radiation is diffuse, then

$$\rho_\lambda = \frac{1}{\pi} \int_0^{2\pi} \int_0^\pi \rho_\lambda(\lambda, \theta, \phi) \cos\theta \sin\theta \, d\theta \, d\phi \tag{12.44}$$

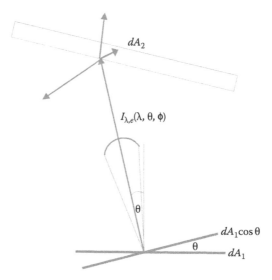

FIGURE 12.13
Directional spectral surface properties.

The total hemispherical reflectivity is the fraction of the total irradiation that is reflected:

$$\rho = \frac{G_{ref}}{G} = \frac{\int_0^\infty \rho_\lambda G_\lambda d\lambda}{\int_0^\infty G_\lambda d\lambda} \tag{12.45}$$

Again, the denominator is the total hemispherical irradiation given by Equation 12.40. Similarly, the directional, spectral absorptivity, $\alpha_\lambda(\lambda, \theta, \phi)$, is the fraction of the incoming energy that is absorbed in a particular direction at a particular wavelength:

$$\alpha_\lambda\left(\lambda,\theta,\phi\right) = \frac{I_{\lambda,i,abs}\left(\lambda,\theta,\phi\right)}{I_{\lambda,i}\left(\lambda,\theta,\phi\right)} \tag{12.46}$$

The spectral hemispherical absorptivity and the total hemispherical absorptivity are obtained by replacing ρ with α in Equations 12.44 and 12.45, respectively.

Finally, the directional, spectral transmissivity, $\tau_\lambda(\lambda, \theta, \phi)$, is the fraction of the incoming energy that is transmitted through the surface in a particular direction at a particular wavelength:

$$\tau_\lambda\left(\lambda,\theta,\phi\right) = \frac{I_{\lambda,i,trans}\left(\lambda,\theta,\phi\right)}{I_{\lambda,i}\left(\lambda,\theta,\phi\right)} \tag{12.47}$$

The spectral hemispherical transmissivity and the total hemispherical transmissivity are obtained by replacing ρ with τ in Equations 12.44 and 12.45, respectively. Again, if the incident radiation is diffuse, then spectral emissivity and spectral transmissivity can be obtained from Equation 12.44 by replacing ρ with α and τ, respectively.

From the definitions of the surface properties,

$$I_{\lambda,i}(\lambda,\theta,\phi) = I_{\lambda,i}(\lambda,\theta,\phi)\rho_\lambda(\lambda,\theta,\phi) + I_{\lambda,i}(\lambda,\theta,\phi)\alpha_\lambda(\lambda,\theta,\phi) + I_{\lambda,i}(\lambda,\theta,\phi)\tau_\lambda(\lambda,\theta,\phi) \tag{12.48}$$

or

$$\rho_\lambda(\lambda,\theta,\phi) + \alpha_\lambda(\lambda,\theta,\phi) + \tau_\lambda(\lambda,\theta,\phi) = 1 \tag{12.49}$$

and in terms of total hemispherical properties

$$\rho + \alpha + \tau = 1 \tag{12.50}$$

For an opaque surface, $\tau = 0$.

12.3.5 Gray Body

Real surfaces emit energy less than that of a black surface, and to assess how close a real surface comes to the ideal one, black surface, the concept of emissivity is used. Its most fundamental form is the spectral directional emissivity is defined as

$$\varepsilon(\lambda,\theta,\phi) = \frac{e_{\lambda,e}(\lambda,\theta,\phi)}{e_{\lambda,b}(\lambda,\theta,\phi)} = \frac{e_{\lambda,e}(\lambda,\theta,\phi)}{\pi I_{\lambda,b}(\lambda)} \tag{12.51}$$

which is the ratio of the amount of energy emitted per unit projected area, in a particular direction, for a given wavelength, at a specified temperature over the same quantity for a blackbody. The denominator is given by Equation 12.28 and is independent of the direction. Multiplying the numerator and denominator by cos θ, emissivity is also the ratio of the appropriate emissive powers. For engineering calculations, we are interested in calculating energy and not intensity, and thus we need average values. The adjectives like total and/or hemispherical used for the surface properties effectively refer to different averages of the basic quantities. In general, the average values are defined as numbers that, when multiplied by the maximum possible energy, provide the actual energy. For example, the total hemispherical reflectivity is a number that, when multiplied by the maximum amount of energy that could be reflected over all wavelengths, in all directions, which is the irradiation, provides the actual amount of energy reflected over all wavelengths in all directions. Similarly, the total hemispherical emissivity can be obtained from

$$\varepsilon = \frac{e}{e_b} = \frac{\int_0^\infty \varepsilon_\lambda e_{b,\lambda} d\lambda}{\sigma T^4} \tag{12.52}$$

An ideal surface that closely approximates real surfaces is a gray surface that is a surface whose properties are independent of the wavelength and direction. Consider a gray body in a large enclosure as shown in Figure 12.14. If the two surfaces are in thermal equilibrium, then the energy absorbed by the gray body must be equal to the energy that it emits in order for it to remain in thermal equilibrium. Let irradiation, G, be the total amount of energy incident per unit area that comes to surface from all other sources, in this case the large enclosure, then

$$G\alpha = \varepsilon e_b \tag{12.53}$$

If we replace the gray surface with a blackbody at the same temperature, then

$$G(1) = (1)e_b \tag{12.54}$$

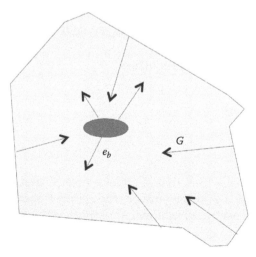

FIGURE 12.14
A small isothermal object surrounded by an isothermal enclosure.

and therefore for a gray surface,

$$\alpha = \varepsilon \qquad (12.55)$$

Although we obtained this for total absorptivity and emissivity when the surfaces are in thermal equilibrium, it also holds for directional and spectral quantities, i.e.,

$$\alpha(\lambda, \theta, \phi) = \varepsilon(\lambda, \theta, \phi) \qquad (12.56)$$

12.3.6 Shape Factor

Consider the two surfaces shown in Figure 12.15. Every point on surface i emits radiation in all directions. Only a portion of the energy that is emitted by surface i reaches surface j. Shape factor, also referred to as view or configuration factor, denoted by F_{ij}, is defined as the fraction of energy leaving surface i and reaching surface j. This is a purely geometrical factor that depends on the shapes of the two surfaces and their relative orientations. It can be shown that

$$F_{ij} = \frac{1}{A_i} \int\limits_{A_i} \int\limits_{A_j} \frac{\cos\theta_i \cos\theta_j}{\pi r^2} dA_j dA_i \qquad (12.57)$$

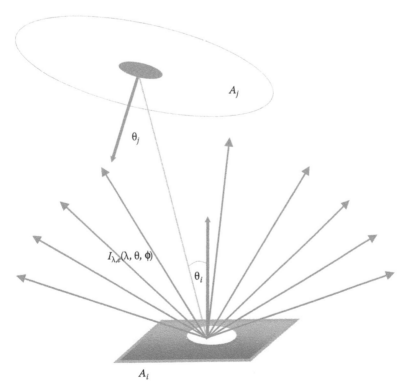

FIGURE 12.15
View factor definition.

and

$$F_{ji} = \frac{1}{A_j} \int\limits_{A_j} \int\limits_{A_i} \frac{\cos\theta_i \cos\theta_j}{\pi r^2} dA_i dA_j \tag{12.58}$$

and since the order of the integration is not important, then

$$A_i F_{ij} = A_j F_{ji} \tag{12.59}$$

This is known as the reciprocity relation. The expressions for shape factors for many geometries have been evaluated and some of those are listed in Table 12.2 [5].

For enclosures, from the definition of shape factors,

$$\sum_{j=1}^{N} F_{ij} = 1 \tag{12.60}$$

Also note that if a surface is concave, i.e., it can see itself, its shape factor is not zero.

Example 12.5

Determine the shape factor between two disks having radii R_1 and R_2:

$$F_{ij} = \frac{1}{A_i} \int\limits_{A_i} \int\limits_{A_j} \frac{\cos\theta_i \cos\theta_j}{\pi r^2} dA_j dA_i$$

$$F_{12} = \frac{1}{A_1} \int\limits_0^{R_1} \int\limits_0^{2\pi} \int\limits_0^{R_2} \int\limits_0^{2\pi} \frac{\cos\theta_1 \cos\theta_2}{\pi r^2} dA_2 dA_1$$

$$\cos\theta_1 = \cos\theta_2 = \frac{D}{r}$$

$$dA_1 = r_1 d\phi_1 dr_1$$

$$dA_2 = r_2 d\phi_2 dr_2$$

$$b^2 = r_1^2 + r_2^2 - 2r_1 r_2 \cos(\phi_1 - \phi_2)$$

$$r^2 = D^2 + b^2$$

After much algebra, the expression given in Table 12.2 is obtained.

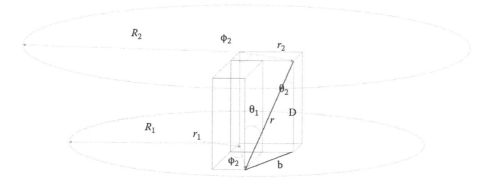

TABLE 12.2

View Factor Expressions

Differential surface parallel to a finite rectangular surface

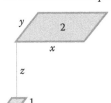

$$F_{12} = \frac{1}{2\pi}\left[\frac{y^*}{\sqrt{1+y^{*2}}}\tan^{-1}\left(\frac{x^*}{\sqrt{1+y^{*2}}}\right) + \frac{x^*}{\sqrt{1+x^{*2}}}\tan^{-1}\left(\frac{y^*}{\sqrt{1+x^{*2}}}\right)\right]$$

$$x^* = \frac{x}{z}, y^* = \frac{y}{z}$$

Differential surface perpendicular to a finite rectangular surface

$$F_{12} = \frac{1}{2\pi}\left[\tan^{-1}(y^*) - \frac{1}{\sqrt{1+x^{*2}}}\tan^{-1}\left(\frac{y^*}{\sqrt{x^{*2}+y^{*2}}}\right)\right]$$

$$x^* = \frac{x}{z}, y^* = \frac{y}{z}$$

Differential spherical surface to a finite rectangular surface

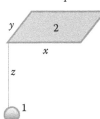

$$F_{12} = \frac{1}{4\pi}\sin^{-1}\left(\frac{x^* y^*}{\sqrt{1+x^{*2}+y^{*2}+x^{*2}y^{*2}}}\right)$$

$$x^* = \frac{x}{z}, y^* = \frac{y}{z}$$

Two parallel disks

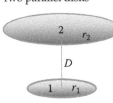

$$F_{12} = \frac{1+r_1^{*2}+r_2^{*2} - \sqrt{\left(1+r_1^{*2}+r_2^{*2}\right)^2 - 4r_1^{*2}r_2^{*2}}}{2r_1^{*2}}$$

$$r_1^* = \frac{r_1}{D}, r_2^* = \frac{r_2}{D}$$

Two parallel rectangles

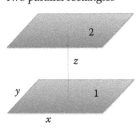

$$F_{12} = \frac{1}{2x^* y^*}\left[\begin{array}{l}\frac{1}{2}\ln\frac{(1+x^{*2})(1+y^{*2})}{1+x^{*2}+y^{*2}} + x^*\sqrt{(1+y^{*2})}\tan^{-1}\frac{x^*}{\sqrt{(1+y^{*2})}} \\ +y^*\sqrt{(1+x^{*2})}\tan^{-1}\frac{y^*}{\sqrt{(1+x^{*2})}} - x^*\tan^{-1}x^* - y^*\tan^{-1}y^*\end{array}\right]$$

$$x^* = \frac{x}{z}, y^* = \frac{y}{z}$$

(continued)

TABLE 12.2 (continued)

View Factor Expressions

Two perpendicular rectangles

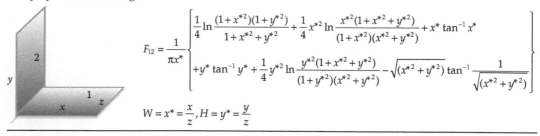

$$F_{12} = \frac{1}{\pi x^*} \left\{ \begin{array}{l} \dfrac{1}{4}\ln\dfrac{(1+x^{*2})(1+y^{*2})}{1+x^{*2}+y^{*2}} + \dfrac{1}{4}x^{*2}\ln\dfrac{x^{*2}(1+x^{*2}+y^{*2})}{(1+x^{*2})(x^{*2}+y^{*2})} + x^* \tan^{-1}x^* \\[2ex] +y^* \tan^{-1}y^* + \dfrac{1}{4}y^{*2}\ln\dfrac{y^{*2}(1+x^{*2}+y^{*2})}{(1+y^{*2})(x^{*2}+y^{*2})} - \sqrt{(x^{*2}+y^{*2})}\,\tan^{-1}\dfrac{1}{\sqrt{(x^{*2}+y^{*2})}} \end{array} \right\}$$

$$W = x^* = \frac{x}{z},\ H = y^* = \frac{y}{z}$$

Example 12.6

Find F_{12} for triangular enclosure that is infinitely long.

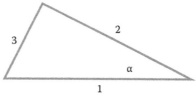

Since none of the surfaces see themselves,

$$F_{12} + F_{13} = 1$$
$$F_{21} + F_{23} = 1$$
$$F_{31} + F_{32} = 1$$

Multiplying each equation with the corresponding area,

$$A_1 F_{12} + A_1 F_{13} = A_1$$
$$A_2 F_{21} + A_2 F_{23} = A_2$$
$$A_3 F_{31} + A_3 F_{32} = A_3$$

Using reciprocity in the last two equations,

$$A_1 F_{12} + A_1 F_{13} = A_1$$
$$A_1 F_{12} + A_2 F_{23} = A_2$$
$$A_1 F_{13} + A_2 F_{23} = A_3$$

Here, we have three equations for three unknowns, resulting in

$$F_{12} = \frac{A_1 + A_2 - A_3}{2A_1}$$

$$F_{13} = \frac{A_1 + A_3 - A_2}{2A_1}$$

$$F_{23} = \frac{A_2 + A_3 - A_1}{2A_2}$$

For an isosceles triangle, $A_1 = A_2$

$$F_{12} = \frac{A_1 + A_2 - A_3}{2A_1} = 1 - \frac{\dfrac{A_3}{2}}{A_1} = 1 - \sin^{-1}\frac{\alpha}{2}$$

where α is the angle between the two sides.

12.3.7 Crossed String Method

For 2D surfaces, i.e., those that are infinitely long in the z direction, the shape factors can be determined by the cross string method. The method works, even if the surfaces can see themselves.

Consider surfaces 1 and 4 shown in Figure 12.16. Imagine that we attach a string from each end of surface 1 to each end of surface 4. We end of with four strings, two of which will cross. From the definition of shape factor, assuming surface 1 does not see itself,

$$F_{12} + F_{14} + F_{12'} = 1 \tag{12.61}$$

Multiplying by A_1 and rearranging,

$$A_1 F_{14} = A_1 - A_1 F_{12} - A_1 F_{12'} \tag{12.62}$$

From the previous example, for the triangular enclosures 123 and 12'3',

$$A_1 F_{12} = \frac{A_1 + A_2 - A_3}{2} \tag{12.63}$$

$$A_1 F_{12'} = \frac{A_1 + A_{2'} - A_{3'}}{2} \tag{12.64}$$

Therefore,

$$A_1 F_{14} = A_1 - \frac{A_1 + A_2 - A_3}{2} - \frac{A_1 + A_{2'} - A_{3'}}{2} \tag{12.65}$$

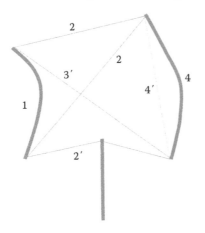

FIGURE 12.16
Crossed string method.

which simplifies to

$$A_1 F_{14} = \frac{(A_3 + A_{3'}) - (A_{2'} + A_2)}{2} \tag{12.66}$$

note that the first parenthesis is the total length of the crossed strings and the second one represents the total length of the uncrossed strings.

Example 12.7

Find F_{12}

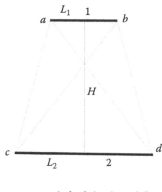

$$A_1 F_{12} = \frac{(ad + bc) - (ac + bd)}{2}$$

and because of symmetry

$$A_1 F_{12} = ad - ac$$

$$ad = \sqrt{H^2 + \left(cd - \frac{cd - ab}{2}\right)^2}$$

$$ac = \sqrt{H^2 + \left(\frac{cd - ab}{2}\right)^2}$$

$$L_1 F_{12} = \sqrt{H^2 + \left(L_2 - \frac{L_2 - L_1}{2}\right)^2} - \sqrt{H^2 + \left(\frac{L_2 - L_1}{2}\right)^2}$$

$$F_{12} = \frac{\sqrt{4 + \left(L_2^* + L_1^*\right)^2} - \sqrt{4 + \left(L_2^* - L_1^*\right)^2}}{2L_1^*}$$

12.4 Radiation Exchange between Surfaces

Radiation exchange is generally broken down between those situations where the exchange is only between surfaces and those where there is a participating medium like a liquid or a gas. We first consider surface energy exchange.

Consider N surfaces participating in exchanging energy radiatively as shown in Figure 12.17. If the surfaces do not form an enclosure, we add a hypothetical surface to turn the system into one. We will later examine the characteristics of this imaginary surface. In case of Figure 12.17, there are five surfaces and surface 6 is added to turn the system into an enclosure. Also note that in radiation heat transfer, if surfaces are at the same temperature and have the same properties, we consider them to be one surface, even if geometrically they are not the same surface. On the other hand, if two surfaces are on the same plane, if they are not at the same temperature or if their properties like emissivity or absorptivity are different, then they should be treated as different surfaces. For example, in Figure 12.17, even though surfaces 3 and 4 are geometrically the same surface, they are treated as two separate surfaces, since they are at different temperatures.

Consider surface 1 of the enclosure as shown in Figure 12.18. The total amount of incident energy per unit area is called irradiation and is denoted by G. It is the energy that reaches surface 1 from all other surfaces. A portion of the energy arriving at the surface will be absorbed, and the rest will be reflected, assuming the surface is opaque. The total amount of energy leaving the surface is sum of the energy reflected and what is emitted by the surface. This is called radiosity, J,

$$J = \varepsilon e_b + \rho G \qquad (12.67)$$

For the temperature of surface 1 to remain constant, it must exchange energy with the surroundings. Let q_1 be the amount of energy that must be supplied to surface 1 to keep its temperature at T_1. Then

$$q_1 = A_1 J_1 - A_1 G_1 \qquad (12.68)$$

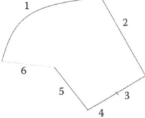

FIGURE 12.17
Radiation exchange in an enclosure.

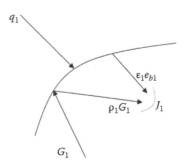

FIGURE 12.18
Surface energy balance.

The total energy that surface 1 receives by radiation is the sum of all the energies leaving the other surfaces and reaching surface 1 or

$$A_1 G_1 = A_1 J_1 F_{11} + A_2 J_2 F_{21} + A_3 J_3 F_{31} + A_4 J_4 F_{41} + \cdots \tag{12.69}$$

Using reciprocity,

$$A_1 G_1 = J_1 A_1 F_{11} + J_2 A_1 F_{12} + J_3 A_1 F_{13} + J_4 A_1 F_{14} + J_5 A_1 F_{15} \tag{12.70}$$

which can be written using summation notation as

$$A_1 G_1 = \sum_{j=1}^{N} J_j A_1 F_{1j} \tag{12.71}$$

since for an enclosure

$$\sum_{j=1}^{N} F_{1j} = 1 \tag{12.72}$$

Substituting Equation 12.71 in Equation 12.68 and multiplying the first term by the left-hand side of Equation 12.72,

$$q_1 = A_1 J_1 \sum_{j=1}^{N} F_{1j} - \sum_{j=1}^{N} J_j A_1 F_{1j} = \sum_{j=1}^{N} A_1 J_1 F_{1j} - \sum_{j=1}^{N} J_j A_1 F_{1j} = \sum_{j=1}^{N} (J_1 - J_j) A_1 F_{1j} \tag{12.73}$$

also

$$J = \varepsilon e_b + (1 - \varepsilon) G \tag{12.74}$$

Solving for irradiation,

$$G = \frac{J - \varepsilon e_b}{(1 - \varepsilon)} \tag{12.75}$$

Substituting from Equation 12.75 into Equation 12.68 to eliminate irradiation,

$$q_1 = A_1 J_1 - A_1 \frac{J_1 - \varepsilon_1 e_{b1}}{(1 - \varepsilon_1)} = \frac{A_1 \varepsilon_1}{1 - \varepsilon_1} (e_{b1} - J_1) \tag{12.76}$$

Therefore, for an enclosure with N surfaces, we get $2N$ equations

$$q_i = \sum_{j=1}^{N} (J_i - J_j) A_i F_{ij} \tag{12.77}$$

$$q_i = \frac{A_i \varepsilon_i}{1 - \varepsilon_i} (e_{bi} - J_i) \tag{12.78}$$

for a maximum of $2N$ unknown combination of $T_i(e_{bi})$ and q_i. If all the temperatures are known, then eliminating qs between the two equations results in N equations for N unknown Js:

$$\sum_{j=1}^{N} (J_i - J_j)A_iF_{ij} = \frac{A_i\varepsilon_i}{1-\varepsilon_i}(e_{bi} - J_i) \tag{12.79}$$

which can then be used in Equation 12.78 to find the heat transfer to each surface.

12.4.1 Radiation Exchange between Two Surfaces

Consider two surfaces that exchange energy radiatively, as shown in Figure 12.19. Starting with Equations 12.77 and 12.78, we get the following four equations:

$$q_1 = (J_1 - J_2)A_1F_{12} \tag{12.80}$$

$$q_2 = (J_2 - J_1)A_2F_{21} \tag{12.81}$$

$$q_1 = \frac{A_1\varepsilon_1}{1-\varepsilon_1}(e_{b1} - J_1) \tag{12.82}$$

$$q_2 = \frac{A_2\varepsilon_2}{1-\varepsilon_2}(e_{b2} - J_2) \tag{12.83}$$

Eliminating q_1 and q_2, we get two equations for the radiosities

$$(J_1 - J_2)F_{12} = \frac{\varepsilon_1}{1-\varepsilon_1}(e_{b1} - J_1) \tag{12.84}$$

$$(J_2 - J_1)F_{21} = \frac{\varepsilon_2}{1-\varepsilon_2}(e_{b2} - J_2) \tag{12.85}$$

After much algebra,

$$J_1 = \frac{\dfrac{1}{\left(F_{21} + \dfrac{\varepsilon_2}{(1-\varepsilon_2)}\right)}\dfrac{\varepsilon_2}{1-\varepsilon_2}e_{b2} + \dfrac{1}{F_{12}}\dfrac{\varepsilon_1}{1-\varepsilon_1}}{\left(1 + \dfrac{\varepsilon_1}{1-\varepsilon_1}\right) - \dfrac{F_{21}}{F_{21} + \dfrac{\varepsilon_2}{1-\varepsilon_2}}} \tag{12.86}$$

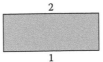

FIGURE 12.19
Radiation exchange between two surfaces.

which can be substituted in Equation 12.84 to find J_2. Knowing the radiosities, the heat transfer to each surface can be found after some algebra resulting in

$$q_1 = \frac{e_{b1} - e_{b2}}{\dfrac{1-\varepsilon_1}{A_1\varepsilon_1} + \dfrac{1}{A_1} + \dfrac{1-\varepsilon_2}{A_2\varepsilon_2}} \tag{12.87}$$

The amount of heat that must be transferred to surface 2 is negative of this value.

12.4.2 Gray Surface Exchange Electrical Analogy

For cases involving three or fewer surfaces, the algebra simplifies greatly by using electrical analogy. Equations 12.77 and 12.78 can be written as

$$q_i = \frac{A_i\varepsilon_i}{1-\varepsilon_i}(e_{bi} - J_i) = \frac{e_{bi} - J_i}{\dfrac{1-\varepsilon_i}{A_i\varepsilon_i}} \tag{12.88}$$

$$q_i = \sum_{j=1}^{N} A_i F_{ij}(J_i - J_j) = \sum_{j=1}^{N} \frac{J_i - J_j}{\dfrac{1}{A_i F_{ij}}} \tag{12.89}$$

Equation 12.88 can be interpreted to indicate that associated with each surface there is a surface resistance equal to

$$\text{Surface resistance} = \frac{1-\varepsilon_i}{A_i\varepsilon_i} \tag{12.90}$$

Equation 12.89 indicates that associated with each pair of surfaces, there is a space resistance given by

$$\text{Space resistance} = \frac{1}{A_i F_{ij}} \tag{12.91}$$

The electrical circuit for radiation exchange between two surfaces is shown in Figure 12.20 and is constructed by starting with the blackbody emissive power of one of the surfaces, for example, surface 1, which is e_{b1}. Associated with surface 1, there is a surface resistance equal to $\dfrac{1-\varepsilon_1}{A_1\varepsilon_1}$. The potential after this resistance drops to J_1. Since there are only two surfaces that exchange energy radiatively, there is one space resistance equal to $\dfrac{1}{A_1 F_{12}}$, which from reciprocity is also equal to $\dfrac{1}{A_2 F_{21}}$. After this resistance the potential drops to J_2, behind which

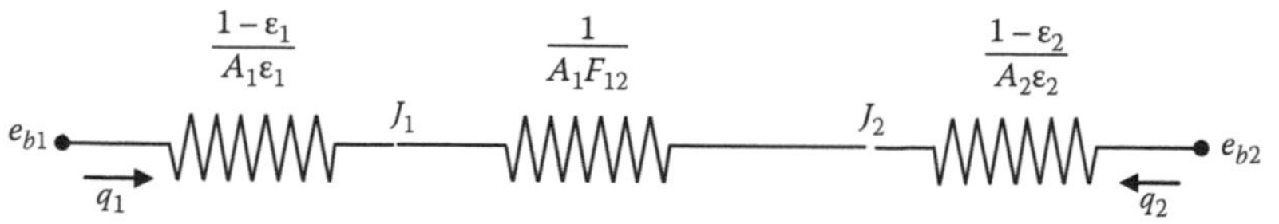

FIGURE 12.20
Equivalent electrical network for radiation exchanging between two surfaces.

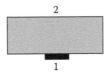

FIGURE 12.21
Small surface 1 exchanging energy with a much larger surface 2.

there is another surface resistance associated with surface 2 that is equal to $\dfrac{1-\varepsilon_2}{A_2\varepsilon_2}$, and the potential after this resistance becomes e_{b2}. The electrical circuit shown in Figure 12.20 consists of three resistances in series. Therefore, the amount of heat transfer is

$$q_1 = \frac{e_{b1} - e_{b2}}{\dfrac{1-\varepsilon_1}{A_1\varepsilon_1} + \dfrac{1}{A_1 F_{12}} + \dfrac{1-\varepsilon_2}{A_2\varepsilon_2}} \tag{12.92}$$

which is the same as Equation 12.87, however, obtained with far less algebra. Dividing both sides of Equation 12.92 by A_1 and rearranging,

$$q_1'' = \frac{e_{b1} - e_{b2}}{\dfrac{1}{\varepsilon_1} - 1 + \dfrac{1}{F_{12}} + \dfrac{A_1}{A_2\varepsilon_2} - \dfrac{A_1}{A_2}} \tag{12.93}$$

If surface 1 does not see itself, then $F_{12} = 1$ and Equation 12.93 simplifies to

$$q_1'' = \frac{e_{b1} - e_{b2}}{\dfrac{1}{\varepsilon_1} + \dfrac{A_1}{A_2\varepsilon_2} - \dfrac{A_1}{A_2}} \tag{12.94}$$

For the special case of a small surface surrounded by a large one (Figure 12.21),

$$q_1'' = \frac{e_{b1} - e_{b2}}{\dfrac{1}{\varepsilon_1} + 0 - 0} \tag{12.95}$$

or

$$q_1 = \varepsilon_1 A_1 \sigma \left(T_1^4 - T_2^4\right) \tag{12.96}$$

Example 12.8

On a clear night, the effective sky temperature is 200 K. Assuming a convective heat transfer coefficient of 25 W/m² K, estimate the maximum air temperature for which frost will form. Assume ground to be black and insulated.

$$T_s = 200$$

$$T_a = ?$$

$$T_g = 0$$

Ground can be considered to be surrounded by a much larger surface

$$q_{conv} = q_{rad}$$

$$hA(T_a - T_g) = \sigma A \varepsilon \left(T_g^4 - T_s^4 \right)$$

$$T_a = T_g + \frac{\sigma \varepsilon}{h} \left(T_g^4 - T_s^4 \right)$$

$$T_a = 273 + \frac{5.67 \times 10^{-8} \times 1}{25} (273^4 - 200^4) = 282\,\text{K} = 9°\text{C} = 48°\text{F}$$

12.4.3 Radiation Shields

Radiation shields are used to control the rate of heat transfer by radiation, and they are analogous to adding insulation to reduce heat transfer by conduction. Consider two infinitely long parallel plates that exchange energy only by radiation, and their equivalent electrical circuit is shown in Figure 12.22. For this geometry, the heat transfer is

$$q_1'' = \frac{e_{b1} - e_{b2}}{\dfrac{1}{\varepsilon_1} + \dfrac{A_1}{A_2 \varepsilon_2} - \dfrac{A_1}{A_2}} \qquad (12.97)$$

The two surfaces have the same area, and if we furthermore assume that they have the same emissivity, then Equation 12.97 reduces to

$$q_1'' = \frac{\sigma \left(T_1^4 - T_2^4 \right)}{\dfrac{2}{\varepsilon} - 1} \qquad (12.98)$$

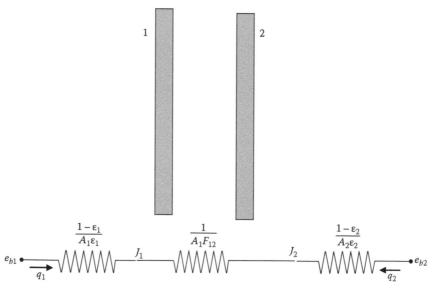

FIGURE 12.22
Electrical network for radiation exchanging between two surfaces.

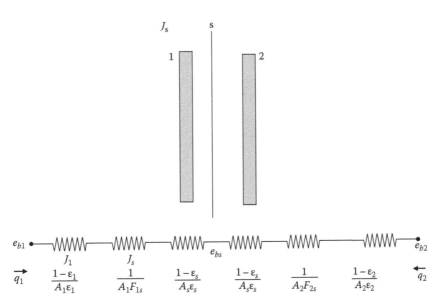

FIGURE 12.23
Electrical network for a shield between two infinitely large surfaces.

Now consider another plate with surface emissivity ε_s is placed between the two plates, as shown in Figure 12.23, along with the electrical circuit. The addition of this shield adds three additional resistances, and therefore the rate of heat transfer becomes

$$q_1 = \frac{\sigma\left(T_1^4 - T_2^4\right)}{\dfrac{1-\varepsilon_1}{A_1\varepsilon_1} + \dfrac{1}{A_1 F_{1s}} + \dfrac{1-\varepsilon_s}{A_1\varepsilon_s} + \dfrac{1-\varepsilon_s}{A_1\varepsilon_s} + \dfrac{1}{A_2 F_{2s}} + \dfrac{1-\varepsilon_2}{A_2\varepsilon_2}} \quad (12.99)$$

The areas are all the same and the shape factors are all equal to one in this case; therefore, the heat flux becomes

$$q_1'' = \frac{\sigma\left(T_1^4 - T_2^4\right)}{\dfrac{1}{\varepsilon_1} + \dfrac{2}{\varepsilon_s} + \dfrac{1}{\varepsilon_2} - 2} \quad (12.100)$$

Again assuming all surfaces to have the same emissivity, the rate of heat transfer becomes

$$q_1'' = \frac{\sigma\left(T_1^4 - T_2^4\right)}{\dfrac{4}{\varepsilon} - 2} = \frac{\sigma\left(T_1^4 - T_2^4\right)}{2\left(\dfrac{2}{\varepsilon} - 1\right)} = \frac{q_{no\ shield}''}{2} \quad (12.101)$$

In general, adding N shields will reduce the heat transfer by a factor of $(N + 1)$ or

$$q_{N\ shields} = \frac{q_{no\ shield}}{N+1} \quad (12.102)$$

12.4.4 Radiation Exchange between Three and More Surfaces

The electrical network for general case of radiation exchange between three surfaces is shown in Figure 12.24. In this case, the circuit is analyzed by using Kirchhoff's law that the sum of currents flowing into each node must add up to zero. For example, for node 1, assuming arbitrary directions for currents, then

$$\frac{e_{b1} - J_1}{\dfrac{1 - \varepsilon_1}{A_1 \varepsilon_1}} = \frac{J_1 - J_2}{\dfrac{1}{A_1 F_{12}}} + \frac{J_1 - J_3}{\dfrac{1}{A_1 F_{13}}} \tag{12.103}$$

and similar expressions can be arrived at for nodes 2 and 3. The three equations are solved for the three values of radiosity, which can then be used to find the current or q in each of the branches. For the general case, there is no advantage in using the electrical analogy or solving the equations directly. However, if one or more of the surfaces are insulated or black, or if any of the shape factors is zero, then using the equivalent electrical network simplifies the algebra.

An insulated surface does not exchange energy with the surroundings directly, and is termed a reradiating surface. No current flows in the surface resistance associated with an insulated surface, or its radiosity is equal to its blackbody emissive power. Furthermore, its surface resistance can be eliminated, simplifying the network by using the equivalent resistances.

If a surface is black, its surface resistance is zero, and even though in this case again radiosity is equal to its blackbody emissive power, the radiosity is fixed by the temperature

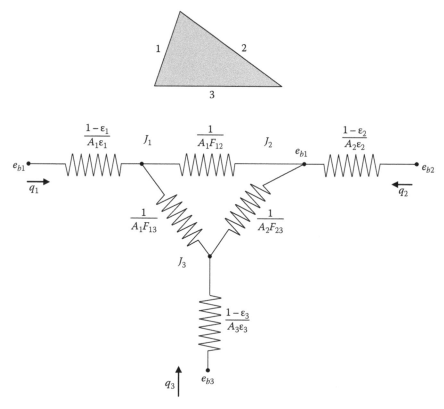

FIGURE 12.24
Electrical network for radiation exchange between three surfaces.

of the surface. The network is acting as if a current source (a battery) is attached to the network at the point, and therefore the resistances cannot be added and the circuit must be analyzed using Kirchhoff's law.

If a shape factor is zero, i.e., two of the surfaces do not see each other, then the space resistance between them is infinity or the network is open, which again simplifies the network.

Example 12.9

Consider the radiation exchange between the three surfaces of the cylindrical enclosure. If surface 3 is insulated, determine T_3 and q_1''.

Starting with the general electrical network, since surface 3 is insulated, $q_3 = 0$, and the surface resistance associated with surface 3 can be removed. Therefore, the electrical circuit becomes as shown in Figure 12.25. In the lower branch, there are two resistances in series; therefore, they can be replaced with one resistance equal to

$$R = \frac{1}{A_1 F_{13}} + \frac{1}{A_2 F_{23}}$$

Then we have two resistances in parallel, and the equivalent resistance becomes

$$\frac{1}{R_{eq}} = \frac{1}{\dfrac{1}{A_1 F_{12}}} + \frac{1}{R} = A_1 F_{12} + \frac{1}{\dfrac{1}{A_1 F_{13}} + \dfrac{1}{A_2 F_{23}}}$$

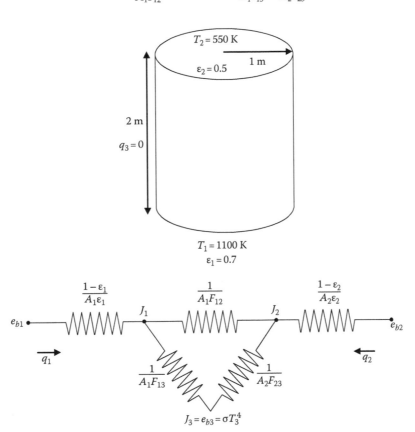

FIGURE 12.25
Electrical network for radiation exchange between three surfaces, with surface 3 insulated.

or

$$R_{eq} = \cfrac{1}{A_1 F_{12} + \cfrac{1}{\cfrac{1}{A_1 F_{13}} + \cfrac{1}{A_2 F_{23}}}}$$

And finally there are three resistances in series, and the heat transfer becomes

$$q_1 = \cfrac{\sigma\left(T_1^4 - T_2^4\right)}{\cfrac{1-\varepsilon_1}{A_1 \varepsilon_1} + R_{eq} + \cfrac{1-\varepsilon_2}{A_2 \varepsilon_2}}$$

To calculate the heat transfer, we need the shape factors. F_{12} can be found from the correlation for two parallel disks:

$$F_{12} = 0.1716$$

since

$$F_{12} + F_{13} = 1$$

also from symmetry

$$F_{13} = F_{23} = 1 - F_{12}$$

$$q_1'' = 24.82 \text{ kW/m}^2$$

σ	5.67E−08
T_1	1100
T_2	550
T_3	974.8
ε_1	0.7
ε_2	0.5
ε_3	0.9
A_1	3.14
A_2	3.14
A_3	12.57
F_{12}	0.172
F_{13}	0.828
F_{23}	0.828
J_1	72,377.4
J_2	30,008.2
J_3	51,192.8
q_1	77,973.8
q_2	−77,973.8
q_3	0.0
e_{b1}	83,014.5
e_{b2}	5188.4
e_{b3}	51,192.8
q_1''	24,819.8
q_2''	−24,819.8
Sum q	0.00

Example 12.10

Solve the previous problem, by using the equations directly:

$$q_i = \sum_{j=1}^{N} (J_i - J_j) A_i F_{ij}$$

$$q_i = \frac{A_i \varepsilon_i}{1 - \varepsilon_i} (e_{bi} - J_i)$$

Eliminating qs and solving for radiosity for each equation result in

$$J_i = e_{bi} - \frac{1 - \varepsilon_i}{\varepsilon_i} \sum_{j=1}^{N} (J_i - J_j) F_{ij}$$

which are three equations for three unknown radiosities that can be solved iteratively as shown in the spreadsheet. In this case,

$$J_3 = e_{b3}$$

σ	5.67E−08
T_1	1100
T_2	550
T_3	
ε_1	0.7
ε_2	0.5
ε_3	0.9
A_1	3.14
A_2	3.14
A_3	12.57
F_{12}	0.172
F_{13}	0.828
F_{11}	0.000
F_{23}	0.828
F_{21}	0.172
F_{22}	0.000
F_{31}	0.207
F_{32}	0.207
F_{33}	0.586
e_{b1}	83,014.5
e_{b2}	5188.4
e_{b3}	51,192.8
J_1	72,377.4
J_2	30,008.2
J_3	51,192.8
q_1	77,973.8
q_2	−77,973.8
q_3	0
q_1''	24,819.8
q_2''	−24,819.8
q_3''	77,973.8
Sum q	0.00

Example 12.11

Consider the previous problem, except that surface 3 is black at 600 K.

The electrical circuit is similar to the previous problem and is shown in Figure 12.26. The difference between the two cases is that in the previous case the potential at node 3 was a floating potential, whereas in this case it is a known fixed potential equal to the blackbody emissive power of surface 3. This is like placing a battery of known potential at node 3 that delivers a current directly at node 3. Thus, in this case, the resistances cannot be considered in series, since the current flowing in them is not the same and we need to use the Kirchhoff's law for currents at nodes 1 and 2, and noting that since 3 is a black surface, $J_3 = e_{b3}$:

$$q_1 = \frac{e_{b1} - J_1}{\frac{1 - \varepsilon_1}{\varepsilon_1}} = \frac{J_1 - J_2}{\frac{1}{F_{12}}} + \frac{J_1 - e_{b3}}{\frac{1}{F_{13}}}$$

$$q_2 = \frac{e_{b2} - J_2}{\frac{1 - \varepsilon_2}{\varepsilon_2}} = \frac{J_2 - J_1}{\frac{1}{F_{12}}} + \frac{J_2 - e_{b3}}{\frac{1}{F_{23}}}$$

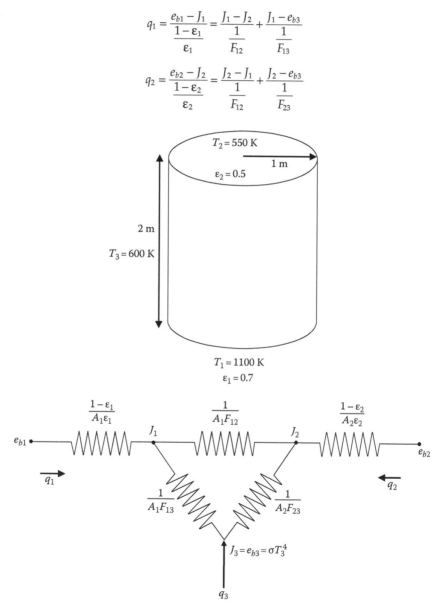

FIGURE 12.26
Electrical network for radiation exchange between three surfaces, with surface 3 being black.

Solving these two equations for J_1 and J_2 results in

$$J_1 = 60{,}493.77$$
$$J_2 = 10{,}828.25$$

which can be used to find

$$q_1 = 165.1 \text{ kW}$$
$$q_2 = -17.7 \text{ kW}$$

and q_3 can be found by using Kirchhoff's law at node 3:

$$q_3 = \frac{J_3 - J_1}{\dfrac{1}{A_1 F_{13}}} + \frac{J_3 - J_2}{\dfrac{1}{A_2 F_{23}}}$$

$$q_3 = -147.4 \text{ kW}$$

Note that the net heat transfer must be zero. The solution can also be found by solving the governing equations given in the previous example. In this case,

$$e_{b3} = \sigma T_3^4$$

$$\varepsilon_3 = 1$$

which automatically results in

$$J_3 = e_{b3}$$

The solution is given in the spreadsheet:

σ	5.67E−08
T_1	1100
T_2	550
T_3	600.0
ε_1	0.7
ε_2	0.5
ε_3	1
A_1	3.14
A_2	3.14
A_3	12.57
F_{12}	0.172
F_{13}	0.828
F_{11}	0.000
F_{23}	0.828
F_{21}	0.172
F_{22}	0.000
F_{31}	0.207
F_{32}	0.207
F_{33}	0.586
e_{b1}	83,014.5
e_{b2}	5188.4
e_{b3}	7348.3
J_1	60,493.8
J_2	10,828.2
J_3	7348.3
q_1	165,085.3
q_2	−17,718.1
q_3	−147,367.3
Sum q	0.00

Problems

12.1 Determine the intensity and emissive power for a blackbody at 500 and 100 K.

12.2 The spectral emissive power for a diffuse emitter is given by

$e_{\lambda b}$ (W/m² μm)	λ (μm)
0	$\lambda < 2$
100	$2 < \lambda < 10$
1000	$10 < \lambda < 100$
200	$100 < \lambda < 1000$
0	$1000 < \lambda$

a. Calculate its total emissive power.
b. Determine its total intensity.
c. Determine the fraction of the emissive power between $60 \leq \theta \leq 90$.

12.3 The spectral emissivity of a diffuse surface at 1500 K is 0.1 for $\lambda \leq 2$ μm and 0.5 at longer wavelengths. Determine
a. The total hemispherical emissivity
b. The emissive power $1 \leq \lambda \leq 8$ μm and $45 \leq \theta \leq 90$

12.4 Calculate the total emissivity of a surface at 500 K, if its spectral emissivity is

$e_{\lambda b}$ (W/m² μm)	λ (μm)
0	$\lambda < 0.1$
0.5	$0.1 < \lambda < 0.7$
0.7	$0.7 < \lambda < 10$
0.6	$10 < \lambda < 50$
0	$50 < \lambda$

12.5 The spectral irradiation incident on a surface is given by

G_{λ} (W/m² μm)	λ (μm)
200	$\lambda < 2$
3000	$2 < \lambda < 10$
1000	$10 < \lambda < 100$
0	$100 < \lambda$

The spectral absorptivity of the surface is

α_{λ} (W/m² μm)	λ (μm)
0.1	$\lambda < 2$
0.74	$2 < \lambda < 10$
0.2	$10 < \lambda$

a. What is the surface's spectral reflectivity?
b. Determine the total hemispherical absorptivity.

12.6 The spectral emissivity of a surface is given by the function $\varepsilon_\lambda = e^{-\frac{(\lambda-\lambda_1)^2}{2\sigma^2}}$ (normal distribution function) with $\lambda_1 = 0.7$ and $\sigma = 0.2$ (shown in the figure below). Assuming the emissivity can be approximated by a seven-step function as shown in the table and plotted in the figure, determine the fraction of sun's radiation that is absorbed by the surface.

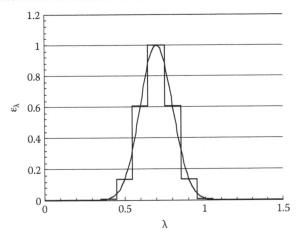

ε_λ	λ (μm)	λ (μm)
1.0	$0.65 < \lambda < 0.7$	$0.7 < \lambda < 0.75$
0.607	$0.55 < \lambda < 0.65$	$0.75 < \lambda < 0.85$
0.135	$0.45 < \lambda < 0.55$	$0.85 < \lambda < 0.95$
0.011	$0.35 < \lambda < 0.45$	$0.95 < \lambda < 1.05$

12.7 The spectral emissivity of a surface is given by the function $\varepsilon_\lambda = e^{-\frac{(\lambda-\lambda_1)^2}{2\sigma^2}}$ (variation of normal distribution function). For $\lambda_1 = 0.7$ and $\sigma = 0.2$ (shown in the figure for Problem 12.6), numerically determine what fraction of sun's radiation is absorbed by the surface.

12.8 The spectral emissivity of a gas is given by the function $\varepsilon_\lambda = e^{-\frac{(\lambda-2)^2}{0.032}} + 0.8e^{-\frac{(\lambda-7)^2}{0.04}}$, as shown in the figure below. Numerically determine the fraction of energy from a blackbody at 300 K that the gas absorbs.

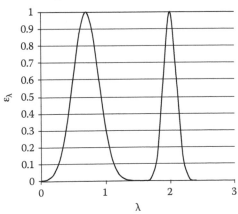

12.9 Two equal-diameter ($D = 0.3$ m) coaxial parallel disks are a distance 0.5 m apart. If $T_1 = 600$ K, $\varepsilon_1 = 0.7$ and $T_2 = 400$ K, $\varepsilon_2 = 0.9$, and the environment is at 300 K, determine the heat transfer to the two disks.

12.10 The ceiling of a cubical pizza oven, shown in the figure below, is embedded with radiant heaters that maintain the ceiling's temperature at 400°C. The floor is well insulated and the sidewalls are maintained at 300°C. All surfaces are diffuse and gray, the ceiling is black, and the other walls have an emissivity of 0.5. Determine
   a. How much heat is dissipated from the heaters.
   b. The temperature of the floor.

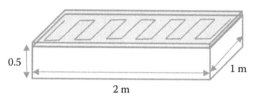

12.11 Two long coaxial cylinders of diameters $D_1 = 0.30$ m and $D_2 = 0.10$ m and emissivities $\varepsilon_1 = 0.7$ and $\varepsilon_2 = 0.4$ are maintained at uniform temperatures of $T_1 = 600$ K and $T_2 = 350$ K, respectively:
   a. Determine the radiation heat loss from surface 1.
   b. How fast should air at 300 K be flowing over surface 2 to remove 350 W/m of energy from the outer cylinder?

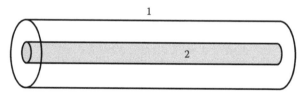

12.12 Two long coaxial cylinders of diameters $D_1 = 0.30$ m and $D_2 = 0.10$ m and emissivities $\varepsilon_1 = 0.7$ and $\varepsilon_2 = 0.4$ are maintained at uniform temperatures of $T_1 = 600$ K and $T_2 = 350$ K, respectively:
   a. Determine the radiation heat loss from surface 1, if a coaxial radiation shield of diameter $D_3 = 0.20$ m having emissivity of 0.3 is placed between the two cylinders.
   b. How fast should air at 800 K be flowing over surface 1 to supply 350 W/m of energy to the outer cylinder?

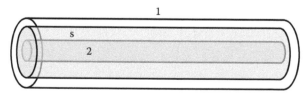

12.13 A flat-bottom hole 6 mm in diameter is drilled to a depth of 22 mm in a diffuse gray material having an emissivity of 0.6 and a uniform temperature of 1000 K:
   a. Determine the radiant power leaving the cavity.
   b. Effective emissivity of a cavity (ε_f) defined as the radiant power leaving the cavity divided by emissive power of a black surface the size of the cavity opening and at the same temperature as that of the cavity surfaces. Determine the effective emissivity of this cavity.

12.14 Two concentric spheres of radii $r_1 = 5$ cm and $r_2 = 15$ cm are maintained at $T_1 = 1000$ K and $T_2 = 300$ K. The emissivity for the surfaces are $\varepsilon_1 = 0.6$ and $\varepsilon_2 = 0.9$:
  a. Calculate the heat loss from the inner sphere.
  b. To reduce the heat loss, a spherical radiation shield having an emissivity of 0.2 on both surfaces and a radius of 10 cm is placed between the two spheres. Determine the heat loss from the inner sphere.

12.15 Consider a cone of height H whose side surface is insulated, having a base diameter D, whose underside has an emissivity ε, and is maintained at temperature T. The base has a hole of diameter $d = D/4$ at its center. Determine the amount of energy leaving the hole.

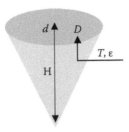

12.16 All surfaces of the tetrahedron enclosure are equal, with three being isothermal and gray, with $T_1 = 600$ K, $\varepsilon_1 = 0.80$; $T_2 = 400$ K, $\varepsilon_2 = 0.50$; and $T_3 = 350$, $\varepsilon_3 = 0.20$. The fourth surface is insulated. Determine its temperature.

12.17 In Problem 12.16, determine heat flux of surface 4, if it is black at 300 K.

References

1. Fishbane, P. M., Gasiorowicz, S., and Thornton, S. T. (2005) *Physics for Scientists and Engineers with Modern Physics*, Upper Saddle River, NJ: Prentice Hall.
2. Siegel, R. and Howell, J. R. (1981) *Thermal Radiation Heat Transfer*, 2nd edn., Washington: Hemisphere Publishing Corporation.
3. McGraw-Hill. (2003) *McGraw-Hill Dictionary of Scientific and Technical Terms*, New York: McGraw-Hill.
4. Mahan, J. R. (2002) *Radiation Heat Transfer: A Statistical Approach*, New York: John Wiley.
5. Özışık, M. N. (1977) *Basic Heat Transfer*, New York: McGraw-Hill.

13

Participating Medium

13.1 Introduction

In the previous chapter, we considered the radiation exchange between solid surfaces, separated by a transparent or nonparticipating medium, vacuum, or a gas that allowed the passage of radiation without impacting it. Gases that allow passage of radiation, generally absorb, emit, and scatter radiation, and the study of radiation heat transfer in participating medium is considerably more challenging, partly due to the strong wavelength dependence. Radiation heat transfer in participating media is important in a number of engineering applications, including combustion, propulsion, climate change, and plasma cutting.

Figure 13.1 shows the absorption coefficient (fraction of energy absorbed) over a broad range of wavelengths by the different gases in the atmosphere [1]. As mentioned before, the sun can be approximated as a blackbody at around 5800 K, and around 12% of the sun's radiation is in the ultraviolet region. Very little radiation in the ultraviolet region (<0.3 m or 300 nm) reaches the surface of the earth, as a result of the presence of oxygen and ozone, shielding us effectively from their harmful effects. Oxygen and ozone are transparent in the visible and mostly transparent in the infrared portion of the spectrum. Also 51% of solar flux is in the infrared region, and carbon dioxide and water vapor in the atmosphere absorb much of the infrared radiation beyond 13 μm.

In between, the absorption spectrum is more complex with many peaks of varying width. For example, carbon dioxide has two wide absorption bands at about 2.8 and 4.5 μm. Water has several absorption bands in the infrared and even has some absorption well into the microwave region. The gases shown are primarily transparent in the visible range, which makes for a brighter world.

Additionally, due to its low surface temperature, much of the energy emitted by the earth is in the infrared region, and absorption of the earth's radiation by the atmospheric gases reduces the heat loss from the earth. There are a few "open windows" for earth to radiate away its energy, and the concern is that manmade gases will narrow these windows and reduce the earth's ability to loose energy.

To better understand the strong spectral dependence of radiation in gases, it is helpful to consider how radiation interacts with molecules and atoms. Atoms possess different, but, discrete energy states, i.e., the amount of energy that they have cannot change continuously. Different atoms/molecules have different allowable energy states that are dependent on the pressure and temperature; therefore, they can absorb radiation only at specific frequencies.

Let us consider a hydrogen atom, which is the simplest atom. Its energy levels are given by Bohr model [2] and are shown in Figure 13.2. The zero energy level is when

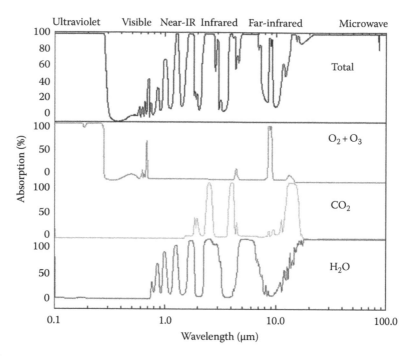

FIGURE 13.1
Absorption of radiation by various atmospheric gases. (With kind permission from Springer Science+Business Media: *Physics of Climate*, 1992, Peixoto, J.P. and Oort, A.H.)

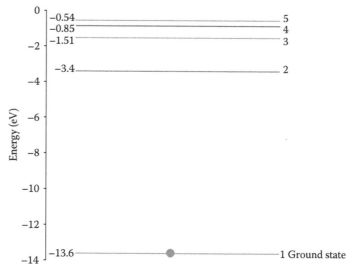

FIGURE 13.2
Hydrogen energy levels.

the hydrogen atom is ionized, i.e., its single electron is removed. Relative to the zero energy state, hydrogen has five allowable energy levels. The ground state is the lowest permissible energy level, which for hydrogen atom is -13.6 eV (1 eV $= 1.602 \times 10^{-19}$ J), or the electron in the ground state has energy -13.6 eV. The energy of the ground state is known as the zero-point energy of the system. If the electron is at an excited state, one with energy greater than the ground state, it can have energy corresponding to only that energy state, for example, if it is in the third state, it must have –1.51 eV of energy.

To increase the energy of an atom from one state (E_i) to another state (E_j), only photons that have energy equal to the difference between the two energy levels, i.e., having the frequency

$$v = \frac{E_j - E_i}{h} \tag{13.1}$$

can be used. If the frequency of the incident radiation corresponds to any one of the allowable energy states, then the radiation at that particular frequency will be absorbed by the electrons. Similarly, emission from atoms can happen at specific frequencies, corresponding to the allowable energy levels; therefore, the emissivity and absorptivity of the gases will have peaks at certain wavelengths.

For the hydrogen electron to jump from the first to the second energy level, it needs to absorb a photon having the frequency

$$v = \frac{(-3.4 - (-13.6))1.602 \times 10^{-19}}{6.63 \times 10^{-34}} = 2.47 \times 10^{15} \text{ Hz}$$

This corresponds to a wavelength of

$$\lambda = \frac{c}{v} = \frac{299,792,458 \times 10^6}{2.47 \times 10^{15}} = 0.122 \, \mu m$$

which is in the ultraviolet range, or for a hydrogen atom to go from the ground state to the second state, it must absorb a photon of ultraviolet light having a wavelength of 0.122 μm. Conversely, if the hydrogen electron drops from the second state to the ground state, it will emit a photon at the same wavelength.

Although one may expect to have sharp peaks, corresponding to specific wavelengths, as can be seen from Figure 13.1, the peaks are not that sharp and in fact have finite width. There are a number of factors that contribute to the broadening. There is uncertainty associated with the exact energy levels E_i and E_j, which is called natural broadening. There is also broadening due to Doppler effect. If the atom that is emitting at frequency v is also moving with a velocity u relative to an observer, then the frequency of the emitted radiation will be

$$v_e = v\left(1 \pm \frac{u}{c}\right) \tag{13.2}$$

and since atoms move with varying velocities in all directions, the emitted or radiated frequencies will be broadened. The higher the temperature of the gas, the broader will be the velocity of the molecules and thus the broader the spectral line. However, most of the broadening is due to collision. It takes a finite amount of time for transition from one state to another. Often this time is much longer than the interval between atomic collisions. Collisions during the transition increase uncertainty in the allowable energy states, broadening the emission line.

13.2 Energy Balance and Equation of Transfer

In dealing with gases, as with solid surfaces, it is also convenient to use the concept of intensity. For gases, intensity in a given direction is defined as the radiation per unit projected area (hypothetical) in that direction, per unit solid angle centered at the point, and per unit wavelength. In a medium that does not scatter and emit radiation, intensity remains constant in a given direction. It was also shown before that the intensity of a blackbody is independent of the direction.

When radiation travels through a participating medium such as a gas in a particular direction, its intensity changes as a result of its interactions with the medium. Now consider a differential control volume of cylindrical shape, Figure 13.3, having the cross-sectional area dA in the s direction in a medium that participates in radiation heat transfer by absorbing, emitting, and scattering radiation. If $I'_\lambda(s)$ is the radiation intensity in the direction s, some of the radiation will be absorbed, some scattered away, and additional energy will be emitted and scattered into the direction of interest, with all strongly dependent on the wavelength. Doing an energy balance, in the direction s, then

Intensity in + Intensity emitted + Intensity gain as a result of scattered radiation
= Intensity out + Intensity absorbed + Intensity loss as a result of scattering

The amount of energy entering the control volume at s is $\left(I'_\lambda dA\right)_s$ and the amount leaving is $\left(I'_\lambda dA\right)_{s+\Delta s}$. If $\dot{q}$ is the net amount of radiative energy generated per unit volume, the energy balance results in

$$\left(I'_\lambda dA\right)_s + \dot{q}dA\Delta s = \left(I'_\lambda dA\right)_{s+\Delta s} \tag{13.3}$$

Simplifying and using Taylor series, the first law becomes

$$\frac{dI'_\lambda}{ds} = \dot{q} \tag{13.4}$$

The prime indicates that in general, intensity is dependent on the direction. The net radiative energy generation per unit volume is also equal to the net intensity production per unit length. The radiative energy in the direction of interest is destroyed as a result of absorption by the gas and loss due to scattering and is produced as a result of emission by the gas and gain by scattering of radiation from other directions, or $\dot{q} = \dot{q}_a + \dot{q}_{s,L} + \dot{q}_e + \dot{q}_{s,G}$ with the first two terms on the right-hand side being negative and the last two positive.

Absorption of radiation intensity has been found experimentally to be proportional to the local value of intensity, or

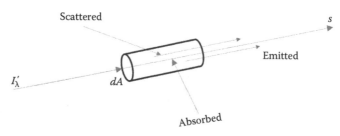

FIGURE 13.3
Components of radiative energy.

$$\dot{q}_a = -a_\lambda I'_\lambda(s) \tag{13.5}$$

where a_λ is the proportionality constant and is called absorption coefficient and has units of inverse length. The gas may attenuate the intensity by scattering it in other directions or out-scattering. Both absorption and scattering cause the intensity to decrease; the difference is that absorption increases internal energy of the gas, while scattering simply changes the direction, adding to intensity in other directions. As will be discussed later, scattering from other directions also increases intensity in a given direction. The reduction in intensity along the path s as a result of scattering is also proportional to the local intensity and thus, the destruction as a result of scattering becomes

$$\dot{q}_{s,L} = -a_{s,\lambda} I'_\lambda(s) \tag{13.6}$$

where $a_{s,\lambda}$ is the scattering coefficient.

Gases emit radiation as a result of their temperature, which will add to the intensity. This is known as the ordinary or spontaneous emission. If the gas is in local thermodynamic equilibrium, using the concept of emissivity for a gas and noting that it is equal to its absorptivity, then the energy generation per unit volume (intensity generation per unit length) as a result of emission is

$$\dot{q}_e = a_\lambda I'_{\lambda,b}(s) \tag{13.7}$$

where $I'_{\lambda,b}$ is the black body intensity. Additionally, when a photon encounters a molecule at an excited state, there is the likelihood that the molecule will emit a photon with the same frequency as the one that it encountered in the same direction, resulting in the molecule's loss of energy and going to a lower energy state. This is called simulated or induced emission.

Finally, scattering from other directions, or in-scattering, can increase the intensity in the direction of interest. Consider radiation intensity in the direction s_i incident on the area dA as shown in Figure 13.4. From Equation 13.6, the total energy scattered in the direction $\vec{s}_i$ is $a_{s,\lambda} I'_\lambda(s_i)$. The probability that the scattered rays from the direction $\vec{s}_i$ are scattered into the direction $\vec{s}$ is given by $\dfrac{\Phi(s_i, s)}{4\pi}$, and therefore, the energy scattered in the $\vec{s}$ direction from $\vec{s}_i$ becomes $a_{s,\lambda} I'_\lambda(s_i) \dfrac{\Phi(s_i, s)}{4\pi}$. The function $\Phi(s_i, s)$ is known as the scattering phase function and represents the ratio of the intensity scattered in a given direction, divided by the intensity for isotropic (uniform) scattering. Note that phase function can be greater

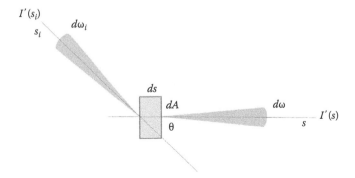

FIGURE 13.4
Scattering intensity into the direction s.

than or smaller than one, with $\Phi(s_i, s) = 1$ for isotropic scattering. In general, scattering phase function is a complicated function of direction.

The total amount of energy generated per unit volume in the s direction as a result of scattering from all the other directions is obtained by integrating over the sphere surrounding the element. Therefore, the energy generation per unit volume (intensity generation per unit length) as a result of scattering is

$$\dot{q}_{s,G} = a_{s,\lambda} \frac{1}{4\pi} \int_{4\pi} I_\lambda'(s_i)\Phi(s_i,s)d\omega_i \tag{13.8}$$

Substituting from Equations 13.5 to 13.8 into 13.4 results in the following for the determination of the radiation intensity change in a gas:

$$\frac{dI_\lambda'(s)}{ds} = -\left(a_\lambda + a_{s,\lambda}\right)I_\lambda'(s) + a_\lambda I_{\lambda b}' + a_{s,\lambda} \frac{1}{4\pi} \int_{4\pi} I_\lambda'(s_i)\Phi(s_i,s)d\omega_i \tag{13.9}$$

In this equation, $I_\lambda'(s)$ is the radiation intensity in the direction s, and $I_{\lambda b}'$ is the blackbody intensity. The prime indicates that, in general, they are dependent on their direction. The coefficients a_λ and $a_{s,\lambda}$ are absorption and scattering coefficients respectively, and Φ is a phase function that describes the angular distribution of the scattered energy. Although it can be a complicated function of direction, for isotropic scattering, $\Phi = 1$.

Often, it is helpful to use the concepts of extinction coefficient that combines the effects of absorption and out-scattering in one parameter defined as

$$K_\lambda(\lambda,T,P) = a_\lambda(\lambda,T,P) + a_{s,\lambda}(\lambda,T,P) \tag{13.10}$$

and optical thickness

$$\kappa_\lambda(s) = \int_0^s K_\lambda ds' = \int_0^s (a_\lambda + a_{s,\lambda})ds \tag{13.11}$$

and albedo for scattering

$$\Omega_{0,\lambda} = \frac{a_{s,\lambda}}{a_\lambda + a_{s,\lambda}} \tag{13.12}$$

which when substituted in the earlier equation results in

$$\frac{dI_\lambda'(\kappa_\lambda)}{d\kappa_\lambda} = -I_\lambda'(\kappa_\lambda) + (1 - \Omega_{0,\lambda})I_{b\lambda} + \frac{\Omega_{0,\lambda}}{4\pi} \int I_\lambda'(\kappa_\lambda,\omega_i)\Phi(\lambda,\omega,\omega_i)d\omega_i \tag{13.13}$$

This equation is known as the equation of radiative transfer that describes the intensity of radiation along a particular direction in an absorbing, scattering, and emitting medium. In a more compact form, Equation 13.13 can be written as

$$\frac{dI_\lambda'(\kappa_\lambda)}{d\kappa_\lambda} + I_\lambda'(\kappa_\lambda) = \tilde{I}_\lambda' \tag{13.14}$$

where

$$\tilde{I}_\lambda' = (1 - \Omega_{0,\lambda})I_{\lambda,b} + \frac{\Omega_{0,\lambda}}{4\pi} \int I_\lambda'(\kappa_\lambda,\omega_i)\Phi(\lambda,\omega,\omega_i)d\omega_i \tag{13.15}$$

is the source function, or the source for intensity that is due to emission and scattering. Note that the term $I_{b\lambda}$ is the blackbody intensity, which will be constant only if the medium is isothermal; otherwise, it will be a function of position. The boundary condition for this equation is

$$s = 0 \quad I_\lambda'(s) = I_\lambda'(0) \tag{13.16}$$

Multiplying both sides of the equation by the integrating factor, e^{κ_λ}, the left-hand side can be written as a single derivative

$$\frac{d\left[e^{\kappa_\lambda}I_\lambda'(\kappa_\lambda)\right]}{d\kappa_\lambda} = \tilde{I}_\lambda' e^{\kappa_\lambda} \tag{13.17}$$

Equation 13.17 can be integrated from $\kappa_\lambda = 0$ to κ_λ

$$I_\lambda'(\kappa_\lambda) = I_\lambda'(0)e^{-\kappa_\lambda} + \int_0^{\kappa_\lambda} \tilde{I}_\lambda' e^{-(\kappa_\lambda - z)}dz \tag{13.18}$$

The term on the left-hand side is the radiation intensity at any location "s" that corresponds to the optical depth κ_λ. This equation indicates that intensity at a point is due to two effects: the first term on the right-hand side is what is left of the intensity entering at $s = 0$, which drops exponentially with increased optical depth due to absorption and scattering. The second term is the intensity increase, due to the gas emission and incoming scattering, which increases exponentially with increasing optical depth.

In the absence of scattering ($\Omega_{0,\lambda} = 0$), for an emitting and absorbing medium, Equation 13.18 becomes

$$I_\lambda'(\kappa_\lambda) = I_\lambda'(0)e^{-\kappa_\lambda} + \int_0^{\kappa_\lambda} I_{\lambda,b}e^{-(\kappa_\lambda - z)}dz \tag{13.19}$$

Additionally, if the medium is isothermal, the blackbody intensity will be constant and $\kappa_\lambda(s) = a_\lambda s$; therefore,

$$I_\lambda'(s) = I_\lambda'(0)e^{-a_\lambda s} + I_{\lambda,b}(1 - e^{-a_\lambda s}) \tag{13.20}$$

Also note that in this case, the radiation intensity varies exponentially between the intensity of the entering radiation, $I_\lambda'(0)$, and the blackbody intensity of the gas, $I_{b\lambda}$, reached in the limit when $\kappa_\lambda \to \infty$ (optically thick limit), or an optically thick medium emits like a blackbody.

13.3 Radiative Properties of Gases

From this solution, we can also define the radiative properties of gases, which, unlike those for a solid, depend on the thickness of the layer. Consider the case where the gas emission is negligible, i.e., the gas is at a relatively low temperature. Then,

$$\tau_\lambda = \frac{I_\lambda(\kappa_\lambda)}{I_\lambda(0)} = e^{-\kappa_\lambda} \tag{13.21}$$

which defines transmissivity of the gas layer, or the fraction of the intensity that has reached location s. Since

$$\tau_\lambda + \alpha_\lambda = 1 \tag{13.22}$$

then from Kirchhoff's law, absorptivity and emissivity must be equal, thus

$$\alpha_\lambda = \varepsilon_\lambda = 1 - e^{-\kappa_\lambda} \tag{13.23}$$

in the optically thick limit, $\kappa_\lambda > 5$, the medium can be treated as a blackbody, absorbing all the incident energy passing through it at the particular wavelength and emitting energy like a blackbody at the same wavelength. The total emittance (emissivity) is obtained from

$$\varepsilon = \frac{\int_0^\infty \varepsilon_\lambda I_{b\lambda} d\lambda}{\int_0^\infty I_{b\lambda} d\lambda} = \frac{\int_0^\infty \varepsilon_\lambda I_{b\lambda} d\lambda}{\dfrac{\sigma T^4}{\pi}} = \frac{\pi \int_0^\infty (1 - e^{-\kappa_\lambda}) I_{b\lambda} d\lambda}{\sigma T^4} \tag{13.24}$$

In the optically thick limit, from Equation 13.24, the spectral emissivity approaches 1 over the absorption bands and the total directional emissivity can be determined from the blackbody fractions.

Example 13.1

From Figure 13.1, CO_2 has four bands centered at 2, 2.8, 4.5, and 15 μm. Assuming the band widths to be 0.5, 0.5, 0.5, and 5, respectively, the following table summarizes the ranges and absorptivities. Determine the absorptivity of carbon dioxide at 300 K and 1 atm.

Band		1		2		3		4		
Range	0	1.75	2.25	2.55	3.05	4.25	4.75	10	20	∞
$\alpha_{\lambda,j}$	0	0.5	0.5	1	1	1	1	1	1	0

$$\alpha = \frac{\int_0^\infty \alpha_\lambda e_{b\lambda} d\lambda}{\sigma T^4} = \alpha_{\lambda,1} \int_{\lambda_1 - \frac{\Delta\lambda_1}{2}}^{\lambda_1 + \frac{\Delta\lambda_1}{2}} \frac{e_{b\lambda}}{\sigma T^4} d\lambda + \alpha_{\lambda,2} \int_{\lambda_2 - \frac{\Delta\lambda_2}{2}}^{\lambda_2 + \frac{\Delta\lambda_2}{2}} \frac{e_{b\lambda}}{\sigma T^4} d\lambda + \alpha_{\lambda,3} \int_{\lambda_3 - \frac{\Delta\lambda_3}{2}}^{\lambda_3 + \frac{\Delta\lambda_3}{2}} \frac{e_{b\lambda}}{\sigma T^4} d\lambda + \alpha_{\lambda,4} \int_{\lambda_4 - \frac{\Delta\lambda_4}{2}}^{\lambda_4 + \frac{\Delta\lambda_4}{2}} \frac{e_{b\lambda}}{\sigma T^4} d\lambda$$

These can be expressed in terms of blackbody fraction

$$\alpha = \sum_{j=1}^{4} \alpha_{\lambda,j} \left[f_{0 - \left(\lambda_j - \frac{\Delta\lambda_j}{2}\right)T} - f_{0 - \left(\lambda_j + \frac{\Delta\lambda_j}{2}\right)T} \right]$$

and the results are shown in the following spreadsheet.

Assuming that the surface of the Earth is at 300 K, this shows that 47.5% of earth's radiation will be absorbed by the carbon dioxide.

Band		1		2		3		4		
Range	0	1.75	2.25	2.55	3.05	4.25	4.75	10	20	∞
$\alpha_{\lambda,j}$	0	0.5	0.5	1	1	1	1	1	1	0
T	300									
λT		525	675	765	915	1275	1425	3000	6000	
$f_{0-\lambda T}$		0	0	0	3E–04	0.004	0.009	0.27	0.74	
α	0.475									

Example 13.2

What fraction of sun's radiation will be absorbed by carbon dioxide?
The solution is similar to the previous case, except that temperature is different. Therefore, carbon dioxide absorbs only 3.7% of sun's radiation and 47.5% of earth's radiation. The only unsaturated range is the first one, which contributes little to the absorptivity of the energy radiated by earth.

Band		1		2		3		4		
Range	0	1.75	2.25	2.55	3.05	4.25	4.75	10	20	∞
$\alpha_{\lambda,j}$	0	0.5	0.5	1	1	1	1	1	1	0
T	5,800									
λT		10,150	13,050	14,790	17,690	24,650	27,550	58,000	116,000	
$f_{0-\lambda T}$		0.92	0.96	0.966	0.98	0.992	0.995	1	1	
α	0.037									

13.3.1 Emissivity and Mean Beam Length

If an isothermal gas is surrounded by a black enclosure, the amount of heat that is radiated from the gas to the chamber walls can be calculated relatively easily. This problem finds applications in combustion chambers. The gas can be considered isothermal and at a constant pressure, and the chamber walls can be considered black. Now consider a small surface, dA, located at the center of a black hemisphere filled with a gas at a temperature T as shown in Figure 13.5. Since there is no emission at the wall, and a_λ is constant, the intensity arriving at dA is given by Equation 13.20 and simplifies to

$$I'_\lambda(R) = I_{\lambda,b}(1 - e^{-a_\lambda R}) \tag{13.25}$$

The spectral heat flux striking area dA is

$$q_\lambda = \int_0^{2\pi} \int_0^{\frac{\pi}{2}} I'_\lambda(R)\cos\theta \sin\theta\, d\theta\, d\phi \tag{13.26}$$

Since the blackbody intensity is independent of the direction, Equation 13.26 simplifies to

$$q_\lambda = (1 - e^{-a_\lambda R})E_{\lambda,b} \tag{13.27}$$

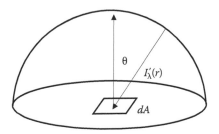

FIGURE 13.5
Radiation received by a surface element from the surrounding gas.

Therefore, the hemispherical spectral emissivity for the gas becomes

$$\varepsilon_\lambda = \frac{q_\lambda}{E_{\lambda b}} = (1 - e^{-a_\lambda R}) \tag{13.28}$$

and from Equation 13.22, the spectral transmissivity becomes

$$\tau_\lambda = e^{-a_\lambda R} \tag{13.29}$$

The total emissivity or transmissivity can be calculated by integrating the earlier two equations over all wavelengths if the relation between absorptivity and wavelength is known.

This equation is valid only for a surface at the center of a hemisphere and is not applicable in general. For example, if we consider an element of area on the surface of the hemisphere, the amount of energy intercepted will be different than that predicted by Equation 13.28.

Siegel and Howell [3] have shown that the transmissivity for a gas inside a sphere to an element on the surface of the sphere is

$$\tau_\lambda = \frac{2}{(a_\lambda D)^2} \left[1 - (1 + a_\lambda D) e^{-a_\lambda D} \right] \tag{13.30}$$

Nevertheless, Equation 13.28 provides an approximate approach for the determination of the heat transfer from an isothermal gas, as long as an equivalent characteristic path length (L_e) can be determined. This characteristic path length is called the mean beam length and is defined by

$$\tau_\lambda = e^{-a_\lambda L_e} \tag{13.31}$$

For example, for a sphere of gas radiating to any area on its surface, equating Equations 13.30 and 13.31 results in

$$\frac{L_e}{D} = -\frac{1}{a_\lambda D} \left\{ \ln \frac{2}{(a_\lambda D)^2} \left[1 - (1 + a_\lambda D) e^{-a_\lambda D} \right] \right\} \tag{13.32}$$

Figure 13.6 is a plot of the ratio of the mean beam length to diameter as a function of $a_\lambda D$. As can be seen in the optically thin limit, the mean beam length is essentially constant. In this limit, the mean beam length is given by

$$L_{e,0} = \frac{4V}{A} \tag{13.33}$$

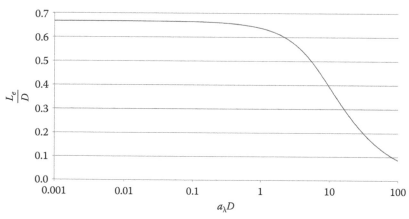

FIGURE 13.6
Mean beam length variation for sphere radiating to its surface.

TABLE 13.1

Mean Beam Length for Radiation from a Gas Volume to Different Surfaces

Geometry	Mean Beam Length, Optically Thin Limit	Mean Beam Length, Optically Thick Limit
Surface of a sphere	$\frac{2}{3}D$	0.65D
Area at the center of hemisphere	$\frac{1}{2}D$	0.5D
Curved area of an infinite cylinder	D	0.95D
Base of a semi-infinite cylinder	0.81D	0.65D
Entire surface of a circular cylinder whose diameter is equal to its height	$\frac{2}{3}D$	0.6D
Infinite parallel plates distance D apart	2D	1.8D
Radiation to any surface of a cube of side D	$\frac{2}{3}D$	0.6D
Arbitrary object of volume V and inside area A	$4\dfrac{V}{A}$	$3.6\dfrac{V}{A}$

which for a sphere becomes

$$L_{e,0} = \frac{4\dfrac{\pi D^3}{6}}{\pi D^2} = \frac{2}{3}D \tag{13.34}$$

which is very close to the value in Figure 13.6.

For an optically thick gas, an average mean beam length can also be calculated. Table 13.1 provides the characteristic path length for several geometries for both optically thin and thick limits, and additional ones can be found in Ref. [3].

13.3.2 Emissivity Charts

The total amount of energy emitted by the gas over all wavelength can be obtained by integrating Equation 13.27

$$q = \int_0^\infty (1 - e^{-a_\lambda L_e}) I_{\lambda,b} d\lambda \tag{13.35}$$

Defining the total gas emittance as

$$\varepsilon_g = \frac{q}{\sigma T^4} \tag{13.36}$$

then

$$\varepsilon_g = \frac{\int_0^\infty (1 - e^{-a_\lambda L_e}) e_{\lambda,b} d\lambda}{\sigma T^4} \tag{13.37}$$

As was shown in Figure 13.1, the spectral emissivity is not a continuous function of the wavelength and only has nonzero value over finite number of bands. Assuming that

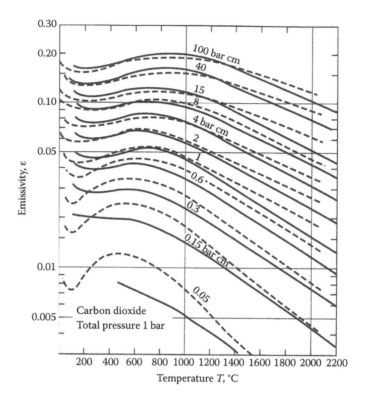

FIGURE 13.7
Emissivity of CO_2 [5] for different P_cL values. (Data from Hottel, H.C., *Heat Transmission*, 3rd edn., chapter 4. New York: McGraw Hill, 1954 [solid lines]; Leckner, B., *Combustion and Flame* 19(1), 33–48, 1972 [dashed lines].)

the gas is at a uniform pressure and temperature and that the band widths are narrow enough so that the blackbody emissive power can be assumed to be constant over each of the N bands, and that the bands do not overlap, Equation 13.37 can be approximately written as

$$\varepsilon = \sum_{i=1}^{N} \frac{e_{\lambda,b,i}}{\sigma T^4} \int_{\lambda_i - \frac{\Delta\lambda_i}{2}}^{\lambda_i + \frac{\Delta\lambda_i}{2}} (1 - e^{-a_i L_e}) d\lambda \tag{13.38}$$

Hottel [4] provides charts for calculation of emittance for a number of gases using experimental measurements, including those for carbon dioxide, Figure 13.7, and water vapor, Figure 13.8 [5]. The charts are a plot of experimentally determined values of emissivity as a function of temperature for different values of the product of the mean distance travelled by radiation, L_e (Table 13.1), and the partial pressure of the participating gas. The charts are for a mixture of the participating gas and a nonabsorbing gas at a total pressure of one bar, when the partial pressure of the absorbing gas is very small [5]. More recent work [6–8] has improved the accuracy of the charts. Leckner [6] provides the following expression [5]

$$\varepsilon_0 = \exp\left[\sum_{i=0}^{M} \sum_{j=0}^{N} c_{ji}\left(\frac{T}{1000}\right)^j \left(\log_{10}\frac{P_a L}{1000}\right)^i\right] \tag{13.39}$$

where
  T is the gas temperature in K
  P_a is the partial pressure of the participating gas in Pascal
  L is the thickness of gas layer in m
  c_{mn} are the correlation constants, which for water are

$$c_{mn} = \begin{pmatrix} -2.2118 & -1.987 & 0.035596 \\ 0.85667 & 0.93048 & -0.14391 \\ -0.10838 & -0.17156 & 0.045915 \end{pmatrix} \qquad (13.40a)$$

and for carbon dioxide [5; pp. 377, 368] are

$$c_{mn} = \begin{pmatrix} -3.9893 & 2.7669 & -2.1081 & 0.39163 \\ 1.271 & -1.1090 & 1.0195 & -0.21897 \\ -0.23678 & 0.19731 & -0.19544 & 0.044644 \end{pmatrix} \qquad (13.40b)$$

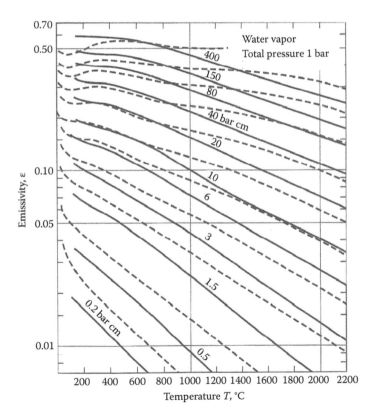

FIGURE 13.8
Emissivity of H_2O [5] for different P_wL values. (Data from Hottel, H.C., *Heat Transmission*, 3rd edn., chapter 4. New York: McGraw Hill, 1954 [solid lines]; Leckner, B., *Combustion and Flame* 19(1), 33–48, 1972 [dashed lines].)

For pressures other than 1 bar, an expression for a correction factor is also provided [5].

If the participating medium is a mixture of different gases, then the different species in the gas may have overlapping bands, as seen from Figure 13.1, for water and carbon dioxide. Hottel and Sarofim [9] have shown that the emissivity of a mixture of two gases 1 and 2 is

$$\varepsilon_{1+2} = C_1\varepsilon_1 + C_2\varepsilon_2 - \Delta\varepsilon \tag{13.41}$$

where the last term is the correction factor if species 1 and 2 have partially overlapping bands. If the bands totally overlap, then $\Delta\varepsilon = \varepsilon_1\varepsilon_2$ [5]. Water and carbon dioxide have overlapping bands, and the correction factor is given by [5]

$$\Delta\varepsilon = \left[\frac{\zeta}{10.7+101\zeta} - 0.0089\zeta^{10.4} \right] \left[\log_{10} \frac{P_{H_2O}L}{1000\zeta} \right]^{2.76} \tag{13.42}$$

$$\text{where } \zeta = \frac{P_{H_2O}}{P_{H_2O}+P_{CO_2}}$$

The values C_1 and C_2 are equal to 1 for the gas total pressure of 1 atm and for other pressures, they can be obtained from Figures 13.9 and 13.10. The correction factor $\Delta\varepsilon$ for a mixture of participating gases that have overlapping bands is given in Figure 13.11 for a mixture of water and carbon dioxide.

Hottel and Egbert [10] have shown that the absorptivity of a mixture of water and carbon dioxide gases can be obtained from

$$\alpha_{c+w} = C_c\alpha_c + C_w\alpha_w - \Delta\alpha \tag{13.43}$$

where the correction factors are given in Figures 13.9 and 13.10, and $\Delta\alpha = \Delta\varepsilon$, determined from Figures 13.11, and the absorptivities are given by

$$\alpha_w = \left(\frac{T_g}{T_s} \right)^{0.45} \times \varepsilon_w\left(T_s, p_wL_e\frac{T_s}{T_g} \right) \tag{13.44a}$$

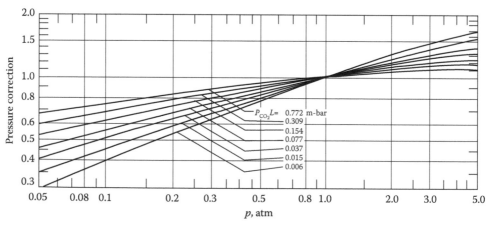

FIGURE 13.9
Pressure correction for CO_2.

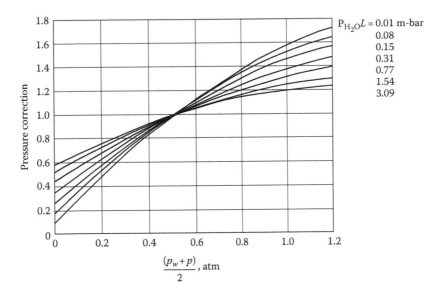

FIGURE 13.10
Pressure correction for H_2O.

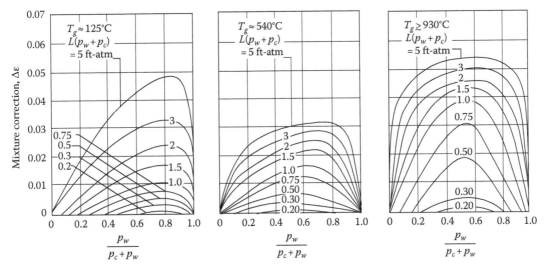

FIGURE 13.11
Correction factor for a mixture of H_2O and CO_2.

$$\alpha_c = \left(\frac{T_g}{T_s}\right)^{0.65} \times \varepsilon_c\left(T_s, p_c L_e \frac{T_s}{T_g}\right) \tag{13.44b}$$

The emissivities in Equations 13.44 are obtained from Figures 13.7 and 13.8 but evaluated at the surface temperature with the equivalent mean beam length modified by the ratio of surface to gas temperatures.

13.4 Mixed-Mode Heat Transfer

Radiation in participating media depends on many factors, including the local temperature of the medium. At the same time, the inclusion of the radiation heat transfer impacts the temperature distribution. To include the impact of radiation, when it is important, the general energy equation needs to be modified (Figure 13.10). For incompressible flow, and in the absence of energy generation, the first law of thermodynamics becomes

$$\rho \frac{D(c_p T)}{Dt} = -\nabla \cdot \left(\vec{q}_c'' + \vec{q}_r'' \right) \tag{13.45}$$

where
$\vec{q}_c''$ is the conduction flux vector and accounts for heat transfer by conduction
$\vec{q}_r''$ is the radiative flux vector, which accounts for the contribution of radiation heat transfer

The conduction flux vector is given by the Fourier law of heat conduction, and thus the divergence of conduction flux is the familiar

$$\nabla \cdot \vec{q}_c'' = -\left[\frac{\partial}{\partial x}\left(k \frac{\partial T}{\partial x} \right) + \frac{\partial}{\partial y}\left(k \frac{\partial T}{\partial y} \right) + \frac{\partial}{\partial z}\left(k \frac{\partial T}{\partial z} \right) \right] \tag{13.46}$$

which for constant thermal conductivity is proportional to the Laplacian of temperature. The development of divergence of radiative flux is presented next.

13.4.1 Divergence Radiative Flux

Consider a hypothetical differential area dA in a participating medium. Let I_λ' represent the spectral directional intensity in a particular direction s arriving at dA (Figure 13.12). Since intensity is the energy per unit projected area per unit solid angle, the amount of radiative energy crossing the area dA from a particular direction at a particular wavelength is

$$dq_{r,\lambda} = I\, dA \cos\theta\, d\omega \tag{13.47}$$

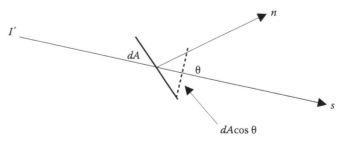

FIGURE 13.12
Intensity incident on a differential surface.

where θ is the angle between the direction of $I'(s)$ and normal to the surface. Therefore, the heat flux crossing dA from a particular direction becomes

$$q''_{r,\lambda} = \frac{dq_{r,\lambda}}{dA} = I' \cos\theta d\omega \tag{13.48}$$

The hemispherical radiative flux in the direction perpendicular to the surface dA can be obtained by integrating over all directions

$$q''_{r,\lambda}(\lambda) = \int_{4\pi} I' \cos\theta d\omega \tag{13.49}$$

The total radiative heat flux in the direction s is obtained by integrating over all wavelengths

$$\vec{q}''_r = \int_0^\infty \int_{4\pi} I' \cos\theta d\omega d\lambda \tag{13.50}$$

The total radiative flux is a vector and can be expressed in terms of its components along the chosen coordinate system. Having already determined the radiative intensity in a particular direction s, then

$$\frac{dI'_\lambda}{ds} = \frac{dI'_\lambda}{dx}\frac{dx}{ds} + \frac{dI'_\lambda}{dy}\frac{dy}{ds} + \frac{dI'_\lambda}{dz}\frac{dz}{ds} \tag{13.51}$$

As can be seen from Figure 13.13,

$$\frac{dI'_\lambda}{ds} = \frac{dI'_\lambda}{dx}\cos\theta_x + \frac{dI'_\lambda}{dy}\cos\theta_y + \frac{dI'_\lambda}{dz}\cos\theta_z \tag{13.52}$$

where θ_x, θ_y, and θ_z are the angles between the x, y, and z axes and intensity, as shown in Figure 13.13. The direction s can be defined by vector $\vec{s}$ given by

$$\vec{s} = \cos\theta_x\vec{i} + \cos\theta_y\vec{j} + \cos\theta_z\vec{k} \tag{13.53}$$

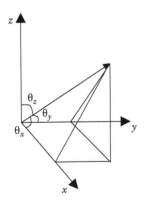

FIGURE 13.13
Components of the intensity flux in Cartesian coordinates.

Therefore,

$$\frac{dI'_\lambda}{ds} = \vec{s} \cdot \nabla I'_\lambda \tag{13.54}$$

and Equation 13.54 becomes

$$\frac{dI'_\lambda(s)}{ds} = \vec{s} \cdot \nabla I'_\lambda = -(a_\lambda + a_{s,\lambda})I'_\lambda(s) + a_\lambda I'_{\lambda b} + \frac{a_{s,\lambda}}{4\pi}\int_{4\pi} I'_\lambda(\kappa_\lambda, \omega_i)\Phi(\lambda, \omega, \omega_i)d\omega_i \tag{13.55}$$

Equation 13.55 can be integrated over all solid angles

$$\int_{4\pi} \vec{s} \cdot \nabla I'_\lambda d\omega = -\int_{4\pi}(a_\lambda + a_{s,\lambda})I'_\lambda(s)d\omega + \int_{4\pi} a_\lambda I'_{\lambda b}d\omega + \int_{4\pi}\int_{4\pi}\frac{a_{s,\lambda}}{4\pi} I'_\lambda(\kappa_\lambda, \omega_i)\Phi(\lambda, \omega, \omega_i)d\omega_i d\omega \tag{13.56}$$

Furthermore,

$$\nabla \cdot \vec{q}''_{r,\lambda} = \nabla \cdot \int_{4\pi} I'_\lambda \vec{s} d\omega = \int_{4\pi} \vec{s} \cdot \nabla I'_\lambda d\omega = \int_{4\pi}\frac{dI'_\lambda(s)}{ds}d\omega \tag{13.57}$$

Therefore, Equation 13.56 becomes

$$\nabla \cdot \vec{q}''_{r,\lambda} = -\int_{4\pi}(a_\lambda + a_{s,\lambda})I'_\lambda(s)d\omega + \int_{4\pi} a_\lambda I'_{\lambda b}d\omega + \int_{4\pi}\int_{4\pi}\frac{a_{s,\lambda}}{4\pi} I'_\lambda(\kappa_\lambda, \omega_i)\Phi(\lambda, \omega, \omega_i)d\omega_i d\omega \tag{13.58}$$

By defining

$$\bar{\Phi}(\lambda, \omega_i) = \frac{1}{4\pi}\int_{\omega=4\pi} \Phi(\lambda, \omega, \omega_i)d\omega \tag{13.59}$$

which represents the extent of the scattering of the incident radiation, and average intensity

$$\bar{I}(s, \lambda) = \frac{1}{4\pi}\int_{\omega=4\pi} I'_\lambda(s)d\omega \tag{13.60}$$

the divergence of the spectral radiative flux becomes

$$\nabla \cdot q''_{r,\lambda} = -(a_\lambda + a_{s,\lambda})4\pi\bar{I}(s, \lambda) + a_\lambda 4\pi I_{\lambda b} + a_{s,\lambda}\int_{\omega_i=4\pi} I'_\lambda(s, \omega_i)\bar{\Phi}(\lambda, \omega_i)d\omega_i \tag{13.61}$$

The local divergence of the total radiative flux is obtained by integrating Equation 13.61 over all wavelengths. For the general case of absorbing, emitting, and scattering media, the divergence becomes [3, p. 462]

$$\nabla \cdot q''_r = 4\int_0^\infty \left[-\pi[a_\lambda(\lambda) + a_{s,\lambda}(\lambda)]\bar{I}(\lambda) + a_\lambda(\lambda)e_{b\lambda}(\lambda) + \frac{a_{s,\lambda}(\lambda)}{4}\int_{\omega_i=4\pi} I'_\lambda(\lambda, \omega_i)\bar{\Phi}(\lambda, \omega_i)d\omega_i \right] d\lambda \tag{13.62}$$

Depending on the medium, the radiative flux takes different forms.

13.4.2 Absorbing–Emitting Medium without Scattering

If scattering can be neglected, then

$$a_{s,\lambda} = \Phi(\lambda, \omega, \omega_i) = \bar{\Phi}(\lambda, \omega_i) = \Omega_{0,\lambda} = 0 \tag{13.63}$$

Then from Equation 13.15

$$\tilde{I}'_\lambda = I'_{b\lambda} \tag{13.64}$$

and Equation 13.18 becomes

$$I'_\lambda(\kappa_\lambda) = I'_\lambda(0)e^{-\kappa_\lambda} + \int_0^{\kappa_\lambda} I'_{b\lambda}e^{-(\kappa_\lambda - \hat{\kappa}_\lambda)}d\hat{\kappa}_\lambda \tag{13.65}$$

from Equation 13.116

$$\nabla \cdot q_r = 4 \int_0^\infty a_\lambda [e_{b\lambda} - \pi \bar{I}(\lambda)]d\lambda \tag{13.66}$$

13.4.3 Absorbing–Emitting Medium with Isotropic Scattering

If radiation is scattered uniformly in all directions, then the scattering is called isotropic. For isotropic scattering,

$$\Phi(\lambda, \omega, \omega_i) = \bar{\Phi}(\lambda, \omega_i) = 1 \tag{13.67}$$

Then

$$\tilde{I}'_\lambda = (1 - \Omega_{0,\lambda})I'_{b\lambda} + \frac{\Omega_{0,\lambda}}{4\pi} \int I'_\lambda(\kappa_\lambda, \omega)d\omega \tag{13.68}$$

And Equation 13.66 simplifies to

$$\nabla \cdot q''_r = 4 \int_0^\infty \left[-\pi[a_\lambda(\lambda) + a_{s,\lambda}(\lambda)]\bar{I}(\lambda) + a_\lambda(\lambda)e_{b\lambda}(\lambda) + \frac{a_{s,\lambda}(\lambda)}{4} \int_{\omega_i = 4\pi} I'_\lambda(\lambda, \omega_i)d\omega_i \right] d\lambda \tag{13.69}$$

and substituting from Equation 13.60, it simplifies to

$$\nabla \cdot q_r = 4 \int_0^\infty a_\lambda [e_{b\lambda} - \pi \bar{I}(\lambda)]d\lambda \tag{13.70}$$

Although similar in form to Equation 13.66, which is for the non-scattering case, the two equations are different.

13.4.4 One-Dimensional Solutions

The derivation of the radiative flux in three dimensions is rather involved, and also the general three-dimensional solutions to the energy equation including radiation contributions are complicated. To get a better insight into the interaction of radiation with the other modes of heat transfer, and developing a better understanding of the salient features of the radiation heat transfer in participating media, we focus on cases where radiation flux is only important in one direction. Starting with the equation of transfer in terms of x

$$\kappa_\lambda(x) = \kappa_\lambda(s)\cos\theta_x \tag{13.71}$$

Therefore, equation of radiative transfer (Equation 13.14) becomes

$$\cos\theta_x \frac{dI'_\lambda(\kappa_\lambda(x))}{d\kappa_\lambda(x)} + I'_\lambda(\kappa_\lambda(x)) = \tilde{I}'_\lambda \tag{13.72}$$

Defining

$$\mu = \cos\theta_x \tag{13.73}$$

then it is convenient to break Equation 13.72 into two equations, one for I^+, which is the intensity in the positive x direction where $0 \le \theta_x \le \frac{\pi}{2}$ $(0 \le \mu \le 1)$ and another I^-, which is the intensity in the negative x direction where $\frac{\pi}{2} \le \theta_x \le \pi$ $(-1 \le \mu \le 0)$.

$$\mu \frac{dI^+_\lambda}{d\kappa_\lambda(x)} + I^+_\lambda(\kappa_\lambda,\mu) = \tilde{I}'(\kappa_\lambda,\mu) \tag{13.74}$$

$$\mu \frac{dI^-_\lambda}{d\kappa_\lambda(x)} + I^-_\lambda(\kappa_\lambda,\mu) = \tilde{I}'(\kappa_\lambda,\mu) \tag{13.75}$$

The boundary conditions for the two equations are

$$\begin{aligned}
\text{at } \kappa_\lambda = 0 \quad & I^+_\lambda(\kappa_\lambda,\mu) = I^+_\lambda(0,\mu) \\
\text{at } \kappa_\lambda = \kappa_{D\lambda} \quad & I^-_\lambda(\kappa_\lambda,\mu) = I^-_\lambda(\kappa_{D\lambda},\mu)
\end{aligned} \tag{13.76}$$

where

$$\kappa_{D\lambda} = \int_0^D (a_\lambda + a_{s,\lambda})dx \tag{13.77}$$

Using the integration factor, the solutions become

$$I^+_\lambda(\kappa_\lambda,\mu) = I^+_\lambda(0,\mu)e^{-\frac{\kappa}{\mu}} + \int_0^{\kappa_\lambda} \tilde{I}'_\lambda(\hat{\kappa}_\lambda,\mu)e^{-\frac{\kappa-\hat{\kappa}_\lambda}{\mu}} \frac{d\hat{\kappa}_\lambda}{\mu} \quad \text{for } 0 \le \mu \le 1 \tag{13.78}$$

$$I_\lambda^-(\kappa_\lambda,\mu) = I_\lambda^-(\kappa_{D\lambda},\mu)e^{-\frac{\kappa_\lambda-\kappa_{D\lambda}}{\mu}} - \int_0^{\kappa_{D\lambda}} \tilde{I}_\lambda'(\hat{\kappa}_\lambda,\mu)e^{-\frac{\kappa-\hat{\kappa}_\lambda}{\mu}}\frac{d\hat{\kappa}_\lambda}{\mu} \quad \text{for } -1 \le \mu \le 0 \tag{13.79}$$

From Equation 13.73 since $d\mu = -\sin\theta_x d\theta_x$,

$$q_{r,\lambda}'' = \int_0^{2\pi}\int_0^\pi I' \cos\theta \sin\theta d\theta d\phi = 2\pi\left[\int_0^1 I^+\mu d\mu - \int_0^1 I^-\mu d\mu\right] \tag{13.80}$$

Substituting from Equations 13.78 and 13.79 into Equation 13.80 and after much algebra, for a plane layer, the radiative flux in an arbitrary direction becomes

$$q_{r,\lambda}'' = 2\pi\int_0^1 I_\lambda^+(0,\mu)e^{-\frac{\kappa_\lambda}{\mu}}\mu d\mu - 2\pi\int_0^1 I_\lambda^-(\kappa_{D\lambda},-\mu)e^{-\frac{\kappa_{D\lambda}-\kappa_\lambda}{\mu}}\mu d\mu$$

$$+ 2\pi\int_0^1\int_0^{\kappa_\lambda} \tilde{I}_\lambda(\hat{\kappa}_\lambda,\mu)e^{-\frac{\kappa_\lambda-\hat{\kappa}_\lambda}{\mu}}d\hat{\kappa}_\lambda d\mu - 2\pi\int_0^1\int_{\kappa_\lambda}^{\kappa_{D\lambda}} \tilde{I}_\lambda(\hat{\kappa}_\lambda,-\mu)e^{-\frac{\hat{\kappa}_\lambda-\kappa_\lambda}{\mu}}d\hat{\kappa}_\lambda d\mu \tag{13.81}$$

The divergence of heat flux is obtained by differentiating Equation 13.81 with respect to κ_λ.

13.4.5 Gray Body with Isotropic Scattering

If the properties of the participating medium can be assumed to be independent of the wavelength, then the medium under consideration is called gray. This is an idealization, since very few real gases have properties that are independent of the wavelength. However, in certain ranges, gas properties can be considered to be independent of wavelength. Integrating the equations over all wavelengths and defining

$$I' = \int_0^\infty I_\lambda' d\lambda \tag{13.82}$$

$$\tilde{I}' = \int_0^\infty \tilde{I}_\lambda' d\lambda \tag{13.83}$$

and noting that

$$\int_0^\infty I_{\lambda b}' d\lambda = \frac{\sigma T^4}{\pi} \tag{13.84}$$

the radiative heat flux becomes [3]

$$q_r'' = 2\pi \int_0^1 I^+(0,\mu)e^{-\frac{\kappa}{\mu}}\mu d\mu - 2\pi \int_0^1 I^-(\kappa_D,-\mu)e^{-\frac{\kappa_D-\kappa}{\mu}}\mu d\mu + 2\pi \int_0^\kappa \tilde{I}(\hat{\kappa})E_2(\kappa-\hat{\kappa})d\hat{\kappa} - 2\pi \int_0^{\kappa_D} \tilde{I}(\hat{\kappa})E_2(\hat{\kappa}-\kappa)d\hat{\kappa}$$

(13.85)

and its divergence becomes

$$\frac{dq_r''}{d\kappa} = -2\pi \int_0^1 I^+(0,\mu)e^{-\frac{\kappa}{\mu}}d\mu - 2\pi \int_0^1 I^-(\kappa_D,-\mu)e^{-\frac{\kappa_D-\kappa}{\mu}}d\mu - 2\pi \int_0^{\kappa_D} \tilde{I}(\hat{\kappa})E_1(|\kappa-\hat{\kappa}|)d\hat{\kappa} + 4\pi\tilde{I}$$ (13.86)

where

$$\tilde{I} = (1-\Omega_0)\frac{\sigma T^4}{\pi} + \frac{\Omega_0}{2}\left[\int_0^1 I^+(0,\mu)e^{-\frac{\kappa}{\mu}}d\mu + \int_0^1 I^-(\kappa_D,-\mu)e^{-\frac{\kappa_D-\kappa}{\mu}}d\mu + \int_0^{\kappa_D} \tilde{I}(\hat{\kappa})E_1(|\hat{\kappa}-\kappa|)d\hat{\kappa}\right]$$ (13.87)

and

$$E_n(x) = \int_0^1 \mu^{n-2}e^{-\frac{x}{\mu}}d\mu$$ (13.88)

is the exponential integral function.

13.4.6 Exponential Integral Function

The exponential integral function is defined as

$$E_n(x) = \int_1^\infty \frac{e^{-xt}}{t^n}dt$$ (13.89)

An alternative form of the integral can be obtained by defining $\mu = \dfrac{1}{t}$

$$E_n(x) = \int_1^\infty \frac{e^{-xt}}{t^n}dt = \int_0^1 \mu^{n-2}e^{-\frac{x}{\mu}}d\mu$$ (13.90)

For $n = 1$, the exponential function can be approximated by

$$E_1(x) = -\left(\gamma + \ln x + \sum_{i=1}^\infty \frac{(-1)^i(x)^i}{i!\,i}\right)$$ (13.91)

where $\gamma = 0.577216$ is the Euler–Mascheroni constant. Another approximation is given by

$$E_1(x) = \Gamma(0, x) \approx \cfrac{e^{-x}}{x+1-\cfrac{1}{x+3-\cfrac{4}{x+5-\cfrac{9}{x+7-\cfrac{16}{x+9-\cdots}}}}} \qquad (13.92)$$

where $\Gamma(0, x)$ is the incomplete gamma function. Note that the earlier series solutions converge slowly, and to get accurate results, many terms may have to be used. Figure 13.14 is a comparison of the two solutions with the exact results of Siegel and Howell [3] using eight terms in each series. As can be seen, Equation 13.91 is more accurate at smaller values of x while Equation 13.92 is more accurate at larger values of x.

The recurrence formula

$$E_{n+1}(x) = \frac{1}{n}\left[e^{-x} - xE_n(x)\right] \qquad (13.93)$$

can be used to obtain the values of other integrals. Also by differentiating Equation 13.93, it is easy to show that the derivative of exponential functions can be obtained from

$$\frac{dE_n(x)}{dx} = -E_{n-1}(x) \qquad (13.94)$$

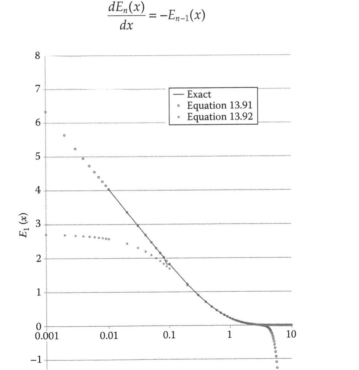

FIGURE 13.14
Comparison of the exact and approximate solutions to the exponential function.

13.4.7 Gray Medium in Radiative Equilibrium

When heat transfer is dominated by radiation and the other modes of heat transfer, i.e., conduction and convection are negligible, the energy balance is between emitted, absorbed, and scattered radiation. Conservation of energy requires that $\dfrac{dq_r''}{d\kappa} = 0$. In addition, if the scattering is negligible, $\tilde{I} = \dfrac{\sigma T^4}{\pi}$ and from Equation 13.85,

$$q_r'' = 2\pi \int_0^1 I^+(0,\mu) e^{-\frac{\kappa}{\mu}} \mu d\mu - 2\pi \int_0^1 I^-(\kappa_D,-\mu) e^{-\frac{\kappa_D-\kappa}{\mu}} \mu d\mu + 2 \int_0^\kappa \sigma T^4 E_2(\kappa - \hat{\kappa}) d\hat{\kappa} - 2 \int_\kappa^{\kappa_D} \sigma T^4 E_2(\hat{\kappa} - \kappa) d\hat{\kappa}$$

$$(13.95)$$

If we further assume that the boundaries are diffuse and gray, and the incident fluxes at the boundaries are $q''^+(0) = \pi I^+(0,\mu)$ and $q''^-(\kappa_D) = \pi I^-(\kappa_D,-\mu)$, then

$$q_r'' = 2q''^+(0)E_3(\kappa) - 2q''^-(\kappa_D)E_3(\kappa_D - \kappa) + 2 \int_0^\kappa \sigma T^4 E_2(\kappa - \hat{\kappa}) d\hat{\kappa} - 2 \int_\kappa^{\kappa_D} \sigma T^4 E_2(\hat{\kappa} - \kappa) d\hat{\kappa} \quad (13.96)$$

For an isothermal gas at temperature T confined between two black walls at temperatures T_1 and T_2, Equation 13.96 simplifies to

$$\frac{q_r''(\kappa)}{2\sigma T^4} = \left(\frac{T_1^4}{T^4} - 1\right)E_3(\kappa) - \left(\frac{T_2^4}{T^4} - 1\right)E_3(\kappa_D - \kappa) \quad (13.97)$$

And the radiative flux at each wall is obtained by evaluationg Equation 13.97 at $\kappa = 0$, and $\kappa = \kappa_D$.

13.4.8 One-Dimensional Conduction Radiation

Consider heat transfer by conduction in a semitransparent medium of thickness D, where the surfaces can be assumed to be isothermal and black at T_1 and T_2. If scattering is neglected, then the energy equation becomes

$$0 = k\frac{d^2T}{dx^2} - \frac{dq_r}{dx} \quad (13.98)$$

For a gray medium, $\kappa = ax$; therefore,

$$ka\frac{d^2T}{d\kappa^2} = \frac{dq_r}{d\kappa} \quad (13.99)$$

Since the boundaries are black,

$$I^+(0,\mu) = \frac{\sigma T_1^4}{\pi} \quad (13.100)$$

$$I^-(\kappa_D,-\mu) = \frac{\sigma T_2^4}{\pi} \quad (13.101)$$

In the absence of scattering, Equation 13.87 reduces to

$$\tilde{I} = \frac{\sigma T^4}{\pi} \tag{13.102}$$

and therefore Equation 13.87 becomes

$$\frac{dq_r}{d\kappa} = -2\sigma T_1^4 \int_0^1 e^{-\frac{\kappa}{\mu}} d\mu - 2\sigma T_2^4 \int_0^1 e^{-\frac{\kappa_D - \kappa}{\mu}} d\mu - 2\int_0^{\kappa_D} \sigma T^4 E_1(|\kappa - \hat{\kappa}|) d\hat{\kappa} + 4\sigma T^4 \tag{13.103}$$

which, when substituted in the energy equation, results in

$$ka \frac{d^2 T}{d\kappa^2} = -2\sigma T_1^4 E_2(\kappa) - 2\sigma T_2^4 E_2(\kappa_D - \kappa) - 2\int_0^{\kappa_D} \sigma T^4 E_1(|\kappa - \hat{\kappa}|) d\hat{\kappa} + 4\sigma T^4 \tag{13.104}$$

This is a second-order integro-differential equation. Defining

$$T^* = \frac{T}{T_1}, \quad \kappa^* = \frac{\kappa}{\kappa_D} \tag{13.105}$$

The nondimensional form of the energy equation reduces to

$$N_{cr} \frac{d^2 T^*}{d\kappa^{*2}} = T^{*4} - \frac{1}{2}\left[E_2(\kappa_D \kappa^*) + T_2^{*4} E_2[\kappa_D(1-\kappa^*)] + \kappa_D \int_0^1 T^{*4} E_1(\kappa_D(|\kappa^* - \hat{\kappa}^*|)) d\hat{\kappa}^* \right] \tag{13.106}$$

where

$$N_{cr} = \frac{ka}{4\sigma T_1^3 \kappa_D^2} \tag{13.107}$$

is the conduction radiation parameter and a measure of the relative importance of conduction to radiation. Note that some authors define the radiation conduction parameter as $\dfrac{ka}{4\sigma T_1^3}$. From Equation 13.106, the dimensionless solution depends on the dimensionless temperature at surface 2, the conduction radiation parameter, and the optical depth. For small values of N_{cr}, radiation dominates, and as it increases, the impact of radiation decreases and the solution approaches that of pure conduction, with temperature varying linearly.

Here we present the numerical solution by using central difference approximation for the second derivative resulting in

$$T_i^* = \frac{T_{i+1}^* + T_{i-1}^*}{2} - \frac{\Delta\kappa^{*2}}{2N_{cr}}\left\{ T_i^{*4} - \frac{1}{2}\left[E_2(\kappa) + T_2^{*4} E_2(\kappa_D - \kappa) + \kappa_D \int_0^1 T^{*4} E_1\left(\kappa_D(|\kappa - \hat{\kappa}|)\right) d\hat{\kappa}^* \right] \right\} \tag{13.108}$$

The numerical solution to this equation involves a few complicating factors. The equation is nonlinear, having T^4 on the right-hand side, requiring an iterative approach. The next challenge is evaluation of the integral that is a function of the unknown temperature. This also requires the solution to be obtained iteratively. The numerical integration of the integral on the right-hand side is performed using the trapezoidal rule, which results in

$$\int_0^1 T_i^{*4} E_1(\kappa_D(|\kappa^* - \hat{\kappa}^*|)) d\hat{\kappa}^* = \frac{\hat{\kappa}^*}{2}\left(2\sum_{j=1}^m f_{i,j} - (f_{i,1} + f_{i,m})\right) \tag{13.109}$$

where

$$f_{i,j} = T_i^{*4} E_1(\kappa_D(|\kappa^* - \hat{\kappa}^*|)) \tag{13.110}$$

This requires the evaluation of the exponential functions that can be done using series solution provided by Equations 13.91 and 13.92. The numerical solution is shown in Figure 13.15 for several values of the radiation conduction parameter. The results of numerical simulations are very close to those of Viskanta and Grosh [11] for different values of N_{cr}.

When radiation effects increase, the temperature distribution becomes flatter near higher temperature boundary and then drops rapidly near the colder boundary.

The total rate of heat transfer in the medium is the combination of the conduction and radiation terms

$$q_T'' = -k\frac{dT}{dx} + q_r = -ka\frac{dT}{d\kappa} + q_r \tag{13.111}$$

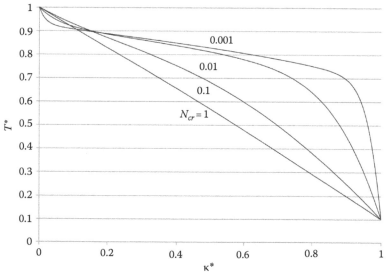

FIGURE 13.15
Solution to one-dimensional conduction radiation problem.

and from first law, it is a constant, and therefore can be determined at any position. At the left wall ($\kappa = 0$) Equation 13.96 simplifies to

$$q_r'' = 2\pi \int_0^1 I^+(0,\mu)\mu d\mu - 2\pi \int_0^1 I^-(\kappa_D,-\mu)e^{-\frac{\kappa_D}{\mu}}\mu d\mu - 2\pi \int_0^{\kappa_D} \tilde{I}(\hat{\kappa})E_2(\hat{\kappa})d\hat{\kappa} \tag{13.112}$$

and in the absence of scattering

$$\tilde{I} = \frac{\sigma T^4}{\pi} \tag{13.113}$$

$$q_r'' = 2\pi \frac{\sigma T_1^4}{\pi}\left(\frac{1}{2}\mu^2\right)_0^1 - 2\pi \frac{\sigma T_2^4}{\pi}E_3(\kappa_D) - 2\pi \int_0^{\kappa_D} \frac{\sigma T^4}{\pi}E_2(\hat{\kappa})d\hat{\kappa} \tag{13.114}$$

Therefore, the total heat flux at the left boundary is

$$q_T''(\kappa = 0) = -ka\frac{dT}{d\kappa}\bigg|_{\kappa=0} + \sigma T_1^4 - 2\sigma T_2^4 E_3(\kappa_D) - 2\int_0^{\kappa_D} \sigma T^4 E_2(\hat{\kappa})d\hat{\kappa} \tag{13.115}$$

Nondimensionalizing this equation

$$\frac{q_T''}{\sigma T_1^4} = -N_{cr}4\kappa_D\frac{dT^*}{d\kappa^*}\bigg|_{\kappa^*=0} + 1 - 2\left[T_2^{*4}E_3(\kappa_D) + \kappa_D\int_0^1 T^{*4}E_2(\hat{\kappa})d\hat{\kappa}^*\right] \tag{13.116}$$

In the limit $N_{cr} \to \infty$, the heat transfer is by pure conduction, and it is easy to show that

$$\lim_{N_{cr}\to\infty}\frac{q}{\sigma T_1^4} = \frac{T_1 - T_2}{\frac{D}{k}\sigma T_1^4} = \frac{1 - T_2^*}{\frac{\kappa_D}{ak}\sigma T_1^3} = N_{CR}4K_D(1 - T_2^*) \tag{13.117}$$

Table 13.2 shows the values obtained by Viskanta and Grosh [12] as reported by Ref. [3]. Two values calculated by the numerical solution provided earlier are shown in the fifth column, and the last column shows the values when heat transfer is by pure conduction, corresponding to large values of N_{cr}.

13.4.9 Boundary Layer Flow

The boundary layer flow with gas radiation serves as an example of the convection–radiation interaction. Several solution techniques have been used to determine the interaction of radiation and convective flow. An excellent textbook on the topic is by Özisik [13], where a thorough literature review and the description of the problem formulation are presented. The continuity and momentum equations are unchanged, and the energy equation includes the radiative flux and becomes

TABLE 13.2

Total Heat Flux at the Left Boundary

K_D	T_2^*	N_{CR}	$\dfrac{q''}{\sigma T_1^4}$ [12]	$\dfrac{q''}{\sigma T_1^4}$ Numerical	$\lim\limits_{N_{cr}\to\infty}\dfrac{q''}{\sigma T_1^4}$
0.1	0.5	0	0.859		0
0.1	0.5	1	1.074		0.2
0.1	0.5	10	2.88		2
0.1	0.5	100	20.88		20
0.1	0.5	1000	200.88		200
1	0.5	0	0.518		0
1	0.5	0.01	0.595	0.579	0.02
1	0.5	0.1	0.798		0.2
1	0.5	1	2.6		2
1	0.5	10	20.6		20
1	0.1	0	0.556		0
1	0.1	0.01	0.658	0.640	0.036
1	0.1	0.1	0.991		0.36
1	0.1	1	4.218		3.6
1	0.1	10	36.6		36
10	0.5	0	0.102		0
10	0.5	0.0001	0.114		0.002
10	0.5	0.001	0.131		0.02
10	0.5	0.01	0.315		0.2
10	0.5	0.1	2.114		2

$$\rho c_p\left(u\frac{\partial T}{\partial x}+v\frac{\partial T}{\partial y}\right)=\frac{\partial}{\partial y}\left[k\frac{\partial T}{\partial y}-q_r\right]\tag{13.118}$$

The widely used approach for treating gas phase radiation is one-dimensional optically thin approximation [14]. Viskanta and Grosh [15] used Roseland approximation, where radiative flux is given by

$$q_r=-\frac{16n^2\sigma T^3}{3a}\frac{\partial T}{\partial y}\tag{13.119}$$

and n is the index of refraction of the medium. This approximation is valid for optically thick media, with strong absorption. The energy equation can then be written as

$$\rho c_p\left(u\frac{\partial T}{\partial x}+v\frac{\partial T}{\partial y}\right)=\frac{\partial}{\partial y}\left[k_{eff}\frac{\partial T}{\partial y}\right]\tag{13.120}$$

where

$$k_{eff}=k+\frac{16n^2\sigma T^3}{3a}\tag{13.121}$$

with the second term on the right-hand side representing an equivalent radiative conductivity. Although an optically thick boundary layer is not a common occurrence, their analysis provides much insight into the physics of the problem.

Cess [16] assumed optically thin approximation and presented a perturbation technique to solve for regions close to the leading edge of the plate. In another study, Cess [17] dropped the optically thin approximation and used linearization of radiation term to obtain results. Lee et al. [18] developed a multilayer solution that is particularly suited for numerical solution. Using similarity variables, the momentum and energy equations for combined radiation–convection for compressible laminar boundary layer flow of gray absorbing–emitting gases over a flat plate reduce to

$$lf''' + ff'' = 0 \tag{13.122}$$

$$\frac{l}{\mathrm{Pr}} \frac{\partial^2 T^*}{\partial \eta^2} + f \frac{\partial T^*}{\partial \eta} = 2\xi \left(f' \frac{\partial T^*}{\partial \xi} + \frac{1}{N_r} \frac{\partial q_r^*}{\partial \kappa} \right) \tag{13.123}$$

where

$$\eta = \left(\frac{U_\infty}{2\rho_\infty \mu_\infty x} \right)^{\frac{1}{2}} \int_0^y \rho\, dy \tag{13.124}$$

$$\xi = \frac{x}{L_r} = \frac{x}{\dfrac{U_\infty}{\rho_\infty \mu_\infty} \dfrac{\rho^2}{a^2}} \tag{13.125}$$

$$T^* = \frac{T}{T_w} \tag{13.126}$$

$$l = \frac{\rho\mu}{\rho_w \mu_w} = cons \tag{13.127}$$

$$\frac{\rho}{a} = cons \tag{13.128}$$

$$\kappa = \int_0^y a\, dy' = ay \tag{13.129}$$

$$q_r^* = \frac{q_r}{\sigma T_w^4} \tag{13.130}$$

$$N_r = \frac{\rho_\infty \mu_\infty c_p}{\sigma T_w^3} \frac{a}{\rho} \tag{13.131}$$

The radiative flux for an isothermal gray gas layer is given by Equation 13.96. In finite difference solution, the domain is divided into small layers (grids), and assuming that the temperature in the integrals of Equation 13.96 can be approximated by the average temperature of the layer, Equation 13.75 simplifies to

$$q_r'' = 2q''^+(0)E_3(\kappa) - 2q''^-(\kappa_D)E_3(\kappa_D - \kappa) + 2\pi\sigma T_m^4 \int_0^\kappa E_2(\kappa - \hat{\kappa})d\hat{\kappa} - 2T_m^4 \int_\kappa^{\kappa_D} F_2(\hat{\kappa} - \kappa)\,d\hat{\kappa}$$

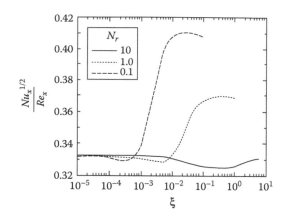

FIGURE 13.16
Impact of radiation on Nu for boundary layer flow.

Evaluating the last two integrals results in the following expression for the radiative flux in each layer [18]

$$q''_r = 2E_3(\kappa)q''^+(0) - 2E_3(\kappa_D - \kappa)q''^-(\kappa_D) - 2\sigma T_m^4\left[E_3(\kappa_D - \kappa) - E_3(\kappa)\right] \qquad (13.132)$$

This equation provides the radiative flux at each layer of the finite difference grid.

Figure 13.16 is a plot of $\dfrac{Nu_x}{\sqrt{Re_x}}$ along the plate. At small values of ξ, the solution must approach to 0.332, which is the similarity solution in the absence of radiation. At larger values of ξ, the local value of Nusselt number increases as a result of gas phase radiation. Lee et al. report that at very large values of ξ, the solution becomes numerically unstable. The numerical results are in good agreement with those of Cess [17].

Problems

13.1 Consider an equimolar mixture of CO_2 and O_2 gases at 600 K and a total pressure of 1 atm. Determine the emissivity of the gas if the path length is 1 m.

13.2 Repeat Problem 13.1, if the concentration of oxygen is three time that of CO_2.

13.3 A spherical shell $D = 1$ m contains a mixture of water and air at 1000°C and 2 bar. The water mass fraction is 0.2 and the inside surface of the sphere is black and maintained at 400°C:

a. Show that the net radiative transfer between the gas and the inner surface is

$$q = \sigma A\left(\varepsilon_g T_g^4 - \alpha_g T_s^4\right)$$

b. Determine the heat transfer rate.

13.4 A gas mixture of air and water at 800°C and 1 bar flows in a 15 cm circular pipe that is at a constant temperature of 400°C. The water's mole fraction is 0.08. Determine the heat transfer between the gas and pipe.

13.5 Hot gases at 1000°C and 2 bar are contained between two parallel plates, 50 cm apart that can be assumed to be black at 200°C. The gas mixture is 10% O_2, 60% N_2, 15% CO_2, and 15% H_2O. Determine the heat transfer between the gas and the pipe.

13.6 A rectangular furnace has dimensions of $1 \times 1 \times 3$ m. The combustion products are at 1200°C. The gas mixture is 10% O_2, 30% CO_2, and 15% H_2O, and the balance is nitrogen. Determine the energy radiated by the gas to the interior furnace walls that can be assumed to be black.

13.7 Steam leaving the boiler of a power plant leaves at 100 bar and 800°C and is sent to the turbine through a 20 cm pipe. Determine the radiative heat transfer to the interior surface of the pipe.

13.8 Starting with Equation 13.96, derive Equation 13.97.

13.9 A gray gas at the uniform temperature of 500 K having an absorption coefficient $\alpha = 0.3$ m^{-1} flows between two isothermal parallel plates, 50 cm apart, that are maintained at 300 K. Determine the radiation heat flux at the walls.

13.10 Show that in the optically thin limit, $\kappa_D \to 0$, for a gas between two infinitely long parallel plates, the radiative heat transfer is given by

$$q_{\kappa_D \to 0} = \sigma A \left(T_1^4 - T_2^4 \right)$$

13.11 For 1D conduction in a semitransparent medium in the optically thin limit,
  a. Show that the energy equation simplifies to

$$N_{cr} \frac{d^2 T^*}{d\kappa^{*2}} = T^{*4} - \frac{1}{2}\left[1 + T_2^{*4} \right]$$

  b. Numerically obtain the solution for $N_{cr} = 0.01$ and $T_2^* = 0.1$.

13.12 Numerically determine temperature distribution for heat transfer by conduction in a semitransparent medium of thickness D, where the surfaces can be assumed to be isothermal and black at T_1 and T_2 for $N_{cr} = 0.1$, as shown in Figure 13.15.

References

  1. Peixoto, J. P. and Oort, A. H. (1992) *Physics of Climate*, New York: Springer.
  2. Fishbane, P. M., Gasiorowicz, S., and Thornton, S. T. (2005) *Physics for Scientists and Engineers with Modern Physics*, Upper Saddle River, NJ: Prentice Hall.
  3. Siegel, R. and Howell, J. R. (1968–1971) *Thermal Radiation Heat Transfer*, Washington, DC: Scientific and Technical Information Division, National Aeronautics and Space Administration (for sale by The Superintendent of Documents, U.S. Government Printing Office).

4. Hottel, H. C. (1954) Radiant heat transmission, in W. H. Mcadams, ed., *Heat Transmission*, 3rd edn., New York: McGraw-Hill, Ch 4.
5. Modest, M. F. (1993) *Radiative Heat Transfer*, New York: McGraw-Hill.
6. Leckner, B. (1972) Spectral and total emissivity of water vapor and carbon dioxide, *Combustion and Flame* 19(1), 33–48.
7. Boynton, F. P. and Ludwig, C. B. (1971) Total emissivity of hot water vapor-II. Semi-empirical charts deduced from long-path spectral data, *International Journal of Heat and Mass Transfer* 14(7), 963–973.
8. Sarofim, A. F., Farag, I. H., and Hottel, H. C. (1978) Radiative heat transfer transmission from non-luminous gases. Computational study of the emissivities of carbon dioxide, *American Society of Mechanical Engineers*, Paper 78-HT-16.
9. Hottel, H. C. and Sarofim, A. F. (1967) *Radiative Transfer*, New York: McGraw-Hill.
10. Hottel, H. C. and Egbert, R. B. (1942) Radiant heat transmission from water vapour, *AIChE Trans.* 38, 531.
11. Viskanta, R. and Grosh, R. J. (1962) Heat transfer by simultaneous conduction and radiation in an absorbing medium, *Journal of Heat Transfer* 84(1), 63–72.
12. Viskanta, R. and Grosh, R. J. (1962) Effect of surface emissivity on heat transfer by simultaneous conduction and radiation, *International Journal of Heat and Mass Transfer* 5(8), 729–734.
13. Özisik, M. N. (1973) *Radiative Transfer and Interactions with Conduction and Convection*. New York: Wiley.
14. Brier, R. A., Pagni, P. J., and Okoh, C. I. (1984) Soot and radiation in combusting boundary layers, *Combustion Science and Technology* 39, 235–262.
15. Viskanta, R. and Grosh, R. J. (1962) Boundary layer in thermal radiation absorbing and emitting media, *International Journal of Heat and Mass Transfer* 5(9), 795–806.
16. Cess, R. D. (1966) The interaction of thermal radiation with free convection heat transfer, *International Journal of Heat and Mass Transfer* 9(11), 1269–1277.
17. Cess, R. D. (1966) The interaction of thermal radiation in boundary layer heat transfer, *Proceedings of the Third International Heat Transfer Conference*, Chicago, IL, 5, 154–163.
18. Lee, H. S., Menart, J. A., and Fakheri, A. (1990) Multilayer radiation solution for boundary-layer flow of gray gases, *Journal of Thermophysics and Heat Transfer* 4(2), 180–185.

14

Phase Change

14.1 Introduction

When a liquid is heated above its saturation temperature, it evaporates, and when a vapor is cooled below its saturation temperature, it condenses. The amount of energy released or absorbed during phase change (latent heat) is relatively large, making phase change an effective heat transfer process, in nature and in engineered systems. During intense exercising, the body could lose as much as 85% of its heat through perspiration [1]. Transpiration from leaves' pores is needed for the flow of moisture and nutrients from the roots to the leaves, helps in photosynthesis, and prevents leaves' overheating. An acre of corn can transpire as much as 4000 gal of water per day [2]. Transpiration cooling is used in a number of applications like turbines and reentry vehicles to remove large amounts of heat.

In an evaporative cooler, air flows over a wet fibrous mat, causing water to evaporate by removing heat from the dry air, cooling as well as humidifying the air. This is an effective and economical method of air-conditioning in dry hot climates. Single-stream heat exchangers, where one fluid goes through a phase change, are widely used when large amounts of heat need to be transferred in a small volume. Evaporators or condensers are examples of such devices and are used in air-conditioning applications. Evaporation and condensation are integral part of many processes used in chemical plants, paper mills, and oil refineries. Rankin cycle, which is based on evaporation and condensation of a working fluid, is the primary cycle used for electric power generation.

Phase change can also play a detrimental role in heat transfer process. For example, the outer surface temperature of the evaporator in an air conditioner is typically below the dew point of the humid air flowing over the coil, leading to accumulation of liquid on the evaporator surface. At low Reynolds numbers, the condensate inhibits heat transfer rate and increases the pressure drop, but at high Reynolds numbers, the condensate will be removed and the droplets on the surface will effectively increase the surface roughness, increasing the heat transfer and pressure drop [3,4].

Similar to heat transfer by convection, evaporation and condensation are broadly categorized depending on whether the fluid flow occurs due to naturally occurring forces or an external mean.

14.2 Condensation

Consider a cold inclined plate exposed to a saturated vapor. There are two different modes of condensation. If the condensate wets the wall, a thin film would be formed as a result of heat transfer from the vapor to the wall. This film slides down along the plate as a result of gravity and the forced flow of the saturated vapor if the vapor is flowing over the surface. This is known as the film condensation. On some surfaces, the condensate will not stick to the wall; instead, small droplets form on the surface imperfections like scratches or pits. The droplets grow, merge, and move down the surface, creating room for the formation of more droplets. This process is known as dropwise condensation.

14.3 Film Condensation

Consider a cold surface at temperature T_w in contact with a superheated vapor at a temperature T_v. As one moves horizontally toward the plate, the temperature decreases, and once the vapor reaches the saturation temperature, it condenses, forming a liquid film that falls along the surface. Since vapor has a much lower density, it will have a much higher velocity normal to the surface and primarily in the horizontal direction toward the surface. The liquid's normal velocity is very small, and it predominantly moves along the surface, and therefore the boundary layer approximation can be invoked.

14.3.1 Vertical Flat Plate

The basic physics of film condensation is demonstrated by considering condensation over a vertical flat plate. In general, the superheated vapor moves toward the plate, and its temperature, T', decreases from T_v, far away from the plate, to the saturation temperature, T_s, at the edge of the condensed film, and then the saturated vapor condenses to saturated liquid and is further cooled down to a compressed liquid state while falling along the wall with a velocity u. Using boundary layer approximation, as was shown in Equation 10.8, the governing equations are

$$\frac{\partial u}{\partial x} + \frac{\partial v}{\partial y} = 0 \tag{14.1}$$

$$\rho u \frac{\partial u}{\partial x} + \rho v \frac{\partial u}{\partial y} = \mu \frac{\partial^2 u}{\partial y^2} - (\rho_g - \rho_f)g \tag{14.2}$$

$$\rho c_p \left(u \frac{\partial T}{\partial x} + v \frac{\partial T}{\partial y} \right) = k \frac{\partial^2 T}{\partial y^2} \tag{14.3}$$

In Equation 14.2, ρ_g and ρ_f are the densities of saturated liquid and vapor respectively $(\rho_f \gg \rho_g)$, and the negative sign in front of the body force is due to the fact that gravity is acting in the direction of the flow, as opposed to the natural convection.

14.3.2 Approximate Solution

The liquid velocity is expected to be very low, and therefore as an approximation, we can neglect inertial forces in the momentum equation and convective terms in the energy equation, simplifying them to

$$\mu \frac{\partial^2 u}{\partial y^2} + (\rho_f - \rho_g)g = 0 \tag{14.4}$$

$$\frac{\partial^2 T}{\partial y^2} = 0 \tag{14.5}$$

The boundary conditions are

$$
\begin{aligned}
y = 0 \quad & u = 0 \quad && T = T_w \\
y = \delta \quad & \frac{du}{dy} = 0 \quad && T = T_s
\end{aligned}
\tag{14.6}
$$

Doing an energy balance at the edge of the boundary layer, as shown in Figure 14.1,

$$-k\frac{\partial T}{\partial y} + \rho_f v_f h_f = -k_v \frac{\partial T'}{\partial y} + \rho_g v_g h_g \tag{14.7}$$

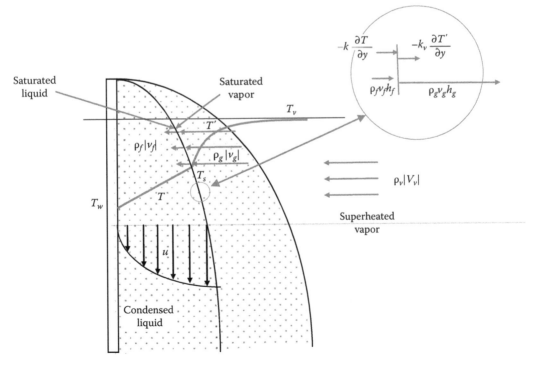

FIGURE 14.1
Condensation of superheated vapor over a vertical flat plate.

and from continuity

$$\rho_g v_g = \rho_f v_f = \rho_v v_v \tag{14.8}$$

then

$$-k\frac{\partial T}{\partial y} = -k_v\frac{\partial T'}{\partial y} + \rho_f v_f h_{fg} \tag{14.9}$$

Note that the velocities of liquid, saturated vapor, and superheated vapor in the y direction (normal to the plate) are all negative, since they all flow toward the plate.

14.3.2.1 Saturated Vapor Solution

We first consider the case where the plate is in contact with saturated vapor as shown in Figure 14.2. The liquid temperature increases from the wall temperature to the saturation temperature at the interface, and the vapor temperature remains constant at T_s past that point, or for $y > \delta$ $\frac{\partial T'}{\partial y} = 0$, and Equation 14.9 simplifies to

$$-k\frac{\partial T}{\partial y} = \rho_f v_f h_{fg} \tag{14.10}$$

14.3.2.2 Nondimensionalization

Defining the nondimensional variables as

$$x^* = \frac{x}{L} \quad y^* = \frac{y}{L} \quad u^* = \frac{u}{u_s} \quad v^* = \frac{v}{u_s} \quad T^* = \frac{T - T_w}{T_s - T_w} \quad \delta^* = \frac{\delta}{L} \tag{14.11}$$

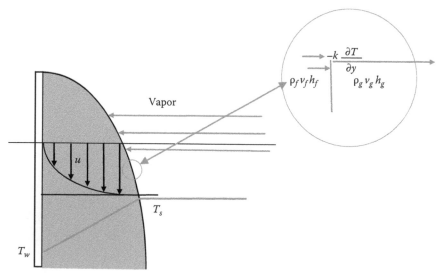

FIGURE 14.2
Condensation of saturated vapor over a vertical flat plate.

the continuity equation simplifies to

$$\frac{\partial u^*}{\partial x^*} + \frac{\partial v^*}{\partial y^*} = 0 \qquad (14.12)$$

and substituting in the momentum equation,

$$\frac{\partial^2 u^*}{\partial y^{*2}} + \frac{L^2}{\mu u_s}(\rho_f - \rho_g)g = 0 \qquad (14.13)$$

The relevant velocity scale is then

$$u_s = \frac{L^2}{\mu}(\rho_f - \rho_g)g \qquad (14.14)$$

The condensate velocity is expected to be of the same order of magnitude as this velocity scale. It also shows that the velocity of the condensate is proportional to the density difference (liquid density primarily), the square of the distance along the plate, and inversely proportional to the fluid viscosity. The nondimensional form of the momentum equation becomes

$$\frac{\partial^2 u^*}{\partial y^{*2}} = -1 \qquad (14.15)$$

$$\frac{\partial^2 T^*}{\partial y^{*2}} = 0 \qquad (14.16)$$

The nondimensional form of the energy balance at the interface, Equation 14.10, becomes

$$\frac{\partial T^*}{\partial y^*} = -\frac{(\rho_f - \rho_g)L^3 g}{\rho_f \alpha v} \frac{h_{fg}}{c_p(T_s - T_w)} v_f^* \qquad (14.17)$$

From Chapter 5, the Jacob number is defined as

$$Ja = \frac{c_p(T_s - T_w)}{h_{fg}} \qquad (14.18)$$

The dimensionless group

$$Ra_{c,L} = \frac{(\rho_f - \rho_g)gL^3}{\rho_f v \alpha} \qquad (14.19)$$

is similar to the Rayleigh number and can be viewed as the condensation Rayleigh number. Equations 14.14 through 14.17 only depend on the ratio of condensation Rayleigh number to Jacob number, or the relevant dimensionless group for film condensation is

$$\frac{Ra_{c,L}}{Ja} = \frac{(\rho_f - \rho_g)gL^3 h_{fg}}{kv(T_s - T_w)} \qquad (14.20)$$

and therefore the boundary condition at the edge of the boundary layer for the energy equation becomes

$$\frac{\partial T^*}{\partial y^*} = -\frac{Ra_{c,L}}{Ja} v_f^*$$

(14.21)

Note that the factor $\frac{(\rho_f - \rho_g)}{\rho_f}$ is typically very close to one, and, for example, the local value of the condensation Rayleigh number simplifies to

$$Ra_{c,x} = \frac{(\rho_f - \rho_g)gx^3}{\rho_f v\alpha} \approx \frac{gx^3}{v\alpha}$$

(14.22)

Integrating the momentum and energy equations twice and evaluating the constants, the solutions become

$$u^* = \delta^{*2}\left(\frac{y^*}{\delta^*} - \frac{y^{*2}}{2\delta^{*2}}\right)$$

(14.23)

$$T^* = \frac{y^*}{\delta^*}$$

(14.24)

And in terms of the dimensional variables

$$u = \frac{(\rho_f - \rho_g)g\delta^2}{\mu}\left(\frac{y}{\delta} - \frac{y^2}{2\delta^2}\right)$$

(14.25)

$$\frac{T - T_w}{T_s - T_w} = \frac{y}{\delta}$$

(14.26)

The mass flow rate of the condensate can be determined from

$$\dot{m} = W\int_0^\delta \rho_f u\, dy = \frac{W\rho_f\left(\rho_f - \rho_g\right)g\delta^3}{3\mu}$$

(14.27)

where W is the plate width. The thickness of the boundary layer is still an unknown and is determined from Equation 14.21. Doing a mass balance on an elemental control volume extending from the plate to the edge of boundary layer and having a width Δx, then

$$\frac{d\dot{m}}{dx} = -\rho_f W v_f$$

(14.28)

Evaluating $\frac{d\dot{m}}{dx}$ from Equation 14.27, substituting in Equation 14.28, and nondimensionalizing, results in an expression for the nondimensional y component of the fluid velocity

$$v_f^* = -\delta^{*2}\frac{d\delta^*}{dx^*}$$

(14.29)

which is then substituted both in Equation 14.21 resulting in

$$\frac{1}{\delta^*} = \frac{Ra_c}{Ja} \delta^{*2} \frac{\partial \delta^*}{\partial x^*} \tag{14.30}$$

which simplifies to the following first-order differential equation:

$$\frac{d\delta^{*4}}{dx^*} = \frac{4}{Ra_c/Ja} \tag{14.31}$$

whose solution is

$$\delta^* = \left(\frac{4}{Ra_{c,L}/Ja} x^* \right)^{1/4} \tag{14.32}$$

and using the local value of condensation Rayleigh number, the boundary layer thickness becomes

$$\frac{\delta}{x} = \left[\frac{4}{Ra_{c,x}/Ja} \right]^{1/4} \tag{14.33}$$

Comparing this equation with the boundary layer equation for laminar flow over a flat plate, it is clear that the ratio of $\dfrac{Ra_{c,x}}{Ja}$ is the relevant nondimensional group for condensation, and its square root $\sqrt{\dfrac{Ra_{c,x}}{Ja}}$ serves the same function in the development of the film condensation thickness, as Reynolds number does in the development of boundary layer. In terms of the dimensional quantities, the thickness of the condensate is

$$\delta = \left[\frac{4k\nu(T_s - T_w)x}{(\rho_f - \rho_g)gh_{fg}} \right]^{1/4} \tag{14.34}$$

The heat transfer coefficient can be defined as

$$h = \frac{-k\dfrac{\partial T}{\partial y}}{T_w - T_s} = \frac{k}{\delta} \tag{14.35}$$

Therefore,

$$Nu_x = \frac{hx}{k} = \left[\frac{Ra_{c,x}}{4Ja} \right]^{1/4} = 0.707 \left[\frac{Ra_{c,x}}{Ja} \right]^{1/4} \tag{14.36}$$

The average heat transfer coefficient is

$$\bar{h} = \frac{1}{L} \int_0^L h\,dx = \frac{4}{3} h_{x=L} \tag{14.37}$$

The average Nusselt number is

$$\overline{Nu} = \frac{\overline{h}L}{k} = 0.943 \left[\frac{Ra_{c,L}}{Ja} \right]^{1/4} \tag{14.38}$$

which in terms of the dimensional variables becomes

$$\overline{Nu} = \frac{\overline{h}L}{k} = 0.943 \left[\frac{\rho_f(\rho_f - \rho_g)gL^3 h_{fg}}{\mu k(T_s - T_w)} \right]^{1/4} \tag{14.39}$$

The earlier derivation is a modification of the original derivation by Nusselt in 1916, who used integral solution.

14.3.3 Exact Solution

The solution can be obtained using similarity method. Tamm et al. [5] by defining the similarity variable as

$$\eta = \left(\frac{\dfrac{(\rho_f - \rho_g)}{\rho_f} g}{4\nu^2 x} \right)^{1/4} y \tag{14.40}$$

and the dimensionless stream function and temperature as

$$f = \frac{\psi}{\left(64 \dfrac{(\rho_f - \rho_g)}{\rho_f} g\nu^2 x^3 \right)^{1/4}} \tag{14.41}$$

$$T^* = \frac{T - T_s}{T_w - T_s} \tag{14.42}$$

provide the following two equations for the momentum and energy equations:

$$f''' + 3ff'' - 2f'^2 + 1 = 0 \tag{14.43}$$

$$T^{*''} + 3\Pr f T^{*'} = 0 \tag{14.44}$$

Boundary conditions

$$\begin{aligned}
\eta = 0 \quad & f = f' = T^* = 0 \\
\eta = \eta_\delta \quad & f'' = 0, \quad T^* = 1
\end{aligned} \tag{14.45}$$

The energy balance at the interface results in [5,6]

$$\frac{3\Pr}{Ja} = \frac{T^{*'}(\eta_\delta)}{f(\eta_\delta)} \tag{14.46}$$

The solution must be obtained iteratively by choosing a value for η_δ and solving the momentum and energy equations simultaneously for the given Prandtl number. The temperature gradient and stream function at the assumed value of η_δ must satisfy Equation 14.46; if not, a different value for η_δ is assumed and the process is repeated until the convergence is obtained.

Tamm et al. [5] provide an estimate for

$$\eta_\delta = \left[\frac{Ja}{Pr} \right]^{1/4} \tag{14.47}$$

which can be used as an initial estimate for the numerical solution.

14.3.4 Horizontal Cylinder

Condensation on horizontal tubes is commonly encountered in condensers, Figure 14.3. As was shown in Chapter 7, for axisymmetric flows, Boltz transformation essentially reduces the boundary layer equations to those for flow over a flat plate, but the gravity term must be modified to account for the angular variation and reduces to

$$\rho u \frac{\partial u}{\partial x} + \rho v \frac{\partial u}{\partial y} = \mu \frac{\partial^2 u}{\partial y^2} + (\rho_g - \rho_f) g \sin \phi \tag{14.48}$$

where

$$\phi = \frac{x}{\dfrac{D}{2}} \tag{14.49}$$

Using the same assumptions as those for the flat plate solution, the momentum equation simplifies to

$$\frac{\partial^2 u}{\partial y^2} + \frac{(\rho_g - \rho_f)}{\mu} g \sin \frac{2x}{D} = 0 \tag{14.50}$$

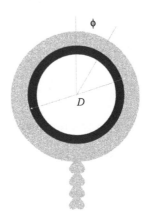

FIGURE 14.3
Condensation over a horizontal cylinder.

and the energy equation will be the same as Equation 14.8. The solution is similar to that for flat plate, but somewhat more involved mathematically, and the results are also very similar. The average Nusselt number for a single tube becomes [7]

$$\overline{Nu} = \frac{\overline{h}D}{k} = 0.728 \left[\frac{Ra_{c,D}}{Ja} \right]^{1/4}$$

(14.51)

For inline tubes, where horizontal tubes are stacked on top of each other in N rows, such that the condensate drips from the top layer to the lower ones, the average heat transfer coefficient becomes [7]

$$\overline{Nu} = 0.728 \left[\frac{Ra_{c,D}/Ja}{N} \right]^{1/4}$$

(14.52)

For a sphere, the constant on the right hand side of Equation 14.51 becomes 0.826.

14.3.5 Vapor Superheat

The previous analyses were based on the assumption that the surface is in contact with saturated vapor. If the vapor is superheated, then adjacent to the condensate layer, there will be another layer, where the superheated vapor cools down from T_v to the saturated vapor state at T_s, as shown also in Figure 14.1. In this layer, the superheated vapor is moving in the y direction, and the energy equation in the superheated layer becomes

$$\rho_v c_{pv} v_v \frac{\partial T'}{\partial y} = k_v \frac{\partial^2 T'}{\partial y^2}$$

(14.53)

The boundary conditions are

$$y \to \infty \qquad T' = T_v$$

(14.54)

$$y = \delta \qquad T' = T_s$$

(14.55)

This is an advection–diffusion problem, and the solution is

$$T' = T_v + (T_s - T_v)e^{\frac{v_v}{\alpha_v}(y-\delta)}$$

(14.56)

Note that the y component of velocity is negative ($v_v < 0$). Evaluating the temperature gradient of the vapor at the edge of boundary layer

$$\left. \frac{\partial T'}{\partial y} \right|_{y=\delta} = (T_s - T_v)\frac{v_v}{\alpha_v}$$

(14.57)

Substituting Equation 14.48 into Equation 14.9 results in

$$-k\frac{\partial T}{\partial y} = \rho_f v_f \left[h_{fg} + c_{pv}(T_v - T_s) \right]$$

(14.58)

Comparing this equation with Equation 14.10, it can be seen that the effect of vapor superheat can be accounted for by replacing h_{fg} in the previous equations, with

$$h'_{fg} = h_{fg} + c_{pv}(T_v - T_s)$$

(14.59)

or Jacob number needs to be modified to

$$Ja = \frac{c_p(T_s - T_w)}{h_{fg} + c_{pv}(T_v - T_s)}$$ (14.60)

14.3.6 Transition to Turbulence

As the distance along the plate increases, more and more vapor will condense, increasing the thickness of the film and the average velocity and therefore the Reynolds number. The condensate flow along the plate is similar to that of an open channel flow; as Reynolds number increases, waves and ripples begin to appear on the surface of the condensed film and contribute to surface mixing. As Reynolds number is further increased, these instabilities intensify and finally the flow becomes turbulent. The relevant Reynolds number is based on the average velocity and the hydraulic diameter of the condensate. The average condensate velocity at each location along the plate is

$$\bar{u} = \frac{1}{\delta} \int_0^\delta \frac{(\rho_f - \rho_g)g\delta^2}{\mu} \left(\frac{y}{\delta} - \frac{y^2}{2\delta^2} \right) dy = \frac{1}{3} \frac{(\rho_f - \rho_g)g\delta^2}{\mu}$$ (14.61)

and for the condensate flow, the hydraulic diameter becomes

$$D_h = \frac{4A}{p} = \frac{4 \times 1 \times \delta}{1 + \delta} \approx 4\delta$$ (14.62)

therefore the Reynolds number becomes

$$Re_{D_h} = \frac{\bar{u}4\delta}{\nu} = \frac{4}{3} \frac{(\rho_f - \rho_g)g\delta^3}{\nu\mu} \approx \frac{4}{3} \frac{g\delta^3}{\nu^2}$$ (14.63)

Note that $\left(\frac{\nu^2}{g} \right)^{1/3}$ has the dimension of length, and Re_{D_h} is therefore a dimensionless boundary layer thickness to the third power. For $Re_{D_h} < 30$ [6], the condensate flow will be laminar and the surface of the liquid is smooth, and therefore the local Nusselt number is given by Equation 14.36.

$$Nu_x = \frac{hx}{k} = 0.707 \left[\frac{Ra_{c,x}}{Ja} \right]^{1/4}$$ (14.64)

As discussed in Mills [7], as Reynolds number increases, initially ripples and waves form on the surface of the condensate. With increasing Reynolds numbers, the wave amplitudes increase and eventually the flow becomes fully turbulent with large surface waves that extend into the condensate. For water, the transition to turbulence happens for Reynolds number

$$Re_{D_h,tr} = \frac{5800}{Pr^{1.06}}$$ (14.65)

Defining a modified Nusselt number based on the length scale $\left(\dfrac{v^2}{g}\right)^{1/3}$,

$$Nu^* = \frac{h\left(\dfrac{v^2}{g}\right)^{1/3}}{k} \tag{14.66}$$

Based on experimental measurements, Chun and Seban [8] provided the following correlations for the modified Nusselt number for wavy laminar and turbulent flows:

$$Nu^* = 0.822 Re_{D_h}^{-0.22} \qquad 30 < Re_{D_h} < \frac{5800}{Pr^{1.06}} \tag{14.67}$$

$$Nu^* = 0.0038 Re_{D_h}^{0.4} Pr^{0.65} \qquad Re_{D_h} > \frac{5800}{Pr^{1.06}} \tag{14.68}$$

14.4 Dropwise Condensation

When a cold surface suddenly comes in contact with a saturated vapor, initially, the water droplets form on the cold surface. The number of droplets formed increases with time, and some droplets will coalesce to form larger ones. The condensate is retained on the surface by surface tension and may slide along the wall due to gravity or vapor motion. Depending on the surface, and the condensation rate, the condensation process may result in the formation of a film that was discussed in the previous sections. If the rate of condensation is low, or the vapor is mixed with noncondensing gases, as in humid air, or the fluid does not stick to the surface, then the condensation mode will remain dropwise condensation. In dropwise condensation, the droplet removal greatly reduces the resistance to heat transfer, and therefore, heat transfer coefficient is very high and could be as much as one order of magnitude higher compared to film condensation.

A good review of the topic is conducted by Garg [9]. There are two theories for explaining dropwise condensation. The first and the original model was proposed by Jakob [10] and postulates that the droplets are formed at the nucleation sites on the surface. If their initial radius is more than the equilibrium radius, they grow in size and join other droplets. Once external forces acting on the droplet, which include gravitational force and/or viscous force exerted by the moving vapor, become sufficiently large, the droplets are removed, and new ones will be formed.

The second model proposed originally by Emonsi [11] assumes that initially a thin layer of condensate is formed on the surface, and as the thickness grows and the layer reaches a critical thickness (around 1 μm), it breaks down into droplets. The condensation process continues over the newly formed bare areas that have a much smaller resistance to heat transfer as well as some over the existing droplets.

Consider a liquid drop on a surface as shown in Figure 14.4. The wettability of the surface is defined by the contact angle. The contact angle, θ, is the angle between the droplet face and the surface, as defined by the angle between the tangent to the droplet and the surface at the point of contact. If the angle of contact is less than 90°, then the surface is called hydrophilic or water loving, and if it is greater than 90, the surface is termed hydrophobic or water fearing. For contact angles greater than 150, the surface is called superhydrophobic. If a droplet moves along a surface, the contact angle at the front is called the advancing angle θ_a

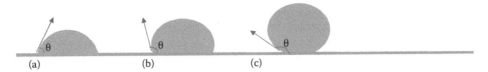

FIGURE 14.4
(a) Hydrophilic surface, (b) hydrophobic surface, and (c) superhydrophobic surface.

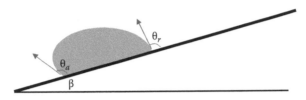

FIGURE 14.5
Droplet motion along a surface.

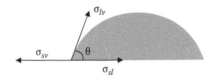

FIGURE 14.6
Different components of surface tension.

and the contact angle at the rear is called the receding angle θ_r as shown in Figure 14.5. The difference between the two angles is called the *contact angle hysteresis* and is an important parameter in characterizing the water retention properties of the surface.

At the interface of two different substances as shown in Figure 14.6, neither substance has like molecules on all sides, and consequently the molecules cohere unevenly, forming a surface "film" that requires a finite force to break through. This is known as surface tension; for example, when water at 20°C is in contact with air, its surface tension is 0.0728 N/m, meaning that in order to break 1 m long film, 0.0728 N must be applied. Small insects can walk on water because their weight is not enough to overcome surface tension. The contact angle is dependent on the surface tension (σ) between the contacting phases and can be calculated from

$$\cos\theta = \frac{\sigma_{sv} - \sigma_{sl}}{\sigma_{lv}} \tag{14.69}$$

Using Gibbs free energy, it can be shown that the minimum droplet diameter above which the droplet continues to grow in size is given by

$$D_{min} = \frac{4\sigma T_s}{\rho_f h_{fg}(T_s - T_w)} \tag{14.70}$$

Example 14.1

What is the minimum droplet diameter formed on a glass window exposed to humid air at 20°C if the glass temperature is 5°C?

$$D_{min} = \frac{4 \times 0.0728 \times (273.15 + 5)}{998 \times 2454.12 \times (20 - 5)} = 2.2 \times 10^{-9} \text{ m} = 0.0022 \text{ μm} \tag{14.71}$$

The higher heat transfer coefficients associated with dropwise condensation have led to extensive research in developing techniques to promote and sustain this mode of condensation. Much of these involve chemical treatment of surfaces or including additives to the vapor. Chemical treatment of the surfaces is effective for limited cycles and is not practical in continuously operating systems. For example, *n*-tetradecanoic acid can be used to create superhydrophobic copper surfaces. Inclusion of additives is generally expensive, and sometimes not feasible. Research is also underway to use nanotechnology to create coatings and also use micromachining for guided flow of the droplets and enhanced heat transfer.

14.5 Pool Boiling

Pool boiling is evaporation where the liquid and vapor motions are caused by natural forces. This is the process that occurs in a pan filed with water on a stove. Consider a pan filled with water at room temperature, which is being heated over an electric stove. As the heating starts, natural convection currents are established, enhancing the conduction of heat. As the temperature increases, small bubbles begin to form on the walls of the pan. These are dissolved air bubbles in water that separate as the water temperature increases. Metallic surfaces typically have surface defects in the form of scratches or pits that are filled with air, and water molecules do not penetrate as a result of surface tension. The dissolved air will then diffuse into these sites, causing the bubbles to grow and eventually detach, further promoting flow and mixing. As the pan temperature is further increased beyond the saturation temperature, water changes phase and vapor bubbles form on the superheated wall. The vapor bubbles detach, and as they move up in the colder fluid, they condense and collapse, further heating the fluid by convection. Eventually, the fluid reaches the saturation temperature, and columns of vapor stream to the top, which is nucleate boiling. The vapor that rises must be replaced with water flowing down from the top. If the surface temperature is too high, it is possible for a vapor film to be formed on the surface, adding a large resistance, and drastically reducing the heat transfer rate, which results in great increase in the surface temperature.

Nukiyama [12] was the first to experimentally investigate pool boiling and established the different boiling regimes. As shown in Figure 14.7, he inserted an electrically heated wire in a saturated liquid and studied boiling on the surface of the wire as a result of the power dissipated. Figure 14.8 shows the variation of heat flux as a function of the difference between the wire and the saturated liquid temperatures. This plot can be used to identify the different boiling regimes. He identified four different boiling regimes: natural convection, nucleate, transition, and film boiling, depending on the excess temperature. Note that the ratio of the y axis to x axis values represents the heat transfer coefficient.

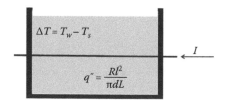

FIGURE 14.7
Pool boiling experimental setup.

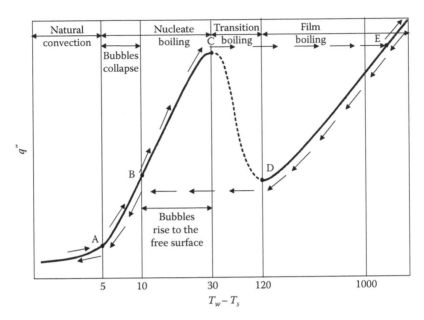

FIGURE 14.8
Boiling regimes of water.

Initially, as the power supplied to the wire is increased (increasing the current), the surface temperature increases; however, no bubbles are observed until point A where the surface temperature is somewhere between 2°C and 6°C above the saturation temperature. Up to point A, the heat transfer is by natural convection, and evaporation takes place from the surface of the liquid. As power is further increased, the bubbles appear first at point A, and nucleate boiling initiates at defects on the surface of the wire. For wire surface temperatures less than 10°C above the saturation temperature of the liquid, the bubbles detach from the wire; however, as a result of heat loss, they condense and collapse. The bubble formation and collapse induce further motion in the liquid and increase the heat transfer coefficient, enabling the dissipation of large amounts of energy over relatively small temperature differences, as a result of evaporation.

Between points B and C, the bubble are hot enough that they will not collapse as they separate from the wire and rise to the free surface, further enhancing the fluid motion and increasing heat transfer coefficient and removing very large amounts of heat. There are two competing effects that impact the heat flux. The bubble formation, detachment, and upward rise enhances convection currents and increases heat transfer coefficient and heat removal. On the other hand, increased bubble formation on the wire, covers the wire by more poorly conducting vapor, increases thermal resistance and tends to decrease the rate of heat transfer, and therefore eventually a maximum point is reached at point C. For water, this maximum heat flux is around 1 MW/m², and the excess temperature difference is about 30°C, or heat transfer coefficient reaches a value of 30,000 W/m² K. The nucleate boiling is the most effective boiling regime that allows for the removal of large amounts of heat over a relatively small temperature difference.

As the current is increased further, the wire will be covered by a vapor film, drastically reducing the heat transfer coefficient. In order to dissipate the supplied heat flux, since the heat transfer coefficient has significantly deteriorated as a result of the wire being covered by a layer of vapor, the surface temperature of the wire will have to increase drastically to point E, to be able to dissipate the supplied energy. It is very likely that the increase in surface temperature will go beyond the melting point of the wire and lead to catastrophic failure. The peak heat flux is also known as the burnout point. In practice, boilers operate at excess temperatures below the burnout point.

The transition region is unstable, with boiling switching between pool and film boiling, leading to surface temperature fluctuation, and is avoided in practice. Past point D, if the wire does not melt, the boiling turns into film boiling where the heater surface gets covered with a stable film of vapor. Film boiling is avoided in applications involving water, as it requires very high surface temperature; however, it is used extensively in cryogenic applications, where the fluid boiling temperature is very low, and the heater excess temperature is still significantly below the melting point of the heater and therefore large amounts of heat across large temperature differences can be transferred.

Starting in the film boiling region (past point E), if the power input to the wire is reduced, the temperature difference and the heat flux continuously decrease and the relation between the two is given by the cooling path along ED portion of the curve. The vapor film is stable, and as the power is reduces, the thickness of the film decreases, but the film does not break up into bubbles and maintains its integrity long after point C. Eventually, the vapor film breaks down when the heat flux reaches a minimum value at point D, which is known as the Leidenfrost point, and then an abrupt temperature drop to point B with convective boiling commencing and bypassing the nucleate boiling.

14.5.1 Nucleate Boiling

As mentioned before, bubbles are initiated at the surface defects. To develop a better understanding of the basic physics of the bubble formation and detachment, consider a spherical bubble in a fluid that is at the saturation temperature T_s and saturation pressure P_s as shown in Figure 14.9. The pressure inside the fluid would change slightly, due to the hydrostatic pressure, and therefore the pressure outside of the bubble is P_∞. Inside of the bubble, there is a mixture of vapor that can be assumed to be at the saturation pressure (P_s) corresponding to the local temperature, and the other gases (air, at least initially) at P_a. The force balance must account for the surface tension, and therefore, if R_c is the critical radius beyond which the bubble will detach, then

$$(P_s + P_a)\pi R_c^2 = P_\infty \pi R_c^2 + 2\pi R_c \sigma \tag{14.72}$$

which can be used to determine the saturation pressure, and therefore temperature that is needed for the bubble to detach:

$$P_s = P_\infty + \frac{2\sigma}{R_c} - P_a \tag{14.73}$$

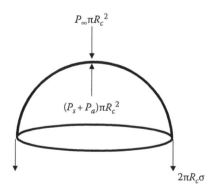

FIGURE 14.9
Force balance on a vapor bubble.

Assuming that there are no dissolved gases in the fluid, and also neglecting hydrostatic pressure, which means assuming $P_\infty = P_s(T_s)$, then T_w, at which bubbles will detach, will be the saturation temperature corresponding to pressure

$$P_s(T_w) = P_s(T_s) + \frac{2\sigma}{R_c} \tag{14.74}$$

The radius at which the bubble will detach is obtained by a balance between buoyancy and surface tension

$$(\rho_f - \rho_g)\frac{1}{2}\frac{4\pi}{3}R_b^3 g = 2\pi R_b \sigma \tag{14.75}$$

which results in

$$R_b = \sqrt{\frac{3\sigma}{g(\rho_f - \rho_g)}} \tag{14.76}$$

This radius is also the relevant length scale for nucleate boiling, which typically is taken without the factor 3,

$$R_b = \sqrt{\frac{\sigma}{g(\rho_f - \rho_g)}} \tag{14.77}$$

The maximum heat flux must be related to the surface's ability to create the bubbles and the bubbles' ability to clear the hot surface. If τ is the time that it takes to create a bubble of radius R_b, then the amount of heat transferred to the bubble is $q''\pi R_b^2 t$. This much thermal energy must be equal to the energy needed to evaporate the mass of saturated liquid needed to fill the bubble; therefore,

$$\rho_g \frac{1}{2}\frac{4\pi}{3}R_b^3 h_{fg} = q''\pi R_b^2 \tau \tag{14.78}$$

and solving for time,

$$\tau = \frac{2}{3}\frac{\rho_g R_b h_{fg}}{q''} \tag{14.79}$$

We can also estimate the time that it takes for the bubbles to detach and create room for the next bubble by calculating the initial acceleration of the bubble. If the bubble has a volume V, then the buoyancy force acting on it is $(\rho_f - \rho_g)Vg$ and the mass of the bubble is $\rho_g V$, which is due to the buoyancy force, acting on the mass of the bubble,

$$\frac{d^2y}{dt^2} = \frac{(\rho_f - \rho_g)g}{\rho_g} \tag{14.80}$$

Therefore, the time that it takes for the bubble to move up in distance R_b is approximately

$$t \approx \sqrt{\frac{R_b}{\frac{(\rho_f - \rho_g)g}{\rho_g}}} \tag{14.81}$$

TABLE 14.1

Value of C_{max} in Equation 14.83 for Different Geometries

Heater Geometry	C_{max}	Range $L^* = \dfrac{L_c}{R_s}$ where $R_s = \left(\dfrac{\sigma}{g(\rho_f - \rho_g)}\right)^{1/2}$
Flat heater (L_c = width or diameter)	$0.15\dfrac{12\pi R_s^2}{\text{Area}}$	$9 < L^* < 20$
	0.15	$L^* > 27$
Horizontal cylinder (L_c = radius)	$0.12\,L^{*-0.25}$	$0.15 < L^* < 1.2$
	0.12	$L^* > 1.2$
Sphere (L_c = radius)	$0.227\,L^{*-0.5}$	$0.15 < L^* < 4.26$
	0.11	$L^* > 4.26$
Arbitrary-shaped object	0.12	

The maximum heat flux is going to take place when these two times are equal, i.e., the frequency by which the bubbles are generated is equal to the frequency by which the bubble can clear the heater, therefore, setting the two times equal:

$$\sqrt{\frac{R_b}{\dfrac{(\rho_f - \rho_g)g}{\rho_g}}} \approx \frac{2}{3}\frac{\rho_g R_b h_{fg}}{q''} \tag{14.82}$$

Substituting for R_b from Equation 14.62, and simplifying and combining all constants in one value, and turning the approximation into an equality, results in

$$q''_{max} = C_{max}h_{fg}\left[\sigma\rho_g^2(\rho_f - \rho_g)g\right]^{1/4} \tag{14.83}$$

This is equation was originally proposed Kutateladze [13] and independently by Zuber [14] as reported in Ref. [7]. The value of the constant depends on the dimensionless heater geometry, and the values are given in Table 14.1, and all properties are evaluated at the saturation temperature, T_s.

The rate of heat transfer during nucleation boiling is strongly dependent on the surface conditions, which are difficult to quantify, making its theoretical predictions very difficult, requiring reliance on experimental data. The most widely used experimentally determined correlation is that developed by Rohsenow, which is given as follows in terms of Nusselt number:

$$\overline{Nu} = \frac{\overline{h}R_s}{k} = \frac{Ja^2}{C_{sl}^3 Pr^{3n-1}} \tag{14.84}$$

where $n = 1$ for water and 1.7 for all other substances. The values of the constant C_{sl} depend on the solid and liquid involved and the values for a combination of different liquid solid combinations are given in Table 14.2. Caution should be exercised in using this equation as it is an approximate equation, and the predicted heat fluxes can be off significantly.

14.5.2 Film Boiling

In film boiling, the surface is separated from the liquid by a relatively thick layer of vapor across which there exists a large temperature difference driving the heat transfer.

TABLE 14.2

Value of C_{sl} in Equation 14.84 for Different Liquid-Surface Combinations

Liquid	Surface	C_{sl}
35% K_2CO_3	Copper	0.0027
Benzene	Chromium	0.010
Carbon tetrachloride	Copper	0.013
Ethanol	Chromium	0.0027
Isopropyl alcohol	Copper	0.0023
n-Butyl alcohol	Copper	0.0030
n-Pentane	Chromium	0.015
Water	Brass	0.006
Water	Copper, scored	0.0068
Water	Copper, polished	0.013
Water	Nickel	0.006
Water	Platinum	0.013
Water	Stainless steel, chemically etched	0.013
Water	Stainless steel, ground and polished	0.008
Water	Stainless steel, mechanically polished	0.013

As mentioned before, due to the large temperature differences involved, film boiling finds much of its application in cryogenic systems. Film boiling and film condensation are very similar but reverse phenomena. An important difference is that the impact of liquid drag on the vapor flow is not negligible. Therefore, with slight modifications, we can use Equation 14.39 to determine the heat transfer coefficient while accounting for superheating [7]:

$$\overline{Nu} = \frac{\overline{h}L}{k_g} = C_{fb}\left[\frac{\rho_g(\rho_f - \rho_g)gL^3}{\mu_g k_g(T_s - T_w)}\left[h_{fg} + 0.35c_{pv}(T_w - T_s)\right]\right]^{1/4} \tag{14.85}$$

where
C_{fb} is 0.71, 0.62, or 0.67, for vertical plate, horizontal cylinder, and sphere, respectively
L is plate length or the diameter

To account for the superheat, the latent heat is modified. The properties of the vapor are evaluated at the film temperature $\frac{T_w + T_s}{2}$, and liquid properties are evaluated at the saturation temperature. Similar to film condensation, for large x values or high evaporation rates, waves begin to form at the interface of liquid and vapor, and at higher values, the flow becomes turbulent.

The film boiling persists until the Leidenfrost point at which point Zuber [15] showed that the minimum heat flux is

$$q''_{min} = 0.09 h_{fg}\rho_g\left[\frac{\sigma g(\rho_f - \rho_g)}{(\rho_f + \rho_g)^2}\right]^{1/4} \tag{14.86}$$

which as mentioned earlier, if the heat flux drops below this value, the vapor film enveloping the surface collapses, the surface temperature drops significantly and the boiling changes from film to nucleate boiling.

14.6 Multiphase Flow

Forced flow condensation and evaporation of fluids, particularly in ducts, serve as examples of a broad class of problems known as multiphase flow problems that have numerous applications in different areas, including power plants, process industries, and air-conditioning systems. Multiphase flow regimes can be gas–liquid; liquid–liquid; gas–solid; liquid–solid flows; and gas–liquid–solid flows.

Theoretical analysis of these problems is rather complicated due to the existence and the flow of the two phases, and a number of models are proposed to deal with the two phase flow problems. Broadly, the available approaches for the theoretical or numerical analysis of multiphase flow problems are either Euler–Lagrange approach or the Euler–Euler approach.

In the Euler–Lagrange approach, Eulerian or control volume approach is used to determine the fluid's velocity and temperature distributions from the Navier–Stokes equations. Lagrangian approach (control mass) is utilized for the second phase, that could be solid particles, vapor bubbles, or liquid droplets, by treating them as discrete particles and determining their position, velocity, and temperature through their interaction with the fluid through exchange momentum, mass, and energy.

The interaction of the two phases requires the particle trajectories to be computed individually, and therefore a large number of particles need to be followed. This approach is practical only when the second phase has a low volume fraction. The approach is also not viable for the modeling of immiscible liquid–liquid mixtures.

In the Euler–Euler approach, all phases are treated as continuum. The volume of a phase cannot be occupied by the other phases; therefore, the concept of volume fraction is introduced for each phase. The volume fractions are then assumed to be a continuous function of space and time, and from the definition, they add up to one. The interaction between the different phases is then modeled using experimental data or theoretical analysis or both.

Problems

14.1 Derive Equations 14.28 and 14.29.

14.2 The dimensionless variable $\frac{y}{\delta}$ in Equation 14.26 and the similarity variable η given by Equation 14.40 essentially represent the same quantity:

a. Show that the $\frac{y}{\delta} = \eta \left[\frac{Pr}{Ja} \right]^{1/4}$. Note that the temperature distribution of Equation 14.24 then becomes $T^* = \left[\frac{Pr}{Ja} \right]^{1/4} \eta$.

b. Show the validity if Equation 14.47.

14.3 Equations 14.43 and 14.44 give the exact solution for condensation of a saturated liquid in contact with a cold vertical plate assuming boundary layer approximation. For $\eta_\delta = 0.36$,

a. Numerically obtain the velocity and temperature distributions for saturated liquid water at 100°C.

b. Determine what Jacob number corresponds to this thickness.

c. What must be the surface temperature?

d. Verify the accuracy of Equation 14.47.

14.4 Equations 14.43 and 14.44 give the exact solution for condensation of a saturated liquid in contact with a cold vertical plate assuming boundary layer approximation. For $\eta_\delta = 0.4$,

a. Numerically obtain the velocity and temperature distributions for saturated liquid 134a at 30°C.

b. Determine the local Nusselt number and compare to Equation 14.36.

c. Determine what Jacob number corresponds to this thickness.

d. What must be the surface temperature?

e. Verify the accuracy of Equation 14.47.

f. Using the results of Problem 14.2, compare the velocity distribution from boundary layer solution with that of the integral solution.

g. Compare the velocity and temperature distributions from boundary layer solution with that of the integral solution.

14.5 Numerically solve for condensation around a circular cylinder.

14.6 Derive an expression for the velocity and temperature distribution for condensation on the outer surface of a vertical cylinder of diameter D.

14.7 Saturated steam at 100°C is being condensed on one side of a 0.5 m by 0.5 m vertical plate at 90°C. Determine the condensation rate.

14.8 Saturated steam at 50 kPa is condensing on the outer surface of a thin horizontal copper pipe carrying water at 30°C:

a. Determine the heat transfer and condensation per unit length of the pipe.

b. Estimate how closely the copper pipes can be placed on a horizontal plane, to not impact each other.

14.9 Saturated refrigerant 134a at 1.40 MPa condenses on a thin 5 mm OD copper tube, carrying water at 20°C. Determine the condensation rate.

14.10 Refrigerant 134a at 1 MPa condenses on the outside of an 18 mm copper tube. Plot the heat transfer coefficient for the copper surface temperature range of –20°C to 20°C.

14.11 A 20 mm steel sphere initially at 80°C is exposed to saturated steam at 100°C. Determine its temperature after 10 min.

14.12 Determine the critical heat flux on a large horizontal surface at 1 atm, for water, and R134a.

14.13 Using the result of Problem 14.12, determine the maximum current that a 0.5 mm wire having a resistance of 0.1 Ω/m can carry if submerged in water at 1 atm.

14.14 Determine the heat flux needed to maintain a polished copper plate at 110°C while boiling water at 1 atm.

14.15 A 30 mm copper sphere initially at 400°C is suddenly immersed in saturated liquid water at one atm. Neglecting radiation loss,

a. Numerically solve for temperature distribution.

b. How long does it take for water to reach the Leidenfrost point?

c. How long does it take for the sphere temperature to reach within 5% of its final value?

14.16 A 30 mm copper sphere initially at 400°C is suddenly immersed in saturated liquid water at 1 atm. Copper has an emissivity of 0.64:
  a. Numerically solve for temperature distribution.
  b. How long does it take for water to reach the Leidenfrost point?
  c. How long does it take for the sphere temperature to reach within 5% of its final value?

References

1. Healthwise. (2011). *WebMD: First Aid & Emergencies: Ways in Which the Body Loses Heat.* Available online: http://firstaid.webmd.com/ways-in-which-the-body-loses-heat. Accessed April 6, 2013.
2. Graham, S., Parkinson, C., and Chahine, M., *The Water Cycle.* Available online: http://earthobservatory.nasa.gov/Features/Water/water_cycle_2000.pdf Accessed April 6, 2013.
3. Kaiser, J. M. and Jacobi, A. M. (2000) Condensate retention effects on the air-side heat transfer performance of automotive evaporator coils, *ACRCCR-32*, July 2000, Air Conditioning and Refrigeration Center, University of Illinois, Urbana, IL.
4. Uv, E. H. and Sonju, O. K. (1992). Heat transfer measurements of circular finned tubes with and without partial condensation. Paper presented at *The Institution of Chemical Engineers Symposium Series*, 1(129), 295–302.
5. Tamm, G., Boettner, D. D., Van Poppel, B. P., Benson, M. J., and Arnas, A. Ö. (2009) On the similarity solution for condensation heat transfer, *Journal of Heat Transfer* 131.11, 1–5.
6. Oosthuizen, P. H. and Naylor, D. (1999) *An Introduction to Convective Heat Transfer Analysis.* New York: WCB/McGraw Hill.
7. Mills, A. F. (1972) *Heat and Mass Transfer.* Burr Ridge, IL: Irwin.
8. Chun, K. R. and Seban, R. A. (1972) Performance prediction of falling-film evaporators, *Journal of Heat Transfer, Transactions ASME* 94 Ser C.4, 432–436.
9. Garg, S. C. (1969) The effect of coatings and surfaces on dropwise condensation, Technical Note, N-1041, http://www.dtic.mil/dtic/tr/fulltext/u2/691394.pdf. Accessed April 6, 2013.
10. Jakob, M. (1936) Heat transfer in evaporation and condensation—II, *Mechanical Engineering* 58, 729–739.
11. Emmons, H. (1939) The mechanism of dropwise condensation, *American Institute of Chemical Engineers, Transactions*, 35, 109–125.
12. Nukiyama, S. (1934) Film boiling water on thin wires, *Japan Society of Mechanical Engineers*, 37, 367 (Translation *International Journal of Heat Mass Transfer*, 9, 1419, 1966).
13. Kutateladze, S. S. (1948) On the transition to film boiling under natural convection, *Kotloturbostroenie* 3, 10.
14. Zuber, N. (1959) Hydrodynamic aspects of nuclear boiling, PhD dissertation, Department of Engineering, University of California, Los Angeles, CA.
15. Zuber, N. (1958) On the stability of boiling heat transfer, *ASME Transactions*, 80, 711–720.

15

Mass Transfer and Chemically Reactive Flows

15.1 Introduction

So far we have treated the medium as having unvarying chemical composition, or at least to behave like one. When we dealt with a mixture, like air, the individual components or species were assumed to behave the same as the mixture. A necessary requirement for this behavior is the absence of concentration gradients or the constituents are uniformly distributed throughout the mixture. Many problems of engineering interest involve changing chemical composition throughout the system and during the process. Air generally contains water vapor and if the operating temperature reaches the saturation temperature of water, some of the water will condense, and a concentration gradient for water will be established. Another example is combustion or the general case of reactive flows. In a combustion process, fuel and oxidizer enter the combustion chamber, react with each other and products are formed, and finally they exit the system. Using the notation

$$b_1 B_1 + b_2 B_2 \rightarrow b_3 B_3 + b_4 B_4 \tag{15.1}$$

which indicates that "b_1" moles of species B_1 react with "b_2" moles of species B_2 and produce "b_3" moles of species B_3 and b_4 moles of species B_4. Therefore, as one moves from the inlet of the combustion chamber toward the exit, the concentrations of B_1 and B_2 will decrease and those of B_3 and B_4 will increase, and therefore there will be concentration gradients within the system. Equation 15.1 represents an ideal mixture where all reactants disappear and only products exit the system.

Analogous to temperature difference causing heat transfer by diffusion, concentration difference causes mass transfer by diffusion. In a multispecies diffusing mixture, different species move with different velocities. Therefore, the velocity of the mixture represents an average velocity, meaning some of the species move faster and others slower than the average velocity. There are a number of ways by which the velocities can be averaged. Before these can be introduced, we need to define some basic concepts.

15.2 Definitions

In dealing with multicomponent systems involving mass diffusion, we need to be able to specify the state of the system, which in general requires the specification of two independent properties as well as the composition of the mixture. There are a number of different ways that the concentrations of different species can be defined, and some of the definitions are reviewed as follows.

15.2.1 Concentrations and Other Mixture Properties

Consider a container having a volume V filled with a mixture made up of N different species or constituents. Let m_i represent the mass of species i, then the total mass of the mixture is

$$m = \sum_{i=1}^{N} m_i \tag{15.2}$$

The mass density of the mixture is

$$\rho = \frac{m}{V} = \frac{\sum_{i=1}^{N} m_i}{V} = \sum_{i=1}^{N} \rho_i \tag{15.3}$$

where ρ_i is the mass concentration (density) of species i and is defined as mass of species i per unit volume of the mixture

$$\rho_i = \frac{m_i}{V} \tag{15.4}$$

Therefore, the density of a mixture is equal to the sum of the densities of its constituents. The molar mass of a substance, M_i, is its mass in grams that is numerically equal to its atomic or molecular mass. For example, 32 g of oxygen is 1 mole of oxygen and contains Avogadro's number ($L = 6.022 \times 10^{23}$) of oxygen molecules. The number of moles of species i, n_i, is

$$n_i = \frac{m_i}{M_i} \tag{15.5}$$

The molar concentration of species i is defined as its number of moles per unit volume

$$c_i = \frac{n_i}{V} = \frac{\rho_i}{M_i} \tag{15.6}$$

and the total number of moles of the mixture, n, and the molar concentration of the system become

$$n = \sum_{i=1}^{N} n_i \tag{15.7}$$

$$c = \frac{n}{V} \tag{15.8}$$

The mass fraction of species i, w_i (or x_i), is

$$w_i = x_i = \frac{m_i}{m} = \frac{\rho_i}{\rho} \tag{15.9}$$

and the mole fraction of species i, y_i, is

$$y_i = \frac{n_i}{n} \tag{15.10}$$

From Equation 15.5, $M_i = \frac{m_i}{n_i}$; therefore, we can use this relation and also define the molar mass for the mixture as

$$M = \frac{m}{n} = \frac{\sum_{i=1}^{N} m_i}{\sum_{i=1}^{N} n_i} = \frac{\sum_{i=1}^{N} n_i M_i}{n} = \sum_{i=1}^{N} y_i M_i \tag{15.11}$$

Note that in a reactive mixture, the mole fractions change, and therefore molecular mass of the system is not constant throughout the system. However, since in many systems, the concentration of reacting species is typically small compared to that of inert gases, M can be treated as a constant in many instances. Also,

$$\sum_{i=1}^{N} w_i = \sum_{i=1}^{N} y_i = 1 \tag{15.12}$$

and

$$w_i = \frac{\rho_i}{\rho} = \frac{c_i M_i}{cM} = \frac{M_i}{M} y_i \tag{15.13}$$

For an ideal gas mixture,

$$PV = n\bar{R}T \tag{15.14}$$

where $\bar{R}$ is the universal gas constant. Using Dalton's model for a mixture of gases, if each species is assumed to be at the same temperature as the mixture temperature and occupies the mixture volume, then each species in the mixture will be at a pressure lower than the total mixture pressure, called its partial pressure P_i:

$$P_i V = n_i \bar{R}T \tag{15.15}$$

Therefore, the partial pressure of species i in an ideal gas mixture

$$\frac{P_i}{P} = y_i \tag{15.16}$$

which is the same as its mole fraction.

Example 15.1

A mole analysis of a gas mixture is shown in column 2 and the other properties are calculated next:

	% by Mole	y	Molecular Weight	m	w/x (%)
CO_2	12	0.12	44	5.28	18.8
O_2	4	0.04	32	1.28	4.6
CO	2	0.02	28	0.56	2.0
H_2O	20	0.20	18	3.6	12.8
N_2	62	0.62	28	17.36	61.8
	$n =$	1	$m =$	28.08	100.0
			$M =$	28.08	

15.2.2 Diffusion Velocities

Different species move with different velocities in a diffusing mixture. Assume v_i to be the absolute velocity of species i, which is its velocity relative to an inertial frame of reference. The mass flux or mass velocity of species i relative to the inertial frame is its mass flow rate per unit area:

$$G_i = \frac{\dot{m}_i}{A} = \rho_i v_i \tag{15.17}$$

The mass flux of the system is

$$G = \rho v \tag{15.18}$$

where v is the local mass average velocity of the mixture; therefore,

$$v = \frac{\sum_{i=1}^{N} \rho_i v_i}{\rho} = \frac{\sum_{i=1}^{N} \rho_i v_i}{\sum_{i=1}^{N} \rho_i} = \sum_{i=1}^{N} w_i v_i \tag{15.19}$$

As a result of mass diffusion, the components whose concentrations decrease in the direction of the flow will be moving faster and those with increasing concentration will be moving slower than the average velocity. Therefore, we can define the diffusion velocity of the species i, $v_{i,d}$, as the difference between its velocity and the average mixture velocity:

$$v_{i,d} = v_i - v \tag{15.20}$$

This is the velocity of species i relative to the mixture velocity, i.e., to an observer moving with the average velocity of the mixture, it would appear that the species i is coming toward the observer with velocity $v_{i,d}$. The relative mass flux of species i

$$j_i = \rho_i(v_i - v) \tag{15.21}$$

represents the migration of species as a result of molecular diffusion. Note that relative to an inertial frame of reference, the mass flux of species i is

$$G_i = \rho_i v_i = j_i + \rho_i v \tag{15.22}$$

If the medium is stationary ($v = 0$), isothermal, and at constant pressure, then the velocity of species i will be its diffusion velocity. Summing all the relative mass fluxes,

$$\sum_{i=1}^{N} j_i = \sum_{i=1}^{N} \rho_i(v_i - v) = \sum_{i=1}^{N} \rho_i v_i - v\sum_{i=1}^{N} \rho_i \tag{15.23}$$

From Equation 15.19, the right-hand side of Equation 15.23 is zero; therefore, for a mixture, the sum of the relative mass fluxes is zero:

$$\sum_{i=1}^{N} j_i = 0 \tag{15.24}$$

This is not surprising in that the net rate of mass flow as a result of diffusion is zero.

15.2.3 Fick's Law

Consider two cylinders having the same volume, being at the same pressure and temperature, but containing two different species A and B. The two cylinders are connected together by a pipe having a valve and assume both gases behave like an ideal gas (Figure 15.1). The initial densities of the two gases are

$$\rho_{A,i} = \frac{M_A P}{\bar{R}T} \tag{15.25}$$

$$\rho_{B,i} = \frac{M_B P}{\bar{R}T} \tag{15.26}$$

Therefore, initially, the two gases have different densities. If the valve opens, there is a density difference (concentration gradient) between the two tanks for both species. If we wait long enough the density of both species will become uniform throughout the two tanks, and at the final state, since the pressures and temperatures will be the same as the initial state, but the available volume to each species will be twice as much as their initial volume, then the final densities will be half of those given by Equations 15.25 and 15.26.

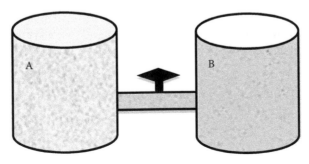

FIGURE 15.1
Mixing two ideal gases initially in two separate tanks A and B.

It is reasonable to expect the diffusion mass flux of species i or the rate of species diffusion to be proportional to the density difference and inversely proportional to the distance that the species has to travel or

$$j_A \propto \frac{\partial \rho_A}{\partial x} \tag{15.27}$$

The proportionality constant is called the binary diffusion coefficient, D_{AB}, or mass diffusivity

$$j_A = -D_{AB} \frac{\partial \rho_A}{\partial x} \tag{15.28}$$

Since the total density is constant, multiplying and dividing the right-hand side of the equation by the mixture density result in

$$j_A = -\rho D_{AB} \frac{\partial w_A}{\partial x} \tag{15.29}$$

The mass flux of species A is then

$$G_A = -\rho D_{AB} \frac{\partial w_A}{\partial x} + \rho v w_A \tag{15.30}$$

Also, since

$$j_A = \rho_A v_{A,d} = -\rho D_{AB} \frac{\partial w_A}{\partial x} \tag{15.31}$$

the diffusion velocity of species A is then given by

$$v_{A,d} = -D_{AB} \frac{1}{w_A} \frac{\partial w_A}{\partial x} \tag{15.32}$$

Equation 15.31 can be generalized, when diffusion is in more than one direction, to

$$\vec{j}_A = -\rho D_{AB} \nabla w_A \tag{15.33}$$

This is Fick's first law of diffusion, which accounts for mass diffusion as a result of concentration gradients. Pressure and temperature gradients also contribute to mass diffusion, although these effects are usually negligible.
  Since

$$j_A + j_B = 0 \tag{15.34}$$

then

$$-\rho D_{AB} \frac{\partial w_A}{\partial x} - \rho D_{BA} \frac{\partial w_B}{\partial x} = 0 \tag{15.35}$$

The gradient of species A and B must have opposite signs, as A and B diffuse in opposite directions.

Since $w_A + w_B = 1$, then

$$\frac{\partial w_A}{\partial x} + \frac{\partial w_B}{\partial x} = 0 \tag{15.36}$$

which when substituted in 15.35 results in

$$D_{AB} = D_{BA} \tag{15.37}$$

for binary diffusion.

Knuth [1] showed that the Fick's law can also be used for the diffusion of different species in a multicomponent system as long as the binary diffusion coefficients of all pairs are equal.

Based on Chapman–Enskog formulas, the following correlation can be obtained for the binary diffusion coefficient [2]:

$$D_{AB} = 0.0018583 \frac{\sqrt{T^3 \left(\dfrac{1}{M_A} + \dfrac{1}{M_B} \right)}}{P \sigma_{AB}^2 \Omega_{D,AB}} \tag{15.38}$$

where
   T is temperature (K)
   P is pressure (atm)
   σ is the collision diameter
   Ω is a parameter that depends on temperature

Based on this equation, D_{AB} is approximately proportional to T^2 at low temperatures, and $T^{1.65}$ at high temperatures. The diffusion coefficient is also dependent on the molecular mass of the species, and therefore the error associated with using the Fick's law for multicomponent systems can be estimated by considering the molecular mass of the species involved.

15.3 Law of Conservation of Species

A transport equation, similar to momentum or energy equations, can be obtained for the conservation of each species in a mixture. Consider a two-dimensional (2D) control volume shown in Figure 15.2. Applying the law of conservation of mass to the species i,

$$\frac{dm_i}{dt} = \sum_{in} \dot{m}_i - \sum_{out} \dot{m}_i + R_i \tag{15.39}$$

where R_i is the rate of production of species i

$$\frac{\partial(\rho_i \Delta x \Delta y \Delta z)}{\partial t} = (G_i \Delta y \Delta z)_x - (G_i \Delta y \Delta z)_{x+\Delta x} + (G_i \Delta x \Delta z)_y - (G_i \Delta x \Delta z)_{y+\Delta y} + (r_i''' \Delta x \Delta y \Delta z) \tag{15.40}$$

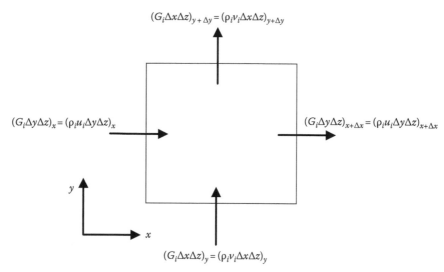

FIGURE 15.2
Two-dimensional differential control volume.

which simplifies to

$$\frac{\partial(\rho_i)}{\partial t} + \frac{\partial G_{i,x}}{\partial x} + \frac{\partial G_{i,y}}{\partial y} = r_i''' \tag{15.41}$$

This is the conservation equation for species i, where r_i''' is the rate of production of species i per unit volume, and G_is are the components of the mass flux vector. Equation 15.41 may be generalized as

$$\frac{\partial \rho_i}{\partial t} + \nabla \cdot \vec{G}_i = r_i''' \tag{15.42}$$

Generalizing Equation 15.30,

$$\vec{G}_i = -\rho D_{AB} \nabla w_A + \rho_i \vec{V} \tag{15.43}$$

and substituting in Equation 15.42,

$$\frac{\partial \rho_i}{\partial t} + \nabla \cdot \rho_i \vec{V} = \nabla \cdot (\rho D_{AB} \nabla w_A) + r_i''' \tag{15.44}$$

A similar equation can be obtained in terms of molar concentrations. These equations are valid for mixtures with variable total density or molar concentration as well as diffusivity. Assuming constant density and thermal diffusivity, the conservation of species i results in

$$\rho \left(\frac{\partial w_i}{\partial t} + u \frac{\partial w_i}{\partial x} + v \frac{\partial w_i}{\partial y} + w \frac{\partial w_i}{\partial z} \right) = \frac{\partial}{\partial x}\left(\rho D_i \frac{\partial w_i}{\partial x} \right) + \frac{\partial}{\partial y}\left(\rho D_i \frac{\partial w_i}{\partial y} \right) + \frac{\partial}{\partial z}\left(\rho D_i \frac{\partial w_i}{\partial z} \right) + r_{i,g} \tag{15.45}$$

where $r_{i,g} = r_i'''$ is the rate of generation of species i in the gas phase per unit volume, usually as a result of chemical reaction.

The boundary conditions for the species equation are obtained by applying the conservation equation at the boundary, or using specified values. For example, for external flows, the mass fraction of the species i will be known in the free stream. At a solid impermeable boundary,

$$-\rho D_i \frac{\partial w_i}{\partial y} = r_{i,s} \tag{15.46}$$

where $r_{i,s}$ is the rate of generation of species i at the boundary, which will be nonzero in cases like solid fuel combustion or catalytic reaction. This boundary condition assumes that there is sufficient time for the adsorption of the reactants to the surface of the catalyst and the formation of the monolayer to eliminate the need for transient boundary condition.

In a reacting mixture containing N species, in addition to continuity, momentum, and energy equations, $(N - 1)$ equations given by Equation 15.45 must be solved for $(N - 1)$ species with the mass faction of the Nth species obtained from Equation 15.12.

15.3.1 Evaporation in a Column

Consider a container partially filled with liquid A, and a binary mixture of gases A and B flow over the top of the container as shown in Figure 15.3. After the steady-state condition has been established, Equation 15.45 appears to simplify to

$$u \frac{\partial w_i}{\partial x} = \frac{\partial}{\partial x}\left(D_i \frac{\partial w_i}{\partial x}\right) \tag{15.47}$$

It may initially be surprising that we retained the convection term on the left-hand side, as a cause for the flow is not readily apparent. We expect the mole fraction of species A to be higher at $x = 0$, where it is evaporating and decreases with increasing x. The reverse is expected to be true for B. Therefore, there will be a density gradient for both species, and

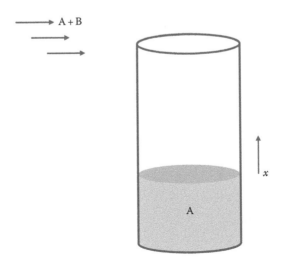

FIGURE 15.3
Evaporation in a column.

both will diffuse in the column. If we assume that species B is not soluble in liquid A, then there must be a convection current to take species B that is diffused in the negative x direction toward the liquid interface, away from this interface, to ensure that the net mass flux of species B is zero. For a binary mixture,

$$D_i = D_{AB} = D_{BA} = D \tag{15.48}$$

This equation can be written as

$$\frac{\partial}{\partial x}\left(D\frac{\partial w_i}{\partial x} - uw_i\right) = 0 \tag{15.49}$$

The term inside the parenthesis is proportional to the mass flux of species i. Integrating this equation once for A,

$$D\frac{\partial w_A}{\partial x} - uw_A = k_1 \tag{15.50}$$

We need an equation to express the fluid velocity in terms of the mole fractions. The net flux of B is zero; therefore, from Equation 15.30,

$$0 = -\rho D\frac{\partial w_B}{\partial x} + \rho u w_B \tag{15.51}$$

Solving for u and substituting in Equation 15.50,

$$D\frac{\partial w_A}{\partial x} - D\frac{w_A}{w_B}\frac{\partial w_B}{\partial x} = k_1 \tag{15.52}$$

Also, noting that $w_B = 1 - w_A$, equation for the mole fraction of A can be arrived at

$$D\frac{\partial w_A}{\partial x} + D\frac{w_A}{1-w_A}\frac{\partial w_A}{\partial x} = k_1 \tag{15.53}$$

which simplifies to

$$D\frac{1}{1-w_A}\frac{\partial w_A}{\partial x} = k_1 \tag{15.54}$$

whose solution is

$$w_A = 1 - k_2 e^{-\frac{k_1}{D}x} \tag{15.55}$$

Using the boundary conditions,

$$x = 0, \quad w_A = w_{A,0}$$
$$x = L, \quad w_A = w_{A,L} \tag{15.56}$$

Evaluating the constants, the solution becomes

$$\frac{1-w_A}{1-w_{A,0}} = \left(\frac{1-w_{A,L}}{1-w_{A,0}}\right)^{\frac{x}{L}} \tag{15.57}$$

and since $w_B = 1 - w_A$

$$\frac{w_B}{w_{B,0}} = \left(\frac{y_{B,L}}{y_{B,0}}\right)^{\frac{x}{L}} \tag{15.58}$$

and the advection velocity becomes

$$u = D\frac{1}{w_B}\frac{\partial w_B}{\partial x} = \frac{D}{L}\ln\frac{y_{B,L}}{y_{B,0}} = \frac{D}{L}\ln\frac{1-w_{A,L}}{1-w_{A,0}} \tag{15.59}$$

which is a positive constant and proportional to the diffusion coefficient.

15.4 Chemically Reactive Flows

Modeling a combustion process is one of the more challenging engineering problems. It involves momentum, heat and mass transfer, and generally turbulence. To gain a basic understanding of the fundamentals of chemically reacting flows, we consider the case of chemically reactive boundary layer flow over a flat plate. The derivation is general to include both gas phase and surface reactions.

Consider an ideal mixture of fuel and oxidizer flowing over an impermeable catalytic flat plate where an exothermic reaction occurs (Figure 15.4). The surface reaction may initiate gas phase reactions as well, and the effects of gas phase reactions are also included. The flow is assumed to be laminar and the body forces and radiation effects are neglected. It is also assumed that the boundary layer approximation can be applied to the momentum, energy, and the species equations. The specific heats of all the species are assumed to be equal and constant.

With the aforementioned assumptions, for a constant pressure flow, the conservation equations for transient, 2D compressible boundary layer flow, involving mass transfer, reduce to the following:

Continuity equation:

$$\frac{\partial \rho}{\partial t} + \frac{\partial \rho u}{\partial x} + \frac{\partial \rho v}{\partial y} = 0 \tag{15.60}$$

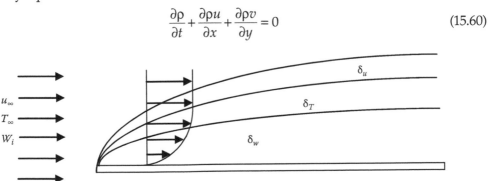

FIGURE 15.4
Combustion of a mixture of fuel and oxidizer flowing over an impermeable catalytic flat plate.

x momentum equation:

$$\rho\left(\frac{\partial u}{\partial t} + u\frac{\partial u}{\partial x} + v\frac{\partial u}{\partial y}\right) = \frac{\partial}{\partial y}\left(\mu\frac{\partial u}{\partial y}\right) \tag{15.61}$$

Energy equation:

$$\rho c_p\left(\frac{\partial T}{\partial t} + u\frac{\partial T}{\partial x} + v\frac{\partial T}{\partial y}\right) = \frac{\partial}{\partial y}\left(k\frac{\partial T}{\partial y}\right) + \sum_{i=1} h_i r_{ig} \tag{15.62}$$

Species equations for the $i = 1$ to N, where N is the number of species:

$$\rho\left(\frac{\partial w_i}{\partial t} + u\frac{\partial w_i}{\partial x} + v\frac{\partial w_i}{\partial y}\right) = \frac{\partial}{\partial y}\left(\rho D_i\frac{\partial w_i}{\partial y}\right) + r_{ig} \tag{15.63}$$

and the equation of state is given by

$$P = \rho R T \tag{15.64}$$

In these equations, r_{ig} is the rate of production of species i in the gas phase, and h_i is the absolute enthalpy of species i. The initial and the boundary conditions are

$$
\begin{aligned}
t < 0, \quad x > 0, \quad & T = T_\infty, \quad w_i = w_{i\infty} \\
t > 0, \quad x = 0, \quad & u = U_\infty, \quad T = T_\infty, \quad w_i = w_{i\infty} \\
x > 0, \quad y \to \infty, \quad & u = U_\infty, \quad T = T_\infty, \quad w_i = w_{i\infty} \\
x > 0, \quad y = 0, \quad & u = 0, \quad v = 0
\end{aligned}
\tag{15.65}
$$

The wall boundary condition is discussed later.

15.4.1 Chemical Reactions

The chemical reaction given by Equation 15.1

$$b_1 B_1 + b_2 B_2 \to b_3 B_3 + b_4 B_4 \tag{15.66}$$

represents the overall reaction between B_1 and B_2. In reality, there would be hundreds of reactions taking place and many more species would be formed, albeit at very low concentrations. It states that b_1 moles ($b_1 M_1$ kg) of species B_1 reacts with b_2 moles ($b_2 M_2$ kg) of species B_2 and produce b_3 moles of species B_3 ($b_3 M_3$ kg) and b_4 moles of species B_4 ($b_4 M_4$ kg) as well as thermal energy. It takes a finite amount of time for the chemical reaction to go to completion and the general form of the reaction rate is given by

$$\bar{r} = K T^m e^{-\frac{E}{RT}} [B_1]^{a_1} [B_2]^{a_2} \tag{15.67}$$

and is known as the Arrhenius rate law. In Equation 15.67, $[B_i]$ is the molar concentration of species i:

$$[B_i] = c_i = \frac{\rho_i}{M_i} = \frac{\rho}{M_i} w_i \tag{15.68}$$

and $n = a_1 + a_2$ is the order of the reaction, E is the activation energy, and K is the pre-exponential factor. In this equation, a_1, a_2, and m are usually determined experimentally. The bar on a quantity implies that it is a molar quantity and thus the reaction rate, $\bar{r}$, has the units of moles/m^3/s and is related to the rate of appearance or disappearance of species i through

$$\bar{r}_i = b_i \bar{r} \tag{15.69}$$

where b_i is negative for the reactants.

In terms of the mass fractions, the rate of generation of species i in kg/m^3/s is

$$r_i = b_i M_i \bar{r} = b_i M_i \frac{K}{M_1^{a_1} M_2^{a_2}} \rho^n T^m e^{-\frac{E}{RT}} w_1^{a_1} w_2^{a_2} \tag{15.70}$$

which is the generation term in Equation 15.63.

15.4.2 Energy Generation Term

Most chemical reactions are exothermic, meaning thermal energy will be released. The generation term in Equation 15.62 accounts for the energy release by the reaction

$$\sum_{i=1}^{N} h_i r_{ig}$$

where

$$h_i = h_{0,i} + \int_{T_0}^{T} c_{pi} dT \tag{15.71}$$

is the total enthalpy
h_i^0 is the enthalpy of formation of species i at the reference state T_0 and P_0

At the standard reference state of 25°C, and 1 atm, stable elements have been assigned the value of zero for their enthalpy of formation. Substituting for the generation of species i, from Equation 15.70,

$$\sum_{i=1}^{N} h_i r_{ig} = \sum_{i=1}^{N} \left[h_i^0 + \int_{T_0}^{T} c_{pi} dT \right] b_i M_i \bar{r}_g \tag{15.72}$$

Expanding the summation into two terms,

$$\sum_{i=1}^{N} h_i r_{ig} = \bar{r}_g \sum_{i=1}^{N} h_i^0 b_i M_i + \bar{r} \sum_{i=1}^{N} b_i M_{ig} \int_{T_0}^{T} c_{pi} dT \tag{15.73}$$

Assuming all the species to have approximately the same specific heat, the integral term can be factored out of the summation, which is zero from conservation of mass

$$\sum_{i=1}^{N} b_i M_i = 0 \tag{15.74}$$

then the second term on the right-hand side of Equation 15.73 drops out, and the generation term becomes

$$\sum_{i=1}^{N} h_i r_{ig} = \bar{r}_g \sum_{i=1}^{N} h_i^0 b_i M_i = \bar{r}_g \sum_{i=1}^{N} b_i \bar{h}_i^0 \tag{15.75}$$

The summation on the right-hand side is the enthalpy of reaction at the reference state.

15.4.3 Wall Boundary Condition

Doing an energy balance and mass balance on the surface of the catalyst, the energy transferred in from the outside plus the energy generated as a result of chemical reactions on the surface of the wall must be equal to the energy leaving the surface. The boundary conditions at $y = 0$ for the energy equation becomes

$$q_w'' + \sum_i h_i r_{i,s} = -k \frac{\partial T}{\partial y} \tag{15.76}$$

where q'' is the energy transferred to the wall from the surroundings, which for an adiabatic wall is zero, h_i is the enthalpy of species i, given by Equation 15.71 that includes the enthalpy of formation, and $r_{i,s}$ is the rate of production of species i as a result of surface reaction and is given by

$$r_{i,s} = b_i M_i \bar{r}_s \tag{15.77}$$

The boundary condition for the species equation is obtained by doing a mass balance at the wall, which requires that the rate of generation of species i at the wall be equal to the rate of its diffusion away from the wall:

$$-\rho D_i \frac{\partial w_i}{\partial y} = r_{i,s} \tag{15.78}$$

where $r_{i,s}$ is the reaction rate at the surface, having units of $kg/m^2/s$, and is different than the gas phase reaction, generally again is given by the Arrhenius rate law.

15.4.4 Nondimensionalization

To get a better understanding of the physics of the problem, we first nondimensionalize the governing equations. If $x^* = \dfrac{x}{L_s}$, $y^* = \dfrac{y}{L_s}$, $t^* = \dfrac{t}{t_s}$, $u^* = \dfrac{u}{U_\infty}$, $\rho^* = \dfrac{\rho}{\rho_\infty}$, $T^* = \dfrac{T - T_\infty}{T_s}$, $w_i^* = \dfrac{w_{i\infty} - w_i}{w_{i,s}}$

$$\frac{\partial \rho^*}{\partial t^*} + \frac{\partial \rho^* u^*}{\partial x^*} + \frac{\partial \rho^* v^*}{\partial y^*} = 0 \tag{15.79}$$

with $t_s = \dfrac{L_s}{U_\infty}$

$$\rho^* \left(\frac{\partial u^*}{\partial t^*} + u^* \frac{\partial u^*}{\partial x^*} + v^* \frac{\partial u^*}{\partial y^*} \right) = \frac{1}{Re} \frac{\partial^2 u^*}{\partial y^{*2}} \tag{15.80}$$

$$\rho^* \left(\frac{\partial T^*}{\partial t^*} + u^* \frac{\partial T^*}{\partial x^*} + v^* \frac{\partial T^*}{\partial y^*} \right) = \frac{1}{RePr} \frac{\partial^2 T^*}{\partial y^{*2}} + \frac{L_s}{U_\infty T_s \rho_\infty c_p} \overline{r}_g \sum_{i=1}^{N} b_i \overline{h}_i^0 \tag{15.81}$$

$$\rho^* \left(\frac{\partial w_i^*}{\partial t^*} + u^* \frac{\partial w_i^*}{\partial x^*} + v^* \frac{\partial w_i^*}{\partial y^*} \right) = \frac{1}{ReSc_i} \frac{\partial^2 w_i^*}{\partial y^{*2}} + \frac{L_s}{U_\infty w_{i,s} \rho_\infty} b_i M_i \overline{r}_g \tag{15.82}$$

where

$$Re = \frac{\rho_\infty U_\infty L_s}{\mu} \tag{15.83}$$

$$Pr = \frac{\mu c_p}{k} = \frac{\nu}{\alpha} \tag{15.84}$$

and the Schmidt number is defined as

$$Sc_i = \frac{\nu}{D_i} \tag{15.85}$$

and is the ratio of momentum diffusion to mass diffusion. As discussed earlier, the ratio of hydrodynamic to thermal boundary layers is proportional to Prandtl number. Similarly, the ratio of hydrodynamic to concentration boundary layers is proportional to Schmidt number. A Schmidt number of greater than one ($Sc > 1$) indicates that velocity boundary layer is thicker than the concentration boundary layer. We still have to determine the scales for mass fractions and temperature.

15.4.5 Scale for Mass Fraction

If we define the scale for mass fraction

$$w_{i,s} = w_{1\infty} \frac{b_i M_i}{b_1 M_1} \tag{15.86}$$

then

$$w_i = w_{i\infty} \left(1 - \frac{w_{1\infty}}{w_{i\infty}} \frac{b_i M_i}{b_1 M_1} w_i^* \right) \tag{15.87}$$

and the species equations become

$$\rho^* \left(\frac{\partial w_i^*}{\partial t^*} + u^* \frac{\partial w_i^*}{\partial x^*} + v^* \frac{\partial w_i^*}{\partial y^*} \right) = \frac{1}{ReSc_i} \frac{\partial^2 w_i^*}{\partial y^{*2}} + \frac{L_s}{U_\infty \rho_\infty} \frac{b_1 M_1}{w_{1\infty}} \overline{r}_g \tag{15.88}$$

The last term on the right-hand side of Equation 15.88 is then the dimensionless species generation term

$$\bar{r}_g^* = \frac{\bar{r}_g}{\dfrac{U_\infty \rho_\infty}{L_s} \dfrac{w_{1\infty}}{b_1 M_1}} \tag{15.89}$$

with the term in the denominator representing the scale for the gas phase reaction rate, simplifying the equation significantly.

15.4.6 Temperature Scale

The boundary condition for the species equation at the wall, including surface reaction becomes

$$-\rho^* \rho_\infty D_i w_{1\infty} \frac{b_i M_i}{b_1 M_1} \frac{1}{L_s} \frac{\partial w_i^*}{\partial y^*} = b_i M_i \bar{r}_s \tag{15.90}$$

which simplifies to

$$-\rho^* \frac{\partial w_i^*}{\partial y^*} = \bar{r}_s^* \tag{15.91}$$

where

$$\bar{r}_s^* = \frac{\bar{r}_s}{\dfrac{\rho_\infty D_i}{L_s} \dfrac{w_{1\infty}}{b_1 M_1}} \tag{15.92}$$

The boundary condition for the energy equation with surface reaction simplifies to

$$q_w'' + \bar{r}_s \sum_{i=1}^{N} b_i \bar{h}_i^0 = -k \frac{T_s}{L_s} \frac{\partial T^*}{\partial y^*} \tag{15.93}$$

Assuming the temperature scale to be

$$T_s = \frac{D_i}{\alpha} \frac{w_{1\infty}}{b_1 M_1} \frac{\sum_{i=1}^{N} b_i \bar{h}_i^0}{c_p} \tag{15.94}$$

then for an adiabatic wall

$$\bar{r}_s^* = -\frac{\partial T^*}{\partial y^*} \tag{15.95}$$

The ratio of mass diffusion to thermal diffusion is called Lewis number, which also becomes the ratio of Prandtl number to Schmidt number:

$$Le = \frac{D}{\alpha} = \frac{Pr}{Sc} \tag{15.96}$$

Note that some authors define Lewis number as inverse of Equation 15.96 or as the ratio of Schmidt number to Prandtl number. For laminar flow, if $Le > 1$, then

$$\frac{\delta_c}{\delta_T} = Le^{1/3} \tag{15.97}$$

The temperature scale becomes

$$T_s = Le \frac{w_{1\infty}}{b_1 M_1} \frac{\sum_{i=1}^{N} b_i \overline{h}_i^0}{c_p}$$

which is the product of Lewis number and the adiabatic flame temperature. Experimentally, local temperatures above adiabatic flame temperatures are observed and this can be explained for $Le > 1$. With this choice of temperature scale, the nondimensional temperature will vary between 0 and 1. The maximum dimensionless temperature then becomes

$$\lambda = \frac{T_s}{T_\infty} = Le \frac{w_{1\infty}}{b_1 M_1} \frac{\sum_{i=1}^{N} b_i \overline{h}_i^0}{c_p T_\infty} \tag{15.98}$$

then

$$T = T_\infty (1 + \lambda T^*) \tag{15.99}$$

Multiplying and dividing the last term in the energy equation with the reaction rate scale and Lewis number, the energy equation becomes

$$\rho^* \left(\frac{\partial T^*}{\partial t^*} + u^* \frac{\partial T^*}{\partial x^*} + v^* \frac{\partial T^*}{\partial y^*} \right) = \frac{1}{RePr} \frac{\partial^2 T^*}{\partial y^{*2}} + Le \frac{\sum_{i=1}^{N} b_i \overline{h}_i^0}{T_s c_p} \frac{w_{1\infty}}{b_1 M_1} \frac{1}{Le} \overline{r}_g^* \tag{15.100}$$

which simplifies to

$$\rho^* \left(\frac{\partial T^*}{\partial t^*} + u^* \frac{\partial T^*}{\partial x^*} + v^* \frac{\partial T^*}{\partial y^*} \right) = \frac{1}{RePr} \frac{\partial^2 T^*}{\partial y^{*2}} + \frac{1}{Le} \overline{r}_g^* \tag{15.101}$$

15.4.7 Length Scale

The length scale, L_s, can be assumed to be the plate length; however, a more appropriate length scale for combustion process can be obtained based on the chemical reactions involved and depending on which one is present or dominates, i.e., has lower activation energy.

Let us consider the case of gas phase combustion alone. Then substituting from Equation 15.70 in Equation 15.92

$$\overline{r}_g^* = \frac{L_s}{U_\infty \rho_\infty} \frac{|b_1| M_1}{w_{1\infty}} \frac{K}{M_1^{a_1} M_2^{a_2}} \rho^n T^m e^{-\frac{E}{RT}} w_1^{a_1} w_2^{a_2}$$

in terms of the dimensionless variables

$$\bar{r}_g^* = \frac{L_s}{U_\infty} \kappa_g \left(1 + \lambda T^*\right)^{m-n} e^{\frac{\beta \lambda T^*}{(1+\lambda T^*)}} \left(1 - \frac{w_{1,s}}{w_{1\infty}} w_1^*\right)^{a_1} \left(1 - \frac{w_{2,s}}{w_{2\infty}} w_2^*\right)^{a_2}$$

where

$$\kappa_g = |b_1| Ke^{-\beta} \rho_\infty^{n-1} T_\infty^m \left(\frac{w_{1\infty}}{M_1}\right)^{a_1-1} \left(\frac{w_{2\infty}}{M_2}\right)^{a_2} = |b_1| Ke^{-\beta} T_\infty^m c_{1\infty}^{a_1-1} c_{2\infty}^{a_2}$$

To eliminate as many parameters as possible, set

$$L_s = \frac{U_\infty}{\kappa_g}$$

then if

$$G = (1 + \lambda T^*)^{m-n} e^{\frac{\beta \lambda T^*}{(1+\lambda T^*)}} \left(1 - w_1^*\right)^{a_1} \left(1 - \omega w_2^*\right)^{a_2} \tag{15.102}$$

where

$$\omega = \frac{b_2 M_2}{b_1 M_1} \frac{w_{1\infty}}{w_{2\infty}} \tag{15.103}$$

then

$$\bar{r}_g^* = G \tag{15.104}$$

For example, for a first-order reaction, where the pre-exponential factor does not depend on temperature, the relevant length scale becomes

$$L_s = \frac{U_\infty}{b_1 Ke^{-\beta}} \tag{15.105}$$

15.4.8 Damkohler Number

With this choice for length scale, then the dimensionless distance along the plate becomes

$$x^* = \frac{x b_1 Ke^{-\beta} T_\infty^m c_{1\infty}^{a_1-1} c_{2\infty}^{a_2}}{U_\infty} \tag{15.106}$$

The Damkohler number is defined as the ratio of the residence time to reaction time. Residence time is the time that the reactants spend in the reactive mixture, and the reaction time is an indication of how quickly the reaction takes place. Rearranging Equation 15.106,

$$x^* = \frac{\dfrac{x}{U_\infty}}{\dfrac{1}{\left(b_1 Ke^{-\beta} T_\infty^m c_{1\infty}^{a_1-1} c_{2\infty}^{a_2}\right)}} \tag{15.107}$$

and examining Equation 15.107 shows that the ratio of $\dfrac{x}{U_\infty}$ is the average time that the reactants have resided over the plate. The terms in the denominator come from the reaction rate and the fraction has the dimension of time. For example, for a first-order reaction with $m = 0$, K must have dimension of second inverse.

At small values of x^*, the residence time is small and the reaction time is long, and therefore there is not sufficient time for the reaction to go to completion. As Damkohler number increases, the reactants remain in the system long enough and the reaction rate is enough, for the process to essentially become diffusion controlled. As soon as the reactants get to the reaction zone, the chemical reaction goes to completion.

15.4.9 Elimination of the Species Equations

The dimensionless form of the governing equations and boundary conditions reduce to

$$\frac{\partial \rho^*}{\partial t^*} + \frac{\partial \rho^* u^*}{\partial x^*} + \frac{\partial \rho^* v^*}{\partial y^*} = 0 \tag{15.108}$$

$$\rho^* \left(\frac{\partial u^*}{\partial t^*} + u^* \frac{\partial u^*}{\partial x^*} + v^* \frac{\partial u^*}{\partial y^*} \right) = \frac{1}{Re} \frac{\partial^2 u^*}{\partial y^{*2}} \tag{15.109}$$

$$\rho^* \left(\frac{\partial T^*}{\partial t^*} + u^* \frac{\partial T^*}{\partial x^*} + v^* \frac{\partial T^*}{\partial y^*} \right) = \frac{1}{RePr} \frac{\partial^2 T^*}{\partial y^{*2}} + \frac{1}{Le} G \tag{15.110}$$

$$\rho^* \left(\frac{\partial w_i^*}{\partial t^*} + u^* \frac{\partial w_i^*}{\partial x^*} + v^* \frac{\partial w_i^*}{\partial y^*} \right) = \frac{1}{ReSc_i} \frac{\partial^2 w_i^*}{\partial y^{*2}} + G \tag{15.111}$$

$$
\begin{aligned}
t^* &< 0, \quad x^* > 0, \quad T^* = 0, \quad w_i^* = 0 \\
t^* &> 0, \quad x^* = 0, \quad u^* = 1, \quad T^* = 0, \quad w_i^* = 0 \\
&\quad\quad x^* > 0, \quad y^* \to \infty, \quad u^* = 1, \quad T^* = 0, \quad w_i^* = 0 \\
&\quad\quad x^* > 0, \quad y^* = 0, \quad\quad u^* = 0, \quad v^* = 0
\end{aligned}
\tag{15.112}
$$

The different combinations of boundary conditions on the plate are given as follows:

$$
\begin{aligned}
x^* > 0, \quad y^* = 0, \quad & \frac{\partial T^*}{\partial y^*} = -\bar{r}_s^*, && \text{Catalytic plate} \\[2mm]
& \rho^* \frac{\partial w_i^*}{\partial y^*} = -\bar{r}_s^*, && \text{Catalytic plate} \\[2mm]
& \frac{\partial T^*}{\partial y^*} = 0, && \text{Adiabatic plate} \\[2mm]
& \frac{\partial w_i^*}{\partial y^*} = 0, && \text{Noncatalytic plate} \\[2mm]
& T^* = T_w^*, && \text{Isothermal plate}
\end{aligned}
\tag{15.113}
$$

In general, for a reactive mixture containing N species, $(N - 1)$ species equations given by Equations 15.111 must be solved, with the mass fraction of the Nth species obtained from the definition of mass fraction. With the choice of the aforementioned references and scales, the modeling of the reactive mixtures is simplified greatly. As mentioned earlier, using Fick's law for a multicomponent system is accurate, as long as the binary diffusion coefficients of all pairs of species are the same, or all species have the same diffusion coefficient, and therefore, the same Schmidt number. This assumption is implicitly used in reactive mixtures.

If the diffusion coefficients are equal, examining Equations 15.111 shows that these equation are the same for all species and does not depend on i. The boundary conditions given by Equations 15.112 and 15.113 are also independent of i. Therefore, the dimensionless form of the mass fraction of all species satisfies the same equation and the same boundary condition, and thus they are all equal or

$$w_1^* = w_2^* = \cdots = w_N^* = w^* \tag{15.114}$$

and

$$G = (1 + \lambda T^*)^{m-n} e^{\frac{\beta \lambda T^*}{(1+\lambda T^*)}} (1 - w^*)^{a_1} (1 - \omega w^*)^{a_2} \tag{15.115}$$

This is a major simplification, resulting in the solution of one transport equation for all species. This simplification also will apply to an impermeable wall in the absence of chemical reactions ($\overline{r}_s^* = 0$).

It is also interesting to note that for a noncatalytic adiabatic plate, for $Pr = Sc$, the energy and species equations and their boundary conditions are identical or

$$T^* = w^* \tag{15.116}$$

and

$$G = (1 + \lambda T^*)^{m-n} e^{\frac{\beta \lambda T^*}{(1+\lambda T^*)}} (1 - T^*)^{a_1} (1 - \omega T^*)^{a_2} \tag{15.117}$$

which again is a major simplification.

15.5 Boundary Layer Flow Transformation

As with all boundary layer flows, solutions are easily obtained using similarity type transformation. Since the governing equations are for compressible flow, Equations 15.79 through 15.82 are transformed into incompressible form using a transformation known as Howarth–Dorodnitzyn transformation [3–6]. The stream function for transient compressible flow is defined by

$$\rho u = \frac{\partial \psi}{\partial y} \tag{15.118}$$

$$\rho v = -\left(\frac{\partial \psi}{\partial x} + \frac{\partial}{\partial t} \int_0^y \rho dy'\right) \tag{15.119}$$

which eliminates the continuity equation. The nondimensional stream function, temperature, and mass fraction are defined as

$$f(\eta) = \frac{\psi}{\sqrt{2\rho_\infty \mu_\infty U_\infty x}} \tag{15.120}$$

and the nondimensional independent variables as

$$\xi = x^* = \frac{x}{L_s} \tag{15.121}$$

$$\eta = \left(\frac{U_\infty}{2\rho_\infty \mu_\infty x}\right)^{\frac{1}{2}} \int_0^y \rho dy' \tag{15.122}$$

$$\tau = t^* = \frac{U_\infty t}{L_s} \tag{15.123}$$

The governing equations are then transformed to

Momentum

$$\frac{\partial}{\partial \eta}\left(l\frac{\partial^2 f}{\partial \eta^2}\right) + f\frac{\partial^2 f}{\partial \eta^2} = 2\xi\left[\frac{\partial^2 f}{\partial \tau \partial \eta} + \frac{\partial f}{\partial \eta}\frac{\partial^2 f}{\partial \eta \partial \xi} - \frac{\partial f}{\partial \xi}\frac{\partial^2 f}{\partial \eta^2}\right] \tag{15.124}$$

Energy

$$\frac{\partial}{\partial \eta}\left(\frac{l}{Pr}\frac{\partial T^*}{\partial \eta}\right) + f\frac{\partial T^*}{\partial \eta} = 2\xi\left[\frac{\partial T^*}{\partial \tau} + \frac{\partial f}{\partial \eta}\frac{\partial T^*}{\partial \xi} - \frac{\partial f}{\partial \xi}\frac{\partial T^*}{\partial \eta} - \frac{1}{Le}G\right] \tag{15.125}$$

Species

$$\frac{\partial}{\partial \eta}\left(\frac{l}{Sc}\frac{\partial w^*}{\partial \eta}\right) + f\frac{\partial w^*}{\partial \eta} = 2\xi\left[\frac{\partial w^*}{\partial \tau} + \frac{\partial f}{\partial \eta}\frac{\partial w^*}{\partial \xi} - \frac{\partial f}{\partial \xi}\frac{\partial w^*}{\partial \eta} - G\right] \tag{15.126}$$

where

$$l = \frac{\rho \mu}{\rho_\infty \mu_\infty} \tag{15.127}$$

The boundary conditions become

$$\tau < 0, \quad \xi > 0, \quad T^* = 0, \quad w^* = 0$$

$$\tau > 0, \quad \xi = 0, \quad f' = 1, \quad T^* = 0, \quad w^* = 0$$

$$\xi > 0, \quad \eta \to \infty, \quad f' = 1, \quad T^* = 0, \quad w^* = 0 \qquad (15.128)$$

$$\xi > 0, \quad \eta = 0, \quad f' = 0, \quad f = 0$$

The different combinations of boundary conditions on the plate are given as follows:

$$\xi > 0, \quad \eta = 0, \quad \frac{\partial T^*}{\partial \eta} = -\bar{r}_s^*, \quad \text{Catalytic plate}$$

$$\frac{\partial w^*}{\partial \eta} = -\bar{r}_s^*, \quad \text{Catalytic plate}$$

$$\frac{\partial T^*}{\partial \eta} = 0, \qquad \text{Adiabatic plate} \qquad (15.129)$$

$$\frac{\partial w^*}{\partial \eta} = 0, \qquad \text{Noncatalytic plate}$$

$$T^* = T_w^* \qquad \text{Isothermal plate}$$

Note that for a catalytic or a noncatalytic adiabatic plate, if Le = 1, then the energy and species equations and their boundary conditions become identical and therefore

$$T^* = w^* \qquad (15.130)$$

or only one transport equation needs to be solved for determining the temperature and the concentrations of all the species.

15.6 Homogeneous Combustion

Consider the flow of a fuel and oxidizer over an isothermal flat plate. The velocity profiles can be assumed to be developed, and independent of time. If we further assume $l = 1$, and Pr and Sc numbers to be constant, then the governing equations simplify to

$$\frac{\partial^3 f}{\partial \eta^3} + f \frac{\partial^2 f}{\partial \eta^2} = 0 \qquad (15.131)$$

$$\frac{1}{Pr} \frac{\partial^2 T^*}{\partial \eta^2} + f \frac{\partial T^*}{\partial \eta} = 2\xi \left[\frac{\partial T^*}{\partial \tau} + \frac{\partial f}{\partial \eta} \frac{\partial T^*}{\partial \xi} - \frac{1}{Le} G \right] \qquad (15.132)$$

Species

$$\frac{1}{Sc}\frac{\partial^2 w^*}{\partial \eta^2} + f\frac{\partial w^*}{\partial \eta} = 2\xi\left[\frac{\partial w^*}{\partial \tau} + \frac{\partial f}{\partial \eta}\frac{\partial w^*}{\partial \xi} - G\right] \qquad (15.133)$$

The equations can be solved numerically using subroutine TRIDI; however, the energy and species equations are coupled together at the boundary, and the boundary condition is highly nonlinear, and more advanced iterative schemes need to be used to obtain converged solutions.

15.7 Catalytic Combustion

A catalyst is a substance in the presence of which chemical reactions take place at lower temperature and higher rates without the catalyst itself being significantly depleted during the process. The catalyst does participate in the reaction chain, but it normally returns back to its original state at the completion of the reaction. There are two types of catalytic process. Homogeneous catalysis is referred to the case where reactants, catalyst, and products are all in the same phase. Heterogeneous catalysis refers to the case where the reactants and catalysts are in different phases, generally reactants are in gas phase and catalyst is on a solid surface. In catalytic reaction, one or more of the reactants are adsorbed on the catalyst in a transitory state and the final products are desorbed from the catalyst, returning it to its original state.

The heterogeneous catalytic reaction is considered to involve five steps:

1. Diffusion of the reactants from the stream to the catalyst
2. Adsorption of the reactants on the catalyst
3. Occurrence of the reactions on the surface
4. Desorption of the products from the catalyst
5. Diffusion of the products from the catalyst to the stream

A catalytic reaction can proceed when all these processes occur continuously. The overall rate of reaction is determined by the slowest step. If diffusion rate (steps 1 and 5) is less than the reaction rate (step 3), then the process is diffusion controlled; otherwise, if reaction rate is lower, then the process is reaction controlled. Since the reaction rate has an exponential dependence on temperature, at low temperatures the problem is reaction controlled and at high temperatures diffusion controlled. A more detailed discussion of the subject can be found in Refs. [7,8].

The transformed wall boundary condition for the species or energy equation becomes

$$\frac{\partial T^*}{\partial \eta} = \frac{\partial w^*}{\partial \eta} = -\delta\sqrt{2\xi}(1+\lambda T^*)^{m-n}e^{\frac{\beta_s \lambda T^*}{(1+\lambda T^*)}}(1-w^*)^{a_{1,s}}(1-\omega w^*)^{a_{2,s}} \qquad (15.134)$$

where

$$\delta = \frac{Sc}{l}\left(\frac{1}{\rho_\infty \mu_\infty}\frac{\kappa_s^2}{\kappa}\right)^{1/2} \qquad (15.135)$$

and

$$\kappa_s = |b_1| K_s e^{-\beta_s} \rho_\infty^{n_s} T_\infty^{m_s} \left(\frac{w_{1\infty}}{M_1}\right)^{a_{1,s}-1} \left(\frac{w_{2\infty}}{M_2}\right)^{a_{2,s}} \tag{15.136}$$

For this case, the energy and species equations are identical and reduce to

$$\frac{1}{Pr}\frac{\partial^2 T^*}{\partial \eta^2} + f\frac{\partial T^*}{\partial \eta} = 2\xi\left[\frac{\partial T^*}{\partial \tau} + \frac{\partial f}{\partial \eta}\frac{\partial T^*}{\partial \xi} - (1+\lambda T^*)^{m-n} e^{\frac{\beta \lambda T^*}{(1+\lambda T^*)}}(1-T^*)^{a_1}(1-\omega T^*)^{a_2}\right] \tag{15.137}$$

and the wall boundary condition is

$$\frac{\partial T^*}{\partial \eta} = -\delta\sqrt{2\xi}\,(1+\lambda T^*)^{m_s-n_s} e^{\frac{\beta_s \lambda T^*}{(1+\lambda T^*)}}(1-T^*)^{a_{1,s}}(1-\omega T^*)^{a_{2,s}} \tag{15.138}$$

The numerical solution of these equations is presented in [4–6,9], and two sample results are shown later. Figures 15.5 and 15.6 show the transient behavior of the temperature of an insulated plate as a function of the dimensionless distance along a noncatalytic and a catalytic plate [10]. The plate temperature increases much faster at the downstream locations, which enhances the combustion and results in a jump in the plate temperature to a value close to the adiabatic flame temperature. The comparison of the two figures shows that the addition of the catalyst greatly enhances the combustion on an adiabatic plate. Both ignition distance and ignition delay times are reduced greatly, explained by the fact that the catalytic reactions have a much lower activation energy so they occur much faster compared to the gas phase reactions. The energy released by the surface reactions then initiates the gas phase reactions or the catalyst lights off the gas phase combustion. Also, note that for Lewis number equal to one, $T^* = w^*$ or Figures 15.5 and 15.6 also show the development of the products concentration along the wall as a function of time.

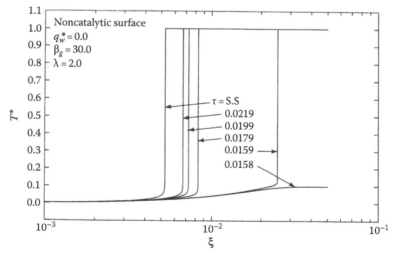

FIGURE 15.5
Transient temperature variation along an adiabatic noncatalytic plate.

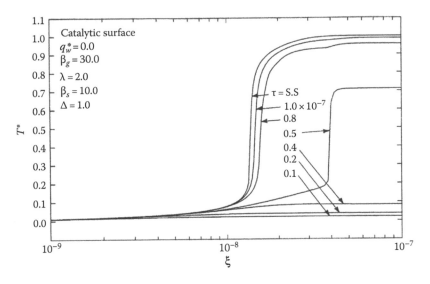

FIGURE 15.6
Transient temperature variation along an adiabatic catalytic plate.

Problems

15.1 Determine the mass fraction, mole fraction, mixture molecular weight of the products for stoichiometric reaction of methane with air, assuming the reaction has gone to completion.

15.2 Derive the species equation for 1D diffusion in cylindrical coordinates.

15.3 Derive the species equation for 1D diffusion in spherical coordinates with gas phase reaction.

15.4 A spherical droplet of radius R of liquid A evaporates into a stationary gas B. Set up the governing equations and boundary conditions for steady-state combustion of a coal spherical particle. Nondimensionalize the governing equations and boundary conditions.

15.5 A spherical steel tank of diameter 50 cm OD and thickness of 3 mm contains helium at 5 MPa and 25°C. If the He diffusion coefficient in steel is 10^{-11} m^2/s, determine the initial rate of mass loss and pressure drop.

15.6 Set up the governing equations and boundary conditions for steady-state combustion at stagnation point of reactants and products impinging on a vertical flat plate. Nondimensionalize the governing equations and boundary conditions.

15.7 A 5 m by 10 m swimming pool is exposed to stationary ambient at 25°C and 30% relative humidity. Estimate the rate of evaporation.

15.8 Determine the concentration change with time in a spherical tank filled with a gas, if the diffusion coefficient is D for the tank wall.

15.9 Set up the governing equations and boundary conditions for methane combustion with air in an insulated circular pipe, assuming that the gas phase reaction is a two-step reaction as shown in the following, and also assuming that both reaction rates are second order and $m = 0$ for both reactions:

$$CH_4 + \frac{3}{2}O_2 \rightarrow CO + 2H_2O$$

$$CO + \frac{1}{2}O_2 \rightarrow CO_2$$

15.10 A stoichiometric mixture of CO and O_2 is flowing over an isothermal catalytic flat plate. Assuming only heterogeneous chemical reactions and further assuming equilibrium at the wall and infinitely fast reactions, such that the concentrations of the reactants are zero at the wall, using equations
  a. Set up the governing equations and the boundary conditions.
  b. Obtain the solution numerically.

References

1. Knuth, E. L. (1959) Multicomponent diffusion and Fick's law, *Physics of Fluids* 2(3), 339–334.
2. Bird, R. B., Stewart, W. E., and Lightfoot, E. N. (1962, c1960) *Transport Phenomena*. New York: Wiley.
3. Schlichting, H. and Gersten, K. (2000) *Boundary-Layer Theory*. Berlin: Springer.
4. Fakheri, A. and Buckius, R. O. (1984) The effects of radiation heat transfer on catalytic combustion. *Combustion and Flame* 57(3), 275–282.
5. Fakheri, A. and Buckius, R. O. (1983) Transient catalytic combustion on a flat plate. *Combustion and Flame* 52C, 169–184.
6. Fakheri, A., Buckius, R. O., and Masel R. I. (1988) Combustion of hydrogen-rich mixtures on a nickel oxide catalyst, *Combustion and Flame* 72(3), 259–269.
7. Wilkinson, F. (1979) *Chemical Kinetics and Reaction Mechanisms*. New York: Van Nostrand Reinhold Co.
8. Satterfield, C. N. (1980) *Heterogeneous Catalysis in Practice*. New York: McGraw-Hill.
9. Fakheri, A. (1985) Analytical and experimental investigations of catalytic combustion, PhD thesis. Mechanical Engineering, UIUC, Urbana, IL.
10. Fakheri, A. and Buckius, R. O. (1987) Transient analysis of heterogeneous and homogeneous combustion in boundary layer flow, *Combustion Science and Technology* 53, 259–275.

16

Computational Fluid Dynamics*

16.1 Introduction

Computers and numerical methods have revolutionized the study of science and engineering problems, and perhaps none more so than fluid mechanics and heat transfer. The field has significantly evolved from its early days when researchers and designers wrote their own codes for solving specific problems, and spent considerable amount of time debugging them. General purpose commercial software packages provide flexibility and broad capabilities. A listing and brief descriptions and links to additional resources for various free and commercial CFD software can be found in [1].

Notable among the available packages are ANSYS Fluent [2] and OpenFOAM [3]. OpenFOAM is an open-source freely distributed CFD code, written in C++ with broad range of capabilities and the user ability to modify and customize the code. ANSYS Fluent is a commercial software with also broad modeling capabilities including for turbulence, heat transfer, chemical reactions, and multiphase systems.

In the discussion later, the general principles are presented first, and then tutorials are presented that walk the readers through the basic process for setting up, solving, and postprocessing the results. This chapter has drawn heavily from the *Fluent User Manual* [2], which is an excellent source for gaining a better understanding of the computational fluid dynamics, as well as understanding the software.

16.2 Governing Equations

The governing equations for fluid flow, heat transfer, phase change, and chemical reactions are derived from the laws of conservation of mass, momentum, energy, and species. The derivations and many different solutions of these equations were presented in the preceding 15 chapters. In shorthand form, the continuity can be written as

$$\frac{\partial}{\partial t}(\rho) + \frac{\partial}{\partial x_k}(\rho u_k) = S_c \qquad (16.1)$$

---

* Co-authored by Dr. Majid Molki, Department Mechanical Engineering, SIU, Carbondale, IL.

and the momentum equations become

$$\frac{\partial}{\partial t}(\rho u_i) + u_k \frac{\partial}{\partial x_k}(\rho u_i) = -\frac{\partial p}{\partial x_i} + \frac{\partial}{\partial x_k}\left(\mu \frac{\partial}{\partial x_k}(\rho u_i)\right) + S_{u_i} \tag{16.2}$$

where S is the source term. As mentioned in Chapter 6, the indices (i, j, k) appearing as subscripts assume the values of 1, 2, and 3, indicating the three coordinates. For example, and x_1, x_2, and x_3 represent x, y, and z coordinates and u_1, u_2, and u_3 represent u, v, and w in Cartesian coordinates. If an index is repeated in a product, summation over the repeated index is implied.

For 2D incompressible constant property flow, the governing equations in conservative form are

$$\frac{\partial u}{\partial x} + \frac{\partial v}{\partial y} = 0 \tag{16.3}$$

$$\frac{\partial u}{\partial t} + \frac{\partial(uu)}{\partial x} + \frac{\partial(vu)}{\partial y} = -\frac{1}{\rho}\frac{\partial p}{\partial x} + \nu\left(\frac{\partial^2 u}{\partial x^2} + \frac{\partial^2 u}{\partial y^2}\right) + g_x \tag{16.4}$$

$$\frac{\partial v}{\partial t} + \frac{\partial(uv)}{\partial x} + \frac{\partial(vv)}{\partial y} = -\frac{1}{\rho}\frac{\partial p}{\partial y} + \nu\left(\frac{\partial^2 v}{\partial x^2} + \frac{\partial^2 v}{\partial y^2}\right) + g_y \tag{16.5}$$

$$\frac{\partial T}{\partial t} + \frac{\partial(uT)}{\partial x} + \frac{\partial(vT)}{\partial y} = \alpha\left(\frac{\partial^2 T}{\partial x^2} + \frac{\partial^2 T}{\partial y^2}\right) + \frac{\dot{q}}{\rho c_p} \tag{16.6}$$

The first term on the LHS of these equations is accumulation (unsteady) term, for example $\frac{\partial u}{\partial t}$ is the rate of x momentum accumulation. The terms like $\frac{\partial(uu)}{\partial x}$ and $\frac{\partial(vu)}{\partial y}$ are advection (convection) terms. The second derivatives on the RHS are the diffusion terms, and terms like $\dot{q}$ are generation, or if negative, are destruction terms. The equations of this form are often referred to as advection–diffusion equations.

The conservation equations are often supplemented by additional transport equations for any additional models such as phase change or turbulence. The objective is to numerically solve the conservation and transport equations for a fluid for a given flow geometry. There are a number of challenges in solving the earlier set of equations. These equations are nonlinear and coupled, which means they all have to be solved together and some type of iterative method must be used to deal with the nonlinearity. Another complicating factor is the absence of an explicit equation for determining the pressure field.

16.3 Spatial Discretization or Grid Generation

The first step in solving a problem using a commercial CFD package is to create the geometry. Once the geometry is created, the different regions of computational interest, like the fluid and solid regions, are divided into a discrete set of cells, with the cell corners

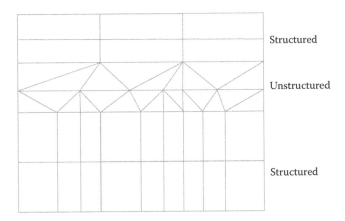

FIGURE 16.1
Structured and unstructured mesh.

defining the nodes. This process and the different components of the mesh are defined in the preprocessor. Depending on the discretization process, the solution to the governing equations is obtained only at the nodes or cell centers.

There are two types of grids: structured and unstructured, as shown in Figure 16.1. A structured grid is one where the grid faces align with the coordinate system, for example, rectangular cubes in Cartesian coordinates. The unstructured or irregular grid, unlike the structured one, does not align with the coordinate system and is generally made of geometries like triangles or tetrahedra, and are arranged in an irregular pattern. Unstructured grids provide more flexibility over structured grids; however, the solution accuracy may suffer. If some portions of flow field are discretized using the structured grid and the rest using unstructured, the resulting grid is called a hybrid grid.

The computational domain generally consists of a large number of cells. A cell is a line segment, an area, or a volume depending on whether the flow is 1D, 2D, or 3D, as shown in Figures 16.2 and 16.3. The boundary of a cell is called a face, which can be a point, a line, or a surface, for 1D, 2D, and 3D cells, respectively. The boundary of a face is called an edge, and nodes or grid points are at the intersection of the edges. To simplify assigning properties or boundary conditions, nodes, faces, and/or cells are often grouped together as zones.

The choices of cell or element types to be used are dependent on the geometry and the problem being considered. Some of the more commonly used types are shown in Figures 16.2 and 16.3.

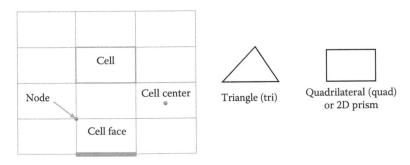

FIGURE 16.2
2D mesh terminology and types.

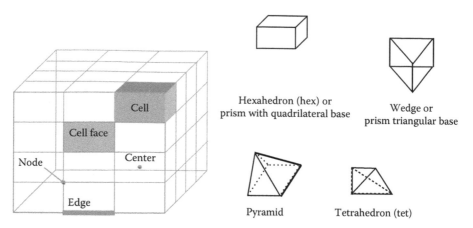

Hexahedron (hex) or
prism with quadrilateral base

Wedge or
prism triangular base

Pyramid

Tetrahedron (tet)

FIGURE 16.3
3D mesh terminology and types.

16.3.1 Geometry and Mesh Generation

The generation of the geometry and the mesh depend on the preprocessor being used. In general, geometry is created as follows:

1. The minimum number of vertices (points) needed to construct the computational domain are determined and then created.
2. Once the vertices are defined, they are connected together to create lines that specify the boundaries of the domain, as well as additional ones that may be used to define the geometry or to mesh the domain properly.
3. The lines are then used to define the surfaces (faces).
4. In the case of 3D geometries, the faces will be used to define the volumes.

Alternatively, if a solid modeler is used, the geometry is created one feature at a time. This approach is used increasingly for creating the geometry. Once the faces/volumes are created, they are meshed. There are a number of methods for creating the mesh. A commonly used method is to mesh the edges, then use those to create face and volume meshes. To develop a better understanding of the meshing process, and the impact of different parameters selected, consider an edge shown in Figure 16.4. If

L is the edge length
N is number of intervals (number of nodes is $N + 1$)
a_1 is length of the first interval
a_N is length of last interval
$d = \dfrac{a_i}{a_{i-1}}$ is the ratio of successive intervals

FIGURE 16.4
Elements of an edge mesh.

Then

$$a_1 = \frac{d-1}{d^N - 1} L \tag{16.7}$$

and

$$a_N = a_1 d^{N-1} = \frac{1 - d^{-1}}{1 - d^{-N}} L \tag{16.8}$$

The ratio of the last interval length to the first interval length is called bias and is

$$\text{Bias} = B = \frac{a_N}{a_1} = d^{N-1} \tag{16.9}$$

If bias is known, the length of the first interval for a given interval ratio or number of nodes can be determined from

$$a_1 = \frac{d-1}{dB - 1} L = \frac{B^{\frac{1}{N-1}} - 1}{B^{\frac{N}{N-1}} - 1} L \tag{16.10}$$

Generally, the number of intervals in a given length L needs to be determined. Usually one has an idea about what the size of the first interval and the ratio of the successive intervals are to be assumed, then the number of intervals can be determined from

$$N = \frac{\ln\left(\dfrac{d-1}{a_1} L + 1\right)}{\ln d} \tag{16.11}$$

These formulas remove the guess work in the selection of parameters like number of intervals or the interval ratios, and are very helpful in creating well-structured meshes.

Example 16.1

Consider boundary layer flow over a flat plate of length l with the computational domain extending to 0.2 in the y direction. Determine

a. The interval ratio, the first and last interval lengths, for a bias of 10 using 50 intervals
b. Keeping the maximum cell aspect ratio below 10, determine the number of nodes needed in the y direction, using the same interval ratio in the x and y directions

Solutions:

a.

$$B = \frac{a_N}{a_1} = d^{N-1}$$

$$10 = d^{50-1}$$

$$d = 1.0481$$

$$a_{1,x} = \frac{d-1}{d^N - 1} L = \frac{1.0481 - 1}{1.0481^{50} - 1} 1 = 0.0051$$

$$a_{N,x} = a_{1,x} d^{N-1} = 0.0507$$

b.

$$\frac{a_{N,x}}{a_{1,y}} = \frac{0.0507}{\dfrac{d_y - 1}{d_y^M - 1} L_y} < 10$$

$$\frac{1.0481 - 1}{1.0481^M - 1} 0.2 > 0.0051$$

which results in

$$M < 22.6$$

Using 22 intervals

$$\frac{a_{N,x}}{a_{1,y}} = \frac{0.254}{\dfrac{d-1}{d^M - 1} L} = \frac{0.254}{0.0266} = 9.555$$

and

$$\frac{a_{N,y}}{a_{1,x}} = \frac{a_{1,y} d^{M-1}}{0.0254} = \frac{0.0712}{0.0254} = 2.808$$

and both of these are <10.

16.3.2 Cell Aspect Ratio and Skewness

The cell aspect ratio is a measure of the relative proportions of cells and an indication of the ratio of the maximum to the minimum distances between the nodes of the cell. Fluent defines it as the ratio of the maximum distance between the cell and face centroids to the minimum distance between cell nodes.

The shape of the cell is also assessed by another parameter called cell skewness, S, defined as

$$S = max\left[\frac{\theta_{max} - \theta_e}{180 - \theta_e}, \frac{\theta_e - \theta_{min}}{\theta_e} \right] \tag{16.12}$$

where
θ_{max} and θ_{min} are the maximum and minimum angles in the face or cell, respectively
θ_e is the angle for an equiangular face or cell which is 60° for a triangle and 90° for a quadrilateral cell

Skewness is a comparison between the cell's shape and that of an ideal cell, with the value of $S = 0$ representing the ideal cell, and $S = 1$ indicating a completely distorted cell that is nearly coplanar. Good meshing will generally have skewness <0.5 and values of $S > 0.8$ are to be avoided. For quadrilateral or hexahedral elements, $S = 0$.

Example 16.2

What is the skewness for a triangle with two interior angles of 120° and 40°?
The third angle of the triangle is 20°; therefore,

$$S = max\left[\frac{120-60}{180-60}, \frac{60-20}{60}\right] = 0.67$$

Some simple rules help in designing meshes that lead to faster convergence and more accurate results. It is a good practice to visually inspect the mesh, to make sure that it appears orderly, and proportionate, where changes in mesh size happen smoothly. Generally, structured grids, aligned with the flow, provide more accurate results. The grid change should not exceed 20%, i.e.,

$$\frac{a_{i+1}}{a_i} < 1.2 \qquad (16.13)$$

and generally it is best to limit it to <5%. Also, high aspect ratios (>10) should be avoided. Although in general high aspect ratio cells are to be avoided, if the flow is mostly aligned with the mesh, the high aspect ratio quadrilateral or hexahedral cells may still be used. This will result in far fewer cells compared to using triangular or tetrahedral cells, since a high aspect ratio in a triangular or tetrahedral cell results in high skewness, which is undesirable, as it may impede accuracy and convergence.

16.4 Discretization of Transport Equations

Numerical solution involves discretization of the transport equations in time and space. The finite difference approximation was covered in Chapter 2, which we used extensively in the subsequent chapters. Finite difference approximation works well for relatively simple problems, resulting in a set of algebraic equations that were solved using spreadsheets either directly, using for example TRIDI, or iteratively. The linear problems, where the domain can be expressed using a structured grid, are good candidates for finite difference approach.

For more general cases, like those involving nonlinear equations or irregular boundaries, the finite difference approximation does not always result in stable or accurate solutions, and special precautions need to be taken. The approach typically used for solution of general NS equations is finite volume approximation. The basic idea behind finite volume approximation is that the conserved quantities, like mass, momentum, and energy, are ensured to be conserved over the discretized regions (cell). This approach has proven to be more stable and accurate than the finite difference approximation.

16.4.1 Finite Difference Solution

Consider a steady 1D incompressible advection–diffusion flow in Cartesian coordinates. The conservation of energy becomes

$$\frac{d(\rho u c_p T)}{dx} = k\frac{d^2 T}{dx^2} \qquad (16.14)$$

Nondimensionalizing Equation 16.14 by defining

$$T^* = \frac{T(x) - T(0)}{T(L) - T(0)} \quad \text{and} \quad x^* = \frac{x}{L} \tag{16.15}$$

results in

$$Pe\frac{dT^*}{dx^*} = \frac{d^2T^*}{dx^{*2}} \tag{16.16}$$

where

$$Pe = \frac{\rho c_p u L}{k} = RePr \tag{16.17}$$

Integrating Equation 16.16 and evaluating the constants, the temperature distribution is given by

$$T^* = \frac{\exp(Pex^*) - 1}{\exp(Pe) - 1} \tag{16.18}$$

Using central difference approximation for the derivatives, the finite difference approximation of Equation 16.16 becomes

$$Pe\frac{T_{i+1}^* - T_{i-1}^*}{2\Delta x^*} = \frac{T_{i+1}^* - 2T_i^* + T_{i-1}^*}{\Delta x^{*2}} \tag{16.19}$$

which simplifies to

$$\left(\frac{1}{2} + \frac{Pe_c}{4}\right)T_{i-1}^* - T_i^* + \left(\frac{1}{2} - \frac{Pe_c}{4}\right)T_{i+1}^* = 0 \tag{16.20}$$

where Pe_c is the cell Peclet number

$$Pe_c = Pe\Delta x^* = \frac{\rho c_p u \Delta x}{k} = Re_c Pr \tag{16.21}$$

The cell Peclet number can be viewed as the ratio of the convective to diffusive fluxes across the cell. At low Peclet number, diffusion is the dominant mechanism for energy transfer, whereas at high cell Peclet number convection dominates. The differencing scheme used must preserve this requirement to arrive at accurate results.

Figure 16.5 shows the solution for $Pe = 20$ and different values of $\Delta x^* = 0.2, 0.1$, and 0.05, resulting in the cell Peclet numbers of 4, 2, and 1, respectively [9,10]. As is seen from the solution, for cell Peclet number greater than 2, the solution oscillates and is physically unrealistic.

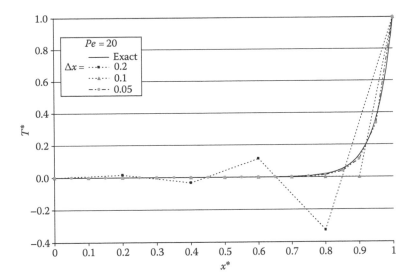

FIGURE 16.5
Numerical solution of advection–diffusion flow for different cell Peclet number.

This is related to the concept of false diffusion or numerical diffusion, where the effective diffusivity becomes more than the actual physical diffusivity due to the grid size and the discretization used. If $Pe_c = 2$, the solution at the interior nodes is identically equal to zero. It can be shown that to get stable solutions, the coefficients of T_{i+1} and T_{i-1} must be positive, which requires $Pe_c < 2$, and as can be seen, the solution approaches the exact solution for $Pe_c = 1$.

16.4.2 False or Numerical Diffusion

A significant source of error in numerical solution is false diffusion (numerical diffusion). False diffusion is caused by the discretization process and the associated truncation errors. False diffusion increases problem's effective diffusion or even introduces diffusion in situations where there is no actual diffusion, e.g., inviscid flow, changing the nature of the problem, resulting in unrealistic solutions. The solutions obtained correspond to a problem that has higher diffusion coefficient than the actual problem [4].

Consider the advection–diffusion problem described earlier. Equation 16.14 can be rearranged as

$$\frac{dT^*}{dx^*} - \frac{d}{dx^{*2}}\left[\frac{T^*}{Pe}\right] = 0 \tag{16.22}$$

with the term, $\dfrac{1}{Pe}$ representing the physical diffusion coefficient. For high values of Peclet number, the diffusion coefficient is low and the problem is dominated by convection, while for low values of Peclet number the diffusion will be the dominant mode of heat transfer. Using the central difference approximation for both derivatives and retaining the higher order terms, we obtain

$$\frac{dT^*}{dx^*} = \frac{T_{i+1}^* - T_{i-1}^*}{2\Delta x^*} - \frac{\Delta x^{*2}}{3!}\left.\frac{d^3T^*}{dx^{*3}}\right|_i + \cdots \tag{16.23}$$

$$\frac{1}{Pe}\frac{d^2T^*}{dx^2} = \frac{1}{Pe}\left[\frac{T_{i+1}^* - 2T_i^* + T_{i+1}^*}{\Delta x^{*2}} - 2\frac{\Delta x^4}{4!}\frac{d^4T^*}{dx^{*4}}\bigg|_i + \cdots\right] \tag{16.24}$$

Subtracting Equation 16.24 from 16.23 and rearranging result in

$$\frac{T_{i+1}^* - T_{i-1}^*}{2\Delta x^*} - \frac{1}{Pe}\frac{T_{i+1}^* - 2T_i^* + T_{i+1}^*}{\Delta x^{*2}} = \frac{dT^*}{dx^*} - \frac{d}{dx^{*2}}\left[\frac{T^*}{Pe}\left(1 + \frac{\Delta x^{*2}}{3!}\frac{Pe}{T^*}\frac{dT^*}{dx^*}\bigg|_i - 2\frac{\Delta x^4}{4!}\frac{1}{T^*}\frac{d^2T^*}{dx^{*2}} + \cdots\right)\right] \tag{16.25}$$

As can be seen, the LHS of Equation 16.25 is the finite difference approximation that we used for the original PDE that we intended to solve (Equation 16.16). However, Equation 16.25 shows that the differencing on the LHS is approximation to a different differential equation, represented by the RHS of Equation 16.25 that has essentially the same form as 16.16, but with an effective diffusivity given by

$$\frac{1}{Pe}\left(1 + \frac{\Delta x^{*2}}{3!}\frac{Pe}{T^*}\frac{dT^*}{dx^*}\bigg|_i - 2\frac{\Delta x^{*4}}{4!}\frac{1}{T^*}\frac{d^2T^*}{dx^{*2}} + \cdots\right) \tag{16.26}$$

that is different than the original value of $\frac{1}{Pe}$. The additional terms in 16.26 are proportional to Δx^*, which is <1, and the temperature gradient, and inversely proportional to temperature. For large Peclet numbers, the second term in the parenthesis is larger than the other higher order terms ($\Delta x^{*2} \gg \Delta x^{*4}$ and $Pe \gg 1$) or the second term will dominate. The impact of these terms is to increase the effective viscosity for positive temperature gradient, or numerical diffusion will become more significant in convection-dominated problems ($Pe \gg 1$) and/or in the regions where the grid size or the gradients are large. For diffusion-dominated problems, where Peclet number is small, the higher order terms in the parenthesis in Equation 16.26 will be small and ($1/Pe$), which is the actual physical diffusion, will dominate and the numerical diffusion will not introduce a large error. Therefore, to reduce instability and numerical diffusion, finer mesh needs to be used, particularly in the regions where one expects to see large gradients. Numerical diffusion can be further reduced by using higher order approximations.

16.4.3 Finite Volume Approximation

This mathematical grounding for the finite volume method (FVM) is the divergence theorem that states the integral of the divergence of a vector field over the volume of a region is equal to the integral of the vector field over the region's boundary. Divergence of velocity, temperature, etc., appears in conservation equations and therefore can be written as surface integrals, simplifying the problem to evaluating fluxes normal to the boundaries of regions of interest. In the FVM, the flux leaving one region enters the adjacent one, and therefore at the cell level, the quantities of interest are conserved. FVM is the most commonly used method in commercial CFD codes.

The FVM approach can also be viewed as ensuring that the conservation equations are satisfied on each grid cell, as shown in Figure 16.6. This is accomplished by integrating the transport equations over a finite volume defined by the cell and then assuming that

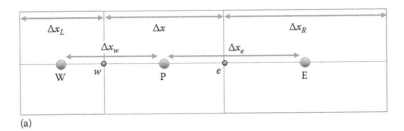

(a)

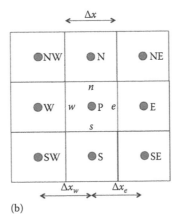

(b)

FIGURE 16.6
Control volumes surrounding nodes.

the mean value of the variable of interest over the cell sufficiently describes the behavior over the cell. For 2D case, the transport equations can be rearranged into the general form

$$\frac{\partial \rho \phi}{\partial t} + \frac{\partial}{\partial x}\left(\rho u \phi - \mu \frac{\partial \phi}{\partial x}\right) + \frac{\partial}{\partial y}\left(\rho v \phi - \mu \frac{\partial \phi}{\partial y}\right) = S \tag{16.27}$$

where ϕ can be any of the variables of interest, like u, v, w, T, etc. The basic approach is demonstrated by considering the energy equation for steady 1D incompressible flow in Cartesian coordinates given by Equation 16.14.

The discretized domain is shown in Figure 16.6a, where node P is the main or present node (node i) and nodes E (east, $i + 1$) and W (west, $i - 1$) are the adjacent nodes. In terms of the dimensional variables and the new notation, the finite difference approximation given by Equation 16.20 becomes

$$\left[\frac{k}{2\Delta x} + \frac{(\rho u c_p)_W}{4}\right] T_W - \frac{k}{\Delta x} T_P + \left[\frac{k}{2\Delta x} - \frac{(\rho u c_p)_E}{4}\right] T_E = 0 \tag{16.28}$$

Here we have used a uniform mesh, namely, $\Delta x_w = \Delta x_e = \Delta x$.

Of course, the backward or forward difference approximation for the first derivative results in alternative representations of the discretized equation.

We now derive the discretized equation using finite volume approximation, which is obtained by applying the law of conservation of energy to a control volume of length Δx surrounding node P, where face w of the control volume is halfway between W and P and face *e* is halfway between nodes *P* and *E*. The first law of thermodynamics for steady, 1D flow, without energy generation requires that

Energy in by conduction + Energy in by convection

= Energy out by conduction + Energy out by convection

$$-k\frac{\partial T}{\partial x}\bigg|_w + (\rho u c_p T)_w = -k\frac{\partial T}{\partial x}\bigg|_e + (\rho u c_p T)_e \tag{16.29}$$

Alternatively, the energy equation (Equation 16.14) can be written as

$$\frac{\partial}{\partial x}\left(\rho u c_p T - k\frac{\partial T}{\partial x}\right) = 0 \tag{16.30}$$

Integrating this equation over the control volume (along *x*) surrounding node *P* results in

$$\left(\rho u c_p T - k\frac{\partial T}{\partial x}\right)_e - \left(\rho u c_p T - k\frac{\partial T}{\partial x}\right)_w = 0 \tag{16.31}$$

which is the same as Equation 16.29.

The values of the dependent variables, in this case *T*, are calculated for nodes like *W*, *P*, and *E*; however, all the terms in Equation 16.29 are evaluated at locations other than the nodes. The derivatives on the cell faces are typically calculated using forward difference approximation; therefore, Equation 16.29 becomes

$$-k\frac{T_P - T_W}{\Delta x_w} + (\rho u c_p T)_w = -k\frac{T_E - T_P}{\Delta x_e} + (\rho u c_p T)_e \tag{16.32}$$

However, the convective terms must still be approximated.

16.4.4 Approximation for Convective Fluxes

The next step is to find approximate values for the terms like $(\rho u c_p T)_w$ and $(\rho u c_p T)_e$, which are the convective fluxes. There are a number of ways by which the convective flux terms can be evaluated [5–10], and the choice depends on a number of factors, including the balance between the desired accuracy and computational expense.

16.4.4.1 Average Values

One simple approach is to use the average values for the cell faces, which in this case becomes

$$-k\frac{T_P - T_W}{\Delta x} + (\rho u c_p)_w \frac{T_W + T_P}{2} = -k\frac{T_E - T_P}{\Delta x} + (\rho u c_p)_e \frac{T_E + T_P}{2} \tag{16.33}$$

which simplifies to

$$\left[\frac{k}{\Delta x} + \frac{(\rho u c_p)_w}{2}\right]T_W - \left[\frac{2k}{\Delta x}\right]T_P + \left[\frac{k}{\Delta x} - \frac{(\rho u c_p)_e}{2}\right]T_E = 0 \tag{16.34}$$

or it reduces to Equation 16.20, which was obtained using central difference approximation. The deficiencies of this approach were discussed earlier.

16.4.4.2 First-Order Upwind Differencing

In the first-order upwind approximation, the value at the face of the cell is taken to be equal to the value in the center of the cell upstream of the face. Since for 1D problem, "upstream" is at smaller x location, then the finite volume approximation becomes

$$-k\frac{T_P - T_W}{\Delta x_w} + (\rho u c_p T)_W = -k\frac{T_E - T_P}{\Delta x_e} + (\rho u c_p T)_P \tag{16.35}$$

which simplifies to

$$k\frac{T_W - 2T_P + T_E}{\Delta x} = (\rho u c_p T)_P - (\rho u c_p T)_W \tag{16.36}$$

Although this equation can be arrived at by finite differencing Equation 16.16 using central difference approximation for the second derivative and backward difference approximation for the first derivative, in general the two approaches do not result in the same differencing equations. In 2D or 3D flows, the velocity will not necessarily be in the positive x direction and therefore the upstream is not always at smaller x location.

16.4.4.3 Higher Order Approximations

A higher order approximation for the convected variables can be calculated by using the values and fitting a curve to more nodes upstream and downstream. This approach is appropriate when the faces are well defined like when structured rectangular (cube for 3D) grids are used. Using the weighted values at two nodes upstream and one node downstream, for 1D case, the value at the east cell can be approximated by

$$\phi_e = \alpha\left[\left(1 - \frac{\Delta x}{2\Delta x_e}\right)\phi_P + \frac{\Delta x}{2\Delta x_e}\phi_E\right] + (1 - \alpha)\left[\left(1 + \frac{\Delta x}{2\Delta x_w}\right)\phi_P - \frac{\Delta x}{2\Delta x_w}\phi_W\right] \tag{16.37}$$

where α is a weighting factor. Different values of α correspond to different higher order schemes.

16.4.4.4 Second-Order Upwind Scheme

In second-order upwind scheme, the value at the cell face is obtained by using the values of the variables at two upstream nodes, using Taylor series expansion. For a 1D case, for example, the value of the variable at any location, using second-order upwind, is obtained by using linear interpolation between the two upstream nodes from e which are P and W, resulting in

$$\phi = \left[\frac{x - x_P}{x_P - x_W} + 1\right]\phi_P + \left[\frac{x_P - x}{x_P - x_W}\right]\phi_W \tag{16.38}$$

on the east face, $x_P = 0$, $x = \dfrac{\Delta x}{2}$, $x_W = -\Delta x_W$, reducing Equation 16.38 to

$$\phi_e = \left[1 + \frac{\Delta x}{2\Delta x_w}\right]\phi_P - \left[\frac{\Delta x}{2\Delta x_w}\right]\phi_W \tag{16.39}$$

The same can also be obtained by setting $\alpha = 0$ in Equation 16.37. Note that $\alpha = 1$ results in another method called central second-order interpolation.

16.4.4.5 QUICK Scheme

The quadratic upwind interpolation (QUICK) [6] is obtained by fitting a second-order polynomial to the cell values. The results can also be obtained by setting $\alpha = 3/4$ in Equation 16.36. For example, for uniform spacing

$$\phi_e = \frac{3-2\alpha}{2}\phi_P + \frac{\alpha}{2}\phi_E - \frac{(1-\alpha)}{2}\phi_W = \frac{3}{4}\phi_P + \frac{3}{8}\phi_E - \frac{1}{8}\phi_W \tag{16.40}$$

Again, the QUICK scheme will be typically more accurate on structured grids aligned with the flow direction.

16.4.4.6 Power Law Approximation

The power law discretization scheme [5] interpolates the face value of a variable, using the exact solution to a 1D convection–diffusion equation (16.14) given by Equation 16.18. Applying this equation between W and P, Figure 16.6a results in the temperature distribution for different values of cell Peclet number,

$$\frac{T(x)-T_W}{T_P-T_W} = \frac{\exp\left(\dfrac{\rho c_p u \Delta x_w}{k}\dfrac{x}{\Delta x_w} - 1\right)}{\exp\left(\dfrac{\rho c_p u \Delta x_w}{k} - 1\right)} = \frac{\exp\left(Pe_c \dfrac{x}{\Delta x_w}\right)-1}{\exp(Pe_c)-1} \tag{16.41}$$

shown in Figure 16.7. Note that x is measured from node W. As can be seen, for large values of cell Peclet number ($Pe_c = 10$), corresponding to high velocity, the value at face w is close to the upstream value (W). Similarly, when the fluid is flowing in the negative x direction ($Pe = -10$), the temperature on the same face is close to upstream value (P). This means that upwinding provides an accurate approximation for high velocity flows. At lower cell Peclet numbers, higher order approximations may be required, making QUICK an appropriate choice.

First-order approximations converge faster, but are less accurate. When the flow is predominantly flowing along the grids, like boundary layer flow over a flat plate, using quad elements, first-order upwind approximation provides sufficient accuracy. When the flow is not perpendicular to the grids, or triangular or tetrahedral elements are used, higher order approximations are needed. To improve convergence and accuracy in more complicated problems, it may be advantageous to initially start with first-order upwinding to get results close to the converged solution and then switch to second-order differencing to improve accuracy.

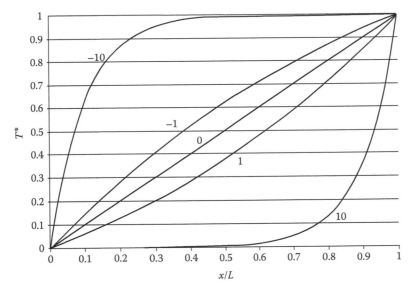

FIGURE 16.7
The exact solution of 1D convection–diffusion equation for different Peclet numbers.

16.5 Pressure Velocity Coupling

The finite volume formulation of the conservation equations results in a set of algebraic equations of the form [5]

$$A_p \phi_P = \sum_{nb} A_{nb} \phi_{nb} + S \tag{16.42}$$

where
ϕ stands for any of the field variables like the different components of velocity, or temperature, etc.
The subscript nb stands for the neighboring points
A is the discretization coefficient
S is the source term

For example for steady 2D incompressible flows, without body forces, the finite volume form of the x momentum equation becomes

$$A_p u_P = A_N u_N + A_S u_S + A_E u_E + A_W u_W - C \frac{\partial P}{\partial x} \tag{16.43}$$

or for velocity components the source term in Equation 16.42 contains the pressure gradients, and therefore the pressure field is needed for solving the momentum equations.

As mentioned earlier, a complicating factor in numerical solution of the NS equations is the absence of an explicit equation for determining the pressure field. The pressure gradient, and not the absolute value of the pressure, is needed for obtaining the velocity field, although for compressible flows, the absolute value of the pressure is needed for determination of density. If the flow can be considered incompressible, i.e., has low Mach number or does not experience large temperature changes, the general approach is to

- Arrive at a guess to the pressure field
- Solve the momentum equation to obtain the velocity distribution
- Check the accuracy of the guessed pressure field, by determining if the continuity equation is satisfied and if not improves the pressure field

This is sometimes referred to as a pressure-based scheme for solving the NS equations. The estimate of the pressure field is arrived at by solving a pressure or a pressure correction equation obtained from continuity and momentum equations. For compressible flows, the continuity is used to determine the density, and the pressure is obtained from the equation of state, such as the ideal gas law.

In the pressure-based approaches, by using guessed values for pressure field, the governing equations are decoupled (segregated from one another) and solved sequentially. Typically, the momentum equations are solved first, one after another, using the recently calculated values of the other variables. Once the velocity field is obtained, it is used to obtain the new pressure field, using either a pressure or a pressure correction equation. The new pressure field is used to correct the velocity field. Since the equations are nonlinear and coupled, each solution is based on the values of the other variables from a previous iteration and therefore the final solution must be obtained iteratively. In this approach, the computer memory requirements are low, but the convergence rate is slow.

The new values of quantities like pressure or velocity are expressed as

$$\phi = \phi^* + \phi' \tag{16.44}$$

where
ϕ^* is the current or the guessed value (not to be confused with nondimensional ϕ)
ϕ' is the correction

The new values obtained from earlier are then used to obtain the other quantities of interest, like turbulent kinetic energy and dissipation, temperature, species, concentrations, etc. The process is continued until the convergence criteria are met.

16.5.1 Pressure Correction Equation

There are a number of ways by which an explicit equation for determining pressure can be arrived at by manipulating the continuity and momentum equations. For example, for 2D incompressible flows, differentiating the x momentum with respect to x and y momentum with respect to y, and adding the two equations, result in

$$\frac{\partial}{\partial x}\left(u\frac{\partial u}{\partial x}+v\frac{\partial u}{\partial y}\right)+\frac{\partial}{\partial y}\left(u\frac{\partial v}{\partial x}+v\frac{\partial v}{\partial y}\right)=-\frac{1}{\rho}\left(\frac{\partial^2 P}{\partial x^2}+\frac{\partial^2 P}{\partial y^2}\right) \tag{16.45}$$

which simplifies to [9]

$$\frac{\partial^2 P}{\partial x^2}+\frac{\partial^2 P}{\partial y^2}=2\rho\left[\frac{\partial u}{\partial x}\frac{\partial v}{\partial y}-\frac{\partial u}{\partial y}\frac{\partial v}{\partial x}\right] \tag{16.46}$$

Along a solid boundary the velocity components are zero, and the pressure satisfies the Poisson equation. This equation is used in a number of algorithms developed for the solution of the NS equations in primitive variables, such as marker and cell (MAC) and simplified marker and cell (SMAC) [11,12], to name a few.

16.5.2 SIMPLE Scheme

The semi-implicit method for pressure-linked equations (SIMPLE) algorithm is a pressure correction method first proposed by Patankar and Spalding in 1972 [13]. Substituting the exact and guessed values of velocity in the discretized form of the momentum equation results in

$$A_p u_P^* = \sum_{nb} A_{nb} u_{nb}^* - C_p \left(P_P^* - P_W^* \right) \tag{16.47}$$

$$A_p u_P = \sum_{nb} A_{nb} u_{nb} - C_p (P_P - P_W) \tag{16.48}$$

Subtracting Equation 16.47 from 16.48 results in

$$A_p u_P' = \sum_{nb} A_{nb} u_{nb}' - C_p \left(P_P' - P_W' \right) \tag{16.49}$$

which is an equation for calculating the velocity corrections, provided pressure correction values are known. For a converged solution, all the corrections are zero. In the SIMPLE algorithm, the summation on the RHS is neglected resulting in

$$u_P' = \frac{C_p}{A_p} (P_W' - P_P') \tag{16.50}$$

This is the correction for the velocity, and therefore the corrected velocity can be obtained from Equation 16.44

$$u = u^* + \frac{C_p}{A_p} (P_W' - P_P') \tag{16.51}$$

Similar expressions can be found at the neighboring nodes and also for other components of velocity. The corrected velocities must satisfy the discretized continuity equation, providing an equation for calculating the pressure correction in the form of Equation 16.42

$$A_p P_P' = \sum_{nb} A_{nb} P_{nb}' + S \tag{16.52}$$

where the source term in this equation results from the fact that guessed velocities like u^* are approximations and as the solution converges the values of P' and therefore S will approach zero. Note that the elimination of the summation in Equation 16.49 will not introduce any errors in the converged solution, since once converged the corrections and therefore the summation will be zero. The numerical solution using the SIMPLE algorithm proceeds as follows:

1. Guess the pressure field P^*.
2. Solve the momentum equations to get the velocity components u^*, v^*, and w^*.
3. Solve Equation 16.52 to get pressure correction values.
4. Get the new velocities from Equation 16.51.
5. Calculate the other scalar values, like temperature, turbulence kinetic energy.
6. Get the new pressure values from Equation 16.44.
7. Check convergence, if not converged go to step 2.

The SIMPLE algorithm and its variations are the primary method currently employed in commercial CFD packages for handling the velocity–pressure coupling.

16.5.3 SIMPLEC Scheme

In the SIMPLE-Consistent (SIMPLEC) algorithm, the summation on the RHS of Equation 16.49 is not neglected, as is done in the SIMPLE approach, rather the summation is approximated by

$$\sum A_{nb} u'_{nb} \approx u'_P \sum A_{nb} \tag{16.53}$$

resulting in

$$u'_P = \frac{C_p}{A_p - \sum A_{nb}} \left(P'_W - P'_P \right) \tag{16.54}$$

The solution procedure is similar to that outlined for the SIMPLE algorithm; however, the solutions typically convergence faster.

16.5.4 SIMPLER Scheme

The SIMPLE algorithm results in an equation for pressure correction but not the pressure field directly. Neglecting the summation on the RHS of Equation 16.47 also results in overestimation of the pressure correction term, and therefore it becomes necessary to adjust the corrections using a factor <1, which as discussed later is called the under-relaxation factor.

The convergence can be improved by also correcting the pressure using an equation that directly solves for pressure. The SIMPLE Revised (SIMPLER) contains a separate equation to correct pressure at every iteration. Again by using the continuity and momentum equations, we can arrive at an equation for P in the form of

$$A_p P_P = \sum_{nb} A_{nb} P_{nb} + S \tag{16.55}$$

The approach is then

1. Guess the velocity field.
2. Solve the pressure equation (Equation 16.55) and obtain the pressure.
3. Solve momentum equations using the initial velocity field and just-computed pressure to find u^* and v^*.
4. Solve the pressure correction equation to determine P'.
5. Use P' to correct the velocities but not pressure.
6. Check for convergence. If not converged, go to step 2.

Unlike SIMPLE that requires good guesses for pressure, SIMPLER arrives at a reasonable estimate for pressure from the guessed velocities and the pressure equation. Since it solves an additional equation, it is computationally more expensive compared to SIMPLE; however, the solutions typically converge faster, requiring fewer iterations.

16.5.5 PISO Scheme

Pressure-implicit with splitting of operators (PISO) is also another variation of SIMPLE algorithm. This scheme is particularly useful for unsteady problems using larger time steps, although it can be used for steady flow problems as well. For steady-state problems, PISO with neighbor correction does not provide any noticeable advantage over SIMPLE or SIMPLEC with optimal under-relaxation factors [2]. The two main features of PISO are that no under-relaxation is used, i.e., the under-relaxation is set to one, and that the pressure correction and therefore velocities are corrected a set number of times during each iteration.

PISO with skewness correction is recommended for both steady-state and transient calculations on meshes with a high degree of distortion. If you use just the PISO skewness correction for highly distorted meshes (without neighbor correction), set the under-relaxation factors for momentum and pressure so that they sum to 1 (e.g., 0.3 for pressure and 0.7 for momentum) [2].

16.6 Under-Relaxation

The discretization of the transport equations results in algebraic equations in the form of Equation 16.42 for different components of velocity, temperature, pressure, species, etc. These equations are coupled and some are nonlinear and therefore are typically solved iteratively. To obtain the solution iteratively, guessed values are used to evaluate the coefficients and source terms, to be able to arrive at the new solution. The value to be used to start the next iteration, ϕ, is generally obtained from

$$\phi = \phi_{old} + \alpha(\phi_{new} - \phi_{old}) \tag{16.56}$$

where
ϕ_{old} is the value that started the previous iteration and provided the latest solution ϕ_{new}
$0 < \alpha < 1$ is the under-relaxation factor

Note that $\alpha = 1$ only uses the latest value calculated, and $\alpha = 0$ will not advance the solution.

The choice of under-relaxation factor used for the different variables is problem-dependent and more of an art than science, which improves with experience. A good practice is to start with small values of the under-relaxation factor initially and increase to values close to one, as the solution approaches the converged values. Patankar [5] and [14] suggested using an under-relaxation value of 0.5 for velocity and 0.8 for pressure, when using SIMPLE algorithm. Different CFD software provide different default under-relaxation factors that generally serve as good initial values to start the solution with.

16.7 Boundary Conditions

All fluid mechanics and heat transfer problems are governed by the same laws of nature that are typically expressed by partial differential equations. The solution of a set of partial differential equations requires boundary conditions for closure. What differentiate the different solutions are primarily the geometry and the boundary conditions. Determining the proper boundary conditions is an important step in numerical solution.

The number of boundary conditions for a variable in each direction is equal to the highest order derivative of the variable appearing in the equation in that direction. The NS equations, energy, species, etc., are all second order in the spatial directions and first order in time; therefore, two boundary conditions are generally needed in each direction for each variable, and one initial condition, when dealing with transient problems.

The boundary conditions are conditions that are imposed at the boundaries. The boundaries separate the domain, over which the solution is to be obtained, from the surroundings. The boundaries are defined by specific values of the independent variables, and therefore boundary conditions are conditions imposed on the dependent variables or their derivatives at specific values of the independent variables. In Cartesian coordinates, the boundary conditions can often be expressed as

$$A(x,y)\phi + B(x,y)\frac{\partial \phi}{\partial z} = C(x,y) \tag{16.57}$$

Equation 16.57 applies to a particular set of x and y that define the particular boundary. The boundary condition is referred to as Dirichlet type for $B = 0$, Neumann type for $A = 0$, and mixed type otherwise. Depending on what ϕ represents, the boundary conditions are arrived at by common sense, or application of conservation laws.

An important point to remember is that the computational domain is often different than the physical domain of interest. For example, although we may only be interested in developing flow in the entrance region of a pipe, the computational domain may extend well into the fully developed region to be able to properly specify the boundary conditions.

The boundary conditions vary greatly depending on the problem and the variable. In what follows, some of the more commonly encountered boundary conditions are reviewed.

16.7.1 Wall

When a viscous fluid is in contact with a solid boundary, it assumes the velocity of the solid. If the boundary is stationary, then all three components of velocity are zero, $u = v = w = 0$, which is known as the no-slip condition in viscous flows. If the wall is moving or rotating, the normal component of velocity is still zero, but the tangential component of fluid velocity is equal to the wall velocity. For turbulent flows, the surface roughness may also have to be specified. It is also possible to specify the shear stress at the boundary, which can be used, when two fluids are in contact or to approximate a free surface.

For the energy equation, the wall boundary conditions include

$$\text{Isothermal wall} \qquad T = \boldsymbol{T_w} \qquad\qquad (16.58)$$

$$\text{Uniform heat flux} \qquad -k\frac{\partial T}{\partial n} = \boldsymbol{q''} \qquad\qquad (16.59)$$

$$\text{Convection} \qquad -k\frac{\partial T}{\partial n} = \boldsymbol{h}(T_w - \boldsymbol{T_\infty}) \qquad\qquad (16.60)$$

$$\text{External radiation} \qquad -k\frac{\partial T}{\partial n} = \boldsymbol{\varepsilon}\sigma\left(T_w^4 - \boldsymbol{T_\infty^4}\right) \qquad\qquad (16.61)$$

Depending on the boundary condition imposed, the bold face variables in the earlier equations must be supplied.

16.7.2 Mass Flow Inlet

A simple inlet boundary condition is the specified mass flow rate. For incompressible flows, the velocity inlet boundary conditions will fix the mass flow, and therefore velocity inlet is more appropriate. When the flow rate is specified, the pressure at the inlet is not known, and is adjusted in response to the internal conditions. This results in slower convergence, and therefore, if the specified mass flow and specified pressure at the inlet are both acceptable as boundary conditions, the inlet pressure condition results in faster convergence [2].

16.7.3 Velocity Inlet

In many problems, the velocity is known at a given boundary. The examples include velocity at the inlet of a pipe or leading edge of a flat plate. This type of boundary condition is known as velocity inlet, where velocity can be specified by magnitude normal to the boundary, magnitude of the velocity components, or magnitude and the direction of velocity. Generally, the velocity profile is assumed uniform, although a nonuniform velocity profile can also be specified. When velocity is specified at a boundary, the static pressure is adjusted to accommodate the specified velocity distribution. This boundary condition is used with incompressible flows and can result in physically unrealistic results, if used for compressible flows. The boundary condition can also be used at an outlet by specifying negative velocity, but caution must be taken that the overall mass flow rate is conserved when multiple velocity inlets are used. When this boundary condition is used, there needs to be at least one outflow or pressure outlet boundary condition. Also, if the flow at the boundary is turbulent, the turbulence kinetic energy and dissipation rate must also be specified.

16.7.4 Outflow

Specifying boundary conditions downstream are usually a challenge, since the flow details or pressure is not known. The typical boundary condition used is to assume the diffusion fluxes normal to the boundary to be zero, i.e., $\dfrac{\partial^2 \phi}{\partial n^2} = 0$ for all variables except pressure. This boundary condition allows gradients in the cross-stream direction at the outflow boundary, and only the diffusion fluxes in the direction normal to the outflow boundary are assumed to be zero. This boundary condition is appropriate when the flow has settled and changes in the normal direction are either small or zero, like fully developed duct flow. Even when the flow is not fully developed, the outflow boundary condition can be used if the assumption of a zero diffusion flux is expected to minimally impact the solution. The outflow boundary condition cannot be used if the problem includes pressure inlet boundaries, is compressible, or is unsteady with varying density.

16.7.5 Pressure Inlet and Outlet

Pressure inlet boundary condition is used in cases when the pressure is known at the boundary but not the flow rate or velocity. This boundary condition arises at the inlet of a duct in which a fluid is drawn by a pump or a fan or in buoyancy-driven flows. Pressure inlet boundary conditions can also be used to define a "free" boundary in an external or unconfined flow.

The pressure outlet boundary condition is used when the pressure at the boundary is known like when the fluid is exhausted into an environment with known pressure. Generally, the pressure is assumed to be constant over the outlet. This boundary condition allows for backflow into the domain. The pressure inlet requires a pressure outlet boundary condition.

16.7.6 Symmetry and Axis Boundaries

When there is a line or plane, where the solution for a variable on one side is the mirror image of the solution on the other side, then that line or plane constitutes a symmetry boundary, and thus only the solution on one side needs to be obtained. Taking advantage of symmetry can greatly reduce the computational effort. Mathematically, the gradients of variables normal to a line of symmetry are zero. Also, velocity components normal to a line of symmetry are zero.

The axis boundary condition takes advantage of symmetry that exists in 2D axisymmetric flows.

16.7.7 Periodic Boundary Condition

The periodic boundary condition refers to situations where there is a repeating pattern in field variables, over fixed intervals L, called periodic length. For example, if all components of velocity satisfy the following

$$u(x,y,z) = u(x+L,y,z) \tag{16.62}$$

then the periodic boundary condition can be used for component u in the x direction. It also follows that the pressure difference is constant over the periodic length L for translationally periodic flows.

16.8 Grid Independence

The solutions obtained numerically must be independent of the grid selected. Generally, finer grids provide more accurate results, but are usually less stable and are computationally more expensive. It is therefore critical to obtain the solution using more than one grid, to ensure the results are independent of the grid. As a general rule, you should start with a relatively coarse mesh to quickly verify the geometry, the boundary condition, and other important features of the problem, and once the solution for this grid is obtained, or if the solution does not converge, to progressively refine the mesh, changes in the variables of interest are below a small threshold.

16.9 Tutorial 1: Laminar Boundary Layer Flow over a Flat Plate

This tutorial walks the user through a step-by-step process of setting up and solving a problem using Fluent, analyzing the results and taking corrective measures, if needed. In this tutorial, you will solve the problem of laminar boundary layer flow over a sharp-edged flat plate and compare the results with the classical similarity solution for boundary layer flow. The specific objectives of this tutorial are to

- Become familiar with obtaining CFD solution using Fluent.
- Compute the flow field.
- Compare the fluid velocity, skin friction coefficient, and thickness of the boundary layer with the Blasius similarity solution to assess accuracy.
- Take corrective measures to improve numerical solution.

16.9.1 Problem Statement

Consider the laminar flow of air over a flat plate of length L, as shown in Figure 16.8. The fluid approaches the plate with a constant free stream velocity, U_∞, which remains constant far from the plate. In this tutorial, the plate length $L = 1$ m, and the air properties are taken the same as the default values in Fluent, namely, density $\rho = 1.225$ kg/m³ and viscosity $\mu = 1.7894 \times 10^{-5}$ Pa s. The solution is obtained for the free stream velocities of $U_\infty = 1$ m/s, resulting in Reynolds number based on the length $Re = U_\infty L/\nu = (1.225 \times 1 \times 1.0)/(1.7894 \times 10^{-5}) = 68{,}459$, which is less than the critical Reynolds number $Re_c = 5 \times 10^5$ and thus the flow remains laminar.

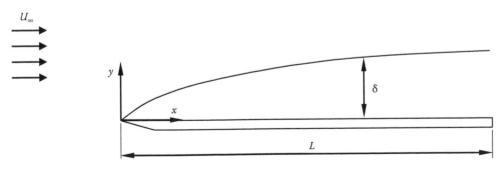

FIGURE 16.8
Laminar flow of air over a flat plate.

16.9.2 Mathematical Formulation

The continuity and NS equations for a 2D incompressible flow are

$$\text{Continuity} \qquad \frac{\partial u}{\partial x} + \frac{\partial v}{\partial y} = 0 \tag{16.63}$$

$$x \text{ momentum} \quad \rho\left(u\frac{\partial u}{\partial x} + v\frac{\partial u}{\partial y}\right) = -\frac{\partial P}{\partial x} + \mu\left(\frac{\partial^2 u}{\partial x^2} + \frac{\partial^2 u}{\partial y^2}\right) \tag{16.64}$$

$$y \text{ momentum} \quad \rho\left(u\frac{\partial v}{\partial x} + v\frac{\partial v}{\partial y}\right) = -\frac{\partial P}{\partial y} + \mu\left(\frac{\partial^2 v}{\partial x^2} + \frac{\partial^2 v}{\partial y^2}\right) \tag{16.65}$$

The solution of these equations provides the x and y components of the velocity in the flow field above the plate. As it is also known, the u and v change over a small distance close to the plate, which is called the boundary layer thickness. Outside of the boundary layer, the velocity field is unaffected by the presence of the plate and is equal to the free stream value. Therefore, one is only interested in obtaining the solution in the boundary layer region. The boundary conditions are

- On the flat plate, $y = 0$: $u = 0$, $v = 0$
- Free stream condition, $y \rightarrow \infty$: $u = U_\infty$, $v = 0$
- Upstream, $u = U_\infty$, $v = 0$
- Downstream, $\frac{\partial u}{\partial x} = 0$, $\frac{\partial v}{\partial x} = 0$

The nondimensional forms of the conservation equations are

$$\text{Continuity} \qquad \frac{\partial u^*}{\partial x^*} + \frac{\partial v^*}{\partial y^*} = 0 \tag{16.66}$$

$$x \text{ momentum} \quad u^*\frac{\partial u^*}{\partial x^*} + v^*\frac{\partial u^*}{\partial y^*} = -\frac{\partial P}{\partial x^*} + \frac{1}{Re}\left(\frac{\partial^2 u^*}{\partial x^{*2}} + \frac{\partial^2 u^*}{\partial y^{*2}}\right) \tag{16.67}$$

$$y \text{ momentum} \quad u^*\frac{\partial v^*}{\partial x^*} + v^*\frac{\partial v^*}{\partial y^*} = -\frac{\partial P^*}{\partial y^*} + \frac{1}{Re}\left(\frac{\partial^2 v^*}{\partial x^{*2}} + \frac{\partial^2 v^*}{\partial y^{*2}}\right) \tag{16.68}$$

We will solve this problem in the nondimensional form using Fluent even though solves the dimensional form of the equations. Comparing Equations 16.63 through 16.65 with Equations 16.66 through 16.68, it can be seen that the two sets become identical, if density of the fluid is one and the viscosity is equal to the inverse of Reynolds number. Caution must be exercised in using this approach, particularly when other effects such as buoyancy or turbulence are included.

16.9.3 Geometry and Mesh Generation

The computational domain is shown in Figure 16.9. The solution to boundary layer flow over a flat plate is arrived at by the boundary layer approximation that reduces to Blasius similarity solution presented in Chapter 7. From the Blasius solution, the boundary layer thickness is

$$\delta_x = \frac{5x}{\sqrt{Re_x}} \tag{16.69}$$

For the given Reynolds number, the dimensionless thickness of the boundary layer at the end of the plate is

$$\delta_L^* = \frac{5}{\sqrt{Re_L}} = \frac{5}{\sqrt{68,459}} = 0.0191 \tag{16.70}$$

or the dimensionless thickness of the boundary layer is around 0.02. Although the velocity is expected to change over a maximum distance of 0.02, we extend the computational domain to about 0.1, or about five times the thickness of the boundary layer at the end of the plate. Therefore, the computational domain is 1×0.2.

There are a number of ways to make the geometry and mesh. GAMBIT was the original preprocessor for Fluent and can still be used, as can other software such as Pro/E or Hypermesh, etc. In this tutorial, the simulation is performed in ANSYS Workbench environment which is an integrated collection of different engineering design and simulation software. For example, the geometry can be generated using DesignModeler, which is a solid modeler, the mesh is generated using ANSYS Meshing, and the solution can be obtained and analyzed using Fluent. It also allows standard files to be imported into the environment for further analysis.

To organize many files that will be generated during the different stages of analysis, it is best to create a directory where all the work will be stored. For example, create a folder on desktop naming it BLflow.

To start ANSYS Workbench

Start > All Programs > Ansys 13.0 > Workbench

Depending on the installation, you will eventually get to a window similar to Figure 16.10 but with only two panels. The left panel is the Toolbox, listing the available analysis tools. The panel on the right is called Project Schematic, which is the work area where geometry and mesh are created and simulations and analyses are performed. Also, in response to actions performed, the informational messages are displayed at the bottom of the window.

Initially, the right panel is blank. To start a Fluent analysis, drag the Fluid Flow (FLUENT) icon from the Toolbox menu into the right panel, Project Schematic area. A new window

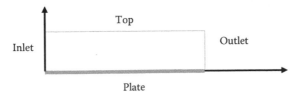

FIGURE 16.9
Computational domain and boundary condition types assigned in mesher.

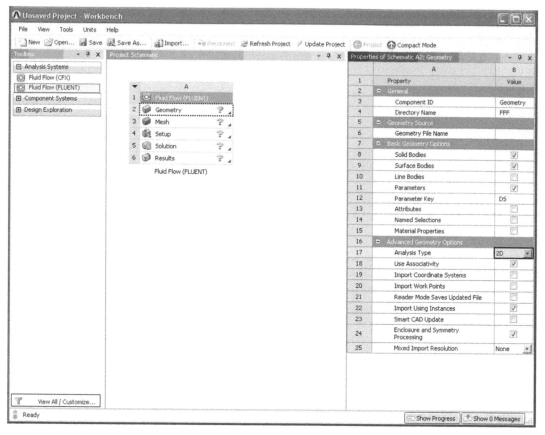

FIGURE 16.10
Initial fluent window.

Fluid Flow (Fluent) appears, listing the steps that you will need to go through. This window is also shown in Figure 16.10.

The geometry is generated using DesignModeler, which is a feature-based parametric solid modeler, using industry CAD standards to create the entire geometry or modify a model, prepared by other standard CAD programs, for analysis by Fluent. It uses 2D sketches for 2D problems and generates volumes by extruding, revolving, sweeping, etc., of 2D sketches for 3D analysis.

Right click on Geometry and select Properties. The properties menu will appear to the right of the Workbench window, shown in Figure 16.10. We need to specify that the geometry to be created will be 2D. In the right panel, under Advance Geometry Options, change the Analysis Type (item 17) to 2D as shown.

At this step, save the project in the directory BLflow. This will create file BLflow.wbpj and a folder called BLflow_files in the working directory BLflow, both are needed in order to edit or to continue the project.

16.9.3.1 Creating the Geometry

In the Fluid Flow panel, double click on Geometry to invoke the DesignModeler to create the model. A window with two panels will appear. The left panel is the Tree Outline and the right panel is the Graphics with a blue background, as also shown in Figure 16.11a.

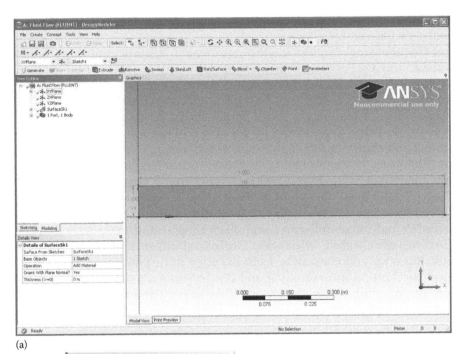

(a)

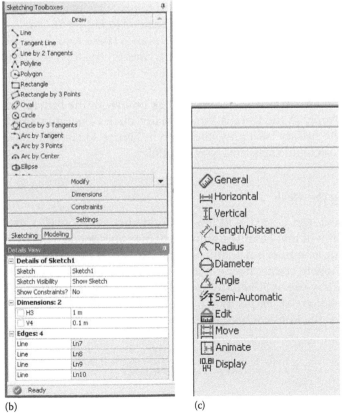

(b)   (c)

FIGURE 16.11

Geometry creation: (a) graphics panel; (b) sketching toolboxes; (c) dimension tab.

Another window will also appear to select the length unit to be used. Use the default unit (meter) and click OK.

The Tree Outline panel displays the features created. It starts with three default datum planes XY, XZ, and YZ. For this 2D problem, the physical model is on the XY plane; therefore, the computational domain will be created on the XY plane. In the Tree Outline, select XY plane, and then click on the Z axis on the bottom right corner of the Graphics window. This will rotate the XY plane and make it parallel to the computer screen.

Switch from Modeling to sketching by clicking on Sketching tab, which is below Tree Outline panel and above the Details View. This will change the Tree Outline panel to the Sketching Toolboxes. In the Sketching Toolboxes, select Rectangle. In the Graphics window, move the cursor to the origin and make sure that it snaps to the origin, click, and then move the cursor to another point on the positive XY quadrant and click again. At this point, a green rectangle will be displayed. If the dimensions are not shown, then click the Dimensions tab and then scroll down and select Display and make sure that both Name and Value boxes are checked. Then select Horizontal dimensioning and click on the two ends of the top edge and click away from the edge to display the dimension of the top edge. Select the Vertical dimensioning and do the same for the left edge. At this point, the dimensions are arbitrary. In the Details View under Dimensions, change the value of horizontal line to 1 and the vertical line to 0.1 and then click on Generate on the horizontal menu bar above the Graphics window.

Use Zoom fit button on the top menu to resize the image. If the dimension lines are not at the proper place, use Move under Dimension to properly place the dimension lines and again zoom to Fit and the geometry will look similar to Figure 16.11a.

So far, we have created four edges on the XY plane. These four lines are used to define a surface. To accomplish this, from the top menu, click on the top horizontal menu

Concept > Surface From Sketches

This will create a new surface SurfaceSK1. Click on any of the four lines of the rectangle, and select click Apply in the Details View. Finally, click Generate to generate the surface. This will make the surface turn gray, as shown in Figure 16.11a. The geometry is now created. Save the project and close the DesignModeler.

16.9.3.2 Creating the Mesh

Meshing is the next step. Go to the BLflow—Workbench window (Figure 16.10) and double click Mesh, which will invoke the meshing program and opens a new window.

The key to accurate numerical solution is the use of the proper mesh. Here, we provide some details on how to create the mesh and what considerations need to go into selecting the mesh. We start by use of 51 equally spaced nodes in the x direction (50 intervals) on the plate in the x direction. At the position of the second node along the plate ($x_2 = 0.02$) from Equation 16.69, the thickness of boundary layer is approximately

$$\delta_x = \frac{5x}{\sqrt{Re_x}} = \frac{5(0.02)}{\sqrt{68,459(0.02)}} = 0.0027 \tag{16.71}$$

We must select the number of nodes and the interval ratio in the y direction such that we will have a reasonable number of nodes inside the boundary layer at all values of x, including even at the second node. Choosing also 51 nodes in the y direction (50 intervals) with

TABLE 16.1

Meshing Parameters Used

Re	68,459
δ at x_2	0.0027
δ at L	0.0191
x direction	
L	1
d	1
N	51
x_2	0.02
Bias	1
y direction	
L	0.1
d	1.1
N	51
y_2	7.80E–05
Bias	117.4
Ny in BL at x_2	15.7
Ny in BL at L	34.0

an interval ratio of 1.1 results in a Bias of around 117, with 15 nodes inside the boundary layer at x_2 and 34 nodes at the end of the plate, as shown in Table 16.1.

Meshing generally falls under two broad categories of free or mapped. A free mesh is not restricted with regards to the type of elements or the pattern used. Mapped meshing places limits on both of these aspects. A mapped area will only have quadrilateral or triangular elements, and a mapped volume only hexahedron. For this particular problem, we will use Mapped Face Meshing. Click Mesh Control > Mapped Face Meshing, then click anywhere on the surface (rectangle), which would then become green. Then Apply under details of "Mapped Face Meshing" and the rectangle turns dark blue.

The edge meshing will be used to specify the mesh. From the pull down Mesh Control menu select Sizing, then Click Edge Selection Filter in the top menu (cube with the top left edge highlighted green). Then click the bottom edge of the rectangle, which should highlight green. Next, hit Apply in the Details of Sizing window to the right of Geometry. Set Type pull down menu to Number of Divisions, set Number of Divisions to 50, and Behavior to Hard. This last step will disable the Mesher from overwriting any of the edge sizing specifications. Repeat the process for the top edge starting by selecting Sizing, etc.

For the y directions, we will use 50 nonuniform divisions by using bias. Again start from Sizing and go through the steps followed for the horizontal edges, except set the bias so that the divisions get larger as we move away, by selecting the bias type. After one of the options is selected, the edge divisions will be shown and if needed it can be reversed by changing the bias type to make sure the intervals grow in the positive y direction on the left and right edges. Choose 117 for the Bias Factor. Select Update and Mesh in the Outline panel and the meshed face would appear as shown in Figure 16.12.

In order to assign boundary conditions in FLUENT, the boundary Edges need to be named. The names to be used are arbitrary and are typically selected for easy identification. The names to be used for this case are shown in Figure 16.9. To assign the name inlet to the left vertical edge, activate the Edge Selection Filter, then left click on the left edge of the rectangle and it should highlight green, then, right click the same edge and choose

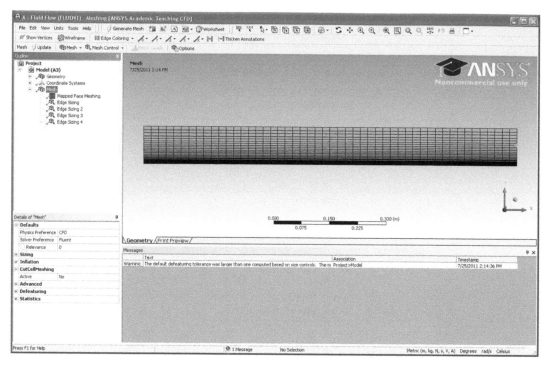

FIGURE 16.12
Computational mesh.

Create Named Selection. The Selection Name window will appear. Enter Inlet and OK. Repeat the same process for the other three edges. Meshing is now completed; Save Project.

16.9.4 Solution

The next step is to do the CFD analysis. We can run the Fluent analysis from within the Workbench, or create a file that can then be read by Fluent. To run from within the Workbench, close the Mesher window, and go to the BLflow—Workbench window and click the Update Project button on the top menu. A check mark will appear next to the Mesh (item 3), then double click on Setup.

Alternatively, to run Fluent independently, you first need to generate the input file Fluent needs by going to the meshing window and File > Export and Save as Fluent Input Files (*.msh) and for file name (*) use BLflow. Close Meshing then Start > All Programs > Ansys 13.0 > Fluent, then File > Read > BLflow.msh

In both cases, when the FLUENT Launcher appears, change the options to "Double Precision," and then click OK. In Double Precision option, each floating point number is represented using 64 bits as opposed to use 32 bits for single precision. The Double Precision mode increases the precision, but requires more memory.

Fluent will read in the geometry and the mesh and will display some information on the main screen. If the geometry and mesh are set up appropriately, it should give no errors, and the word "Done" should appear. It is also a good idea to check the grid by clicking Check. Some information about the mesh will be displayed, including a warning about high aspect ratios, followed by Done, as shown in Figure 16.13. You should see no errors in the Command pane. Note if you have trouble reading your grid, you probably made a mistake somewhere. Go back and try to either modify your grid or remake it following the module earlier.

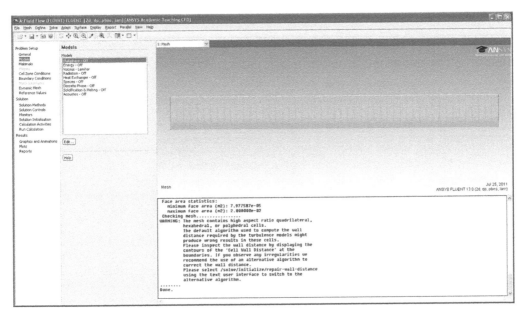

FIGURE 16.13
Launching Fluent.

16.9.4.1 Setup

We will only solve for the velocity distribution, so the Model should show Viscous–Laminar with all others off. Select Materials in Define/Materials and observe the property values. Since we are solving the problem in the nondimensional plane, we name the fluid "fictitious," having density of 1 whose viscosity is the inverse of Reynolds number being considered which in this case becomes 1.7607e–5, as shown in Figure 16.14. Click Change/Create and then say Yes then Close.

Next step is to define the boundary conditions for the four named boundaries. The inlet will be specified first. Boundary Conditions > inlet. Note that the Boundary Condition Type

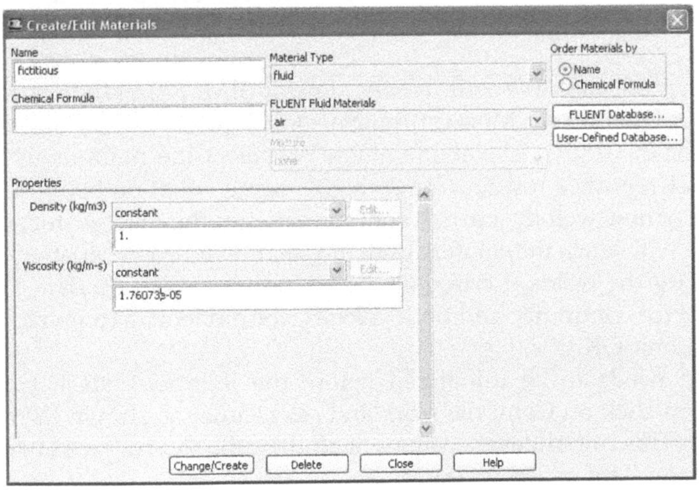

FIGURE 16.14
Defining material properties.

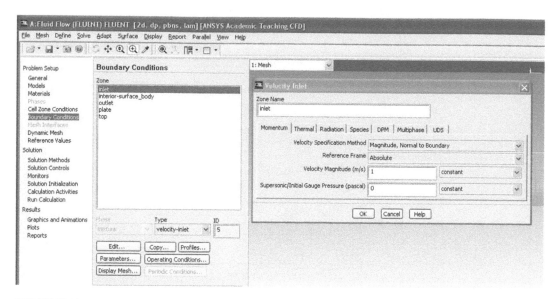

FIGURE 16.15
Setting boundary conditions.

should have been automatically set to velocity inlet. FLUENT guesses boundary conditions based on the label of the named selections. Now, the velocity at the inlet will be specified. Click Edit, etc., and in the Velocity Inlet menu set the Velocity Specification Magnitude, Normal to Boundary, and set the Velocity Magnitude (m/s) to 1 m/s, as shown in Figure 16.15.

Next select outlet and note that the panel is automatically set to pressure outlet. For the outlet boundary condition, the gauge pressure needs to be set to zero. The default gauge pressure is zero, thus no changes need to be made. OK.

Next select Plate and the Type should have been automatically set to wall. If the Boundary Condition Type is not set to wall, then set it to wall. We will use the default setting for the wall boundary condition, thus no changes are needed.

Then Select top. Change the type to wall and then select Specified shear and set both component to zero. At this point, save your work in the FLUENT Window by clicking the save button.

Click on Solution Methods. We will use the default SIMPLE, Standard for Pressure, but use Second-order Upwind for Momentum. Next Click on Solution Control and retain the default values for the under-relaxation values. We select the parameters for the solution method. FLUENT reports a residual for each governing equation being solved. The residual is a measure of how well the current solution satisfies the discrete form of each governing equation. We will iterate the solution until the residual for each equation falls below 10^{-12}. In order to specify the residual criteria (Click) Monitors > Residuals > Edit, etc., enter a value of 1×10^{-12} for continuity and both velocity components to ensure full convergence during the iterations. OK.

The flow field needs to be initialized before the iterations start. Click on Solution Initialization then click on Compute from and select inlet as shown later. Then click the Initialize button. This initializes the values of all the cells to 1 m/s and 0 for x and y velocity components and 0 for gauge pressure.

The maximum number of iterations is set next by clicking on Run Calculation and entering 1000 for the Number of Iterations.

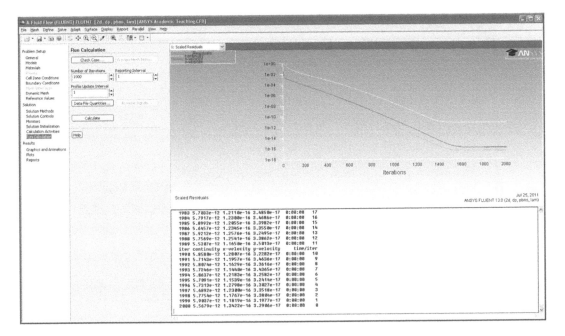

FIGURE 16.16
Solution iteration and convergence.

Save the project now, and click on Calculate in order to run the calculation. The residuals for the iterations are printed and plotted on the screen. After running the calculation for another 1000 iterations, you should obtain the residual plot as shown in Figure 16.16. At this point, save the project once again.

16.9.5 Results

In this section, we examine the x component of the velocity profiles across the boundary layer at two x locations along the flat plate, namely at $x = \mathbf{0.5\ m}$ and $x = \mathbf{1.0\ m}$, so a line needs to be defined within Fluent at $x = 0.5$. In the main *Fluent* window, Surface > Line/ Rake. Type in the desired starting and ending x and y locations of the vertical line, i.e., a vertical line going from (0.5,0) to (0.5,0.1). The *New Surface Name* should be assigned at this point. It is suggested that this line be called "profile0.50" or something descriptive of its intended purpose. Click on Create to create the line, and then Close.

To view this newly created line, return to the main *Fluent* window, and Display > Mesh > Display. Unselect (by left mouse click) the default interior, and select the newly created lines instead. Display. The line should be visible at the appropriate location. If not, create them again more carefully. In order to have an xy plot of variables, select XY Plot Set Up. Set the parameters as shown in Figure 16.17, and the x component of velocity along two lines will be plotted as shown in Figure 16.18.

16.9.6 Velocity Overshoot

As seen in Figure 16.18, the velocity that is expected to approach asymptotically to the free stream velocity actually goes above 1.0 and there is a velocity overshoot. This behavior makes no physical sense and has not been observed experimentally.

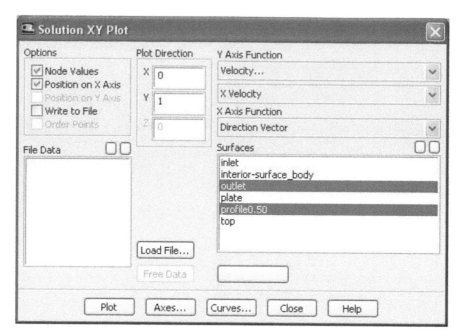

FIGURE 16.17

Setting up the plot of the x component of velocity in the y direction in the middle and end of the plate.

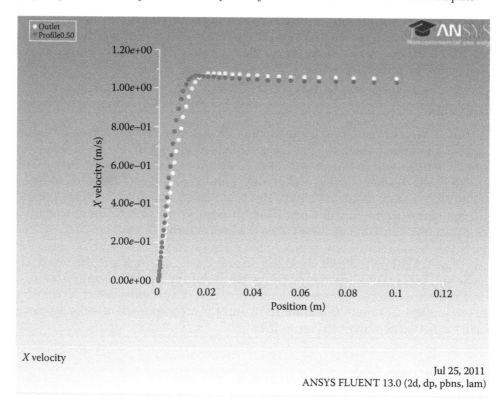

FIGURE 16.18

The xy plot of the x component of velocity along two vertical lines.

As can be seen in the plot, there is about 7% overshoot in velocity along $x = 1.0$ line which corresponds to the end of the plate. The reason for the overshoot is that mass must be conserved over the computational domain. For the free stream, we used the zero shear boundary condition. We could have alternatively used the symmetry condition. For both choices, the vertical component of velocity on the top boundary will be set to zero, or no mass leaves the computational domain from the top. Therefore, the mass flow rate at the left vertical edge must be equal to the mass flow rate at each vertical plane, including at the right boundary. If h is the height of the computational domain, then the conservation of mass requires that the mass flow at the left edge be the same as the mass flow at any x location at which the boundary layer thickness is δ

$$\rho U_\infty h = \int_0^\delta \rho u \, dy + \int_\delta^h \rho u \, dy \tag{16.72}$$

solving for h,

$$h = \int_0^\delta \frac{u}{U_\infty} \, dy + \int_\delta^h \frac{u}{U_\infty} \, dy \tag{16.73}$$

The first integral can be expressed in terms of the similarity variables

$$h = \frac{1}{\sqrt{\dfrac{U_\infty}{vx}}} \int_0^\delta f' \sqrt{\frac{U_\infty}{vx}} \, dy + \int_\delta^h \frac{u}{U_\infty} \, dy \tag{16.74}$$

and simplifies to

$$h = \sqrt{\frac{vx}{U_\infty}} \int_0^5 f' d\eta + \int_\delta^h \frac{u}{U_\infty} \, dy \tag{16.75}$$

Note that the upper limit of the first integral becomes 5 since $\eta = 5$ corresponds to the thickness of the boundary layer. The first integral can be easily evaluated, and assuming the velocity outside of the boundary layer is constant, then

$$h = \sqrt{\frac{vx}{U_\infty}} f(5) + \frac{u}{U_\infty} (h - \delta) \tag{16.76}$$

From the Blasius solution

$$\delta = 5 \sqrt{\frac{vx}{U_\infty}} \tag{16.77}$$

$$h = \delta \frac{f(5)}{5} + \frac{u}{U_\infty} (h - \delta) \tag{16.78}$$

Solving for the dimensionless velocity outside of the boundary layer results in

$$\frac{u}{U_\infty} = \frac{h - \delta \dfrac{f(5)}{5}}{h - \delta} \tag{16.79}$$

The velocity overshoot is then how much this quantity exceeds 1, or

$$\text{Overshoot} = \frac{u}{U_\infty} - 1 = \frac{h - \delta \dfrac{f(5)}{5}}{h - \delta} - 1 \tag{16.80}$$

For $Re = 68,459$, $h = 0.1$, $f(5) = 3.284$, and at $x = 0.5$, or in the middle of the plate, $\delta = 0.0135$, and the earlier analysis predicts an overshoot of about 5.36% at which it is close to the results observed. This analysis also shows that the velocity overshoot can be reduced by increasing h. For example if h is moved to 1, the maximum overshoot reduces to 0.4%, since the mass flow deficit is now distributed over a much larger area.

16.9.7 Modifications

Seeing the error in the obtained results, and having identified a possible explanation for the error, the necessary modifications can now be made. The new computational domain is shown in Figure 16.19. The geometry is generated as before. The plate AB is assigned the Wall boundary condition. This condition sets the velocity components on the plate equal to zero. The outlet boundary BC is selected as Pressure Outlet, which sets the velocity gradients in the x direction equal to zero. Boundary CD is taken as Wall with zero shear. The inlet boundary DE is considered Velocity Inlet with $u = U_\infty$ and $v = 0$. The boundary EA is selected as Symmetry line.

Using 100 nodes along EA, AB, ED, and BC edges with a bias of 100 results in the mesh shown in Figure 16.20. The rest of the solution process is the same as earlier.

16.9.7.1 Results

First, we examine the velocity field. Figure 16.21 shows the x component of velocity along several vertical lines at $x = 0.2, 0.4, 0.5, 0.6, 0.8$, and exit ($x = 1$). As seen in Figure 16.21, the velocity approaches asymptotically to the free stream velocity. The x–y coordinates of the plot given in Figure 16.21 can be written into a file by selecting the Write to File option in the Solution XY Plot panel, Figure 16.17. Then this file may be opened in the spreadsheet and analyzed.

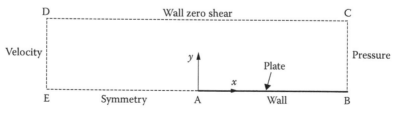

FIGURE 16.19
New computational domain and boundary condition types.

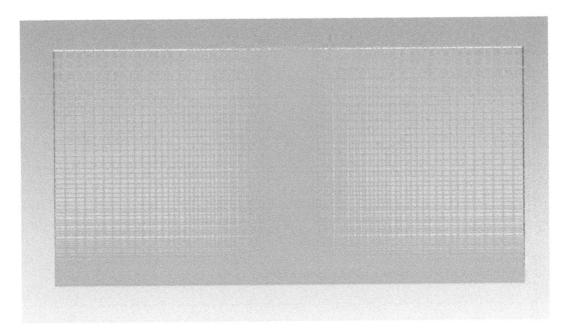

FIGURE 16.20
New mesh.

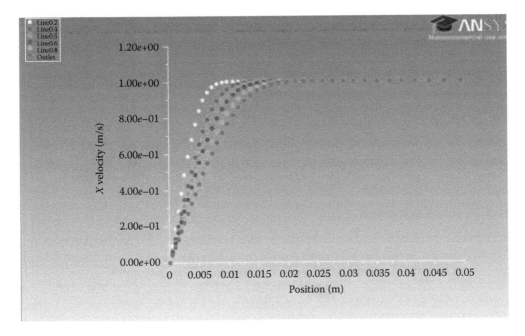

FIGURE 16.21
The *xy* plot of the *x* component of velocity on several vertical lines.

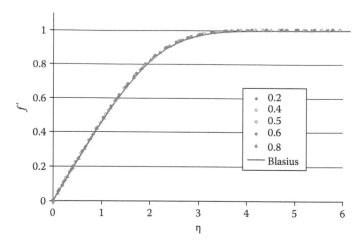

FIGURE 16.22
Velocity distribution and comparison with the Blasius similarity solution.

It is easy to show that the similarity variable for flow over flat plate is

$$\eta = y\sqrt{\frac{u_\infty}{2\nu x}} = \frac{y^*}{\sqrt{x^*}}\sqrt{\frac{1}{2}Re} \tag{16.81}$$

Therefore, the velocity at a given x^*, y^*, for a given Re, can be expressed in terms of the Blasius profile, and if the solution obtained is accurate, all of the data points in Figure 16.21 must collapse on the same curve and match the Blasius solution. Figure 16.22 is a plot of data in Figure 16.21, in terms of the similarity variable, as well as the Blasius solution, and shows that the modified mesh produces results (the symbols) that are in excellent agreement with the Blasius similarity solution.

16.10 Tutorial 2: Turbulent Pipe Flow*

The objective of this tutorial is also to go through a step-by-step process of setting up and solving turbulent flow in a circular pipe using Fluent, analyzing the results and comparing the numerical solution with the available analytical and experimental results.

16.10.1 Problem Statement

Consider the turbulent flow of air in a circular pipe of diameter 5 cm and length L, as shown in Figure 16.23. Air enters the pipe with a uniform inlet velocity U and temperature T_{in}, and the pipe wall is maintained at a constant temperature T_w. The air properties are taken to be the same as the default values in Fluent, namely, $\rho = 1.225$ kg/m³, $\mu = 1.7894 \times 10^{-5}$ kg/m s, $c_p = 1006.43$ J/kg, and $k = 0.0242$ W/m K, resulting in $Pr = 0.744$. The solution is obtained for Reynolds number of 20,000, corresponding to an inlet velocity $U = 5.84$ m/s. The $Re = 20,000$ is more than the critical Reynolds number of $Re_c = 2300$, thus the flow is turbulent.

---

* The author acknowledges Mr. Matthew West for preparing this tutorial.

FIGURE 16.23
Geometry for the turbulent flow in a circular pipe.

16.10.2 Mathematical Formulation

The continuity, momentum, and energy equations for a 2D incompressible turbulent flow in cylindrical coordinates are

Continuity
$$\frac{\partial \bar{u}}{\partial x} + \frac{1}{r}\frac{\partial(r\bar{v})}{\partial r} = 0 \tag{16.82}$$

x momentum $\bar{u}\dfrac{\partial \bar{u}}{\partial x} + \bar{v}\dfrac{\partial \bar{u}}{\partial r} = -\dfrac{1}{\rho}\dfrac{\partial \bar{P}}{\partial x} + \dfrac{\partial}{\partial x}\left[(\nu+\nu_t)\dfrac{\partial \bar{u}}{\partial x}\right] + \dfrac{1}{r}\dfrac{\partial}{\partial r}\left[r(\nu+\nu_t)\dfrac{\partial \bar{u}}{\partial r}\right]$ (16.83)

y momentum $\bar{u}\dfrac{\partial \bar{v}}{\partial x} + \bar{v}\dfrac{\partial \bar{v}}{\partial r} = -\dfrac{1}{\rho}\dfrac{\partial \bar{P}}{\partial r} + \dfrac{\partial}{\partial x}\left[(\nu+\nu_t)\dfrac{\partial \bar{v}}{\partial x}\right] + \dfrac{1}{r}\dfrac{\partial}{\partial r}\left[r(\nu+\nu_t)\dfrac{\partial \bar{v}}{\partial r}\right] - (\nu+\nu_t)\dfrac{\bar{v}}{r^2}$

$$\tag{16.84}$$

Energy $\bar{u}\dfrac{\partial \bar{T}}{\partial x} + \bar{v}\dfrac{\partial \bar{T}}{\partial r} = \dfrac{\partial}{\partial x}\left[\left(\dfrac{\nu}{Pr} + \dfrac{\nu_t}{Pr_t}\right)\dfrac{\partial \bar{T}}{\partial x}\right] + \dfrac{1}{r}\dfrac{\partial}{\partial r}\left[r\left(\dfrac{\nu}{Pr} + \dfrac{\nu_t}{Pr_t}\right)\dfrac{\partial \bar{T}}{\partial r}\right]$ (16.85)

The solution of these equations provides the x and r components of the velocity, pressure, and temperature in the flow field in the pipe. The axial and radial components of the velocity change axially over a distance close to the pipe inlet, which is called the entrance length. Beyond the entrance length, the r component of velocity disappears, and the velocity field becomes independent of axial position. The boundary conditions are

$$r = 0 \quad \frac{\partial \bar{u}}{\partial r} = \bar{v} = \frac{\partial \bar{T}}{\partial r} = 0 \tag{16.86}$$

$$r = R = 0.025, \quad \bar{u} = \bar{v} = 0, \quad \bar{T}_w = 600 \text{ K} \tag{16.87}$$

$$x = 0, \quad \bar{u} = 5.84, \quad \bar{v} = 0, \quad \bar{T}_{in} = 300 \text{ K} \tag{16.88}$$

$$x = L, \quad \frac{\partial \bar{u}}{\partial x} = \bar{v} = 0, \quad \frac{\partial^2 \bar{T}}{\partial x} = 0 \tag{16.89}$$

16.10.3 Geometry and Mesh Generation

For turbulent flow, the hydrodynamic and thermal entrance lengths become

$$x_{fd,h} = 4.4Re^{1/6}D_h = 1.15 \text{ m} \tag{16.90}$$

$$x_{fd,t} = 4.4Re^{1/6}PrD_h = 0.85 \text{ m} \tag{16.91}$$

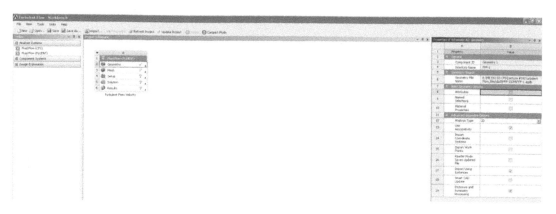

FIGURE 16.24
Geometry options inside of Workbench.

Therefore, the right boundary must extend beyond these values to ensure the fully developed condition is reached. The accuracy of these values will also be verified by the numerical solution. The problem is axisymmetric and the computational domain is taken to be a rectangle of 0.025 m ($D/2$) tall and 2.1 m long, with 0.1 m in front of the entrance allowing for the adjustment of the flow and avoiding the singularity on the surface of the pipe at the entrance. The remaining 2 m are in the pipe to ensure fully developed condition at the exit.

To start ANSYS Workbench:

1. Start > All Programs > Ansys 13.0 > Workbench.
2. Drag Fluid Flow (FLUENT) icon from the Toolbox menu into the right panel, Project Schematic area.
3. Right click on Geometry and select Properties. In the right panel, under Advance Geometry Options, change the Analysis Type (item 12) to 2D as shown in Figure 16.24.
4. Save the project in the directory Turbulent Flow. This will create file Turbulentflow. wbpj and a folder called Turbulentflow_files in the working directory.

16.10.3.1 Creating the Geometry

1. In the Fluid Flow panel, double click on Geometry. In the ANSYS Workbench window, select the default unit (meter) and click OK.
2. In the Tree Outline, select XY plane, then click on the Z axis on the bottom right corner of the Graphics window. This will rotate the XY plane. Make it parallel to the computer screen.
3. Switch from Modeling to sketching by clicking on Sketching tab. In the Sketching Toolboxes, select line.
4. In the Graphics window, move the cursor to the origin and make sure that it snaps to the origin, click, and then move the cursor to another point on the positive X axis and click again. Click, move up vertically, and click again, move back horizontally in the negative x direction and click once intersecting the Y axis. Continue moving horizontally to some negative X point and click. Move vertically down and click on the X axis, and move along the positive X axis and click on the origin again. At this point, you have created four horizontal and two vertical line segments that form a rectangle.

5. Make sure that the lines are either horizontal or vertical, by using Constraints.

6. If the dimensions are not shown, then click the Dimensions tab and then scroll down and select Display and make sure that both Name and Value boxes are checked.

7. Select Horizontal dimensioning and click on the two ends of the top left edge and click away from the edge to display the dimension of the top edge.

8. Do the same for the top right edge.

9. Select the Vertical dimensioning and do the same for the left edge.

10. At this point, the dimensions are arbitrary. In the Details View under Dimensions, as shown in Figure 16.25, change the value of top left horizontal line to 0.1, the top right horizontal line to 2.0, the vertical line to 0.025, and then click on Generate on the horizontal menu bar above the Graphics window.

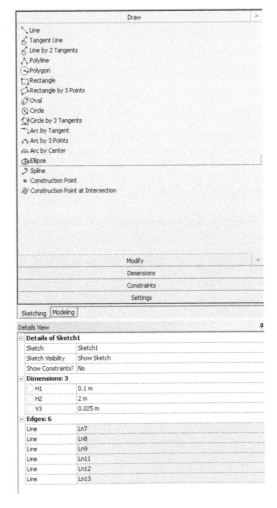

FIGURE 16.25
Dimensions of the sketched geometry.

11. Use Zoom fit button on the top menu to resize the image. If the dimension lines are not at the proper place, use Move under Dimension menu as seen in Figure 16.26 to place the dimension lines. The final geometry with dimensions is shown in Figure 16.27.

12. Click on the top horizontal menu Concept > Surface From Sketches. This will create a new surface SurfaceSK1. Click on any of the six lines of the rectangle, and select click Apply in the Details View. You should see a turquoise rectangle.

13. Finally, click Generate to generate the surface. This will make the surface turn gray, and the rectangle dark blue. The geometry is now created. Save the project and close the DesignModeler.

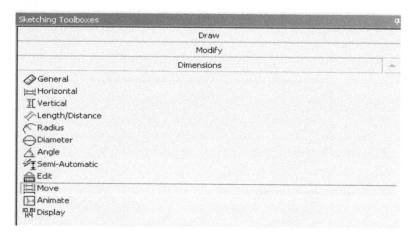

FIGURE 16.26
Dimension menu inside geometry editor.

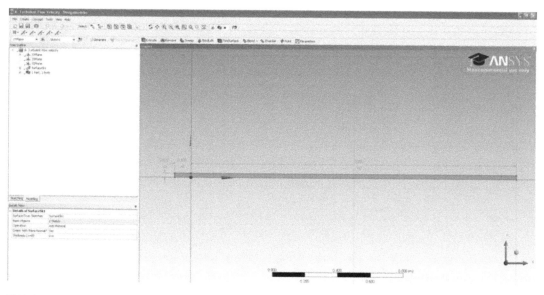

FIGURE 16.27
Final geometry with dimensions.

16.10.3.2 Creating the Mesh

The mesh details are shown in Table 16.2. In the pipe, 3300 uniformly spaced nodes are used in the x direction and 150 nodes with a bias of 10 are used in the r direction with higher density close to the wall. Outside of the pipe, 50 nodes using variable spacing, clustered close to the inlet are used in the x direction.

1. Go to the Turbulent Flow Workbench window, and double click Mesh which will invoke the meshing program and opens a new window.
2. Select Mesh in the Outline pane, click Mesh Control > Mapped Face Meshing, and then click anywhere on the surface (rectangle), which would then become green. Then Apply under details of "Mapped Face Meshing," and the rectangle turns dark blue.
3. The edge meshing will be used to specify the mesh. From the pull down Mesh Control menu select Sizing, and then Click Edge Selection Filter in the top menu (cube with the top left edge highlighted green). Then click the right edge of the rectangle, which should highlight green. Next, hit Apply in the Details of Edge Sizing window, which is below Outline window. In the same window, set Type pull down menu to Number of Divisions, set Number of Divisions to 150, and Behavior to Hard. This last step will disable the Mesher from overwriting any of the edge sizing specifications. Repeat the process for the left edge starting by selecting Sizing. Also, a bias of 10 needs to be added such that the nodes get closer and closer to the wall. After one of the options is selected, the edge divisions will be shown and if needed can be reversed by changing the bias type to make sure the intervals decrease in the positive r direction on the left and right edges.
4. For the x direction, use 3300 uniform divisions for the wall and axis section and 50 nodes with a bias of 5 for the section before the wall. Again, start from Sizing and go through the steps followed for the vertical edges and change the bias as needed, ensuring that the nodes are clustered near the wall. Select Update and Mesh in the Outline pane, and the meshed face will appear as shown in Figure 16.28.
5. The names assigned to be used are shown in Table 16.3. To assign the name inlet to the left vertical edge, activate the Edge Selection Filter up on the toolbar; then

TABLE 16.2

Mesh Specifications

Re	20,000
x direction	
Before pipe	
L	0.1
N	50
Bias	5
In the pipe	
L	2
N	3300
Bias	None
r direction	
R	0.025
N	150
Bias	10

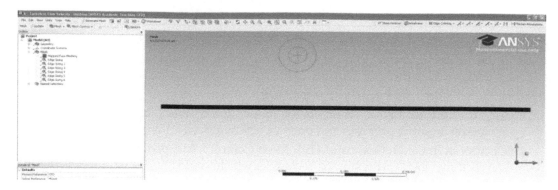

FIGURE 16.28
Turbulent flow mesh.

TABLE 16.3

Naming Guide

Name	Location
Inlet	Left vertical line
Symmetry	Top left horizontal line
Wall	Top right horizontal line
Outlet	Right vertical line
Axis	Bottom right horizontal line
Axis_inlet	Bottom left horizontal line

left click on the left edge of the rectangle, and it should highlight green. Then right click the same edge and choose Create Named Selection. The Selection Name window will appear. Enter Inlet and OK. Repeat the same process for the other five edges using Table 16.3 as a guide. Meshing is now completed; Save Project.

16.10.4 Solution

1. Generate the input file Fluent needs by going to the meshing window and File > Export and Save as Fluent Input Files (*.msh) and for file name (*) use Turbulent Flow. Close Meshing.

2. Start > All Programs > Ansys 13.0 > Fluent

3. FLUENT Launcher appears as shown in Figure 16.29. Change the options to "Double Precision" 2D, and parallel processing, if your computer is multiprocessor and enter the specific number of processors it has and then click OK.

4. Click File > Read > Turbulent Flow.msh

5. Fluent will read in the geometry and the mesh, and the word "Done" should appear. Note that two warnings will appear on the screen because of the two sections that were named axis when defining the mesh in the earlier steps.

6. Select the Axisymmetric 2D Space condition under the Solver section of the General Tab as shown in Figure 16.30.

7. Check the grid by clicking Check. Some information about the mesh will be displayed, followed by Done, as shown in Figure 16.30. You should see no errors in the Command pane.

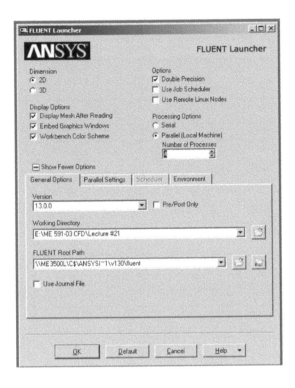

FIGURE 16.29
Fluent luncher.

16.10.4.1 Setup

1. We will solve for the velocity and temperature distributions, so the Model should show Viscous–Standard k–ε, Enhanced Wall Fn, and Energy On.

2. Select Materials in Define/Materials and ensure that air appears in the window. The default values used by Fluent can be seen by clicking on the Change/Create button. When finished examining the properties, close the window by clicking Close. The material properties for air showing should be the ones given earlier.

3. Define the boundary conditions by selecting Boundary Conditions > inlet. Note that the Boundary Condition Type should have been automatically set to velocity inlet. FLUENT guesses boundary condition to be set based on the label of the named selections. Now, the velocity at the inlet will be specified. Click Edit, etc. and in the Velocity Inlet menu set the Velocity Specification Method, Components, and set the Axial Velocity (m/s) to 5.482939, the Radial Velocity (m/s) to 0. Under Turbulence, change the Specification Method to Intensity and Hydraulic Diameter with a Turbulent Intensity of 10% and Hydraulic Diameter of 0.05 m as shown in Figure 16.31. Under Thermal tab, set the temperature equal to 300 K as shown in Figure 16.32.

4. Next select outlet and note that the panel is automatically set to outflow. For the outflow condition, the Flow Rate Weighting should be set to 1 and then click OK.

5. Next select Wall and the Boundary Condition Type should have been automatically set to wall, if set to wall. The default setting will be used for the wall boundary condition for the momentum. Set a constant temperature equal to 600 K using the thermal tab near the top of the window.

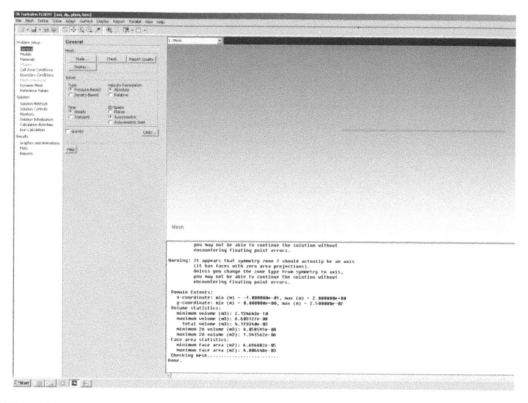

FIGURE 16.30
Mesh loaded into Fluent with no errors.

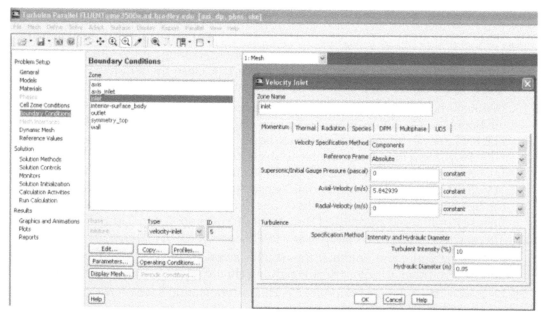

FIGURE 16.31
Inlet velocity boundary condition window.

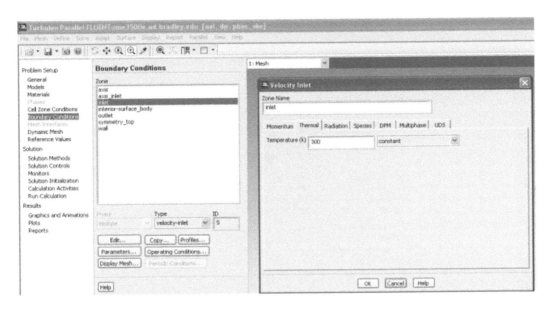

FIGURE 16.32
Inlet temperature boundary condition window.

6. Select both the axis and axis_inlet boundaries and specify both as axis boundary condition.

7. Save the work in the FLUENT Window by clicking the save button. Click on Solution Methods. Leave the default SIMPLE algorithm for the Pressure–Velocity Coupling and Standard for the Pressure. Change the Momentum, Turbulent Kinetic Energy, and Turbulent Dissipation Rate to Second Order Upwind, as shown in Figure 16.33.

8. Next Click on Solution Control, and retain the default values for the under-relaxation values.

9. Click Monitors > Residuals > Edit, etc., and enter a value of 1×10^{-10} for continuity, both velocity components, k, and ε while setting the energy equation residuals to 1×10^{-14} in the panel shown in Figure 16.34 to ensure full convergence. Click OK.

10. The flow field needs to be initialized before the iterations start. Click on Solution Initialization, and then click on Compute from and select Inlet. Then, click the Initialize button. This initializes the values of all the cells to 5.48 m/s and 0 for x and r velocity components and 0 for gauge pressure.

11. The maximum number of iterations is set next by clicking on Run Calculation and entering 1000 for the Number of Iterations.

12. Save the project now, and click on Calculate in order to run the calculation. The residuals for the iterations are printed and plotted on the screen. After running the calculation for another 8760 iterations, the residual plot as shown in Figure 16.35 is obtained. The solution has now converged to the specified convergence criterion. At this point, save the project once again.

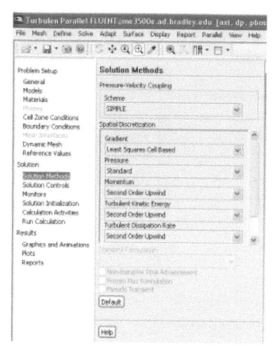

FIGURE 16.33
Solution methods for turbulent flow problem.

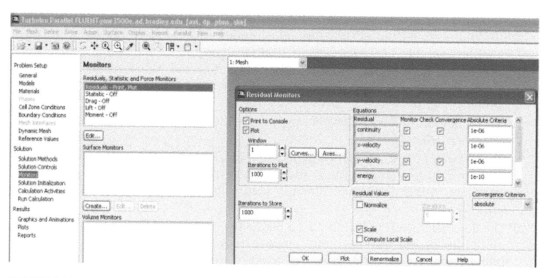

FIGURE 16.34
Residual window.

16.10.5 Results

1. In this section, the x component of the velocity profiles is examined in the developing and hydrodynamically fully developed regions. The velocity profiles will be examined at several cross sections starting at $x = 0.125$ m to $x = 1.00$ m with a spacing of 0.125 m and also lines at 1.1, 1.5, 1.8, and 1.85 m. In the main *Fluent*

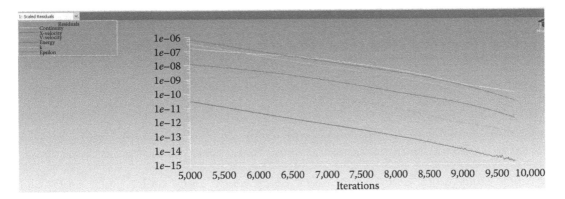

FIGURE 16.35
Residual plot at 9760 iterations.

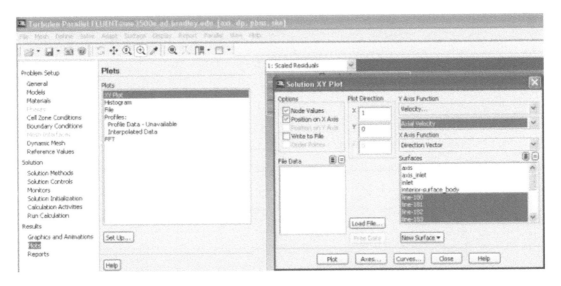

FIGURE 16.36
XY plot window.

window near the top, select Surface > Line/Rake. Type in the desired starting and ending x and y locations of the vertical line. For example, the first vertical line is drawn from (0.125,0) to (0.125,0.025). The New Surface Name assigned for this line is "line0.125." Click on Create to create the line, and repeat for the other lines. When finished, click Close.

2. To view the newly created lines, return to the main *Fluent* window, and click Display > Mesh > Display. Unselect (by left mouse click) the default interior, and select the newly created lines instead. Click Display. The lines should be visible at the appropriate location. If not, create the line again more carefully.

3. In order to have an xy plot of variables, select XY Plot Set Up. Set the parameters as shown in Figure 16.36, and the x component of velocity along the newly created lines will be plotted as shown in Figure 16.37. As seen in Figure 16.37, the velocity increases quickly and remains fairly flat at nearly 6.9 m/s.

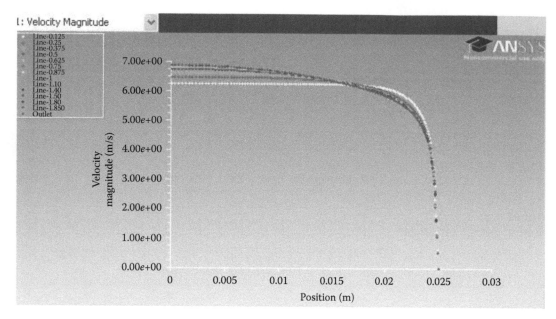

FIGURE 16.37
Velocity profiles at several *x* locations.

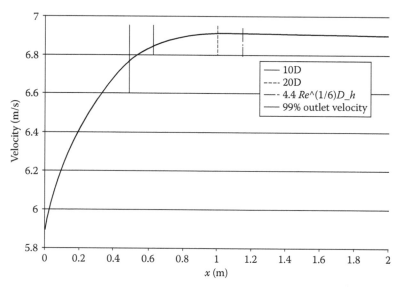

FIGURE 16.38
Centerline velocity along the pipe.

4. Figure 16.38 is a plot of the centerline velocity from pipe inlet. As seen, the velocity increases and asymptotes to a constant value which is about 15% more than the inlet or the mean velocity. A number of different criteria are given for the fully developed condition, and they all are also plotted in Figure 16.38 and can used to ascertain the efficacy of each criterion.

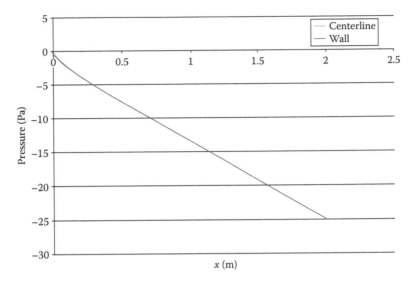

FIGURE 16.39
Pressure at centerline and wall.

5. Pressure at the wall and the centerline were outputted to a file and plotted in Figure 16.39. As can be seen, pressure drops linearly along the pipe soon after the pipe entrance. It also shows that the pressure variation across the pipe is negligible.

6. The friction factor can be calculated from

$$f = \frac{-\dfrac{dP}{dx} D_h}{\dfrac{1}{2} \rho U^2}$$

(16.92)

Since pressure at different locations is known, f can easily be calculated. The results from the numerical simulations are compared with one of the many available correlations for fully developed turbulent flow in a circular pipe like

$$f = \left[-1.8\log\left(\frac{6.9}{Re}\right) \right]^{-2}$$

(16.93)

The results are shown in Figure 16.40, and there is reasonable agreement (<5%) between the numerical results and the available correlation in the fully developed region.

7. In order to obtain the wall shear stress, Select Wall Fluxes from the Y Axis Function inside of the XY plot window and then select Wall Shear Stress in the box underneath of the Y Axis Function. The values can be outputted to a file and are useful in calculating the friction velocity and length scales needed later.

8. Figure 16.41 is a plot of dimensionless velocity profile at $x = 1.8$ m, using the nondimensional variables defined by Equations 9.42 and 9.43 and compared with the results from Law of the Wall and Spalding's expression. Again, the results are reasonably close.

9. Another important note in solving turbulent flow problems comes from approximations made through experimental models implemented in Fluent. While there

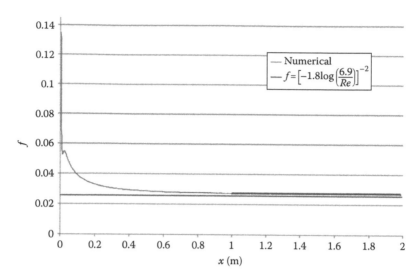

FIGURE 16.40
Friction factor.

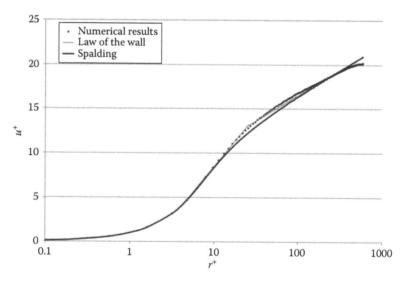

FIGURE 16.41
u^+ versus r^+ for actual data, Law of the Wall, and Spalding at $x = 1.80$ m.

are several different models that have been proposed, the three that will be examined are Law of the Wall, Power Law, and Spalding. These equations are readily available and thus are not presented in this tutorial. Also, important to note is that the shear stress at the wall quickly approaches a constant value, and the velocity profile fully develops around 0.60 m, the equations for r^+ and u^+ do not change because both u^+ and r^+ are functions of the shear stress and velocity. Therefore, only one calculation for u^+ and r^+ must be completed. Figure 16.41 is a log plot of u^+ versus r^+. The model that most closely approximates the actual data is the Law of the Wall. This is expected because of the Enhanced Wall Treatment selected in Fluent uses the Law of the Wall to create the solution.

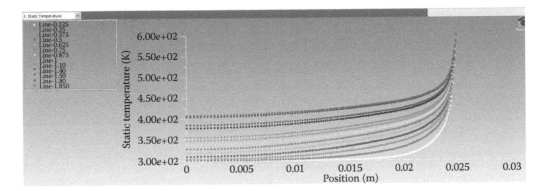

FIGURE 16.42
Temperature profiles at several different x locations.

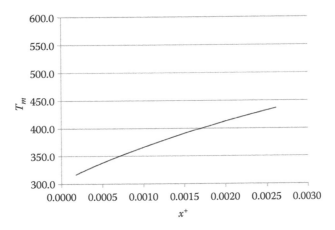

FIGURE 16.43
Mean temperature.

10. Other useful quantities to examine are the temperature along the vertical lines, Figure 16.42, as well as velocity and pressure along the centerline. Finally, the shear stress at the wall is important. Figure 16.43 is a plot of the radial temperature profiles at several different x locations. Again temperature drops significantly over a short distance close to the wall and then varies slowly radially toward the value at the center.

11. The flow becomes thermally fully developed when

$$T^+ = \frac{T(x,r) - T_w(x)}{\overline{T}(x) - T_w(x)} \tag{16.94}$$

becomes independent of axial location. However, in order to calculate T^+ and other quantities of interest like Nusselt number the value of mean temperature is needed. The temperature profiles at a number of rakes were exported and the mean temperature is calculated by numerically from

$$\overline{T} = \frac{2}{R^2 \overline{u}} \int_0^R ruT\,dr = \frac{2}{R^2 \overline{u}} \sum_{j=2}^M (r_j u_j T_j + r_{j-1} u_{j-1} T_{j-1}) \frac{(r_j - r_{j-1})}{2} \tag{16.95}$$

and the results are shown in Figure 16.43.

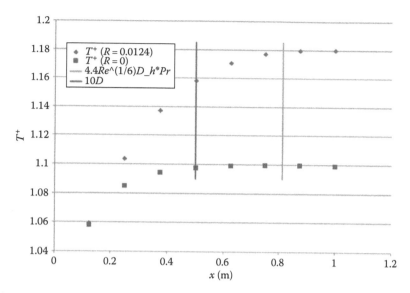

FIGURE 16.44
T^+ versus x.

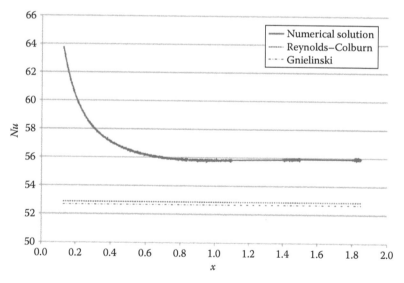

FIGURE 16.45
Nusselt number versus correlations.

12. Once mean temperature is known, T^+ can be calculated from Equation 16.94 and the results for two locations, along the pipe center line, $T^+(x,0)$, and almost half-way between wall and center, $T^+(x,0.0124)$, are shown in Figure 16.44. Again, T^+, although a function of r, becomes independent of axial location. Two criteria for establishment of thermally fully developed flow in the pipe are also shown in the figure, and as can be seen, $10D$ underestimates the thermal entrance length, and Equation 16.91 is a better predictor of the entrance length in this case.

The Nusselt number was calculated using Equation 8.86

$$Nu_x = \frac{1}{4} \frac{1}{T_w^* - \overline{T}^*} \frac{d\overline{T}^*}{dx^+} \qquad (16.96)$$

and is plotted in Figure 16.45 along with the correlations for fully developed turbulent flow given by Gnielinski (Equation 9.114), and Reynolds–Colburn analogy with f calculated from Equation 9.94

$$f = 0.316 Re_d^{-1/4} \qquad (16.97)$$

The numerical result in the fully developed region is within 6% of the values from the correlations.

16.11 Concluding Remarks

Computational fluid mechanics is an evolving field, and still as much an art as it is science. The two tutorials earlier are intended to help the readers with the process of setting up two different problems, obtaining results and analyzing them, detecting errors, and comparing results with the available solutions, and taking corrective actions. As can be seen, the effort involved in setting up a problem is trivial compared to what is needed to obtain accurate solutions, even for the two simple problems considered earlier, for which analytical and experimental results are available and verification of the results is not that difficult.

In numerical analysis, the prudent course of action is to always assume that the initial solution obtained is incorrect, and then try to prove otherwise. This becomes particularly challenging for problems where other analytical and experimental results are unavailable and in such cases, building the model in stages of increasing complexity, verifying the accuracy of simpler problems, and adding more details in subsequent steps may be the approaches to consider. Another method for checking accuracy is to check for conservation of energy, by using uniform heat flux boundary condition in one or more of the boundaries, and insulating the rest, and making sure that the energy is conserved, even for cases where temperature distribution is not needed.

Problems

16.1 Consider laminar flow between two plates at a distance D apart. The fluid approaches the plate with a uniform velocity U_{in} and temperature T_{in} with the walls maintained at a constant temperature T_w. Nondimensionalize the governing equations and the boundary conditions. Solve the problem using Fluent for arbitrary Reynolds and Prandtl numbers and present the results. Compare the numerical results with the available analytical and experimental results to prove that you have correct results.

16.2 War air enter a rectangular A/C duct at 50°C and $Re = 1500$. The duct is 8 cm high, 50 cm wide. After traveling in the duct for 1 m, the duct makes a sharp 90° turn. The duct is in a basement and its surface can be assumed isothermal at 20°C. Determine the pressure drop and heat transfer between two points 50 cm before and after the turn. Approximate the flow in this high aspect ratio duct, as that between two parallel plates.

16.3 Numerically obtain the solution for laminar flow in a square duct subjected to isothermal walls.

16.4 Numerically obtain the solution for laminar flow in a rectangular duct of aspect ratio 0.5, subjected to a uniform heat flux.

16.5 Numerically obtain the solution for turbulent flow between concentric cylinders with the inner cylinder specified as isothermal and the outer cylinder as insulated.

16.6 Numerically study turbulent flow perpendicular to a circular disk for $Re = 1000$.

16.7 Repeat problem 6 for $Re = 100,000$.

16.8 Consider the heat conduction in a rectangular plate with $k = 1$ W/m K subject to the boundary conditions shown in the figure:
a. Write the governing differential equation and boundary conditions
b. Solve the problem using Fluent
c. Plot the contours of temperature
d. Plot the temperature along the vertical and horizontal lines passing through the center point
e. Discuss your findings and trends

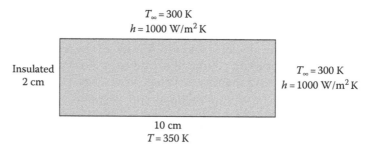

16.9 Solve for transient temperature distribution in a short cylinder of aspect ratio 5.

16.10 Solve the same problem, replacing the top boundary condition with air flowing on the top, with a $Re_L = 50,000$.

16.11 Determine the velocity and pressure drop in an isothermal toroid, with a pipe diameter of 2 cm and a coil diameter of 20 cm.

16.12 Determine the flow and temperature distribution over a circuit board 10 cm long, 1 cm thick, with a 2 cm wide, 0.5 cm thick computer chip imbedded 4 cm from the leading edge, kept at a constant 50°C. Air flows over the circuit board with a Reynolds number of 1000. The materials chosen were copper for the chip and aluminum for the circuit board.

16.13 Solve for the velocity distribution over a square cylinder perpendicular to the flow field.

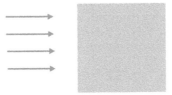

16.14 Solve the same problem except the cylinder is rotated 45°.

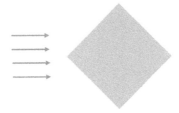

16.15 Model the flow over a tube bank, by considering flow over two inline cylinders. The tubes horizontal and vertical pitch are 2 × 2, and Reynolds numbers of 1000. Assume air at 300 K flows over isothermal tubes maintained at $T_w = 350$ K.

16.16 Solve for turbulent flow over a flat plate.

16.17 Model turbulent axisymmetric jet with a Reynolds number of 1000.

16.18 A semicylinder initially at 95°C is cooled by exposing it to an array of air jets at 25°C and providing an effective heat transfer coefficient of 150 W/m² K. The bottom (flat) side is exposed to air at 25°C and a convective coefficient of 5. Using Fluent, solve for transient temperature distribution if the diameter is 50 cm and it is 80 cm long. Take $\rho = 1200$ kg/m³, $c = 1700$ k/kg, and $k = 0.4$ W/m K.

References

1. http://www.cfd-online.com/Wiki/Codes, Accessed April 7, 2013.
2. http://www.ansys.com/Support/Documentation, Accessed April 7, 2013.
3. http://www.openfoam.org/docs/user/index.php, Accessed April 7, 2013.
4. Ferziger, J. H. and Perić, M. (1996). *Computational Methods for Fluid Dynamics*. Berlin, Germany: Springer-Verlag.
5. Patankar, S. V. (1980) *Numerical Heat Transfer and Fluid Flow*, Washington, Hemisphere Publishing Corporation.
6. Minkowycz, W. J. (ed.) (1988) *Handbook of Numerical Heat Transfer*. New York: Wiley-Interscience.
7. Shih, T. M. (1984) *Numerical Heat Transfer*. Washington, DC: Hemisphere Pub. Corp.
8. Cebeci, T. and Bradshaw, P. (1988) *Physical and Computational Aspects of Convective Heat Transfer*. New York: Springer-Verlag.

9. Jaluria, Y. and Torrance, K. E. (2003) *Computational Heat Transfer*. New York: Taylor & Francis Group.
10. Fletcher, C. A. and Srinivas, J. K. (1991) *Computational Techniques for Fluid Dynamics*. Berlin: Springer-Verlag, c1992 [Print].
11. Harlow, F. H. and Welch, J. E. (1965) Numerical calculation of time-dependent viscous incompressible flow of fluid with free surface. *Physics of Fluids* 8(12), 2182–2189.
12. Amsden, A. A. and Harlow, F. H. (1988) A simplified MAC technique for incompressible fluid flow calculations. *Journal of Computational Physics* 6(2), 322–325.
13. Patankar, S. V. and Spalding, D. B. (1972) A calculation procedure for heat, mass and momentum transfer in three-dimensional parabolic flows. *International Journal of Heat and Mass Transfer* 15(10), 1787–1806.
14. Patankar, S. V. (1981) Calculation procedure for two-dimensional elliptic situations. *Numerical Heat Transfer* 4(4), 409–425.

Appendix: Thermodynamic Properties of Different Substances and Table of Exponential Integral Functions

TABLE A.1

Properties of Metals

Element	Density, ρ (kg/m³)	Specific Heat, c_p (J/kg K)	Thermal Conductivity, k (W/m K)	Diffusivity, α ($\alpha \times 10^6$ m²/s) =	Melting Temperature (K)
Aluminum					
Pure	2702	896	236	97.5	933
2024, temper T351	2800	795	143	64	
2024, temper T4	2800	795	121	54	
5052, temper H32	2680	963	138	53	
5052, temper O	2690	963	144	56	
6061, temper O	2710	1256	180	53	
6061, temper T4	2710	1256	154	45	
6061, temper T6	2710	1256	167	49	
7075, temper T6	2800	1047	130	44	
A356, temper T6	2760	900	128	52	
Beryllium	1850	1750	205	63.3	1550
Brass					
Yellow	8470	402	120	35	1213
Red: 85% Cu, 15% Zn	8800	380	151	45	
Yellow: 65% Cu, 35% Zn	8800	380	119	36	
Chromium	7160	440	91.4	29	2118
Copper	8913	397.8	399.8	113	1353
Pure	8954	380	386	113	
Alloy, 1100	8933	385	388	113	
Aluminum bronze: 95% Cu, 5% Al	8666	410	83	23	
Brass: 70% Cu, 30% Zn	8522	385	111	34	
Bronze: 75% Cu, 25% Sn	8666	343	26	9	
Constantan: 60% Cu, 40% Ni	8922	410	22.7	6	
Drawn wire	8800	376	287	87	
German silver: 62% Cu, 15% Ni, 22% Zn	8618	394	24.9	7	
Red brass: 85% Cu, 9% Sn, 6% Zn	8714	385	61	18	
Copper	8933	383	399	116.6	1356
Gold	19,320	134.003	316.7	122	1336
Gold	19,300	129	316	126.9	1336
Incoloy 800	8027	544.4			1644
Inconel 600	8414	527.6			1644
Invar, 64%Fe 35%Ni	8130	480	13.8	4	
Iron, pure	7897	452	71.8	20	

TABLE A.1 (continued)

Properties of Metals

Element	Density, ρ (kg/m³)	Specific Heat, c_p (J/kg K)	Thermal Conductivity, k (W/m K)	Diffusivity, α ($\alpha \times 10^6$ m²/s) =	Melting Temperature (K)
Iron	7870	452	31.1	22.8	1810
Iron, cast	7197	502.5	80.2	22	1450
Iron, cast	7920	456	55	15	
Iron, wrought, 0.5% C	7849	460	59	16	
Kovar, 54% Fe 029% Ni 017% Co	8360	432	16.3	5	
Lead, pure	11,373	130	35	24	
Lead, liquid	10,712	154.9			255
Lead, solid	11,348	134	35.3	23	600
Lead	11,340	129	35.3	24.1	601
Magnesium	1744	1,130.7			923
Magnesium, Mg 0Al, electrolytic, 8% Al0 2% Zn	1810	1000	66	36	
Magnesium, pure	1746	1013	171	97	
Magnesium	1740	1017	156	88.2	923
Manganese	7290	486	7.78	2.2	1517
Molybdenum	10,214	297.3	130	43	2894
Molybdenum	10,240	251	138	53.7	2883
Monel 400	8830	460.6			1589
Nichrome (80% Ni 20% Cr)	8359	460.6	12	3	1672
Nickel	8885	502.5	90.7	20	1723
Nickel, Ni Cr, 90% Ni 10%Cr	8666	444	17	4	
Nickel, pure	8906	445.9	99	25	
Nickel	8900	446	91	22.9	1726
Platinum	21,451	146.6	71.6	23	2047
Platinum	21,450	133	71.4	25	2042
Potassium	860	741	103	161.6	337
Silicon	2330	703	153	93.4	1685
Silver	10,490	238.7	429	171	1233
Silver, pure	10,510	230	418	173	
Solder					
Solder (50% Pb, 50% Sn)	8940	213.6			456
Hard: 80% Au, 20% Sn	15,000	15	57	253	
Hard: 88% Au, 12% Ge	15,000		88		
Hard: 95% Au, 3% Si	15,700	147	94	41	
Soft: 60% Sn, 40% Pb	9290	180	50	30	
Soft: 63% Sn, 37% Pb	9250	180	51	31	
Soft: 92.5% Pb, 2.5% Ag, 5% In	12,000		39		
Soft: 95% Pb, 5% Sn	11,000	134	32.3	22	
Steel					
Carbon: 0.5% C	7833	465	54	15	
Carbon: 1.0% C	7801	473	43	12	

(*continued*)

TABLE A.1 (continued)

Properties of Metals

Element	Density, ρ (kg/m³)	Specific Heat, c_p (J/kg K)	Thermal Conductivity, k (W/m K)	Diffusivity, α ($\alpha \times 10^6$ m²/s) =	Melting Temperature (K)
Carbon: 1.5% C	7753	486	36	10	
Chrome: Cr 0%	7897	452	73	20	
Chrome: Cr 1%	7865	460	61	17	
Chrome: Cr 20%	7689	460	22	6	
Chrome: Cr 5%	7833	460	40	11	
Chrome and nickel: 18%Cr, 8%Ni	7817	460	16.3	5	
Invar: 36% Ni	8137	460	10.7	3	
Mild	7861	510.9	45	11	1683
Nickel: Ni 0%	7897	452	73	20	
Nickel: Ni 20%	7933	460	19	5	
Nickel: Ni 40%	8169	460	10	3	
Nickel: Ni 80%	8618	460	35	9	
SAE 1010	7832	434	59	17	
SAE 1010, sheet	7832	434	63.9	19	
Stainless 304	7916	502.5	14	4	1672
Stainless 430	7612	460.6	14	4	1728
Stainless, 316	8027	502.1	16.3	4	
Tungsten: W 0%	7897	452	73	20	
Tungsten: W 1%	7913	448	66	19	
Tungsten: W 10%	8314	419	48	14	
Tungsten: W 5%	8073	435	54	15	
Tantalum	16,607	146.6			3269
Tin					
Pure	7304	226.5	64	39	
Tin	5750	227	67	51.3	505
Cast, hammered	7352	226	62.5	38	
Liquid	7003	217.8			255
Solid	7280	272.2	66.6	34	505
Titanium	4510	544	15.6	6	
Titanium 99.0%	4539	544.4	21.9	9	1941
Titanium	4500	611	22	8	1953
Tungsten	19,292	167.5	174	54	3683
Tungsten	19,300	134	179	69.2	3653
Type metal (85% Pb015% Sb)	10,712	167.5			533
Uranium	19,070	113	27.4	12.7	1407
Vanadium	6100	502	31.4	10.3	2192
Zinc	7141	402	116	40	692
Zinc	7140	385	121	44	693
Zirconium	6477	280.6	251	138	2116

TABLE A.2

Thermal Properties of Selected Building Materials and Insulations at 293 K (20°C) or 528°R (65°F)

Element	Density, ρ (kg/m³)	Specific Heat, c_p (J/kg K)	Thermal Conductivity, k (W/m K)	Diffusivity α ($\alpha \times 10^6$ (m²/s) =
Asbestos	383	816	0.113	0.362
Asphalt	2120		0.698	
Bakelite	1270		0.233	
Brick				
Carborundum (50% SiC)	2200		5.82	
Common	1800	840	0.38	0.251
Magnesite (50% MgO)	2000		2.68	
Masonry	1700	837	0.658	0.462
Silica (95% SiO_2)	1900		1.07	
Cardboard			0.14	
Cement (hard)			1.047	
Clay (48.7% moist)	1545	880	1.26	0.927
Coal (anthracite)	1370	1260	0.238	0.138
Concrete (dry)	500	837	0.128	0.306
Cork board	150	1880	0.042	0.149
Cork (expanded)	120		0.036	
Earth (diatomaceous)	466	879	0.126	0.308
Earth (clay with 28% moist)	1500		1.51	
Earth (sandy with 8% moist)	1500		1.05	
Glass fiber	220		0.035	
Glass (window pane)	2800	800	0.81	0.362
Glass (wool)	200	670	0.04	0.299
Granite	2750		3	
Ice at 0°C	913	1830	2.22	1.329
Kapok	25		0.035	
Linoleum	535		0.081	
Mica	2900		0.523	
Pine bark	342		0.08	
Plaster	1800		0.814	
Plywood	590		0.109	
Polystyrene	1050		0.157	
Rubber				
Buna	1250		0.465	
Ebonite	1150	2009	0.163	0.071
Spongy	224		0.055	
Sand				
Dry			0.582	
Moist	1640		1.13	
Sawdust	215		0.071	

(continued)

TABLE A.2 (continued)

Thermal Properties of Selected Building Materials and Insulations at 293 K (20°C) or 528°R (65°F)

Element	Density, ρ (kg/m^3)	Specific Heat, c_p (J/kg K)	Thermal Conductivity, k (W/m K)	Diffusivity α ($\alpha \times 10^6$ (m^2/s) =
Wood				
Fir, pine, and spruce	444	2720	0.15	0.124
Oak	705	2390	0.19	0.113
Celotex	400		0.055	
Fiber sheets	200		0.047	
Wool	200		0.038	

TABLE A.3

Properties of Saturated Liquids

Temperature, T	Density, ρ (kg/m³)	Specific Heat, c_p (J/kg K)	Kinematic Viscosity, ν	Thermal Conductivity, k (W/m K)	Diffusivity, α ($\alpha \times 10^7$ m²/s) =	Prandtl Number, Pr	β ($\beta \times 10^3$ 1/K) =
Ammonia (NH₃)							
−50	703	4463	0.435	0.547	1.742	2.60	
−40	691	4467	0.406	0.547	1.775	2.28	
−30	679	4476	0.387	0.549	1.801	2.15	
−20	666	4509	0.381	0.547	1.819	2.09	
−10	653	4564	0.378	0.543	1.825	2.07	
0	640	4635	0.373	0.540	1.819	2.05	
10	626	4714	0.368	0.531	1.801	2.04	
20	611	4798	0.359	0.521	1.775	2.02	2.45
30	596	4890	0.349	0.507	1.742	2.01	
40	580	4999	0.34	0.493	1.701	2.00	
50	564	5116	0.33	0.476	1.654	1.99	
Carbon dioxide (CO₂) [1]							
−50	1156	1840	0.119	0.0655	0.4021×10^{-7}	2.96	
−40	1117	1880	0.118	0.1011	0.481	2.46	
−30	1076	1970	0.117	0.1116	0.5272	2.22	
−20	1032	2050	0.115	0.1151	0.5445	2.12	
−10	983	2180	0.113	0.1099	0.5133	2.20	
0	926	2470	0.108	0.1045	0.4578	2.38	
10	860	3140	0.101	0.0971	0.3608	2.80	
20	772	5000	0.091	0.0872	0.2219	4.10	14.0
30	597	36,400	0.08	0.0703	0.0279	28.70	
Dichlorodifluoromethane (Freon-12) (CCl₂F₂) [1]							
−50	1546	875	0.31	0.067	0.501	6.2	2.63
−40	1518	884.7	0.279	0.069	0.514	5.4	
−30	1489	895.6	0.253	0.069	0.526	4.3	
−20	1460	907.3	0.235	0.071	0.539	4.4	
−10	1429	920.3	0.221	0.073	0.55	4.0	
0	1397	934.5	0.214	0.073	0.557	3.8	
10	1364	949.6	0.203	0.073	0.56	3.6	
20	1330	965.9	0.198	0.073	0.56	3.5	
30	1295	983.5	0.194	0.071	0.56	3.5	
40	1257	1001.9	0.191	0.069	0.555	3.5	
50	1215	1021.6	0.19	0.067	0.545	3.5	
Engine oil (unused) [1]							
0	899	1796	0.00428	0.147	0.911	47,100	
20	888	1880	0.0009	0.145	0.872	10,400	0.70
40	876	1964	0.00024	0.144	0.834	2870	
60	864	2047	0.839	0.140	0.8	1050	
80	852	2131	0.375	0.138	0.769	490	
100	840	2219	0.203	0.137	0.738	276	

(continued)

TABLE A.3 (continued)

Properties of Saturated Liquids

Temperature, T	Density, ρ (kg/m³)	Specific Heat, c_p (J/kg K)	Kinematic Viscosity, ν	Thermal Conductivity, k (W/m K)	Diffusivity, α ($\alpha \times 10^7$ m²/s) =	Prandtl Number, Pr	β ($\beta \times 10^3$ 1/K) =
120	828	2307	0.124	0.135	0.71	175	
140	816	2395	0.08	0.133	0.686	116	
160	805	2483	0.056	0.132	0.663	84	
Ethylene glycol (C₂H₄(OH₂)) [1]							
0	1130	2294	57.53	0.242	0.934	615.0	
20	1116	2382	19.18	0.249	0.939	204.0	0.65
40	1101	2474	8.69	0.256	0.939	93.0	
60	1087	2562	4.75	0.26	0.932	51.0	
80	1077	2650	2.98	0.261	0.921	32.4	
100	1058	2742	2.03	0.263	0.908	22.4	
Eutectic calcium chloride solution (29.9%CaCl₂) [1]							
−50	1319	2608	36.35	0.402	1.166	312.0	
−40	1314	2635.6	24.97	0.415	1.2	208.0	
−30	1310	2661.1	17.18	0.429	1.234	139.0	
−20	1305	2688	11.04	0.445	1.267	87.1	
−10	1300	2713	6.96	0.459	1.3	53.6	
0	1296	2738	4.39	0.472	1.332	33.0	
10	1291	2763	3.35	0.485	1.363	24.6	
20	1286	2788	2.72	0.498	1.394	19.6	
30	1281	2814	2.27	0.511	1.419	16.0	
40	1277	2839	1.92	0.523	1.445	13.3	
50	1272	2868	1.65	0.535	1.468	11.3	
Glycerin (C₃H₅(OH)₃)							
0	1276	2261	0.00831	0.282	0.983	84.7×10^3	
10	1270	2319	0.003	0.284	0.965	31.00	
20	1264	2386	0.00118	0.286	0.947	12.50	0.50
30	1258	2445	0.0005	0.286	0.929	5.38	
40	1252	2512	0.00022	0.286	0.914	2.45	
50	1244	2583	0.00015	0.287	0.893	1.63	
Mercury (Hg)							
0	13,628	140.3	0.124	8.20	42.99	0.0288	
20	13,579	139.4	0.144	8.69	46.06	0.0249	182
50	13,505	138.6	0.104	9.40	50.22	0.0207	
100	13,384	137.3	0.0928	10.51	57.16	0.0162	
150	13,264	136.5	0.0853	11.49	63.54	0.0134	
200	13,144	157	0.0802	12.34	69.08	0.0116	
250	13,025	135.7	0.0765	13.07	74.06	0.0103	
315.5	12,847	134	0.0673	14.02	81.5	0.0083	

TABLE A.3 (continued)

Properties of Saturated Liquids

Temperature, T	Density, ρ (kg/m³)	Specific Heat, c_p (J/kg K)	Kinematic Viscosity, ν	Thermal Conductivity, k (W/m K)	Diffusivity, α ($\alpha \times 10^7$ m²/s) =	Prandtl Number, Pr	β ($\beta \times 10^3$ 1/K) =
Methyl chloride (CH₃Cl)							
−50	1052	1475.9	0.32	0.215	1.388	2.31	
−40	1033	1482.6	0.318	0.209	1.368	1.32	
−30	1016	1492.2	0.314	0.202	1.337	2.35	
−20	999	1504.3	0.309	0.196	1.301	2.38	
−10	981	1519.4	0.306	0.187	1.257	2.43	
0	962	1537.8	0.302	0.178	1.213	2.49	
10	942	1560	0.297	0.171	1.166	2.55	
20	923	1586	0.293	0.163	1.112	2.63	
30	903	1616.1	0.288	0.154	1.058	2.72	
40	883	1650.4	0.281	0.144	0.996	2.83	
50	861	1689	0.274	0.133	0.921	2.97	
Sulfur dioxide (SO₂) [1]							
−50	1560	1359.5	0.484	0.242	1.141	4.24	
−40	1536	1360.7	0.424	0.235	1.13	3.74	
−30	1520	1361.6	0.371	0.23	1.117	3.31	
−20	1488	1362.4	0.324	0.225	1.107	2.93	
−10	1463	1362.8	0.288	0.218	1.097	2.62	
0	1438	1363.6	0.257	0.211	1.081	2.38	
10	1412	1364.5	0.232	0.204	1.066	2.18	
20	1386	1365.3	0.21	0.199	1.05	2.00	1.94
30	1359	1366.2	0.19	0.192	1.035	1.83	
40	1329	1367.4	0.173	0.185	1.019	1.70	
50	1299	1368.3	0.162	0.177	0.999	1.61	

TABLE A.4

Properties of Saturation Liquid Water

T (°C)	P (kPa)	ρ (kg/m³)	h (kJ/kg)	c_p (kJ/kg)	$k \times 10^3$ (W/m K)	$\nu \times 10^6$ (m²/s)	$\alpha \times 10^6$ (m²/s)	Pr	Coefficient of Thermal Expansion $\beta \times 10^4$ (1/K)	Surface Tension σ N/m)
0.01	0.6	999.8	0.0006	4.220	561.0	1.7920	0.1330	13.4700	−0.6796	0.0757
5	0.9	999.9	21.0	4.205	570.5	1.5180	0.1357	11.1900	0.1573	0.0749
10	1.2	999.7	42.0	4.196	580.0	1.3060	0.1383	9.4470	0.8769	0.0742
15	1.7	999.1	63.0	4.189	589.3	1.1390	0.1408	8.0860	1.5064	0.0735
20	2.3	998.2	83.9	4.184	598.4	1.0030	0.1433	7.0040	2.0667	0.0727
25	3.2	997.0	104.8	4.182	607.2	0.8928	0.1456	6.1300	2.5717	0.0720
30	4.2	995.6	125.7	4.180	615.5	0.8009	0.1479	5.4150	3.0333	0.0712
35	5.6	994.0	146.6	4.180	623.3	0.7237	0.1500	4.8230	3.4588	0.0704
40	7.4	992.2	167.5	4.180	630.6	0.6581	0.1521	4.3280	3.8541	0.0696
45	9.6	990.2	188.4	4.180	637.3	0.6020	0.1540	3.9100	4.2264	0.0688
50	12.4	988.0	209.3	4.182	643.6	0.5535	0.1558	3.5530	4.5779	0.0679
55	15.8	985.7	230.3	4.183	649.2	0.5113	0.1575	3.2470	4.9122	0.0671
60	20.0	983.2	251.2	4.185	654.3	0.4744	0.1590	2.9830	5.2329	0.0662
65	25.0	980.5	272.1	4.187	659.0	0.4419	0.1605	2.7530	5.5421	0.0654
70	31.2	977.7	293.1	4.190	663.1	0.4131	0.1619	2.5520	5.8402	0.0645
75	38.6	974.8	314.0	4.193	666.8	0.3875	0.1631	2.3760	6.1305	0.0636
80	47.4	971.8	335.0	4.197	670.0	0.3646	0.1643	2.2200	6.4139	0.0627
85	57.9	968.6	356.0	4.201	672.8	0.3441	0.1654	2.0810	6.6921	0.0618
90	70.2	965.3	377.0	4.205	675.3	0.3257	0.1663	1.9580	6.9667	0.0608
95	84.6	961.9	398.1	4.210	677.3	0.3091	0.1673	1.8480	7.2378	0.0599
100	101.4	958.3	419.2	4.216	679.1	0.2940	0.1681	1.7490	7.5070	0.0589
105	120.9	954.7	440.3	4.222	680.5	0.2803	0.1688	1.6600	7.7742	0.0579
110	143.4	950.9	461.4	4.228	681.7	0.2678	0.1695	1.5800	8.0408	0.0570
115	169.2	947.1	482.6	4.236	682.6	0.2565	0.1702	1.5070	8.3085	0.0560
120	198.7	943.1	503.8	4.244	683.2	0.2461	0.1707	1.4410	8.5781	0.0550
125	232.2	939.0	525.1	4.252	683.6	0.2365	0.1712	1.3810	8.8488	0.0540
130	270.3	934.8	546.4	4.261	683.7	0.2277	0.1716	1.3270	9.1239	0.0529
135	313.2	930.5	567.7	4.272	683.6	0.2197	0.1720	1.2770	9.4014	0.0519
140	361.5	926.1	589.2	4.283	683.3	0.2122	0.1723	1.2320	9.6836	0.0509
145	415.7	921.6	610.6	4.294	682.8	0.2053	0.1725	1.1900	9.9718	0.0498
150	476.2	917.0	632.2	4.307	682.0	0.1990	0.1727	1.1520	10.2661	0.0487
155	543.5	912.3	653.8	4.321	681.1	0.1931	0.1728	1.1170	10.5678	0.0477
160	618.2	907.4	675.5	4.335	680.0	0.1876	0.1728	1.0850	10.8783	0.0466
165	700.9	902.5	697.2	4.351	678.6	0.1825	0.1728	1.0560	11.1911	0.0455
170	792.2	897.5	719.1	4.368	677.0	0.1778	0.1727	1.0290	11.5209	0.0444
175	892.6	892.3	741.0	4.386	675.3	0.1734	0.1726	1.0050	11.8682	0.0433
180	1003.0	887.0	763.1	4.405	673.3	0.1693	0.1723	0.9822	12.2210	0.0422
185	1123.0	881.6	785.2	4.425	671.1	0.1654	0.1720	0.9616	12.5907	0.0411
190	1255.0	876.1	807.4	4.447	668.8	0.1618	0.1716	0.9429	12.9666	0.0400
195	1399.0	870.4	829.8	4.471	666.1	0.1585	0.1712	0.9258	13.3732	0.0388
200	1555.0	864.7	852.3	4.496	663.3	0.1553	0.1706	0.9104	13.7851	0.0377
205	1724.0	858.8	874.9	4.523	660.3	0.1524	0.1700	0.8965	14.2292	0.0365

TABLE A.4 (continued)

Properties of Saturation Liquid Water

T (°C)	P (kPa)	ρ (kg/m3)	h (kJ/kg)	c_p (kJ/kg)	$k \times 10^3$ (W/m K)	$\nu \times 10^6$ (m²/s)	$\alpha \times 106$ (m²/s)	Pr	Coefficient of Thermal Expansion $\beta \times 10^4$ (1/K)	Surface Tension σ (N/m)
210	1908.0	852.7	897.6	4.551	657.0	0.1496	0.1693	0.8840	14.6945	0.0354
215	2106.0	846.5	920.5	4.582	653.4	0.1471	0.1685	0.8729	15.1802	0.0342
220	2320.0	840.2	943.6	4.615	649.7	0.1446	0.1676	0.8632	15.6986	0.0331
225	2550.0	833.7	966.8	4.650	645.6	0.1423	0.1665	0.8547	16.2409	0.0319
230	2797.0	827.1	990.2	4.688	641.3	0.1402	0.1654	0.8476	16.8178	0.0307
235	3063.0	820.3	1014.0	4.728	636.7	0.1382	0.1642	0.8418	17.4448	0.0296
240	3347.0	813.4	1038.0	4.772	631.8	0.1363	0.1628	0.8372	18.0969	0.0284
245	3651.0	806.2	1062.0	4.819	626.7	0.1345	0.1613	0.8339	18.8167	0.0272
250	3976.0	798.9	1086.0	4.870	621.2	0.1328	0.1597	0.8319	19.5769	0.0260
255	4323.0	791.4	1110.0	4.925	615.4	0.1312	0.1579	0.8313	20.4069	0.0249
260	4692.0	783.6	1135.0	4.986	609.2	0.1298	0.1559	0.8321	21.2991	0.0237
265	5085.0	775.7	1160.0	5.051	602.8	0.1284	0.1538	0.8344	22.2767	0.0225
270	5503.0	767.5	1185.0	5.123	595.9	0.1270	0.1516	0.8382	23.3355	0.0213
275	5946.0	759.0	1211.0	5.202	588.7	0.1258	0.1491	0.8437	24.5191	0.0202
280	6417.0	750.3	1237.0	5.289	581.1	0.1246	0.1465	0.8510	25.8163	0.0190
285	6915.0	741.3	1263.0	5.385	573.2	0.1235	0.1436	0.8603	27.2494	0.0178
290	7442.0	731.9	1290.0	5.493	565.0	0.1225	0.1405	0.8717	28.8564	0.0167
295	7999.0	722.2	1317.0	5.614	556.3	0.1215	0.1372	0.8856	30.6840	0.0155
300	8588.0	712.1	1345.0	5.750	547.4	0.1206	0.1337	0.9023	32.7342	0.0144
305	9209.0	701.6	1373.0	5.906	538.2	0.1198	0.1299	0.9221	35.1055	0.0132
310	9865.0	690.7	1402.0	6.085	528.7	0.1190	0.1258	0.9456	37.8312	0.0121
315	10560.0	679.2	1432.0	6.293	519.1	0.1182	0.1215	0.9734	41.0483	0.0110
320	11280.0	667.1	1462.0	6.537	509.2	0.1175	0.1168	1.0070	44.8508	0.0099
325	12050.0	654.3	1494.0	6.830	499.2	0.1169	0.1117	1.0470	49.4574	0.0088
330	12860.0	640.8	1526.0	7.186	489.1	0.1163	0.1062	1.0950	55.1342	0.0077
335	13710.0	626.3	1559.0	7.632	478.8	0.1158	0.1002	1.1560	62.3184	0.0067
340	14600.0	610.7	1595.0	8.208	468.5	0.1153	0.0935	1.2340	71.7537	0.0056
345	15540.0	593.6	1631.0	8.988	458.0	0.1149	0.0858	1.3390	84.7540	0.0046
350	16530.0	574.7	1671.0	10.120	447.4	0.1146	0.0770	1.4900	103.9325	0.0037
355	17570.0	553.1	1714.0	11.890	436.5	0.1144	0.0664	1.7240	134.9304	0.0027
360	18670.0	527.6	1762.0	15.000	425.7	0.1143	0.0538	2.1260	191.2434	0.0019
365	19820.0	495.7	1818.0	21.410	416.4	0.1145	0.0392	2.9200	311.2770	0.0011
370	21040.0	451.4	1891.0	45.160	425.0	0.1153	0.0209	5.5320	763.8458	0.0004

Source: Lemmon, E.W. et al., Thermophysical properties of fluid systems, in NIST *Chemistry WebBook*, P.J. Linstrom and W.G. Mallard (eds.), NIST Standard Reference Database Number 69, National Institute of Standards and Technology, Gaithersburg MD, http://webbook.nist.gov (retrieved on June 18, 2012).

TABLE A.5

Properties of Saturation Water Vapor

T (°C)	P (kPa)	ρ (kg/m³)	h (kJ/kg)	c_p (kJ/kg)	$k \times 10^3$ (W/m K)	$\nu \times 10^6$ (m²/s)	$\alpha \times 10^6$ (m²/s)	Pr	β (1/K)
0.01	0.6117	0.0049	2501	1.884	17.07	1898.0	1866.0000	1.0170	0.0037
5	0.8726	0.0068	2510	1.889	17.34	1372.0	1349.0000	1.0170	0.0036
10	1.228	0.0094	2519	1.895	17.62	1006.0	988.6000	1.0170	0.0036
15	1.706	0.0128	2528	1.900	17.92	747.00	734.3000	1.0170	0.0035
20	2.339	0.0173	2537	1.906	18.23	561.80	552.4000	1.0170	0.0034
25	3.17	0.0231	2547	1.912	18.55	427.60	420.5000	1.0170	0.0034
30	4.247	0.0304	2556	1.918	18.89	329.10	323.8000	1.0170	0.0033
35	5.629	0.0397	2565	1.925	19.24	256.00	251.9000	1.0160	0.0033
40	7.385	0.0512	2574	1.931	19.6	201.20	198.0000	1.0160	0.0033
45	9.595	0.0656	2582	1.939	19.97	159.50	157.1000	1.0150	0.0032
50	12.35	0.0832	2591	1.947	20.36	127.70	125.8000	1.0150	0.0032
55	15.76	0.1046	2600	1.955	20.77	103.00	101.6000	1.0140	0.0031
60	19.95	0.1304	2609	1.965	21.19	83.840	82.6800	1.0140	0.0031
65	25.04	0.1615	2618	1.975	21.62	68.730	67.8000	1.0140	0.0031
70	31.2	0.1984	2626	1.986	22.07	56.750	55.9900	1.0130	0.0030
75	38.6	0.2422	2635	1.999	22.53	47.180	46.5500	1.0130	0.0030
80	47.41	0.2937	2643	2.012	23.01	39.470	38.9400	1.0140	0.0030
85	57.87	0.3539	2651	2.027	23.51	33.230	32.7700	1.0140	0.0030
90	70.18	0.4239	2660	2.043	24.02	28.140	27.7400	1.0150	0.0029
95	84.61	0.5049	2668	2.061	24.55	23.960	23.5900	1.0160	0.0029
100	101.4	0.5982	2676	2.080	25.1	20.510	20.1700	1.0170	0.0029
105	120.9	0.7050	2683	2.101	25.66	17.650	17.3200	1.0190	0.0029
110	143.4	0.8269	2691	2.124	26.24	15.250	14.9400	1.0210	0.0029
115	169.2	0.9654	2699	2.150	26.85	13.240	12.9400	1.0240	0.0029
120	198.7	1.1220	2706	2.177	27.47	11.550	11.2400	1.0270	0.0029
125	232.2	1.2990	2713	2.207	28.11	10.110	9.8080	1.0310	0.0029
130	270.3	1.4970	2720	2.239	28.76	8.885	8.5820	1.0350	0.0029
135	313.2	1.7190	2727	2.274	29.44	7.8380	7.5330	1.0400	0.0029
140	361.5	1.9670	2733	2.311	30.14	6.9390	6.6320	1.0460	0.0029
145	415.7	2.2420	2740	2.351	30.86	6.1630	5.8530	1.0530	0.0029
150	476.2	2.5480	2746	2.394	31.6	5.4910	5.1800	1.0600	0.0029
155	543.5	2.8860	2752	2.440	32.35	4.9070	4.5950	1.0680	0.0029
160	618.2	3.2600	2757	2.488	33.13	4.3980	4.0850	1.0770	0.0030
165	700.9	3.6710	2763	2.540	33.93	3.9530	3.6390	1.0860	0.0030
170	792.2	4.1220	2768	2.594	34.75	3.5620	3.2490	1.0960	0.0030
175	892.6	4.6170	2773	2.652	35.59	3.2170	2.9060	1.1070	0.0031
180	1003	5.1590	2777	2.713	36.45	2.9130	2.6040	1.1180	0.0031
185	1123	5.7500	2781	2.777	37.33	2.6430	2.3380	1.1300	0.0032
190	1255	6.3950	2785	2.844	38.24	2.4030	2.1020	1.1430	0.0032
195	1399	7.0980	2789	2.915	39.16	2.1900	1.8930	1.1570	0.0033
200	1555	7.8610	2792	2.990	40.11	1.9990	1.7070	1.1710	0.0033
205	1724	8.6900	2795	3.068	41.09	1.8280	1.5410	1.1860	0.0034
210	1908	9.5880	2797	3.150	42.09	1.6750	1.3930	1.2020	0.0035
215	2106	10.560	2799	3.237	43.11	1.5370	1.2610	1.2190	0.0036

TABLE A.5 (continued)

Properties of Saturation Water Vapor

T (°C)	P (kPa)	ρ (kg/m³)	h (kJ/kg)	c_p (kJ/kg)	$k \times 10^3$ (W/m K)	$\nu \times 10^6$ (m²/s)	$\alpha \times 10^6$ (m²/s)	Pr	β (1/K)
220	2320	11.620	2801	3.329	44.17	1.4130	1.1420	1.2370	0.0036
225	2550	12.750	2802	3.426	45.26	1.3000	1.0360	1.2560	0.0037
230	2797	13.990	2803	3.528	46.38	1.1990	0.9398	1.2760	0.0038
235	3063	15.310	2803	3.638	47.53	1.1060	0.8532	1.2970	0.0040
240	3347	16.750	2803	3.754	48.73	1.0220	0.7751	1.3190	0.0041
245	3651	18.300	2802	3.878	49.97	0.9460	0.7043	1.3430	0.0042
250	3976	19.970	2801	4.011	51.26	0.8762	0.6402	1.3690	0.0044
255	4323	21.770	2799	4.153	52.61	0.8124	0.5819	1.3960	0.0045
260	4692	23.710	2797	4.308	54.03	0.7539	0.5290	1.4250	0.0047
265	5085	25.810	2793	4.475	55.53	0.7003	0.4808	1.4560	0.0049
270	5503	28.070	2790	4.656	57.11	0.6510	0.4369	1.4900	0.0051
275	5946	30.520	2785	4.855	58.8	0.6057	0.3968	1.5260	0.0054
280	6417	33.160	2780	5.073	60.61	0.5638	0.3603	1.5650	0.0057
285	6915	36.030	2774	5.314	62.57	0.5252	0.3268	1.6070	0.0060
290	7442	39.130	2767	5.582	64.71	0.4895	0.2962	1.6520	0.0063
295	7999	42.500	2759	5.882	67.05	0.4564	0.2682	1.7010	0.0067
300	8588	46.170	2750	6.220	69.65	0.4256	0.2426	1.7550	0.0072
305	9209	50.170	2739	6.604	72.55	0.3971	0.2190	1.8130	0.0077
310	9865	54.540	2728	7.045	75.84	0.3705	0.1974	1.8770	0.0083
315	10560	59.340	2715	7.557	79.58	0.3457	0.1775	1.9480	0.0090
320	11280	64.640	2701	8.159	83.91	0.3225	0.1591	2.0270	0.0099
325	12050	70.510	2684	8.878	88.96	0.3008	0.1421	2.1160	0.0109
330	12860	77.050	2666	9.753	94.94	0.2804	0.1263	2.2190	0.0122
335	13710	84.410	2645	10.84	102.1	0.2612	0.1116	2.3410	0.0138
340	14600	92.760	2622	12.24	110.9	0.2431	0.0977	2.4880	0.0159
345	15540	102.40	2595	14.09	121.9	0.2260	0.0845	2.6750	0.0187
350	16530	113.60	2564	16.69	135.9	0.2097	0.0717	2.9250	0.0227
355	17570	127.10	2527	20.63	154.7	0.1940	0.0590	3.2870	0.0288
360	18670	143.90	2481	27.36	181.5	0.1788	0.0461	3.8770	0.0394
365	19820	166.30	2423	41.80	224.9	0.1635	0.0323	5.0560	0.0625
370	21040	201.80	2335	96.60	323.8	0.1470	0.0166	8.8530	0.1517

Source: Lemmon, E.W. et al., Thermophysical properties of fluid systems, in *NIST Chemistry WebBook*, P.J. Linstrom and W.G. Mallard (eds.), NIST Standard Reference Database Number 69, National Institute of Standards and Technology, Gaithersburg MD, http://webbook.nist.gov (retrieved on June 18, 2012).

TABLE A.6

Properties of Saturation Liquid Ethane, 1,1,1,2-Tetrafluoro-(R134a)

T (°C)	P (kPa)	ρ (kg/m³)	h (kJ/kg)	c_p (kJ/kg)	$k \times 10^3$ (W/m K)	$\nu \times 10^6$ (m²/s)	$\alpha \times 10^6$ (m²/s)	Pr	Coefficient of Thermal Expansion $\beta \times 10^4$ (1/K)	Surface Tension σ (N/m)
−60	0.01591	1474	123.4	1.223	120.7	0.448	0.06694	6.692	18.853	208.0
−55	0.02183	1460	129.5	1.23	118.1	0.4121	0.06574	6.269	19.212	199.9
−50	0.02945	1446	135.7	1.238	115.6	0.3809	0.06454	5.902	19.592	191.8
−45	0.03912	1432	141.9	1.246	113.1	0.3536	0.06335	5.581	20.007	183.9
−40	0.05121	1418	148.1	1.255	110.6	0.3294	0.06218	5.298	20.444	176.0
−35	0.06614	1403	154.4	1.264	108.2	0.308	0.06101	5.048	20.941	168.1
−30	0.08438	1388	160.8	1.273	105.8	0.2888	0.05985	4.825	21.470	160.4
−25	0.1064	1373	167.2	1.283	103.4	0.2715	0.05869	4.626	22.039	152.7
−20	0.1327	1358	173.6	1.293	101.1	0.2559	0.05755	4.447	22.666	145.1
−15	0.1639	1343	180.1	1.304	98.77	0.2417	0.05641	4.285	23.343	137.6
−10	0.2006	1327	186.7	1.316	96.49	0.2287	0.05527	4.139	24.099	130.2
−5	0.2433	1311	193.3	1.328	94.24	0.2168	0.05413	4.006	24.928	122.9
0	0.2928	1295	200	1.341	92.01	0.2058	0.05299	3.884	25.830	115.6
5	0.3497	1278	206.8	1.355	89.81	0.1957	0.05185	3.774	26.854	108.4
10	0.4146	1261	213.6	1.37	87.62	0.1863	0.05071	3.673	27.994	101.4
15	0.4884	1243	220.5	1.387	85.44	0.1775	0.04955	3.582	29.284	94.4
20	0.5717	1225	227.5	1.405	83.28	0.1692	0.04838	3.498	30.727	87.6
25	0.6654	1207	234.5	1.425	81.13	0.1615	0.0472	3.422	32.353	80.8
30	0.7702	1187	241.7	1.446	78.99	0.1542	0.04599	3.353	34.263	74.2
35	0.887	1168	249	1.471	76.85	0.1473	0.04475	3.292	36.421	67.7
40	1.017	1147	256.4	1.498	74.72	0.1408	0.04348	3.238	38.971	61.3
45	1.16	1125	263.9	1.53	72.57	0.1346	0.04217	3.191	42.000	55.0
50	1.318	1102	271.6	1.566	70.43	0.1286	0.04079	3.153	45.626	48.9
55	1.492	1078	279.5	1.609	68.27	0.1229	0.03935	3.123	50.037	43.0
60	1.682	1053	287.5	1.66	66.09	0.1174	0.03781	3.105	55.508	37.2
65	1.89	1026	295.8	1.723	63.89	0.1121	0.03615	3.101	62.524	31.6
70	2.117	996.2	304.3	1.804	61.67	0.1069	0.03432	3.115	71.863	26.1
75	2.364	964.1	313.1	1.911	59.42	0.1018	0.03224	3.158	84.825	21.0
80	2.633	928.2	322.4	2.065	57.15	0.09679	0.02982	3.246	104.159	16.0
85	2.926	887.2	332.2	2.306	54.88	0.09172	0.02682	3.42	136.046	11.4
90	3.244	837.8	342.9	2.756	52.75	0.08647	0.02285	3.785	198.854	7.1
95	3.591	772.7	355.2	3.938	51.45	0.08072	0.01691	4.775	373.625	3.3
100	3.972	651.2	373.3	17.59	58.88	0.07284	0.00514	14.17	2542.998	0.37

Source: Lemmon, E.W. et al., Thermophysical properties of fluid systems, in *NIST Chemistry WebBook*, P.J. Linstrom and W.G. Mallard (eds.), NIST Standard Reference Database Number 69, National Institute of Standards and Technology, Gaithersburg MD, http://webbook.nist.gov (retrieved on June 18, 2012).

TABLE A.7

Properties of Saturation Vapor Ethane, 1,1,1,2-Tetrafluoro-(R134a)

T (°C)	P (kPa)	ρ (kg/m³)	h (kJ/kg)	c_p (kJ/kg)	$k \times 10^3$ (W/m K)	$\nu \times 10^6$ (m²/s)	$\alpha \times 10^6$ (m²/s)	Pr	β (1/K)
−60	0.01591	0.9268	361.3	0.692	6.555	9.1930	10.2200	0.8998	0.0049
−55	0.02183	1.246	364.5	0.706	6.959	6.9880	7.9100	0.8834	0.0049
−50	0.02945	1.65	367.7	0.720	7.363	5.3930	6.2010	0.8697	0.0048
−45	0.03912	2.152	370.8	0.734	7.767	4.2210	4.9170	0.8585	0.0048
−40	0.05121	2.769	374	0.749	8.174	3.3470	3.9400	0.8494	0.0048
−35	0.06614	3.521	377.2	0.765	8.581	2.6850	3.1880	0.8423	0.0048
−30	0.08438	4.426	380.3	0.781	8.991	2.1770	2.6010	0.8369	0.0048
−25	0.1064	5.506	383.4	0.798	9.402	1.7830	2.1400	0.8332	0.0048
−20	0.1327	6.784	386.6	0.816	9.816	1.4740	1.7740	0.8310	0.0049
−15	0.1639	8.287	389.6	0.835	10.230	1.2280	1.4800	0.8302	0.0049
−10	0.2006	10.04	392.7	0.854	10.660	1.0320	1.2420	0.8308	0.0050
−5	0.2433	12.08	395.7	0.875	11.080	0.8730	1.0480	0.8327	0.0051
0	0.2928	14.43	398.6	0.897	11.510	0.7434	0.8894	0.8358	0.0052
5	0.3497	17.13	401.5	0.921	11.950	0.6369	0.7580	0.8403	0.0054
10	0.4146	20.23	404.3	0.946	12.400	0.5488	0.6486	0.8461	0.0055
15	0.4884	23.76	407.1	0.972	12.860	0.4752	0.5569	0.8533	0.0057
20	0.5717	27.78	409.7	1.001	13.330	0.4135	0.4797	0.8621	0.0060
25	0.6654	32.35	412.3	1.032	13.820	0.3614	0.4142	0.8726	0.0062
30	0.7702	37.54	414.8	1.065	14.340	0.3172	0.3585	0.8849	0.0065
35	0.887	43.42	417.2	1.103	14.870	0.2794	0.3107	0.8995	0.0069
40	1.017	50.09	419.4	1.145	15.450	0.2470	0.2695	0.9168	0.0074
45	1.16	57.66	421.5	1.192	16.060	0.2191	0.2338	0.9373	0.0079
50	1.318	66.27	423.4	1.246	16.730	0.1949	0.2026	0.9618	0.0086
55	1.492	76.1	425.2	1.310	17.480	0.1739	0.1754	0.9915	0.0094
60	1.682	87.38	426.6	1.387	18.330	0.1555	0.1512	1.0280	0.0104
65	1.89	100.4	427.8	1.482	19.310	0.1394	0.1297	1.0750	0.0117
70	2.117	115.6	428.6	1.605	20.470	0.1253	0.1104	1.1350	0.0134
75	2.364	133.5	429	1.771	21.900	0.1128	0.0926	1.2170	0.0159
80	2.633	155.1	428.8	2.012	23.740	0.1017	0.0761	1.3370	0.0195
85	2.926	181.9	427.8	2.397	26.200	0.0919	0.0601	1.5290	0.0254
90	3.244	216.8	425.4	3.121	29.820	0.0832	0.0441	1.8860	0.0366
95	3.591	267.1	420.7	5.020	36.080	0.0753	0.0269	2.7990	0.0665
100	3.972	373	407.7	25.350	58.980	0.0682	0.0062	10.9300	0.3906

Source: Lemmon, E.W. et al., Thermophysical properties of fluid systems, in *NIST Chemistry WebBook*, P.J. Linstrom and W.G. Mallard (eds.), NIST Standard Reference Database Number 69, National Institute of Standards and Technology, Gaithersburg MD, http://webbook.nist.gov (retrieved on June 18, 2012)

TABLE A.8

Ideal Gas Properties of Air

T (°C)	ρ (kg/m³)	c_p (J/kg K)	$k \times 10^3$ (W/m² K)	$\mu \times 10^6$ (kg/s m)	$\nu \times 10^6$ (m²/s)	$\alpha \times 10^6$ (m²/s)	Pr
−50	1.582	1003.0	19.790	14.740	9.317	12.480	0.747
−45	1.547	1003.0	20.180	15.010	9.698	13.010	0.746
−40	1.514	1003.0	20.570	15.270	10.080	13.550	0.744
−35	1.482	1003.0	20.960	15.530	10.480	14.100	0.743
−30	1.452	1003.0	21.340	15.790	10.880	14.660	0.742
−25	1.423	1003.0	21.730	16.050	11.280	15.230	0.741
−20	1.394	1003.0	22.110	16.300	11.690	15.810	0.739
−15	1.367	1003.0	22.500	16.550	12.100	16.400	0.738
−10	1.341	1003.0	22.880	16.800	12.520	17.000	0.737
−5	1.316	1003.0	23.260	17.050	12.950	17.600	0.736
0	1.292	1004.0	23.640	17.290	13.380	18.220	0.734
5	1.269	1004.0	24.010	17.540	13.820	18.850	0.733
10	1.247	1004.0	24.390	17.780	14.260	19.480	0.732
15	1.225	1004.0	24.760	18.020	14.710	20.130	0.731
20	1.204	1004.0	25.140	18.250	15.160	20.780	0.729
25	1.184	1005.0	25.510	18.490	15.610	21.450	0.728
30	1.164	1005.0	25.880	18.720	16.080	22.120	0.727
35	1.146	1005.0	26.250	18.950	16.540	22.800	0.726
40	1.127	1005.0	26.620	19.180	17.020	23.490	0.724
45	1.110	1006.0	26.990	19.410	17.490	24.180	0.723
50	1.092	1006.0	27.350	19.630	17.970	24.890	0.722
55	1.076	1006.0	27.720	19.860	18.460	25.600	0.721
60	1.060	1007.0	28.080	20.080	18.950	26.330	0.720
65	1.044	1007.0	28.450	20.300	19.450	27.060	0.719
70	1.029	1007.0	28.810	20.520	19.950	27.790	0.718
75	1.014	1008.0	29.170	20.740	20.450	28.540	0.717
80	1.000	1008.0	29.530	20.960	20.970	29.290	0.716
85	0.986	1009.0	29.880	21.170	21.480	30.060	0.715
90	0.972	1009.0	30.240	21.390	22.000	30.820	0.714
95	0.959	1010.0	30.600	21.600	22.520	31.600	0.713
100	0.946	1010.0	30.950	21.810	23.050	32.380	0.712
105	0.934	1011.0	31.300	22.020	23.590	33.180	0.711
110	0.921	1011.0	31.650	22.230	24.120	33.970	0.710
115	0.910	1012.0	32.010	22.430	24.670	34.780	0.709
120	0.898	1012.0	32.350	22.640	25.210	35.590	0.708
125	0.887	1013.0	32.700	22.840	25.760	36.410	0.708
130	0.876	1014.0	33.050	23.050	26.320	37.240	0.707
135	0.865	1014.0	33.400	23.250	26.880	38.070	0.706
140	0.854	1015.0	33.740	23.450	27.440	38.910	0.705
145	0.844	1016.0	34.080	23.650	28.010	39.750	0.705
150	0.834	1016.0	34.430	23.850	28.590	40.600	0.704
155	0.825	1017.0	34.770	24.040	29.160	41.460	0.703
160	0.815	1018.0	35.110	24.240	29.740	42.320	0.703

TABLE A.8 (continued)

Ideal Gas Properties of Air

T (°C)	ρ (kg/m³)	c_p (J/kg K)	$k \times 10^3$ (W/m² K)	$\mu \times 10^6$ (kg/s m)	$\nu \times 10^6$ (m²/s)	$\alpha \times 10^6$ (m²/s)	Pr
165	0.806	1019.0	35.450	24.440	30.330	43.190	0.702
170	0.797	1019.0	35.790	24.630	30.920	44.070	0.702
175	0.788	1020.0	36.120	24.820	31.510	44.950	0.701
180	0.779	1021.0	36.460	25.010	32.110	45.840	0.701
185	0.771	1022.0	36.790	25.200	32.710	46.730	0.700
190	0.762	1023.0	37.130	25.390	33.320	47.630	0.700
195	0.754	1023.0	37.460	25.580	33.930	48.540	0.699
200	0.746	1024.0	37.790	25.770	34.540	49.440	0.699
250	0.675	1034.0	41.040	27.600	40.900	58.820	0.695
300	0.616	1045.0	44.180	29.340	47.640	68.660	0.694
350	0.567	1056.0	47.210	31.010	54.730	78.910	0.694
400	0.524	1068.0	50.150	32.610	62.180	89.520	0.695
450	0.488	1080.0	52.980	34.150	69.950	100.500	0.696
500	0.457	1092.0	55.720	35.630	78.040	111.700	0.698
550	0.429	1104.0	58.370	37.070	86.440	123.300	0.701
600	0.404	1115.0	60.930	38.460	95.120	135.200	0.704
650	0.382	1126.0	63.410	39.810	104.100	147.300	0.707
700	0.363	1136.0	65.810	41.110	113.300	159.700	0.710
750	0.345	1145.0	68.120	42.390	122.800	172.400	0.713
800	0.329	1154.0	70.370	43.620	132.600	185.400	0.715
850	0.314	1162.0	72.540	44.830	142.600	198.500	0.718
900	0.301	1170.0	74.650	46.000	152.900	212.000	0.721
950	0.289	1178.0	76.700	47.150	163.400	225.600	0.724
1000	0.277	1185.0	78.680	48.260	174.100	239.500	0.727

TABLE A.9

Ideal Gas Properties of Carbon Monoxide (CO)

T (°C)	ρ (kg/m³)	c_p (J/kg K)	$k \times 10^3$ (W/m² K)	$\mu \times 10^6$ (kg/s m)	$\nu \times 10^6$ (m²/s)	$\alpha \times 10^6$ (m²/s)	Pr
−50	1.530	1081.0	19.010	13.790	9.012	11.500	0.784
−45	1.496	1076.0	19.390	14.050	9.387	12.040	0.779
−40	1.464	1071.0	19.770	14.300	9.769	12.600	0.775
−35	1.433	1067.0	20.150	14.560	10.160	13.170	0.771
−30	1.404	1064.0	20.530	14.810	10.550	13.750	0.767
−25	1.376	1060.0	20.910	15.060	10.950	14.330	0.764
−20	1.348	1057.0	21.290	15.310	11.360	14.930	0.761
−15	1.322	1055.0	21.660	15.560	11.770	15.530	0.758
−10	1.297	1052.0	22.030	15.810	12.190	16.140	0.755
−5	1.273	1050.0	22.410	16.050	12.610	16.760	0.752
0	1.250	1048.0	22.780	16.290	13.040	17.390	0.750
5	1.227	1047.0	23.150	16.540	13.470	18.020	0.748
10	1.206	1045.0	23.520	16.770	13.910	18.670	0.746
15	1.185	1044.0	23.880	17.010	14.360	19.320	0.744
20	1.164	1043.0	24.250	17.250	14.810	19.970	0.742
25	1.145	1042.0	24.610	17.480	15.270	20.640	0.740
30	1.126	1041.0	24.980	17.720	15.730	21.310	0.738
35	1.108	1040.0	25.340	17.950	16.200	21.990	0.737
40	1.090	1040.0	25.700	18.180	16.670	22.680	0.735
45	1.073	1039.0	26.060	18.400	17.150	23.370	0.734
50	1.056	1039.0	26.410	18.630	17.640	24.070	0.733
55	1.040	1039.0	26.770	18.850	18.120	24.770	0.732
60	1.025	1039.0	27.120	19.080	18.620	25.490	0.731
65	1.009	1039.0	27.480	19.300	19.120	26.210	0.730
70	0.995	1039.0	27.830	19.520	19.620	26.930	0.729
75	0.981	1039.0	28.180	19.740	20.130	27.670	0.728
80	0.967	1039.0	28.530	19.950	20.640	28.400	0.727
85	0.953	1040.0	28.880	20.170	21.160	29.150	0.726
90	0.940	1040.0	29.230	20.380	21.680	29.900	0.725
95	0.927	1040.0	29.570	20.590	22.210	30.660	0.725
100	0.915	1041.0	29.920	20.800	22.740	31.420	0.724
105	0.903	1042.0	30.260	21.010	23.280	32.190	0.723
110	0.891	1042.0	30.600	21.220	23.820	32.960	0.723
115	0.879	1043.0	30.940	21.430	24.360	33.740	0.722
120	0.868	1044.0	31.280	21.630	24.910	34.530	0.722
125	0.857	1044.0	31.620	21.830	25.470	35.320	0.721
130	0.847	1045.0	31.960	22.040	26.030	36.110	0.721
135	0.836	1046.0	32.300	22.240	26.590	36.920	0.720
140	0.826	1047.0	32.630	22.440	27.150	37.720	0.720
145	0.816	1048.0	32.960	22.630	27.720	38.540	0.719
150	0.807	1049.0	33.300	22.830	28.300	39.360	0.719
155	0.797	1050.0	33.630	23.020	28.880	40.180	0.719
160	0.788	1051.0	33.960	23.220	29.460	41.010	0.718

TABLE A.9 (continued)

Ideal Gas Properties of Carbon Monoxide (CO)

T (°C)	ρ (kg/m³)	c_p (J/kg K)	$k \times 10^3$ (W/m² K)	$\mu \times 10^6$ (kg/s m)	$\nu \times 10^6$ (m²/s)	$\alpha \times 10^6$ (m²/s)	Pr
165	0.779	1052.0	34.290	23.410	30.050	41.840	0.718
170	0.770	1053.0	34.620	23.600	30.640	42.680	0.718
175	0.762	1054.0	34.940	23.790	31.230	43.530	0.718
180	0.753	1055.0	35.270	23.980	31.830	44.370	0.717
185	0.745	1056.0	35.590	24.160	32.430	45.230	0.717
190	0.737	1057.0	35.920	24.350	33.040	46.090	0.717
195	0.729	1059.0	36.240	24.530	33.650	46.950	0.717
200	0.721	1060.0	36.560	24.720	34.260	47.820	0.716
250	0.653	1072.0	39.720	26.480	40.580	56.780	0.715
300	0.596	1085.0	42.770	28.120	47.220	66.190	0.713
350	0.548	1098.0	45.730	29.660	54.150	76.030	0.712
400	0.507	1111.0	48.600	31.110	61.360	86.280	0.711
450	0.472	1123.0	51.390	32.490	68.820	96.930	0.710
500	0.442	1135.0	54.120	33.790	76.530	108.000	0.709
550	0.415	1147.0	56.770	35.030	84.480	119.400	0.708
600	0.391	1157.0	59.370	36.230	92.680	131.200	0.706
650	0.370	1168.0	61.910	37.400	101.100	143.400	0.705
700	0.351	1178.0	64.420	38.540	109.900	156.000	0.705
750	0.334	1187.0	66.890	39.670	118.900	168.900	0.704
800	0.318	1196.0	69.330	40.800	128.300	182.300	0.704
850	0.304	1204.0	71.740	41.950	138.000	196.100	0.704
900	0.291	1212.0	74.150	43.120	148.200	210.300	0.705
950	0.279	1219.0	76.550	44.320	158.800	224.900	0.706
1000	0.268	1226.0	78.940	45.570	170.000	240.100	0.708

TABLE A.10

Ideal Gas Properties of Carbon Dioxide (CO_2)

T (°C)	ρ (kg/m³)	c_p (J/kg K)	$k \times 10^3$ (W/m² K)	$\mu \times 10^6$ (kg/s m)	$\nu \times 10^6$ (m²/s)	$\alpha \times 10^6$ (m²/s)	Pr
−50	2.403	746.0	10.510	11.290	4.699	5.860	0.802
−45	2.351	753.0	10.910	11.540	4.910	6.165	0.797
−40	2.300	759.9	11.320	11.790	5.126	6.475	0.792
−35	2.252	766.7	11.730	12.040	5.346	6.791	0.787
−30	2.206	773.3	12.130	12.290	5.570	7.111	0.783
−25	2.161	779.9	12.540	12.530	5.799	7.437	0.780
−20	2.119	786.3	12.940	12.780	6.031	7.768	0.777
−15	2.078	792.6	13.350	13.020	6.268	8.104	0.774
−10	2.038	798.9	13.750	13.270	6.509	8.445	0.771
−5	2.000	805.0	14.150	13.510	6.754	8.790	0.768
0	1.964	811.0	14.560	13.750	7.003	9.141	0.766
5	1.928	816.9	14.960	13.990	7.256	9.497	0.764
10	1.894	822.8	15.360	14.230	7.513	9.858	0.762
15	1.861	828.5	15.770	14.470	7.775	10.220	0.761
20	1.830	834.2	16.170	14.710	8.040	10.590	0.759
25	1.799	839.8	16.570	14.950	8.309	10.970	0.758
30	1.769	845.3	16.970	15.180	8.582	11.350	0.756
35	1.741	850.7	17.380	15.420	8.860	11.730	0.755
40	1.713	856.1	17.780	15.660	9.141	12.120	0.754
45	1.686	861.4	18.180	15.890	9.426	12.520	0.753
50	1.660	866.5	18.580	16.120	9.714	12.920	0.752
55	1.634	871.7	18.980	16.360	10.010	13.320	0.751
60	1.610	876.7	19.380	16.590	10.300	13.730	0.751
65	1.586	881.7	19.780	16.820	10.600	14.140	0.750
70	1.563	886.6	20.180	17.050	10.910	14.560	0.749
75	1.541	891.5	20.580	17.280	11.220	14.980	0.749
80	1.519	896.3	20.980	17.510	11.530	15.410	0.748
85	1.498	901.0	21.380	17.740	11.840	15.840	0.748
90	1.477	905.7	21.770	17.960	12.160	16.280	0.747
95	1.457	910.3	22.170	18.190	12.490	16.720	0.747
100	1.437	914.8	22.570	18.410	12.810	17.160	0.746
105	1.418	919.3	22.970	18.640	13.140	17.610	0.746
110	1.400	923.8	23.360	18.860	13.480	18.070	0.746
115	1.382	928.2	23.760	19.090	13.810	18.530	0.746
120	1.364	932.5	24.160	19.310	14.150	18.990	0.745
125	1.347	936.8	24.550	19.530	14.500	19.460	0.745
130	1.330	941.0	24.950	19.750	14.850	19.930	0.745
135	1.314	945.2	25.340	19.970	15.200	20.400	0.745
140	1.298	949.3	25.740	20.190	15.550	20.880	0.745
145	1.283	953.3	26.130	20.410	15.910	21.370	0.745
150	1.267	957.4	26.520	20.630	16.270	21.860	0.745
155	1.253	961.4	26.920	20.840	16.640	22.350	0.744
160	1.238	965.3	27.310	21.060	17.010	22.850	0.744

TABLE A.10 (continued)

Ideal Gas Properties of Carbon Dioxide (CO_2)

T (°C)	ρ (kg/m³)	c_p (J/kg K)	$k \times 10^3$ (W/m² K)	$\mu \times 10^6$ (kg/s m)	$\nu \times 10^6$ (m²/s)	$\alpha \times 10^6$ (m²/s)	Pr
165	1.224	969.2	27.700	21.280	17.380	23.350	0.744
170	1.210	973.0	28.100	21.490	17.760	23.860	0.744
175	1.197	976.8	28.490	21.700	18.140	24.370	0.744
180	1.184	980.6	28.880	21.920	18.520	24.880	0.744
185	1.171	984.3	29.270	22.130	18.900	25.400	0.744
190	1.158	987.9	29.660	22.340	19.290	25.930	0.744
195	1.146	991.6	30.050	22.550	19.690	26.450	0.744
200	1.134	995.2	30.440	22.760	20.080	26.980	0.744
250	0.675	1034.0	41.040	27.600	40.900	58.820	0.695
300	0.616	1045.0	44.180	29.340	47.640	68.660	0.694
350	0.567	1056.0	47.210	31.010	54.730	78.910	0.694
400	0.524	1068.0	50.150	32.610	62.180	89.520	0.695
450	0.488	1080.0	52.980	34.150	69.950	100.500	0.696
500	0.457	1092.0	55.720	35.630	78.040	111.700	0.698
550	0.429	1104.0	58.370	37.070	86.440	123.300	0.701
600	0.404	1115.0	60.930	38.460	95.120	135.200	0.704
650	0.382	1126.0	63.410	39.810	104.100	147.300	0.707
700	0.363	1136.0	65.810	41.110	113.300	159.700	0.710
750	0.345	1145.0	68.120	42.390	122.800	172.400	0.713
800	0.329	1154.0	70.370	43.620	132.600	185.400	0.715
850	0.314	1162.0	72.540	44.830	142.600	198.500	0.718
900	0.301	1170.0	74.650	46.000	152.900	212.000	0.721
950	0.289	1178.0	76.700	47.150	163.400	225.600	0.724
1000	0.277	1185.0	78.680	48.260	174.100	239.500	0.727

TABLE A.11

Ideal Gas Properties of Helium (He)

T (°C)	ρ (kg/m³)	c_p (J/kg K)	$k \times 10^3$ (W/m² K)	$\mu \times 10^6$ (kg/s m)	$\nu \times 10^6$ (m²/s)	$\alpha \times 10^6$ (m²/s)	Pr
−50	0.219	5193.0	127.200	16.300	74.620	112.100	0.666
−45	0.214	5193.0	129.100	16.550	77.440	116.400	0.665
−40	0.209	5193.0	131.100	16.790	80.300	120.700	0.665
−35	0.205	5193.0	133.000	17.030	83.210	125.100	0.665
−30	0.201	5193.0	134.900	17.270	86.160	129.600	0.665
−25	0.197	5193.0	136.800	17.510	89.150	134.100	0.665
−20	0.193	5193.0	138.700	17.750	92.180	138.700	0.665
−15	0.189	5193.0	140.600	17.990	95.260	143.400	0.664
−10	0.185	5193.0	142.500	18.230	98.380	148.100	0.664
−5	0.182	5193.0	144.300	18.460	101.500	152.900	0.664
0	0.179	5193.0	146.200	18.690	104.700	157.700	0.664
5	0.175	5193.0	148.000	18.930	108.000	162.600	0.664
10	0.172	5193.0	149.900	19.160	111.300	167.600	0.664
15	0.169	5193.0	151.700	19.390	114.600	172.600	0.664
20	0.166	5193.0	153.500	19.620	118.000	177.700	0.664
25	0.164	5193.0	155.300	19.850	121.400	182.900	0.664
30	0.161	5193.0	157.100	20.070	124.800	188.100	0.664
35	0.158	5193.0	158.900	20.300	128.300	193.400	0.663
40	0.156	5193.0	160.700	20.520	131.800	198.700	0.663
45	0.153	5193.0	162.400	20.750	135.400	204.100	0.663
50	0.151	5193.0	164.200	20.970	139.000	209.600	0.663
55	0.149	5193.0	166.000	21.190	142.600	215.100	0.663
60	0.146	5193.0	167.700	21.410	146.300	220.700	0.663
65	0.144	5193.0	169.400	21.630	150.000	226.300	0.663
70	0.142	5193.0	171.200	21.850	153.800	232.000	0.663
75	0.140	5193.0	172.900	22.070	157.600	237.700	0.663
80	0.138	5193.0	174.600	22.290	161.400	243.500	0.663
85	0.136	5193.0	176.300	22.510	165.300	249.400	0.663
90	0.134	5193.0	178.000	22.720	169.200	255.300	0.663
95	0.132	5193.0	179.700	22.940	173.200	261.300	0.663
100	0.131	5193.0	181.400	23.150	177.200	267.300	0.663
105	0.129	5193.0	183.100	23.370	181.200	273.400	0.663
110	0.127	5193.0	184.800	23.580	185.300	279.600	0.663
115	0.126	5193.0	186.400	23.790	189.400	285.800	0.663
120	0.124	5193.0	188.100	24.000	193.500	292.000	0.663
125	0.123	5193.0	189.800	24.210	197.700	298.400	0.663
130	0.121	5193.0	191.400	24.420	201.900	304.700	0.663
135	0.120	5193.0	193.100	24.630	206.200	311.200	0.663
140	0.118	5193.0	194.700	24.840	210.500	317.600	0.663
145	0.117	5193.0	196.300	25.050	214.800	324.200	0.663
150	0.115	5193.0	198.000	25.260	219.200	330.800	0.663
155	0.114	5193.0	199.600	25.460	223.600	337.400	0.663
160	0.113	5193.0	201.200	25.670	228.000	344.100	0.663

TABLE A.11 (continued)

Ideal Gas Properties of Helium (He)

T (°C)	ρ (kg/m³)	c_p (J/kg K)	$k \times 10^3$ (W/m² K)	$\mu \times 10^6$ (kg/s m)	$\nu \times 10^6$ (m²/s)	$\alpha \times 10^6$ (m²/s)	Pr
165	0.111	5193.0	202.800	25.870	232.500	350.900	0.663
170	0.110	5193.0	204.400	26.080	237.000	357.700	0.663
175	0.109	5193.0	206.000	26.280	241.500	364.600	0.663
180	0.108	5193.0	207.600	26.490	246.100	371.500	0.663
185	0.106	5193.0	209.200	26.690	250.700	378.500	0.663
190	0.105	5193.0	210.800	26.890	255.400	385.500	0.663
195	0.104	5193.0	212.400	27.090	260.100	392.600	0.663
200	0.103	5193.0	213.900	27.290	264.800	399.700	0.663
250	0.093	5193.0	229.400	29.270	314.000	473.900	0.663
300	0.085	5193.0	244.500	31.200	366.700	553.300	0.663
350	0.078	5193.0	259.200	33.080	422.700	637.600	0.663
400	0.072	5193.0	273.500	34.920	482.000	726.900	0.663
450	0.067	5193.0	287.500	36.720	544.500	820.900	0.663
500	0.063	5193.0	301.200	38.490	610.200	919.600	0.664
550	0.059	5193.0	314.700	40.230	679.000	1023.000	0.664
600	0.056	5193.0	328.000	41.940	750.800	1131.000	0.664
650	0.053	5193.0	341.000	43.620	825.700	1243.000	0.664
700	0.050	5193.0	353.800	45.280	903.400	1359.000	0.665
750	0.048	5193.0	366.400	46.910	984.100	1480.000	0.665
800	0.045	5193.0	378.900	48.530	1068.000	1605.000	0.665
850	0.043	5193.0	391.100	50.120	1154.000	1734.000	0.665
900	0.042	5193.0	403.200	51.690	1243.000	1868.000	0.666
950	0.040	5193.0	415.200	53.240	1335.000	2005.000	0.666
1000	0.038	5193.0	427.000	54.780	1430.000	2146.000	0.666

TABLE A.12

Ideal Gas Properties of Diatomic Hydrogen (H$_2$)

T (°C)	ρ (kg/m³)	c_p (J/kg K)	k × 10³ (W/m² K)	μ × 10⁶ (kg/s m)	ν × 10⁶ (m²/s)	α × 10⁶ (m²/s)	Pr
−50	0.110	12,635.0	140.400	7.293	66.240	100.900	0.656
−45	0.108	12,834.0	143.000	7.406	68.780	103.500	0.665
−40	0.105	13,013.0	145.500	7.518	71.350	106.100	0.672
−35	0.103	13,173.0	148.100	7.630	73.960	108.900	0.679
−30	0.101	13,317.0	150.600	7.741	76.610	111.900	0.685
−25	0.099	13,446.0	153.000	7.851	79.290	115.000	0.690
−20	0.097	13,563.0	155.500	7.960	82.020	118.100	0.694
−15	0.095	13,667.0	158.000	8.068	84.780	121.400	0.698
−10	0.093	13,761.0	160.400	8.176	87.580	124.800	0.702
−5	0.092	13,845.0	162.800	8.284	90.410	128.300	0.705
0	0.090	13,920.0	165.200	8.391	93.290	131.900	0.707
5	0.088	13,988.0	167.600	8.497	96.190	135.600	0.709
10	0.087	14,048.0	169.900	8.602	99.140	139.400	0.711
15	0.085	14,102.0	172.200	8.707	102.100	143.200	0.713
20	0.084	14,151.0	174.600	8.812	105.100	147.200	0.714
25	0.082	14,194.0	176.900	8.916	108.200	151.200	0.716
30	0.081	14,233.0	179.100	9.019	111.300	155.300	0.717
35	0.080	14,267.0	181.400	9.122	114.400	159.500	0.717
40	0.078	14,298.0	183.700	9.224	117.600	163.700	0.718
45	0.077	14,325.0	185.900	9.326	120.800	168.000	0.719
50	0.076	14,349.0	188.100	9.427	124.000	172.400	0.719
55	0.075	14,370.0	190.300	9.528	127.300	176.900	0.719
60	0.074	14,389.0	192.500	9.628	130.600	181.400	0.720
65	0.073	14,406.0	194.700	9.728	133.900	186.000	0.720
70	0.072	14,420.0	196.800	9.828	137.300	190.700	0.720
75	0.071	14,432.0	199.000	9.927	140.700	195.400	0.720
80	0.070	14,443.0	201.100	10.030	144.100	200.200	0.720
85	0.069	14,453.0	203.200	10.120	147.600	205.000	0.720
90	0.068	14,461.0	205.300	10.220	151.100	209.900	0.720
95	0.067	14,467.0	207.400	10.320	154.600	214.800	0.720
100	0.066	14,473.0	209.500	10.420	158.200	219.900	0.720
105	0.065	14,478.0	211.600	10.510	161.800	224.900	0.719
110	0.064	14,482.0	213.600	10.610	165.400	230.000	0.719
115	0.063	14,485.0	215.700	10.700	169.100	235.200	0.719
120	0.062	14,487.0	217.700	10.800	172.800	240.400	0.719
125	0.062	14,489.0	219.700	10.900	176.600	245.700	0.719
130	0.061	14,491.0	221.700	10.990	180.300	251.100	0.718
135	0.060	14,491.0	223.700	11.080	184.100	256.400	0.718
140	0.059	14,492.0	225.700	11.180	188.000	261.900	0.718
145	0.059	14,492.0	227.600	11.270	191.800	267.300	0.718
150	0.058	14,492.0	229.600	11.370	195.700	272.900	0.717
155	0.057	14,491.0	231.500	11.460	199.700	278.400	0.717
160	0.057	14,491.0	233.500	11.550	203.700	284.100	0.717

TABLE A.12 (continued)

Ideal Gas Properties of Diatomic Hydrogen (H_2)

T (°C)	ρ (kg/m³)	c_p (J/kg K)	$k \times 10^3$ (W/m² K)	$\mu \times 10^6$ (kg/s m)	$\nu \times 10^6$ (m²/s)	$\alpha \times 10^6$ (m²/s)	Pr
165	0.056	14,490.0	235.400	11.640	207.700	289.700	0.717
170	0.055	14,489.0	237.300	11.740	211.700	295.400	0.717
175	0.055	14,488.0	239.200	11.830	215.700	301.200	0.716
180	0.054	14,487.0	241.100	11.920	219.800	307.000	0.716
185	0.054	14,486.0	243.000	12.010	224.000	312.800	0.716
190	0.053	14,485.0	244.900	12.100	228.100	318.700	0.716
195	0.052	14,484.0	246.800	12.190	232.300	324.600	0.716
200	0.052	14,482.0	248.600	12.280	236.500	330.600	0.716
250	0.653	1072.0	39.720	26.480	40.580	56.780	0.715
300	0.596	1085.0	42.770	28.120	47.220	66.190	0.713
350	0.548	1098.0	45.730	29.660	54.150	76.030	0.712
400	0.507	1111.0	48.600	31.110	61.360	86.280	0.711
450	0.472	1123.0	51.390	32.490	68.820	96.930	0.710
500	0.442	1135.0	54.120	33.790	76.530	108.000	0.709
550	0.415	1147.0	56.770	35.030	84.480	119.400	0.708
600	0.391	1157.0	59.370	36.230	92.680	131.200	0.706
650	0.370	1168.0	61.910	37.400	101.100	143.400	0.705
700	0.351	1178.0	64.420	38.540	109.900	156.000	0.705
750	0.334	1187.0	66.890	39.670	118.900	168.900	0.704
800	0.318	1196.0	69.330	40.800	128.300	182.300	0.704
850	0.304	1204.0	71.740	41.950	138.000	196.100	0.704
900	0.291	1212.0	74.150	43.120	148.200	210.300	0.705
950	0.279	1219.0	76.550	44.320	158.800	224.900	0.706
1,000	0.268	1226.0	78.940	45.570	170.000	240.100	0.708

TABLE A.13

Ideal Gas Properties of Diatomic Nitrogen (N₂)

T (°C)	ρ (kg/m³)	c_p (J/kg K)	$k \times 10^3$ (W/m² K)	$\mu \times 10^6$ (kg/s m)	$\nu \times 10^6$ (m²/s)	$\alpha \times 10^6$ (m²/s)	Pr
−50	1.530	957.3	20.010	13.910	9.091	13.660	0.666
−45	1.496	972.9	20.400	14.170	9.467	14.010	0.676
−40	1.464	986.0	20.790	14.420	9.848	14.400	0.684
−35	1.434	996.9	21.180	14.670	10.240	14.820	0.691
−30	1.404	1006.0	21.570	14.930	10.630	15.270	0.696
−25	1.376	1013.0	21.950	15.180	11.030	15.740	0.701
−20	1.349	1020.0	22.330	15.420	11.440	16.240	0.704
−15	1.322	1025.0	22.710	15.670	11.850	16.760	0.707
−10	1.297	1029.0	23.090	15.920	12.270	17.300	0.709
−5	1.273	1032.0	23.460	16.160	12.690	17.860	0.711
0	1.250	1035.0	23.840	16.400	13.120	18.430	0.712
5	1.227	1037.0	24.210	16.640	13.560	19.020	0.713
10	1.206	1039.0	24.580	16.880	14.000	19.630	0.713
15	1.185	1040.0	24.950	17.120	14.450	20.250	0.714
20	1.165	1041.0	25.310	17.360	14.910	20.890	0.714
25	1.145	1041.0	25.680	17.590	15.360	21.530	0.714
30	1.126	1042.0	26.040	17.830	15.830	22.190	0.713
35	1.108	1042.0	26.400	18.060	16.300	22.860	0.713
40	1.090	1042.0	26.750	18.290	16.780	23.550	0.712
45	1.073	1042.0	27.110	18.520	17.260	24.240	0.712
50	1.056	1042.0	27.460	18.740	17.740	24.940	0.711
55	1.040	1042.0	27.820	18.970	18.230	25.650	0.711
60	1.025	1042.0	28.170	19.190	18.730	26.380	0.710
65	1.010	1042.0	28.510	19.420	19.230	27.110	0.710
70	0.995	1042.0	28.860	19.640	19.740	27.850	0.709
75	0.981	1042.0	29.210	19.860	20.250	28.590	0.708
80	0.967	1042.0	29.550	20.080	20.770	29.350	0.708
85	0.953	1041.0	29.890	20.300	21.290	30.110	0.707
90	0.940	1041.0	30.230	20.510	21.820	30.880	0.707
95	0.927	1041.0	30.570	20.730	22.350	31.650	0.706
100	0.915	1041.0	30.900	20.940	22.890	32.440	0.706
105	0.903	1041.0	31.230	21.150	23.430	33.230	0.705
110	0.891	1041.0	31.570	21.360	23.980	34.020	0.705
115	0.880	1041.0	31.900	21.570	24.530	34.820	0.704
120	0.868	1042.0	32.230	21.780	25.080	35.630	0.704
125	0.857	1042.0	32.550	21.990	25.640	36.440	0.704
130	0.847	1042.0	32.880	22.190	26.210	37.260	0.703
135	0.836	1042.0	33.200	22.400	26.780	38.080	0.703
140	0.826	1043.0	33.520	22.600	27.350	38.910	0.703
145	0.816	1043.0	33.850	22.800	27.930	39.750	0.703
150	0.807	1043.0	34.160	23.000	28.510	40.580	0.703
155	0.797	1044.0	34.480	23.200	29.100	41.430	0.702
160	0.788	1044.0	34.800	23.400	29.690	42.270	0.702

TABLE A.13 (continued)

Ideal Gas Properties of Diatomic Nitrogen (N_2)

T (°C)	ρ (kg/m³)	c_p (J/kg K)	$k \times 10^3$ (W/m² K)	$\mu \times 10^6$ (kg/s m)	$\nu \times 10^6$ (m²/s)	$\alpha \times 10^6$ (m²/s)	Pr
165	0.779	1045.0	35.110	23.600	30.290	43.130	0.702
170	0.770	1046.0	35.430	23.790	30.890	43.980	0.702
175	0.762	1046.0	35.740	23.990	31.490	44.840	0.702
180	0.753	1047.0	36.050	24.180	32.100	45.710	0.702
185	0.745	1047.0	36.360	24.370	32.710	46.580	0.702
190	0.737	1048.0	36.660	24.560	33.330	47.450	0.702
195	0.729	1049.0	36.970	24.750	33.950	48.330	0.702
200	0.722	1050.0	37.270	24.940	34.570	49.210	0.703
250	0.653	1059.0	40.240	26.770	41.020	58.220	0.705
300	0.596	1070.0	43.090	28.490	47.830	67.580	0.708
350	0.548	1083.0	45.830	30.120	54.980	77.260	0.712
400	0.507	1095.0	48.480	31.660	62.420	87.270	0.715
450	0.472	1108.0	51.050	33.120	70.160	97.620	0.719
500	0.442	1120.0	53.580	34.510	78.160	108.300	0.722
550	0.415	1132.0	56.060	35.840	86.420	119.400	0.724
600	0.391	1143.0	58.520	37.110	94.920	130.900	0.725
650	0.370	1154.0	60.970	38.340	103.700	142.900	0.725
700	0.351	1164.0	63.440	39.510	112.600	155.400	0.725
750	0.334	1173.0	65.930	40.650	121.800	168.400	0.724
800	0.318	1182.0	68.470	41.760	131.300	182.000	0.721
850	0.304	1191.0	71.060	42.840	140.900	196.300	0.718
900	0.291	1199.0	73.740	43.890	150.800	211.400	0.714
950	0.279	1206.0	76.510	44.920	160.900	227.200	0.708
1000	0.268	1213.0	79.380	45.940	171.300	244.000	0.702

TABLE A.14

Ideal Gas Properties of Diatomic Oxygen (O_2)

T (°C)	ρ (kg/m³)	c_p (J/kg K)	$k \times 10^3$ (W/m² K)	$\mu \times 10^6$ (kg/s m)	$\nu \times 10^6$ (m²/s)	$\alpha \times 10^6$ (m²/s)	Pr
−50	1.748	984.4	20.670	16.160	9.246	12.020	0.769
−45	1.709	974.8	21.080	16.470	9.636	12.650	0.762
−40	1.673	966.3	21.490	16.780	10.030	13.300	0.754
−35	1.637	958.9	21.900	17.090	10.430	13.950	0.748
−30	1.604	952.5	22.310	17.390	10.840	14.610	0.743
−25	1.571	946.8	22.720	17.690	11.260	15.270	0.738
−20	1.540	942.0	23.120	17.990	11.680	15.930	0.733
−15	1.511	937.8	23.520	18.290	12.110	16.600	0.729
−10	1.482	934.2	23.920	18.580	12.540	17.280	0.726
−5	1.454	931.2	24.320	18.870	12.980	17.960	0.723
0	1.428	928.7	24.720	19.160	13.420	18.650	0.720
5	1.402	926.6	25.120	19.450	13.870	19.340	0.717
10	1.377	924.9	25.520	19.740	14.330	20.030	0.715
15	1.353	923.5	25.920	20.020	14.790	20.730	0.713
20	1.330	922.5	26.310	20.300	15.260	21.440	0.712
25	1.308	921.8	26.700	20.580	15.730	22.150	0.710
30	1.286	921.4	27.100	20.850	16.210	22.860	0.709
35	1.266	921.1	27.490	21.130	16.690	23.580	0.708
40	1.245	921.1	27.880	21.400	17.180	24.310	0.707
45	1.226	921.3	28.280	21.670	17.680	25.040	0.706
50	1.207	921.7	28.670	21.940	18.180	25.770	0.705
55	1.188	922.2	29.060	22.200	18.680	26.510	0.705
60	1.171	922.9	29.450	22.470	19.190	27.260	0.704
65	1.153	923.7	29.840	22.730	19.710	28.010	0.704
70	1.136	924.6	30.220	22.990	20.230	28.770	0.703
75	1.120	925.6	30.610	23.250	20.750	29.530	0.703
80	1.104	926.7	31.000	23.500	21.280	30.290	0.703
85	1.089	927.8	31.390	23.760	21.820	31.070	0.702
90	1.074	929.1	31.770	24.010	22.360	31.850	0.702
95	1.059	930.4	32.160	24.260	22.910	32.630	0.702
100	1.045	931.8	32.540	24.510	23.460	33.420	0.702
105	1.031	933.2	32.930	24.760	24.010	34.220	0.702
110	1.018	934.7	33.310	25.010	24.570	35.020	0.702
115	1.005	936.2	33.700	25.260	25.140	35.820	0.702
120	0.992	937.7	34.080	25.500	25.710	36.640	0.702
125	0.979	939.3	34.460	25.740	26.280	37.460	0.702
130	0.967	940.9	34.840	25.980	26.860	38.280	0.702
135	0.955	942.6	35.220	26.220	27.450	39.110	0.702
140	0.944	944.2	35.610	26.460	28.040	39.950	0.702
145	0.933	945.9	35.990	26.700	28.630	40.790	0.702
150	0.922	947.6	36.370	26.940	29.230	41.640	0.702
155	0.911	949.3	36.750	27.170	29.830	42.500	0.702
160	0.900	951.0	37.130	27.400	30.440	43.360	0.702

TABLE A.14 (continued)

Ideal Gas Properties of Diatomic Oxygen (O$_2$)

T (°C)	ρ (kg/m³)	c_p (J/kg K)	$k \times 10^3$ (W/m² K)	$\mu \times 10^6$ (kg/s m)	$\nu \times 10^6$ (m²/s)	$\alpha \times 10^6$ (m²/s)	Pr
165	0.890	952.7	37.500	27.640	31.050	44.230	0.702
170	0.880	954.4	37.880	27.870	31.670	45.100	0.702
175	0.870	956.1	38.260	28.100	32.290	45.990	0.702
180	0.861	957.8	38.640	28.330	32.910	46.870	0.702
185	0.851	959.6	39.010	28.550	33.540	47.770	0.702
190	0.842	961.3	39.390	28.780	34.180	48.660	0.702
195	0.833	963.0	39.760	29.000	34.820	49.570	0.702
200	0.824	964.7	40.140	29.230	35.460	50.480	0.703
250	0.745	981.4	43.850	31.410	42.140	59.950	0.703
300	0.680	997.1	47.510	33.500	49.230	70.030	0.703
350	0.626	1011.0	51.100	35.500	56.740	80.740	0.703
400	0.579	1025.0	54.630	37.440	64.630	92.040	0.702
450	0.539	1037.0	58.090	39.320	72.910	103.900	0.702
500	0.504	1048.0	61.480	41.140	81.560	116.400	0.701
550	0.474	1058.0	64.800	42.910	90.580	129.300	0.700
600	0.447	1067.0	68.050	44.640	99.960	142.800	0.700
650	0.422	1075.0	71.240	46.330	109.700	156.800	0.699
700	0.401	1083.0	74.360	47.990	119.800	171.300	0.699
750	0.381	1091.0	77.430	49.610	130.200	186.300	0.699
800	0.363	1097.0	80.430	51.200	140.900	201.700	0.699
850	0.347	1104.0	83.390	52.770	152.000	217.600	0.699
900	0.332	1110.0	86.300	54.310	163.400	233.900	0.699
950	0.319	1116.0	89.160	55.820	175.100	250.700	0.699
1000	0.306	1121.0	91.980	57.320	187.100	267.800	0.699

TABLE A.15

Surface Emissivity

Material	Temperature °C	Emissivity
Alloys		
20-Ni, 24-CR, 55-FE, oxidized	200	0.9
20-Ni, 24-CR, 55-FE, oxidized	500	0.97
60-Ni, 12-CR, 28-FE, oxidized	270	0.89
60-Ni, 12-CR, 28-FE, oxidized	560	0.82
80-Ni, 20-CR, oxidized	100	0.87
80-Ni, 20-CR, oxidized	600	0.87
80-Ni, 20-CR, oxidized	1300	0.89
Aluminum		
Unoxidized	25	0.02
Unoxidized	100	0.03
Unoxidized	500	0.06
Oxidized	199	0.11
Oxidized	599	0.19
Heavily oxidized	93	0.2
Heavily oxidized	504	0.31
Highly polished	100	0.09
Roughly polished	100	0.18
Commercial sheet	100	0.09
Highly polished plate	227	0.04
Highly polished plate	577	0.06
Bright rolled plate	170	0.04
Bright rolled plate	500	0.05
Alloy A3003, oxidized	316	0.4
Alloy A3003, oxidized	482	0.4
Alloy 1100-0	93–427	0.05
Alloy 24ST	24	0.09
Alloy 24ST, polished	24	0.09
Alloy 75ST	24	0.11
Alloy 75ST, polished	24	0.08
Bismuth, bright	80	0.34
Bismuth, unoxidized	25	0.05
Brass		
Polished	247	0.03
83% Cu, 17% Zn, polished	277	0.03
Matte	20	0.07
Burnished to brown color	20	0.4
Unoxidized	25	0.04
Unoxidized	100	0.04
Cadmium	25	0.02
Carbon		
Lampblack	25	0.95
Unoxidized	25	0.81

TABLE A.15 (continued)

Surface Emissivity

Material	Temperature °C	Emissivity
Unoxidized	100	0.81
Unoxidized	500	0.79
Candle soot	121	0.95
Filament	260	0.95
Graphitized	100	0.76
Graphitized	300	0.75
Graphitized	500	0.71
Chromium	38	0.08
Chromium	538	0.26
Chromium, polished	150	0.06
Cobalt, unoxidized	500	0.13
Cobalt, unoxidized	1000	0.23
Columbium, unoxidized	816	0.19
Columbium, unoxidized	1093	0.24
Copper		
Cuprous oxide	38	0.87
Cuprous oxide	260	0.83
Cuprous oxide	538	0.77
Black, oxidized	38	0.78
Matte	38	0.22
Polished	38	0.03
Rolled	38	0.64
Rough	38	0.74
Nickel plated	38–260	0.37
Gold		
Enamel	100	0.37
Polished	38–260	0.02
Polished	538–1093	0.03
Inconel sheet	538	0.28
Iron		
Oxidized	100	0.74
Oxidized	499	0.84
Oxidized	1199	0.89
Unoxidized	100	0.05
Rusted	25	0.65
Cast iron		
Oxidized	199	0.64
Oxidized	599	0.78
Unoxidized	100	0.21
Wrought iron		
Dull	25	0.94
Smooth	38	0.35
Polished	38	0.28

(continued)

TABLE A.15 (continued)

Surface Emissivity

Material	Temperature °C	Emissivity
Lead		
Polished	38–260	0.06–0.08
Rough	38	0.43
Oxidized	38	0.43
Magnesium	38–260	0.07–0.13
Magnesium oxide	1027–1727	0.16–0.20
Mercury	25	0.1
Molybdenum	38	0.06
Monel, Ni-Cu	200	0.41
Monel, Ni-Cu	400	0.44
Monel, Ni-Cu	600	0.46
Nickel		
Polished	38–260	0.31–0.46
Oxidized	25	0.05
Unoxidized	500	0.12
Electrolytic	260	0.06
Platinum	38	0.05
Platinum, black	38	0.93
Silver		
Plate (0.0005 on Ni)	93–371	0.06–0.07
Polished	38	0.01
Steel		
Cold rolled	93	0.75–0.85
Ground sheet	938–1099	0.55–0.61
Polished sheet	38	0.07
Polished sheet	260	0.1
Polished sheet	538	0.14
Mild steel, polished	24	0.1
Mild steel, smooth	24	0.12
Mild steel, Liquid	1599–1793	0.28
Steel, unoxidized	100	0.08
Steel, oxidized	25	0.8
Steel alloys		
Type 301, polished	24	0.27
Type 303, oxidized	316–1093	0.74–0.87
Type 310, rolled	816–1149	0.56–0.81
Type 316, polished	24	0.28
Type 321	93–427	0.27–0.32
Type 347, oxidized	316–1093	0.87–0.91
Type 350	93–427	0.18–0.27
Stellite, polished	20	0.18
Tantalum, unoxidized	727	0.14
Tin, unoxidized	25	0.04
Tinned iron, bright	24	0.05

TABLE A.15 (continued)

Surface Emissivity

Material	Temperature °C	Emissivity
Titanium		
Polished	149–649	0.08–0.19
Filament (aged)	38	0.03
Filament (aged)	2760	0.35
Zinc		
Bright, galvanized	38	0.23
Commercial 99.1%	260	0.05
Galvanized	38	0.28
Polished	38	0.02
Non-metals		
Asbestos board	38	0.96
Basalt	20	0.72
Brick		
Red, rough	21	0.93
Gault cream	1371–2760	0.26–0.30
Fire clay	1371	0.75
Light buff	538	0.8
Lime clay	1371	0.43
Fire brick	1000	0.75–0.80
Ceramic		
Alumina on inconel	427–1093	0.69–0.45
Earthenware, glazed	21	0.9
Porcelain	22	0.92
Clay	20	0.39
Fired	70	0.91
Shale	20	0.69
Tiles, light red	1371–2760	0.32–0.34
Concrete		
Rough	0–1093	0.94
Tiles, natural	1371–2760	0.63–0.62
Cotton cloth	20	0.77
Glass		
Convex D	100	0.8
Nonex	100	0.82
Smooth	0–93	0.92–0.94
Granite	21	0.45
Gravel	38	0.28
Gypsum	20	0.80–0.90
Ice, smooth	0	0.97
Ice, rough	0	0.98
Paints		
Blue, Cu_2O_3	24	0.94
Black, CuO	24	0.96
Green, Cu_2O_3	24	0.92
Red, Fe_2O_3	24	0.91

(*continued*)

TABLE A.15 (continued)

Surface Emissivity

Material	Temperature °C	Emissivity
White, Al_2O_3	24	0.94
White, Y_2O_3	24	0.9
White, ZnO	24	0.95
White, $MgCO_3$	24	0.91
White, ZrO_2	24	0.95
White, ThO_2	24	0.9
White, MgO	24	0.91
White, $PbCO_3$	24	0.93
Yellow, PbO	24	0.9
Yellow, $PbCrO_4$	24	0.93
Red lead	100	0.93
Rubber, hard	23	0.94
Rubber, soft, gray	24	0.86
Sand	20	0.76
Sandstone	38	0.67
Sandstone, red	38	0.60–0.83
Sawdust	20	0.75
Snow, fine particles	−7	0.82
Soil		
Surface	38	0.38
Black loam	20	0.66
Plowed field	20	0.38
Soot		
Acetylene	24	0.97
Camphor	24	0.94
Candle	121	0.95
Coal	20	0.95
Stonework	38	0.93
Water	38	0.67
Wood		0.80–0.90
Beech, planed	70	0.94
Oak, planed	38	0.91
Spruce, sanded	38	0.89

Source: http://www.omega.com/temperature/z/pdf/z088-089.
pdf (retrieved on June 18, 2012).

TABLE A.16

Diffusion Coefficients of Binary Mixtures of Air and Another Gas
at 1 atm

T(K)	D—Binary Diffusion Coefficient (m²/s × 10⁴)							
	O_2	CO_2	Co	C_7H_{16}	H_2	NO	SO_2	He
200	0.095	0.074	0.098	0.036	0.375	0.088	0.058	0.363
300	0.188	0.157	0.202	0.075	0.777	0.18	0.126	0.713
400	0.325	0.263	0.332	0.128	1.25	0.303	0.214	1.14
500	0.475	0.385	0.485	0.194	1.71	0.443	0.326	1.66
600	0.646	0.537	0.659	0.27	2.44	0.603	0.44	2.26
700	0.838	0.684	0.854	0.354	3.17	0.782	0.576	2.91
800	1.05	0.857	1.06	0.442	3.93	0.978	0.724	3.64
900	1.26	1.05	1.28	0.538	4.77	1.18	0.887	4.42
1000	1.52	1.24	1.54	0.641	5.69	1.41	1.06	5.26
1200	2.06	1.69	2.09	0.881	7.77	1.92	1.44	7.12
1400	2.66	2.17	2.7	1.13	9.9	2.45	1.87	9.2
1600	3.32	2.75	3.37	1.41	12.5	3.04	2.34	11.5
1800	4.03	3.28	4.1	1.72	15.2	3.7	2.85	13.9
2000	4.8	3.94	4.87	2.06	18	4.48	3.36	16.6

Source: Mills, A.F., *Heat and Mass Transfer*, Burr Ridge, IL, Irwin, 1972.

TABLE A.17

Schmidt Number for Binary Mixtures of Air and Low Concentration
of Another Gas at 1 atm

T[K]	$\nu \times 10^6$ (m²/s) [1] Air	Sc—Schmidt Number							
		O_2	CO_2	Co	C_7H_{16}	H^2	NO	SO_2	He
200	7.66	0.81	1.04	0.78	2.13	0.20	0.87	1.32	0.21
300	16.00	0.85	1.02	0.79	2.13	0.21	0.89	1.27	0.22
400	26.56	0.82	1.01	0.80	2.07	0.21	0.88	1.24	0.23
500	39.03	0.82	1.01	0.80	2.01	0.23	0.88	1.20	0.24
600	53.22	0.82	0.99	0.81	1.97	0.22	0.88	1.21	0.24
700	68.99	0.82	1.01	0.81	1.95	0.22	0.88	1.20	0.24
800	86.23	0.82	1.01	0.81	1.95	0.22	0.88	1.19	0.24
900	104.88	0.83	1.00	0.82	1.95	0.22	0.89	1.18	0.24
1000	124.86	0.82	1.01	0.81	1.95	0.22	0.89	1.18	0.24
1200	168.67	0.82	1.00	0.81	1.91	0.22	0.88	1.17	0.24
1400	217.37	0.82	1.00	0.81	1.92	0.22	0.89	1.16	0.24
1600	270.78	0.82	0.98	0.80	1.92	0.22	0.89	1.16	0.24
1800	328.76	0.82	1.00	0.80	1.91	0.22	0.89	1.15	0.24
2000	391.23	0.82	0.99	0.80	1.90	0.22	0.87	1.16	0.24

Source: Mills, A.F., *Heat and Mass Transfer*, Burr Ridge, IL, Irwin, 1972.

TABLE A.18

Exponential Integral Functions

x	E1	E2	E3
0	∞	1	0.5
0.01	4.0379	0.9497	0.4903
0.02	3.3547	0.9131	0.4810
0.03	2.9591	0.8817	0.4720
0.04	2.6813	0.8535	0.4633
0.05	2.4679	0.8278	0.4549
0.06	2.2953	0.8040	0.4468
0.07	2.1508	0.7818	0.4388
0.08	2.0269	0.7610	0.4311
0.09	1.9187	0.7412	0.4236
0.1	1.8229	0.7225	0.4163
0.15	1.4645	0.6410	0.3823
0.2	1.2227	0.5742	0.3519
0.25	1.0443	0.5177	0.3247
0.3	0.9057	0.4691	0.3000
0.35	0.7942	0.4267	0.2777
0.4	0.7024	0.3894	0.2573
0.45	0.6253	0.3562	0.2387
0.5	0.5598	0.3266	0.2216
0.55	0.5034	0.3001	0.2059
0.6	0.4544	0.2762	0.1916
0.65	0.4115	0.2546	0.1783
0.7	0.3738	0.2349	0.1661
0.75	0.3403	0.2171	0.1548
0.8	0.3106	0.2009	0.1443
0.85	0.2840	0.1860	0.1347
0.9	0.2602	0.1724	0.1257
0.95	0.2387	0.1599	0.1174
1	0.2194	0.1485	0.1097
1.1	0.1860	0.1283	0.0959
1.2	0.1584	0.1111	0.0839
1.3	0.1355	0.0964	0.0736
1.4	0.1162	0.0839	0.0646
1.5	0.1000	0.0731	0.0567
1.6	0.0863	0.0638	0.0499
1.7	0.0747	0.0558	0.0439
1.8	0.0647	0.0488	0.0387
1.9	0.0562	0.0428	0.0341
2	0.0489	0.0375	0.0301
2.1	0.0426	0.0330	0.0266
2.2	0.0372	0.0290	0.0235
2.3	0.0325	0.0255	0.0208
2.4	0.0284	0.0225	0.0184
2.5	0.0249	0.0198	0.0163
2.6	0.0218	0.0175	0.0144

TABLE A.18

Exponential Integral Functions

x	E1	E2	E3
2.7	0.0192	0.0154	0.0128
2.8	0.0169	0.0136	0.0113
2.9	0.0148	0.0120	0.0101
3	0.0130	0.0106	0.0089
3.1	0.0115	0.0094	0.0079
3.2	0.0101	0.0083	0.0070
3.3	0.0089	0.0074	0.0063
3.4	0.0079	0.0065	0.0056
3.5	0.0070	0.0058	0.0049
3.6	0.0062	0.0051	0.0044
3.7	0.0054	0.0046	0.0039
3.8	0.0048	0.0041	0.0035
3.9	0.0043	0.0036	0.0031
4	0.0038	0.0032	0.0028
4.1	0.0033	0.0028	0.0025
4.2	0.0030	0.0025	0.0022
4.3	0.0026	0.0022	0.0020
4.4	0.0023	0.0020	0.0017
4.5	0.0021	0.0018	0.0015
4.6	0.0018	0.0016	0.0014
4.7	0.0016	0.0014	0.0012
4.8	0.0015	0.0013	0.0011
4.9	0.0013	0.0011	0.0010
5	0.0011	0.0010	0.0009

References

1. Lemmon, E. W., McLinden, M. O., and Friend, D. G. (2012) Thermophysical properties of fluid systems, in *NIST Chemistry WebBook*, NIST Standard Reference Database Number 69, P. J. Linstrom and W. G. Mallard (eds.). Gaithersburg, MD: National Institute of Standards and Technology. http://webbook.nist.gov (retrieved on June 18, 2012).
2. http://www.omega.com/temperature/z/pdf/z088-089.pdf (retrieved on June 18, 2012).
3. Mills, A. F. (1972) *Heat and Mass Transfer*, Burr Ridge, IL: Irwin.

Index

Milton Keynes UK
Ingram Content Group UK Ltd.
UKHW051900071024
449327UK00025B/2031